VII A	VIII A
					1 **H** 1.0079	2 **He** 4.00260
III A	IV A	V A	VI A			
5 **B** 10.81	6 **C** 12.011	7 **N** 14.0067	8 **O** 15.9994	9 **F** 18.998403	10 **Ne** 20.179	
13 **Al** 26.98154	14 **Si** 28.0855	15 **P** 30.97376	16 **S** 32.06	17 **Cl** 35.453	18 **Ar** 39.948	

I B	II B					
28 **Ni** 58.70	29 **Cu** 63.546	30 **Zn** 65.38	31 **Ga** 69.72	32 **Ge** 72.59	33 **As** 74.9216	34 **Se** 78.96	35 **Br** 79.904	36 **Kr** 83.80
46 **Pd** 106.4	47 **Ag** 107.868	48 **Cd** 112.41	49 **In** 114.82	50 **Sn** 118.69	51 **Sb** 121.75	52 **Te** 127.60	53 **I** 126.9045	54 **Xe** 131.30
78 **Pt** 195.09	79 **Au** 196.9665	80 **Hg** 200.59	81 **Tl** 204.37	82 **Pb** 207.2	83 **Bi** 208.9804	84 **Po** (209)	85 **At** (210)	86 **Rn** (222)

Metals ← → Nonmetals

63 **Eu** 151.96	64 **Gd** 157.25	65 **Tb** 158.9254	66 **Dy** 162.50	67 **Ho** 164.9304	68 **Er** 167.26	69 **Tm** 168.9342	70 **Yb** 173.04	71 **Lu** 174.967
95 **Am** (243)	96 **Cm** (247)	97 **Bk** (247)	98 **Cf** (251)	99 **Es** (252)	100 **Fm** (257)	101 **Md** (258)	102 **No** (259)	103 **Lr** (260)

INTRODUCTION TO GENERAL, ORGANIC, AND BIOLOGICAL CHEMISTRY

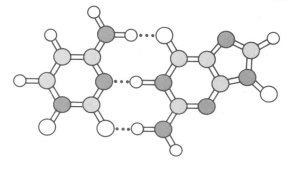

To my mother

INTRODUCTION TO GENERAL, ORGANIC, AND BIOLOGICAL CHEMISTRY

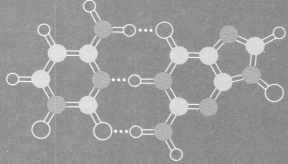

Martha J. Gilleland
California State College, Bakersfield

WEST PUBLISHING COMPANY
St. Paul • New York • Los Angeles • San Francisco

Copy editor: Pamela S. McMurry
Interior text design: Larry Layton
Illustrations: Etc. Graphics
Composition: Etc. Graphics
Production Management: Etc. Graphics
Cover design: Jack Deskin
Cover photograph: Manfred Kage/Peter Arnold, Inc.

The cover photograph is a photomicrograph of aspirin.

A study guide has been developed to assist you in mastering the concepts presented in this text. The study guide reinforces concepts by presenting them in condensed, concise form. Additional problems and examples are also included. The study guide is available from your local bookstore under the title *Study Guide to Accompany Introduction to General, Organic, and Biological Chemistry* prepared by Gloria Mumford.

In addition, a laboratory manual with 43 class-tested experiments accompanies this text. The manual is also available from your local bookstore under the title *Basic Experiments for General, Organic, and Biological Chemistry* prepared by Thomas I. Pynadath.

If you cannot locate the study guide or the lab manual in your bookstore, ask the manager to order them for you.

Credits and acknowledgments appear on page **G-19**.

COPYRIGHT © 1982 By WEST PUBLISHING CO.
50 West Kellogg Boulevard
P.O. Box 3526
St. Paul, Minnesota 55165

All rights reserved
Printed in the United States of America

Library of Congress Cataloging in Publication Data
Gilleland, Martha J.
Introduction to general, organic, and biological chemistry.

Includes index.
1. Chemistry. I. Title.
QD31.2.G524   540   81-19733
ISBN 0-314-63173-9   AACR2

2nd Reprint—1984

CONTENTS

Preface *xi*

1 THE BASIS OF CHEMISTRY 2

Alchemy and Chemistry 3

PERSPECTIVE: Paracelsus—Father of Chemotherapy 4

Scientific Measurements 5

Systems of Measurements 5 *Conversions* 7 *Precision, Accuracy, and Significant Figures* 11 *Mass and Weight* 14 *Density* 15 *Specific Gravity* 17 *Energy* 17 *Heat* 18 *Temperature* 18 *Specific Heat* 21

Summary 23

Study Questions and Problems 24

2 ELEMENTS AND ATOMS 26

The Essence of Chemistry 27

Elements 27 *Atoms* 29 PERSPECTIVE: Origin of the Concept of Element 29 *The Periodic Table* 32

Atomic Structure 33

Subatomic Particles 33 *Atomic Number* 34 *Mass Number* 34 *Isotopes* 35 *Atomic Mass Units* 35 *Atomic Weight* 36 *Electron Configuration* 36 *Chemical Periodicity* 41 *The Periodic Table Revisited* 44 *Elements and Life* 47 *Properties of Some Elements Important to Life* 48

Radioactivity 50

Alpha Decay 51 *Beta Decay* 52 *Gamma Decay* 52 *Nuclear Reactions* 53 PERSPECTIVE: Madame Curie 53

Summary 55

Study Questions and Problems 56

3 CHEMICAL BONDS 58

Covalent Bonds and Molecular Compounds 60

Covalent Bonds 60 *Molecular Compounds* 61 *Practice in Writing Electronic Formulas* 62 *Naming Binary Molecular Compounds* 64 *Properties of Molecular Elements and Compounds* 65

Ionic Bonds and Ionic Compounds 66

Ionic Bonds 66 *Electron Configurations of Ions* 67 PERSPECTIVE: Lithium, Ions, and Depression 68 *Ionic Compounds* 68 *Naming Binary Ionic Compounds* 70 *Naming Complex Ionic Compounds* 71 *Properties of Ionic Compounds* 72

Additional Concepts of Bonding 74

Shapes of Molecules 74 *Electronegativity* 77 *Polar Covalent Bonds* 78 PERSPECTIVE: Linus Pauling, Crusading Scientist 80 *Multiple Covalent Bonds* 80 *Resonance* 81 PERSPECTIVE: Nitrates and Nitrites in Meat 82

Summary 83

Study Questions and Problems 84

4 CHEMICAL REACTIONS 86

Chemical Equations and Mass Relationships 88

The Language of Chemical Equations 88 *Conservation of Mass* 89 *Balancing Chemical Equations* 90 *Molecular Weights and Formula Weights* 93 *The Chemical Mole* 95 *Calculations and Chemical Equations* 98

Types of Chemical Reactions 101

Single Replacement Reactions 101 *Double Replacement Reactions* 101 *Combination Reactions* 102 *Decomposition Reactions* 102 *Isomerization Reactions* 102 *Oxidation-Reduction Reactions* 105 PERSPECTIVE: Chemical Reactions and Life Processes 108

Summary 109

Study Questions and Problems 109

5 GASES 112

Properties of Gases 113

Mass and Volume 113 *Pressure* 114 *Compressibility* 115 *Gaseous Expansion and Contraction* 115 *Diffusion* 116

Relationships of Volume, Temperature, and Pressure 116

Volume and Pressure 116 *Volume and Temperature* 119 *Pressure and Temperature* 122 *More Calculations with the Gas Laws* 124 PERSPECTIVE: The Earth's Atmosphere 125

Interpretation of Gaseous Behavior 127

Avogadro's Hypothesis 127 *The Kinetic Molecular Theory* 131

Gas Mixtures *132*
Real Gases *134*
Summary *135*
Study Questions and Problems *136*

6 LIQUIDS AND SOLIDS *138*

Intermolecular Forces *140*

Dipole Attractions **140** *London Dispersion Forces* **141** *Hydrogen Bonds* **142**

Liquids *144*

Density **144** *Surface Tension* **145** *Viscosity* **145**

Vaporization *146*

Dynamic Equilibrium **146** *Boiling Point* **147** *Heat of Vaporization* **148**

Solids *150*

Crystalline Solids **150** *Amorphous Solids* **154** PERSPECTIVE: The Manufacture of Glass **155**

Fusion *156*

The Melting Process **156** *Heat of Fusion* **156** *Melting Point* **157**

Phase Changes *158*

Liquid Crystals *160*

Summary *160*

Study Questions and Problems *161*

7 SOLUTIONS, DISPERSIONS, AND SUSPENSIONS *164*

Water *165*

Review of Structure and Physical Properties **165** *Solvent Properties* **167** *Hydrates* **167** *Water in the Human Body* **168**

Solution Formation *169*

The Dissolution Process **171** *Rates of Dissolution* **172** *Solubility* **172** *Factors Affecting Solubility* **174**

Preparation of Solutions *176*

Concentrations **176** *Dilutions* **180**

Properties of Solutions *182*

Homogeneity **183** *Solute Particles* **183** *Colligative Properties* **184**

Colloids *187*

Suspensions *189*

Dialysis *189*

Transport of Body Fluids and Solutes *190*

PERSPECTIVE: Hemodialysis *190*
Summary *191*
Study Questions and Problems *192*

8 CHEMICAL EQUILIBRIUM AND RATES OF REACTIONS *196*

Chemical Equilibrium *197*

The Dynamic Equilibrium of Reversible Reactions **197** *Equilibrium Constants* **199** *Utility of the Equilibrium Constant* **201** *LeChatelier's Principle* **202** *Energy and Chemical Change* **205** PERSPECTIVE: Chemical Energy from Food **207** *Energy and the Biosphere* **210**

Rates of Chemical Reactions *211*

Energy of Activation **211** *Factors Affecting Reaction Rates* **213**

The Relationship Between Chemical Equilibrium and Reaction Rates *215*

Summary *216*

Study Questions and Problems *217*

9 ACIDS, BASES, AND SALTS *220*

Acids and Bases *221*

Naming Acids and Bases **225** *Properties of Acids and Bases* **225** *Common Acids and Bases* **226** *Strong and Weak Acids* **229** PERSPECTIVE: Acid Rain **230** *Examples of Strong and Weak Acids* **233** *Strong and Weak Bases* **236** *Normality* **238** PERSPECTIVE: The Chemistry of Antacids **240** *Acid-Base Titrations* **243**

Salts *245*

Summary *248*

Study Questions and Problems *248*

10 ELECTROLYTES, pH, AND BUFFERS *252*

Electrolytes *253*

Strong Electrolytes, Weak Electrolytes, and Nonelectrolytes **254** *Concentrations of Electrolytes* **255** *Electrolysis* **255** PERSPECTIVE: Charles Martin Hall and the Production of Aluminum **258**

Ionization of Water *258*

The Concept of pH *259*

The pH Scale **262** *Measurement of pH* **263**

Buffers *265*

The Henderson-Hasselbalch Equation **267** *Buffers in Blood* **271** *Acidosis and Alkalosis* **271**

PERSPECTIVE: The High Solubility of Carbon Dioxide in Water *272*

Summary *274*

Study Questions and Problems *274*

11 AN INTRODUCTION TO ORGANIC CHEMISTRY: HYDROCARBONS AND THEIR HALOGEN DERIVATIVES *278*

Organic Chemistry *279*

The Hydrocarbons *281*

Occurrence and General Uses **281** *Physical Properties* **281** *Types of Hydrocarbons* **284**

Alkanes *284*

Structure **284** *Nomenclature* **286** *Physical Properties* **291** *Chemical Properties* **291** *Sources and Uses* **293**

Alkenes *293*

Structure **293** *Nomenclature* **294** PERSPECTIVE: *Petroleum Refining* **295** *Physical Properties* **297** *Chemical Properties* **298** *Sources and Uses* **301**

Alkynes *301*

Structure **302** *Nomenclature* **303** *Physical Properties* **303** *Chemical Properties* **303** *Sources and Uses* **304**

Cyclic Aliphatic Hydrocarbons *304*

Aromatic Hydrocarbons *305*

Benzene Structure **306** *Benzene Derivatives* **307** *Polycyclic Aromatic Hydrocarbons* **307**

Halogen Derivatives of Hydrocarbons *307*

PERSPECTIVE: Halogenated Hydrocarbons in the Environment *311*

Summary *312*

Study Questions and Problems *313*

12 ALCOHOLS, ETHERS, AND RELATED COMPOUNDS *318*

Alcohols *319*

Structure **319** *Nomenclature* **321** *Physical Properties* **323** *Chemical Properties* **325** *Sources of Alcohols* **331** *Uses of Alcohols* **332**

Ethers *334*

Structure **335** *Nomenclature* **335** *Physical Properties* **336** *Chemical Properties* **337** *Sources and Uses of Ethers* **337**

Phenols *337*

Structure and Nomenclature **338** *Properties* **338** *Sources and Uses of Phenols* **340**
PERSPECTIVE: Joseph Lister, Founder of Modern Surgery *342*

Thiols *342*

Summary *343*

Study Questions and Problems *344*

13 ALDEHYDES, KETONES, AND CARBOHYDRATES *348*

Aldehydes and Ketones *349*

Structure and Physical Properties **350** *Nomenclature* **351** *Chemical Properties* **354** *Sources and Uses* **358**

Carbohydrates *359*

Optical Isomerism **361** *Examples of Monosaccharides* **366** *Disaccharides* **374** *Examples of Disaccharides* **375** PERSPECTIVE: *Lactose Intolerance* **376** *Polysaccharides* **377**

Summary *380*

Study Questions and Problems *380*

14 CARBOXYLIC ACIDS, ESTERS, AND LIPIDS *384*

Carboxylic Acids *385*

Structure **386** *Nomenclature* **386** *Physical Properties* **388** *Chemical Properties* **389** *Examples of Carboxylic Acids* **392**

Esters *394*

Nomenclature **395** *Physical Properties and Uses* **396** *Chemical Properties* **397** *Examples of Esters* **399** PERSPECTIVE: Alfred Nobel, Inventor of Dynamite *402*

Lipids *403*

Classification of Lipids **403** *Fatty Acids* **404** *Waxes* **405** *Triacylglycerols* **405** *Chemical Properties of Triacylglycerols* **407** *Phosphoglycerides* **409** *Sphingolipids* **411** *Terpenes* **412** *Steroids* **414** *Prostaglandins* **416**

Summary *416*

Study Questions and Problems *417*

15 ORGANIC NITROGEN COMPOUNDS *420*

Amines *421*

Structure **422** *Nomenclature* **423** *Physical Properties* **425** *Chemical Properties* **427** *Examples of Biologically Important Amines* **430**

Amino Acids *432*

Structure 432   Classification 433   Physical Properties 436   Chemical Properties 437   Unusual Amino Acids 441

Amides *442*

Structure 442   Nomenclature 442   Physical Properties 444   Chemical Properties 446   Examples of Amides 447   PERSPECTIVE: Nylon, A Synthetic Polyamide *448*

Heterocyclic Nitrogen Compounds *448*

Five-Membered Rings 448   Six-Membered Rings 449   Fused-Ring Nitrogen Heterocyclics 449

Summary *450*

Study Questions and Problems *450*

16 PROTEINS *454*

Functions of Proteins *455*

Protein Structure *457*

Primary Structure 457   PERSPECTIVE: Endorphins, The Morphine from Within *462   Secondary Structure 463   Tertiary Structure 467   Quaternary Structure 471*

Classes of Proteins *474*

Properties of Proteins *475*

PERSPECTIVE: The Glycoproteins, a Diverse Group of Biomolecules *476*

Separation of Protein Mixtures *477*

Liquid Chromatography 477   Electrophoresis 479

Summary *481*

Study Questions and Problems *482*

17 ENZYMES AND COENZYMES *484*

Names and Classes of Enzymes *485*

Properties and Functions of Enzymes *491*

Enzyme Catalysis *491*

Rates and Equilibria 491   Enzyme Specificity 492   Theories of Enzyme Action 493

Factors Affecting Enzyme Activity *494*

Cofactor Concentration 494   Substrate Concentration 495   Enzyme Concentration 495   Temperature 496   The Effect of pH 496

Coenzymes *497*

The Pyridine Coenzymes 498   The Flavin Coenzymes 499   Coenzyme A 500

Regulation of Enzyme Activity *502*

Activation of Zymogens 502   PERSPECTIVE: Pernicious Anemia—A Vitamin Deficiency? *502   Covalent Modification 504   Effect of Inhibitors 504*

Uses of Enzymes *506*

PERSPECTIVE: Recent Advances in the Use of Enzymes in Medicine *506*

Summary *508*

Study Questions and Problems *509*

18 BODY FLUIDS AND HORMONES *512*

Body Fluids in Digestion *513*

Saliva 514   Gastric Juice 514   Pancreatic Juice 516   Intestinal Juice 516   Bile 517   The Large Intestine 520

Blood *520*

Composition of Blood 521   Blood Plasma 522   Red Blood Cells 524   Hemoglobin 524   Blood Clot Formation 528

Urine *530*

PERSPECTIVE: Hemophilia and the Fall of the Russian Monarchy *531   Diuretics 532   Homones Affecting Urine Composition 532   The Role of the Kidneys in Acid-Base Balance 533   The Role of the Kidneys in Electrolyte Balance 533   The Role of the Kidneys in Regulation of Blood Pressure 533*

Hormones *534*

Classes of Hormones 537   Prohormones 538   The Molecular Basis of Hormone Action 539

Summary *539*

Study Questions and Problems *540*

19 AN INTRODUCTION TO METABOLISM: BIOENERGETICS AND THE CITRIC ACID CYCLE *542*

The Cell and an Overview of Metabolism *543*

Features of Eucaryotic Cells 544   Overview of Metabolism 546

Membranes *548*

Membrane Lipids 548   Lipid Bilayers 548   Membrane Permeability 550   Membrane Proteins 550   Membrane Fluidity 551   The Fluid Mosaic Model 553   Membrane Transport 553

Bioenergetics *555*

PERSPECTIVE: Obesity—An Enzyme Deficiency? *556   The Central Figure—ATP 556*

The Pathway of Electron Transport 558
Oxidative Phosphorylation 560

The Citric Acid Cycle 562

Reactions of the Citric Acid Cycle 562 *Regulation of the Citric Acid Cycle* 568 *Summary and Major Features of the Citric Acid Cycle* 569 *Energetics of the Citric Acid Cycle* 569

Summary 570

Study Questions and Problems 571

20 METABOLISM OF CARBOHYDRATES 574

Glucose—The Central Figure 575

Blood Sugar 576

Glycogen Metabolism 577

Glycogen Synthesis 577 *Glycogen Breakdown* 578 *Inborn Errors of Glycogen Metabolism* 580

Glycolysis 580

Reactions 581 *Energy Yield* 586

Fates of Pyruvate 587

Reduction to Lactate 587 *Reduction to Ethanol* 588 *Conversion to Acetyl CoA* 589

Complete Oxidation of Glucose 590

Gluconeogenesis 592

The Pentose Phosphate Pathway 593

PERSPECTIVE: Calories from Alcohol 594

Regulation of Carbohydrate Metabolism 594

Inborn Errors of Carbohydrate Utilization and Glycolysis 598

Summary 599

Study Questions and Problems 600

21 METABOLISM OF LIPIDS AND AMINO ACIDS 602

Metabolism of Lipids 603

Absorption of Fats 603 *Blood Lipids* 604 *Mobilization of Fatty Acids* 606 *Fatty Acid Oxidation* 608 *Energetics of Fatty Acid Oxidation* 611 *Ketone Bodies* 612 *Fatty Acid Synthesis* 614 *Synthesis of Triacylglycerols and Other Lipids* 617 *Cholesterol Metabolism* 617 *Relationships Between Carbohydrate and Fatty Acid Metabolism* 620 *Inborn Errors of Lipid Metabolism* 621

Metabolism of Amino Acids 622

The Amino Acid Pool 622 PERSPECTIVE: Ridding the Body of Toxic Materials 623 *Amino Acid Degradation* 623 *The Urea Cycle* 626 *Fates of the Carbon Atoms from Amino Acids* 629 *Amino Acid Synthesis* 630 *Inborn Errors of Amino Acid Metabolism* 630

Metabolic Interrelationships 632

Summary 633

Study Questions and Problems 634

22 NUCLEIC ACIDS AND HEREDITY 636

Nucleic Acids 637

Composition of Nucleic Acids 638 *Nucleosides* 639 *Nucleotides* 639 *The Primary Structure of Nucleic Acids* 641

DNA 643

Secondary Structure of DNA 644 *The Role of DNA* 645 *Replication of DNA* 646 *DNA Repair* 646

RNA 647

Messenger RNA 648 *Transfer RNA* 648 *Ribosomal RNA* 648

Protein Biosynthesis 648

Genetic Mutations 652

Viruses 653

PERSPECTIVE: Slow Viruses 654

Carcinogens 656

Chemical Carcinogens 656 *Radiation* 657 *Viruses* 657

Genetic Engineering 658

Enzymes Used in Genetic Engineering 658 *Plasmids* 659 *Recombinant DNA* 660

Summary 662

Study Questions and Problems 663

23 NUTRITION 666

Human Nutritional Requirements 667

Energy Provided by the Macronutrients 668

The Macronutrients 670

Proteins 670 PERSPECTIVE: Liquid Protein Diets 673 *Fats* 673 *Carbohydrates* 674 PERSPECTIVE: Sugar—Is It Really Harmful? 675

The Micronutrients 675

The Water-Soluble Vitamins 675 *The Fat-Soluble Vitamins* 683 *Minerals* 687

Balanced Diets 689

Diet Therapy *690*
   Restorative Diets **690**   *Restrictive Diets* **691**

Summary *692*

Study Questions and Problems *693*

24 CONSUMER PRODUCTS *696*

Food Preservation and Food Products *697*

   Food Preservation **697**   *Dairy Products* **700**
   Fats and Oils **703**   *Food Additives* **704**
   PERSPECTIVE: The Chemistry of Leavening Agents **707**

Drugs *708*

   Drug Names **708**   *Drugs for Treating Symptoms* **709**
   Drugs for Curing Disease **711**

Home Products *717*

   Dental Chemistry and Fluoride Treatment **717**
   Soaps and Detergents **718**   *Drain Cleaners and Oven Cleaners* **719**

Summary *720*

Study Questions and Problems *721*

25 RADIOACTIVITY AND NUCLEAR PROCESSES *724*

Radioactivity *725*

   Natural Radioactivity **725**   *Artificial Radioactivity* **726**

Radiation: Detection and Measurement *727*

   Detection of Radiation **729**   *Half-Life* **731**
   Measurement of Radiation **732**

Radiation Safety *735*

   The Basis for Biological Damage **735**   *Radiation Exposure* **736**   *Protection Against Radiation* **737**

Applications of Radiochemistry *739*

   Archaeological Dating **739**   *Isotope Tracers* **740**
   Radiation Therapy **741**   *Medical Diagnosis* **742**

Nuclear Power *744*

   Nuclear Binding Energy **745**   *Nuclear Fission* **746**
   PERSPECTIVE: Development of the Atomic Bomb **748**   *Nuclear Fusion* **750**

Summary *751*

Study Questions and Problems *752*

Appendix A *Basic Mathematics* **A-1**

Appendix B *Table of Logarithms* **A-7**

Appendix C *Decimal pH Values* **A-9**

Appendix D *Solutions to Exercises and Answers to Selected Study Questions and Problems* **A-11**

Glossary *G-1*

Illustration Sources *G-19*

Index *I-1*

PREFACE

Introduction to General, Organic, and Biological Chemistry is a textbook designed for students who have no formal background in chemistry but who need a working understanding of the science in order to achieve their career goals. Students majoring in nursing, allied health, nutrition, home economics, agriculture, and the life sciences will find the textbook especially helpful because of its emphasis on relationships between chemistry and these fields. In addition, liberal arts students wishing a comprehensive introduction to chemistry will benefit from this text. Historical developments in chemistry are presented, particularly in the early chapters and in some of the biological chemistry chapters, and numerous chemical applications to health and everyday life are thoroughly integrated throughout the text. Because the students using this textbook are not likely to have developed mathematical skills beyond elementary algebra, ways of solving mathematical problems are presented clearly, based on dimensional analysis and logical reasoning. Examples of both mathematical and reasoning problems are abundant, with detailed solutions following each. In addition, exercises appear with most examples to give students an opportunity to practice newly acquired problem-solving skills. Detailed solutions to these appear in Appendix D.

Each chapter begins with a brief introduction, followed by a logical sequence of topics; a topic outline is given on the opening page of each chapter. Margin notes in each chapter provide definitions of new terms, pronunciation of technical terms, and supplementary information. Also included are boxed inserts, called Perspectives, that present historical developments, practical applications, current issues, or human involvement in chemistry. As the name implies, these are designed to help each student develop an individual perspective of chemistry. At the conclusion of each chapter is a summary and an extensive set of study questions and problems. Answers to selected study questions and problems appear in Appendix D. Additional appendices include a review of basic mathematics (exponents, scientific notation, logarithms, and simple algebra), a table of logarithms, and the calculation of decimal pH values. The textbook concludes with a comprehensive glossary and a detailed index.

Chapter Sequence

The text begins with a treatment of the historical basis of chemistry and the fundamental nature of units and measurements. The early chapters deal with elements and atoms, chemical bonds, and chemical reactions in order to introduce students to the nature of chemistry as soon as possible. Following this sequence are the three states of matter, solutions, chemical equilibrium and reaction rates, acids and bases, and electrolytes, pH, and buffers. Thus, chapters 1–10 are devoted to general chemistry. Chapter 11 is an introduction to organic chemistry and a discussion of hydrocarbons and their halogen derivatives, and chapter 12 is a treatment of alcohols, ethers, and related compounds. I have included only the reactions of organic chemistry that are needed for a basic understanding of biological chemistry. Chapter 13 includes the organic chemistry of aldehydes and ketones, but blends these topics with a consideration of carbohydrate structure and function. In this way, carbohydrates, usually treated as a biochemical topic, appear in conjunction with their simpler organic relatives. Chapter 14 also blends organic and biochemical topics; carboxylic acids, esters, and lipids are discussed together, once again to emphasize the close relationship between organic and biological chemistry. To continue this theme, organic nitrogen compounds and amino acids are discussed together in chapter 15. Chapter 16, Proteins, is the first chapter devoted exclusively to biological chemistry, and it is followed by chapters on enzymes and coenzymes, body fluids and hormones, metabolism, and nucleic acids and heredity. Chapters 23 and 24 emphasize the chemistry of nutrition and consumer products, respectively, and include information regarding nutrients, diet, food preservation and food products, drugs, and home products. The last chapter, Radioactivity and

Nuclear Processes, consists of topics introduced in chapter 2 but amplified with applications in chapter 25. For those who wish to include these topics earlier in their courses, chapter 25 can be taken up immediately after chapter 2, provided that a brief preliminary explanation of the nature of ions and molecules is given.

Throughout the text, the emphasis is on application of principles and understanding of concepts. Principles and concepts are introduced and then reiterated in different contexts wherever possible in order to have the students consider them from different viewpoints. An example of such a recurring topic is the concept of equilibrium, introduced as dynamic equilibrium in chapter 6, extended to reversible reactions, chemical equilibria, and reaction rates in chapter 8, used to develop the concepts of weak acids and weak bases in chapter 9, used to develop the concepts of pH and buffers in chapter 10, applied to the formation and hydrolysis of esters in chapter 14, used in explaining the formation of enzyme-substrate complexes in chapter 17, and appearing in many instances in the chapters on metabolism. Other recurring ideas are shapes of molecules (introduced in chapter 3), types of chemical reactions (introduced in chapter 4), intermolecular forces (introduced in chapter 6), solution properties (introduced in chapter 7), energetics (introduced in chapter 8), reaction rates (introduced in chapter 8), and Bronsted-Lowry acid-base theory (introduced in chapter 9).

Alternate Chapter Sequences and Coverages

For those colleges and universities that offer a two-semester sequence of introductory chemistry, this text can be used in its entirety; however, instructors of such courses may wish to take up chapter 25, Radioactivity and Nuclear Processes, immediately after chapter 2, where radioactivity is introduced. Other alterations in chapter sequence and coverage are possible, as indicated by chapter correlations given later in the preface.

For institutions that must restrict their course sequences to two quarters, the textbook offers a number of possible topic arrangements, depending on the emphasis desired. Two such arrangements are outlined below.

Two-quarter course sequence with emphasis on general chemistry:

Chapter	Recommended Parts
1	Entire chapter
2	Entire chapter
25	Entire chapter
3	Entire chapter
4	Entire chapter
5	Entire chapter
6	Entire chapter
7	Water; Solution Formation; Preparation of Solutions; Properties of Solutions
8	Entire chapter
9	Entire chapter
10	Entire chapter
11	Entire chapter
12	Entire chapter
13	Entire chapter
14	Entire chapter
15	Entire chapter
16	Entire chapter
17	Entire chapter
19	Entire chapter

Additional coverage, such as chapter 7 in its entirety, chapter 22, and chapter 23, is recommended if time permits.

Two-quarter course sequence with emphasis on biological chemistry:

Chapter	Recommended Parts
1	Entire chapter
2	Entire chapter
3	Entire chapter
4	Entire chapter
5	Properties of Gases
6	Entire chapter
7	Entire chapter
8	Entire chapter
9	Entire chapter
10	Entire chapter
11	Entire chapter
12	Entire chapter
13	Entire chapter
14	Entire chapter
15	Entire chapter

Chapter	Recommended Parts
16	Entire chapter
17	Entire chapter
18	Entire chapter
19	Entire chapter

Additional coverage, such as chapters 20, 21, and 22, is recommended if time permits.

For colleges and universities that offer a one-semester course in introductory chemistry, the following selection of chapters is recommended.

Chapter	Recommended Parts
1	Entire chapter
2	The Essence of Chemistry; Atomic Structure
3	Entire chapter
4	Entire chapter
5	Properties of Gases
6	Entire chapter
7	Water; Solution Formation; Preparation of Solutions; Properties of Solutions
8	Entire chapter
9	Entire chapter
10	Entire chapter
11	Entire chapter
12	Entire chapter
13	Entire chapter
14	Entire chapter
15	Entire chapter
16	Entire chapter

Additional coverage, such as chapters 17, 18, and 19, is recommended if time permits.

Chapter Correlations

The following table correlates each chapter in the text with preceding chapters that contain some essential background.

Chapter	Prerequisite Chapters
1	–
2	1
3	1, 2
4	1–3
5	1–4[a]
6	1–4[a]
7	1–4[a], 6
8	1–4
9	1–4, 6–8
10	1–4, 6–9
11	1–4, 6
12	1–4, 6, 8[b], 11
13	1–4, 6, 8[b], 11, 12
14	1–4, 6–8[b], 9, 11-13
15	1–4, 6–8[b], 9–14
16	1–4, 6–8[b], 9–15
17	1–4, 6–16
18	1–4, 6–17
19	1–4, 6–17
20	1–4, 6–9, 11–17, 19
21	1–4, 6–17, 19
22	1–4, 6–9, 11–17, 19
23	1–4, 6–17, 19
24	1–4, 6–17, 19
25	1–3[c]

[a]Mole concept
[b]Reversible reactions
[c]Ions and molecules

Textbook Supplements

Two excellent supplements have been prepared for use with this text. The student study guide, by Gloria M. Mumford of Central Piedmont Community College, contains an outstanding collection of study aids, such as learning objectives, enrichment discussions, additional problems and self-tests (both with solutions), key word lists, and solutions to the even-numbered study questions/problems in the text. An accompanying laboratory manual, by Thomas I. Pynadath of Kent State University, provides a rich assortment of student-tested experiments in the areas of general, organic, and biological chemistry. Instructor's manuals are available for both the laboratory manual and the textbook.

Acknowledgments

This textbook was written for students, and it was my goal to write it in a manner that students would find interesting, informative, and readable. While working toward this goal, I was fortunate in receiving thoughtful and incisive comments from many reviewers around the country. In particular, I wish to thank Joseph F. Becker for his generous efforts in reviewing almost the entire manuscript. I am deeply grateful to him and all of the reviewers listed below.

Demir A. Barker
Orange County Community College—New York

Joseph F. Becker
Montclair State College—New Jersey

Jerry A. Bell
Simmons College—Massachusetts

Larry F. Bray
Miami Dade Community College—Florida

William E. Bull
University of Tennessee

William H. Day
Fresno City College—California

J. Lindsley Foote
Western Michigan University

Howard H. Guyer
Fullerton College—California

Florence Haimes
San Francisco State University

Delwin D. Johnson
St. Louis Community College

Doris K. Kolb
Illinois Central University

Gloria M. Mumford
Central Piedmont Community College—North Carolina

Pat O'Day
Pasadena City College—California

Thomas Onak
California State University, Los Angeles

Robert K. Ono
Los Angeles Pierce College

Thomas I. Pynadath
Kent State University—Ohio

Don Roach
Miami Dade Community College—Florida

Miriam Smith
Pasadena City College—California

Edward C. Stark
Los Angeles City College

Mary S. Vennos
Essex Community College—Maryland

Les Wynston
California State University, Long Beach

I would also like to thank Denise Simon for her superb work in coordinating this project, John Lindley for masterfully overseeing production, Pamela McMurry for editing the manuscript and checking solutions to the problems, and Patricia Weedon for typing the first draft of the manuscript. Their contributions were invaluable in arriving at this finished product. In addition, I wish to thank California State College, Bakersfield, for granting me a sabbatical leave to work on this project. I am especially grateful to my colleagues and students for fruitful discussions and encouragement during various stages in the preparation of this textbook. Finally, I invite students and instructors who use the text to make comments and criticisms; I look forward to receiving them.

INTRODUCTION TO GENERAL, ORGANIC, AND BIOLOGICAL CHEMISTRY

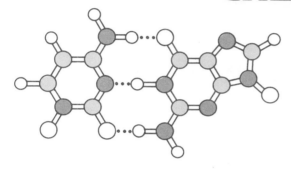

CHAPTER 1

THE BASIS OF CHEMISTRY

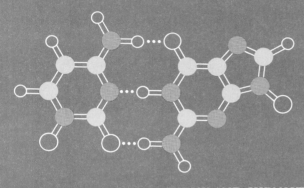

ALCHEMY AND CHEMISTRY
SCIENTIFIC MEASUREMENTS
- Systems of Measurements
- Conversions
- Precision, Accuracy, and Significant Figures
- Mass and Weight
- Density
- Specific Gravity
- Energy
- Heat
- Temperature
- Specific Heat

ALCHEMY AND CHEMISTRY

The science of chemistry had its beginnings in the ancient mystical art of *alchemy* (see Figure 1-1), thought by many to have originated in China around the fourth century B.C. Early Chinese alchemy had as its goal the prolongation of life; it was believed that immortality could be achieved by drinking the "elixir of life," an imaginary solution of gold. As time went by, alchemy developed in the Hindu, Greek, Arabic, and Latin cultures. It was brought to the Western world by Christian scholars during the twelfth century A.D. During its evolution, alchemy developed a second goal—converting, or *transmuting,* common metals such as lead and iron into gold. This latter goal was especially popular in the Western world, and even Sir Isaac Newton conducted transmutation experiments in the seventeenth century. In the eighteenth century, however, alchemists grew weary of searching for the secret of gold making, and they devoted themselves to religious aims. It was then that modern chemistry began to emerge as a science.

Many scientists of the eighteenth century became curious about the gaseous substances which could be generated by heating such things as blood, tallow, horn, wood, and chalk. They found that gases could combine with other materials to form liquids and solids, and the idea evolved that composition of substances is fundamental to understanding matter. In this way, modern chemistry developed into the study of matter in its many forms and transformations.

Today chemistry is an important influence on our quality of life. The early alchemists who dreamed of prolonging life would be pleased to know of chemical advances in the treatment and prevention of disease. For example, rickets was once a common childhood disease until the chemical compound called vitamin D was added to milk and other

alchemy (AL-ke-me): an ancient art concerned principally with finding an elixir of life and with discovering a method for transmuting base metals into gold.

Fig. 1-1 An Alchemist at Work

vitamin D: a vitamin occurring naturally in fish-liver oils; childhood deficiencies result in rickets, a disease in which softening of the bones causes deformities.

PERSPECTIVE: Paracelsus—Father of Chemotherapy

Paracelsus

One of the most colorful alchemists was Theophrastus Bombastus von Hohenheim, better known as Paracelsus, a name he gave himself to indicate to the world that he was greater than Celsus, a celebrated Roman writer on medicine. Paracelsus was born in 1493 in Switzerland, and he was taught chemistry and alchemy by his father, a physician. Paracelsus' restless nature led him to travel about Europe as a young man, associating with physicians, alchemists, astrologers, apothecaries, miners, gypsies, and various practitioners of the occult. After receiving an M.D. degree in Italy, he settled in Germany in 1526, where he was quickly acclaimed as a physician.

Paracelsus practiced rather unorthodox medicine for his time, and his original ideas got him into trouble more than once. He openly criticized the use of old-fashioned herbal remedies, advocating medicines made from minerals. In addition, he made no secret of his love for alcoholic beverages. His venomous attacks on critics and his scandalous behavior caused Paracelsus to fall out of favor, and once again he set out on a life of wandering. His tramplike appearance led to his being refused entry to at least one town, but at last his ideas were appreciated by a Bavarian prince who invited Paracelsus to live a peaceful life in Salzburg, Austria. Although he had finally found a congenial atmosphere, Paracelsus died in 1541 at the age of 48, physically worn out by his restless and strenuous life. Paracelsus lived during the transition of alchemy from an art to a science, and he is remembered today for originating the practice of chemical medicine, or in modern terminology, chemotherapy.

foods. Chemical science addresses the problems of world nutrition by synthesis and study of herbicides and pesticides which have dramatically increased the supply and availability of food. Chemistry plays vital roles in the manufacture of commercial products. Development of an almost infinite variety of plastics has provided us with egg cartons, telephones, bath tubs, boat hulls, and automobile components, to name a few. As our concern over dwindling energy supplies heightens, we look to chemistry to develop ways to store solar energy in batteries, convert low-quality crude oil into high-quality gasoline, and create storage cartridges for use in hydrogen-powered automobiles. Indeed, the science of chemistry is the basis of our very lives, as our bodies degrade the chemical components of food during digestion and use those products to build tissue, provide energy for life processes, synthesize hormones, and register nerve impulses.

Chemisty is an all-pervasive influence whose development was accomplished, in part, by the ability of scientists to make accurate measurements of the properties of matter (Figure 1-2). *Matter* is anything that

Fig. 1-2 A Laboratory Worker Uses a Laboratory Balance to Measure Mass Precisely and Accurately.

occupies space and has mass. It exists in three forms—solid, liquid, and gas. Figure 1-3 illustrates the three forms of water. Solid objects have rigid shapes and fixed volumes, while liquids flow to assume the shapes of their containers. Like solids, liquids also have fixed volumes. Gases resemble liquids in their ability to flow, but the most distinctive property of gases is that they expand or contract to fill the volume of their containers. Thus a gas has neither fixed shape nor volume.

Consistent and accurate measurements of properties of matter such as mass, volume, and temperature have helped us to extend our knowledge of the causes and consequences of changes in matter. Thus, we will begin our study of chemistry with a consideration of scientific measurements.

SCIENTIFIC MEASUREMENTS

Systems of Measurement

The two most popular systems of measurement are the *British system* and the *metric system*. The British system is traditionally used in the United States, the United Kingdom, and Canada; it has such familiar units as feet, pounds, and gallons. Scientists use the metric system, and recently a standard metric system called the International System of Measurements, or SI (from the French *Systeme International*), was adopted for use worldwide (see Table 1-1).

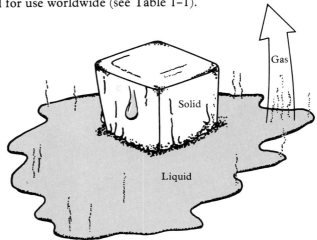

Fig. 1-3 Water in Its Three States: Solid (ice), Liquid, and Gaseous (water vapor)

Table 1-1 Units in Measurement Systems

Property	Metric	British	Unit Relationships
length	kilometer (km)	mile (mi)	1.61 km = 1 mi
	meter (m)	yard (yd)	1 m = 39.37 in
			1000 m = 1 km
	centimeter (cm)	foot (ft)	30.48 cm = 1 ft
		inch (in)	2.54 cm = 1 in
			100 cm = 1 m
	millimeter (mm)		1000 mm = 1 m
	micrometer (μm)		10^6 μm = 1 m
	nanometer (nm)		10^9 nm = 1 m
volume	**cubic meter (m^3)**	quart (qt)	1 m^3 = 1057 qt
	liter (L)		1 L = 1.057 qt
			1 L = 0.001 m^3
	deciliter (dL)		10 dL = 1 L
	milliliter (mL)		1000 mL = 1 L
	cubic centimeter (cm^3)		1 cm^3 = 1 mL
	microliter (μL)		10^6 μL = 1 L
mass	**kilogram (kg)**	pound (lb)	1 kg = 2.2 lb
	gram (g)		453.6 g = 1 lb
			1000 g = 1 kg
	milligram (mg)		1000 mg = 1 g
	microgram (μg)		10^6 μg = 1 g
	nanogram (ng)		10^9 ng = 1 g
energy	**joule (J)**		1 J = 10^7 erg
	kilocalorie (kcal)		1 kcal = 4184 J
	calorie (cal)		1000 cal = 1 kcal
			1 cal = 4.184 J
	erg (erg)		4.184 $\times$ 10^7 erg = 1 cal
temperature	**Kelvin (K)**	Fahrenheit (°F)	K = °C + 273
	Celsius (°C)		°F = 9/5 °C + 32
			°C = (°F − 32) (5/9)

(Bold face print signifies SI standard units.)

The basic unit of length in SI is the *meter* (also spelled metre), abbreviated as m. One meter is equivalent to 39.4 inches (or 1.09 yards). The metric system is based on the decimal system, and standard prefixes are used to denote powers of ten (for help with exponents, see Appendix A):

$$1\ \textit{nano}\text{meter (nm)} = \frac{1}{1,000,000,000}\ \text{meter (m)} = 10^{-9}\ \text{meter (m)}$$

$$1\ \textit{micro}\text{meter (}\mu\text{m)} = \frac{1}{1,000,000}\ \text{meter (m)} = 10^{-6}\ \text{meter (m)}$$

$$1\ \textit{milli}\text{meter (mm)} = \frac{1}{1,000}\ \text{meter (m)} = 10^{-3}\ \text{meter (m)}$$

$$1\ \textit{centi}\text{meter (cm)} = \frac{1}{100}\ \text{meter (m)} = 10^{-2}\ \text{meter (m)}$$

$$1\ \textit{kilo}\text{meter (km)} = 1000\ \text{meter (m)} = 10^{3}\ \text{meter (m)}$$

The SI standard unit of volume, the cubic meter (m^3), is a large volume (about 274 gallons), and it has not yet become a popular unit. Most scientists use the unit of volume called the *liter* (L). This unit is defined as 0.001 cubic meter, and thus 1000 liters = 1 m^3. Subunits of the liter such as milliliters (mL) and deciliters (dL) are often used to measure small volumes. These subunits are defined in Table 1-1.

liter (LEE-ter)

The SI standard unit of mass, the *kilogram* (kg), is also a large quantity, slightly more than two pounds, and commonly used subunits are the gram (g), milligram (mg), and microgram (μg). These subunits are also defined in Table 1-1. We will return to the topic of mass for a more complete discussion later in this chapter.

kilogram (KIH-lo-gram)

Conversions

British units may be converted to metric units by using conversion factors. In making conversions and in working other problems, we will rely on *dimensional analysis,* the use of dimensions (units) associated with a quantity as an aid in setting up the solution to a problem. Consider the relationship between meters and yards:

$$1\ \text{m} = 1.09\ \text{yd}$$

If we divide the expression by 1.09 yd, we obtain

$$\frac{1\ \text{m}}{1.09\ \text{yd}} = 1, \qquad (1)$$

a statement of unity. Similarly, if we divide the relationship by 1 m, we also obtain an expression of unity:

$$1 = \frac{1.09\ \text{yd}}{1\ \text{m}} \qquad (2)$$

Now we have two factors which are each equal to 1 (unity). We can choose from these *unit factors* to make a conversion.

For example, if we wish to convert 2.00 yards to meters, we must choose the unit factor which, when multiplied by the quantity of yards, will leave only meter as the final unit. We choose unit factor (1), because the units of yards will cancel, and we will have meters as our final units:

$$2.00 \text{ yd} \times \frac{1 \text{ m}}{1.09 \text{ yd}} = \frac{2.00}{1.09} \text{ m} = 1.83 \text{ m}$$

We can use the same approach to convert meters to yards. Suppose we wish to calculate the number of yards corresponding to 3.00 meters. We choose unit factor (2) and multiply it by the quantity of meters:

$$3.00 \text{ m} \times \frac{1.09 \text{ yd}}{1 \text{ m}} = 3.27 \text{ yd}$$

By using dimensional analysis to set up our problem, we have arrived at the *unit-factor method* for making conversions. It employs four simple steps:

1. Write down the fundamental relationship between units and derive the two unit factors.

2. Write down the quantity to be converted, making sure to include units. Referring to the last example, this would be

$$3.00 \text{ m}$$

3. Multiply that quantity by the unit factor which allows cancellation of the unwanted units. Again, in the last example,

$$3.00 \text{ m} \times \frac{1.09 \text{ yd}}{1 \text{ m}}$$

4. Complete the calculations,

$$\frac{3.00 \times 1.09}{1} \text{ yd} = 3.27 \text{ yd}$$

This is a fail-proof approach for converting units from one system into those of another as well as for making conversions within one system.

EXAMPLE 1.1
Convert 2.7 yards to feet.

SOLUTION

Unit relationship: $\quad 1 \text{ yd} = 3 \text{ ft}$

Unit factors: $\quad \dfrac{1 \text{ yd}}{3 \text{ ft}} = 1$

$\quad\quad\quad\quad\quad\quad \dfrac{3 \text{ ft}}{1 \text{ yd}} = 1$

Choose unit factor and cancel units:

$$2.7 \text{ yd} \times \frac{3 \text{ ft}}{1 \text{ yd}}$$

Complete calculations:

$$\frac{2.7 \times 3}{1} \text{ ft} = 8.1 \text{ ft}$$

EXAMPLE 1.2
Convert 4.2 yards to meters.

SOLUTION

Unit relationship: $1 \text{ m} = 1.09 \text{ yd}$

Unit factors:

$$\frac{1 \text{ m}}{1.09 \text{ yd}} = 1$$

$$\frac{1.09 \text{ yd}}{1 \text{ m}} = 1$$

Choose unit factor and cancel units:

$$4.2 \text{ yd} \times \frac{1 \text{ m}}{1.09 \text{ yd}}$$

Complete calculations:

$$\frac{4.2 \times 1}{1.09} \text{ m} = 3.9 \text{ m}$$

EXAMPLE 1.3
Convert 3.3 meters to millimeters.

SOLUTION

Unit relationship: $1 \text{ mm} = 10^{-3} \text{ m}$

Unit factors:

$$\frac{1 \text{ mm}}{10^{-3} \text{ m}} = 1$$

$$\frac{10^{-3} \text{ m}}{1 \text{ mm}} = 1$$

Choose unit factor and cancel units:

$$3.3 \text{ m} \times \frac{1 \text{ mm}}{10^{-3} \text{ m}}$$

Complete calculations:

$$\frac{3.3 \times 1}{10^{-3}} \text{ mm} = 3.3 \times 10^{3} \text{ mm}$$

EXERCISE 1.1 Make the indicated conversions.
(a) 17 milliliters to liters
(b) 10 decigrams to grams
(c) 2 liters to milliliters
(d) 1.68 kilograms to grams

The unit factor method is especially helpful when two or more conversion factors are needed. In these cases, the unit factors are used in series when there is not a direct unit relationship between initial units and final units. One unit factor converts initial units to a second set of units, and then one or more unit factors are used to convert the second set of units to the desired final units. The following examples illustrate this use of unit factors.

EXAMPLE 1.4

If 1 kilometer is equal to 0.621 miles, convert 1.20 kilometers to feet.

SOLUTION

Unit relationships:
$$1 \text{ km} = 0.621 \text{ mi}$$
$$1 \text{ mi} = 5280 \text{ ft}$$

Unit factors:

$$\frac{1 \text{ km}}{0.621 \text{ mi}} = 1$$

$$\frac{0.621 \text{ mi}}{1 \text{ km}} = 1$$

$$\frac{1 \text{ mi}}{5280 \text{ ft}} = 1$$

$$\frac{5280 \text{ ft}}{1 \text{ mi}} = 1$$

Choose unit factors and cancel units:

$$1.20 \text{ k\cancel{m}} \times \frac{0.621 \cancel{\text{mi}}}{1 \cancel{\text{km}}} \times \frac{5280 \text{ ft}}{1 \cancel{\text{mi}}}$$

Complete calculations:

$$\frac{1.20 \times 0.621 \times 5280}{1} \text{ ft} = 3930 \text{ ft}$$

EXAMPLE 1.5

Convert 1750 micrograms to kilograms.

SOLUTION

Unit relationships:
$$1 \text{ µg} = 10^{-6} \text{ g}$$
$$1 \text{ kg} = 10^{3} \text{ g}$$

Unit factors:

$$\frac{1\ \mu g}{10^{-6}\ g} = 1$$

$$\frac{10^{-6}\ g}{1\ \mu g} = 1$$

$$\frac{1\ kg}{10^{3}\ g} = 1$$

$$\frac{10^{3}\ g}{1\ kg} = 1$$

Choose unit factors and cancel units:

$$1750\ \mu g \times \frac{10^{-6}\ g}{1\ \mu g} \times \frac{1\ kg}{10^{3}\ g}$$

Complete calculations:

$$\frac{1750 \times 10^{-6} \times 1}{10^{3}}\ kg = 1750 \times 10^{-9}\ kg = 1.75 \times 10^{-6}\ kg$$

EXERCISE 1.2 Make the indicated conversions.
(a) 4230 feet to kilometers
(b) 3.65 kilograms to micrograms
(c) 4.30 meters to feet
(d) 1.67 cubic meters to milliliters

Precision, Accuracy, and Significant Figures

Accuracy and precision are terms used to describe quality of measurement. For example, you might measure the length of a sheet of paper with a ruler and find it to be 11.0 inches. You might find other sheets to have lengths of 10.9 and 11.1 inches. You would be able to measure these different lengths only if you used a ruler calibrated to 0.1 inch. Figure 1-4 illustrates the difference between a ruler calibrated to 0.1-inch intervals and one calibrated to 1-inch intervals. Thus, the calibration of the ruler determines how closely you can measure the paper. This fineness of measurement is *precision*. Precision also refers to the extent of reproducibility of the measurement. In other words, if you

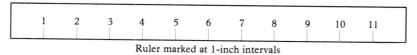
Ruler marked at 1-inch intervals

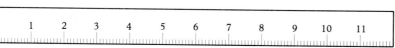
Ruler marked at 0.1-inch intervals

Fig. 1-4 Precision: Fineness of Measurement

accuracy: freedom from error.

Fig. 1-5 The Precision and Accuracy of this Medical Thermometer Depend on the Care Put into Its Construction.

significant figures: those digits in a numerical value which have actual physical meaning.

were to measure the same sheet of paper several times, and each time the length of the paper measures the same, then you have measured with good precision.

Accuracy refers to the amount of error in the measurement. Errors in measurement can be caused by such things as misreading the measuring scale, misplacement of the measuring device, poor choice or design of the measuring device, and other human errors. The medical thermometer in Figure 1-5, for example, must be designed well for its intended purpose, constructed carefully, and used properly to obtain accurate measurements. Measurements which are found to have very small errors are said to be highly accurate. Both precision and accuracy are important in making measurements.

In recording and reporting measurements, we must be careful that the numerical data signify the precision of our measurements. Suppose we measure a sheet of paper with a ruler calibrated to 0.1-inch intervals and find the length to be slightly greater than 11.1 inches. We know with certainty that the length is closer to 11.1 inches than to 11.0 inches or 11.2 inches. Since the length is just slightly longer than 11.1 inches, we estimate it to be 11.12 inches. In this measurement, we acknowledge that the last digit is an estimate. Thus, in our number the first three digits are highly reliable, but the last digit is uncertain. In using scientific data, we always assume that the last digit is an estimate; the amount of uncertainty is frequently one unit of measure at that position. In other words, our measurement of length is 11.12 inches with an uncertainty of 0.01 inches. If we were to use a more precisely calibrated ruler, we might find the paper to be 11.123 inches long with an uncertainty of 0.001 inches.

In the examples given above, each digit in the numbers was significant; that is to say, each digit had physical meaning. If we had measured with a ruler calibrated only to 1-inch intervals, a reported length of 11.123 inches would have no meaning, as it would be physically impossible to estimate the last two digits.

But what if we were interested in finding the area of the sheet of paper? Assume the length and width were 11.12 inches and 10.42 inches, respectively. Multiplying these dimensions together gives the area:

$$A = 11.12 \text{ in} \times 10.42 \text{ in}$$

Using a calculator, we find the answer to be 115.8704 in^2. Are we justified in reporting this seven-digit value? No, because our measurements contained only four digits each. Multiplying those numbers together gives an area whose value is only as precise as the measurements from which it is derived. In other words, we must report a value for the area which contains no more than the least number of *significant figures* in the measurements. Since both length and width were reported to a precision of four significant figures, our answer would report the area to be 115.9 in^2. The following rules should be used in mathematical operations to be sure that calculated values have the correct number of significant figures:

1. In addition and subtraction, the answer must contain the *same* number of *decimal places* as that of the term with the *least* number of decimal places.

273.05	
16.1	The answer is reported
<u> 8.32</u>	correctly as 297.5.
297.47	

2. In multiplication and division, the answer must contain the *same* number of *significant figures* as the term with the *least* number of *significant figures*.

   $(23.48)(10.1) = 237.148$ The answer is reported correctly as 237.

   The number of zeros in a reported value can be a problem, because they represent digits in a number. However, the number of digits is not always the same as the number of significant figures. For example, the values of 4 mg and 0.004 g are both the same number. Even though they have different numbers of digits (one and four, respectively), they both have only one significant figure, because the zeros in 0.004 g are used only for locating the decimal point. The following rules are used to deal with the problem of zeros and significant figures:

1. Zeros used to locate a decimal point are not significant. For example, we can express the length of an object as 5 cm or as 0.05 m. Both values denote the same length because 1 cm is 0.01 m. However, both values contain only *one* significant figure, since the zeros in 0.05 m serve only to locate the decimal point.

2. Zeros which are part of the measurement are significant. The number 0.00309 has three significant figures. The zero between 3 and 9 is significant because it is a part of the measurement. The zeros before 3 are not significant because they serve only to locate the decimal point.

3. In large numbers containing zeros, it is sometimes difficult to interpret the number of significant figures. For example, in the number 4,000, how many zeros are significant? This dilemma can only be solved by using exponential notation. Then the number can be written in any one of the following ways, depending on how precisely the measurement was made:

4.000×10^3	four significant figures
4.00×10^3	three significant figures
4.0×10^3	two significant figures
4×10^3	one significant figure

4. Values within the same measurement system which are used as definitions are exact and may be considered to have an infinite number of

significant figures (zeros) following the decimal point. For example, 1 meter has, by definition, *exactly* 1000 (1000.000 . . .) millimeters.

5. Similarly, values obtained by counting individual objects are also exact. For example, a deck of playing cards contains *exactly* 52 cards, not 51.7 or 52.4.

One last bit of information about significant figures is important. When we calculate values, we often must reduce the number of figures to comply with the rules on significant figures. In cases where a calculated value contains more figures than are significant, we round off the number. The following rules describe how to round off properly:

1. If the nonsignificant figure is *less* than 5, it is *dropped,* and the significant figure remains the *same.* Thus, 8.173 is equal to 8.17 when rounded off to three significant figures.

2. If the nonsignificant figure is *more* than 5, it is *dropped,* and the significant figure is *increased* by 1.

   8.176 rounded off to three significant figures is 8.18.
   8.1757 rounded off to three significant figures is 8.18.

3. If the nonsignificant figure is *exactly* 5 (or 5 followed by only zeros), the 5 is *dropped* and the significant figure is *increased* by 1 if it is *odd.* Thus, 8.1750 is 8.18 when rounded off to three significant figures.

4. If the nonsignificant figure is *exactly* 5 (or 5 followed by only zeros), the 5 is *dropped* and the significant figure remains the *same* if it is *even.* Thus, 8.185 is 8.18 when rounded off to three significant figures.

EXERCISE 1.3 Give the number of significant figures in each quantity.
(a) 7.04×10^3
(b) 2.367
(c) 17,000
(d) 0.00037
(e) 6.085
(f) 0.1834

EXERCISE 1.4 Round off each number to three significant figures.
(a) 2076
(b) 1.893×10^5
(c) 1665
(d) 0.035850
(e) 1.9857
(f) 3.3333

EXERCISE 1.5 Perform the indicated operations and round off the answer to the correct number of significant figures.
(a) (17)(575)(2) =
(b) 1.75 + 2.3 + 3.02 =
(c) 19 − 6.76 − 2.35 =
(d) 2.046/1.58 =
(e) (1757)(2.7)(3.082) =
(f) [(62)(85)(1.04)]/(2.33) =

Mass and Weight

The amount of matter contained in an object is its *mass*. For example, the average mass of protein required per day in a normal adult diet is 50 grams. In this expression, gram is the unit of mass. The mass of an object is a constant value, as contrasted to weight. The *weight* of an object is the force exerted on the mass of the object by the earth's gravitational attraction. Weight varies with location, and the farther an object is away from the center of the earth, the less its weight. Thus, a person weighs less in the mile-high city of Denver than at sea level, even though the person's mass is the same at both locations. However, the weight variation with altitude is so slight that we usually ignore it.

The British units of ounces and pounds are often used in English-speaking countries for measuring weight. Even though mass and weight clearly are different properties of matter, you will find the terms used interchangeably. Just remember that if we mention weight in chemistry, we usually mean mass.

Density

Since the mass and volume of a given type of material can be changed, they are considered to be variable quantities. However, they are used to derive density, a fixed property for a particular substance that can be used to identify or characterize the substance. For example, the density of urine is determined in the hospital laboratory as a means of assessing kidney function and aiding a physician in making a diagnosis.

Density is defined as the amount of mass in a volume of one unit. In other words, density is mass divided by volume:

$$\text{density} = \frac{\text{mass}}{\text{volume}}$$

or,

$$d = \frac{m}{v}$$

Thus, measurement of mass and volume leads to a calculation of density, usually expressed as grams per milliliter, g/mL. Table 1-2 shows the densities of some common solids, liquids, and gases.

EXAMPLE 1.6

A sample of urine containing 50.00 mL was found to weigh 51.23 g. Calculate the density of the urine sample.

Table 1-2 Densities of Common Substances
(0 °C, 1 atmosphere pressure)

SOLIDS		LIQUIDS		GASES	
Substance	Density (g/mL)	Substance	Density (g/mL)	Substance	Density (g/L)
Lead	11.4	Mercury	13.6	Air	1.3
Iron	7.9	Water	0.999	Oxygen	1.43
Cement	3	Butter	0.9	Nitrogen	1.25
Sand	2.3	Oil	0.8	Helium	0.18
Ice	0.917	Gasoline	0.7	Hydrogen	0.09
Balsa wood	0.11				

SOLUTION

$$d = \frac{m}{v}$$

$$= \frac{51.23 \text{ g}}{50.00 \text{ mL}}$$

$$= 1.025 \text{ g/mL}$$

EXAMPLE 1.7

An iron object has a density of 7.86 g/mL and a mass of 10.92 g. What is the volume of the iron object?

SOLUTION

First we must solve the density expression for volume:

$$d = \frac{m}{v}$$

$$v = \frac{m}{d}$$

Then, substituting values for m and d, and canceling units gives us

$$v = \frac{10.92 \text{ g}}{7.86 \text{ g/mL}}$$

$$= 1.39 \text{ mL}$$

EXAMPLE 1.8

A sample of mercury has a volume of 5.0 mL. If the density of mercury is 13.6 g/mL, what is the mass of the mercury sample?

SOLUTION

First we must solve the density expression for mass:

$$d = \frac{m}{v}$$

$$m = dv$$

Then, substituting values for d and v, and canceling units gives us

$$m = (13.6 \text{ g/mL})(5.0 \text{ mL})$$

$$= 68 \text{ g}$$

EXERCISE 1.6 A sample of seawater containing 21.2 mL weighs 21.7 g. Calculate the density of the seawater.

EXERCISE 1.7 A sample of seawater has a density of 1.02 g/mL and a mass of 27.6 g. What is its volume?

EXERCISE 1.8 An aluminum object has a volume of 38 mL. If the density of aluminum is 2.70 g/mL, what is the mass of the aluminum object?

Specific Gravity

A quantity very closely related to density is *specific gravity*, defined as the density of a material relative to that of water at 4°C. Since this standard density of water is 1.0000 g/mL, specific gravity can be expressed as follows:

$$\text{specific gravity} = \frac{\text{density of substance in g/mL}}{\text{density of water at 4°C in g/mL}}$$

$$= \frac{\text{density of substance in } \cancel{g/mL}}{1.000 \ \cancel{g/mL}}$$

$$= \text{density of substance (no units)}$$

As you can see, for all practical purposes, specific gravity is a measurement which shows how many times more dense the substance is than water at 4°C. Note that specific gravity has no units.

EXAMPLE 1.9

What is the specific gravity of a urine sample whose density is 1.025 g/mL?

SOLUTION

$$\text{specific gravity} = \frac{1.025 \ \cancel{g/mL}}{1.0000 \ \cancel{g/mL}} = 1.025$$

EXERCISE 1.9 Determine the specific gravity of seawater, density 1.02 g/mL.

Energy

Energy is usually defined as the capacity to do work, to move an object over a distance. In this sense, gasoline contains energy because burning gasoline can cause an automobile to move. Animals possess energy because they are able to move themselves or their body parts. In both cases, work is performed.

In the broadest sense, energy can be divided into two classes: kinetic and potential. *Kinetic energy* is possessed by an object if it is moving. A moving car and a thrown baseball are examples of objects having kinetic energy; both have the capacity for doing work if they strike other objects. The car can displace a light post, a tree, or almost any other object in its path, and a misdirected throw of a baseball can move glass from a window pane. *Potential energy* is energy that is stored in an object because of the object's position, condition, or composition. A car parked on top of a hill has stored energy which represents a potential for doing work. If the car begins to roll downhill, some of its potential energy is transformed into kinetic energy. A tightly drawn bowstring has potential energy because of its condition; when released, the drawn bowstring imparts kinetic energy to an arrow. An example of potential energy due to composition is dynamite, a mixture of chemicals that explodes with great force when detonated. Its potential energy is transformed to kinetic energy as parts of its storage container fly apart.

Energy exists in many intangible yet familiar forms. It is a principle of nature called the *law of conservation of energy* that energy cannot be created or destroyed, but it can be changed from one form to another. The following forms of energy are the most well-known:

1. Heat energy is the form of energy that causes water to evaporate and that warms us in cold weather. It can be obtained from the sun and from burning wood, petroleum, and natural gas.

2. Mechanical energy is possessed by moving machinery and even by moving animal body parts.

3. Electrical energy is the electricity carried by wires from power plants to homes and factories and from a battery to an automobile ignition system.

4. Chemical energy is energy stored in a substance which can be released by chemical processes. Chemical energy stored in plants provides fuel for animals that eat the plants.

5. Light energy is the energy that radiates as light from the sun and from light bulbs. However, light that is invisible to humans, such as ultraviolet light, also possesses light energy.

joule (jool)

The SI unit of energy is the *joule* (J), but a more traditional unit, the *calorie* (cal), is very popular. A calorie is that quantity of energy that will raise the temperature of 1 g of water from 14.5°C to 15.5°C (C stands for the Celsius scale, to be explained shortly). For measuring larger quantities of energy, *kilocalories* (kcal) are used. This is the same unit as the Calorie used by nutritionists. Unit relationships for the various energy units are given in Table 1-1

Heat

exothermic: heat-releasing; from the Greek **exo-**, meaning "outside," and **therm**, meaning "heat."

endothermic: heat-absorbing; from the Greek **endo-**, meaning "within," and **therm**, meaning "heat."

Heat is a familiar form of energy that can be transferred between objects. If two objects of different temperatures are brought together, heat flows (energy is transferred) from the warmer object to the cooler one. The release of heat is called an *exothermic* process, and absorption of heat is an *endothermic* process.

Temperature

temperature: a measure of the warmth or coldness of an object.

Fahrenheit scale: named for Daniel Fahrenheit (1686-1736), the German physicist who invented the alcohol and mercury thermometers, and consequently, the Fahrenheit temperature scale.

Temperature is our way of measuring how hot an object is. An abnormally high body temperature is an indicator of infection in a human. We often measure body temperature in Fahrenheit degrees, the temperature scale traditionally used in medicine and meteorology. The *Fahrenheit* scale, as with other temperature scales, utilizes two reference points, the freezing point and the boiling point of pure water. In Fahrenheit these values are 32° and 212°, respectively.

The *Celsius* scale, formerly called the centigrade scale, is used for scientific work, and as the United States converts to the metric system, use of the Celsius scale will become widespread in this country. The Celsius scale sets the freezing point of pure water at 0° and the boiling point at 100°.

A comparison of temperature scales (Figure 1-6) shows that there are 180 Fahrenheit degrees between the freezing and boiling points of water but only 100 Celsius degrees for the interval between these reference points. The ratio 100/180, which reduces to 5/9, is important in converting measurements made on one temperature scale to readings on the other. It is also important to note that the interval of 0°–32° on the Fahrenheit scale is below zero on the Celsius scale. Thus, when we make temperature conversions, we must subtract 32 degrees from the Fahrenheit reading. If we first make the subtraction and then multiply the remainder by 5/9, to reduce the number of Fahrenheit degrees to the right number of Celsius degrees, we arrive at the following relationship:

$$°C = (°F-32)(5/9)$$

For conversion of Celsius temperature to Fahrenheit, we need only rearrange the equation as follows:

$$9/5 \, °C = °F - 32$$

$$32 + 9/5 \, °C = °F$$

Celsius scale: named for Anders Celsius (1701-1744), a Swedish astronomer who devised the temperature scale that bears his name.

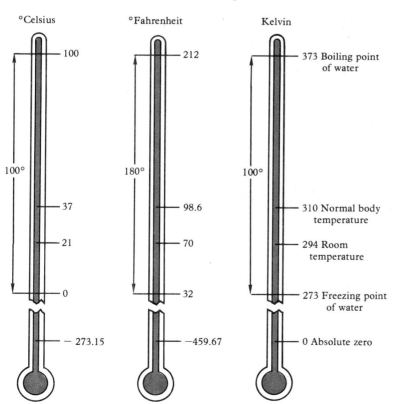

Fig. 1-6 Comparison of the Three Temperature Scales

This relationship is usually written as

$$°F = 9/5 \, °C + 32$$

We can keep the factors 5/9 and 9/5 straight by remembering that for our reference interval, there are more Fahrenheit degrees than Celsius degrees. Hence, when we want to convert to Celsius temperature, we must reduce the number of Fahrenheit degrees by multiplying by the factor 5/9. The opposite reasoning is true for conversion of Celsius to Fahrenheit. (You can practice converting from Celsius to Fahrenheit using the temperature in Figure 1-7.)

Fig. 1-7 (© 1980 by Sidney Harris/*Science* 80)

"*Let's go over to Celsius' place. I hear it's only 36° there.*"

Kelvin scale: named for Lord Kelvin (1824–1901), an English physicist and mathematician who developed the absolute temperature scale.

A third temperature scale, the *Kelvin* (or absolute) scale, is used in SI. This scale is derived from the theoretical behavior of gases at very low temperatures. The freezing point of water on the Kelvin scale is 273.15, and the boiling point is 373.15. Units on the Kelvin scale are called Kelvins; they have the same magnitude as Celsius degrees, but we do not use the degree sign when writing Kelvins. Kelvin and Celsius scales may be interconverted by the relationship

$$K = °C + 273$$

(Note that we have rounded off 273.15 to 273.)

EXAMPLE 1.10

Convert the body temperature of a healthy person, 98.6°F, to the Celsius temperature.

SOLUTION

$$\begin{aligned} °C &= 5/9 \, (°F - 32) \\ &= 5/9 \, (98.6 - 32) \\ &= 5/9 \, (66.6) \\ &= 37.0 \end{aligned}$$

(Since 32 is an exact number, we round off to three significant figures.)

EXAMPLE 1.11

Convert a reading of 69.0°C to Fahrenheit temperature.

SOLUTION

$$\begin{aligned} °F &= 9/5°C + 32 \\ &= 9/5 \, (69.0) + 32 \\ &= 124 + 32 \\ &= 156 \end{aligned}$$

EXAMPLE 1.12

Convert a reading of 34°C to Kelvin temperature.

SOLUTION

$$\begin{aligned} K &= °C + 273 \\ &= 34 + 273 \\ &= 307 \end{aligned}$$

EXERCISE 1.10
(a) 17.3°F to °C
(b) 25°C to K
(c) 43°C to °F
(d) 0°F to °C
(e) 0°C to K
(f) 78°F to K

Specific Heat

The addition of heat usually causes the temperature of a material to increase. However, each material has a different capacity for absorbing heat. Water must absorb 1.00 calorie per gram in order for its temperature to increase by 1°C. On the other hand, it takes only 0.031 calories to raise the temperature of 1 g of gold 1°C. Thus, water has a high heat capacity relative to gold. We should be grateful for this property of water because it allows us to maintain a rather constant body temperature. The term *specific heat* (short for specific heat capacity) is used to describe the amount of heat, measured in calories, required to raise the temperature of 1 g of a substance 1°C. Thus,

$$\text{specific heat} = \text{cal per g per °C}$$

$$\text{specific heat} = \frac{\text{cal}}{\text{g °C}}$$

Table 1-3 shows the specific heats of various solids and liquids.

Table 1-3 Specific Heats of Selected Materials (cal/g °C)

Solids		Liquids	
ice	0.492	water	1.00
aluminum	0.215	ethyl alcohol	0.581
iron	0.108	glycerine	0.540
copper	0.092	ethyl ether	0.529
silver	0.057	mineral oil	0.50
lead	0.038	olive oil	0.471
gold	0.031		

EXAMPLE 1.13

It takes 12,000 calories of heat to raise the temperature of a 1000-gram cast-iron pot from 25°C to 125°C. Calculate the specific heat of cast iron.

SOLUTION

$$\text{specific heat} = \frac{\text{cal}}{\text{g °C}}$$

$$= \frac{12000 \text{ cal}}{(1000 \text{ g})(100°C)}$$

$$= 0.120 \text{ cal/g °C}$$

EXAMPLE 1.14

Calculate the amount of heat in calories needed to increase the temperature of 2.0 g of water by 3.0°C.

SOLUTION

We can solve this problem using the known specific heat of water, 1.00 cal/g °C, but first we must solve the specific heat expression for cal, where °C represents the temperature change in Celsius degrees:

$$\text{specific heat} = \frac{\text{cal}}{\text{g °C}}$$

$$\text{cal} = \text{specific heat} \times \text{g} \times °C$$

Then, by substituting values for specific heat, g, and °C, we find that 6.0 cal are required:

$$cal = (1.00 \text{ cal/g °C})(2.0 \text{ g})(3.0°C)$$

$$cal = 6.0 \text{ cal}$$

EXERCISE 1.11 The specific heat of aluminum is 0.217 cal/g °C. Calculate the amount of heat in calories required to raise the temperature of 10.0 g of aluminum by 13°C.

The high specific heat of water accounts for the fact that small amounts of water absorb large amounts of heat. Organisms make use of this property of water to control their temperature and thus provide a favorable internal environment for life-sustaining chemical processes. Another consequence of the high specific heat of water is that large bodies of water take a long time for cooling. Hence the warm waters of the Gulf Stream keep much of western Europe warmer than the corresponding region of eastern North America. We shall return to the topic of specific heat in chapter 6.

SUMMARY

Alchemy and Chemistry: Since its beginnings in ancient alchemy, the science of chemistry has developed into the study of matter and its transformations. The development of chemistry was largely dependent on the ability of scientists to make accurate and precise measurements of such properties as mass, volume, and temperature.

Scientific Measurements: Two popular systems of measurements exist, the British system and the metric system. A standard metric system, SI, was recently adopted for international use. Most scientific measurements are made in the metric system. Conversions between or within measurement systems may be made by the unit-factor method.

Precision refers to the fineness of measurement and also to the reproducibility of measurement. Accuracy is an indication of the amount of error in a measurement. Significant figures are those digits in a reported measurement which have actual physical meaning. Scientific measurements must be made with accuracy and precision, and numerical values should be reported in proper significant figures.

The mass of an object is the amount of matter contained in the object. Mass is a constant value for any particular object, whereas weight varies with the distance of the object from the center of the earth. Weight is a consequence of the earth's gravitational attraction for matter; the farther an object is away from the center of the earth, the less will be its weight. Density is a property used to characterize matter. It is determined by dividing the mass of material by its volume. The density of a material relative to that of water at 4°C is called specific gravity. Specific gravity has no units.

Energy, the capacity to do work, is classified as kinetic or potential. Kinetic energy is possessed by an object if it is moving, while potential energy is energy stored in an object by virtue of the object's position, condition, or composition. Some familiar forms of energy are heat energy, mechanical energy, electrical energy, chemical energy, and light energy. Temperature is a measurement of how hot an object is. The three temperature scales are Fahrenheit, Celsius, and Kelvin. Temperature measurements may be converted from one scale to another by use of standard conversion formulas. Specific heat, the capacity of a material for absorbing heat, is expressed as the amount of heat (in calories) required to raise the temperature of one gram of material one degree Celsius.

STUDY QUESTIONS AND PROBLEMS

1. Define the following terms:
   (a) alchemy
   (b) matter
   (c) British system
   (d) metric system
   (e) SI
   (f) unit-factor method
   (g) precision
   (h) accuracy
   (i) significant figures
   (j) mass
   (k) weight
   (l) density
   (m) specific gravity
   (n) energy
   (o) kinetic energy
   (p) potential energy
   (q) heat
   (r) temperature
   (s) specific heat

2. Give the number of significant figures for each quantity.
   (a) 7.32×10^4
   (b) 0.0052
   (c) 8050
   (d) 603.42
   (e) 700.3
   (f) 10.09
   (g) 310
   (h) 2.000×10^{-5}
   (i) 0.325
   (j) 30.79

3. Round off each number to four significant figures.
   (a) 65.3785
   (b) 7.4892
   (c) 759,350
   (d) 0.00073428
   (e) 0.059763
   (f) 0.050631
   (g) 9.99976
   (h) 83.3333
   (i) 6.666667
   (j) 1,056,793

4. Perform the indicated operations and round off the answer to the correct number of significant figures.
   (a) $(216.5)(0.018) =$
   (b) $4.15/2.077 =$
   (c) $(8.1)(1.23)(0.47) =$
   (d) $75 - 0.76 - 1.49 =$
   (e) $[(19)(1.059)(75)]/[(6.2)(1.86)] =$
   (f) $39.0 + 42.6 + 1.39 =$
   (g) $[(1.008)(32.06)]/149.1 =$
   (h) $55.841 - 4.10 =$
   (i) $(67.9 - 12.73)/0.5152 =$

5. Make conversions as indicated.
   (a) 0.394 m to mm
   (b) 0.457 mL to L
   (c) 52.3 mm to cm
   (d) 84.92 g to kg
   (e) 0.059 g to ng
   (f) 7.6 g to mg
   (g) 20.6 m to mm
   (h) 14 mg to g
   (i) 72 kg to g
   (j) 3.0 L to cm^3

6. Make conversions as indicated.
   (a) 17.32 inches to m
   (b) 24.833 quarts to L
   (c) 39 cm to inches
   (d) 13.0 yards to m
   (e) 166 km to miles
   (f) 145 g to pounds
   (g) 250 mL to quarts
   (h) 2.63 m to feet
   (i) 12.6 inches to cm
   (j) 1.65 quarts to L

7. Distinguish between accuracy and precision.

8. A block of ice, temperature $-10°C$, has a volume of 1.00186 L and a mass of 1.000 kg. Calculate its density.

9. Calculate the volume occupied by a 20.0-gram sample of mercury, density 13.5939 g/mL.

10. A metal object has a volume of 3.0 mL and weighs 16.6 g. Calculate the density of the object.

11. What volume is occupied by 736 g of water, density 1.00 g/mL?

12. An alcohol has a density of 0.79 g/mL. What is the mass of 109 mL of the alcohol?

13. A sample of marble, specific gravity 2.70, has a volume of 7.4 mL. What is the mass of the marble?

14. Aluminum and molybdenum have specific gravities of 2.70 and 10.2, respectively. Which occupies the greater volume, 85 g of aluminum or 25 g of molybdenum?

15. Copper has a specific gravity of 8.96. What volume of copper has a mass of 74.3 g?

16. A small glass container weighs 3.764 g when empty. After 2.0 mL of urine are added to the container, it weighs 5.966 g. What is the specific gravity of the urine?

17. The specific gravity of air at $25°C$ is 11.843×10^{-4}. At this temperature, how many grams of air can the human lungs hold if the average lung capacity of an adult is 6,000 mL?

18. An average healthy person excretes 1.50 L of urine each day, corresponding to a mass of 1.530×10^3 g. What is the specific gravity of normal urine?

19. Classify each as having kinetic or potential energy.
    (a) water at the top of a waterfall
    (b) a person playing ping pong
    (c) a runner in a track race
    (d) a sugar cube
    (e) a gopher digging a hole
    (f) a stretched rubber band
    (g) a ripe apple in a tree

20. An air conditioner transfers heat from the cooler interior of a house to the warmer exterior. The direction of heat transfer is the reverse of what would occur spontaneously, and the air conditioner has to do work to bring about the reversal. What is the energy source for the work done by the air conditioner?

21. Make the following temperature conversions:
    (a) 98.6°F to K
    (b) 32°F to °C
    (c) 37°C to °F
    (d) 45°C to K
    (e) 78°F to °C
    (f) 0°F to °C
    (g) −21°C to °F
    (h) −150°F to °C
    (i) 220°F to °C
    (j) 600°C to °F

22. How much heat, in calories, is needed to raise the temperature of 100 mL of water from 25°C to 100°C?

23. The specific heat of an alcohol is 0.581 cal/g °C. How many calories are needed to raise the temperature of 100 g of the alcohol from 29°C to 78°C?

24. Freon is a gas used in refrigeration systems. It has a specific heat of 0.1297 cal/g °C. Calculate the amount of heat required to change the temperature of 400 g of freon from 0°C to 25°C.

25. A beaker contains 250 mL of water at a temperature of 78°F. The beaker is heated until the water temperature is 200°F. How many calories of heat were necessary to bring about the temperature change?

26. What is the mass of 5 lbs of tomatoes in kg?

27. A person weighs 115 lbs. What volume of mercury (density 13.5939 g/mL), in liters, would have the same mass?

28. A person has a fever of 102.3°F. By how many degrees is his temperature elevated on the Celsius scale? (Normal body temperature is 98.6°F.)

29. In some parts of the U.S., gasoline is measured in liters. How many gallons of gasoline are in 12 liters?

30. How many calories must be removed from 50 g of water to lower its temperature from 78°F to 32°F?

31. A drug is to be administered by intravenous injection to a patient who weighs 200 lb. The recommended dosage is 50 mg per kg of body weight. The drug is supplied in the form of a solution containing 20 mg of drug per mL. How many mL of the solution will be required for one dose?

32. For one particular brand of cigarettes, each cigarette contains 40 mg of tar. If a person smokes 20 of the cigarettes each day, how many grams of tar does the person inhale each week?

33. An aspirin tablet contains 5 grains of aspirin. If 15 grains equal 1 gram, how many grams of aspirin are in a dose of 2 tablets?

34. If the average density of bone is 1.8 g/mL, what mass of bone is needed for a bone graft requiring 4 cm^3 of bone tissue?

CHAPTER 2
ELEMENTS AND ATOMS

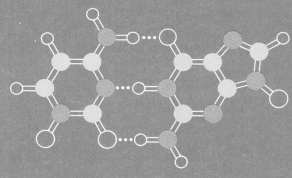

THE ESSENCE OF CHEMISTRY
- Elements
- Atoms
- The Periodic Table

ATOMIC STRUCTURE
- Subatomic Particles
- Atomic Number
- Mass Number
- Isotopes
- Atomic Mass Units
- Atomic Weight
- Electron Configuration
- Chemical Periodicity
- The Periodic Table Revisited
- Elements and Life
- Properties of Some Elements Important to Life

RADIOACTIVITY
- Alpha Decay
- Beta Decay
- Gamma Decay
- Nuclear Reactions

Elements are the fundamental substances of which all matter is composed. Robert Boyle first recognized the nature of a chemical element in 1661, and by 1789 the first list of elements (Figure 2-1) was published. As the idea of elements developed, scientists were also trying to understand the composition of matter. The English chemist John Dalton was the first to explain laws of chemical combination. In doing so, he postulated the existence of atoms with unique sets of properties. Today Dalton's atomic hypothesis is universally accepted, although modern scientists have made a few modifications. In order to understand the roles of atoms and elements in the science of chemistry, we will study three major topics in this chapter, each subdivided as indicated in the topic outline.

THE ESSENCE OF CHEMISTRY

Elements

Ancient Greek philosophy held that all matter is composed of four elements: air, earth, fire, and water (see Figure 2-2). This concept was accepted for almost two thousand years, but the science of chemistry prevailed, and by the end of the eighteenth century, 23 elements were known. Scientists now identify *elements* as pure substances which cannot be separated into simpler substances by ordinary processes. What the ancient Greeks thought to be elements are actually mixtures or chemical combinations of simpler substances, the true elements. For example, air is composed primarily of oxygen and nitrogen, both elements in the true sense. Earth is composed of far too many elements to list here, while water is a chemical compound formed by the elements hydrogen and oxygen. Fire is a chemical reaction between oxygen and

Fig. 2-1 The First List of Elements (Reproduction from A.L. Lavoisier, *Traite elementaire de chimie,* 1789)

Noms nouveaux.	Noms anciens correspondans.
Substances simples qui appartiennent aux trois règnes & qu'on peut regarder comme les élémens des corps.	
Lumière.........	Lumière.
Calorique.........	Chaleur. Principe de la chaleur. Fluide igné. Feu. Matière du feu & de la chaleur.
Oxygène.........	Air déphlogistiqué. Air empiréal. Air vital. Base de l'air vital.
Azote...........	Gaz phlogistiqué. Mofete. Base de la mofete.
Hydrogène.......	Gaz inflammable. Base du gaz inflammable.
Substances simples non métalliques oxidables & acidifiables.	
Soufre..........	Soufre.
Phosphore.......	Phosphore.
Carbone.........	Charbon pur.
Radical muriatique.	Inconnu.
Radical fluorique .	Inconnu.
Radical boracique,.	Inconnu.
Substances simples métalliques oxidables & acidifiables.	
Antimoine........	Antimoine.
Argent..........	Argent.
Arsenic.........	Arsenic.
Bismuth.........	Bismuth.
Cobolt..........	Cobolt.
Cuivre..........	Cuivre.
Etain...........	Etain.
Fer............	Fer.
Manganèse.......	Manganèse.
Mercure.........	Mercure.
Molybdène.......	Molybdène.
Nickel..........	Nickel.
Or.............	Or.
Platine..........	Platine.
Plomb..........	Plomb.
Tungstène.......	Tungstène.
Zinc...........	Zinc.
Substances simples salifiables terreuses.	
Chaux..........	Terre calcaire, chaux.
Magnésie........	Magnésie, base du sel d'Epsom.
Baryte..........	Barote, terre pesante.
Alumine.........	Argile, terre de l'alun, base de l'alun.
Silice...........	Terre siliceuse, terre vitrifiable.

LAVOISIER'S LIST OF THE ELEMENTS (1789).

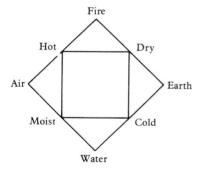

Fig. 2-2 The Four Greek Elements

compound: a pure substance composed of two or more elements and having constant composition.

other elements. Other examples of elements are the metals gold, silver, tin, and lead, and nonmetals sulfur, neon, and phosphorus. A total of 106 elements have been identified to date.

Elements can combine with one another to form a wide variety of more complex substances called *compounds*. The number of compounds is almost infinite; millions are known, and more are being discovered or created each day. When two or more elements combine to form a compound, they lose their separate identities, and the compound has characteristics quite different from those of the constituent elements. A familiar example of a compound is water. Two gaseous elements, hydrogen and oxygen, combine to form water.

$$\text{hydrogen} + \text{oxygen} \longrightarrow \text{water}$$

Since water can be decomposed into its constituent elements, we know it is not an element. Hydrogen and oxygen, however, are elements because they cannot be decomposed into simpler substances by ordinary processes.

Most naturally occuring substances are mixtures of compounds. Seawater, for example, is a mixture of water and a large number of other compounds, the most common of which is sodium chloride, or table salt. *Mixtures* differ from pure substances in that they can be separated into their component parts by physical processes. For example, the simple process of evaporation separates water from other compounds in seawater. In fact, table salt can be obtained in this way.

mixture: a collection of two or more substances which are not chemically united and which do not exist in fixed proportion to each other.

Atoms

As scientists began to understand the concept of elements in the seventeenth and eighteenth centuries, they became more concerned with matter and its properties. Democritus, a Greek philosopher, put forth a theory of discontinuous matter in the fifth century B.C. His theory was

Democritus (deh-MOCK-rih-tus)

PERSPECTIVE: Origin of the Concept of Element

Robert Boyle

Antoine Laurent Lavoisier

In the seventeenth century, a philosopher and chemist by the name of Robert Boyle (1627-1691) recognized the fundamental nature of elements. Boyle came from a wealthy Irish family, and as a young man, he spent several years touring Europe with a tutor. He then settled in England where he did scientific work and wrote moral essays, one of which is said to have inspired Jonathan Swift to write *Gulliver's Travels*. Boyle suffered from poor health and often concocted various medicines using recipes from miscellaneous sources for himself and his friends. A quiet, modest man, Boyle preferred to spend his time in scholarly work, particularly applications of science to metallurgy, medicine, and the manufacture of chemicals, dyes, and glass.

In the eighteenth century another chemist, Antoine Laurent Lavoisier (1743-1794), became known for his study of the elements. Lavoisier received an excellent education in France, where he studied mathematics, astronomy, chemistry, and botany. By 1789 he had learned enough to put forth the first list of elements (Figure 2-1). He included in his list lime, alumina, and silica, which are now known to be very stable compounds but at the time seemed indestructible. Because of his role as a tax collector, Lavoisier was put to death at the guillotine after the French Revolution. The judge who pronounced sentence commented that the Republic had no need for scientists. Lavoisier made many important contributions to the development of chemistry, and he is regarded as the father of modern chemistry.

Aristotle (AIR-is-tot-ul)

hypothesis (high-PAH-theh-sis): a proposition set forth as an explanation for a set of observations.

atomic hypothesis: the proposition that the smallest representative unit of an element is an atom.

that matter is divisible only up to a point, the point at which certain basic units are reached. Another Greek philosopher, Aristotle, considered matter to be continuous, a concept in which matter was believed to be infinitely divisible, each part exhibiting identical properties regardless of size. Aristotle's theory, more popular than that of Democritus, generally prevailed until experiments in the sixteenth century forced a return to the discontinuous theory.

Democritus used the word *atom* to refer to fundamental bits of matter which he believed to be the ultimate carriers of the properties associated with bulk matter. Numerous experiments, including work by Dalton, have demonstrated the validity of Democritus' concept, and we now call it the *atomic hypothesis*.

Atoms are so small that it is impossible to see them with ordinary microscopes. Scientists have developed an indirect technique for photographing atoms, but even these pictures (Figure 2-3, for example) are not detailed enough to allow us to distinguish the internal structure of atoms. However, other evidence suggests that atoms are highly complex structures composed of even more fundamental particles. Within each atom is a small nucleus which contains protons and neutrons. The nucleus occupies only a negligible portion of the total volume of an atom; it is surrounded by very small particles called electrons. The space in which electrons move about the nucleus accounts for most of the volume of the atom.

Fig. 2-3 Images of Uranium Atoms Photographed Against a Dark Background (Courtesy of Dr. Albert Crewe)

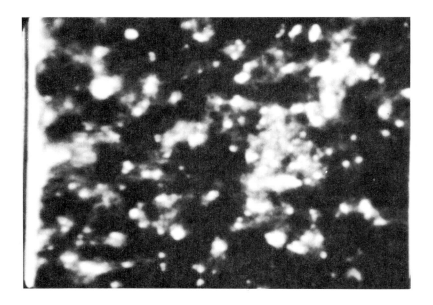

I A	II A	III B	IV B	V B	VI B	VII B	VIII B			I B	II B	III A	IV A	V A	VI A	VII A	VIII A
1 H 1.0079																1 H 1.0079	2 He 4.00260
3 Li 6.941	4 Be 9.01218											5 B 10.81	6 C 12.011	7 N 14.0067	8 O 15.9994	9 F 18.998403	10 Ne 20.179
11 Na 22.98977	12 Mg 24.305											13 Al 26.98154	14 Si 28.0855	15 P 30.97376	16 S 32.06	17 Cl 35.453	18 Ar 39.948
19 K 39.0983	20 Ca 40.08	21 Sc 44.9559	22 Ti 47.90	23 V 50.9415	24 Cr 51.996	25 Mn 54.9380	26 Fe 55.847	27 Co 58.9332	28 Ni 58.70	29 Cu 63.546	30 Zn 65.38	31 Ga 69.72	32 Ge 72.59	33 As 74.9216	34 Se 78.96	35 Br 79.904	36 Kr 83.80
37 Rb 85.4678	38 Sr 87.62	39 Y 88.9059	40 Zr 91.22	41 Nb 92.9064	42 Mo 95.94	43 Tc (98)	44 Ru 101.07	45 Rh 102.9055	46 Pd 106.4	47 Ag 107.868	48 Cd 112.41	49 In 114.82	50 Sn 118.69	51 Sb 121.75	52 Te 127.60	53 I 126.9045	54 Xe 131.30
55 Cs 132.9054	56 Ba 137.33	57 *La 138.9055	72 Hf 178.49	73 Ta 180.9479	74 W 183.85	75 Re 136.207	76 Os 190.2	77 Ir 192.22	78 Pt 195.09	79 Au 196.9665	80 Hg 200.59	81 Tl 204.37	82 Pb 207.2	83 Bi 208.9804	84 Po (209)	85 At (210)	86 Rn (222)
87 Fr (223)	88 Ra 226.0254	89 'Ac 227.0278	104 § (261)	105 § (262)	106 § (263)												

Metals ← | → Nonmetals

§ The International Union for Pure and Applied Chemistry has not adopted official names or symbols for these elements.

Note: Atomic masses shown here are 1977 IUPAC values.

★ Lathanoid Series

58 Ce 140.12	59 Pr 140.9077	60 Nd 144.24	61 Pm (145)	62 Sm 150.4	63 Eu 151.96	64 Gd 157.25	65 Tb 158.9254	66 Dy 162.50	67 Ho 164.9304	68 Er 167.26	69 Tm 168.9342	70 Yb 173.04	71 Lu 174.967

' Actinoid Series

90 Th 232.0381	91 Pa 231.0359	92 U 238.029	93 Np 237.0482	94 Pu (244)	95 Am (243)	96 Cm (247)	97 Bk (247)	98 Cf (251)	99 Es (252)	100 Fm (257)	101 Md (258)	102 No (259)	103 Lr (260)

Fig. 2-4 The Periodic Table

The Periodic Table

As the science of chemistry developed, scholars made several attempts to classify the elements on the basis of physical features. It was not until the latter part of the nineteenth century that our current classification scheme was developed. This remarkably simple arrangement is called the *periodic table* (Figure 2-4).

Notice that the elements are arranged in horizontal rows, called *periods,* which are layered one on top of another so that vertical columns exist as well. These vertical columns are called *groups* or *families*. The format of the periodic table allows us to recognize elements having similar properties easily, because those elements in any group (vertical column) behave very much alike. Thus, as we examine a list of elements, we find similar elements recurring periodically in the list (see Figure 2-5). This predictable occurrence of similar elements is the basis for the name periodic table.

Elements are represented in the periodic table by *symbols*, used as a kind of convenient shorthand notation. Early alchemists first used symbols such as those in Figure 2-6 to represent elements, and the practice was handed down to modern chemists.

Most names for elements are derived from the Latin and Greek languages, and the chemical symbols are abbreviations of these names. For example, the name carbon is derived from the Latin word *carbo*, meaning "coal;" the symbol for carbon is C. A few symbols taken from Latin names do not resemble English names, such as K for potassium and Pb for lead. The more recently discovered elements have been given pseudo-Latin names such as Berkelium (Bk), Californium (Cf), and Einsteinium (Es) to indicate places of discovery or to honor famous scientists. Some elements have a single capital letter as their symbols (for example, H, C,

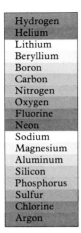

Fig. 2-5 A Partial List of the Elements; Shading Indicates Elements with Similar Properties

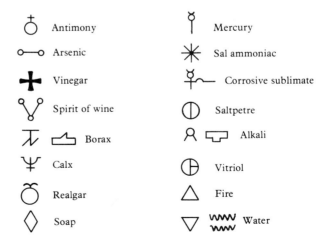

Fig. 2-6 Alchemical Symbols

O, N), while others have two-letter symbols. In those cases, only the first of the two letters is capitalized. A complete list of elements and their symbols is given inside the back cover of this book.

ATOMIC STRUCTURE

Subatomic Particles

An atom is composed of a compact *nucleus* surrounded by moving electrons. The nucleus itself seems to contain perhaps a dozen different kinds of particles, but the most important ones are protons and neutrons. A *proton*, symbolized as p, is a particle of matter having a positive electrical charge and a mass of 1.673×10^{-24} grams. *Neutrons*, symbolized as n, have a similar mass, 1.675×10^{-24} grams, but they have no electrical charge. This electrical neutrality is the basis for the name neutron. Protons and neutrons are packed tightly together to form the dense core of an atom.

atomic nucleus: the dense core of an atom which contains neutrons and protons.

Atoms do not have an overall electrical charge. This means that the positive charges of protons are balanced by exactly the same number of negative charges elsewhere in the atom. These negative charges are possessed by *electrons*, symbolized as e^-, which move continuously about the nucleus. Thus, each atom has exactly the same number of electrons as protons. Electrons, with a mass of 9.110×10^{-28} grams, are much smaller than protons and neutrons. Figure 2-7 shows simplified atomic structures for several elements; the electrons are represented by dots surrounding each nucleus.

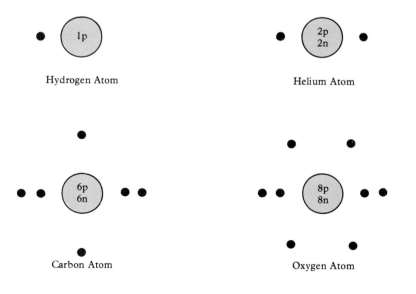

Fig. 2-7 Simplified Representations of Atoms

Atomic Number

atomic number: the number of protons in an atom and also the number assigned to each element in the periodic table.

The basis for ordering elements in the periodic table is the number of protons in each atom. All atoms of any element have the same number of protons; this number is the *atomic number* of the element. Elements appear in the periodic table in order of increasing atomic number. Since protons cannot exist in fractional amounts, atomic numbers are always whole, exact quantities.

Mass Number

Another quantity useful for characterizing an atom is its *mass number*, defined as the sum of protons and neutrons in its nucleus:

$$\text{mass number} = \text{number of protons} + \text{number of neutrons}$$

or,

$$\text{mass number} = \text{atomic number} + \text{number of neutrons}$$

The number of neutrons is the difference between the mass number and the atomic number:

$$\text{number of neutrons} = \text{mass number} - \text{atomic number}$$

EXAMPLE 2.1

A carbon atom has six protons and six neutrons. Find its mass number.

SOLUTION

$$\text{mass number} = \text{number of protons} + \text{number of neutrons}$$
$$= 6 + 6$$
$$= 12$$

EXAMPLE 2.2

An atom having a mass number of 32 has 16 neutrons. How many protons does the atom have? Identify the element.

SOLUTION

$$\text{mass number} = \text{number of protons} + \text{number of neutrons}$$
$$\text{number of protons} = \text{mass number} - \text{number of neutrons}$$
$$= 32 - 16$$
$$= 16$$

Since the number of protons is the atomic number, this is an atom of element 16, sulfur.

EXERCISE 2.1 Each entry below describes the nucleus of an atom. Identify each element.

(a) mass number 16; 8 neutrons
(b) 17 protons and 18 neutrons
(c) mass number 23; 12 neutrons
(d) 13 protons and 14 neutrons
(e) mass number 20; 10 neutrons

Isotopes

As mentioned earlier, all atoms of an element have the same number of protons. This is the atomic number which identifies each element. But atoms of an element can have different numbers of neutrons and thus different mass numbers. Hydrogen atoms, for example, may have 0, 1, or 2 neutrons. Each of these types of hydrogen atoms is an *isotope*. Isotopes are forms of the same element having different mass numbers. The hydrogen isotope having 0 neutrons, called *protium*, has a mass number of 1; the one having 1 neutron, called *deuterium*, has a mass number of 2. The third isotope of hydrogen, *tritium*, has a mass number of 3 and is radioactive. All three isotopes are hydrogen atoms because each has one proton. In nature, protium is the most abundant form and comprises 99.9% of all hydrogen. Deuterium exists to the extent of about 0.1%, and tritium is found only in trace amounts in our environment. All elements have isotopes, and a total of about fourteen hundred are known to exist. For a given element, all isotopes show the same chemical behavior.

isotope (EYE-sah-tope)

protium (PRO-tih-um)
deuterium (doo-TEH-rih-um)
tritium (TRIH-tih-um)

Atomic Mass Units

The masses of subatomic particles are so exceedingly small that they have very little physical meaning to us. In addition, use of exponential notation for expressing masses of subatomic particles is very cumbersome. For convenience, scientists define an *atomic mass unit* (amu) as 1/12 the mass of the carbon isotope with mass number 12. This corresponds to a mass of 1.660×10^{-24} grams for an atomic mass unit. Atomic mass units are used to express masses of subatomic particles as follows:

$$\text{mass of proton} = 1.673 \times 10^{-24} \text{ g} = 1.007 \text{ amu}$$
$$\text{mass of neutron} = 1.675 \times 10^{-24} \text{ g} = 1.009 \text{ amu}$$
$$\text{mass of electron} = 9.110 \times 10^{-28} \text{ g} = 5.486 \times 10^{-4} \text{ amu}$$

By expressing the masses in amu, we have eliminated the need for exponents in the masses of protons and neutrons. In fact, we often round off their masses and say that both protons and neutrons have masses of 1 amu. In the case of electrons, the mass is an exponential quantity even when expressed in amu, and it is so small compared to the mass of the proton or neutron that it is usually ignored altogether. Table 2-1 summarizes the properties of the proton, neutron, and electron.

Table 2-1 Subatomic Particles

Particle	Symbol	Mass (amu)	Charge
proton	p	1.007	+1
neutron	n	1.009	0
electron	e^-	0	+1

Atomic Weight

The elements are listed in the periodic table by atomic number. However, a second number also appears for each element, and it is called the *atomic weight*. The atomic weight is the average mass of all isotopes of the element as they occur in nature. Atomic weights are expressed in atomic mass units.

Let us examine how the atomic weight of carbon is calculated. Carbon exists primarily as a mixture of two isotopes. One has a mass number of 12 (carbon-12) and makes up 98.89% of all carbon found in nature, and the isotope with mass number 13 (carbon-13) accounts for 1.108% of natural carbon. Thus, of every 100.00 carbon atoms, 98.89 have a mass of 12 amu and 1.108 have a mass of 13 amu. The atomic weight of carbon is calculated as follows:

$$99.89 \text{ atoms} \times 12 \text{ amu/atom} = 1187 \text{ amu}$$
$$1.108 \text{ atoms} \times 13 \text{ amu/atom} = 14 \text{ amu}$$

Total mass of 100.00 atoms = 1201 amu

$$\text{Average mass per atom} = \frac{1201 \text{ amu}}{100.00 \text{ amu}} = 12.01 \text{ amu}$$

EXERCISE 2.2 As it occurs in nature, boron contains 80.30% boron-11 and 19.70% boron-10. Calculate the atomic weight of boron.

Electron Configuration

Unlike protons and neutrons, which reside quietly in the nucleus of an atom, electrons move about constantly at high rates of speed. It is not possible to pinpoint the exact position of an electron at any one time, but modern theory can calculate the probability of finding an electron in a certain region of space. That region of space near an atomic nucleus where an electron is most likely to be found is called an *atomic orbital*.

Electrons have negative charges and thus are attracted to the positively charged nucleus of an atom. The fact that electrons remain outside the nucleus is a consequence of their movement. Since electrons are constantly moving, they possess energy of motion (kinetic energy). Thus a certain amount of energy is associated with each atomic orbital. Electrons move about in orbitals at various energy levels; the higher the energy level, the greater the energy of the electron, and the farther away the electron is likely to be found from the nucleus.

Atomic orbitals are of various sizes and shapes, but we will be concerned primarily with just two types: *s*- and *p*-orbitals (see Figure 2-8). (The origins of the *s* and *p* designations are historical and need not concern us.) An *s*-orbital is shaped like a sphere, but with a rather indefinite outer boundary. The boundary is not well defined because it represents only the region where an electron is most likely to be found. A *p*-orbital is shaped like a dumbbell, and an electron moving within it is likely to be found in either lobe of the dumbbell with equal probabil-

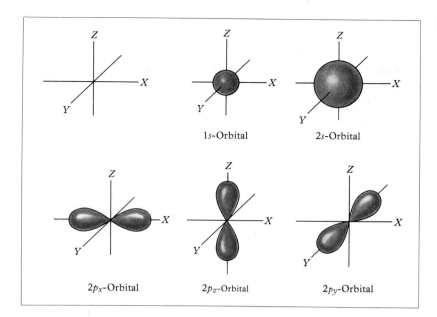

Fig. 2-8 s- and p-Orbitals

ity. The boundaries of a *p*-orbital are also rather indefinite. An electron in an *s*-orbital is likely to be anywhere around the nucleus (at an average distance corresponding to the *s*-orbital radius), but electrons in *p*-orbitals are found only along three mutually perpendicular axes. Thus, *p*-orbitals exist in sets of three, one for the *x*-axis (p_x), one for the *y*-axis (p_y), and one for the *z*-axis (p_z), as shown in Figure 2-8.

Now let us consider *electron configurations* by examining one atom at a time in the periodic table. The simplest atom, hydrogen, has only one electron, and it occupies the lowest principal energy level. The term *principal energy level* is used to indicate a range of energy within which there may be several sublevels slightly different from each other in energy content. Only an *s*-orbital may be located in the lowest principal energy level, also called the first principal energy level. The notation 1*s* is used for this orbital. To indicate the presence of one electron in the 1*s*-orbital, we write $1s^1$; this is the electron configuration for hydrogen.

The next atom, helium, has one more proton than hydrogen, and thus it also has one more electron, giving a total of two protons and two electrons for helium. The spinning motion of each electron creates a tiny magnetic field, and the two electrons spin in opposite directions. Because of their opposite spins, they are similar to two opposite magnetic poles (see Figure 2-9), and they are consequently slightly attracted to each other in spite of their identical electrical charges. This small magnetic attraction allows them to occupy the same atomic orbital. We write the electron configuration for helium as $1s^2$, where the superscript 2 indicates that two electrons are in the 1*s*-orbital.

Next we come to lithium, atomic number 3. As you would expect, two lithium electrons occupy the 1*s*-orbital, but atomic theory maintains that an orbital can hold a maximum of two electrons, and these electrons

electron configuration: the designation of electron distribution in atomic orbitals in an atom.

principal energy level: a region of space located within an atom, occupied by one or more atomic orbitals having similar energy content.

Fig. 2-9 Magnetic Fields of Spinning Electrons

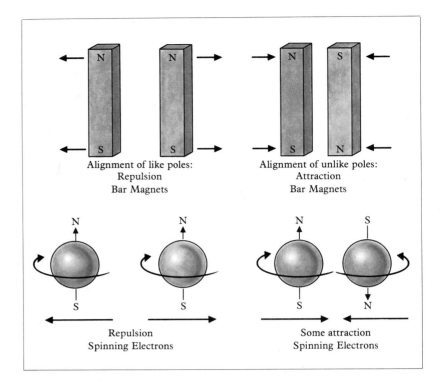

must have opposite spins. Now we are faced with finding another orbital for the third electron of lithium. Since only the *s*-orbital is found in the first principal energy level, we must look to the second principal energy level for another orbital, even though it will be farther away from the nucleus and higher in energy content. The second principal energy level contains an *s*-orbital and three *p*-orbitals. However, the *s*-orbital is just slightly lower in energy content than the *p*-orbitals; this is usually true for *s*- and *p*-orbitals within the same principal energy level. Hence, lithium's third electron occupies the *s*-orbital in the second principal energy level. We write the configuration as $1s^2 2s^1$. Electron configurations for the first 36 elements are shown in Table 2-2.

As you will notice in Table 2-2, element 4, beryllium, has two electrons at the 2*s* level. Continuing on to elements 5, 6, and 7, you can see that electrons exist singly in each of the three 2*p*-orbitals. The following diagrams illustrate these configurations. By using small arrows to indicate electron spin direction, we can easily tell which electrons are "paired."

Boron, atomic number 5

$$\frac{\downarrow\uparrow}{2s} \quad \frac{\downarrow}{2p_x} \quad \frac{}{2p_y} \quad \frac{}{2p_z}$$

$$\frac{\downarrow\uparrow}{1s}$$

$$1s^2 2s^2 2p^1$$

ATOMIC STRUCTURE   39

Table 2-2 Electron Configurations of the First Thirty-Six Elements

Atomic Number	Symbol	Electronic Configuration				
1	H	$1s^1$				
2	He	$1s^2$				
3	Li	$1s^2\ 2s^1$				
4	Be	$1s^2\ 2s^2$				
5	B	$1s^2\ 2s^2\ 2p^1$				
6	C	$1s^2\ 2s^2\ 2p^2$				
7	N	$1s^2\ 2s^2\ 2p^3$				
8	O	$1s^2\ 2s^2\ 2p^4$				
9	F	$1s^2\ 2s^2\ 2p^5$				
10	Ne	$1s^2\ 2s^2\ 2p^6$				
11	Na	$1s^2\ 2s^2\ 2p^6$	$3s^1$			
12	Mg	$1s^2\ 2s^2\ 2p^6$	$3s^2$			
13	Al	$1s^2\ 2s^2\ 2p^6$	$3s^2$	$3p^1$		
14	Si	$1s^2\ 2s^2\ 2p^6$	$3s^2$	$3p^2$		
15	P	$1s^2\ 2s^2\ 2p^6$	$3s^2$	$3p^3$		
16	S	$1s^2\ 2s^2\ 2p^6$	$3s^2$	$3p^4$		
17	Cl	$1s^2\ 2s^2\ 2p^6$	$3s^2$	$3p^5$		
18	Ar	$1s^2\ 2s^2\ 2p^6$	$3s^2$	$3p^6$		
19	K	$1s^2\ 2s^2\ 2p^6$	$3s^2$	$3p^6$		$4s^1$
20	Ca	$1s^2\ 2s^2\ 2p^6$	$3s^2$	$3p^6$		$4s^2$
21	Sc	$1s^2\ 2s^2\ 2p^6$	$3s^2$	$3p^6\ 3d^1$		$4s^2$
22	Ti	$1s^2\ 2s^2\ 2p^6$	$3s^2$	$3p^6\ 3d^2$		$4s^2$
23	V	$1s^2\ 2s^2\ 2p^6$	$3s^2$	$3p^6\ 3d^3$		$4s^2$
24	Cr	$1s^2\ 2s^2\ 2p^6$	$3s^2$	$3p^6\ 3d^5$		$4s^1$
25	Mn	$1s^2\ 2s^2\ 2p^6$	$3s^2$	$3p^6\ 3d^5$		$4s^2$
26	Fe	$1s^2\ 2s^2\ 2p^6$	$3s^2$	$3p^6\ 3d^6$		$4s^2$
27	Co	$1s^2\ 2s^2\ 2p^6$	$3s^2$	$3p^6\ 3d^7$		$4s^2$
28	Ni	$1s^2\ 2s^2\ 2p^6$	$3s^2$	$3p^6\ 3d^8$		$4s^2$
29	Cu	$1s^2\ 2s^2\ 2p^6$	$3s^2$	$3p^6\ 3d^{10}$		$4s^1$
30	Zn	$1s^2\ 2s^2\ 2p^6$	$3s^2$	$3p^6\ 3d^{10}$		$4s^2$
31	Ga	$1s^2\ 2s^2\ 2p^6$	$3s^2$	$3p^6\ 3d^{10}\ 4s^2$		$4p^1$
32	Ge	$1s^2\ 2s^2\ 2p^6$	$3s^2$	$3p^6\ 3d^{10}\ 4s^2$		$4p^2$
33	As	$1s^2\ 2s^2\ 2p^6$	$3s^2$	$3p^6\ 3d^{10}\ 4s^2$		$4p^3$
34	Se	$1s^2\ 2s^2\ 2p^6$	$3s^2$	$3p^6\ 3d^{10}\ 4s^2$		$4p^4$
35	Br	$1s^2\ 2s^2\ 2p^6$	$3s^2$	$3p^6\ 3d^{10}\ 4s^2$		$4p^5$
36	Kr	$1s^2\ 2s^2\ 2p^6$	$3s^2$	$3p^6\ 3d^{10}\ 4s^2$		$4p^6$

Transition Metals Period 4 (Z = 21–30)

Carbon, atomic number 6

$$\frac{\uparrow\downarrow}{2s} \quad \frac{\downarrow}{2p_x} \quad \frac{\downarrow}{2p_y} \quad \frac{}{2p_z}$$

$$\frac{\uparrow\downarrow}{1s}$$

$1s^2 2s^2 2p^2$

Nitrogen, atomic number 7

$$\frac{\uparrow\downarrow}{2s} \quad \frac{\downarrow}{2p_x} \quad \frac{\downarrow}{2p_y} \quad \frac{\downarrow}{2p_z}$$

$$\frac{\uparrow\downarrow}{1s}$$

$1s^2 2s^2 2p^3$

Note that the three 2p-orbitals are shown separately in the diagrams. However, all p-orbitals in any given principal energy level have equal amounts of energy associated with them, and we do not distinguish among them in writing electron configurations.

Despite the small attractions due to opposite spins, electrons also have a certain amount of repulsion for each other because of their like charges. Since this repulsion is rather strong, they tend to exist singly in atomic orbitals whenever possible. Thus, boron, carbon, and nitrogen have only unpaired p-electrons. But in the case of oxygen, element 8, two electrons pair in a 2p-orbital, because the only alternative arrangement would be for one of those electrons to occupy the next higher energy level. When faced with this high-cost alternative, the electrons pair at the 2p level; of course, this coexistence in the same orbital is facilitated by their opposite spins.

$$\text{Oxygen, atomic number 8}$$

$$\underline{\downarrow\uparrow}\quad \underline{\downarrow\uparrow}\quad \underline{\downarrow}\quad \underline{\downarrow}$$
$$2s\quad 2p_x\quad 2p_y\quad 2p_z$$

$$\underline{\downarrow\uparrow}$$
$$1s$$

$$1s^2 2s^2 2p^4$$

As we move progressively through atoms in the periodic table, more and more electrons appear in the 2p-orbitals until we reach element 10, neon. At this point, both the first and second principal energy levels are full, and the next element, sodium (atomic number 11), begins filling the third principal energy level. As can be seen in Table 2-2, the third level has three types of orbitals: s, p, and d. However, the 4s-orbital (in the fourth principal energy level) is of lower energy content than the 3d-orbitals, so the 4s-orbital fills before the 3d-orbitals. Thus, potassium (atomic number 19) and calcium (atomic number 20) have one and two 4s-electrons, respectively, but no 3d-electrons. Since the 3d-orbitals, of which there are five, can only be occupied after the 4s-orbital is filled, scandium (atomic number 21) is the first element to have d-electrons.

A fourth type of orbital, f, appears in the fourth principal energy level. Although we will not be concerned with capacities of f-orbitals, notice that the number of each principal energy level (1, 2, 3, and so on) is also the number of orbital types in the level.

It is important that you practice writing electron configurations for energy levels containing s- and p-orbitals because these will form the basis for understanding material in later chapters. The following steps can be used to determine electron configurations in a systematic way:

1. Use the atomic number to obtain the number of electrons in the atom.

2. Remember the following orbital locations:

First principal energy level	1s
Second principal energy level	2s, 2p
Third principal energy level	3s, 3p, 3d

3. Remember the following electron capacities:

   First principal energy level: 2 electrons
   　　$1s$ holds 2 electrons

   Second principal energy level: 8 electrons
   　　$2s$ holds 2 electrons
   　　$2p$ holds 6 electrons

   Third principal energy level: 18 electrons
   　　$3s$ holds 2 electrons
   　　$3p$ holds 6 electrons
   　　$3d$ holds 10 electrons

4. Use small arrows to assign electrons in energy level diagrams. Distribute electrons one at a time to orbitals in the same principal energy level, starting at the first principal energy level and proceeding in order of increasing energy levels. Pair electrons before making assignments at higher principal energy levels. Remember that p-orbitals are slightly higher in energy content than s-orbitals at the same principal energy level; therefore, p-orbitals will generally not start to fill until the s-orbitals are full.

5. Write the electron configuration in proper notation.

EXAMPLE 2.3

Write the electron configuration for magnesium, atomic number 12.

SOLUTION

Since the atomic number of magnesium is 12, there are 12 electrons.

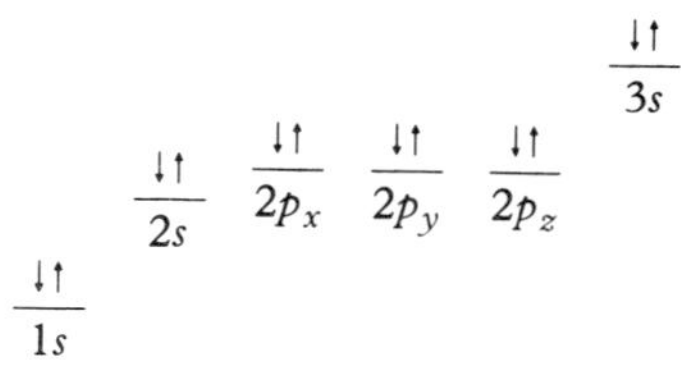

$$\begin{array}{cccccc}
 & & & & & \underline{\uparrow\downarrow} \\
 & & & & & 3s \\
 & & \underline{\uparrow\downarrow} & \underline{\uparrow\downarrow} & \underline{\uparrow\downarrow} & \\
 & \underline{\uparrow\downarrow} & 2p_x & 2p_y & 2p_z & \\
 & 2s & & & & \\
\underline{\uparrow\downarrow} & & & & & \\
1s & & & & &
\end{array}$$

The electron configuration is $1s^2 2s^2 2p^6 3s^2$

EXERCISE 2.3 Write the electron configuration for phosphorus, atomic number 15.

Chemical Periodicity

Perhaps you have wondered what constitutes a period of elements in the periodic table. What determines the end of one period of elements and the beginning of another? If you examine electron configurations, you will find that a new period begins when a new principal energy level contains an electron. Thus, each element in Group IA has one electron in its outermost energy level. Immediately preceding each element in Group IA is one from Group VIIIA with eight electrons in its outermost energy level. As will be explained later, the presence of eight electrons in outer energy levels correlates with lowered chemical reactivity, a property illustrated by the Group VIIIA elements. In contrast, a single electron in the outermost energy level causes high chemical reactivity, a distinguishing feature of Group IA elements.

As you scan the periodic table, you will see that all elements in any chemical group (family) have the same number of electrons in their outermost energy levels. In fact, the group number signifies the number of outer electrons for each member. These outer electrons are called *valence electrons*; they are highly significant in the ability of one element to combine chemically with another. Because members of a chemical group possess the same number of valence electrons, they are very much alike in their chemical properties.

Members of a chemical family also exhibit similarity in physical properties. For example, Figure 2-10 illustrates how boiling points of

valence electrons: electrons occupying the highest principal energy level in an atom.

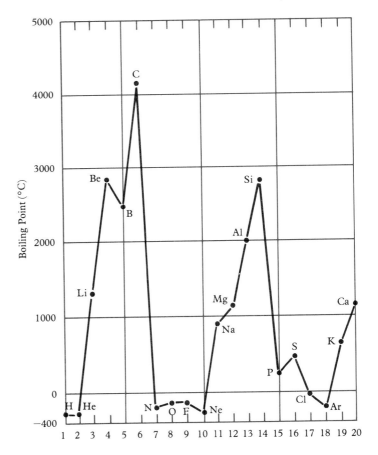

Fig. 2-10 The Boiling Points of the Elements as a Function of Atomic Number

elements vary with atomic number. Notice the regular intervals between peaks and troughs on the graph. The periodic recurrence of similar physical and chemical properties as a function of atomic number is called the *periodic law*.

Another periodic property is atomic size. We can think of atoms as being small spheres. Because of the small size, atomic radii are usually given in units of angstroms (1 Å = 10^{-10} m). Sizes of atomic radii are given for a number of atoms in Figure 2-11. Notice that atomic size decreases from left to right across a period but increases from top to bottom going down a group. The decrease in size across a period occurs because the same outer energy level is being filled with electrons, but each successive nucleus has an additional proton. The increased positive charge draws the valence electrons closer and closer to the nucleus, causing the atoms to get smaller within a period. In contrast, atomic radii increase going down a family because outermost electrons occupy successively higher energy levels. Even the greater attraction for electrons by the increased number of protons in going from one family member to the next does not offset the size effect of additional energy levels.

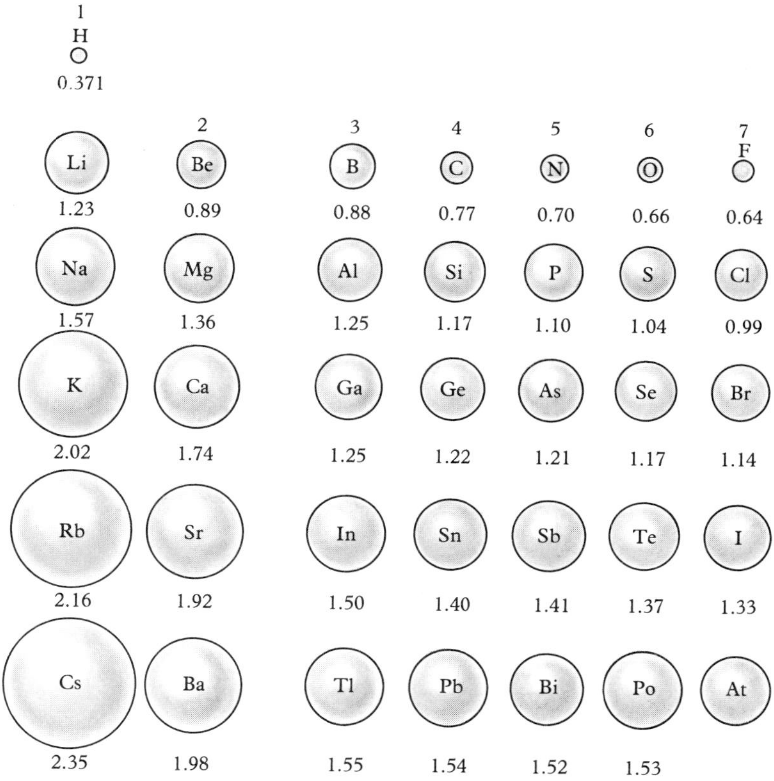

Fig. 2-11 Atomic Radii, in Angstroms

	1	2	3	4	5	6	7
	H						
	0.371						
	Li	Be	B	C	N	O	F
	1.23	0.89	0.88	0.77	0.70	0.66	0.64
	Na	Mg	Al	Si	P	S	Cl
	1.57	1.36	1.25	1.17	1.10	1.04	0.99
	K	Ca	Ga	Ge	As	Se	Br
	2.02	1.74	1.25	1.22	1.21	1.17	1.14
	Rb	Sr	In	Sn	Sb	Te	I
	2.16	1.92	1.50	1.40	1.41	1.37	1.33
	Cs	Ba	Tl	Pb	Bi	Po	At
	2.35	1.98	1.55	1.54	1.52	1.53	

The Periodic Table Revisited

Our newly acquired knowledge of electron configurations allows us to divide the periodic table into sections according to type of orbital in which outer electrons exist. Figure 2-12 summarizes this analysis.

Groups IA and IIA contain the *active metals*. In general, metals have hard, lustrous surfaces, can be pounded into sheets and drawn into wires, and are good conductors of heat and electricity. The active metals are chemically reactive because they have only one or two valence electrons. These occupy *s*-orbitals.

The *transition elements* are those whose *d*-orbitals are incomplete and being filled. In the periodic table, transition elements are usually designated by the letter B following the group number. The transition elements are metals, and as a group, they are relatively unreactive. For example, the coinage metals (copper, silver, and gold) can be melted and molded to form coins, and they can be drawn out into wires. Their low reactivity accounts for their suitability as durable coins and wires.

The *representative elements* are those in which the outermost *p*-orbitals are being filled. Half of these elements are metals, while the other half are nonmetals. The zigzag line separates the two types of elements, with

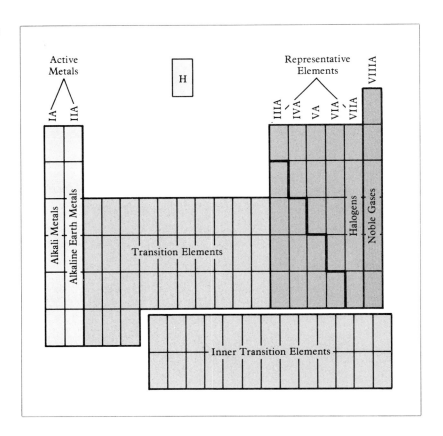

Fig. 2-12 A Block Diagram of the Periodic Table

metals to the left and nonmetals to the right. In general, nonmetals tend to crumble easily, if they are solids, and they are poor conductors of heat and electricity. Elements bordering the zigzag line possess some properties of metals and some of nonmetals. For this reason they are sometimes called *metalloids* (meaning "metal-like").

The *noble gases* (also called rare gases and inert gases) constitute another section of the periodic table. Until 1962, these nonmetals were thought to be incapable of forming chemical compounds. In fact, they came to be called the "noble" gases because of their aloofness from other elements. However, in 1962 two compounds of xenon were prepared, and since then a number of xenon compounds have been synthesized, as well as some compounds of krypton and radon. The outer orbitals for the noble gases are s- and p-orbitals (except for He, which has only an s outer orbital), and these are filled with electrons.

Since the combined electron capacity of s- and p-orbitals in any one principal energy level past the first is eight, the noble gases (except for He) have an octet of valence electrons. Because the noble gases are rather unreactive, a statement called the *octet rule* has been formulated. It says that an atom is least reactive when it contains eight electrons in its highest occupied principal energy level. An exception to this rule is helium, which can have a maximum of only two electrons in its outer level. Thus, helium satisfies the octet rule with two valence electrons.

The final section of the periodic table is composed of two series of elements for which f-orbitals are being filled. These elements are called the *inner transition elements*.

Four chemical families have names which refer to the entire group. Group IA, the *alkali metals*, derives its name from the Arabic word *al-qili*, which means "plant ashes"—a traditional source for obtaining these elements on a small scale. The alkali metals (or the alkalies) are found in greatest quantity as dissolved salts in the ocean; sodium (Na) and potassium (K) are the sixth and seventh most abundant elements, respectively, in the earth's crust. These two metals are also present in physiological fluids as dissolved salts.

Group IIA elements are called the *alkaline earth metals* because they resemble the alkalies and because, in chemical combination with other elements, they fit the old-fashioned definition of "earths"—substances insoluble in water and unchanged by fire. Two members of this group, magnesium and calcium, are the basis for bone structures in the human body.

The family name for Group VIIA, the *halogens*, comes from the fact that they are found in seawater in combination with other elements, frequently the alkalies. Thus the group name comes from the Greek word *hals*, meaning "from the sea." A compound of iodine, the element necessary for proper function of the thyroid gland, is an ingredient of iodized salt found in grocery stores.

Lastly, there is Group VIIIA, the *noble gases*, as described earlier. Neon, a member of this group, is familiar for its use in the orange red electric lights used for advertising.

The sources and uses of some familiar elements are presented in Table 2-3.

Table 2-3 Some Well Known Elements

Element	Source	Uses
Hydrogen	Various compounds containing hydrogen	Manufacture of ammonia; conversion of oils to margarine and shortening; rocket fuel
Helium	Underground gas deposits	Inert component of welding gases; weather balloons; cryogenics
Boron	Underground deposits of ore (borax)	Hardening agent in steel; component of semiconductors, glass, and washing powders
Carbon	Underground deposits (diamond, graphite, coal)	Jewels (diamonds); lubricant (graphite); coloring agent (carbon black)
Nitrogen	Earth's atmosphere	Manufacture of ammonia; liquid coolant
Oxygen	Earth's atmosphere	Metallurgical processes; manufacture of chemicals; medical respirators
Neon	Earth's atmosphere	Electric signs
Aluminum	Underground deposits of ore (bauxite)	Aircraft construction; building materials; containers and wrapping materials
Sulfur	Underground deposits	Processing dried fruits; manufacture of gunpowder and rubber; fertilizers
Chromium	Underground deposits (chromite)	Manufacture of alloys (chrome, stainless steel)
Iron	Underground deposits of iron ore	Construction of all types; machines and parts; tools; stainless steel
Copper	Underground deposits	Electrical wires; manufacture of alloys (brass, bronze); coins
Silver	Underground deposits	Silver bullion and coins; silverware; jewelry; photographic industry
Tin	Natural deposits of tin oxide	Plating steel cans; metal bearings; alloys (soft solder, pewter, bronze)
Platinum	Underground deposits	Electrical apparatus; jewelry; dental alloys
Gold	Underground deposits	Gold bullion; jewelry; electrical devices
Mercury	Natural deposits of mercury sulfide (cinnebar)	Thermometers; barometers; electrical switches; dental fillings
Lead	Natural deposits of lead sulfide	Roofing materials; underwater coverings; linings for water pipes; storage batteries; ammunition; radiation shielding

Elements and Life

Recent space explorations have revealed exciting information about the natural abundance of elements in the solar system. Table 2-4 summarizes this information. As you can see, hydrogen is by far the most abundant element. It is thought to be the basic material from which all other elements in the universe are synthesized.

In contrast to its abundance in the solar system, hydrogen is present in only small amounts on earth. It and helium, the second most abundant element in the solar system, are very light gases. Even though they may have been present in large amounts when the earth was formed, they have since escaped into space. Table 2-5 shows that iron is the

Table 2-4 The Most Abundant Elements in the Solar System

Element	Relative Abundance (% of Total Atoms)
Hydrogen	91
Helium	9
Oxygen, Nitrogen, Carbon, Neon, Iron, Silicon, Magnesium, Sulfur, Nickel, Aluminum, Calcium	0.1

Table 2-5 Elemental Composition of the Earth and Its Atmosphere (Percent by Weight)

Element	%
Iron	35.4
Oxygen	27.8
Magnesium	17.0
Silicon	12.6
Sulfur	2.7
Nickel	2.7
Calcium	0.6
Aluminum	0.4
Cobalt	0.2
Sodium	0.1
Manganese	0.1
Potassium	0.1
Titanium, Phosphorus, Chromium, Hydrogen, Chlorine, Nitrogen, Argon	0.3

most plentiful element on earth, followed by oxygen, magnesium, and silicon. Most of the iron is in the hot, molten core of the earth, while magnesium is found in various minerals in the earth's crust. Silicon and oxygen are especially plentiful in the crust and on the surface of the earth in silicate compounds such as sand. Oxygen is also present as an uncombined element in the earth's atmosphere.

At least 24 elements are essential for life, as illustrated in Table 2-6. The four most abundant elements found in all living organisms are hydrogen (63%), oxygen (25.5%), carbon (9.5%), and nitrogen (1.4%). Twenty-one other elements account for the remaining 0.6%.

Hydrogen, oxygen, carbon, and nitrogen are also among the most abundant elements in the human body, as shown in Table 2-7. These elements are found in the principal biological compounds—fats, carbohydrates, and proteins. Hydrogen and oxygen are also the constituent elements of water, as mentioned earlier.

Properties of Some Elements Important to Life

Hydrogen is a colorless, odorless, tasteless gas which is highly flammable. As mentioned above, hydrogen is very light, and it was once used in blimps; however, this practice was ended in 1937 when the Hindenburg dirigible caught fire and crashed. Since then helium has been used for such purposes because of its nonflammability. Hydrogen is used to

Table 2-6 Elements Essential for Life

Element	Percent of Mass in Living Organisms
Hydrogen	63.0
Oxygen	25.5
Carbon	9.5
Nitrogen	1.4
Fluorine	
Sodium	
Magnesium	
Silicon	
Phosphorus	
Sulfur	
Chlorine	
Potassium	
Calcium	
Vanadium	
Chromium	0.6
Manganese	
Iron	
Cobalt	
Nickel	
Copper	
Zinc	
Selenium	
Molybdenum	
Tin	
Iodine	

make industrial chemicals and to convert vegetable oils to margarine and shortening.

Oxygen is also a colorless, odorless, and tasteless gas. It is the reaction of flammable materials with oxygen that causes fire. Thus, although oxygen is not said to be flammable (because it does not react with itself), its presence in large amounts poses a fire hazard. It is necessary to eliminate flames and sparks when oxygen is in use, because even a slowly burning cigarette will flare up suddenly into an intense flame in the presence of pure oxygen.

An adult uses approximately five hundred liters of oxygen each day in respiration; the gas is used in the body to combine with food substances to provide energy. In cases of respiratory illness, it is sometimes necessary to use respirators, which increase oxygen content from the normal 20% in the air to as much as 60%.

Carbon is found naturally as diamond and graphite, forms of the pure element. Thus, it can exist as both hard, colorless, clear crystals and as a black, slippery, solid material. Because it forms more compounds than any other element, an entire branch of chemistry—organic chemistry—is devoted to the study of carbon compounds.

Nitrogen, another colorless, odorless, tasteless gas, is the most plentiful component in our atmosphere; it makes up 78% of the air we breathe.

Table 2-7 Elemental Composition of the Human Body

Element	Percent of Total Body Mass
Oxygen	65.0
Carbon	18.0
Hydrogen	10.0
Nitrogen	3.0
Calcium	2.0
Phosphorus	1.0
Potassium	0.35
Sulfur	0.25
Chlorine	0.15
Sodium	0.15
Magnesium	0.05
Iron	0.004
Molybdenum	
Manganese	
Aluminum	
Iodine	
Copper	
Nickel	
Fluorine	
Zinc	0.046
Arsenic	
Bromine	
Tin	
Selenium	
Rubidium	
Cobalt	
Others	

Although it is ordinarily not very reactive chemically, it forms compounds at high temperatures with many other elements.

RADIOACTIVITY

radioactivity: the spontaneous emission of particles and/or energy or the capture of electrons by unstable atomic nuclei.

Radioactivity was discovered by a French physicist, Henri Becquerel, in 1896. Since then scientists have found that radioactivity is caused by transformations occurring within the nuclei of atoms. When radioactive rays are given off by an atom, the atomic nucleus usually changes into the nucleus of another element. Indeed, this process of radioactive change might seem to fulfill the dreams of alchemists who searched for the secret of transmutation, but unfortunately the products of radioactivity are usually not gold. However, a number of radioactive isotopes do have important medical uses (see Table 2-8).

All isotopes of elements beyond bismuth, atomic number 83, are radioactive. For most of these, the atomic weights shown in parentheses in the periodic table are the mass numbers for the most stable isotope. In addition to elements 84–106, some elements with smaller atomic numbers have radioactive isotopes (radioisotopes), such as hydrogen-3 (tritium), carbon-14, and cobalt-60. (The numbers 3, 14, and 60 are the mass numbers of the radioisotopes.) Although our understanding of the causes of radioactivity is still developing, we do know that radioactivity is a way for unstable nuclei to transform themselves into more stable nuclei. They do so by emitting particles and/or energy or by capturing electrons. The ultimate result of this nuclear transformation, called *radioactive decay*, is conversion of the radioisotope to a nonradioactive isotope.

radioactive decay: the change in the nucleus of an atom resulting from radioactivity.

Since all isotopes of atomic number greater than 83 are radioactive, scientists have studied the nuclear composition of these elements in an attempt to discover causes of radioactivity. An important observation is that atoms of low atomic number have approximately the same number of neutrons as protons in the nucleus; that is, the *neutron/proton ratio* in most of these cases is very close to 1. On the other hand, atoms of high

Table 2-8 Radioisotopes in Medicine

Radioisotope	Use
Radium-226	Destruction of tumors
Cobalt-60	Destruction of tumors
Phosphorus-32	Treatment of leukemia
Iodine-131	Treatment of overactive thyroid gland
Technetium-99m	Various organ scans
Xenon-133	Lung scans
Copper-64	Diagnosis of Wilson's disease (copper storage disease)
Selenium-75	Pancreas scan
Barium-131	Detection of bone tumors

atomic number (greater than 83) have neutron/proton ratios approaching 1.5. In other words, radioisotopes tend to have 1.5 times as many neutrons as protons. It is thought that the role of neutrons is to stabilize the nucleus by helping to insulate the positively charged protons so that they do not repel each other strongly enough to leave the nucleus. Apparently, however, when there are as many as 84 or more protons in a nucleus, even a greater number of neutrons is not sufficient to overcome proton-proton repulsions, and nuclear decay occurs. If a particular nucleus is unstable, it will undergo radioactive decay to form a new nucleus. If the product nucleus is stable, no further change will occur; but if the product nucleus is also unstable, additional decay will take place until a stable nucleus is formed. The three major types of nuclear decay are summarized in Table 2-9 and discussed in the following sections.

Alpha Decay

When nuclear decay was first discovered, three types were found. They were named alpha (α), beta (β), and gamma (γ), for the first three letters of the Greek alphabet. Alpha decay occurs when unstable isotopes attempt to stabilize themselves by emitting what is called an *alpha particle*. This particle contains two protons and two neutrons. It has an atomic number of 2 and a mass number of 4, and it is identical to a helium nucleus. Its symbol is $^{4}_{2}\alpha$, where the subscript is the atomic number and the superscript is the mass number. Emission of an alpha particle reduces the atomic number of the unstable nucleus by 2 and the mass number by 4. The natural process of alpha decay was first recognized in 1902 when Sir Ernest Rutherford observed the transformation of radium-226 to radon-222. An equation can be written for this reaction as follows:

$$^{226}_{88}Ra \longrightarrow ^{4}_{2}\alpha + ^{222}_{86}Rn$$

Note that the sum of atomic numbers on the left side of the arrow is equal to that on the right side. The same is true for mass numbers. These equalities reflect the fact that the total number of protons and neutrons in the products of the nuclear process is the same as the total number of protons and neutrons in the starting material.

Table 2-9 The Three Types of Radioactive Decay

Type of Decay	Nuclear Change	Particle Mass (amu)	Particle Charge	Particle Description
Alpha	Emission of $^{4}_{2}\alpha$	4	+2	Helium nucleus
Beta	Emission of $^{0}_{-1}\beta$	0	−1	Electron
	Emission of $^{0}_{1}\beta$	0	+1	Positron
	Capture of $^{0}_{-1}\beta$	0	−1	Electron
Gamma	Emission of γ-rays	no particle	no particle	High-energy radiation

Beta Decay

There are three types of *beta decay*, which will be discussed in the following paragraphs. All involve either electrons or a similar particle called a positron.

Electron Emission

One form of beta decay results when electrons are emitted by the nucleus. In a nuclear reaction, electrons are called *beta particles*, symbolized as $_{-1}^{0}\beta$. As you recall, there are no electrons in atomic nuclei, and the question arises, how can a nucleus emit an electron if it does not have one? The answer is that the electron is created at the instant of emission by conversion of a neutron to an electron and a proton. Thus, emission of a beta particle increases the atomic number of the unstable nucleus by 1 but has no effect on its mass number. In the following example, carbon-14 is converted to nitrogen-14 by electron emission. Notice once again that the sum of the atomic numbers and the sum of the mass numbers are the same for both sides of the equation.

$$^{14}_{6}C \longrightarrow {}^{0}_{-1}\beta + {}^{14}_{7}N$$

Positron Emission

A *positron* is identical to an electron except that the positron has a positive charge. Positrons are created by conversion of a proton to a neutron and a positron, symbolized as $_{1}^{0}\beta$. Positron emission reduces the atomic number of the unstable nucleus by 1 but has no effect on its mass number. One such transformation occurs when antimony-116 is converted to tin-116:

$$^{116}_{51}Sb \longrightarrow {}^{0}_{1}\beta + {}^{116}_{50}Sn$$

Electron Capture

Radioisotopes are unstable because of an excessive number of protons in the nucleus. The number of protons can be reduced by a process in which the nucleus captures an electron from outside. Such electron capture results in a decrease in atomic number with no effect on mass number. In other words, entry of an electron into the nucleus changes a proton into a neutron. An example of electron capture is the decay of radioactive gold to platinum:

$$^{195}_{79}Au + {}^{0}_{-1}\beta \longrightarrow {}^{195}_{78}Pt$$

Gamma Decay

gamma ray: a highly penetrating form of energy emitted by an unstable nucleus.

Frequently α or β decay produces unstable products which then release energy in the form of *gamma* (γ) *rays*. In most cases, gamma rays are given off only when other types of radiation occur. Gamma decay causes no change in atomic number or mass number, but other accompanying types of decay have their own effects on these quantities. Gamma emission accompanies the decay of cobalt-60, a radioisotope used in radiation treatment of cancer:

$$^{60}_{27}\text{Co} \longrightarrow {}^{60}_{28}\text{Ni} + {}^{0}_{-1}\beta + \gamma$$

Nuclear Reactions

It is possible to predict the atomic product of radioactive decay if the type of decay is known. For example, tritium decays by giving off one β particle from each atom. Thus we can write

$$^{3}_{1}\text{H} \longrightarrow {}^{0}_{-1}\beta + {}^{3}_{2}\text{?}$$

The product with atomic number 2 corresponds to the helium isotope of mass number 3. The completed equation is then

$$^{3}_{1}\text{H} \longrightarrow {}^{0}_{-1}\beta + {}^{3}_{2}\text{He}$$

EXAMPLE 2.4

Potassium-40 emits a beta particle and gamma radiation. Write a complete equation for the decay.

SOLUTION

By looking up the atomic number for potassium, we can write the following partial equation:

$$^{40}_{19}\text{K} \longrightarrow {}^{0}_{-1}\beta + \gamma + \text{?}$$

PERSPECTIVE: Madame Curie

Marie Curie in Her Laboratory

Marie Sklodowski Curie (1867–1934) was a pioneer in the study of radioactivity. She was a brilliant high school student in her native Poland, and by working as a governess for several years, she saved enough money to begin studies at the Sorbonne, the famous Paris university. Marie often studied late into the night in her garret in the students' quarter where she lived on a diet of mostly bread, butter, and tea. Her academic devotion was rewarded two years later with a master's degree in physics and, after another year, a master's degree in mathematics.

Near the end of those studies, Marie met a young French physicist, Pierre Curie, and they were married in 1895. As Marie searched for a research project for her doctorate, she became intrigued with Henri Becquerel's newly discovered "radioactivity," a term originated by Marie. Pierre joined her in the study of radioactive minerals, and in 1898 they announced their discovery of two new elements, polonium and radium. In 1903, the Curies shared the Nobel Prize with Becquerel for investigations of radioactivity. Marie was awarded her doctorate that same year.

One year after the joint Nobel Prize, in a tragic accident, Pierre Curie was run down and killed by a horse-drawn wagon on the streets of Paris. Marie Curie then devoted all of her energy to completing the scientific work they had started. In 1911 she was awarded a second Nobel Prize, an unprecedented achievement, for her earlier discovery of polonium and radium.

Since the sum of atomic numbers on the right side of the equation must be 19, the same as the left side, then ? has an atomic number of 20 ($-1 + 20 = 19$). Similarly, we deduce the mass number of ? to be 40. Since the element with atomic number 19 is calcium, the complete equation is

$$^{40}_{19}K \longrightarrow\ ^{0}_{-1}\beta + \gamma +\ ^{40}_{20}Ca$$

EXERCISE 2.4 Write a complete equation for each nuclear process.
(a) positron emission by nitrogen-13
(b) alpha decay by polonium-198

If the type of decay and the atomic product are known, the starting material can be deduced. For example, the element polonium, discovered by Madame Curie and named for her native Poland, is produced by alpha decay. The partial equation is

$$^{222}_{86}? \longrightarrow\ ^{4}_{2}\alpha +\ ^{218}_{84}Po$$

Since the atomic number of the starting material is 86, the element must be radon, and the complete equation is

$$^{222}_{86}Rn \longrightarrow\ ^{4}_{2}\alpha +\ ^{218}_{84}Po$$

EXAMPLE 2.5

What radioisotope produces indium-109 by positron emission?

SOLUTION

By looking up the atomic number for indium, we can write the following partial equation:

$$? \longrightarrow\ ^{0}_{1}\beta +\ ^{109}_{49}In$$

Since the atomic number of ? must be the sum of 1 and 49, then ? has an atomic number of 50. Similarly, the atomic mass must be 109. Since the element with atomic number 50 is tin, the complete equation is

$$^{109}_{50}Sn \longrightarrow\ ^{0}_{1}\beta +\ ^{109}_{49}In$$

EXERCISE 2.5 Write a complete equation for each nuclear process.
(a) alpha decay to form osmium-186
(b) electron emission to form bismuth-210

It is also possible to predict the type of radioactive decay if starting material and product are identified. Radioactive nitrogen-13 decays to produce carbon-13. A partial equation for the transformation is

$$^{13}_{7}N \longrightarrow\ ^{0}_{1}? +\ ^{13}_{6}C$$

This reaction occurs by emission of a particle of atomic number 1 and mass number 0, corresponding to a positron. The final equation is therefore

$$^{13}_{7}N \longrightarrow\ ^{0}_{1}\beta +\ ^{13}_{6}C$$

EXAMPLE 2.6

Uranium-238, the predominant radioisotope in uranium ore, decays to produce thorium-234. What type of radiation is produced?

SOLUTION

The partial equation

$$^{238}_{92}U \longrightarrow ? + {}^{234}_{90}Th$$

indicates that the nuclear particle produced has atomic number 2 and mass number 4. This corresponds to an alpha particle, and the complete equation is

$$^{238}_{92}U \longrightarrow {}^{4}_{2}\alpha + {}^{234}_{90}Th$$

EXERCISE 2.6 Write a complete equation for each nuclear process.
(a) formation of rhodium-99 from palladium-99
(b) formation of osmium-188 from platinum-192

The principles of radioactivity discussed in this section are the basis for numerous applications in medicine and production of nuclear power. These applications will be explored in chapter 25.

SUMMARY

The Essence of Chemistry: Elements, pure substances which cannot be decomposed into simpler substances, combine with one another to form compounds. A total of 106 elements are now known. Elements are made of atoms, the smallest units of an element to possess properties of the bulk element. Atoms contain a dense nucleus of protons and neutrons surrounded by moving electrons. Elements are arranged in the periodic table in a format which allows us to recognize easily those with similar properties.

Atomic Structure: The nucleus of an atom contains neutrons and protons. These have electrical charges of 0 and +1, respectively. Atoms do not have overall electrical charges because their protons are balanced by exactly the same number of electrons circulating about the nucleus. Each electron has a charge of −1. The basis for ordering elements in the periodic table is the number of protons in each atom, the atomic number. Another quantity useful for characterizing an atom is its mass number, the sum of neutrons and protons in its nucleus. Atoms having the same atomic number but different mass numbers are called isotopes. The average mass of all isotopes of an element as they occur in nature is called the atomic weight of the element. Atomic weights are expressed in atomic mass units. Electron configuration, the distribution of electrons in atomic orbitals, is the basis for chemical reactivity and chemical periodicity. The octet rule states that an atom is least reactive when it contains eight electrons in its highest principal energy level.

Radioactivity: Radioactivity occurs when an unstable atomic nucleus attempts to stabilize itself by emitting particles and/or energy or by capturing an electron. Nuclear instability results from proton-proton repulsions; all elements with atomic numbers greater than 83 are radioactive as well as some isotopes of lower atomic number. The three kinds of radiation are alpha, beta, and gamma. Alpha decay occurs when a nucleus emits a particle of atomic number 2 and mass number 4. Beta decay occurs as a nuclear emission of an electron or a positron, or as capture of an electron by the nucleus. Gamma decay arises when energy is released from a nucleus. Equations may be written to describe the three types of radiation processes.

STUDY QUESTIONS AND PROBLEMS

1. Define the following terms:
   - (a) element
   - (b) compound
   - (c) mixture
   - (d) atom
   - (e) periodic table
   - (f) chemical period
   - (g) chemical group
   - (h) chemical symbol
   - (i) proton
   - (j) neutron
   - (k) electron
   - (l) atomic number
   - (m) mass number
   - (n) isotope
   - (o) atomic mass unit
   - (p) atomic weight
   - (q) atomic orbital
   - (r) electron configuration
   - (s) principal energy level
   - (t) valence electron
   - (u) periodic law
   - (v) metals
   - (w) nonmetals
   - (x) active metals
   - (y) transition elements
   - (z) representative elements
   - (aa) metalloids
   - (bb) noble gases
   - (cc) octet rule
   - (dd) alkali metals
   - (ee) alkaline earth metals
   - (ff) halogens
   - (gg) radioactivity
   - (hh) radioactive decay
   - (ii) alpha particle
   - (jj) beta particle
   - (kk) positron
   - (ll) gamma rays

2. How are mixtures different from pure substances?

3. How does the theory of discontinuous matter differ from the theory of continuous matter? Which theory is currently accepted?

4. In the periodic table, what is the difference between periods and groups?

5. List the three important subatomic particles with their charges and masses in amu.

6. Where are the three important subatomic particles found in atoms?

7. How is atomic number different from mass number?

8. Each of the following describes the nucleus of an atom. Identify each element.
   - (a) mass number 56; 30 neutrons
   - (b) 56 protons and 81 neutrons
   - (c) mass number 197; 118 neutrons
   - (d) 15 protons and 16 neutrons
   - (e) mass number 84; 48 neutrons
   - (f) mass number 201; 121 neutrons
   - (g) mass number 238; 146 neutrons
   - (h) 17 protons and 18 neutrons
   - (i) mass number 207; 125 neutrons
   - (j) 47 protons and 61 neutrons

9. Give the number of protons, neutrons, and electrons in each atom.
   - (a) boron-11
   - (b) carbon-13
   - (c) nitrogen-14
   - (d) oxygen-16
   - (e) sodium-23
   - (f) aluminum-27
   - (g) phosphorus-30
   - (h) sulfur-32
   - (i) chlorine-35
   - (j) argon-40

10. Naturally occurring neon, the gas used in brightly colored electric signs, is composed of 90.92% neon-20, 0.26% neon-21, and 8.82% neon-22. Calculate the atomic weight of neon.

11. Magnesium, a lightweight metal used in aircraft construction, occurs naturally as 78.60% magnesium-24, 10.11% magnesium-25, and 11.29% magnesium-26. Calculate the atomic weight of magnesium.

12. Sulfur, the yellow, solid nonmetal, occurs naturally as 95.02% sulfur-32, 0.76% sulfur-33, and 4.22% sulfur-34. Calculate the atomic weight of sulfur.

13. Iron, the strong, heavy metal, occurs naturally as 5.84% iron-54, 91.68% iron-56, 2.17% iron-57, and 0.31% iron-58. Calculate the atomic weight of iron.

14. Draw an energy level diagram and write an electron configuration for each of the following atoms:
    - (a) fluorine
    - (b) argon
    - (c) magnesium
    - (d) silicon
    - (e) sodium
    - (f) chlorine
    - (g) aluminum
    - (h) phosphorus
    - (i) neon
    - (j) sulfur

15. What information does the group number in the periodic table convey regarding electronic structure of atoms in the group?

16. How many valence electrons are present in members of the following chemical groups?
    - (a) Group IA
    - (b) Group VIIIA
    - (c) Group IIIA
    - (d) Group IVA
    - (e) Group IIA
    - (f) Group VA
    - (g) Group VIIA
    - (h) Group VIA

17. Are boiling points of elements a periodic property? Explain your answer.

18. How does atomic size change within a chemical period? Why does this trend occur?

19. How does atomic size change within a chemical group? Why does this trend occur?

20. Arrange members of each set in order of decreasing atomic size (largest one first and smallest one last).

    (a) C, Li, B
    (b) F, Li, N
    (c) Sb, I, Rb
    (d) Si, Na, Cl
    (e) Tl, Bi, Ba
    (f) K, Se, Ga
    (g) Sb, Sr, Te
    (h) P, Na, Al
    (i) Po, Ba, Pb
    (j) N, O, Be

21. Arrange members of each set in order of decreasing atomic size (largest one first and smallest one last).

    (a) O, Se, Po
    (b) Al, In, B
    (c) Li, Cs, K
    (d) I, F, Cl
    (e) Mg, Sr, Ca
    (f) Fr, Na, Rb
    (g) Tl, B, Ga
    (h) As, Sb, N
    (i) Te, Se, O
    (j) Ga, In, Tl

22. Classify each of the following elements as a metal or a nonmetal.

    (a) argon
    (b) molybdenum
    (c) calcium
    (d) carbon
    (e) nickel
    (f) bromine
    (g) potassium
    (h) chlorine
    (i) sodium
    (j) oxygen

23. Give three ways in which metals differ from nonmetals.

24. What is the major difference in electron configuration between the active metals and the representative elements?

25. Give an example for each of the following:

    (a) an alkali metal
    (b) an alkaline earth
    (c) a transition element
    (d) a representative element
    (e) a nonmetal
    (f) a halogen
    (g) a noble gas

26. What is the relationship between the noble gases and the octet rule?

27. List the four most abundant elements found in living organisms, and classify each as metal or nonmetal and as solid, liquid, or gas.

28. How does radioactivity result in transmutation?

29. How do radioisotopes attempt to achieve stability?

30. Explain how the neutron/proton ratio is significant to radioactivity.

31. What is the apparent role of neutrons in an atomic nucleus?

32. Name the three types of nuclear decay, and give the distinguishing feature of each.

33. What are the two types of beta particles? How do they differ from each other?

34. Describe the process of electron capture.

35. How is it that an unstable nucleus can give off an electron?

36. How are positrons created?

37. Give the changes in atomic number and mass number brought about by each type of nuclear decay.

38. Write a complete equation for each nuclear reaction:

    (a) alpha decay of polonium-205
    (b) positron emission and gamma decay by cobalt-56
    (c) decay of radium-223 to produce radon-219
    (d) electron emission by neptunium-238
    (e) electron capture by plutonium-237
    (f) electron emission by helium-6
    (g) formation of beryllium-8 from lithium-8
    (h) formation of boron-10 by positron emission
    (i) formation of oxygen-18 by electron emission
    (j) formation of osmium-190 by electron capture

CHAPTER 3
CHEMICAL BONDS

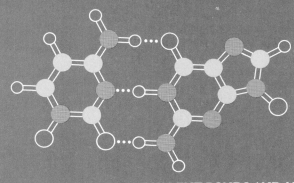

COVALENT BONDS AND MOLECULAR COMPOUNDS
- Covalent Bonds
- Molecular Compounds
- Practice in Writing Electronic Formulas
- Naming Binary Molecular Compounds
- Properties of Molecular Elements and Compounds

IONIC BONDS AND IONIC COMPOUNDS
- Ionic Bonds
- Electron Configurations of Ions
- Ionic Compounds
- Naming Binary Ionic Compounds
- Naming Complex Ionic Compounds
- Properties of Ionic Compounds

ADDITIONAL CONCEPTS OF BONDING
- Shapes of Molecules
- Electronegativity
- Polar Covalent Bonds
- Multiple Covalent Bonds
- Resonance

Early scientists were aware that atoms link themselves together to form compounds, but their ideas of how the linkage comes about were sometimes fanciful when compared to modern concepts. For example, it was once thought that atoms had barbs, and each barb was capable of hooking to a barb on another atom (see Figure 3-1). Thus, atoms simply hooked together, with each atom having its own particular number of hooks.

Fig. 3-1 Two Barbed Atoms Link Together

The notion of barbed atoms led to the idea of valence—the number of hydrogen atoms any other atom could combine with. Hydrogen was selected as the reference element because a single atom of hydrogen could never be found to combine with more than one atom of any other element. Thus, atoms of elements that combine with only one hydrogen atom were said to have a valence of 1 (chlorine, for example), while others such as oxygen, nitrogen, and carbon have valences of 2, 3, and 4, respectively. The compounds formed with hydrogen could then be represented quite simply by the number of atoms involved for each element: hydrogen chloride, HCl; water, H_2O; ammonia, NH_3; and methane, CH_4. Our modern definition of *valence*, the number of bonds an atom can form, is based on this earlier concept of combining with hydrogen.

Early in the nineteenth century it became clear that attractive forces join atoms together to form compounds. We call these attractive forces *chemical bonds*. There are two types of chemical bonds: covalent and ionic. The types of bonds in a compound determine many of the compound's properties, including how it will interact with other compounds.

COVALENT BONDS AND MOLECULAR COMPOUNDS

Covalent Bonds

In order to develop an understanding of what is meant by a covalent bond, let us compare the first two elements in the periodic table—hydrogen and helium. A hydrogen atom consists of a positively charged nucleus and one electron, whereas a helium atom has two electrons and a nucleus holding two protons. Hence, the electronic structures of hydrogen and helium atoms differ only by one electron (see Figure 3-2). Keep in mind, however, that helium, a noble gas, forms no compounds; yet hydrogen, so similar in structure, is found in countless compounds. As the element, hydrogen exists as two atoms bonded together to form a *molecule* (Figure 3-3). Helium does not even bond to itself.

molecule: a group of two or more atoms connected by covalent bonds.

Hydrogen molecules apparently form from pairs of hydrogen atoms as a result of attractions of both nuclei for both electrons. If two hydrogen nuclei were forced together, we would then have a helium nucleus. But even though the two nuclei remain separated in the hydrogen molecule, each still has an attraction for two electrons. Thus, in the hydrogen molecule, two nuclei attract two electrons, and vice versa. These mutual attractions between electrons and protons allow the hydrogen molecule to exist. Furthermore, by sharing a pair of electrons, each hydrogen atom now has the electronic configuration of the noble gas helium.

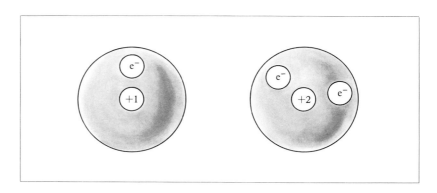

Fig. 3-2 Hydrogen and Helium Atoms

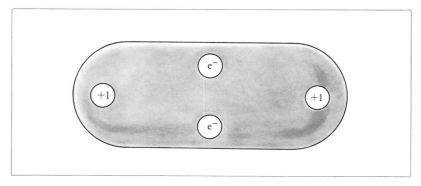

Fig. 3-3 A Hydrogen Molecule

We can represent the hydrogen atom, the helium atom, and the hydrogen molecule by *electronic formulas* (also called electron dot structures or Lewis structures):

H·	He:	H:H	H—H
hydrogen atom	helium atom	hydrogen molecule	hydrogen molecule

electronic formula: a formula which shows covalent bonds, nonbonded valence electrons, and the arrangement in which atoms are bonded together in a molecule.

The electronic formulas reflect the fact that a shared pair of electrons binds the hydrogen molecule together. As a matter of convenience, we usually designate a shared electron pair by a line drawn between the attached atoms. The attractive force between two atoms that arises from sharing a pair of electrons is called a *covalent bond*. Compounds in which all of the bonds are covalent are composed of molecules.

Since each hydrogen atom has one electron, it is logical to expect that hydrogen atoms may form covalent bonds with other atoms that can contribute a single electron toward bond formation. Furthermore, we can anticipate that many atoms are capable of forming covalent bonds; their compounds will be stable if bond formation gives each atom a noble gas configuration.

After hydrogen and helium, we will next examine fluorine and neon. Once again we are contrasting an element with the immediately following noble gas. In some ways, fluorine atoms are rather similar to hydrogen atoms. For example, a fluorine atom always bonds to just one other atom. A second similarity is that a fluorine atom can also achieve a noble gas electron configuration by sharing a pair of electrons. The electronic formula for the fluorine atom is shown below. (Note that we show only valence electrons in an electronic formula.)

$$:\!\ddot{F}\!\cdot$$

Fluorine is another example of a diatomic molecule:

$$:\!\ddot{F}\!-\!\ddot{F}\!:$$

Counting valence electrons in the molecule confirms that each atom, by sharing the bonding electron pair, obeys the octet rule. In fact, each fluorine atom in the molecule has the electron configuration of the noble gas neon. Those valence electrons not involved in the covalent bond in the fluorine molecule are called *unshared pairs, nonbonding electrons,* or *lone pairs*.

Molecular Compounds

Now that you can write electronic formulas for H_2 and F_2 molecules, let's proceed to a *molecular compound* composed of those two elements. Hydrogen fluoride, HF, is a gas formed when hydrogen and fluorine react. We can construct an electronic formula for the molecule by first

molecular compound: a compound having only covalent bonds and thus composed entirely of molecules.

hydrogen fluoride: a colorless gas which causes painful irritation to the eyes and skin and which is extremely poisonous if inhaled.

writing electronic formulas for the individual atoms. We see that hydrogen has one valence electron available for covalent bond formation, while fluorine has seven valence electrons. We then draw the covalent bond between the two atoms, and indicate the nonbonded pairs on the fluorine atom.

$$\text{H}\cdot \qquad \cdot\ddot{\underset{\cdot\cdot}{\text{F}}}: \qquad \text{H}-\ddot{\underset{\cdot\cdot}{\text{F}}}:$$

Note once again that each atom in the molecule, by sharing the bonding electron pair, has a noble gas configuration.

Now let's try the water molecule, H_2O. First, we write electronic formulas for each atom. (Note oxygen's two unpaired $2p$ electrons.) Then we arrange the atoms symmetrically, and draw in the covalent bonds. Finally we indicate the nonbonded electrons on oxygen.

$$\text{H}\cdot \qquad \cdot\ddot{\underset{\cdot\cdot}{\text{O}}}\cdot \qquad \text{H}-\ddot{\underset{\cdot\cdot}{\text{O}}}-\text{H}$$

Practice in Writing Electronic Formulas

It is possible to construct electronic formulas for molecules on the basis of *molecular formulas* and common valences of the atoms. The following steps can be used in writing electronic formulas:

molecular formula: a formula which gives the number of atoms of each element in one molecule of a substance.

1. Draw the atomic skeleton of the molecule, remembering the valence of each element, and join the atoms with single covalent bonds. For example, the molecule H_2O would be arranged as $H-O-H$. Use the following common valences as guides:
   Carbon usually has a valence of 4.
   Nitrogen usually has a valence of 3.
   Oxygen usually has a valence of 2.
   Hydrogen usually has a valence of 1.

2. Count the total number of valence electrons of all atoms in the molecule. (You can use the group number of each atom as a guide for this step.)

3. Subtract two valence electrons from the total for each single covalent bond written in step 1. Then distribute the remaining electrons as unshared pairs to give each atom an octet of electrons (except for hydrogen, which can only have two electrons).

Let's take ammonia, NH_3, as an example. The atomic skeleton might be drawn in a number of ways, but only one arrangement satisfies the usual valences of nitrogen (3) and hydrogen (1). The total number of valence electrons is eight (one each from three hydrogen atoms and five from nitrogen). Since we have drawn three covalent bonds, we must subtract six electrons from the total of eight (step 3), leaving two electrons as an unshared pair on nitrogen.

$$\text{H}\cdot \qquad \cdot \ddot{\text{N}}\cdot \qquad \text{H}-\overset{\displaystyle \ddot{}}{\underset{\displaystyle \text{H}}{\text{N}}}-\text{H}$$

We can double-check our electronic formula by confirming that each atom has a noble gas configuration. Since each hydrogen atom has two electrons (like helium) and nitrogen has an octet in its outer shell (like neon), we know that our electronic formula represents a stable structure and is correct.

EXAMPLE 3.1

Write an electronic formula for hydrogen bromide, HBr.

SOLUTION

Only one skeleton is possible for HBr:

$$\text{H}\cdot \qquad \cdot \ddot{\underset{\displaystyle ..}{\text{Br}}}: \qquad \text{H}-\text{Br}$$

Then, adding valence electrons, we get

$$\text{total number of valence electrons} = 1 + 7 = 8$$

Subtracting two electrons for the covalent bond leaves six electrons to be distributed around the Br atom:

$$\text{H}-\ddot{\underset{\displaystyle ..}{\text{Br}}}:$$

EXAMPLE 3.2

Write an electronic formula for methanol, CH_3OH, the simplest alcohol.

SOLUTION

Electronic formulas for the individual atoms are

$$\cdot \dot{\text{C}}\cdot \qquad \text{H}\cdot \qquad \cdot \ddot{\text{O}}\cdot$$

After trying various arrangements of atoms and bonds, we find only one skeleton which satisfies the usual valences of carbon, hydrogen, and oxygen:

$$\text{H}-\overset{\displaystyle \text{H}}{\underset{\displaystyle \text{H}}{\text{C}}}-\text{O}-\text{H}$$

Counting valence electrons,

$$\text{total number of valence electrons} = 4 + 4 + 6 = 14$$

and subtracting two electrons for each of the five covalent bonds,

$$14 - 10 = 4$$

we find that there are four electrons to be distributed around the oxygen atom as unshared pairs:

$$\begin{array}{c} H \\ | \\ H-C-\ddot{O}-H \\ | \\ H \end{array}$$

EXERCISE 3.1 Write an electronic formula for each molecule.
(a) CH_4, methane (major constituent of natural gas)
(b) H_2S, hydrogen sulfide (the gas that smells like rotten eggs)
(c) CH_3CH_2OH, ethanol (the next higher alcohol after methanol)
 Clue: two carbon atoms are bonded together in this compound.

Naming Binary Molecular Compounds

binary (BYE-nary): composed of two things.

Binary molecular compounds are composed of only two elements. Although more complex molecular compounds exist, we need only be concerned with naming binary ones at this time. The following rules apply:

1. The name of the more metallic element (the element farther to the left in the periodic table) is written first, and its symbol is first in the formula.

2. The ending of the name of the more nonmetallic element (the element farther to the right in the periodic table) is changed to "-ide."

3. Greek prefixes are used to signify the number of each kind of atom in the formula, except that for single atoms, the prefix "mono-" is usually dropped. (Greek prefixes are listed in Table 3-1.)

4. The binary molecular compound is named by writing the name of the more metallic element (with appropriate prefix) followed by the name of the more nonmetallic element with its "-ide" ending (and appropriate prefix).

Table 3-1 Greek Prefixes

Prefix	Meaning
Mono-	One
Di-	Two
Tri-	Three
Tetra-	Four
Penta-	Five
Hexa-	Six
Hepta-	Seven
Octa-	Eight
Nona-	Nine
Deca-	Ten

COVALENT BONDS AND MOLECULAR COMPOUNDS 65

Many binary molecular compounds are composed of a nonmetal and oxygen or halogen. Examples of these are given below. Older names still often used are given in parentheses, along with various comments.

CO	carbon monoxide (an exception to the use of "mono-", Rule 3)
CO_2	carbon dioxide
SO_2	sulfur dioxide
SO_3	sulfur trioxide
N_2O	dinitrogen oxide (nitrous oxide)
NO	nitrogen oxide (nitric oxide)
N_2O_3	dinitrogen trioxide
NO_2	nitrogen dioxide
N_2O_4	dinitrogen tetroxide ("Tetroxide" is a combination of "tetra" and "oxide.")
N_2O_5	dinitrogen pentoxide ("Pentoxide" is a combination of "penta" and "oxide.")
CCl_4	carbon tetrachloride
CBr_4	carbon tetrabromide

Some binary molecular compounds have carried names and formulas which originated before nomenclature rules were created. In these cases, we still use common names and formulas, even though more proper ones exist. Some examples which you have already seen are water (H_2O), ammonia (NH_3), and methane (CH_4).

EXERCISE 3.2 Name the following binary molecular compounds.
(a) CS_2 (b) HI (c) PCl_3 (d) CF_4

Properties of Molecular Elements and Compounds

Many nonmetals exist in their elemental states as molecules. For example, the gaseous elements hydrogen, nitrogen, oxygen, fluorine, and chlorine exist as the diatomic molecules H_2, N_2, O_2, F_2, and Cl_2. Fluorine, pale yellow in color, and chlorine, greenish yellow, have pungent, irritating odors. The other gases are colorless and odorless. The element bromine is a dark reddish brown liquid whose irritating odor is caused by its vapor. Both liquid and vapor are composed of Br_2 molecules. Another halogen, iodine, is a purple solid which gives off purple vapors of I_2.

Nonmetals combine with one another to form molecular compounds, many of which are gases. The oxides of carbon, CO and CO_2, are odorless, colorless gases, whereas oxides of sulfur are colorless gases with irritating odors. Most nitrogen oxides are colorless and odorless gases, except for nitrogen dioxide, NO_2, the reddish brown constituent of smog.

hydrocarbons: binary molecular compounds composed of hydrogen and carbon; the naming system for hydrocarbons will be discussed in Chapter 11.

Molecular compounds also exist as liquids and solids. For example, petroleum is composed of hundreds of molecular compounds called *hydrocarbons*; some are gases (propane, C_3H_8, and butane, C_4H_{10}) and others are liquids (octane, C_8H_{18}, and decane, $C_{10}H_{22}$). Still others, such as naphthalene, $C_{10}H_8$, are solids. However, the liquids tend to have relatively low boiling points (50–200°C), and the solids have relatively low melting points (100–300°C).

The fact that some molecular compounds exist as liquids and solids is evidence that their molecules have attractive forces for each other. Attractions are very weak among gas molecules; nevertheless, some attractions do exist, and if gases are cooled sufficiently, they condense into liquids or solids.

IONIC BONDS AND IONIC COMPOUNDS

Ionic Bonds

The discussion of chemical bonds thus far has considered only nonmetals and their compounds. Now let us turn to metals and some of their compounds.

In contrast to hydrogen and helium, lithium has its only valence electron in the second principal energy level; its first principal energy level is filled with two electrons. When lithium forms a compound with fluorine, one electron is *transferred* from lithium to fluorine. Lithium then takes on the electron configuration of the noble gas (He) which immediately precedes it in the periodic table, and fluorine takes on the electron configuration of neon, the noble gas immediately following fluorine in the periodic table. The formula for the compound is LiF. When lithium loses its valence electron, it becomes a positively charged *ion*; the gain of one electron by fluorine transforms that atom to a negatively charged ion.

ion (EYE-on): an atom or a group of bonded atoms with an electrical charge.

$$\text{Li·} \qquad \text{·}\ddot{\underset{..}{F}}\text{:} \qquad \left[\text{Li}\right]^+ \left[\text{:}\ddot{\underset{..}{F}}\text{:}\right]^-$$

lithium atom   fluorine atom   lithium fluoride

ionic compound: a compound composed entirely of ions.

cation (CAT-eye-on)

anion (AN-eye-on)

Ions are held together in a compound not by shared electrons but by the electrical attraction of a positive ion for a negative ion. Special names are used: a positive ion is called a *cation*, and a negative ion is called an *anion*. The strong electrical attraction between oppositely charged ions is called an *ionic bond*. As we might expect, lithium forms ionic compounds with all of the halogens.

EXAMPLE 3.3 ───────────────

Write an electronic formula for lithium chloride, LiCl.

SOLUTION

The electronic formulas for lithium and chlorine atoms are

$$\text{Li} \cdot \qquad \cdot \ddot{\underset{\cdot\cdot}{\text{Cl}}}:$$

Loss of one electron gives lithium a noble gas configuration, and gain of one electron gives chlorine a noble gas configuration. After the electron transfer, the electronic formula is written as

$$\left[\text{Li}\right]^{+} \quad \left[:\ddot{\underset{\cdot\cdot}{\text{Cl}}}:\right]^{-}$$

EXERCISE 3.3 Write an electronic formula for each ionic compound.
(a) NaCl, sodium chloride
(b) LiF, lithium fluoride
(c) KBr, potassium bromide
(d) RbI, rubidium iodide

Calcium, an alkaline earth metal, also forms ionic compounds with halogens. In this case, calcium has two valence electrons which are transferred to a halogen during compound formation. However, two halogen atoms are necessary to accept the two valence electrons from calcium. Thus the electronic formula for calcium fluoride, CaF_2, is as follows:

$$\text{Ca}: \qquad \cdot\ddot{\underset{\cdot\cdot}{\text{F}}}: \qquad \left[:\ddot{\underset{\cdot\cdot}{\text{F}}}:\right]^{-} \left[\text{Ca}\right]^{2+} \left[:\ddot{\underset{\cdot\cdot}{\text{F}}}:\right]^{-}$$

EXERCISE 3.4 Write an electronic formula for each ionic compound.
(a) $MgBr_2$, magnesium bromide
(b) $BaCl_2$, barium chloride
(c) CaI_2, calcium iodide
(d) SrF_2, strontium fluoride

Electron Configurations of Ions

Ions are formed by transfer of electrons between atoms. As a result of the transfer, negatively charged ions have more electrons than protons, and positively charged ions have fewer electrons than protons. Since the electron transfer involves only valence electrons, the electron configuration of an ion is quite similar to that of the parent atom. The only differences appear in the highest principal energy level. We write electron configurations for ions in the same way as for atoms. For example, Li^+ is formed when a lithium atom, atomic number 3, loses a valence electron. Thus the electron configurations for Li and Li^+ are

$$\begin{array}{cc} \text{Li} & \text{Li}^+ \\ 1s^2 2s^1 & 1s^2 \end{array}$$

Similarly, the electron configurations for F and F^- are

$$\begin{array}{cc} \text{F} & \text{F}^- \\ 1s^2 2s^2 2p^5 & 1s^2 2s^2 2p^6 \end{array}$$

PERSPECTIVE: Lithium, Ions, and Depression

Approximately thirty years ago, Australian physician John Cade discovered that lithium compounds could control the violent mood swings of his manic-depressive patients. The ionic compound lithium carbonate is now used for treatment of a large portion of manic-depressives, but scientists are still trying to understand how the treatment works.

It has been known for some time that manic-depressive illness is an inheritable biochemical defect that can affect the brain. Recent evidence indicates that the manic-depressive's extreme mood changes are the symptoms of a breakdown in the brain's chemical communication system, a complicated interplay of complex ions called neurotransmitters. It now appears that lithium ions probably intereact with the transport system in cell membranes, the same system that controls natural levels of physiologically important ions such as sodium, potassium, calcium, and magnesium.

Barbara Ehrlich, a young graduate student at the University of California, Los Angeles, was intrigued by the action of lithium ions on brain function, and she proceeded to study the effects of lithium compounds on rats and frogs. By 1979 she had found that lithium ions slow the transport out of cells of a compound needed by the brain to synthesize a neurotransmitter called acetylcholine. Thus, lithium may correct a defect in brain cells which causes them to release compounds necessary for the synthesis of acetylcholine. Slower release of these compounds allows synthesis of more normal levels of acetylcholine, and the manic symptoms are alleviated.

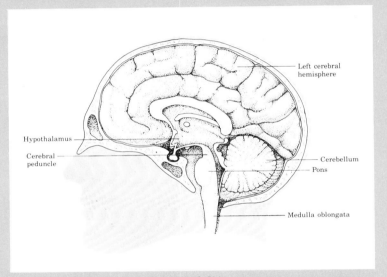

The Human Brain as Viewed from the Left Side

EXERCISE 3.5 Write electron configurations for the following pairs of atoms and ions: Na and Na^+, Cl and Cl^-.

Ionic Compounds

Metal atoms such as those in Groups IA and IIA tend to lose valence electrons to nonmetals, particularly the halogens, oxygen, and sulfur. This transfer of electrons is the basis for chemical reactions between metals and nonmetals. For example, sodium metal reacts violently with chlorine gas to produce sodium chloride, the white, crystalline solid known as table salt. The compound has no physical resemblance to its constituent elements, a fact true for all chemical compounds.

Crystals of sodium chloride are composed of Na^+ and Cl^- arranged in an orderly array called a *crystal lattice* (Figure 3-4). The nearest neigh-

crystal: a bit of solid matter having a regular geometric shape and composed of ions or molecules.

crystal lattice: the orderly array of ions or molecules in a crystal.

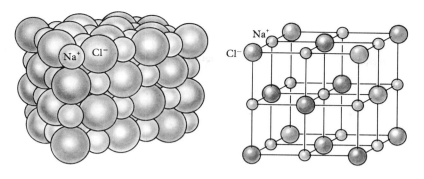

Fig. 3-4 Two Representations of the Crystal Structure of Sodium Chloride. The structure on the left shows the ions in their correct relative sizes, while the structure on the right emphasizes the location of the ions in the crystal lattice.

bors of each sodium cation are chloride anions. Conversely, the nearest neighbors of each chloride anion are sodium cations. In theory, a crystal of sodium chloride is composed of an infinite number of ions; however, crystals are easily fractured by mechanical forces in handling, and we usually observe a collection of small crystals rather than large ones. Nevertheless, each crystal is electrically neutral; this means that in every crystal there is exactly the same number of positive charges as negative charges.

The octet rule is used to predict charges on simple ions and how these ions are combined in compounds. Each atom in Group IA has a single valence electron, and loss of this electron gives an ion with a 1+ (usually written without the 1) charge and a noble gas configuration. Similarly, each atom in Group IIA will form an ion by losing its two valence electrons; each of these ions will have a 2+ charge.

Elements in groups VIA and VIIA are often the acceptors of electrons lost by the active metals. For example, an oxygen atom (Group VIA) has six valence electrons; by accepting two electrons, oxygen takes on a 2− charge and achieves an octet of valence electrons. The ion then has the electron configuration of the noble gas neon.

$$\begin{array}{cc} O & O^{2-} \\ 1s^2 2s^2 2p^4 & 1s^2 2s^2 2p^6 \end{array}$$

Similarly, each atom in Group VIIA will accept a single electron and become an anion with a 1− charge, such as F^-, Cl^-, and Br^-.

An ionic compound has no net charge, even though it is composed of charged particles. Thus, by knowing the charges on cations and anions, we can deduce the ratios in which they combine. For example, Na^+ forms an ionic compound with O^{2-} in the ratio of two Na^+ for each O^{2-}, and there is no overall charge for the compound. Thus, the formula for the ionic compound must be Na_2O. In the case of Mg^{2+} and Br^-, the formula must be $MgBr_2$.

EXERCISE 3.6 Predict the formula for the ionic compound formed by each pair.

(a) potassium and sulfur
(b) sodium and iodine
(c) magnesium and fluorine
(d) calcium and oxygen

Naming Binary Ionic Compounds

Up to this point, we have discussed binary ionic compounds, those which are composed of two elements. These compounds can be named from their formulas by use of the following rules:

1. The name of the metal (the positively charged ion) is written first, and its symbol is first in the formula.

2. The name ending of the nonmetal (the negatively charged ion) is changed to "-ide." Thus common ions from Groups VIA and VIIA are named as follows:

Group VIA	Group VIIA
Oxide	Fluoride
Sulfide	Chloride
	Bromide
	Iodide

3. The binary ionic compound is named by writing the name of the metal and then that of the nonmetal with its "-ide" ending. Some examples are:

    Ionic Compounds from Groups IA and VIA

Na_2S	Sodium sulfide
K_2O	Potassium oxide
Li_2S	Lithium sulfide
Cs_2O	Cesium oxide
Rb_2S	Rubidium sulfide

    Ionic Compounds from Groups IA and VIIA

LiI	Lithium iodide
NaF	Sodium fluoride
KBr	Potassium bromide
RbCl	Rubidium chloride
CsI	Cesium iodide

    Ionic Compounds from Groups IIA and VIA

MgS	Magnesium sulfide
CaO	Calcium oxide
SrS	Strontium sulfide
BaO	Barium oxide

    Ionic Compounds from Groups IIA and VIIA

MgF_2	Magnesium fluoride
$CaCl_2$	Calcium chloride
$SrBr_2$	Strontium bromide
BaF_2	Barium fluoride

    Ionic Compounds from Groups IIIA and VIA

Al_2O_3	Aluminum oxide
Al_2S_3	Aluminum sulfide

    Ionic Compounds from Groups IIIA and VIIA

$AlCl_3$	Aluminum chloride
$AlBr_3$	Aluminum bromide

4. Each transition metal has several possible charges. In these cases, the name of the binary ionic compound includes a roman numeral in parentheses after the name of the metal. This numeral represents the charge on the metal. The most common examples in this category are compounds of iron and copper, some of which are shown below. Older names still often used are shown in parentheses.

$FeBr_2$	iron(II) bromide	(ferrous bromide)
$FeBr_3$	iron(III) bromide	(ferric bromide)

FeO	iron(II) oxide	(ferrous oxide)
Fe_2O_3	iron(III) oxide	(ferric oxide)
CuCl	copper(I) chloride	(cuprous chloride)
$CuCl_2$	copper(II) chloride	(cupric chloride)
Cu_2O	copper(I) oxide	(cuprous oxide)
CuO	copper(II) oxide	(cupric oxide)

Naming Complex Ionic Compounds

Many anions are composed of two or more elements. For example, the principal anion in bone tissue is phosphate, PO_4^{3-}, and one of the most important anions in blood is bicarbonate, HCO_3^-. In these polyatomic anions the atoms are covalently bonded together so that they behave as one unit; the charges arise from an excess of electrons in the unit. Names used to designate polyatomic anions are given in Table 3-2.

poly- (PAH-ly): a Greek prefix meaning "many."

We can refer to ionic compounds containing polyatomic anions as complex ionic compounds. In naming these, we must remember that an ionic compound always contains a cation and an anion; however, the compound may contain more than one of each as needed to balance opposite charges. The name of the positive ion is followed by the name of the negative ion, just as with binary compounds. Examples of some

Table 3-2 Common Polyatomic Ions

Element	Polyatomic Ion	Name
H	OH^-	hydroxide
C	HCO_3^-	hydrogen carbonate or bicarbonate
	CO_3^{2-}	carbonate
	CN^-	cyanide
N	NH_4^+	ammonium
	NO_3^-	nitrate
	NO_2^-	nitrite
P	$H_2PO_4^-$	dihydrogen phosphate
	HPO_4^{2-}	hydrogen phosphate
	PO_4^{3-}	phosphate
S	HSO_4^-	hydrogen sulfate or bisulfate
	SO_4^{2-}	sulfate
	HSO_3^-	hydrogen sulfite or bisulfite
	SO_3^{2-}	sulfite
Cl	ClO^-	hypochlorite
	ClO_3^-	chlorate
Cr	CrO_4^-	chromate
	$Cr_2O_7^{2-}$	dichromate
Mn	MnO_4^-	permanganate
As	AsO_4^-	arsenate
Br	BrO^-	hypobromite
	BrO_3^-	bromate

common complex ionic compounds (with older names in parentheses) are:

Na_2SO_4	sodium sulfate
$MgSO_4$	magnesium sulfate
$CaSO_4$	calcium sulfate
$Al_2(SO_4)_3$	aluminum sulfate
$NaHCO_3$	sodium hydrogen carbonate (sodium bicarbonate)
$Mg(HCO_3)_2$	magnesium hydrogen carbonate (magnesium bicarbonate)
Na_2CO_3	sodium carbonate
$CaCO_3$	calcium carbonate
$Al_2(CO_3)_3$	aluminum carbonate
$KMnO_4$	potassium permanganate
Na_2CrO_4	sodium chromate
$Mg(NO_3)_2$	magnesium nitrate
$NaNO_2$	sodium nitrite
NaH_2PO_4	sodium dihydrogen phosphate
Na_2HPO_4	sodium hydrogen phosphate
Na_3PO_4	sodium phosphate
$Ca_3(PO_4)_2$	calcium phosphate

EXERCISE 3.7 Give the formula and name the compound formed by each pair of ions.
(a) Na^+ and S^{2-}
(b) Li^+ and CO_3^{2-}
(c) Al^{3+} and NO_3^-
(d) Ba^{2+} and F^-
(e) Mg^{2+} and PO_4^{3-}
(f) K^+ and HPO_4^{2-}

Properties of Ionic Compounds

As we have seen, ionic compounds are formed as a result of electron transfer, and the oppositely charged ions have electrical attractions for each other. These attractions are much stronger than the attractive forces which exist between the molecules of molecular compounds. As a result, ionic compounds tend to be solids with very high melting points. When a solid melts, its particles become much more mobile, moving randomly through the liquid. It requires much heat energy, hence high temperatures, to disrupt ionic bonds and cause the ions to move about. However, once the ionic compound has been melted, it has all the properties of a liquid. Much higher temperatures are required to separate the ions completely and thus form a gas. Figure 3-5 illustrates the melting and vaporizing processes for a crystalline solid. Examples of the high melting and boiling points of ionic compounds are shown in Table 3-3.

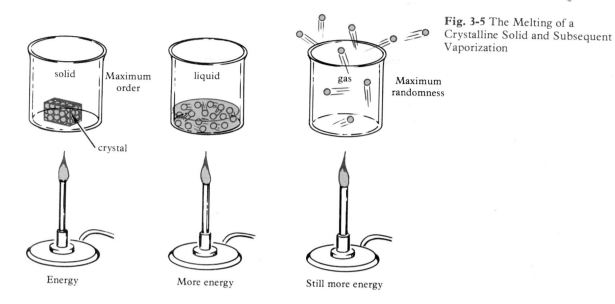

Fig. 3-5 The Melting of a Crystalline Solid and Subsequent Vaporization

Table 3-3 Melting Points and Boiling Points of Ionic Compounds

Compound	Formula	Melting Point (°C)	Boiling Point (°C)
Ammonium chloride	NH_4Cl	340	520
Barium chloride	$BaCl_2$	963	1560
Barium oxide	BaO	1920	2000
Calcium fluoride	CaF_2	1418	2500
Calcium oxide	CaO	2590	2850
Lead chloride	$PbCl_2$	500	950
Lithium chloride	$LiCl$	610	1360
Magnesium fluoride	MgF_2	1263	2230
Magnesium oxide	MgO	2800	3580
Potassium chloride	KCl	772	1411
Sodium chloride	$NaCl$	808	1473
Sodium fluoride	NaF	1450	2490

Table 3-4 Physiologically Important Ions

Cations	Anions
Sodium, Na^+	Chloride, Cl^-
Potassium, K^+	Bicarbonate, HCO_3^-
Calcium, Ca^{2+}	Hydrogen phosphate, HPO_4^-
Magnesium, Mg^{2+}	Sulfate, SO_4^{2-}

Many ionic compounds dissolve in water. When this happens, the crystals are broken apart, and the individual ions are scattered throughout the solution. In such an arrangement, we often refer to names of individual ions instead of names of compounds. Table 3–4 lists a selection of ions commonly found in physiological fluids.

Ionic compounds are often found as natural deposits in the earth; some examples are shown in Table 3-5. For a long time, the study of these compounds was restricted to the field of *inorganic chemistry*, the chemistry of nonliving substances. However, modern research has demonstrated the importance of ionic compounds in living organisms, from single cells to complex organisms, so ionic compounds are now considered to be active participants in the chemistry of life.

ADDITIONAL CONCEPTS OF BONDING

Shapes of Molecules

We have learned to write electronic formulas of molecules showing covalent bonds, unshared pairs of electrons, and the arrangement of the atoms, but these representations do not tell much about the three-dimensional shapes of molecules. To learn about molecular shapes, we must consider the angles at which atoms are bonded together.

Since each covalent bond is composed of a pair of electrons, we can consider the bond to be a region of space occupied by two negative charges. In this sense, covalent bonds repel each other because of their like charges, and the covalent bonds to any one atom will stay as far away from each other as possible. Using this concept of bond repulsion, we can make predictions about bond angles and molecular shapes. For example, methane has the electronic formula:

$$\begin{array}{c} H \\ | \\ H-C-H \\ | \\ H \end{array}$$

How can its bonds get as far away from each other as possible? The answer is that the bonds to carbon fan out into space to give the largest possible bond angles. If we were to build a model of the carbon atom

Table 3-5 Some Ionic Compounds in the Environment

Mineral Name	Chemical Name	Formula
Alumina	Aluminum oxide	Al_2O_3
Barite	Barium sulfate	$BaSO_4$
Borax	Sodium tetraborate decahydrate	$Na_2B_4O_7 \cdot 10H_2O$
Calamine	Zinc carbonate	$ZnCO_3$
Chalk	Calcium carbonate	$CaCO_3$
Cinnebar	Mercury(II) sulfide	HgS
Hematite	Iron(III) oxide	Fe_2O_3
Lime	Calcium oxide	CaO
Magnesia	Magnesium oxide	MgO
Saltpeter	Potassium nitrate	KNO_3
Talc	Magnesium silicate	$Mg_3Si_4O_{10}(OH)_2$

and its four bonds using a gumdrop and four toothpicks, we would find that the largest possible equal bond angles are 109.5°. Indeed, this is the value for all the bond angles in methane. If we consider that a hydrogen atom is also attached to each bond in methane, we find that these bond angles put each hydrogen atom in the corner of a geometric figure called a regular *tetrahedron* (see Figure 3-6). This figure has four equivalent sides and equal corner angles.

The concept of bond repulsion can also be applied to deduce the shape of the ammonia molecule, NH_3. The electronic formula for ammonia is shown in Figure 3-7. In this case there is an unshared pair of electrons on nitrogen that we must take into consideration. This electron pair occupies a region of space just as the bonding electron pairs do, and if we allow for repulsions among the total four pairs of valence electrons in the ammonia molecule, we would expect the same bond

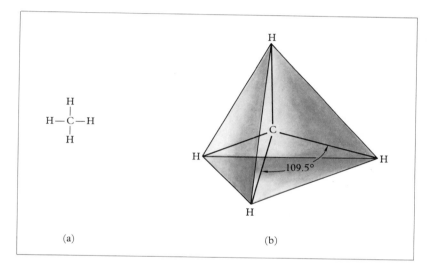

Fig. 3-6 (a) The Electronic Formula of Methane (b) The Geometry of the Methane Molecule

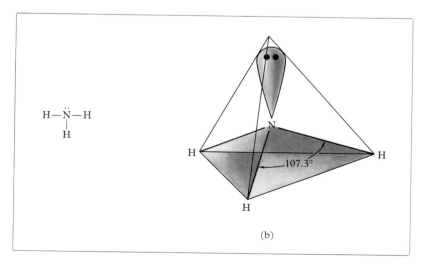

Fig. 3-7 (a) The Electronic Formula for Ammonia (b) The Geometry of the Ammonia Molecule

angles in ammonia as in methane. Since ammonia is only one of many molecules having unshared electron pairs, let us rename our concept as valence electron pair repulsion. Measurements have shown the bond angles in ammonia to be slightly less than the 109.5° we predicted. The small difference is explained by the fact that the unshared electron pair occupies a slightly larger region of space than bonding pairs, and this has the effect of squeezing the bonds slightly closer together. Thus, the ammonia molecule has bond angles of 107.3°, and it is shaped like a pyramid (see Figure 3-7). In this case, the three hydrogen atoms are located at corners of a slightly irregular tetrahedron, and the fourth corner is occupied by the unshared electron pair.

Let's examine one more molecule. The electronic formula for water is given in Figure 3-8. In this case, we have two unshared pairs of electrons to consider, and each occupies its own region of space. Once again, there are four pairs of mutually repulsive valence electrons, and we would predict tetrahedral geometry. However, the two pairs of unshared electrons on oxygen compress the bond angles even more than the single pair in ammonia and distort the tetrahedron even more. Thus, the bond angle to oxygen in the water molecule is 104.5°. This gives the water molecule a bent shape as illustrated in Figure 3-8.

From reasoning based on the valence electron pair repulsion concept, we have deduced that when there are four separate negatively charged regions of space about an atom, the bond angles to the atoms will be very close to 109.5°. These bond angles then determine the overall shape of the molecule. For our examples, we can now write somewhat more representative electronic formulas:

$$
\begin{array}{ccc}
\quad\quad\text{H} & & \\
\quad\quad | & \ddot{\text{N}} & \ddot{\text{O}}\!\!: \\
\text{H}\!-\!\underset{|}{\text{C}}\!-\!\text{H} & \text{H}\!\diagup\ \underset{|}{}\ \diagdown\text{H} & \text{H}\diagup\ \diagdown\text{H} \\
\quad\quad\text{H} & \quad\text{H} &
\end{array}
$$

Fig. 3-8 (a) The Electronic Formula for Water
(b) The Geometry of the Water Molecule

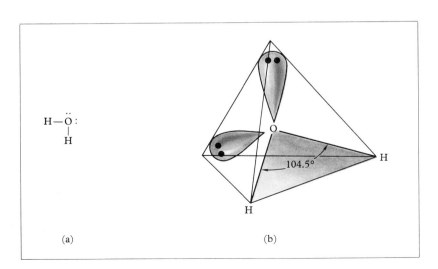

EXERCISE 3.8 Predict the shape of each of the following molecules.
(a) CCl_4, carbon tetrachloride
(b) PH_3, phosphine
(c) H_2S, hydrogen sulfide

Electronegativity

Up to this point we have classified chemical bonds as covalent or ionic. In fact, these are two extremes, and there are many bonds intermediate between these two types. True covalent bonds result from equal sharing of an electron pair. This can only happen in a bond between two identical atoms. Our picture of ionic bonds is also a bit oversimplified, in that transfer of an electron is never totally complete, even with atoms which have extremely different tendencies to gain and lose electrons. Thus, we must modify our concepts by introducing the idea of *electronegativity*, the ability of an atom to draw electrons in a covalent bond toward itself. For example, fluorine atoms in molecular compounds tend to draw bonding electrons toward themselves strongly; thus, fluorine atoms have a high electronegativity. On the other hand, carbon atoms do not pull bonding electrons toward themselves as strongly as fluorine atoms do, and carbon thus has a lower electronegativity than fluorine. These tendencies seem reasonable when we remember that carbon and fluorine atoms have the same number of principal energy levels but that a fluorine atom has more protons in its nucleus. Thus, a fluorine atom has a stronger electrical attraction for valence electrons than does a carbon atom.

The first scientist to assign numerical electronegativity values to elements was the American chemist Linus Pauling. (His name may be familiar because of his recent research on vitamin C.) In the early 1930s, Pauling suggested that electronegativity could be determined for any element by measuring the energy required to break chemical bonds in several of its compounds. Thus, he derived an electronegativity scale, as shown in Figure 3-9.

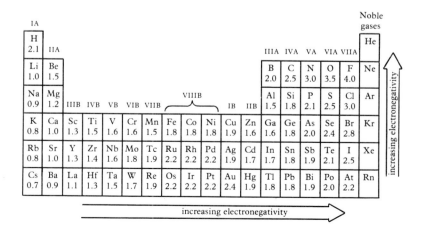

Fig. 3-9 Pauling Electronegativity Values

Pauling's method for calculating electronegativity gives relative values. Fluorine, the most electronegative element, was chosen as the reference element, and all other elements are rated relative to fluorine. (Total nuclear charge is only one of several factors which determine electronegativity; for this reason, we can only compare nuclear charges of atoms in the same chemical period.) Figure 3-9 illustrates that electronegativity increases within each chemical period as atomic number (and thus nuclear charge) increases. Within a chemical family, electronegativity decreases with atomic number. In general, metals have low electronegativity values and nonmetals have high values.

The concept of electronegativity becomes very important in bonds between two different kinds of atoms. For example, the difference in electronegativities between lithium and fluorine is so large that, for all practical purposes, electron transfer is complete. On the other hand, there are many molecules in which covalent bonds represent unequal *sharing* of bonding electrons.

Polar Covalent Bonds

A *polar covalent bond* is one in which a shared pair of electrons is shifted toward an atom of high electronegativity. Because of this shift, the more electronegative atom has a higher density of electrons than the less electronegative atom. Hydrogen fluoride, HF, is such a molecule. The difference in electronegativity between hydrogen and fluorine $(4.0 - 2.1 = 1.9)$ is not great enough to give ionic bonding, but it is large enough that the bonding electrons are shifted toward fluorine. We can indicate this shift as

$$\overset{+}{H}\longrightarrow\overset{-}{F}$$

where the arrow represents the direction of electron shift and the charges indicate relative electron density. However, this representation is somewhat misleading because neither atom has a whole charge residing on it. A better representation is

$$\overset{\delta+}{H}\longrightarrow\overset{\delta-}{F}$$

where the small Greek letter delta (δ) indicates a partial charge of less than one full unit.

The difference in electronegativity between two atoms bonded together can be used to predict the nature of the bond. If the electronegativity difference is less than 0.4, the bond will be nonpolar covalent, but if the electronegativity difference is greater than 2.0, the bond will be ionic. Electronegativity differences in the range of 0.4–2.0 generally lead to polar covalent bonds.

Because a polar covalent bond has a positive end (pole) and a negative end (pole), this bond can also be called a *dipolar bond*. It is in fact an *electric dipole*. Because this is the only bond in HF, then the molecule itself is an electric dipole, and it possesses a *dipole moment*. In this case, we say that the molecule is polar. Polar molecules align themselves as

electric dipole: a localized separation of positive and negative electrical charge.

dipole moment: a measurable physical property possessed by an electric dipole.

shown in Figure 3-10, with the partially positive end of one near the partially negative end of another to form a network of interacting molecules.

On the basis of the discussion above, we can predict that hydrogen atoms will form polar covalent bonds with more electronegative elements such as halogens, oxygen, and nitrogen. An important example of such a compound is water, H_2O. The dipole moment of water is consistent with the following *structural formula*:

$$\overset{\delta-}{O}$$
$$H\diagup\quad\diagdown H$$
$$\delta+\qquad\quad\delta+$$

structural formula: a formula which indicates the geometry of a molecule.

A molecule containing two or more polar covalent bonds can be either polar (water, for example) or nonpolar, depending on the bonding electron distribution over the molecule. Carbon tetrachloride, for example, has four polar C—Cl bonds, but they are arranged symmetrically so that the molecule does not have a positive end and a negative end (see Figure 3-11).

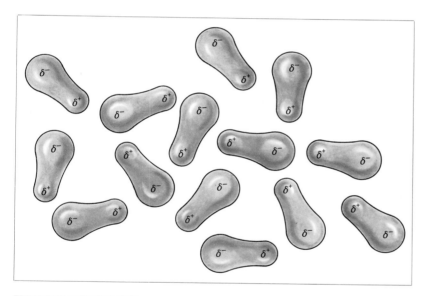

Fig. 3-10 Interacting Polar Molecules

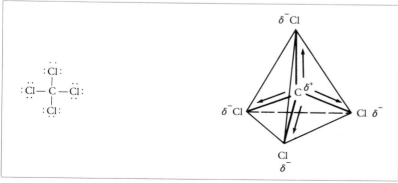

Fig. 3-11 Carbon Tetrachloride, A Nonpolar Molecule Containing Polar Bonds

PERSPECTIVE: Linus Pauling, Crusading Scientist

Linus Carl Pauling, born in Portland, Oregon, in 1901, is a world-renowned chemist and crusader for peace. In his early work on molecular structure, he originated the idea of polar covalent bonds, and his concept of electronegativity helped clarify properties of covalent bonds.

By 1934, Pauling had begun to study the complex molecules of living tissues, and in the late 1940s, he suggested that sickle-cell anemia is a genetic defect associated with hemoglobin formation. He also developed a molecular model for the explanation of anesthesia and introduced ideas toward the understanding of memory processes. His latest work has dealt with vitamin C treatment for various illnesses, including cancer.

Linus Pauling is known for his intuitive ideas aided by a phenomenal memory of chemical facts. Because of his intuitive approach to science and his sometimes unorthodox ideas, Pauling has created much controversy. Following the development of nuclear weapons, his deep concern about the hazards of radiation prompted him to write a book entitled *No More War* in 1958. In January of the same year, he brought to the United Nations a petition signed by 11,021 scientists from all over the world urging an end to nuclear weapons testing. His pacifist views estranged him from many scientists with whom he had once been closely associated, and his loyalty to the United States was questioned in conservative political circles.

Linus Pauling has received two Nobel Prizes, an honor conferred only once before when Marie Curie received two of the awards. His first Nobel Prize came in 1954 for recognition of his research into the nature of chemical bonds. He was honored again in 1962, this time with the Nobel Peace Prize. No official reason was given for the award, but it is widely assumed that he received it for his efforts on behalf of banning nuclear testing.

Linus Pauling

Multiple Covalent Bonds

The molecular compounds known as hydrocarbons contain only hydrogen and carbon. We can draw electronic formulas for two members of the group, methane (CH_4) and ethane (C_2H_6), as shown:

$$\begin{array}{c} \text{H} \\ | \\ \text{H}-\text{C}-\text{H} \\ | \\ \text{H} \end{array} \qquad \begin{array}{c} \text{H} \ \ \text{H} \\ | \ \ \ | \\ \text{H}-\text{C}-\text{C}-\text{H} \\ | \ \ \ | \\ \text{H} \ \ \text{H} \end{array}$$

methane   ethane

Note that in ethane two carbon atoms are bonded together. These two formulas demonstrate that each carbon atom in a hydrocarbon forms bonds with hydrogen atoms or carbon atoms. And yet, there are many hydrocarbon molecules in which there seems to be too little hydrogen; examples are ethene, C_2H_4, and ethyne, C_2H_2 (commonly called acetylene). We must ask the question, with so few hydrogen atoms, how can the carbon atoms form enough bonds to obtain stable electron structures (octets) in their outer shells? The answer is that carbon often forms multiple covalent bonds to other atoms. In the case of ethene, the following electronic formula illustrates a multiple bond—in this case, a *double bond*:

$$\begin{array}{c} H \quad\quad H \\ \diagdown\;\;\;\diagup \\ C=C \\ \diagup\;\;\;\diagdown \\ H \quad\quad H \end{array}$$

ethene

In a double bond, two atoms share four bonding electrons.

In predicting the shape of ethene, we can consider the double bond to behave as if it were a single electron pair. Thus, there are three regions of high electron density around each carbon atom in ethene, and the best arrangement for these gives bond angles of 120° about each carbon atom. This arrangement results in a planar (flat) shape for ethene, as shown in Figure 3-12.

You probably have guessed that ethyne, having still fewer hydrogen atoms, has even more bonding electrons between the two carbon atoms. If so, you are correct. Ethyne is an example of a compound having a *triple bond*:

$$H-C\equiv C-H$$

For both ethene and ethyne, notice that there are eight valence electrons assigned to each carbon atom and two valence electrons assigned to each hydrogen atom.

For ethyne, we can consider the triple bond to behave as another single region of high electron density. Since there are two such regions about each of the carbon atoms, ethyne has a linear shape, as Figure 3-13 illustrates.

As you continue your study of chemistry, you will encounter many compounds having double or triple bonds. The following are examples of common multiple bonds:

$\diagup C=C\diagdown$	carbon-carbon double bond
$-C\equiv C-$	carbon-carbon triple bond
$-\ddot{N}=\ddot{N}-$	nitrogen-nitrogen double bond
$:N\equiv N:$	nitrogen-nitrogen triple bond
$\diagup C=\ddot{N}-$	carbon-nitrogen double bond
$-C\equiv N:$	carbon-nitrogen triple bond
$\diagup C=\ddot{O}$	carbon-oxygen double bond

double bond: a covalent bond composed of two pairs of shared electrons.

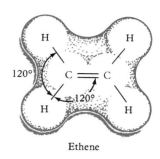

Ethene

Fig. 3-12 Geometry of the Ethene Molecule

ethyne (EH-thine)

triple bond: a covalent bond composed of three pairs of shared electrons.

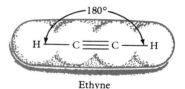

Ethyne

Fig. 3-13 Geometry of the Ethyne Molecule

Resonance

There are many molecules whose bonds can be arranged to give more than one satisfactory electronic formula. One of these is ozone, the molecule composed of three oxygen atoms. The molecular formula for ozone is O_3; structural studies of the molecule indicate that it has a bent

PERSPECTIVE: Nitrates and Nitrites in Meat

Sodium nitrate, NaNO$_3$, is an ionic compound which has been used for centuries for curing ham, bacon, and corned beef. A closely related compound, sodium nitrite (NaNO$_2$), came into use in the curing process in 1925. Shown here is an example of a label showing the presence of NaNO$_2$ in a cured meat. These compounds function to preserve meat and to stabilize the familiar pink color of ham and bacon. As we eat the meats, these compounds dissolve to form nitrate ions, NO$_3^-$, and nitrite ions, NO$_2^-$. Furthermore, it appears that nitrates are converted to nitrites under certain conditions in the stomach.

The use of nitrites in food has been of concern since the early 1960s because nitrites are known to react with other food compounds usually present in the digestive tract to form nitrosamines. These are known to cause cancer in certain laboratory animals and are considered to be potential human carcinogens. Nitrosamines can also be formed while the meat is cooking, so that they may be ingested with fried bacon, for example.

Nitrates and nitrites occur naturally in foods, particularly dark green leafy vegetables like spinach and kale, and also in drinking water. Thus, food processors claim that the small amounts added to meat are inconsequential. Nevertheless, the U.S. Department of Agriculture is presently conducting tests to determine whether meats can be cured successfully in the absence, or with lowered amounts, of nitrates and nitrites.

Typical ingredients in cured ham.

shape, and that all bonds are identical. As we attempt to draw its electronic structure, we find that two structures can be drawn:

The two alternative structures for ozone are equivalent except for the placement of electrons. Equivalent electronic structures, as in this example, are called *resonance forms*. However, neither structure describes the molecule adequately, because each has one single bond and one double bond, and yet we know that both bonds are identical. But if we were to take photographs of the two structures and then put one photograph on top of the other, we would obtain a new structure as shown below.

The new structure with identical bonds implies that each oxygen-oxygen bond has partial double-bond character.

Another method of representing the structure of ozone follows. This representation means that the real structure is a mixture of the two resonance forms.

resonance (REZ-o-nance) **forms:** two or more alternative electronic formulas for a molecule or ion which differ only in the placement of electrons.

$$\left[\begin{array}{c}\ddot{\text{O}}\\ \diagup\diagdown\\ :\ddot{\text{O}}\cdot\quad\cdot\ddot{\text{O}}:\end{array}\right] \longleftrightarrow \left[\begin{array}{c}\ddot{\text{O}}\\ \diagup\diagdown\\ :\ddot{\text{O}}\cdot\quad\cdot\ddot{\text{O}}:\end{array}\right]$$

The double-headed arrow is a symbol reserved for use in drawing resonance forms, and the brackets emphasize that these are resonance forms, not true structures. Please do not think that the molecule oscillates back and forth between the resonance forms, because only one true structure exists for the ozone molecule. Since we are unable to represent the true structure of the molecule on paper, we must mentally average the resonance forms to imagine the real structure.

Another example of a structure for which there are resonance forms is the nitrite ion, NO_2^-. This ion has a shape similar to that of ozone. It has a negative charge because it has an excess of one electron. Resonance forms for the nitrite ion are

$$\left[\begin{array}{c}\ddot{\text{N}}\\ \diagup\diagdown\\ :\ddot{\text{O}}\cdot\quad\cdot\ddot{\text{O}}:\end{array}\right]^- \longleftrightarrow \left[\begin{array}{c}\ddot{\text{N}}\\ \diagup\diagdown\\ :\ddot{\text{O}}\cdot\quad\cdot\ddot{\text{O}}:\end{array}\right]^-$$

The nitrate ion, NO_3^-, is an example of a structure that has three resonance forms:

$$\left[\begin{array}{c}:\ddot{\text{O}}:\\ \|\\ \text{N}\\ \diagup\diagdown\\ :\ddot{\text{O}}\cdot\quad\cdot\ddot{\text{O}}:\end{array}\right]^- \longleftrightarrow \left[\begin{array}{c}:\ddot{\text{O}}:\\ |\\ \text{N}\\ \diagup\diagdown\\ :\ddot{\text{O}}\cdot\quad\cdot\ddot{\text{O}}:\end{array}\right]^- \longleftrightarrow \left[\begin{array}{c}:\ddot{\text{O}}:\\ |\\ \text{N}\\ \diagup\diagdown\\ :\ddot{\text{O}}\cdot\quad\cdot\ddot{\text{O}}:\end{array}\right]^-$$

When you are drawing resonance forms, you should remember one rule: the arrangement of nuclei must be the same in each form. This means that the same atoms must be bonded to each other in all of the forms, so that the only differences are in the arrangement of electrons. Those molecules and ions for which two or more resonance forms can be drawn are said to possess *resonance*. We shall explore this topic in more detail in later chapters dealing with carbon compounds.

resonance: the quality possessed by a molecule or ion having resonance forms.

SUMMARY

Covalent Bonds and Molecular Compounds: Covalent bonds are forces which join atoms together to form molecules. A covalent bond forms when two atoms share a pair of electrons. Any remaining valence electrons not involved in a covalent bond are called unshared pairs, nonbonding electrons, or lone pairs. A molecular compound is composed only of molecules. Molecules can be represented by electronic formulas which show atoms, covalent bonds, and nonbonded electrons. Most molecular compounds are composed of nonmetals. Molecular compounds exist as gases, liquids, and solids, but their boiling and melting points are low compared to those of ionic compounds.

Ionic Bonds and Ionic Compounds: Ionic bonds are the forces of electrical attraction between oppositely charged ions. Ions are created when one or more electrons are transferred from one atom to another. Electron configurations may be written for ions just as for atoms. Ionic compounds are composed of ions arranged in a crystal lattice. Each crystal is electrically neutral. Ionic compounds are usually solids, and much energy is required to separate oppositely charged ions to form liquids and gases. For this reason, ionic compounds have very high melting and boiling points. Ionic compounds are often water soluble.

Additional Concepts of Bonding: Shapes of molecules can be predicted on the basis of the concept of valence electron pair repulsion. Electronegativity, a measure of the ability of an atom to draw electrons in a covalent bond toward itself, accounts for the fact that most covalent bonds are polar. Polar bonds confer dipole moment to a molecule if the bonding electrons are distributed unevenly over the molecule. Multiple bonds are common in hydrocarbon molecules, and carbon forms multiple bonds to many kinds of atoms. Resonance is the quality possessed by a molecule or ion for which two or more equivalent electronic formulas can be drawn. These alternative structures are called resonance forms. The true structure of a molecule or ion having resonance is a mixture of all of the resonance forms.

STUDY QUESTIONS AND PROBLEMS

1. Define the following terms:

   (a) valence
   (b) chemical bond
   (c) molecule
   (d) electronic formula
   (e) covalent bond
   (f) unshared electron pair
   (g) molecular formula
   (h) hydrocarbon
   (i) ion
   (j) anion
   (k) cation
   (l) crystal lattice
   (m) polyatomic anion
   (n) tetrahedron
   (o) electronegativity
   (p) polar covalent bond
   (q) electric dipole
   (r) structural formula
   (s) double bond
   (t) triple bond
   (u) resonance forms

2. Write an electronic formula for each atom.

   (a) Ne
   (b) Al
   (c) N
   (d) Mg
   (e) K
   (f) Be
   (g) C
   (h) P
   (i) H
   (j) Cl

3. Write an electronic formula for each molecular compound or element.

   (a) H_2O
   (b) Br_2
   (c) H_2S
   (d) N_2
   (e) ClBr
   (f) CH_3CH_2OH
   (g) PCl_3
   (h) NH_3
   (i) CH_4
   (j) H_2

4. Using only the rules for naming molecular compounds given in this chapter, name the following:

   (a) CO
   (b) H_2S
   (c) Cl_4
   (d) PCl_3
   (e) HBr
   (f) N_2O
   (g) CO_2
   (h) SO_3
   (i) BF_3
   (j) P_2O_5

5. Write an electronic formula for the ionic compound formed from each of the following pairs of elements:

   (a) sodium and bromine
   (b) magnesium and oxygen
   (c) aluminum and sulfur
   (d) potassium and oxygen
   (e) calcium and chlorine
   (f) aluminum and fluorine

6. Distinguish between a covalent bond and an ionic bond.

7. Write the electron configuration for each cation.

   (a) potassium ion
   (b) magnesium ion
   (c) aluminum ion
   (d) calcium ion
   (e) lithium ion
   (f) hydrogen ion

8. Write the electron configuration for each anion.

   (a) oxide ion
   (b) fluoride ion
   (c) sulfide ion
   (d) chloride ion

9. Give the simple formula for each ionic compound in question 5.

10. Name each of the ionic compounds in question 5.

11. Name each of the following complex ionic compounds:

    (a) $Al_2(SO_4)_3$
    (b) NaOH
    (c) $Ca_3(PO_4)_2$
    (d) $KHSO_4$
    (e) Li_2CO_3
    (f) Na_2SO_3
    (g) $K_2Cr_2O_7$
    (h) $NaHCO_3$
    (i) NaOCl
    (j) KH_2PO_4

12. Why do ionic compounds have extremely high melting points?

13. Draw the three-dimensional structure of each molecule and explain the basis for predicting the structure.

    (a) CH_4 (b) NH_3 (c) H_2O

14. Predict the shape of each molecule.

    (a) H_2S (b) PCl_3 (c) CF_4

15. Describe the trend in electronegativity values from left to right across a period of elements and then from top to bottom in a group of elements.

16. Classify the bonds in each of the following as ionic, polar covalent, or nonpolar covalent.

    (a) H_2
    (b) MgO
    (c) H_2O
    (d) NaI
    (e) CF_4
    (f) HCl
    (g) S_2
    (h) CCl_4
    (i) AlF_3
    (j) NH_3

17. Indicate the direction of polarity in each bond in the following molecules:

    (a) HF
    (b) NH_3
    (c) $ClBr$
    (d) H_2S
    (e) PCl_3

18. Classify each of the following molecules as polar or nonpolar.

    (a) CBr_4
    (b) H_2S
    (c) Br_2
    (d) PH_3

19. Describe the geometry of the ethene and ethyne molecules, and explain the electronic basis of their shapes.

20. Each of the following molecules has at least one multiple bond. Write an electronic formula for each, and predict its geometry.

    (a) CO_2
    (b) N_2
    (c) HCN
    (d) $HCHO$
    (e) C_2H_2
    (f) CH_2NH

21. Draw resonance forms for each molecule. The total number of resonance forms is given in parentheses after each entry.

    (a) 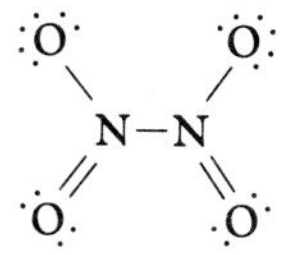
    $$\ddot{\underset{\ddot{O}}{O}}\diagdown\underset{\diagdown\underset{\ddot{O}}{\ddot{O}}}{N-N}\diagup\overset{\ddot{O}}{\diagup}$$
    (4)

    (b) 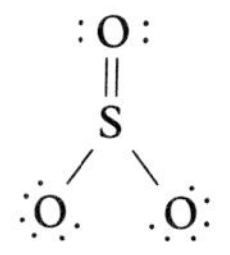
    $$\underset{\overset{\ddot{O}}{\diagdown}\overset{}{}\overset{}{}\overset{\ddot{O}}{\diagup}}{\overset{\overset{:O:}{\|}}{S}}$$
    (3)

    (c) 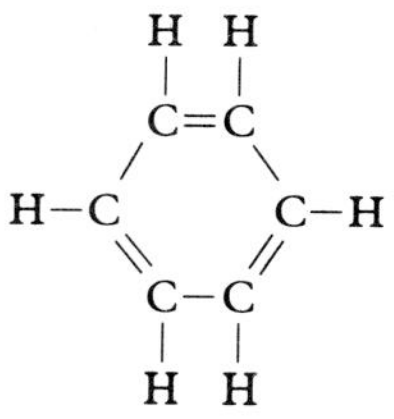
    $$\begin{array}{c} \text{H} \quad \text{H} \\ | \quad | \\ \text{C}=\text{C} \\ \diagup \qquad \diagdown \\ \text{H}-\text{C} \qquad \text{C}-\text{H} \\ \diagdown \qquad \diagup \\ \text{C}-\text{C} \\ | \quad | \\ \text{H} \quad \text{H} \end{array}$$
    (2)

CHAPTER 4
CHEMICAL REACTIONS

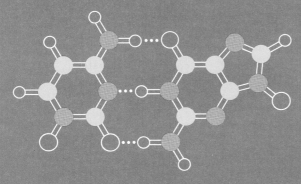

CHEMICAL EQUATIONS AND MASS RELATIONSHIPS
- The Language of Chemical Equations
- Conservation of Mass
- Balancing Chemical Equations
- Molecular Weights and Formula Weights
- The Chemical Mole
- Calculations and Chemical Equations

TYPES OF CHEMICAL REACTIONS
- Single Replacement Reactions
- Double Replacement Reactions
- Combination Reactions
- Decomposition Reactions
- Isomerization Reactions
- Oxidation-Reduction Reactions

The idea of change is one which we have learned to accept as a primary fact of life in the modern world. Social and environmental changes occur at an increasingly rapid pace, and we have difficulty absorbing all of the new information which confronts us in the newpapers. Chemistry plays an important role in our continuously changing environment. Since chemistry is the study of matter and its changes, it is necessary for us to consider types of chemical changes (reactions) and the equations which describe them.

In the broad sense, matter can undergo three kinds of change, as shown in Figure 4-1: change of position, change of form, and change of substance. A change of position occurs when matter moves from one place to another, as when an airplane and its passengers fly from Los Angeles to New York. A change of form takes place when matter changes from one physical state to another, as when solid water (ice) melts and forms liquid water. But a change of substance corresponds to a *chemical reaction*; the very identity of a substance changes to produce a new substance with its own particular characteristics and features. In this process, the original substance disappears and a new substance is formed.

chemical reaction: a transformation of one substance into another.

We encounter chemical reactions everywhere we turn. When we burn wood to create warmth, the original substances (oxygen from the air and wood) are transformed in a chemical reaction to produce water vapor, carbon dioxide, and ash. When we observe rust accumulating on iron, we are actually witnessing a slow chemical reaction which converts oxygen from the air and iron to the reddish brown material called iron(III) oxide. Other familiar examples of chemical reactions are the conversion of sugar to alcohol and carbon dioxide in the fermentation process, the tarnishing of silver and copper, the digestion of food, and the incorporation of fluoride into tooth enamel.

Fig. 4-1 The Three Kinds of Change

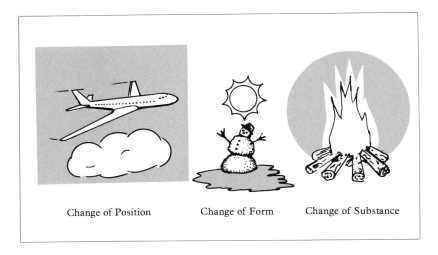

Change of Position Change of Form Change of Substance

CHEMICAL EQUATIONS AND MASS RELATIONSHIPS

The Language of Chemical Equations

Chemical equations represent concise and accurate descriptions of chemical change. An equation for the process of rust formation follows (O_2 indicates that gaseous oxygen exists as molecules containing two atoms each):

$$4\ Fe\ +\ 3\ O_2\ \longrightarrow\ 2\ Fe_2O_3$$
iron   oxygen   iron(III) oxide

The fermentation process can also be summarized by an equation:

$$C_6H_{12}O_6\ \longrightarrow\ 2\ C_2H_5OH\ +\ 2\ CO_2$$
glucose   ethanol   carbon dioxide

In each of these equations, formulas for the starting substances are shown on the left side of the arrow; these materials are called reactants. New substances produced by the chemical change are called products; their formulas are located to the right of the arrow. The arrow itself stands for words such as "produce" and "yield." The numbers in front of the formulas are *coefficients*, which indicate the smallest number of atoms or other chemical units that can participate in the reaction. The equation for rusting of iron can be expressed in words: Four atoms of iron react with three molecules of oxygen to produce two formula units of iron(III) oxide. (Note that iron(III) oxide is an ionic compound and thus does not exist as molecules; therefore we must refer to Fe_2O_3 as a formula unit of iron(III) oxide.)

Similarly, we can use words to paraphrase the equation for fermentation: One molecule of glucose yields two molecules of ethanol and two molecules of carbon dioxide.

coefficient: a small whole number placed in front of a formula to balance a chemical equation.

Occasionally you may see various other kinds of notations in chemical equations which are meant to indicate observations or conditions. Some of these are shown below.

    Solid   (s) or ↓
    Liquid (l)
    Gas   (g) or ↑
    Heat Δ

As you read through the chemical equations in this chapter, practice putting them into words. By doing this you will see that chemical equations actually represent complete sentences that are statements of fact.

Conservation of Mass

Through centuries of observations of chemical reactions, one principle of nature has always held true. This principle, the *law of conservation of mass*, is stated as follows: matter is neither created nor destroyed in a chemical reaction. Another way of stating this is that the total mass of substance which undergoes chemical change is equal to the total mass of new substance produced (see Figure 4–2).

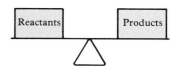

Fig. 4-2 The Mass of Reactants is Always Equal to the Mass of Products in a Chemical Reaction

Let us consider the law of conservation of mass in the chemical change that occurs when charcoal is burned. Although charcoal contains other minor ingredients, for most purposes it can be considered to be composed of elemental carbon. We can write an equation for the *combustion* of carbon as follows:

combustion: the process of burning.

$$C + O_2 \longrightarrow CO_2$$

In this equation, carbon and oxygen are the reactants and carbon dioxide is the product. If we were to measure the mass of reactants, we would find this total mass to be equal to the mass of carbon dioxide produced.

EXAMPLE 4.1

The elements hydrogen and fluorine exist as molecules containing two atoms each. Hydrogen and fluorine react to form hydrogen fluoride as shown in the following equation:

$$H_2 + F_2 \longrightarrow 2\ HF$$

If 2 grams of hydrogen and 38 grams of fluorine react, how much hydrogen fluoride is formed?

SOLUTION

In this reaction, hydrogen and fluorine are the reactants, and hydrogen fluoride is the product. Since the total mass of reactants is 40 grams, the mass of the product must be 40 grams:

$$\text{total mass of reactants} = \text{total mass of products}$$
$$2\,g + 38\,g = 40\,g$$

EXERCISE 4.1 When 5.00 grams of carbon is burned, 13.3 grams of oxygen is consumed. How many grams of carbon dioxide are produced?

$$C + O_2 \longrightarrow CO_2$$

EXERCISE 4.2 A sample of 8.50 grams of sulfur burns to produce 17.0 grams of sulfur dioxide. How many grams of oxygen are used in the reaction?

$$S + O_2 \longrightarrow SO_2$$

Balancing Chemical Equations

In order for an equation to describe a reaction accurately, the equation must be balanced. For example, when carbon burns, carbon dioxide is formed:

$$\text{carbon} + \text{oxygen} \longrightarrow \text{carbon dioxide}$$

Measurements have demonstrated that for every atom of carbon used up, one molecule of oxygen is also used, and one molecule of carbon dioxide is formed. We can now write a chemical equation based on this knowledge in which we show one atom of C reacting with one molecule of O_2 to produce one molecule of CO_2:

$$C + O_2 \longrightarrow CO_2$$

Notice that our equation is balanced as it stands. That is, there is the same number of like atoms on each side of the arrow, and the equation satisfies the law of conservation of mass (see Figure 4-3).

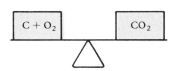

Fig. 4-3 The Balanced Equation: $C + O_2 \longrightarrow CO_2$

Let's try another example. The elements hydrogen and chlorine, which exist as diatomic molecules, react to form hydrogen chloride. We can write an initial equation containing formulas of reactants and product,

$$H_2 + Cl_2 \longrightarrow HCl$$

but we see that the equation is not balanced. There are two hydrogen atoms on the left and only one on the right; the same is true for chlorine. If we leave the equation this way, we are saying that one hydrogen atom and one chlorine atom disappear from the reaction; this is a violation of the law of conservation of mass (see Figure 4-4) and thus an impossible situation. We can rectify things by inserting the number 2 as a coefficient in front of HCl, and the equation now states the facts accurately: two molecules of hydrogen chloride are formed from one molecule of hydrogen and one molecule of chlorine:

$$H_2 + Cl_2 \longrightarrow 2\,HCl$$

The equation is now balanced, illustrating that atoms are neither created nor destroyed in the chemical reaction. In fact, the number of atoms is the same before and after the reaction; only the bonds have been rearranged to form a new compound. Figure 4-4 also shows the effect of balancing the equation, in terms of conservation of mass.

As you can see, balancing equations is only a matter of accurate bookkeeping. We need only know formulas to write the initial equation,

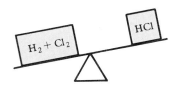

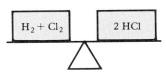

Fig. 4-4 Balancing the Equation $H_2 + Cl_2 \longrightarrow 2\,HCl$

and we can usually deduce these from our knowledge of valence and the periodic table. We can then balance the equation on the basis of the following rules:

1. Write an initial equation using correct formulas for reactants and products. Use + signs when there are two or more reactants or products, and separate reactants from products with an arrow.
2. Count the number of atoms of each element in the reactants and products.
3. Insert coefficients for one element at a time to balance the number of atoms of each element in the reactants and products.
4. Recount the number of atoms of each element in the reactants and products.
5. If you unbalance one element while balancing another, change the coefficients until you achieve a balanced equation for all elements.

Let's practice using these rules by writing a balanced equation for the formation of ammonia from nitrogen and hydrogen.

Step 1. Initial equation:

$$N_2 + H_2 \longrightarrow NH_3$$

Step 2. Count atoms:

Element	Reactants	Products
N	2	1
H	2	3

Step 3. Insert coefficients for nitrogen:

$$N_2 + H_2 \longrightarrow 2\,NH_3$$

Insert coefficients for hydrogen:

$$N_2 + 3\,H_2 \longrightarrow 2\,NH_3$$

Step 4. Recount atoms:

Element	Reactants	Products
N	2	2
H	6	6

Step 5. Change coefficients if necessary:

This step is not necessary in this case.

EXAMPLE 4.2

Write a balanced equation for the combustion of methane, CH_4, producing water and carbon dioxide. (Combustion means reaction with molecular oxygen, or in simpler terms, burning.)

SOLUTION

Step 1. Initial equation:

$$CH_4 + O_2 \longrightarrow H_2O + CO_2$$

Step 2. Count atoms:

Element	Reactants	Products
C	1	1
O	2	3
H	4	2

Step 3. Insert coefficients for oxygen:

$$CH_4 + 2\,O_2 \longrightarrow 2\,H_2O + CO_2$$

Now hydrogens are also balanced.

Step 4. Recount atoms:

Element	Reactants	Products
C	1	1
O	4	4
H	4	4

Step 5. Change coefficients if necessary:

This step is not necessary in this case.

EXAMPLE 4.3

Sodium reacts vigorously with water to form sodium hydroxide (NaOH) and hydrogen (H_2). Write a balanced equation for this reaction.

SOLUTION

Step 1. Initial equation:

$$Na + H_2O \longrightarrow NaOH + H_2$$

Step 2. Count atoms:

Element	Reactants	Products
Na	1	1
H	2	3
O	1	1

Step 3. Insert coefficients:

Since Na and O are balanced, we need only insert coefficients for H. Insertion of ½ before H_2 in the product will balance the H atoms:

$$Na + H_2O \longrightarrow NaOH + \tfrac{1}{2} H_2$$

However, we usually try to find the smallest whole-number coefficients, and therefore use of ½ is not permitted. We can multiply the entire equation by 2 to get a balanced equation with whole-number coefficients:

$$2\,Na + 2\,H_2O \longrightarrow 2\,NaOH + H_2$$

EXERCISE 4.3 Balance the following equations.

(a) $S + O_2 \longrightarrow SO_3$
(b) $C_2H_6 + O_2 \longrightarrow CO_2 + H_2O$
(c) $Na + Cl_2 \longrightarrow NaCl$
(d) $Na + O_2 \longrightarrow Na_2O$

Molecular Weights and Formula Weights

Chemical equations also provide information about masses of reactants and products. In order to understand those relationships, it is first necessary to understand some basic definitions. The *molecular weight* (MW) is the sum of atomic weights of all of the elements in a molecular element or compound. If two or more atoms of the same element are present in the molecule, the atomic weight of the element is added as many times as atoms of the element appear in the formula. Thus, the molecular weight of the O_2 molecule is twice the atomic weight of oxygen (2×16.0 amu), or 32.0 amu. The molecular weight of CO_2 is calculated as follows:

$$MW = \left(1\,\text{atom C} \times \frac{12.0\text{ amu}}{\text{atom C}}\right) + \left(2\,\text{atoms O} \times \frac{16.0\text{ amu}}{\text{atom O}}\right)$$

$$= 12.0\text{ amu} + 32.0\text{ amu}$$

$$= 44.0\text{ amu}$$

The idea of molecular weight cannot be applied to ionic compounds because they do not exist as molecules. To circumvent this problem, we use the more general term *formula weight* (FW), defined as the sum of atomic weights of all of the elements in a *formula unit* of the compound, each atomic weight being added as many times as atoms of that element appear in the formula unit. A formula unit is considered to be a unit of the ionic compound corresponding to its simplest formula. For example, the ionic compound NaCl is composed of many sodium ions and chloride ions, and one sodium ion is not joined to one specific chlorine ion, but a formula unit of NaCl contains one sodium ion and one chloride ion. Although formula weights can be used for both molecular and ionic compounds, they are usually applied only to ionic compounds. For NaCl, the formula weight is calculated as follows:

$$FW = \left(1\,\text{atom Na} \times \frac{23.0\text{ amu}}{\text{atom Na}}\right) + \left(1\,\text{atom Cl} \times \frac{35.5\text{ amu}}{\text{atom Cl}}\right)$$

$$= 23.0\text{ amu} + 35.5\text{ amu}$$

$$= 58.5\text{ amu}$$

It is common practice, but improper, to use the term molecular weight with ionic compounds.

EXAMPLE 4.4

Find the molecular weight for methane, CH_4, the principal constituent of natural gas.

SOLUTION

$$\begin{aligned} MW &= \left(1\,\cancel{\text{atom C}} \times \frac{12.0 \text{ amu}}{\cancel{\text{atom C}}}\right) + \left(4\,\cancel{\text{atoms H}} \times \frac{1.0 \text{ amu}}{\cancel{\text{atom H}}}\right) \\ &= 12.0 \text{ amu} + 4.0 \text{ amu} \\ &= 16.0 \text{ amu} \end{aligned}$$

EXAMPLE 4.5

Find the formula weight for calcium carbonate, $CaCO_3$, the ionic compound that is the major component of limestone and chalk.

SOLUTION

$$\begin{aligned} FW &= \\ &\left(1 \text{ atom Ca} \times \frac{40.1 \text{ amu}}{\cancel{\text{atom Ca}}}\right) + \left(1\,\cancel{\text{atom C}} \times \frac{12.0 \text{ amu}}{\cancel{\text{atom C}}}\right) + \left(3\,\cancel{\text{atoms O}} \times \frac{16.0 \text{ amu}}{\cancel{\text{atom O}}}\right) \\ &= 40.1 \text{ amu} + 12.0 \text{ amu} + 48.0 \text{ amu} \\ &= 100.1 \text{ amu} \end{aligned}$$

EXAMPLE 4.6

Find the formula weight for magnesium sulfate, $Mg(SO_4)_2$.

SOLUTION

This is a case where two polyatomic anions appear in the simplest formula unit; thus we must add atomic weights twice for all atoms appearing inside the parentheses.

$$\begin{aligned} FW &= \\ &\left(1\,\cancel{\text{atom Mg}} \times \frac{24.3 \text{ amu}}{\cancel{\text{atom Mg}}}\right) + \left(2\,\cancel{\text{atoms S}} \times \frac{32.1 \text{ amu}}{\cancel{\text{atom S}}}\right) + \left(8\,\cancel{\text{atoms O}} \times \frac{16.0 \text{ amu}}{\cancel{\text{atom O}}}\right) \\ &= 24.3 \text{ amu} + 64.2 \text{ amu} + 128.0 \text{ amu} \\ &= 216.5 \text{ amu} \end{aligned}$$

EXERCISE 4.4 Calculate molecular weights for the following compounds.

(a) CH_3CH_2OH, ethanol
(b) C_2H_4, ethene
(c) CO, carbon monoxide
(d) N_2O_3, dinitrogen trioxide
(e) CCl_4, carbon tetrachloride
(f) N_2H_4, hydrazine

EXERCISE 4.5 Calculate formula weights for the following compounds.
(a) $Ca(OH)_2$, calcium hydroxide
(b) $AlPO_4$, aluminum phosphate
(c) Na_2SO_4, sodium sulfate
(d) $Al(OH)_3$, aluminum hydroxide
(e) $Mg_3(PO_4)_2$, magnesium phosphate
(f) $NaHCO_3$, sodium bicarbonate

Molecular weights and formula weights can be used for many practical purposes, one of which is to calculate the percentage of an element in a compound. For example, an important concern to many people is the amount of sodium in their diets. Sodium chloride, NaCl (table salt), is usually the major dietary source of sodium, and we can calculate the percentage of sodium in the compound by using the formula weight of 58.5 amu calculated earlier:

$$\% \text{ Na} = \frac{\text{total mass of Na in NaCl}}{\text{formula weight of NaCl}} \times 100$$

$$= \frac{23.0 \text{ amu}}{58.5 \text{ amu}} \times 100$$

$$= 39.3 \%$$

Our calculation shows that sodium constitutes 39.3% of our dietary intake of table salt.

EXERCISE 4.6 Calculate the percentage of each element present in magnesium hydroxide, $Mg(OH)_2$, the main ingredient in milk of magnesia.

The Chemical Mole

Chemical equations tell us how many atoms and/or molecules react, but these are such small particles that we cannot count them out individually. However, we can weigh a sample of a substance composed of atoms or molecules and then calculate how many particles are present. To do this, we rely on the concept of the chemical mole.

A *mole* is defined as 6.02×10^{23} particles. This very large number, known as *Avogadro's number*, is named after the nineteenth century Italian scientist Amedeo Avogadro. We use the term *mole* to refer to any kind of chemical particles: ions, atoms, or molecules. It always stands for 6.02×10^{23} particles just as "dozen" always stands for 12 units or things. A unique feature of the chemical mole, and what makes it so important to chemistry, is that a mole always has a weight that corresponds to the formula weight or molecular weight of a substance expressed in grams. For example, the simplest unit of elemental carbon is a carbon atom, and thus the formula for carbon is C. The atomic weight for carbon is 12.0 amu, but if we express the atomic weight (in this case also the formula weight) of carbon in grams, 12.0 grams, then this quantity contains 6.02×10^{23} carbon atoms, or a mole of carbon. For the molecular compound HCl, the molecular weight expressed in grams is 36.5 grams. This is also the weight of 6.02×10^{23} molecules of HCl, or one mole. In the case of the ionic compound KCl, the formula weight expressed in grams, 74.6 grams, is the weight of 6.02×10^{23} formula units of KCl, or one mole of KCl. Since each formula unit of KCl contains two ions, K^+ and Cl^-, a mole of KCl contains a mole of K^+ ions and a mole of Cl^- ions.

As mentioned before, the concept of a mole as 6.02×10^{23} objects is quite similar to the concept of a dozen as 12 objects. By specifying the mole as a standard measure for a large number of small objects such as molecules, atoms, and ions, we have created a larger, more practical unit to work with. Additional examples of mole relationships are given in Table 4-1.

Table 4-1 Examples of Mole Relationships

Particles	One Mole Contains	One Mole Weighs
C atoms	6.02×10^{23} atoms	12.0 g
H_2 molecules	6.02×10^{23} molecules	2.0 g
Na atoms	6.02×10^{23} atoms	23.0 g
NaCl units	6.02×10^{23} NaCl units	58.5 g
Na^+ ions	6.02×10^{23} Na^+ ions	23.0 g
Cl^- ions	6.02×10^{23} Cl^- ions	35.5 g

EXAMPLE 4.7

Fructose, a molecular compound, is a sugar found in many fruits. Its formula is $C_6H_{12}O_6$. How many grams does one mole of fructose weigh?

SOLUTION

From its formula, we can find the molecular weight of fructose:

$$MW = (6 \text{ atoms C} \times \frac{12.0 \text{ amu}}{\text{atom C}}) + (12 \text{ atoms H} \times \frac{1.0 \text{ amu}}{\text{atom H}})$$

$$+ (6 \text{ atoms O} \times \frac{16.0 \text{ amu}}{\text{atom O}})$$

$$= 72.0 \text{ amu} + 12.0 \text{ amu} + 96.0 \text{ amu}$$

$$= 180.0 \text{ amu}$$

The molecular weight expressed in grams, 180.0 g, is the weight of one mole of fructose.

EXERCISE 4.7 Find the weight in grams of one mole of each of the following substances.
(a) the element hydrogen, H_2
(b) carbon dioxide, CO_2
(c) magnesium sulfate, $MgSO_4$
(d) potassium ions, K^+

By using the mole concept and Avogadro's number, we can interconvert masses and number of particles. For example, let's calculate the number of gold atoms in a pure gold nugget weighing 2.0 grams. We can solve this problem by using the unit-factor method. Here are our unit relationships:

$$1 \text{ mole Au} = 197.0 \text{ g Au}$$

$$1 \text{ mole Au} = 6.02 \times 10^{23} \text{ Au atoms}$$

Our mathematical expression is then

$$\text{Number of Au atoms} = 2.0 \text{ g Au} \times \frac{1 \text{ mole Au}}{197.0 \text{ g Au}} \times \frac{6.02 \times 10^{23} \text{ Au atoms}}{1 \text{ mole Au}}$$

$$= 6.1 \times 10^{21} \text{ Au atoms}$$

Figure 4-5 compares mole quantities of lead, gold, and water in terms of mass, density, and volume.

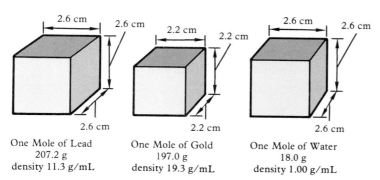

One Mole of Lead
207.2 g
density 11.3 g/mL

One Mole of Gold
197.0 g
density 19.3 g/mL

One Mole of Water
18.0 g
density 1.00 g/mL

Fig. 4-5 Mole Quantities of Lead, Gold, and Water

EXAMPLE 4.8

Sodium bicarbonate, $NaHCO_3$, is the chemical name of ordinary baking soda. This ionic compound is also an important component of human blood. How many sodium ions are in 17.0 grams of sodium bicarbonate?

SOLUTION

To solve this problem, we must first find the formula weight of sodium bicarbonate.

$$FW = \left(1 \text{ atom Na} \times \frac{23.0 \text{ amu}}{\text{atom Na}}\right) + \left(1 \text{ atom H} \times \frac{1.0 \text{ amu}}{\text{atom H}}\right)$$

$$+ \left(1 \text{ atom C} \times \frac{12.0 \text{ amu}}{\text{atom C}}\right) + 3 \text{ atoms O} \times \frac{16.0 \text{ amu}}{\text{atom O}}$$

$$= 23.0 \text{ amu} + 1.0 \text{ amu} + 12.0 \text{ amu} + 48.0 \text{ amu}$$

$$= 84.0 \text{ amu}$$

The formula weight expressed in grams, 84.0 grams, is the weight of one mole of sodium bicarbonate.

$$1 \text{ mole NaHCO}_3 = 84.0 \text{ g NaHCO}_3$$

Using our other unit relationships,

$$1 \text{ mole Na}^+ \text{ ions} = 6.02 \times 10^{23} \text{ Na}^+ \text{ ions}$$

$$1 \text{ mole NaHCO}_3 = 1 \text{ mole Na}^+ \text{ ions}$$

we can set up the following expression:

$$\text{number of Na}^+ \text{ ions} = 17.0 \text{ g NaHCO}_3 \times \frac{1 \text{ mole NaHCO}_3}{84.0 \text{ g NaHCO}_3}$$

$$\times \frac{6.02 \times 10^{23} \text{ Na}^+ \text{ ions}}{1 \text{ mole NaHCO}_3} = 1.22 \times 10^{23} \text{ Na}^+ \text{ ions}$$

Calculations and Chemical Equations

The numerical coefficients in a chemical equation tell us not only how many atoms or molecules react with each other; they also indicate how many moles of substances react. This latter information is of practical importance because we must work with mole collections of atoms and molecules, and the coefficients can be used to arrive at weight relationships. For example, two moles of hydrogen and one mole of oxygen react to form two moles of water according to the following equation:

$$2 H_2 + O_2 \longrightarrow 2 H_2O$$

The mole ratio of the reactants is 2:1. If we state this on a weight basis by converting moles to grams, we then have

$$4.0 \text{ g } H_2 + 32.0 \text{ g } O_2 \longrightarrow 36.0 \text{ g } H_2O$$

Thus, 32.0 grams of oxygen are needed to react with 4.0 grams of hydrogen in the formation of water. Notice once again that our weights are consistent with the law of conservation of mass.

Continuing with the example, the weight relationships may be used as ratios of reacting masses. This means that the reacting mass ratio of hydrogen to oxygen is 4:32. We could just as easily state this mass ratio as 1:8, 2:16, or even 16:128. The point is that the simplest mass ratio of the two reacting elements is 1:8. We are not restricted to mixing 4.0 grams of hydrogen with 32.0 grams of oxygen; we can mix any amounts of the two, but the amounts that react will have a mass ratio of 1:8. Stating this in terms of moles, we can mix any amounts of the two, but the reacting mole ratio will be 2:1. If we should mix more of either reactant than is needed, the excess will remain unreacted.

$2 H_2$	+	O_2	$\longrightarrow$	$2 H_2O$
2 moles		1 mole		2 moles
4.0 g		32.0 g		36.0 g

EXAMPLE 4.9

Magnesium metal reacts with oxygen to form magnesium oxide:

$$2 Mg + O_2 \longrightarrow 2 MgO$$

(a) What is the simplest mole ratio of reactants?
(b) What is the simplest mass ratio of reactants?

SOLUTION

(a) The mole ratio indicated by the balanced equation is 2:1. This is also the simplest mole ratio.
(b) Using the unit-factor method, we convert moles to grams:

$$2 \text{ moles Mg} \times \frac{24.3 \text{ g Mg}}{\text{mole Mg}} = 48.6 \text{ g Mg}$$

$$1 \text{ mole } O_2 \times \frac{32.0 \text{ g } O_2}{\text{mole } O_2} = 32.0 \text{ g } O_2$$

Thus, the mole ratio of 2:1 for the reactants gives a mass ratio of 48.6:32.0. The simplest mass ratio is found by dividing both numbers by the smaller of the two:

$$\text{simplest mass ratio} = \frac{48.6}{32.0} : \frac{32.0}{32.0}$$

$$= 1.52 : 1.00$$

EXERCISE 4.8 Chlorine, a gas which has diatomic molecules, reacts with sodium to form sodium chloride:

$$Cl_2 + 2 \text{ Na} \longrightarrow 2 \text{ NaCl}$$

(a) Find the simplest mole ratio of reactants.
(b) Find the simplest mass ratio of reactants.

The numerical coefficients in a balanced equation also allow us to predict the amount of product which will be formed. For example, carbon burns to form carbon dioxide as shown below.

$$C + O_2 \longrightarrow CO_2$$

Let's suppose that we intend to burn 8.00 grams of carbon. How many grams of oxygen will be needed, and how many grams of carbon dixoide will be produced? We can solve this problem in more than one way, but we will focus on the mole concept. The balanced equation says that the mole ratio of reactants is 1:1. But we have only 8.00/12.0 of a mole of carbon to react. From our mole ratio, we know that the same mole quantity of oxygen is required—8.00/12.0 mole. How many grams of oxygen does that quantity correspond to? Using a unit factor, we get

$$\frac{8.00}{12.0} \text{ moles } O_2 \times \frac{32.0 \text{ g } O_2}{\text{mole } O_2} = 21.3 \text{ g } O_2$$

Thus, 21.3 grams of oxygen are required to burn 8.00 grams of carbon.

And now we turn to the second question—how many grams of carbon dioxide will be produced? Once again, there are several ways to solve this problem, but we will use the mole concept. Since the mole ratio of carbon to carbon dioxide in the balanced equation is 1:1, if we start with

8.00/12.0 moles of carbon, we will produce 8.00/12.0 moles of carbon dioxide. Now all we have to do is convert that mole quantity to grams:

$$\frac{8.00}{12.0} \text{ moles CO}_2 \times \frac{44.0 \text{ g CO}_2}{\text{mole CO}_2} = 29.3 \text{ g CO}_2$$

We can check our answer by the law of conservation of mass:

$$\text{total mass of reactants} = \text{total mass of products}$$

$$8.00 \text{ g} + 21.3 \text{ g} = 29.3 \text{ g}$$

$$29.3 \text{ g} = 29.3 \text{ g}$$

EXAMPLE 4.10

Sodium reacts with diatomic molecules of bromine to form sodium bromide:

$$2 \text{ Na} + \text{Br}_2 \longrightarrow 2 \text{ NaBr}$$

(a) If 3.40 g of sodium are used in the reaction, how many grams of bromine will be needed?
(b) How many grams of sodium bromide will be produced?

SOLUTION

(a) The balanced equation says that the mole ratio of reactants is two to one. Thus, for every mole of sodium present, one-half mole of bromine is needed. First we need to find the number of moles of sodium available by the unit-factor method:

$$3.40 \text{ g Na} \times \frac{1 \text{ mole Na}}{23.0 \text{ g Na}} = 0.147 \text{ mole Na}$$

According to our reasoning,

$$0.147 \text{ mole Na} \times \frac{1 \text{ mole Br}_2}{2 \text{ mole Na}} = 0.0735 \text{ mole Br}_2$$

To find the number of grams of bromine needed, we convert moles of bromine to grams:

$$0.0735 \text{ mole Br}_2 \times \frac{159.8 \text{ g Br}_2}{\text{mole Br}_2} = 11.7 \text{ g Br}_2$$

(b) To find grams of sodium bromide produced, we refer to the balanced equation and find that the mole ratio for sodium and sodium bromide is 1:1. Thus the reaction will produce 0.147 mole of product. Converting moles of sodium bromide to grams, we get

$$0.147 \text{ mole NaBr} \times \frac{102.9 \text{ g NaBr}}{\text{mole NaBr}} = 15.1 \text{ g NaBr}$$

We can check our answer by the law of conservation of mass:

$$\text{total mass of reactants} = \text{total mass of products}$$

$$3.40 \text{ g} + 11.7 \text{ g} = 15.1 \text{ g}$$

$$15.1 \text{ g} = 15.1 \text{ g}$$

EXERCISE 4.9 Sodium and oxygen react to form sodium oxide:

$$4\ Na + O_2 \longrightarrow 2\ Na_2O$$

(a) If 7.69 grams of sodium are used in the reaction, how many grams of oxygen will be needed?
(b) How many grams of sodium oxide will be produced?

TYPES OF CHEMICAL REACTIONS

Although there is a seemingly infinite number of chemical reactions, study and analysis over a long period of time indicate that only a few basic types of reactions exist. Of course a balanced equation must be written, but at that point simple features of the equation allow us to classify the reaction. The following discussion summarizes those basic reaction types.

Single Replacement Reactions

Single replacement reactions occur when an ion, atom, or group of atoms replaces an ion, atom, or group of atoms in a compound. The general pattern is

$$A + BC \longrightarrow AC + B$$

in which B is replaced in the compound BC by A. An example is the reaction of copper with silver nitrate:

$$Cu + 2\ AgNO_3 \longrightarrow Cu(NO_3)_2 + 2\ Ag$$

In this case copper replaces silver in the nitrate compound.

single replacement reaction: also called a substitution or single displacement reaction.

Double Replacement Reactions

Double replacement reactions occur when two parts of two different reacting compounds replace each other. The general pattern for this reaction is

$$AB + CD \longrightarrow AC + BD$$

Examples are the reaction of barium chloride with sodium sulfate,

$$BaCl_2 + Na_2SO_4 \longrightarrow BaSO_4\downarrow + 2\ NaCl$$

and the reaction of potassium carbonate with lead nitrate,

$$K_2CO_3 + Pb(NO_3)_2 \longrightarrow PbCO_3\downarrow + 2\ KNO_3$$

As you can see from the general pattern, double replacement reactions are a kind of trading of ionic or atomic partners between compounds.

double replacement reaction: also called a double displacement reaction or metathesis reaction.

Combination Reactions

combination reaction: also called an addition reaction.

This type of reaction occurs when molecules are formed by combination of two or more atoms or simple compounds:

$$A + B \longrightarrow C$$

Some combustion reactions are examples of combination reactions:

$$C + O_2 \longrightarrow CO_2$$

$$S + O_2 \longrightarrow SO_2$$

$$2\,H_2 + O_2 \longrightarrow 2\,H_2O$$

Other examples are the formation of ammonia,

$$N_2 + 3\,H_2 \longrightarrow 2\,NH_3$$

the reaction of ethene with hydrogen chloride.

$$\underset{\underset{H}{|}}{\overset{\overset{H}{|}}{C}}=\underset{\underset{H}{|}}{\overset{\overset{H}{|}}{C}} \;+\; HCl \longrightarrow H-\underset{\underset{H}{|}}{\overset{\overset{H}{|}}{C}}-\underset{\underset{Cl}{|}}{\overset{\overset{H}{|}}{C}}-H$$

and the reaction of phosphorus with iodine,

$$2\,P + 3\,I_2 \longrightarrow 2\,PI_3$$

Decomposition Reactions

decomposition reaction: a chemical reaction in which a reactant is decomposed into two or more atoms or simpler compounds.

Decomposition reactions are easily recognized because they are just reversals of combination reactions:

$$C \longrightarrow A + B$$

Thus, all of the reaction products in the previous section on combination reactions can be decomposed into smaller units:

$$CO_2 \longrightarrow C + O_2$$

$$SO_2 \longrightarrow S + O_2$$

$$2\,H_2O \longrightarrow 2\,H_2 + O_2$$

Isomerization Reactions

isomers (EYE-sah-mers)

These reactions occur when a compound is converted to an *isomer*. The word isomer is derived from the Greek language and literally means "same" (iso) "parts" (mer). Isomers are compounds having the same numbers and kinds of atoms (parts), but the atoms are arranged in different structural forms.

Many examples of isomers are found in hydrocarbon molecules. Remember that hydrocarbons are binary compounds composed of hydrogen and carbon. Methane, CH_4, is the simplest hydrocarbon. All

Table 4-2 Some Hydrocarbons and Their Names

Structure	Name	Formula
H–C(H)(H)–H	methane	CH_4
H–C(H)(H)–C(H)(H)–H	ethane	C_2H_6
H–C(H)(H)–C(H)(H)–C(H)(H)–H	propane	C_3H_8
H–[C(H)(H)]$_4$–H	butane	C_4H_{10}
H–[C(H)(H)]$_5$–H	pentane	C_5H_{12}
H–[C(H)(H)]$_6$–H	hexane	C_6H_{14}
H–[C(H)(H)]$_7$–H	heptane	C_7H_{16}
H–[C(H)(H)]$_8$–H	octane	C_8H_{18}
H–[C(H)(H)]$_9$–H	nonane	C_9H_{20}
H–[C(H)(H)]$_{10}$–H	decane	$C_{10}H_{22}$

others contain more than one carbon atom, and the carbons are bonded to each other. We can visualize hydrocarbon molecules as being chains of carbon atoms to which hydrogen atoms are attached. Examples of hydrocarbons are given in Table 4-2.

The smallest hydrocarbon molecule to have isomers is butane.

$$\begin{array}{c} \text{H H H H} \\ | \;\; | \;\; | \;\; | \\ \text{H}-\text{C}-\text{C}-\text{C}-\text{C}-\text{H} \\ | \;\; | \;\; | \;\; | \\ \text{H H H H} \end{array} \qquad CH_3-CH_2-CH_2-CH_3$$

butane (C_4H_{10})

As you can see, butane has four carbon atoms. Those on the ends of the chain are bonded to three hydrogens and one carbon. The two interior carbons are each bonded to two hydrogens and two carbons.

Another molecule exists which has the same formula as butane. It is an isomer of butane because it has the same atoms, but they are arranged so that the structure is different from that of butane:

$$\begin{array}{c} \text{H H H} \\ | \;\; | \;\; | \\ \text{H}-\text{C}-\text{C}-\text{C}-\text{H} \\ | \;\; | \;\; | \\ \text{H} \;\; | \;\; \text{H} \\ \text{H}-\text{C}-\text{H} \\ | \\ \text{H} \end{array} \qquad \begin{array}{c} CH_3-CH-CH_3 \\ | \\ CH_3 \end{array}$$

isomer of butane (C_4H_{10})

The isomer resembles butane, but notice that there are now three end carbons, each bonded to three hydrogens and one carbon. Also, there is only one interior carbon, and it is bonded to three other carbons and only one hydrogen. The isomer is an entirely different compound from butane and has its own identifying physical characteristics.

Butane can be converted to its isomer at high temperature. This reaction is then referred to as *isomerization reaction*:

$$CH_3-CH_2-CH_2-CH_3 \xrightarrow{\Delta} \begin{array}{c} CH_3-CH-CH_3 \\ | \\ CH_3 \end{array}$$

butane   isomer of butane
C_4H_{10} C_4H_{10}

The general pattern for an isomerization reaction is A ⟶ A★, where A and A★ are isomers.

Another example of an isomerization reaction is the conversion of urea to ammonium cyanate:

$$\begin{array}{c} \text{O} \\ \| \\ H_2N-C-NH_2 \end{array} \longrightarrow NH_4OCN$$

urea   ammonium cyanate
CH_4N_2O CH_4N_2O

This reaction has been suggested to be the basis for the use of urea to treat sickle-cell anemia. It is thought that when urea is administered to patients, it isomerizes to ammonium cyanate. This compound appears to be effective in preventing red blood cells from taking on the characteristic crescent shape typical of the disease.

EXERCISE 4.10 Classify each of the following reactions as single replacement, double replacement, combination, decomposition, or isomerization.

(a) $2\,P + 3\,H_2 \longrightarrow 2\,PH_3$

(b)
```
    H H                    H   H
    | |                    |   |
  H-C-C-C≡C-H  ⟶    H-C-C≡C-C-H
    | |                    |   |
    H H                    H   H
```

(c) $2\,Na + 2\,HOH \longrightarrow 2\,NaOH + H_2$

(d) $H_2CO_3 \longrightarrow CO_2 + H_2O$

(e) $HCl + NaOH \longrightarrow NaCl + HOH$

Oxidation-Reduction Reactions

The *oxidation process* has long been associated with reactions of various substances with oxygen. In this sense, combustion reactions are oxidation processes. However, in more recent times the term oxidation has taken on a more general usage based on the idea of *oxidation numbers*. These are a bookkeeping device used to keep track of covalent bond formation and electron transfers between ions and compounds during oxidation processes. In order to be consistent in the use of oxidation numbers, we must use the following set of rules:

oxidation number: a signed whole number assigned to an element in a compound to indicate the relative state of oxidation.

1. Uncombined (pure) elements have an oxidation number of zero. Thus, H_2, O_2, Na, S, and N_2 have oxidation numbers of zero.

2. Hydrogen in compounds usually has an oxidation number of $+1$; under certain conditions the oxidation number of hydrogen may be different, but we are not likely to encounter those cases.

3. Except for the transition metals, metallic elements in compounds have positive oxidation numbers corresponding to their valences. Thus, alkali metals (Group IA) have oxidation numbers of $+1$, and alkaline earth metals (Group IIA) have oxidation numbers of $+2$. Transition elements have variable oxidation numbers which must be deduced by Rule 5 below.

4. Oxygen in compounds usually has an oxidation number of -2; under certain conditions the oxidation number for oxygen may be different, but we are not likely to encounter those cases.

5. The sum of oxidation numbers for all atoms or ions in a compound is zero, and the sum of oxidation numbers for all atoms in a polyatomic ion is equal to the charge on the ion. Oxidation numbers for elements not included in Rules 1-4 may be deduced in the context of their compounds and polyatomic ions.

We can use these rules to find oxidation numbers for the elements in the compound sodium sulfate, Na_2SO_4. From Rule 3, we know that sodium has an oxidation number of $+1$; thus the sum of oxidation numbers for sodium is $+2$. Rule 4 tell us that oxygen has an oxidation number of -2; therefore the sum of oxidation numbers for oxygen is -8. At this point we have assigned oxidation numbers to all elements in the compound except sulfur, and we have a total of $+2$ and -8 for positive and negative oxidation numbers, or a net oxidation number of -6. Since Rule 5 states that the sum of oxidation numbers in a compound is zero, then the oxidation number for sulfur must be a $+6$ to balance the -6. We can now write oxidation numbers for each element in the compound,

Compound Na_2SO_4

Oxidation numbers $+1$ $+6$ -2

and we can summarize our calculations in the following table:

Na_2SO_4

Ions or Atoms	Individual Oxidation Numbers	Sum of Oxidation Numbers = 0
2 Na	$+1$	$+2$
S	$+6$	$+6$
4 O	-2	-8
		0

EXAMPLE 4.11

Find oxidation numbers for each element in magnesium chloride, $MgCl_2$.

SOLUTION

Rule 3 tells us that magnesium has an oxidation number of $+2$. Since there are two chloride ions, each must have an oxidation number of -1, to give a sum of zero.

$MgCl_2$

Ions or Atoms	Individual Oxidation Numbers	Sum of Oxidation Numbers = 0
Mg	$+2$	$+2$
2 Cl	-1	-2
		0

EXERCISE 4.11 Find the oxidation numbers for each element in potassium permanganate, $KMnO_4$.

Now that you have learned to assign oxidation numbers, let's see how they apply to oxidation processes. Our first illustration is the combustion of carbon, a combination reaction which is also an oxidation process. The oxidation numbers are shown below the formulas in the equation:

$$C + O_2 \longrightarrow CO_2$$
$$0 \quad\quad 0 \quad\quad\quad +4\;-2$$

(Note that we show a -2 under oxygen in CO_2. Even though there are two oxygen atoms, we show only the oxidation number of the element.) We were able to deduce the oxidation number for carbon from Rule 5. Two oxygen atoms give a total of -4; therefore carbon must be $+4$.

In this example, the oxidation number for carbon increased from 0 to +4. This increase corresponds to the four valence electrons of carbon which are used to form covalent bonds in CO_2:

$$\ddot{\underset{..}{O}}=C=\ddot{\underset{..}{O}}$$

When the oxidation number for an element increases, the element undergoes *oxidation*. The element is said to be oxidized, and the element or compound responsible for oxidation is called the *oxidizing agent*, or *oxidant*. In this case the oxidizing agent is oxygen.

In contrast to carbon, the oxidation number for oxygen decreases from 0 to −2, corresponding to utilization of two valence electrons on each oxygen atom in forming covalent bonds with carbon. Elements whose oxidation numbers decrease in the course of a reaction are said to be reduced. Oxygen therefore undergoes *reduction* in this reaction. The *reducing agent*, or *reductant*, is carbon.

The changes in oxidation number during the oxidation and reduction processes are illustrated in Figure 4-6.

The oxidation process can be thought of as a combination of oxidation and reduction reactions. For this reason the term *oxidation-reduction* (or *redox*) *reaction* is used, and in this context, every oxidation is accompanied by a reduction. You will see that oxidation-reduction reactions often fit into one of the other five reaction categories also. Another illustration is the reaction of iron with oxygen to form ionic iron(III) oxide, Fe_2O_3. The changes in oxidation numbers are shown below.

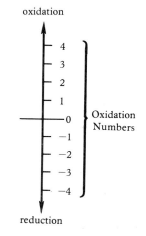

Fig. 4-6 A Positive Change in Oxidation Number Corresponds to Oxidation, and A Negative Change in Oxidation Number Corresponds to Reduction.

$$4\ Fe\ +\ 3\ O_2\ \longrightarrow\ 2\ Fe_2O_3$$

Reactants		Products
Fe	O_2	Fe_2O_3
0	0	+3 −2

In this case, iron is oxidized and oxygen is reduced. (Note that this reaction could be classified also as a combination reaction.)

Oxidation-reduction reactions are essential processes in living organisms. For example, animal cells are dependent on the oxidation of glucose, $C_6H_{12}O_6$, as a source of energy. (For now we will restrict ourselves to oxidation-reduction, but in later chapters we will learn about the energy aspects of the reaction.) In this reaction glucose is oxidized and oxygen is reduced; the products are carbon dioxide and water:

$$C_6H_{12}O_6\ +\ 6\ O_2\ \longrightarrow\ 6\ CO_2\ +\ 6\ H_2O\ +\ \text{energy}$$

Glucose is an example of a molecule composed of many atoms, and assigning oxidation numbers to each atom could become very tedious. For this reason, alternate descriptions for oxidation and reduction are used with large molecules. In these cases, the following statements apply:

Oxidation occurs when a substance either gains oxygen atoms or loses hydrogen atoms.

Reduction occurs when a substance either loses oxygen atoms or gains hydrogen atoms.

> **PERSPECTIVE: Chemical Reactions and Life Processes**
>
> Every living organism can be thought of as a chemical factory. Within the cells of plants and animals a multitude of chemical reactions occur continuously to furnish the organism with energy and molecules needed to synthesize cellular components. These chemical reactions are organized into intricate systems in which products of some reactions become reactants in other reactions. The entire organization of chemical reactions in living cells is called metabolism. In metabolic processes, a vast spectrum of reaction types can be observed.
>
> Animals are dependent on their dietary supplies of food to furnish molecules for metabolic processes. Food is degraded to small molecules by a series of decomposition reactions in the digestive process:
>
> $$\text{food} \longrightarrow \text{small molecules}$$
>
> Then these small molecules can be oxidized to generate energy, or they can be used to synthesize cellular components through a series of isomerization, replacement, and combination reactions:
>
> $$\text{small molecules} + O_2 \longrightarrow CO_2 + H_2O + \text{energy}$$
>
> or
>
> $$\text{small molecules} \longrightarrow \text{cellular components}$$
>
> In addition, living organisims continuously degrade and reform cellular components. A steady supply of food provides animals with energy and raw materials with which to continue growth and development.

Thus glucose is oxidized because it gains oxygen to form carbon dioxide, and oxygen is reduced because it gains hydrogen to form water. For reactions of large molecules, it is convenient to use these alternate, but equally valid, descriptions of oxidation and reduction.

We can summarize our descriptions of oxidation and reduction in the following table:

Oxidation	Reduction
increase in oxidation number	decrease in oxidation number
loss of electrons	gain of electrons
gain of oxygen atoms	loss of oxygen atoms
loss of hydrogen atoms	gain of hydrogen atoms

You can see that regardless of how they are described, oxidation and reduction are always opposites of each other. The general equation for an oxidation-reduction reaction is

$$A\,(\text{ox}) + B\,(\text{red}) \longrightarrow A\,(\text{red}) + B\,(\text{ox})$$

In this equation, reactant A is in the oxidized form and reactant B is in the reduced form. In the reaction A (ox) becomes reduced to A (red), and B (red) is oxidized to B (ox).

EXERCISE 4.12 In each of the following redox equations, indicate which reactant is oxidized and which reactant is reduced.

(a) $Zn + 2\,HCl \longrightarrow ZnCl_2 + H_2$
(b) $2\,NaBr + Cl_2 \longrightarrow 2\,NaCl + Br_2$
(c) $2\,C_2H_4O + O_2 \longrightarrow 2\,C_2H_4O_2$
(d) $C_6H_{12} \longrightarrow C_6H_6 + 3\,H_2$

SUMMARY

Chemical Equations and Mass Relationships: Chemical equations are concise and accurate descriptions of chemical change; they depict the conversion of reactants to products. Chemical reactions obey the law of conservation of mass: as substances undergo chemical change, no change in total mass is observed. Chemical equations must contain correct formulas and must be balanced so that the number of atoms neither increases nor decreases. Balanced equations indicate the number of atoms and molecules involved in a reaction.

Molecular weights are the sum of atomic weights for all atoms in a molecule. Formula weights are the sum of atomic weights for all ions and atoms in a formula unit of a compound. Although formula weights can be used for both ionic and molecular compounds, they are usually applied to ionic compounds. A molecular weight or a formula weight expressed in grams contains one mole, or 6.02×10^{23} molecules or formula units. Coefficients in a balanced equation can be used to devise mole ratios and mass ratios for reactants and products.

Types of Chemical Reactions: Chemical reactions may be classified according to the following types:

Single replacement reaction	$A + BC \longrightarrow AC + B$
Double replacement reaction	$AB + CD \longrightarrow AC + BD$
Combination reaction	$A + B \longrightarrow C$
Decomposition reaction	$C \longrightarrow A + B$
Isomerization reaction	$A \longrightarrow A^*$
Oxidation-Reduction reaction	$A(ox) + B(red) \longrightarrow A(red) + B(ox)$

Oxidation numbers are used to determine which substance is oxidized and which is reduced in an oxidation-reduction reaction. Oxidation corresponds to an increase in oxidation number, a loss of electrons, a gain of oxygen atoms, or a loss of hydrogen atoms. Reduction is characterized as the opposite of oxidation. Oxidation-reduction reactions often fit into one of the other five reaction categories.

STUDY QUESTIONS AND PROBLEMS

1. Define the following terms:
   (a) chemical change
   (b) chemical equation
   (c) combustion
   (d) molecular weight
   (e) formula unit
   (f) formula weight
   (g) mole
   (h) Avogadro's number
   (i) mole ratio
   (j) single replacement reaction
   (k) double replacement reaction
   (l) combination reaction
   (m) decomposition reaction
   (n) isomerization reaction
   (o) oxidation
   (p) reduction
   (q) oxidation-reduction reaction

2. In your own words, summarize the law of conservation of mass.

3. Balance the following equations:
   (a) $Mg + HCl \longrightarrow MgCl_2 + H_2$
   (b) $CH_4 + Cl_2 \longrightarrow CCl_4 + HCl$
   (c) $CH_3OH + O_2 \longrightarrow CO_2 + H_2O$
   (d) $PCl_3 + H_2O \longrightarrow P(OH)_3 + HCl$
   (e) glucose $(C_6H_{12}O_6) \longrightarrow$ fructose $(C_6H_{12}O_6)$
   (f) $HCl + NH_3 \longrightarrow NH_4Cl$
   (g) $KClO_3 \longrightarrow KCl + O_2$
   (h) $KBr + Cl_2 \longrightarrow KCl + Br_2$
   (i) $HgO \longrightarrow Hg + O_2$
   (j) $H_2SO_4 + NaCN \longrightarrow HCN + Na_2SO_4$

4. Distinguish between molecular weight and formula weight.

5. Calculate the molecular weight for each of the following molecular compounds.
   (a) $C_6H_{12}O_6$, glucose
   (b) CF_4, carbon tetrafluoride
   (c) N_2O_5, dinitrogen pentoxide
   (d) PCl_3, phosphorus trichloride
   (e) C_6H_{14}, hexane
   (f) $C_{17}H_{20}O_6N_4$, riboflavin
   (g) $C_6H_6O_6$, ascorbic acid (vitamin C)
   (h) $C_6H_5O_2N$, niacin
   (i) $C_{15}H_{11}I_4NO_4$, thyroxine (thyroid hormone)
   (j) $C_8H_8O_3$, oil of wintergreen

6. Calculate the formula weight for each of the following ionic compounds:
   (a) $Al(BrO_3)_3$, aluminum bromate
   (b) $Ba(CN)_2$, barium cyanide
   (c) $CaCr_2O_4$, calcium chromite
   (d) $Cr_2(SO_4)_3$, chromium(III) sulfate
   (e) $Co(C_2H_3O_2)_3$, cobalt(III) acetate
   (f) $Cu(NO_3)_2$, copper(II) nitrate
   (g) $FeBr_3$, iron(III) bromide
   (h) Li_2SO_4, lithium sulfate
   (i) Na_2HPO_4, sodium hydrogen phosphate
   (j) $Mg(ClO_4)_2$, magnesium perchlorate

7. Calculate the number of moles in each of the following:
   (a) 64.7 g of $MnCl_3$, manganese(III) chloride
   (b) 437 g of Hg_2CO_3, mercury(I) carbonate
   (c) 72.5 g of Ne (exists as single atoms)
   (d) 729 g of $Ni(NO_3)_2$, nickel nitrate
   (e) 58.6 g of NCl_3, nitrogen trichloride
   (f) 100 g of $C_6H_{12}O_6$, glucose
   (g) 943 g of $HC_2H_3O_2$, acetic acid
   (h) 835 g of H_2O, water
   (i) 46.8 g of NaCN, sodium cyanide

8. It is estimated that automobiles in a large city produce 9.1×10^6 kg of carbon monoxide (CO) in a year. How many moles of this toxic gas are formed each year in a typical large city?

9. Aspirin, $C_9H_8O_4$, is the most popular analgesic (pain reliever) in the United States. How many moles are there in a 325-mg tablet of aspirin?

10. How many molecules are present in a drop of water weighing 0.050 g?

11. A sealed container holds 50.0 g of hydrogen gas (H_2). Calculate the number of moles and the number of molecules of hydrogen in the container.

12. Calculate the percentage of each element in each of the following compounds.
    (a) Fe_2O_3, iron(III) oxide
    (b) $CaCl_2$, calcium chloride
    (c) $C_3H_5(NO_3)_3$, nitroglycerin
    (d) $Al_2(SO_4)_3$, aluminum sulfate
    (e) $Ca(CN)_2$, calcium cyanide

13. For the following reaction, how many grams of oxygen are required by 3.10 g of phosphorus?
    $$4 P + 5 O_2 \longrightarrow 2 P_2O_5$$

14. When a sample of $KClO_3$, potassium chlorate, was heated, 0.96 g of oxygen was liberated. How much $KClO_3$ was present?
    $$2 KClO_3 \xrightarrow{\Delta} 2 KCl + 3 O_2$$

15. The following reaction is used for the commercial production of hydrogen chloride gas. How many grams of HCl may be obtained by heating 234 g of NaCl with excess H_2SO_4?
    $$2 NaCl + H_2SO_4 \xrightarrow{\Delta} Na_2SO_4 + HCl$$

16. Calculate the number of moles each of CO_2 and H_2O that are produced by burning 104 g of C_2H_2, ethyne (acetylene).

17. The following reaction can be used to produce AgBr, silver bromide, a substance used in photography. Calculate the number of grams of each reactant needed to produce 93.3 g of AgBr.
    $$AgNO_3 + NaBr \longrightarrow AgBr + NaNO_3$$

18. The leavening action of baking powder (a mixture of cream of tartar and baking soda) is described by the following equation:

    $$KHC_4H_4O_6 + NaHCO_3 \longrightarrow KNaC_4H_4O_6 + H_2O + CO_2$$

potassium hydrogen tartrate	sodium bicarbonate
"cream of tartar"	"baking soda"

    How many grams of baking soda would be required to make a sample of baking powder containing 50.0 g of cream of tartar?

19. When a piece of magnesium weighing 0.32 g was burned, the product weighed 0.53 g.
    (a) Write a balanced equation for the reaction.
    (b) Calculate the moles of oxygen required for the reaction.
    (c) Calculate the number of grams of oxygen required for the reaction.

20. Sodium metal (Na) reacts vigorously with water to form sodium hydroxide (NaOH) and hydrogen (H_2). Write a balanced equation for this reaction and calculate the number of grams of hydrogen produced by 15.0 g of sodium.

21. Calcium oxide (CaO), commonly called lime, reacts with water to produce calcium hydroxide, $Ca(OH)_2$, sometimes referred to as "slaked" lime. Write a balanced equation for this reaction and calculate how many grams of slaked lime would be produced by 100 g of lime.

22. How many grams of aluminum (Al) are needed to react with sulfur (S) to form 600 g of aluminum sulfide, Al_2S_3?

23. Heating sodium nitrate, $NaNO_3$, decomposes it to sodium nitrite, $NaNO_2$, and oxygen, O_2. How many grams of $NaNO_3$ are needed to produce 1.50 g of O_2?

24. Solid calcium oxide, CaO, will absorb water vapor, H_2O, from the air to change completely to calcium hydroxide, $Ca(OH)_2$. If a sample of CaO showed a weight gain of 0.289 g upon being exposed to moist air, how much $Ca(OH)_2$, in moles, was formed?

25. How many grams of O_2 would be needed to burn 500 g of gasoline, C_8H_{18}? The combustion products are carbon dioxide, CO_2, and water, H_2O.

26. Aluminum carbide, Al_4C_3, reacts with water, H_2O, to produce methane, CH_4, and aluminum hydroxide, $Al(OH)_3$. How many grams of $Al(OH)_3$ can be produced from the reaction of 1.00 g of Al_4C_3 with an excessive amount of water?

27. Butane, C_4H_{10}, is the fuel of disposable cigarette lighters. How many grams of carbon dioxide, CO_2, will be produced from burning 7.62 g of butane? The other combustion product is water, H_2O.

28. Classify the reactions in question 3 by type. (Some of the reactions may fit into more than one category; in those cases, specify all pertinent reaction types.)

29. Assign oxidation numbers to each element in the following compounds.

   (a) H_3PO_4
   (b) KNO_3
   (c) N_2O_3
   (d) H_2S
   (e) MnO_2
   (f) CaF_2
   (g) HNO_2
   (h) Na_2SO_3
   (i) $NaHCO_3$
   (j) KI

30. Identify the oxidizing agent and the reducing agent in each of the following reactions.

   (a) $Ca + H_2SO_4 \longrightarrow CaSO_4 + H_2$
   (b) $2 K_2O \longrightarrow 4 K + O_2$
   (c) $C_2H_4 + H_2 \longrightarrow C_2H_6$
   (d) $4 Na + O_2 \longrightarrow 2 Na_2O$
   (e) $Fe + 2 HCl \longrightarrow FeCl_2 + H_2$
   (f) $2 CH_3CHO + O_2 \longrightarrow 2 CH_3CO_2H$
   (g) $Mg + 2 H_2O \longrightarrow Mg(OH)_2 + H_2$
   (h) $2 CH_3OH + O_2 \longrightarrow 2 CH_2O + 2 H_2O$
   (i) $Fe_2O_3 + 3 H_2 \longrightarrow 2 Fe + 3 H_2O$
   (j) $2 S + 3 O_2 \longrightarrow 2 SO_3$

CHAPTER 5
GASES

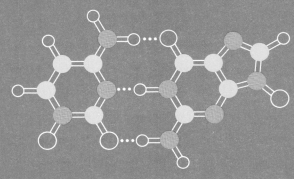

PROPERTIES OF GASES
- Mass and Volume
- Pressure
- Compressibility
- Gaseous Expansion and Contraction
- Diffusion

RELATIONSHIPS OF VOLUME, TEMPERATURE, AND PRESSURE
- Volume and Pressure
- Volume and Temperature
- Pressure and Temperature
- More Calculations with the Gas Laws

INTERPRETATION OF GASEOUS BEHAVIOR
- Avogadro's Hypothesis
- The Kinetic Molecular Theory

GAS MIXTURES

REAL GASES

Each of the three states of matter—gas, liquid, and solid—has its own set of properties that are independent of the particular substance being studied. Thus, water vapor has certain properties because of its existence as a gas; if water vapor is cooled so that it condenses into liquid water, the identical substance, water, will have different properties as a liquid. If liquid water is frozen, it then takes on the properties of a solid.

Several properties are typical of gases. For example, a gas can be compressed, as when we force air into a tire with a pump. When a gas is heated, it expands and rises, as hot air rises over a heater. If a gas is cooled sufficiently, it condenses to a liquid, as many of us have observed by breathing onto the cold surface of a window in winter. The earth's weather is largely the result of changes in the air's gaseous properties, such as pressure and temperature. Clouds, precipitation (rain [Figure 5-1], sleet, hail, snow), storms, wind, and clear skies are all weather occurrences which can be explained by the properties of gases.

Fig. 5-1 A Rainstorm Over Farmland in Missouri

The ability of gases to be compressed and to expand is used in many ways in industry and everyday life. The internal-combustion engine, steam engine, jet engine, and rocket all utilize the properties of gases. So do cooking, welding, deep-sea diving, aircraft, windmills, and mechanical respirators.

PROPERTIES OF GASES

Mass and Volume

Although our earth is surrounded by a blanket of gaseous atmosphere, scientists did not begin to investigate the properties of gases until the seventeenth century, and even then there were gross misconceptions.

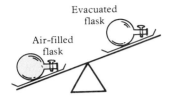

Fig. 5-2 A Flask Filled with Air Weighs More Than An Identical Flask That Is Evacuated

Theories of gaseous behavior began to develop in the nineteenth century and are still evolving. Even so, each of us is aware that gases are a form of matter, and as such, they have mass and occupy space. The fact that gases have *volume* permitted the invention of inflatable rubber tires to replace the solid rubber ones used on early automobiles. We can demonstrate that gases have *mass* by evacuating a rigid container and weighing it; we find that the air-filled container weighs more than the evacuated container (see Figure 5-2). Although we take these properties of gases for granted now, much controversy brewed over them in the seventeenth century.

Pressure

Another characteristic property of gases is that they exert *pressure*. The term pressure implies the idea of a force, something that tends to move an object in a given direction. In scientific terms, pressure is a force that acts on a given surface area. One of the common units of force is the pound, and we often measure pressure (for example, tire pressure) in pounds per square inch (psi).

The most famous demonstration of the pressure exerted by gases was the experiment performed in 1654 by Otto von Guericke with his "Magdeburg hemispheres." Von Guericke placed two large metal cups (hemispheres) together to make a sphere; then he evacuated the sphere with an air pump (Figure 5-3). A team of horses attached to each hemisphere could not pull the two halves apart (Figure 5-4), yet the only force holding the two hemispheres together was the pressure of the atmosphere.

Fig. 5-3 Von Guericke's Experimental Equipment. On the right at the top are the famous "Magdeburg Hemispheres."

Gas pressure is often measured with a barometer much like the first one invented by Evangelista Torricelli in 1643. Torricelli filled a four-foot glass tube with mercury and inverted the tube into a dish (see Figure 5-5). He observed that some of the mercury did not flow out. After much study, he concluded that the variation of the height of the mercury from day to day was caused by changes in atmospheric pressure. We now know that the mercury was prevented from running out of the tube by the pressure that the atmosphere exerted against the surface of the mercury in the dish.

Gas pressure is often expressed in millimeters of mercury, corresponding to the height of a column of mercury supported by the gas pressure.

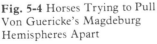

Fig. 5-4 Horses Trying to Pull Von Guericke's Magdeburg Hemispheres Apart

One millimeter of mercury is called one *torr*, in honor of Torricelli. The normal pressure of the atmosphere at sea level supports a mercury column 760 millimeters high, so a pressure of 1 atmosphere (atm) is 760 torr.

Compressibility

A fourth property of gases is their *compressibility*. When we fill an automobile tire, we actually compress a large volume of air into the tire, and by doing so, we create pressure on the tire walls from the air within. Gases commonly used in hospitals, such as oxygen and anesthetics, are stored by compressing large volumes into strong metal cylinders such as those shown in Figure 5-6.

torr: the pressure necessary to support a column of mercury 1 millimeter high at 0°C.

compressibility: the ability to be forced into less space.

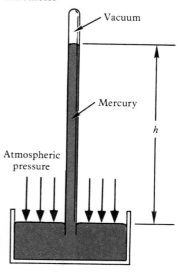

Fig. 5-5 Torricelli's Mercury Barometer

Fig. 5-6 Compressed Gases Are Stored in Strong Metal Cylinders. These gases are used in chemical analysis; the gauges on the top of each cylinder register the pressure exerted by the gas inside.

Gaseous Expansion and Contraction

Temperature influences a fifth property of gases, the ability to *expand* and *contract*. When gases are heated, they expand and become less dense. When fire heats air (Figure 5-7), the less dense hot air rises and the more dense cool air flows downward to replace it.

Because of their tendency to expand when heated, gases confined in containers exert more pressure when warmed. Thus, we are cautioned not to incinerate aerosol spray cans; the intense heat of incineration causes great pressure increases inside the cans, and they can easily explode. A related example is the increase in automobile tire pressure during high-speed driving. The friction created by the tire on the road

Fig. 5-7 Hot Air Rises Over a Fire Because It is Less Dense Than the Surrounding Air

surface generates heat, which in turn causes the air pressure to increase inside the tire. After driving has stopped and the tires have cooled down, the tire pressure returns to its original value.

Diffusion

Another property of gases is *diffusion*, the intermingling of one type of gas with another to form a *homogeneous* mixture. Two or more gases form homogeneous mixtures in all proportions, regardless of the identities of the gases. Anyone who has ever noticed the odor "essence of skunk" in the woods can confirm the ability of one gas to diffuse through other gases. In addition, gases not only diffuse through each other, but also through some solid substances. Thus, even a tight plastic wrapper cannot contain the odor of certain strong cheeses (see Figure 5–8).

homogeneous (hoe-moe-GEE-nee-us): uniform throughout.

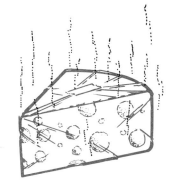

Fig. 5-8 A Tight Plastic Wrap Cannot Contain the Odor of Strong Cheese

RELATIONSHIPS OF VOLUME, TEMPERATURE, AND PRESSURE

We have seen that gases have mass, occupy volume, exert pressure, and are influenced by temperature. Thus, there are four variables to consider in working with gases: mass, volume, pressure, and temperature. To facilitate the study of gases, two of the variables may be held constant while the relationship between the other two is studied.

Volume and Pressure

The English chemist Robert Boyle performed the first careful studies of gases. In fact, it was his work with gases in the seventeenth century that led him to the concept of chemical elements. In 1662 Boyle investigated the relationship between the volume and pressure in a given mass of gas at constant temperature. Boyle's experiments consisted of measuring the volume of a mass of gas as he increased the pressure of the gas. Boyle found that the pressure and volume of a gas are inversely proportional.

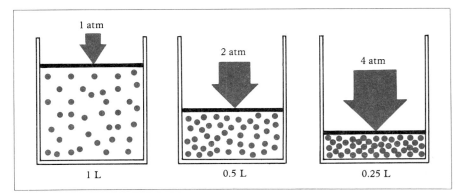

Fig. 5-9 The Effect of Pressure on Gas Volume

As one increases, the other decreases by the same proportion, as illustrated in Figure 5-9. We can state *Boyle's law* mathematically as

$$P \propto \frac{1}{V},$$

Boyle's law: the volume of a gas varies inversely with the pressure applied to the gas at constant temperature.

where the symbol $\propto$ means "proportional to."

Let us consider an illustration of Boyle's law. A gas occupies a volume of 2.0 liters and exerts a pressure of 1.0 atmosphere. What will be the final pressure of the gas if its volume is increased to 4.0 liters? We begin the solution of the problem by writing down the information given:

$$\text{initial volume} = V_i = 2.0 \text{ liters}$$
$$\text{initial pressure} = P_i = 1.0 \text{ atm}$$
$$\text{final volume} = V_f = 4.0 \text{ liters}$$
$$\text{final pressure} = P_f = ?$$

We see that our task is to calculate the final pressure after the gas has been allowed to expand from a volume of 2.0 liters to one of 4.0 liters, at constant temperature and mass. Since the volume of the gas increases, we know (from Boyle's law) that the pressure of the gas must decrease. Thus, we must multiply the original pressure by a fraction whose value is less than one; since the pressure change depends on the volume change, the fraction is made up of the initial and final volumes. Hence, our fraction is 2.0 L/4.0 L, and our calculation is as follows:

$$P_f = 1.0 \text{ atm} \times \frac{2.0 \cancel{L}}{4.0 \cancel{L}}$$

$$= \frac{(1.0)(2.0)}{4.0} \text{ atm}$$

$$= 0.50 \text{ atm}$$

Notice that we have not memorized any formulas; we solved the problem by reasoning. (Note also that our units of atmosphere are in agreement with the quantity, P_f, we are seeking.) In this problem and all others involving pressure-volume relationships, we need only remember that an increase in pressure causes a decrease in volume, and vice versa. Then we simply make a fraction of the initial and final values for one variable; the value of the fraction must be greater than 1 when we wish to increase the other variable and less than 1 when we wish to decrease the other variable.

EXAMPLE 5.1

A sample of gas occupies a volume of 95.2 mL at a pressure of 710 torr. What will be its volume in mL at a pressure of 1 atmosphere?

SOLUTION

$$P_i = 710 \text{ torr} \qquad P_f = 760 \text{ torr}$$
$$V_i = 95.2 \text{ mL} \qquad V_f = ?$$

Since the pressure increases, the volume must decrease. In order to make the volume decrease we must multiply V_i by a fraction composed of pressure terms whose value is less than 1:

$$V_f = 95.2 \text{ mL} \times \frac{710 \text{ torr}}{760 \text{ torr}}$$

$$= \frac{(95.2)(710)}{760} \text{ mL}$$

$$= 88.9 \text{ mL}$$

As a check, we compare the final volume to the initial volume. We see that there was a volume decrease as we had reasoned there should be.

EXERCISE 5.1 A sample of gas occupies 75.0 mL at a pressure of 720 torr. What will be its volume in mL if the pressure is decreased to 700 torr?

EXAMPLE 5.2

A sample of gas occupies 75.0 mL at a pressure of 720 torr. What will be its pressure in torr if the gas volume is decreased to 65.0 mL?

SOLUTION

$V_i = 75.0$ mL $\qquad V_f = 65.0$ mL
$P_i = 720$ torr $\qquad P_f = ?$

Since the volume decreases, the pressure must increase. In order to make the pressure increase we must multiply P_i by a fraction composed of volume terms whose value is greater than 1:

$$P_f = 720 \text{ torr} \times \frac{75.0 \text{ mL}}{65.0 \text{ mL}}$$

$$= \frac{(720)(75.0)}{65.0} \text{ torr}$$

$$= 831 \text{ torr}$$

As a check, we compare the final pressure to the initial pressure. We see that there was an increase as we had reasoned there should be.

EXERCISE 5.2 A sample of gas occupies 140 mL at a pressure of 760 torr. What will be its pressure in torr if the gas volume is increased to 200 mL?

Boyle's law is demonstrated every day of our lives as we conduct the life-sustaining exercise known as breathing, illustrated in Figure 5–10. When it is time to inhale, our lungs expand. This increase in lung volume causes a lower pressure within the lungs, and external air, at higher pressure, flows into the lungs until the internal pressure is equal to that of the external atmosphere. We exhale by contracting lung volume to increase pressure within the lungs. Air then flows from the high-pressure lung space into the lower-pressure external environment until the two pressures are equalized.

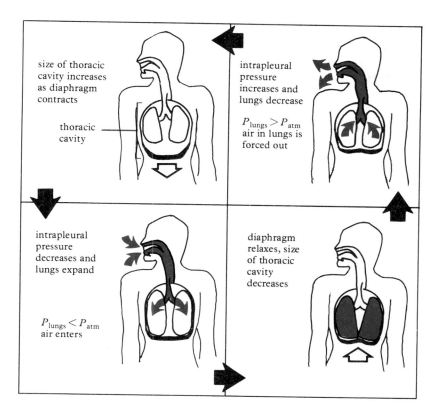

Fig. 5-10 Pressure Changes in Breathing

Volume and Temperature

A second gas law was worked out in 1787 by the French physicist J. A. C. Charles. His studies were subsequently confirmed and published in 1802 by Joseph L. Gay-Lussac, a French chemist. *Charles's law* states that the volume of gas varies directly as its absolute temperature, if its pressure and mass are kept constant (Figure 5-11). Thus, if the temperature of a gas is increased, the volume will increase proportionately, as

Charles's law: the volume of a gas varies directly with the temperature of the gas at constant pressure.

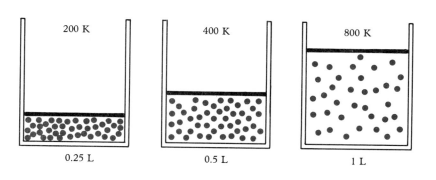

Fig. 5-11 The Effect of Temperature on Gas Volume

long as the pressure within the gas does not change. We can state Charles's law mathematically as

$$V \propto T$$

where T represents the Kelvin temperature of the gas.

Lord Kelvin, an English physicist, continued Charles's observations and reasoned that if a gas could be cooled 273° below 0°C, it would have no volume (see Figure 5-12) and the energy of its molecules would become zero. (In actual fact, a gas will condense to a liquid long before its temperature reaches −273°C.) Since the hypothetical temperature signified the lowest possible temperature, it was given the name *absolute zero*. Thus, the Kelvin (absolute) temperature scale was created in 1848. Because of the inherent dependence of gas volume on absolute temperature, the Kelvin scale must be used in all gas law calculations.

Volume-temperature problems can be solved much like pressure-volume problems. Let's suppose that 2.5 liters of a gas at 25°C are heated to 300°C at constant pressure. What is the final volume of the gas? To solve the problem, we first write down the given information:

$$T_i = 25°C \qquad T_f = 300°C$$
$$V_i = 2.5 \text{ L} \qquad V_f = ?$$

Before we can proceed, we must convert Celsius temperature to the Kelvin scale (see chapter 1):

$$T_i = 25 + 273 = 298 \text{ K}$$
$$T_f = 300 + 273 = 573 \text{ K}$$

Now we can derive a fraction composed of the initial and final temperatures; this fraction must have a value of more than 1, since the tempera-

absolute zero: the hypothetical point of −273°C at which a gas volume would shrink to zero, and the gas molecules would have no energy; this temperature has never been achieved.

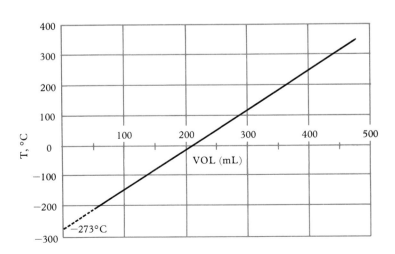

Fig. 5-12 Plot of Temperature (°C) vs. Gas Volume

ture increases and therefore the volume must also increase. Hence our fraction is 573 K/298 K, and our calculation is

$$V_f = 2.5 \text{ L} \times \frac{573 \text{ K}}{298 \text{ K}}$$

$$= \frac{(2.5)(573)}{298} \text{ L}$$

$$= 4.8 \text{ L}$$

EXAMPLE 5.3

A gas occupies a volume of 4.50 L at 27°C. At what temperature in °C would the volume be 6.00 L?

SOLUTION

$V_i = 4.50$ L $\qquad V_f = 6.00$ L
$T_i = 27°C$ $\qquad T_f = ?$

The first thing we must do is convert Celsius temperature to absolute temperature:

$$K = °C + 273$$
$$T_i = 27 + 273 = 300 \text{ K}$$

Now we reason that if volume increases, temperature must also increase. In order to make the temperature increase we must multiply T_i by a fraction composed of volume terms whose value is greater than 1:

$$T_f = 300 \text{ K} \times \frac{6.00 \text{ L}}{4.50 \text{ L}}$$

$$= \frac{(300)(6.00)}{4.50} \text{ K}$$

$$= 400 \text{ K}$$

As a check, we compare the final temperature to the initial temperature. We see that there was a temperature increase as we had reasoned there should be.

Finally, we must convert our final absolute temperature to the Celsius scale:

$$°C = K - 273$$
$$= 400 - 273$$
$$= 127$$

EXERCISE 5.3 A gas occupies a volume of 10.0 mL at 100°C. At what temperature in °C will its volume be 8.00 mL?

EXAMPLE 5.4

A gas occupies a volume of 100 mL at 50°C. What will be its volume in mL at 30°C?

SOLUTION

$$T_i = 50°C \qquad T_f = 30°C$$
$$V_i = 100 \text{ mL} \qquad V_f = ?$$

We must first convert Celsius temperature to absolute temperature:

$$K = °C + 273$$
$$T_i = 50 + 273 = 323 \text{ K}$$
$$T_f = 30 + 273 = 303 \text{ K}$$

Now we reason that if temperature decreases, volume must also decrease. In order to make the volume decrease we must multiply V_i by a fraction composed of temperature terms whose value is less than 1:

$$V_f = 100 \text{ mL} \times \frac{303 \text{ K}}{323 \text{ K}}$$

$$= \frac{(100)(303)}{323} \text{ mL}$$

$$= 93.8 \text{ mL}$$

As a check, we compare the final volume to the original volume. We see that there was a volume decrease as we had reasoned there should be.

EXERCISE 5.4 A gas occupies a volume of 1.00 L at 25°C. What will be its volume in L at 50°C?

Pressure and Temperature

The volume-temperature relationship of gases carries with it an implication about the effect of temperature on gas pressure. When we heat popcorn kernels, steam forms in the seed cores. The pressure created by this steam causes the kernels to explode, and the popcorn "pops." This observation illustrates that when a confined gas is heated at constant volume, the pressure increases. Thus, gas pressure is directly proportional to temperature:

$$P \propto T$$

This statement is illustrated by Figure 5-13 and represents another form of *Charles's law:* at constant volume, the pressure of a given quantity of gas is proportional to absolute temperature.

Charles's law (alternate statement): the pressure of a gas varies directly with its temperature at constant volume.

EXAMPLE 5.5

The temperature of a gas at 1.00 atm pressure is changed from 0°C to 200°C while the gas volume is held constant. What is the final pressure of the gas in torr?

SOLUTION

$$T_i = 0°C \qquad T_f = 200°C$$
$$P_i = 1.00 \text{ atm} \qquad P_f = ?$$

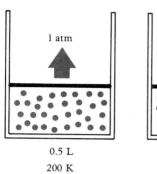  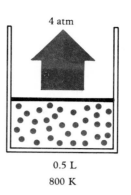

Fig. 5-13 The Effect of Temperature on Gas Pressure

0.5 L 0.5 L 0.5 L
200 K 400 K 800 K

We must first convert Celsius temperature to absolute temperature:

$$K = °C + 273$$
$$T_i = 0 + 273 = 273 \text{ K}$$
$$T_f = 200 + 273 = 473 \text{ K}$$

Now we reason that if temperature increases, pressure must also increase. Thus we must multiply P_i by a fraction composed of temperature terms whose value is greater than 1:

$$P_f = 1 \text{ atm} \times \frac{473 \text{ K}}{273 \text{ K}}$$

$$= \frac{(1.00)(473)}{273} \text{ atm}$$

$$= 1.73 \text{ atm}$$

As a check, we compare the final pressure to the original pressure. We see that there was a pressure increase as we had reasoned there should be.

Finally we must convert our pressure to units of torr:

$$1 \text{ atm} = 760 \text{ torr}$$

$$\frac{1 \text{ atm}}{760 \text{ torr}} = 1$$

$$\frac{760 \text{ torr}}{1 \text{ atm}} = 1$$

We choose the unit factor which allows us to cancel the unwanted units of atm. Thus

$$P_i = 1.73 \text{ atm} \times \frac{760 \text{ torr}}{1 \text{ atm}}$$

$$= \frac{(1.73)(760)}{1} \text{ torr}$$

$$= 1314.8 \text{ torr}$$

Rounding off to three significant figures,

$$P_i = 1310 \text{ torr}$$

EXERCISE 5.5 The temperature of a gas at 2.50 atm pressure is changed from 100°C to 50°C while the gas volume is held constant. What is the final pressure of the gas in torr?

More Calculations with the Gas Laws

You have learned to make calculations based on Boyle's law, in which temperature was constant, and Charles's law, in which either pressure or volume was constant. But what if all of the variables—P, V, and T—change at once? For example, let's assume that we have 1.00 liter of a gas at 1.50 atm pressure and 24°C, and we cause it to expand to a volume of 2.00 liter in a cold room at 10°C. What is the final pressure of the gas? We can solve this problem easily by breaking it into two parts. First, consider the temperature change to occur at constant pressure. Then,

$$T_i = 24°C = 297 \text{ K} \qquad T_f = 10°C = 283 \text{ K}$$
$$V_i = 1.00 \text{ L} \qquad V_f = ?$$

$$V_f = 1.00 \text{ L} \times \frac{283 \text{ K}}{297 \text{ K}}$$

$$= 0.953 \text{ L}$$

We have now calculated the new volume caused only by the change in temperature. Next we must calculate the effect of the volume change on pressure. At this point our given information is

$$V_i = 0.953 \text{ L} \qquad V_f = 2.00 \text{ L}$$
$$P_i = 1.50 \text{ atm} \qquad P_f = ?$$

We calculate the final pressure as follows:

$$P_f = 1.50 \text{ atm} \times \frac{0.953 \text{ L}}{2.00 \text{ L}}$$

$$= 0.715 \text{ atm}$$

Thus the combined effects of volume increase and temperature decrease caused a resulting decrease in gas pressure.

Notice that we could have made the entire calculation in one step:

$$P_f = 1.50 \text{ atm} \times \frac{283 \text{ K}}{297 \text{ K}} \times \frac{1.00 \text{ L}}{2.00 \text{ L}}$$

$$= 0.715 \text{ atm}$$

In the last calculation, we combined Boyle's law and Charles's law to produce a one-step solution to a problem we had previously broken down into two steps. This one-step approach is preferred because it allows fewer chances for making errors in writing down numbers, and it presents a more compact, easy-to-read solution.

When we combine Boyle's law and Charles's law, we are simply making a more compact mathematical expression to replace a two-step solution to a problem. We can summarize the combination of the two laws as follows:

PERSPECTIVE: The Earth's Atmosphere

The earth's atmosphere is the gaseous environment in which we live. Its chemical composition at sea level is shown in the accompanying table. In many ways it is like the liquid sea of water which covers 75% of the earth's surface. Plants and animals living deep in the ocean experience a much different environment than those near the surface. We humans who live at the bottom of our atmospheric sea are seldom aware of the conditions higher up in the atmosphere, and we take for granted the fact that the atmosphere determines the environment we live in.

The pressure of the atmosphere decreases consistently with increasing altitude, as shown in the accompanying figure. Any area on the surface of the earth experiences a pressure resulting from the entire column of atmosphere above it. This pressure, a consequence of the earth's gravitational attraction for the gases in the atmosphere, has an average value of 760 torr at sea level. At higher elevations, however, the pressure of the atmosphere is lower, because there is a smaller total mass of gas in the column above the ground. Thus, the normal atmospheric pressure at Denver, Colorado (elevation 1 mile), is usually about 700 torr. Also, at higher elevations air is less dense, and oxygen is less abundant. For this reason, aircraft flying above 10,000 feet are pressurized so that those in the aircraft will be able to breathe properly.

Average Composition of the Atmosphere (Dry Air at Sea Level)

Gas	Percent by Volume
N_2	78.08
O_2	20.95
Ar	0.93
CO_2	0.033
Ne	0.0018
He	
CH_4	
Kr	trace of each
Xe	
H_2	
N_2O	

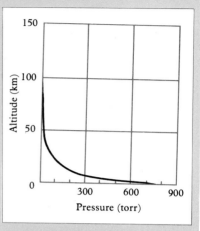

Variation of Atmospheric Pressure with Altitude

$$V_f = V_i \times P \text{ factor} \times T \text{ factor} \quad (1)$$

$$P_f = P_i \times V \text{ factor} \times T \text{ factor} \quad (2)$$

$$T_f = T_i \times V \text{ factor} \times P \text{ factor} \quad (3)$$

In equation (1), the final volume is equal to the initial volume multiplied by a pressure factor and a temperature factor. If the pressure increases, the pressure factor must be less than 1, since increasing the pressure would decrease the initial volume. If the pressure decreases, the pressure factor must be greater than 1, since a decrease in pressure would increase the initial volume. And finally, if the temperature decreases, the temperature factor must be less than 1, and if temperature increases, the temperature factor must be greater than 1. By applying similar reasoning to equations (2) and (3), we can solve for the new pressure and temperature.

EXAMPLE 5.6

A gas occupies 400 mL at 760 torr and 0°C. What volume in mL will it occupy at 850 torr and 100°C?

SOLUTION

$P_i = 760$ torr $\qquad P_f = 850$ torr
$T_i = 0°C = 273$ K $\qquad T_f = 100°C = 373$ K
$V_i = 400$ mL $\qquad V_f = ?$

From equation (1),

$$V_f = V_i \times P \text{ factor} \times T \text{ factor}$$

Since the pressure increase would cause V_i to decrease, the pressure factor must be less than 1. Since the temperature increase would cause V_i to increase, the temperature factor must be greater than 1. Hence, our expression is

$$V_f = 400 \text{ mL} \times \frac{760 \text{ torr}}{850 \text{ torr}} \times \frac{373 \text{ K}}{273 \text{ K}}$$

$$= \frac{(400)(760)(373)}{(850)(273)} \text{ mL}$$

$$= 489 \text{ mL}$$

EXERCISE 5.6 A gas occupies 1.50 L at 800 torr and 150°C. What volume in L will it occupy at 700 torr and 50°C?

EXAMPLE 5.7

A gas occupies 20.0 L at 40°C and 780 torr. Under what pressure in torr would this gas occupy a volume of 75.0 L at 0°C?

SOLUTION

$V_i = 20.0$ L $\qquad V_f = 75.0$ L
$T_i = 40°C = 313$ K $\qquad T_f = 0°C = 273$ K
$P_i = 780$ torr $\qquad P_f = ?$

From equation (2),

$$P_f = P_i \times V \text{ factor} \times T \text{ factor}$$

Since the volume increase would cause a pressure decrease, the volume factor must be less than 1. Since the temperature decrease would cause a pressure decrease, the temperature factor must be less than 1. Hence, our expression is

$$P_f = 780 \text{ torr} \times \frac{20.0 \text{ L}}{75.0 \text{ L}} \times \frac{273 \text{ K}}{313 \text{ K}}$$

$$= \frac{(780)(20.0)(273)}{(75.0)(313)} \text{ torr}$$

$$= 181 \text{ torr}$$

EXERCISE 5.7 A gas occupies 250 mL at 25°C and 760 torr. Under what pressure in torr would this gas occupy a volume of 200 mL at 50°C?

EXAMPLE 5.8

A gas occupies 10.0 L at 10°C and 600 torr. What will the temperature be in Celsius degrees when the gas has a volume of 5.0 L and a pressure of 700 torr?

SOLUTION

$V_i = 10.0$ L $\qquad V_f = 5.0$ L

$P_i = 600$ torr $\qquad P_i = 700$ torr

$T_i = 10°C = 283$ K $\qquad T_f = ?$

From equation (3),

$$T_f = T_i \times V \text{ factor} \times P \text{ factor}$$

Since the volume decrease would cause a temperature decrease, the volume factor must be less than 1. Since the pressure increase would cause a temperature increase, the pressure factor must be greater than 1. Hence, our expression is

$$T_f = 283 \text{ K} \times \frac{5.0 \text{ L}}{10.0 \text{ L}} \times \frac{700 \text{ torr}}{600 \text{ torr}}$$

$$= \frac{(283)(5.0)(700)}{(10.0)(600)} \text{ K}$$

$$= 165 \text{ K}$$

Since the problem asks for the final temperature in Celsius degrees, we must convert absolute temperature to Celsius:

$$°C = K - 273$$
$$= 165 - 273$$
$$= -108$$

EXERCISE 5.8 A gas occupies 7.50 L at 20°C and 720 torr. What will the temperature be in °C when the gas has a volume of 10.0 L and a pressure of 650 torr?

INTERPRETATION OF GASEOUS BEHAVIOR

We have seen that Boyle, Charles, and Gay-Lussac were able to describe the behavior of gases in very concrete terms according to the influences of temperature, pressure, and changes in volume. Their laws are statements derived from experimental observations. Now it is our task to develop an interpretation of the behavior of gases so that we can understand the basis of gas laws.

Avogadro's Hypothesis

Amedeo Avogadro was an Italian physicist who made the first attempt to explain the behavior of gases. Avogadro made two basic assumptions:

1. Gases are composed of molecules.

2. Each molecule of any gas occupies a certain volume of space at a given temperature and pressure. (It was later found that this is a greater volume of space than is actually filled by the nuclei and electrons of the molecule.)

From these assumptions, Avogadro formulated the following hypothesis: equal volumes of gases contain equal numbers of molecules at the same conditions of temperature and pressure. Since Avogadro first proposed his hypothesis, it has been tested many times over, and we now accept it as one of the fundamental truths in chemistry.

Avogadro's hypothesis tells us that a mole of one gas will occupy the same volume as a mole of any other gas under the same conditions of temperature and pressure. Measurements of gas volumes are often converted to *standard temperature and pressure* (*STP*) to make it easy to compare measurements made under different sets of conditions. Standard temperature and pressure (STP) are defined as 1 atmosphere and 273 K (0°C). Countless experiments have shown that one mole of any gas occupies 22.4 liters at STP. This volume of gas contains 6.02×10^{23} molecules (we can regard noble gases to be composed of one-atom molecules); this is Avogadro's number, which is familiar from chapter 4.

We can now summarize each gas law by expressing it in a form which shows the proportionality between volume and one other variable, with the remaining variables held constant (the symbol n stands for number of moles of gas):

Boyle's law:	$V \propto 1/P$	T, n constant
Charles's law:	$V \propto T$	P, n constant
Avogadro's hypothesis:	$V \propto n$	P, T constant

These three relationships can be combined into a single expression,

$$V \propto \frac{nT}{P}$$

To change the proportionality to an equation we insert a constant of proportionality, R, called the molar gas constant:

$$V = \frac{nRT}{P}$$

or $\quad PV = nRT$

Our expression says that the product of pressure and volume of a gas is equal to the product of the number of moles of the gas, the molar gas constant, and the temperature of the gas. The molar gas constant, R, has the value of 0.0820 liter atm/mole K. This expression is called the *combined gas equation*, or the *ideal gas equation*. It can be used for calculations of pressure, volume, number of moles, or temperature, provided that three of the four variables are known.

INTERPRETATION OF GASEOUS BEHAVIOR **129**

EXAMPLE 5.9

A gas sample occupies 7.0 liters at 25°C and 1.0 atmosphere pressure. How many moles of gas are present?

SOLUTION

In this problem there are no changes in variables. Three of the variables (P, V, and T) are given, and we are asked to calculate the fourth, n. To do so, we must first solve the combined gas equation for n:

$$PV = nRT$$

$$n = \frac{PV}{RT}$$

Substituting numerical quantities for P, V, R, and T,

$$n = \frac{(1.0 \text{ atm})(7.0 \text{ L})}{(0.0820 \text{ L atm/mole K})(25 + 273)\text{K}}$$

$$= \frac{(1.0)(7.0)}{(0.0820)(298)} \text{ mole}$$

$$= 0.29 \text{ mole}$$

EXERCISE 5.9 A gas sample occupies 1.25 L at 30°C and 2 atm pressure. How many moles of gas are present?

EXAMPLE 5.10

A gas sample weighs 142 grams and occupies a volume of 24.7 liters at 2.00 atm pressure and 28°C. What is the molecular weight of the gas?

SOLUTION

Values are given for V, P, and T. By using the combined gas equation, we can calculate n, the number of moles of gas. We can then calculate the weight of one mole, and that numerical value expressed in amu is the molecular weight. We start by solving the combined gas equation for n:

$$PV = nRT$$

$$n = \frac{PV}{RT}$$

Substituting,

$$n = \frac{(2.00 \text{ atm})(24.7 \text{ L})}{(0.0820 \text{ L atm/mole K})(28 + 273)\text{K}}$$

$$= \frac{(2.00)(24.7)}{(0.0820)(301)} \text{ mole}$$

$$= 2.00 \text{ mole}$$

At this point we know that 142 grams of the gas contain 2.00 moles. We proceed to calculate the weight of one mole:

$$\text{weight of 1 mole} = \frac{142 \text{ g}}{2.00 \text{ mole}}$$

$$= 71.0 \text{ g/mole}$$

Hence the molecular weight is 71.0 amu.

EXERCISE 5.10 A gas sample weighs 220 g and occupies a volume of 80.1 L at 1.50 atm pressure and 20°C. What is the molecular weight of the gas?

EXAMPLE 5.11

A gas sample containing 2.35 moles occupies a volume of 40.0 L at 1.00 atm pressure. What is the temperature of the gas in °C?

SOLUTION

We must first solve the combined gas equation for T:

$$PV = nRT$$

$$T = \frac{PV}{nR}$$

Substituting numerical quantities,

$$T = \frac{(1.00 \text{ atm})(40.0 \text{ L})}{(2.35 \text{ mole}) \, 0.0820 \text{ L atm/mole K})}$$

$$= \frac{(1.00)(40.0)}{(2.35)(0.0820)} \text{ K}$$

$$= 208 \text{ K}$$

Converting absolute temperature to the Celsius scale,

$$°C = K - 273$$
$$= 208 - 273$$
$$= -65$$

EXERCISE 5.11 A gas sample containing 7.25 moles occupies a volume of 140 L at 1.2 atm pressure. What is the temperature of the gas in °C?

The combined gas equation can be manipulated to still another form when the number of moles (mass) of gas does not change. In this case, under an initial set of conditions,

$$P_i V_i = nRT_i$$

and

$$\frac{P_i V_i}{T_i} = nR$$

Since n and R are constant for any set of conditions, for a final set of values for P, V, and T,

$$\frac{P_f V_f}{T_f} = nR$$

and

$$\frac{P_i V_i}{T_i} = \frac{P_f V_f}{T_f}$$

This last equation is convenient for solving problems in which P, V, T, or any combination of the three varies. In essence, this equation is a summary of equations (1), (2), and (3) developed earlier in this chapter.

EXAMPLE 5.12 _____

A constant mass of a gas occupies a volume of 1.00 L at 1.30 atm pressure and 25°C. What volume will it occupy at 2.00 atm pressure and 30°C?

SOLUTION _____

$$\frac{P_i V_i}{T_i} = \frac{P_f V_f}{T_f}$$

$$V_f = \frac{P_i V_i T_f}{T_i P_f}$$

$$V_f = \frac{(1.30 \text{ atm})(1.00 \text{ L})(303 \text{ K})}{(298 \text{ K})(2.00 \text{ atm})}$$

$$= 0.661 \text{ L}$$

EXERCISE 5.12 A constant mass of gas occupies a volume of 1.35 L at 2.00 atm pressure and 0°C. What will be its temperature in °C when it occupies a volume of 2.00 L at 1.50 atm pressure?

The Kinetic Molecular Theory

Avogadro's hypothesis is the basis for a general theory of gaseous matter called the *kinetic molecular theory*. The main features of the theory are as follows:

1. All gases consist of separate and distinct particles: we call these particles molecules (hence the kinetic *molecular* theory).

2. Individual gas molecules are very small. (The diameter of a gas molecule is estimated to be about 3×10^{-10} meters.)

3. Molecules in a gas are very far apart (roughly 10 times the diameter of one molecule apart).

4. Molecules are in random, straight-line motion and thus they possess kinetic energy.

kinetic: derived from the Greek word *kinetikos*, meaning "moving" or "pertaining to motion."

kinetic molecular theory: the proposed explanation for the behavior of gases which suggests that gas particles move freely and rapidly along straight lines, with frequent collisions causing variations in velocity and direction.

5. Molecules collide with each other and the walls of the container. These collisions are perfectly elastic; this means that the gas molecules do not stick together and they do not spring apart when they collide with each other or with the walls of the container. They behave as very hard billiard balls.

6. The average kinetic energy of each molecule is constant for a given temperature and is proportional to the temperature. (The molecules of a gas do not all have the same kinetic energy but rather have a wide range of kinetic energy because they have different velocities. Thus, the *average* kinetic energy is related to temperature.)

We can now relate the kinetic molecular theory to the observed properties of gases.

1. Gases have mass because they are composed of particles of matter (molecules).

2. Gases occupy volume because their molecules are continuously moving over a wide range of space.

3. Gases exert pressure because the gas molecules collide with the walls of their container.

4. Gases are compressible because there are large empty spaces between molecules.

5. Gases diffuse to fill any vessel because the individual molecules are in rapid, random motion, and they exert no attraction for each other.

Pressure is due to collisions of molecules with the walls of the container (see Figure 5–14). Increasing the number of molecules in a container will increase the number of collisions, and thus the pressure. This statement is in agreement with Boyle's law. If a gas is heated, each molecule has a higher velocity; this increases the number of collisions with the container walls and gives each collision greater impact. These factors cause the pressure to increase (Charles's law).

GAS MIXTURES

Because there are large spaces between gas molecules, we might expect a gas in a mixture of gases to act as if the others were not present. Our expectation is usually true. For example, the oxygen in our atmosphere (air is about 20% oxygen) behaves as if it were pure oxygen with a pressure of 0.2 atmospheres (20% of 1 atmosphere). The fact that gases in a mixture behave independently of each other was discovered by John Dalton (1766–1844). His *law of partial pressures* states that each component in a gas mixture exerts its own pressure (its *partial pressure*) independent of others and that the total pressure is the sum of the partial pressures.

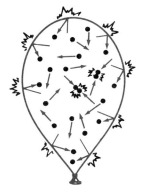

Fig. 5-14 Gas Molecules in Rapid, Random Motion. The pressure on the walls of the balloon is created by the forces of individual impacts.

We can express Dalton's law of partial pressures in mathematical terms as

$$P_{total} = P_1 + P_2 + P_3 + \ldots$$

where P_{total} is the total pressure exerted by the gas mixture, and P_1, P_2, P_3, etc., are the partial pressures of the individual components of the mixture. The partial pressure is directly related to the percentage of each component in a gas mixture.

If we know the partial pressures of gases in a mixture, we can find the total pressure easily by adding together all of the partial pressures. For example, the total pressure exerted by our atmosphere averages 760 torr at sea level. This value is the sum of the partial pressures of the atmospheric components: nitrogen, oxygen, argon, carbon dioxide, water vapor, and traces of other gases.

EXAMPLE 5.13

The partial pressure of oxygen in dry air is 159 torr. Assuming the partial pressures of argon, carbon dioxide, and traces of other gases to be negligible, calculate the partial pressure of nitrogen when the total pressure is 760 torr.

SOLUTION

Dry air is composed of nitrogen, oxygen, argon, carbon dioxide, and traces of other gases. Since only the partial pressures of nitrogen and oxygen need be considered, we can write

$$P_{total} = P_{nitrogen} + P_{oxygen}$$

Then,

$$\begin{aligned} P_{nitrogen} &= P_{total} - P_{oxygen} \\ &= 760 \text{ torr} - 159 \text{ torr} \\ &= 601 \text{ torr} \end{aligned}$$

EXERCISE 5.13 In a laboratory experiment, a mixture of hydrogen chloride gas and water vapor was collected in a sealed container. If the partial pressure of the water vapor was 24 torr and the total pressure was 750 torr, what was the partial pressure of hydrogen chloride?

Table 5-1 Partial Pressures of Gases in the Atmosphere and in the Lungs

Gas	Atmosphere	Inspired Air	Expired Air	Alveolar Air*
O_2	159 torr	149 torr	116 torr	100 torr
CO_2	0 torr	0 torr	28 torr	40 torr
H_2O	6 torr	47 torr	47 torr	47 torr
N_2	595 torr	564 torr	569 torr	573 torr
Total	760 torr	760 torr	760 torr	760 torr

*Air inside the air sacs in the lungs

capillary (CAAH-pih-larry): a tiny blood vessel.

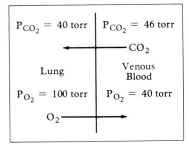

Fig. 5-15 Movement of Oxygen and Carbon Dioxide between Capillaries and Lung Air

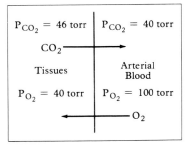

Fig. 5-16 Movement of Oxygen and Carbon Dioxide between Blood and Tissue Cells

Body cells continuously use oxygen in chemical reactions to produce energy; a byproduct of these reactions is carbon dioxide. Thus, blood constitutes a transportation system to carry oxygen to cells and carbon dioxide from cells to the lungs for expiration. Gases are exchanged between lung air and capillary blood by diffusion in response to differences in partial pressure. Table 5-1 compares the partial pressures of gases in the atmosphere to those in the lung. Since the partial pressure of oxygen in lung air is higher than it is in blood entering the lungs, oxygen enters the blood until its partial pressure there equals its partial pressure in lung air. And since carbon dioxide has a higher partial pressure in blood returning from the cells and entering the lungs, it diffuses from the capillaries into the lungs until there is an equalization of pressures. Figures 5-15 and 5-16 summarize the movement of oxygen from lung air to blood to tissue cells and the reverse movement of carbon dioxide.

Because oxygen and carbon dioxide are transported by the blood, they are often referred to as *blood gases*. The quantity of blood gases transported is determined by how rapidly the blood circulates and by how much gas is removed from blood by the tissue cells.

REAL GASES

Even though the kinetic molecular theory helps us to understand the gas laws and to predict the behavior of gases, we must keep in mind that this theory is a scientific model of a gas. The model actually is our conception of an ideal gas whose behavior is described exactly by the combined (ideal) gas equation

$$PV = nRT$$

Real gases are those we encounter in the world about us. Experiments with real gases indicate that they do not behave exactly as ideal gases; instead real gases deviate from the ideal behavior described by the combined gas equation. For example, we can rearrange the combined gas equation to another form by dividing both sides of the equation by nRT:

$$\frac{PV}{nRT} = 1$$

In this form the equation tells us that PV divided by nRT has a value of 1 for all pressures and temperatures. However, if we plot PV/nRT as a function of increasing pressure, as in Figure 5-17, we see that real gases depart dramatically from ideal behavior.

The departures of real gases from ideal behavior can be accounted for by two factors which were ignored by the kinetic molecular theory:

1. Real gases actually possess attractive forces between molecules, and these forces operate over short distances. When real gas molecules are pressed closer together by intermediate applied pressure, the

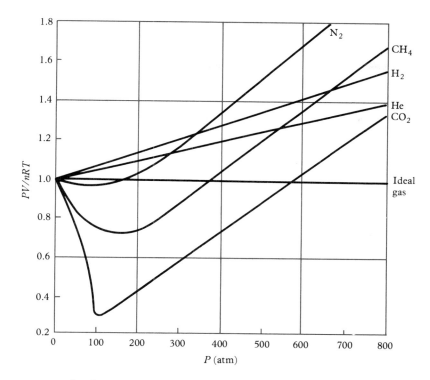

Fig. 5-17 A Graph of PV/nRT vs. Pressure Illustrates How Real Gases Deviate from Ideal Gas Behavior at Intermediate and Very High Pressures

attractive forces cause the real gases to be more compressible than an ideal gas and to exert lower pressures on the walls of their containers than would be predicted by the ideal gas law.

2. Every molecule in a real gas has a real volume. Thus the total volume of the molecules in a container is significant when the space occupied by the gas becomes relatively small, as under conditions of very high pressure.

Figure 5-17 illustrates how real gases deviate from ideal behavior at intermediate and at very high pressures. However, these pressure conditions are somewhat out of the ordinary; at more normal pressures, in the range of one to ten atmospheres, deviations from ideal behavior are not so large, and the ideal gas equation can be used without serious errors.

SUMMARY

Properties of Gases: Gases represent one of the three states of matter, and as such, they possess mass and occupy space. Gases exert pressure, often measured with a barometer and expressed in units of torr. Gases may be compressed by applying external pressure. Gas volume can also be decreased by lowering the temperature; increased temperatures cause gases to expand. Gases can also diffuse, i.e., two or more gases can intermix completely.

Relationships of Volume, Temperature, and Pressure: Boyle found in 1662 that the pressure and volume of a given quantity of gas at constant temperature are inversely proportional. A second gas law, worked out by Charles in 1867, states that the volume of a gas varies directly with its absolute temperature if its pressure and mass are held constant. A variation of Charles's law may be stated: at constant volume, the pressure of a given quantity of gas is proportional to absolute temperature.

Interpretation of Gaseous Behavior: Avogadro advanced a hypothesis in 1811 based on the ideas that gases are composed of molecules and that each molecule occupies a certain amount of space at given temperature and pressure. He concluded that equal volumes of gases contain equal numbers of molecules at the same temperature and pressure. A mole of gas occupies 22.4 L at standard temperature and pressure (STP). The mole concept allows us to combine Boyle's law, Charles's law, and Avogadro's hypothesis to give the combined (or ideal) gas equation, $PV = nRT$. Avogadro's hypothesis is the basis for a general theory of gaseous matter, the kinetic molecular theory, which provides a conceptual foundation for understanding the observed properties of gases.

Gas Mixtures: Each gas in a mixture behaves independently of the others present. Dalton's law of partial pressures states that each component in a gas mixture exerts its own partial pressure independent of others, and that the total pressure is the sum of the partial pressures.

Real Gases: Real gases, those we encounter in the world around us, deviate from ideal behavior because gas molecules are attracted to each other and they have real volumes. However, these deviations usually occur at extraordinary pressures. At pressures of one to ten atmospheres, the combined (ideal) gas equation can be used without serious errors.

STUDY QUESTIONS AND PROBLEMS

1. Define the following terms:
   (a) pressure
   (b) torr
   (c) compressibility
   (d) diffusion
   (e) absolute zero
   (f) standard temperature and pressure
   (g) combined (ideal) gas equation
   (h) kinetic molecular theory
   (i) partial pressure
   (j) blood gases
   (k) ideal gas
   (l) real gas

2. In your own words, describe how gas volume varies with pressure.

3. In your own words, describe how gas volume varies with temperature.

4. In your own words, describe how gas pressure varies with temperature.

5. Describe the six major properties of gases.

6. Summarize Avogadro's hypothesis.

7. Summarize the relationship between the kinetic molecular theory and observed properties of gases.

8. Summarize Dalton's law of partial pressures.

9. Account for deviations of real gases from ideal behavior.

10. Use the kinetic molecular theory to explain Boyle's law.

11. Use the kinetic molecular theory to explain Charles's law.

12. A sample of gas occupies 360 mL at a pressure of 0.750 atm. If the temperature is held constant, what volume in mL will the gas occupy at 1.00 atm pressure?

13. The volume of a sample of gas is 4.00 L at 4.00 atm pressure. If the temperature is held constant, what will be the volume in L of the sample at each of the following pressures?
    (a) 1.00 atm   (b) 0.400 atm   (c) 10.0 atm

14. The volume of a sample of gas is 200 mL at 1.00 atm pressure. If the temperature is held constant, what will be the pressure in atm of the gas at each of the following volumes?
    (a) 250 mL   (b) 100 mL   (c) 1.00 L

15. The volume of a sample of gas is 25.0 L at 1.75 atm pressure. If the temperature is held constant, what will be the pressure in atm of the gas at each of the following volumes?
    (a) 10.0 L   (b) 35.0 L   (c) 2.50 L

16. A sample of gas has a volume of 79.5 mL at 45°C. If the pressure is held constant, what will be the volume in mL at 0°C?

17. The volume of a sample of gas is 100 mL at 90°C. If the pressure is held constant, what will be the volume in mL of the gas at each of the following temperatures?
    (a) 50°C   (b) 140°C   (c) 0°C

18. A container is filled with gas at a pressure of 1.80 atm at 0°C. At what temperature in 0°C will the pressure of the gas be 2.50 atm?

19. A container is filled with a gas at a pressure of 3.00 atm at 25°C.
    (a) What will be the gas pressure in atm at 100°C?
    (b) At what temperature in °C will the gas pressure be 2.50 atm?
    (c) At what temperature in °C will the gas pressure be 3.50 atm?

20. The volume of a sample of gas is 450 mL at 35°C and 1.10 atm pressure. Calculate the volume in mL of the gas at STP.

21. A gas sample was heated from −5°C to 90°C, and the volume increased from 1.00 L to 3.00 L. If the initial pressure was 0.800 atm, what was the final pressure in atm?

22. The pressure in an automobile tire when measured in the winter at 32°F was 2280 torr. The same tire was used during the summer when the temperature was 110°F. What was the tire pressure in the summer?

23. What pressure is required to compress 10 liters of carbon dioxide at 1.0 atmosphere pressure to 1.0 liter at constant temperature?

24. An Eskimo collected 2.0 liters of air in an outdoor container at −40°F. What volume would the air occupy indoors at a temperature of 68°F?

25. A typical weather balloon has a volume of 10 liters. How many weather balloons could be filled at 27°C and 1.0 atmosphere pressure from a tank containing 12 liters of helium at 150 atmospheres pressure and 27°C?

26. Many aerosol spray cans will explode if their internal pressure exceeds 3.0 atmospheres. One such can had a pressure of 2.0 atmospheres at 27°C. Would it be safe to leave the can in an automobile on a hot summer day when the temperature of the automobile interior is 120°F?

27. The average adult human lung capacity is 6,000 milliliters. What volume of air, temperature 68°F, can be inhaled by an average adult human being whose body temperature is 98.6°F?

28. At what pressure will 0.300 moles of N_2 occupy 10.0 L at 90°C?

29. How many moles of CO are present in a 600-mL sample of CO collected at 60°C and 1.25 atm pressure?

30. A sample of nitrogen occupies 47.3 liters at 27°C and 1.50 atmospheres pressure. How many moles of nitrogen are there? What is the mass of the nitrogen?

31. A neon sign contains 0.100 moles of neon gas at a pressure of 2.00 torr and a temperature of 25°C. What is the volume of neon in the neon sign?

32. What is the volume of 4.16×10^{20} molecules of carbon monoxide gas at STP?

33. What volume will 10.0 g of CO_2 occupy at 25°C and 1.75 atm pressure?

34. Cyclopropane is a gas that is used as a general anesthetic. If 1.56 g of the gas occupies a volume of 1.00 L at 0.984 atm pressure and 50°C, what is the molecular weight of cyclopropane?

35. A sample of gas weighs 4.08 g and occupies 2.00 L at 0.850 atm pressure and 32°C. What is the molecular weight of the gas?

36. A sample of air occupies a volume of 5.00 L and has a density of 1.80 g/L. If the air is compressed to a volume of 1.00 L, what will its density be?

37. In a cyclopropane-oxygen mixture used as a general anesthetic, the partial pressure of oxygen is 570 torr. If the total pressure of the mixture is 1.00 atm, what is the partial pressure of cyclopropane?

38. The air inside the lungs of a typical adult has partial pressures as given in Table 5-1. What is the percentage of each gas inside the lungs?

39. If the respiration rate is slowed in a human, the partial pressure of CO_2 in the lungs is increased from 40 torr to 60 torr. Assuming that the partial pressures of other lung gas components (H_2O vapor and N_2) do not change, what will be the effect on the partial pressure of oxygen in the lungs? The normal partial pressure of oxygen in the lungs is 100 torr.

CHAPTER 6
LIQUIDS AND SOLIDS

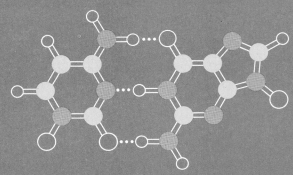

INTERMOLECULAR FORCES
- Dipolar Attractions
- London Dispersion Forces
- Hydrogen Bonds

LIQUIDS
- Density
- Surface Tension
- Viscosity

VAPORIZATION
- Dynamic Equilibrium
- Boiling Point
- Heat of Vaporization

SOLIDS
- Crystalline Solids
- Amorphous Solids

FUSION
- The Melting Process
- Heat of Fusion
- Melting Point

PHASE CHANGES

LIQUID CRYSTALS

In addition to the gaseous state, matter exists in the liquid state and the solid state. Liquids and gases are often called *fluids* because they have the ability to flow and assume the shapes of their containers. When a liquid, water, for example, is poured from a container of one shape into a container of another shape, the liquid takes on the shape of the new container into which it is poured. This behavior contrasts with that of gases, which diffuse freely to fill a container of any size.

A solid object has a definite shape that cannot be easily changed. A bar of soap, for example, has a definite shape and also a definite volume, another characteristic of solid matter. (We all know that a bar of soap does not flow, at least not until we soak it in water for a while.)

Properties of each of the three states of matter can be attributed to the space between particles and the ease with which the particles move about, as illustrated in Figure 6-1. In contrast to gases, the particles in liquids and solids are closer together and they interact with each other to

fluid: a substance which easily changes its shape and is capable of flowing.

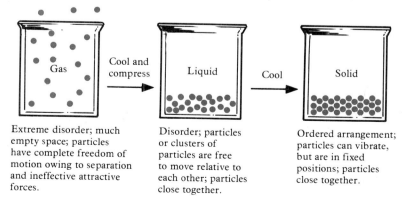

Fig. 6-1 The Three States of Matter

a greater extent. Particles in solids are very close together and tightly packed, allowing only for minimal movement. In this chapter we will study the liquid and solid states of matter, and in particular we will explore the attractive forces which exist in these condensed states.

INTERMOLECULAR FORCES

The kinetic molecular theory applies not only to gases but also to liquids. Recall that gases are composed of widely separated particles undergoing rapid, random motion. Temperature is a measure of the average kinetic energy of the particles; thus, when a gas is cooled, the kinetic energy of the particles is reduced, and they move more slowly. If the cooling is continued, the molecules of a real gas will coalesce into a liquid. This is because real gas molecules exert small attractions for each other as discussed in chapter 5, and if enough kinetic energy is lost, these intermolecular attractive forces will cause the molecules to stick together and form a liquid. We visualize a liquid as consisting of particles clustered closely together but moving freely about throughout the volume of the liquid. As in the case of gases, the average kinetic energy of the particles is proportional to temperature. Since virtually all liquids at ordinary temperatures are composed of molecules, we will consider the types of forces which attract one molecule to another.

intermolecular: between molecules.

Dipolar Attractions

dipolar attractions: attractive forces existing between polar molecules.

For molecules of approximately equal molecular weight, intermolecular forces increase with increasing polarity of the molecule. The strength of the intermolecular forces is reflected in the boiling point of a liquid. The stronger the attractions between molecules, the higher will be the boiling point. Table 6-1 illustrates this trend.

The attractive forces between polar molecules can be explained in terms of the attractions between dipoles. A simple example is hydrogen chloride, HCl. The positive end of one molecule is attracted to the negative end of another molecule because of their opposite partial charges, as illustrated below and in Figure 6-2.

$$\begin{array}{ccc} & & \delta+ \delta- \\ & & H-Cl \\ \delta+ \delta- & \delta+ \delta- & \\ H-Cl & H-Cl & Cl-H \\ & & \delta- \delta+ \end{array}$$

Table 6-1 Boiling Points and Dipole Moments

Substance	Formula	Molecular Weight	Boiling Point (°C)	Dipole Moment (D)
Methane	CH_4	16.0	−161.5	0
Ammonia	NH_3	17.0	−33.4	1.49
Bromine	Br_2	159.8	58.8	0
Iodine Chloride	ICl	162.4	97.8	0.65

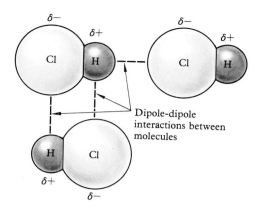

Fig. 6-2 Dipolar Attractions Between Hydrogen Chloride Molecules

Because these are only partial charges, the attractions are not as strong as those in ionic compounds, and they operate over shorter distances than ionic attractions.

London Dispersion Forces

For molecules of approximately equal polarity, intermolecular attractions increase with increasing molecular weight. This trend is illustrated in Table 6-2. Notice that these molecules are nonpolar, yet the boiling points seem to be related to molecular weight. Because the molecules are nonpolar, there can be no dipolar attractions, but there must be a reason for the molecules to group together to form a liquid. The explanation for attraction among nonpolar molecules is based on the idea that the distribution of electrons in any molecule is always in a state of fluctuation (Figure 6-3). Consequently, the density of electrons in a nonpolar molecule changes rapidly over time, so that at any one instant, the molecule

Table 6-2 Boiling Points and Molecular Weights for Nonpolar Molecular Compounds

Compound	Formula	Molecular Weight	Boiling Point (°C)
Methane	CH_4	16.0	−161.5
Ethane	C_2H_6	30.0	−88.6
Propane	C_3H_8	44.0	−42.1
Butane	C_4H_{10}	58.0	−0.5
Pentane	C_5H_{12}	72.0	36.1

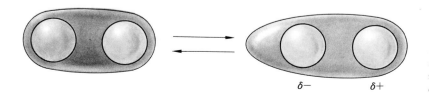

Fig. 6-3 Formation of an Instantaneous Dipole in a Diatomic Molecule. (The spheres are atomic nuclei, and the shaded areas represent electron densities.)

exists as an instantaneous dipole. Because these instantaneous dipoles are constantly changing, a nonpolar molecule does not have a permanent dipole moment which can be measured. However, nonpolar molecules in a liquid synchronize the fluctuations of instantaneous dipoles so that there is a net attraction among molecules, as shown in Figure 6-4. These instantaneous dipolar attractions are known as *London dispersion forces*, named after Fritz London, a German scientist who first developed the theory for this attraction in 1930.

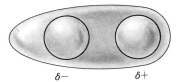 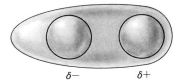

Fig. 6-4 Synchronized Instantaneous Dipoles of Two Diatomic Molecules. (The spheres are atomic nuclei, and the shaded areas represent electron densities.)

The strength of London dispersion forces depends on how easily the electron distribution in the molecules can fluctuate. In general, the larger the molecule, the farther its outer electrons are from the atomic nuclei, and consequently, the easier it is for the electron distribution to shift around in space. Therefore, these intermolecular attractions increase with molecular weight and size. Note that London dispersion forces are the weakest of all intermolecular forces. They exist in all molecules, but their effects are overshadowed if other attractions, such as dipolar forces, exist in the molecules.

Hydrogen Bonds

The attractive forces between molecules in which hydrogen is bonded to highly electronegative atoms such as N, O, and F are much higher than those in other compounds of similar molecular weight and polarity. Table 6-3 illustrates this phenomenon. These unusually large intermolecular attractions always involve compounds which have polar bonds to hydrogen, such as H—F. The explanation is that when hydrogen is

Table 6-3 The Effect of Hydrogen Bonds on Boiling Point

Compound	Formula	Molecular Weight	Boiling Point (°C)	Dipole Moment (D)
Methane	CH_4	16.0	−161.5	0
Ammonia*	NH_3	17.0	−33.4	1.49
Hydrogen Fluoride*	HF	20.0	19.5	1.9
Water*	H_2O	18.0	100	1.84
Ethane	C_2H_6	30.0	−88.6	0
Fluoromethane	CH_3F	34.0	−78.4	1.81
Methyl amine*	CH_3NH_2	31.0	−6.7	1.23
Methanol*	CH_3OH	32.0	64.7	1.66

*compounds capable of forming hydrogen bonds.

bonded to a highly electronegative atom, the bond is very polar, and the bonding electrons are drawn much more closely to the electronegative atom. The hydrogen atom seems almost like a nucleus stripped of its electron, very nearly like a cation, but not quite as positively charged. The hydrogen atom thus experiences a strong attraction to the nonbonded electrons on electronegative atoms of nearby molecules. Because hydrogen is always involved in these interactions, they are called *hydrogen bonds* (H-bonds). In HF, where hydrogen bonding is especially strong (remember that fluorine is the most electronegative element), the hydrogen behaves almost as if it were bound to two fluorine atoms (Figure 6-5).

hydrogen bond: an attractive force existing between molecules that contain hydrogen bonded to nitrogen, oxygen, or fluorine; the attractions are exerted through the hydrogen atoms.

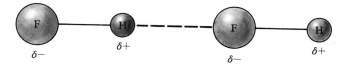

Fig. 6-5 Hydrogen-Bond Formation in Hydrogen Fluoride

Hydrogen bonds are an important feature of liquid water, as shown in Figure 6-6. Each water molecule in the liquid is surrounded by other water molecules with their hydrogen and oxygen atoms oriented in such a way as to provide a maximum number of hydrogen bonds; this is shown in Figure 6-6. For this reason, water has an unusually high boiling point for a molecule of its size. In addition, the polarity of the water molecule allows it to interact favorably with both cations and anions, so that many ionic compounds are soluble in water. Finally, many biological molecules, such as proteins and nucleic acids, have polar and charged groups as parts of their structures. Hydrogen-bond interactions between water and these molecules play a significant role in the biological function of the molecules.

Fig. 6-6 Hydrogen Bonds in Liquid Water

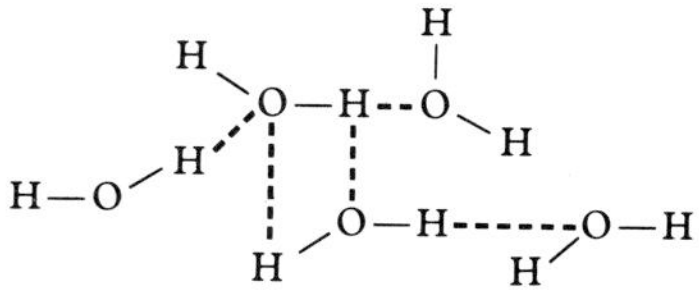

EXAMPLE 6.1

Classify each molecule by the type(s) of intermolecular attraction it would exert for other molecules of the same kind: CH_4, HBr, NH_3, CO_2.

SOLUTION

To solve this problem we must remember that there are three kinds of intermolecular forces: dipolar attractions, London dispersion forces, and hydrogen bonds.

We can determine the kind of intermolecular attraction exerted by a molecule by examining its bonds. If the molecule does not have polar bonds, or if it has symmetrically arranged polar bonds so that it does not have a dipole moment, then the molecule can exert only London dispersion forces. The molecules CH_4 and CO_2 fit this category.

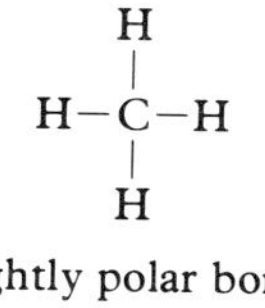

$$\begin{array}{c} H \\ | \\ H-C-H \\ | \\ H \end{array}$$

slightly polar bonds but no dipole moment

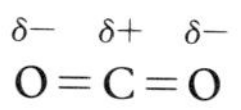

$$\overset{\delta-}{O}=\overset{\delta+}{C}=\overset{\delta-}{O}$$

polar bonds but no dipole moment

If the molecule has polar bonds not involving hydrogen attached to N, O, or F, then the molecule will exert both London dispersion forces and dipolar attractions. HBr is such a molecule.

$$\overset{\delta+}{\text{H}}-\overset{\delta-}{\text{Br}}$$

If the molecule has polar bonds involving hydrogen attached to N, O, or F, then the molecule will exert London dispersion forces and form hydrogen bonds. NH_3 fits this category.

EXERCISE 6.1 Classify each molecule by the type(s) of intermolecular attraction it would exert for other molecules of the same kind.

(a) CO
(b) H_2S (bent molecule)
(c) H_2
(d) HF

LIQUIDS

In general there are two types of pure liquids: nonpolar and polar. Nonpolar liquids are held together by London dispersion forces and usually have low boiling points. Polar liquids, those held together by dipolar forces, tend to have higher boiling points than nonpolar liquids of similar molecular weight. The most extreme kind of polar liquid is one in which there are hydrogen bonds; thus hydrogen-bonded polar liquids show even higher boiling points than polar liquids of similar molecular weight which do not have hydrogen bonds.

Density

Although we know a great deal about the properties of liquids, the structure of a liquid is a much harder feature to study. Thus our knowledge of liquid structure is somewhat limited. We do know that liquids probably contain small regions that have a fairly definite arrangement of molecules and other regions which are completely disordered. The regions of definite structure change as the molecules move through the liquid. Because the molecules cannot pack very close together in the disordered regions, liquids almost always have lower densities than solids. Water is unusually dense in comparison to other liquids. The explanation lies in the fact that each water molecule in the liquid forms an average of three hydrogen bonds. Most other liquid molecules can form only one hydrogen bond apiece at best; many cannot form any hydrogen bonds at all. Hence, the molecules in water are packed together much more tightly than molecules in other liquids, and water has a rather high density for a liquid. Because of their smaller densities, a number of liquids insoluble in water float on its surface. This is why oil floats on the water-based vinegar solution in some salad dressings. Differences in density also account for the surface formation of oil slicks when there are oil spills in the ocean.

Surface Tension

The molecular structure of a liquid has a direct influence on two properties which we have not yet considered. The first of these is *surface tension*, a property we observe on many occasions. It is surface tension that causes water to form beads on a freshly waxed automobile. Surface tension allows us to overfill slightly a glass of water (Figure 6-7), and it allows small insects to walk on water. Surface tension is caused by an imbalance of forces at the surface of a liquid, as shown in Figure 6-8. The molecules in the interior of a liquid are attracted to other molecules in all directions. But those molecules on the surface of a liquid can experience attractions only by molecules on the sides and underneath; there are no attractions from above the surface of the liquid because there are no liquid molecules there. Because of this imbalance, there is a net inward pull on the surface that contracts it almost as if it were a skinlike coating. Thus, a small amount of water on a waxy surface will form beads to minimize its surface area, and drops of a liquid are spherical for the same reason.

Viscosity

Another property which is greatly influenced by liquid molecular structure is *viscosity*, the resistance to flow. The larger the viscosity of a liquid, the more slowly it flows. Liquids with small, compact molecules have low viscosities. Water, for example, has a rather low viscosity. In contrast, liquids having large, unsymmetrical molecules tend to flow slowly. Molasses and cooking oil are examples of viscous liquids. Motor oils are classified by viscosity (20 wt., 30 wt., 40 wt.). Since viscosity increases with decreasing temperature, people who live in cold geographical regions must use the less viscous motor oils in their automobiles. Viscosity is related to the ease with which individual molecules can move by other molecules, and large, unsymmetrical molecules are more likely to become entangled with each other than are small, compact molecules. Table 6-4 lists viscosities of common liquids.

Fig. 6-7 Because of surface tension, this slightly overfilled glass of water has a bulge on the top.

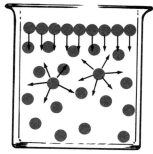

Fig. 6-8 Surface tension is caused by the downward attraction of surface molecules by molecules in the interior of the liquid.

Table 6-4 Viscosities of Common Liquids at 20°C

Substance	Viscosity (kg/m-sec)
Benzene	0.65×10^{-3}
Castor Oil	1027.2×10^{-3}
Ethanol	1.20×10^{-3}
Ether	0.23×10^{-3}
Glycerol (Glycerin)	1490×10^{-3}
Mercury	1.55×10^{-3}
Olive Oil	100.8×10^{-3}
Water	1.00×10^{-3}

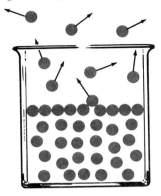

Fig. 6-9 Evaporation of a Liquid

VAPORIZATION

Dynamic Equilibrium

As mentioned earlier, the molecules of a liquid are continuously moving at various speeds. As these molecules move about, every now and then one has enough kinetic energy to break away from the surface of the liquid into the atmosphere (see Figure 6-9). If the liquid is in an open container, this process will continue until all of the liquid has been converted to vapor. Consider a sample of water in an open container. As time passes, the volume of water grows smaller and smaller, until the liquid water disappears. The process of *evaporation* has converted all of the liquid water to water vapor, and the vapor molecules are dispersed throughout the atmosphere.

We have all observed that if we store a liquid in a tightly closed container it will not evaporate. In this case, those molecules that escape from the surface of the liquid cannot disperse through the atmosphere, and some of them return to the liquid, aided by the intermolecular attractive forces. Thus, in the closed container, molecules are always moving back and forth between the vapor and the liquid. When the liquid is first placed in the container, the number of molecules escaping from the liquid to the vapor is higher than the number going in the other direction. But as time goes on, the number of vapor molecules increases. As this build-up occurs, more and more molecules start to return to the liquid. Eventually, the number of molecules escaping from the liquid each second will be the same as the number of vapor molecules captured by the liquid each second, and the liquid and its vapor will be in a state of equilibrium. This process is illustrated in Figure 6-10.

The liquid-vapor equilibrium is a *dynamic equilibrium*, in that molecules are constantly moving between liquid and vapor, but the overall

evaporation: the process in which a liquid is converted to a vapor.

equilibrium: a state of rest or balance due to equal action of opposing forces.

dynamic equilibrium: a state of equilibrium in which there is continuous movement of particles in opposite directions.

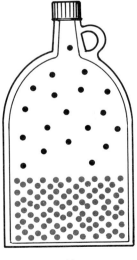
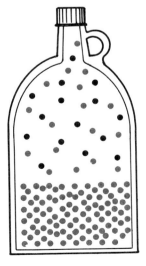

Fig. 6-10 Vaporization of a Liquid in a Closed Container (a) Before Equilibrium and (b) After Equilibrium. The black spheres represent air molecules, and the blue spheres represent liquid and its vapor molecules.

(a)   (b)

number of molecules in each state remains constant. This dynamic liquid-vapor equilibrium occurs whenever a liquid is stored in a closed container. The molecules which have escaped into the vapor behave as any gaseous substance and exert a given pressure if restricted to a certain volume. This pressure is called the equilibrium *vapor pressure*.

vapor pressure: the pressure exerted by a vapor in equilibrium with its corresponding liquid.

Boiling Point

As illustrated in Figure 6-11, the equilibrium vapor pressure of a substance increases with temperature. As temperature increases, more and more liquid molecules acquire enough energy to become vapor molecules. This vapor formation is opposed by the pressure exerted by the atmosphere; consequently, a liquid must overcome the effects of atmospheric pressure in order to boil. In other words, the *boiling point* is the temperature at which the vapor pressure exerted by the liquid is the same as that of the atmosphere. For example, boiling water exerts a vapor pressure of 760 torr at sea level. But because atmospheric pressure varies with altitude, the boiling point of a liquid will also change with altitude. Thus water boils at 95°C in Denver, Colorado, elevation 1 mile above sea level. Table 6-5 shows the decrease in the boiling point of water at higher altitudes.

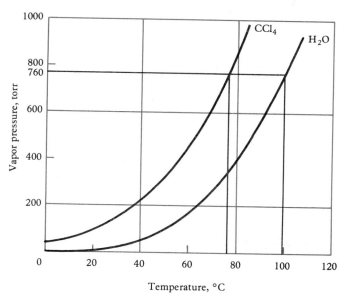

Fig. 6-11 The Increase in Vapor Pressure with Temperature. The boiling points of H_2O and CCl_4 are the temperatures at which their vapor pressures equal atmospheric pressure, 100°C for H_2O and 77°C for CCl_4.

Table 6-5 Variation of the Boiling Point of Water with Pressure

Altitude Above Sea Level (ft)	Atmospheric Pressure	Boiling Point of Water (°C)
0	760 torr	100
5,280 (Denver, Colo.)	630 torr	95
14,500 (Mt. Whitney, Calif.)	450 torr	86
29,000 (Mt. Everest)	253 torr	71

Fig. 6-12 A Pressure Cooker

When a liquid is boiled in a sealed container, the pressure inside the container increases as the liquid vaporizes. The resulting increased pressure raises the boiling point of the liquid. For example, water boils at 120°C at a pressure of 2 atmospheres (1520 torr). This variation of boiling point with pressure is the operating principle of a pressure cooker (Figure 6-12). When water is boiled in a pressure cooker, a valve on top of the sealed pot is adjusted so that the vapor pressure inside the cooker is allowed to build up higher than the atmospheric pressure. Under these conditions, the boiling water reaches a higher temperature than it would otherwise, and this higher temperature cooks food faster than would normally happen. The same principle is utilized in the operation of an autoclave (Figure 6-13), a piece of equipment used for sterilizing hospital equipment. The higher temperatures of steam inside the pressurized autoclave kill bacteria more rapidly than normal boiling water.

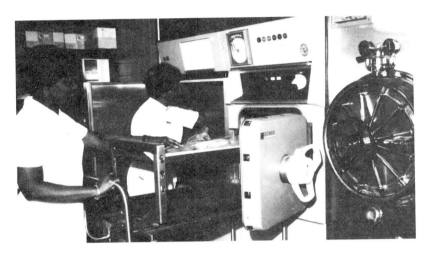

Fig. 6-13 Hospital workers load supplies into an autoclave for sterilization.

Heat of Vaporization

As we heat water at its boiling point in an open container, boiling continues until all of the liquid water is converted to water vapor. If we measure the temperature of the boiling water at any time in the process, it is always the same, even though we are continually adding heat. Where has the heat gone, and why does the temperature remain constant? The answer is that heat is required to change the liquid to vapor at the boiling temperature; since the heat energy is utilized in this way, the temperature of the liquid does not change. The amount of heat required to vaporize one mole of a liquid at its normal boiling point is called the *molar heat of vaporization* of the liquid and is represented by ΔH_{vap}.

(The symbol Δ is the capital Greek letter delta, and it means change, in this case "change in heat.") Molar heats of vaporization are usually expressed in units of kilocalories per mole.

The heat of vaporization is characteristic of a given liquid. Its magnitude is related to the type of intermolecular force in the liquid. For example, water, with its extensive network of hydrogen bonds, has a molar heat of vaporization of 9.72 kcal/mole. On the other hand, methane (CH_4) molecules are held together in the liquid by weak London dispersion forces; thus methane has a molar heat of vaporization of only 1.95 kcal/mole. Table 6-6 lists molar heats of vaporization for various liquids.

Our bodies make practical use of the high heat of vaporization of water when we perspire. Perspiration, being mostly water, evaporates from our skin and in the process absorbs large quantities of heat from our bodies. This loss of body heat results in cooling. It is this same evaporative cooling effect which accounts for the fact that the air temperature in the summer is lower near lakes and oceans.

Although heat of vaporization is defined as the heat required to vaporize a liquid, ΔH_{vap} is also the amount of heat lost by one mole of vapor when it condenses to the liquid state. When heat energy is removed from a vapor by cooling, the molecules lose kinetic energy and form a liquid. For a given mass of vapor, the amount of heat energy lost in this process is identical to the amount of heat energy gained by the same mass of liquid when it vaporizes.

Table 6-6 Molar Heats of Vaporization

Substance	Formula	Molecular Weight	Boiling Point (°C)	ΔH_{vap} (kcal/mole)
Chloroform	$CHCl_3$	119.5	61.7	7.50
Ethanol	C_2H_5OH	46.0	78.5	9.67
Glycerol (glycerin)	$C_3H_8O_3$	92.0	290	18.2
Mercury	Hg	200.6	356	13.6
Methane	CH_4	16.0	−161.5	1.95
Water	H_2O	18.0	100	9.72

EXAMPLE 6.2

Calculate the amount of heat in calories required to vaporize 1.00 gram of water at 100°C.

SOLUTION

The molar heat of vaporization for water is 9.72 kcal/mole (Table 6-6). In order to find the amount of heat needed to vaporize 1.00 g of water, we must first find the number of moles of water in 1.00 g:

$$\text{number of moles } H_2O = \frac{1.00 \text{ g}}{18.0 \text{ g/mole}}$$

$$= 0.0556 \text{ mole}$$

Then we can calculate calories of heat needed to vaporize 0.0556 moles of water at its boiling point:

$$\text{heat required} = \text{moles H}_2\text{O} \times \Delta H_{vap}$$
$$= (0.0556 \text{ mole})(9.72 \text{ kcal/mole})$$
$$= 0.540 \text{ kcal or } 540 \text{ cal}$$

Note that we could have combined the two separate steps as

$$\text{heat required} = \text{moles H}_2\text{O} \times \Delta H_{vap}$$
$$= \left(\frac{1.00 \text{ g}}{18.0 \text{ g/mole}}\right)(9.72 \text{ kcal/mole})$$
$$= 0.540 \text{ kcal or } 540 \text{ cal}$$

EXERCISE 6.2 Calculate the amount of heat in calories needed to vaporize 2.50 g of mercury at 356°C.

SOLIDS

If a liquid is cooled, enough energy can be removed to cause the particles to slow down their movement, and the liquid will freeze into a solid. In the solid state of matter, particles occupy fixed positions in space, with their only movement being slow, gentle vibration about these fixed positions. This limited movement of particles is the reason that solid matter maintains a rigid shape and constant volume. The following discussion will describe the two major types of solid matter, crystalline solids and amorphous solids.

Crystalline Solids

Crystals are composed of ions, molecules, or atoms located at definite positions in the crystal lattice. Although we tend to think of crystals as perfectly ordered arrays of particles, real crystals contain imperfections and defects that occur during crystal formation, such as the defect

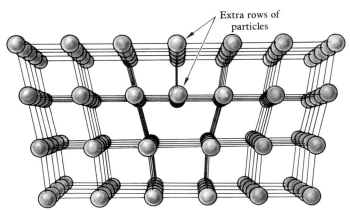

Fig. 6-14 One Type of Crystal Defect

shown in Figure 6-14. If a crystal is allowed to form slowly, the number of defects will be minimal and the crystal size will be large. But the faster a crystal forms, the greater will be the number of imperfections incorporated into the structure and the smaller it will be. The situation is not unlike a crowd of people finding their reserved seats in a theatre. If the people are admitted all at once just before the performance is to begin, many will be misplaced originally. But if the people are allowed to take their seats over a longer period of time, confusion will be minimized.

The fact that small, imperfect crystals are formed as a result of rapid cooling is used to great advantage by the frozen food industry. Food is quick-frozen so that the ice crystals which form are very small. Slow freezing would cause the growth of large ice crystals which rupture the food cells, spilling their contents and reducing the flavor quality and appearance of the product. Quick-freezing avoids this problem and produces frozen food of better taste and appearance.

The most common types of crystals are composed of ions or molecules, but some crystals have single atoms located at points in their lattices. We will discuss five kinds of crystals.

Ionic Crystals

Ionic crystals are held together by electrical forces between cations and anions. Since ionic bonds are very strong, ionic crystals have very high melting points (see Table 3-3).

Potassium chloride is an example of an ionic crystal; its structure is illustrated in Figure 6-15. Notice that each potassium ion is surrounded by six chloride ions; each chloride ion is similarly surrounded by six potassium ions. Because the ions are arranged in the seemingly endless

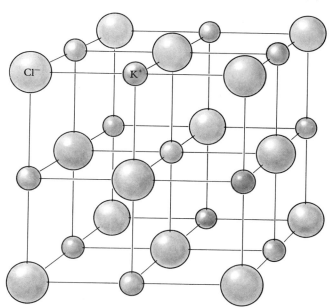

Fig. 6-15 The Crystal Structure of Potassium Chloride

array of crystal lattice, there are no discrete molecules of KCl in the crystal. As you shall see, this situation is in contrast to a molecular crystal in which the crystal is composed of individual molecules. An ionic crystal may be thought of structurally as a single molecule of enormous molecular weight and infinite structure. Of course, the size of the crystal is determined by the conditions under which the crystal is formed. Most ionic crystals that we see each day are rather small because crystals are fragile and hence easily broken.

Hydroxyapatite, $Ca_5(PO_4)_3(OH)$, is an ionic crystalline solid found in bone and teeth. In bone tissue hydroxyapatite makes up 50% of the bone volume and 70% of the bone dry weight. The crystals of hydroxyapatite are responsible for hardness and rigidity in bones and teeth.

Molecular Crystals

Most molecular compounds form *molecular crystals* when they solidify. For example benzene, (C_6H_6), a liquid hydrocarbon, can be solidified by lowering its temperature to 5°C. Benzene crystals are composed of benzene molecules located at lattice points in the crystal (see Figure 6-16). The forces which hold molecular crystals together are the same

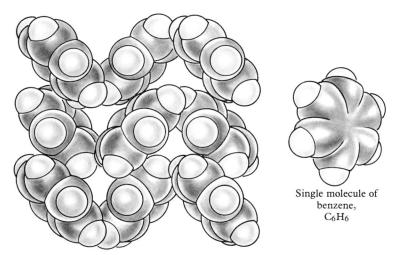

Fig. 6-16 The Crystal Structure of Benzene in the Solid State

Molecular crystal structure of benzene

Single molecule of benzene, C_6H_6

Table 6-7 **Melting Points of Selected Molecular Crystalline Compounds**

Compound	Formula	Melting Point (°C)
Acetylsalicylic acid (aspirin)	$C_9H_8O_4$	135–6
Benzoic acid	$C_7H_6O_2$	122
β-Carotene	$C_{40}H_{56}$	184
Cholesterol	$C_{27}H_{46}O$	149–51
Histamine	$C_5H_9N_3$	86
Urea	CH_4N_2O	133

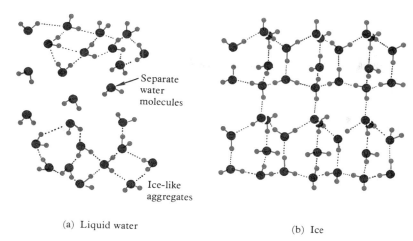

Fig. 6-17 The Arrangement of Molecules in (a) Liquid Water and (b) Ice

(a) Liquid water (b) Ice

ones which are present in the liquid state for the molecular compounds. Since intermolecular forces are weaker than ionic forces, melting points of molecular crystals (examples are shown in Table 6–7) are low in comparison to ionic crystals.

Water forms a molecular crystal when it freezes to ice. Usually the solid form of a substance is more dense than the liquid, but in the case of water, the situation is reversed. Ice is less dense than water because the highly ordered molecules in ice are arranged in such a way that large spaces exist in the solid which are not present in the liquid (see Figure 6–17). Thus, ice floats on water. In cold weather this arrangement permits aquatic life to survive in lakes and oceans. Ice floating on top insulates the water below and thus maintains a temperature suitable for life in the depths of the water.

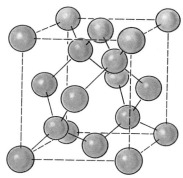

Fig. 6-18 The Crystal Structure of Diamond

Covalent Crystals

Covalent crystals are a network of atoms at lattice points held together by covalent bonds. In a structural sense, each crystal is a giant molecule composed of many atoms. A familiar example is diamond, which is composed entirely of carbon atoms. A model of the diamond structure is shown in Figure 6–18. Because each carbon is bonded to four others at the corners of a tetrahedron, the geometry of this bonding arrangement is the same as that of carbon in methane (CH_4) and in carbon tetrachloride (CCl_4).

Graphite, another form of elemental carbon, is also a covalent crystal. Graphite, whose structure is shown in Figure 6–19, is composed of planes of hexagons, with carbon atoms at each corner. Because of its layered planes, graphite makes an effective solid lubricant. It has a slippery feeling because it flakes easily as the planes slip across one another. Graphite is also used for making pencil "lead" because its black color rubs onto paper easily.

Due to their extensive arrangement of covalent bonds, covalent crystals are often extremely hard and have very high melting points. For example, diamond melts above 3500°C, and graphite passes directly from the solid state to the vapor state at 4000°C.

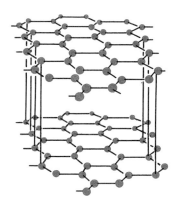

Graphite (C_n)

Fig. 6-19 The Structure of Graphite

Fig. 6-20 Metallic Crystals: Cations Embedded in a Sea of Electrons

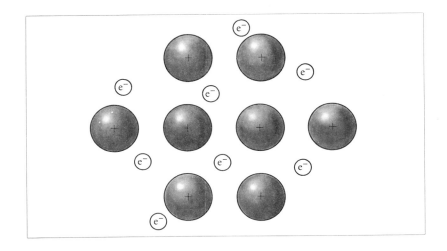

Metallic Crystals

Metallic crystals are composed of metal atoms arranged in a crystal lattice. The forces which hold the metal atoms in place are unlike any which we have yet encountered. Because there are not enough valence electrons in a metal to allow a network of covalent bonds, the electrons do not form definite bonds between any two atoms. Instead they tend to bind atoms to many of their neighbors. We can consider metallic crystals as consisting of cations embedded in a sea of electrons (Figure 6-20). Metals conduct electricity because an electron can move easily to a nearby site. Under the influence of an electric current, electrons in a metallic crystal jump from one atom to another, carrying the electric current through the crystal.

Atomic Crystals

The noble gases can be cooled to sufficiently low temperatures so that they freeze into solids. The crystals of these solids are composed of atoms attracted to each other by the weakest of all forces, London dispersion forces. These are the only known atomic crystals.

Amorphous Solids

amorphous (ah-MOR-fus): lacking definite form or shape.

Extremely fast cooling of a liquid often results in the formation of *amorphous solids*. These materials do not have ordered arrays of particles, and thus they are not crystals. One familiar example of an amorphous solid is glass. Glass is composed largely of silicon dioxide (SiO_2) obtained from sand. When glass is manufactured, purified sand is heated to very high temperatures, causing it to melt. It is then cooled quickly into sheets or other shapes. You will notice that when glass breaks, it does not form any particular geometric pattern as a crystalline material would. Because of its disordered internal structure (shown in Figure 6-21), glass will not maintain its shape completely over a long period of

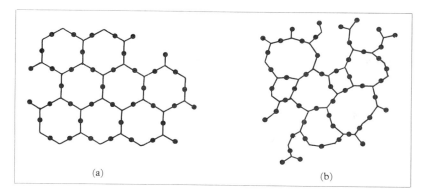

Fig. 6-21 A Comparison of the Structures of (a) Crystalline Silicon Dioxide, SiO_2 (quartz) and (b) Silicon Dioxide in Glass

time. Thus, windowpanes in buildings constructed in the eighteenth century show evidence of glass flowing from the top to produce thick bottoms on the panes.

Other examples of amorphous solids are the synthetic plastics. These materials are formed by causing small molecules to react together to form enormous molecules called *polymers*. Thus, polyethylene is a long, chainlike molecule formed from hundreds or even thousands of ethene units.

polymer (PAH-lih-mur): a large molecule composed of many repeating units; *poly-* means "many" and *-mer* means "parts."

PERSPECTIVE: The Manufacture of Glass

Glass is produced in large furnaces by melting silicon dioxide (from sand), lime, and an alkali at temperatures of 1,500–1,600°C. At these temperatures, glass is as fluid as water. During the melting process, other ingredients are added to give special properties to the glass product. Recycled glass is also added. After thorough mixing and melting, the glass is cooled to 1,000°C so that it can be shaped.

Flat glass is formed by allowing the viscous material to pour vertically out of large containers as sheets which quickly harden. Bottles and other containers are molded from the viscous glass, and light bulbs are made in a machine which blows puffs of compressed air into a continuous ribbon of hot glass. By this method, a machine can produce up to 33 light bulb glasses per second.

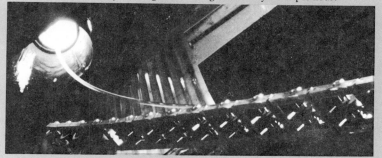

Glass rods are formed by drawing off molten glass at forty miles per hour.

Heat-resistant glass (Pyrex) contains boric oxide (B_2O_3), and it is sometimes called borosilicate glass. Addition of boric oxide causes glass to expand only about one-third as much as ordinary glass for equal increases in temperature. Since it is the rapid expansion produced by heat that causes ordinary glass to break, borosilicate glass can be used for making cooking utensils.

An interesting new glass technology is the manufacture of light-sensitive eyeglasses which darken in bright sunlight. They are constructed of glass which contains silver chloride (AgCl) and silver iodide (AgI). These compounds make the glass lenses turn dark in the presence of bright light in a chemical reaction very much like the one which causes photographic film to turn black when exposed to light. However, unlike photographic film, the glass lenses lose their dark color when the bright light is removed.

$$\begin{array}{c}\text{H}\text{H}\\\diagdown\diagup\\\text{C}=\text{C}\\\diagup\diagdown\\\text{H}\text{H}\end{array} \longrightarrow \ldots \text{C}-\text{C}-\text{C}-\text{C}-\text{C}-\text{C}-\text{C}-\text{C} \ldots$$

ethene (ethylene) → polyethylene (a polymer)

(Ethylene is an older, common name of ethene.) The result is that these large molecules become tangled and are unable to pack themselves into an orderly arrangement in the solid. The disorder of molecules in the solid causes plastics like polyethylene to be deformable, allowing them to be molded.

FUSION

The Melting Process

fusion: the process of particles blending together by melting.

If heat is added to a crystalline solid, the particles within the solid acquire more and more kinetic energy until they have enough motion to disrupt the crystal lattice. This is the process of melting (another term is fusion), in which a solid is converted to a liquid. After melting, the particles have more kinetic energy than they had in the solid state but less than they would have in the vapor state.

When solid and liquid phases of a substance are maintained at the melting point, the particles are in dynamic equilibrium, similar to the situation for liquid and vapor phases at the boiling point. Particles are condensing from the liquid onto the crystal surfaces at the same rate that molecules are passing from the crystal into the liquid.

Heat of Fusion

As a crystalline solid is heated, its temperature increases until the melting point is reached. At that point, the temperature remains constant until all of the solid is converted to liquid, even though heat is being added all the time. This situation is similar to the conversion of liquid to vapor. The amount of heat needed for melting one mole of the crystalline solid at its melting point is called the *molar heat of fusion* (ΔH_{fus}).

Table 6-8 Molar Heats of Fusion

Substance	Formula	Molecular Weight	Melting Point (°C)	ΔH_{fus} (kcal/mole)
Benzene	C_6H_6	78.0	5.5	2.35
Decane	$C_{10}H_{22}$	142.0	−29.7	6.8
Iodine	I_2	253.8	113.6	4.0
Naphthalene	$C_{10}H_8$	128.0	80.2	4.5
Water	H_2O	18.0	0	1.44

The heat of fusion is usually much smaller than the heat of vaporization for a substance. The reason for this is that the heat of fusion is the amount of energy needed to disrupt the crystal lattice but still leave the particles in contact with one another. A much larger amount of energy is required to vaporize the liquid because the attraction between particles must be completely overcome. As in the case of heats of vaporization, heats of fusion also have application to the reverse process, i.e., cooling a liquid until it solidifies (freezes). The freezing point of a pure liquid is the same as the melting point of the corresponding solid. Thus, when a liquid is cooled to its freezing point, an additional amount of heat (equal to ΔH_{fus}) must be lost by the liquid before it will solidify. Heats of fusion for a variety of substances are given in Table 6–8.

EXAMPLE 6.3

Calculate the amount of heat (in calories) necessary to melt 62.0 grams of ice at 0°C.

SOLUTION

The molar heat of fusion for ice is given in Table 6–8 as 1.44 kcal/mole. Thus we can solve the problem as follows:

$$\text{heat required} = \text{moles of water (ice)} \times \Delta H_{fus}$$

$$= \left(\frac{62.0}{18.0} \text{ mole}\right) (1.44 \text{ kcal/mole})$$

$$= 4.96 \text{ kcal or } 4{,}960 \text{ cal}$$

EXERCISE 6.3 Calculate the amount of heat in calories needed to melt 15.0 g of benzene at its melting point.

Melting Point

In the case of pure crystalline solids, conversion to the liquid state occurs over a very narrow temperature range, usually one or two Celsius degrees, and this narrow temperature range is called the melting point. The reason for such sharp melting is that the forces holding a crystal together are approximately equal, and all of the particles achieve the amount of energy necessary to disrupt these forces at very nearly the same temperature. The relative melting points of a series of compounds are related to the strength of the forces holding the crystals together. Thus molecular crystals have much lower melting points than ionic crystals.

The story is somewhat different for amorphous solids. Because the forces which hold these solids together are of many different types and strengths, the particles require a large range of energy to fuse together to form a liquid. For these reasons, amorphous solids do not have sharp melting points. Instead, they slowly soften as they are heated until eventually a free-flowing liquid is produced. In some cases, as with ordinary glass, the temperature range over which softening begins and melting is complete may be several hundred Celsius degrees.

Fig. 6-22 The Heating Curve for Water

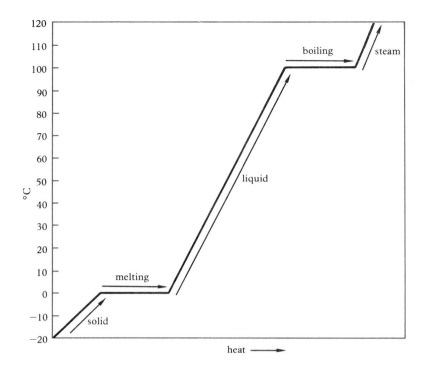

phase (of matter): a state of matter.

PHASE CHANGES

When a substance is heated, its temperature increases unless it is undergoing a phase change such as from solid to liquid. However, during the time a pure substance is melting or boiling its temperature remains constant. A plot of the temperature of a sample of water as a function of time as heat is added at a uniform rate is shown in Figure 6-22. This diagram is called a heating curve. As the ice is heated its temperature rises until the melting point is reached. At that point, the temperature levels off and remains constant until all of the ice is melted. Then, as heating continues, the temperature of water rises until the boiling point is reached. Once again we see a temperature plateau, this time until all of the water is converted to steam.

EXAMPLE 6.4

Calculate the amount of heat in calories required to convert 1.00 mole of ice at −10°C to steam at 100°C.

SOLUTION

In this problem we are asked to determine the total heat which must be added to 1.00 mole of ice at −10°C to

  (1) heat the ice from −10°C to its melting point, 0°C
  (2) convert the ice to water at 0°C

(3) heat the water from 0°C to its boiling point, 100°C
(4) and convert the water to steam at 100°C.

We must begin by consulting Table 1-3 for the specific heat of ice. Then we can complete step (1):

heat required = specific heat of ice × grams of ice × temperature change
= (0.492 cal/g°C) (18.0 g) (10°C)
= 89 cal

Next, we calculate the amount of heat necessary to melt the ice at its melting point, step (2):

heat required = moles of ice × ΔH_{fus}
= (1.00 mole) (1.44 kcal/mole)
= 1.44 kcal

In step (3), we calculate the calories needed to heat the water to its boiling point. The specific heat of liquid water is given in Table 1-3 as 1.00 cal/g°C.

heat required = specific heat of water × grams of water
× temperature change
= (1.00 cal/g°C) (18.0 g) (100°C)
= 1800 cal

At step (4) we must calculate the amount of heat needed to vaporize the water at its boiling point:

heat required = moles of water × ΔH_{vap}
= (1.00 mole) (9.72 kcal/mole)
= 9.72 kcal

Finally, we must convert all of our heat requirements to calories and add them together. Thus

Step (1)		89 cal
Step (2) 1.44 kcal	=	1440 cal
Step (3)		1800 cal
Step (4) 9.72 kcal	=	9720 cal
		13049 cal

We have calculated that the entire process requires 13,049 calories of heat.

EXERCISE 6.4 Calculate the amount of heat in calories required to convert 10.0 g of ice at −5°C to steam at 100°C.

An understanding of the underlying principles of phase changes has allowed applications in the fields of medicine and food preservation. One example is the technique of freeze-drying. In this process, a water-containing substance is frozen, and then its container is evacuated to very low pressures, in the range of 0.1 torr. This extremely low pressure causes the solid water (ice) to pass directly into the vapor state. In this way, water can be removed from a substance at low temperature, thus avoiding possible heat damage. Freeze-drying is used extensively to prepare vaccines and antibiotics and to store skin and blood plasma. It has also gained wide acceptance as a technique for producing instant coffee and nonperishable, light-weight foods for backpacking.

LIQUID CRYSTALS

liquid crystal: a state of matter intermediate between liquid and solid, possessing some properties of liquid matter and others of solid matter.

Certain molecular crystals will not melt directly when heated but instead will turn from a crystalline solid to what is called a *liquid crystal*. This state is a temporary one, and with further heating the liquid crystal completely melts to the liquid state. The liquid crystal state has some characteristics of a crystal and some of a liquid. For example, liquid crystals reflect light like crystals, but they are also able to flow like liquids. Because of this blend of crystalline and liquid characteristics, liquid crystals have unique properties.

One of the more interesting properties of certain liquid crystals is that they react to an electric field by turning from clear to opaque. These types of crystals have recently been incorporated into optical displays for clocks, wristwatches, and calculators. Other liquid crystals react to small changes in temperature by changing color. These liquid crystals are used in color-indicating thermometers to measure air temperature. They are also components of a temperature-indicating paste used to locate blood vessels. When the paste is applied to the skin over a blood vessel, the image of the blood vessel appears in the paste because the blood vessel is warmer than the tissue surrounding it. In this way, medical personnel can locate the vein more easily to give an injection or to withdraw a blood sample.

SUMMARY

Intermolecular Forces: Molecules are attracted to each other by three kinds of forces. Dipolar attractions occur between molecules possessing permanent dipole moments. London dispersion forces are very weak attractions resulting from alignment of instantaneous dipoles which exist in all molecules. Hydrogen bonds, the strongest of the intermolecular attractions, involve molecules which have hydrogen atoms bonded to nitrogen, oxygen, or fluorine atoms.

Liquids: In general, there are two types of pure liquids: polar and nonpolar. It is thought that liquids contain small regions of rather definite structure, while other regions are completely disordered. These regions of order and disorder change as molecules move through a liquid. Because of the relatively loose packing of molecules in the disordered regions, liquids are usually less dense than their corresponding solids. The structure of liquid molecules has a direct influence on the properties of surface tension and viscosity.

Vaporization: Vaporization occurs because some liquid molecules have enough kinetic energy to break away from the liquid surface and become vapor molecules. If a liquid is stored in a closed container, dynamic equilibrium will result. The pressure of vapor molecules in equilibrium with liquid is called the equilibrium vapor pressure. When the temperature of a liquid reaches the boiling point, additional heat of vaporization must be provided in order for the liquid to be converted to vapor. A boiling liquid has a vapor pressure that is equal to atmospheric pressure.

Solids: Solid matter consists of particles packed very closely together and occupying fixed positions in space. When these particles occupy positions in a highly ordered arrangement, the solid is a crystalline solid. If the particles are in disarray, the solid is said to be amorphous. The forces which hold solid matter together may be ionic attractions, dipolar attractions, London dispersion forces, hydrogen bonds, covalent bonds, or electrical forces.

Fusion: Addition of enough heat to disrupt the attractive forces of solid matter causes melting (fusion). When the solid and liquid phases of a substance are maintained at the melting point, dynamic equilibrium occurs. When the temperature of a solid reaches the melting point, additional heat of fusion must be provided in order for the solid to be completely converted to liquid.

Phase Changes: When a substance is heated or cooled, its temperature changes until a phase change occurs. At the melting and boiling points, the temperature remains constant until the phase change is complete, even though heat is being added to or taken away from the substance continually.

Liquid Crystals: The liquid crystal state, a temporary state exhibited by certain molecular crystals when heated, possesses some characteristics of a crystal and some of a liquid.

STUDY QUESTIONS AND PROBLEMS

1. Define the following terms:
   (a) liquid
   (b) dipolar attraction
   (c) London dispersion forces
   (d) hydrogen bond
   (e) surface tension
   (f) viscosity
   (g) evaporation
   (h) dynamic equilibrium
   (i) vapor pressure
   (j) boiling point
   (k) molar heat of vaporization
   (l) solid
   (m) ionic crystal
   (n) molecular crystal
   (o) covalent crystal
   (p) metallic crystal
   (q) atomic crystal
   (r) amorphous solid
   (s) polymer
   (t) melting
   (u) molar heat of fusion
   (v) melting point
   (w) liquid crystal

2. Describe the three kinds of intermolecular forces.

3. Predict the kind(s) of intermolecular forces which would be exerted by each of the following molecules for others of the same kind.
   (a) CCl_4
   (b) CH_3OH
   (c) Ne (a one-atom molecule)
   (d) CH_3-CH_3
   (e) N_2
   (f) $CH_3-N\begin{smallmatrix}H\\H\end{smallmatrix}$

4. Why are liquids usually less dense than solids?

5. Why is water unusually dense in comparison to other liquids?

6. Why is the surface tension of a liquid affected by the nature of intermolecular forces in the liquid?

7. How is the viscosity of a liquid affected by the molecular structure of the liquid?

8. Describe the effect of temperature on viscosity.

9. Describe the dynamic equilibrium that exists in a closed container between a liquid and its vapor.

10. How does the boiling point of a liquid vary with atmospheric pressure?

11. Explain why the boiling point of water is lower than 100°C on a mountain where the atmospheric pressure is 610 torr.

12. Water has a higher boiling point than most other liquids of similar molecular weight. Why?

13. Describe the effect of temperature on vapor pressure?

14. Explain why autoclaves are effective sterilizers.

15. How is the molar heat of vaporization related to intermolecular forces in a liquid?

16. Explain how the human body is cooled by evaporation.

17. Calculate the amount of heat in calories needed to vaporize 3.75 g of water at 100°C.

18. Calculate the amount of heat in calories needed to vaporize 3.00×10^{23} molecules of liquid water at 100°C.
19. Vaporization of 10.0 g of liquid ammonia, NH_3, at its boiling point requires 3.28 kcal of heat. Calculate the molar heat of vaporization for ammonia.
20. Why do slowly formed crystals contain fewer defects than quickly formed crystals?
21. Explain how quick-freezing helps to preserve texture and taste of foods.
22. Why do molecular crystals usually have lower melting points than ionic crystals?
23. Why is ice less dense than water?
24. Why do covalent crystals have such high melting points?
25. Name four forces which exist in crystalline solids and give an example of each.
26. Give two causes for the disordered internal structure of amorphous solids.
27. Explain why glass is classified as an amorphous solid.
28. Describe the dynamic equilibrium that exists between a liquid and its corresponding solid at the melting point of the solid.
29. How is the molar heat of fusion of a solid related to interparticle forces in the solid?
30. Why is the molar heat of fusion for a solid usually lower than the molar heat of vaporization for the corresponding liquid?
31. Calculate the amount of heat in calories required to melt 50.0 g of ice at 0°C.
32. Melting 15.0 g of benzene at 5.5°C requires 452 calories of heat. Calculate the molar heat of fusion for benzene.
33. Why do most pure crystalline solids have sharp melting points?
34. Calculate the amount of heat in calories needed to melt 1.00×10^{23} molecules of I_2 at 114°C.
35. Why do amorphous solids usually show broad melting ranges?
36. Sketch a curve to describe heating of one mole of water from −25°C to 125°C.
37. Calculate the amount of heat in calories required to convert 25.0 g of ice at −15°C to steam at 100°C.
38. Calculate the amount of heat in calories required to convert 4.60×10^{23} molecules of ice at −8°C to steam at 100°C.
39. On the basis of the following information for oxygen, calculate the heat liberated (in calories) when 100 grams of oxygen are cooled from 25°C to −200°C.

melting point	−219°C
boiling point	−183°C
ΔH_{vap}	1630 cal/mole
ΔH_{fus}	106 cal/mole
specific heat of O_2 (gas)	0.228 cal/g °C
specific heat of O_2 (liquid)	0.35 cal/g °C

40. Many of us have learned from experience that steam burns are more severe than burns from boiling water, even though both heat sources are at 100°C. The following calculation will illustrate this point. Calculate the amount of heat released (in calories) when one mole of steam at 100°C condenses and cools to 35.0°C. Compare this value to the amount of heat (in calories) released when one mole of liquid water at 100°C cools to 35.0°C.
41. Explain how the process of freeze-drying is important to the preservation of blood plasma.
42. Some hand-held calculators are advertised as having liquid crystal displays. What does this mean?
43. Compare liquids and solids with respect to the following:
    (a) volume
    (b) shape
    (c) attractive forces between particles
    (d) distances between particles
    (e) mobility of particles
    (f) organization of particles
    (g) types of particles

CHAPTER 7

SOLUTIONS, DISPERSIONS, AND SUSPENSIONS

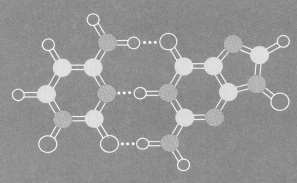

WATER
- Review of Structure and Physical Properties
- Solvent Properties
- Hydrates
- Water in the Human Body

SOLUTION FORMATION
- The Dissolution Process
- Rates of Dissolution
- Solubility
- Factors Affecting Solubility

PREPARATION OF SOLUTIONS
- Concentrations
- Dilutions

PROPERTIES OF SOLUTIONS
- Homogeneity
- Solute Particles
- Colligative Properties

COLLOIDS

SUSPENSIONS

DIALYSIS

TRANSPORT OF BODY FLUIDS AND SOLUTES

The unique and complex substance called water is probably the most familiar of all chemical compounds. Its importance was recognized early in history, and it was considered to be the principal "element" by the sixth century B.C. It was much later that Greek philosophers added fire, earth, and air as fundamental substances of the universe.

Water has the extraordinary ability to mix with countless substances to form solutions, dispersions, and suspensions. These mixtures span the range from homogeneity to heterogeneity. At one extreme is the true solution in which the mixture is uniform throughout. At the other extreme is the suspension—a heterogeneous mixture in which the particles are usually large enough to settle out by gravity. Between these extremes are colloidal dispersions, which are homogeneous mixtures containing particles larger than those in solutions but smaller than those in suspensions.

Protoplasm, the viscous jellylike fluid inside cells, is at once a true solution (of inorganic substances, amino acids, and simple sugars), a colloidal dispersion (of large protein molecules and fat globules), and a suspension (of cellular organelles). The behavior of protoplasm and other body fluids is based on principles that apply to all solutions, dispersions, and suspensions. In this chapter we will explore those principles and some of their applications.

homogeneity (ho-mo-jen-AY-ity): the quality of being uniform throughout.

heterogeneity (HEH-teh-row-jen-AY-ity): the quality of being nonuniform throughout.

WATER

Review of Structure and Physical Properties

The water molecule (Figure 7-1) has a bent shape, with an angle of 104.5° between the two hydrogen atoms. It is thought that liquid water consists of aggregates of hydrogen-bonded molecules with interspersed

Fig. 7-1 The Water Molecule

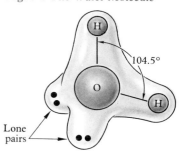

single molecules in dynamic equilibrium with the aggregates (see Figure 7-2). Thus the aggregates are continuously forming, disintegrating, and reforming, and the structure of liquid water is in a state of constant change.

Fig. 7-2 The Structure of Liquid Water

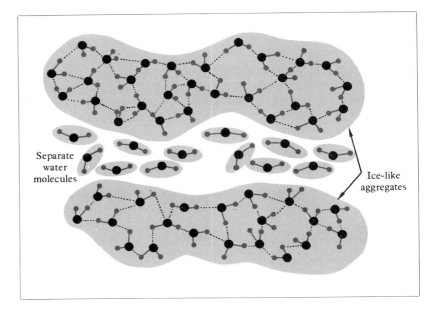

The structure of ice (solid water) is a rigid, more highly ordered arrangement in which each oxygen atom is connected to four hydrogen atoms by two covalent bonds and two hydrogen bonds. This arrangement contains more open spaces than liquid water, and ice is therefore less dense than liquid water. Water molecules occupy their smallest volume at 3.98°C, and at this temperature the density of water is at a maximum of 1.00 g/mL.

Extensive networks of hydrogen bonds in liquid and solid water result in relatively high values for the freezing point, boiling point, heat of vaporization, and heat of fusion. In addition, the geometry of the water molecules and the effect of hydrogen bonding cause liquid water to have a higher density than would be expected. These properties of water are summarized in Table 7-1.

Table 7-1 Physical Properties of Water

Appearance	Colorless liquid
Melting Point	0.00°C (1 atm)
Boiling Point	100.00°C (1 atm)
Density of liquid	0.999 g/mL (0°C)
Density of solid	0.917 g/mL (0°C)
Molar heat of fusion	1.44 kcal/mole
Molar heat of vaporization	9.72 kcal/mole
Dipole moment	1.85 D

Solvent Properties

One of the most remarkable features of water is its versatility as a *solvent*. More substances will dissolve in water than in any other existing liquid. Polar covalent and ionic substances are most likely to dissolve in water because their polar and charged parts are attracted to the polar water molecules. These interactions allow water molecules to surround dissolved particles, as shown in Figure 7-3, forming a uniformly homogeneous mixture.

solvent: the component present in greatest amount in a solution.

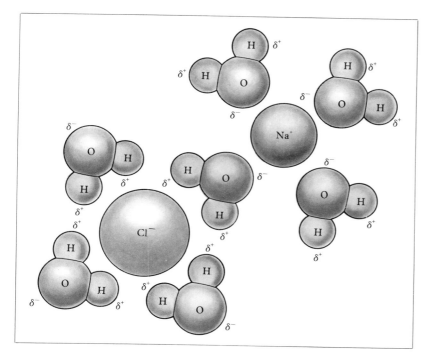

Fig. 7-3 Water Molecules Surround Dissolved Particles

The unique structure of water confers ideal properties to the liquid for its biological role. Its high boiling point and heat of vaporization help to insure that enough water is present in the liquid state to support life. These two properties combined with its high specific heat allow water to cool our external environment as well as the internal environment of our bodies. Because of its excellent solvent qualities, water is able to transport compounds needed in plant and animal metabolism. Finally, the high surface tension of water allows it to be drawn up in plant stems and roots by *capillary action*, which is illustrated in Figure 7-4.

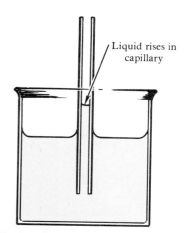

Fig. 7-4 Capillary Action

capillary action: the rise of a liquid within small-diameter tubes.

Hydrates

Some ionic compounds that were formerly dissolved in water have water molecules in their crystals after the water is evaporated. The compounds appear to be dry, with no visible moisture, but each contains a fixed

water of hydration: the fixed amount of water associated with certain ionic crystals.

hydrate: a solid ionic compound that contains a fixed number of water molecules within its crystals.

amount of water and has a definite composition. The attached water molecules are referred to as *water of hydration*, and the compounds are called *hydrates*. Water of hydration is usually released from the compound by heating. Formulas for hydrates are written to show water of hydration as illustrated below. Examples of hydrates are given in Table 7-2.

$$CuSO_4 \cdot 5H_2O$$

copper(II) sulfate pentahydrate

Table 7-2 Commonly Occurring Hydrates

Formula	Name*
$CuSO_4 \cdot 5H_2O$	Copper (II) sulfate pentahydrate (blue vitriol)
$Na_2SO_4 \cdot 10H_2O$	Sodium sulfate decahydrate (Glauber's salt)
$KAl(SO_4)_2 \cdot 12H_2O$	Potassium aluminum sulfate dodecahydrate (alum)
$Na_2B_4O_7 \cdot 10H_2O$	Sodium tetraborate decahydrate (borax)
$FeSO_4 \cdot 7H_2O$	Iron(II) sulfate heptahydrate (green vitriol)
$H_2SO_4 \cdot H_2O$	Sulfuric acid hydrate (m.p. 8.6°C)

*Common names are given in parentheses.

Water in the Human Body

Water provides the medium in which nearly all of the body's chemical reactions take place. In addition, it transports nutrients to cells and takes end products of respiration and metabolism away from cells. Water accounts for 60-70% of human body weight, distributed among various body fluids as indicated in Table 7-3. The total amount of body water remains relatively constant at about 35-50 liters in an adult, depending on the weight of the person. Internal mechanisms regulate our intake and output of water, so that the two are balanced in a healthy human being.

Table 7-3 Distribution of Water in the Human Body

Location	Amount
Intracellular fluid	50% of body weight
Extracellular fluid	15% of body weight
Blood plasma	5% of body weight

Water Intake

An important factor in the amount of water taken into the body is the external temperature, since water is used as a source of cooling in the perspiration process. Thus, at high external temperatures we are likely to drink more water than in cool weather in order to compensate for the loss from perspiration. In addition to drinking it, we get some of our water from food. Metabolic processes also generate a portion of our water needs. A normal adult takes in an average of about 2.5 liters of water each day (see Table 7-4).

Table 7-4 Water Balance in the Human Body

Water Intake		Water Output	
Liquids:	1,200–1,500 mL	Kidneys:	900–1,400 mL
Food:	700–1,000 mL	Lungs:	350 mL
Metabolic water:	200–300 mL	Feces:	150 mL
		Skin:	450–900 mL
Total	2,100–2,800 mL		1,850–2,800 mL

Water Output

Loss of water occurs by way of the skin, lungs, and the intestinal tract, but the most important water loss is brought about by the kidneys (Table 7-4). In this instance, waste materials dissolved in water are excreted. Normal loss of water through urine may be in the range of 1.0–1.5 liters per day. Excretion by this route is highly flexible, and if excessive amounts of water are ingested, the kidney promptly eliminates the excess. Thus, there is no need to be cautious about drinking water. But if there is a shortage of body water, *hormones* secreted by the *pituitary* and *adrenal* glands decrease the amount of water excreted by the kidneys.

In cases of illness, such factors as high body temperature, kidney malfunction, blood loss, vomiting, and diarrhea can cause greater than normal loss of water. If the loss is severe, *dehydration* sets in, and medical attention is necessary.

hormone: a chemical messenger produced by an endocrine gland and secreted directly into the blood stream to exert a specific effect on a distant part of the body.

pituitary (pih-TU-ih-terry) **gland:** a small oval endocrine gland attached to the base of the brain; also called the hypophysis.

adrenal gland: one of a pair of endocrine glands located just above the kidneys.

dehydration: loss of water; in physiology, a condition in which body water is severely depleted.

SOLUTION FORMATION

A *solution* is a mixture of two or more components evenly distributed in each other. Thus, solutions are homogeneous mixtures. Solutions may be composed of gases, liquids, or solids. Examples of various types of solutions are shown in Table 7-5. The earth's atmosphere is an example of a gaseous solution. Dental amalgam, consisting of mercury, silver,

Table 7-5 Types of Solutions

Type	Examples
gas in a gas	air, some anesthetic gases
gas in a liquid	carbonated beverages (CO_2 in H_2O)
gas in a solid	— —
liquid in a gas	humid air
liquid in a liquid	rubbing alcohol, antifreeze
liquid in a solid	dental amalgam (tooth fillings)
solid in a gas	— —
solid in a liquid	salt water
solid in a solid	brass, alloys, jewelry gold and silver

aqueous: containing water.

tin, copper, and zinc, is a solid solution. Liquid solutions, particularly those having water as the solvent, are the ones most familiar to us, and we will focus our attention on *aqueous* solutions in the following sections.

Solutions are made up of a major component, the *solvent*, and one or more minor components, the *solutes*. The liquid solution existing in greatest quantity on earth is seawater. The solvent, of course, is water, and sodium chloride is the most abundant solute. Table 7-6 gives the average composition of seawater, excluding dissolved gases. The oceans are separated from each other by land masses, and the composition of a sample of seawater will vary with its origin and also the depth from which it was taken.

Table 7-6 Average Composition of Seawater (Excluding Dissolved Gases)

Constituent	g/kg of Seawater	Proportion of Total Salt Content (%)
Chloride (Cl^-)	18.980	55.044
Sulfate (SO_4^{2-})	2.649	7.682
Bicarbonate (HCO_3^-)	0.140	0.406
Bromide (Br^-)	0.065	0.189
Fluoride (F^-)	0.001	0.003
Boric acid (H_3BO_3)	0.026	0.075
Sodium (Na^+)	10.556	30.613
Magnesium (Mg^{2+})	1.272	3.689
Calcium (Ca^{2+})	0.400	1.160
Potassium (K^+)	0.380	1.102
Strontium (Sr^{2+})	0.013	0.038
Total	34.482	100.000
Water (with traces of other substances	965.518	
Total	1,000.000	

The Dissolution Process

Consider what happens when a solute dissolves in a liquid solvent. The solute particles must separate from each other and disperse themselves throughout the solvent. At the same time, solvent molecules must separate so that solute particles can be dispersed among them. In essence, there must be very close interaction between solute and solvent, and this interaction becomes favorable if there are attractive forces between solute and solvent particles.

An example of the dissolving process (*dissolution*) is the formation of a solution of sodium chloride in water. Figure 7-5 illustrates the process.

Fig. 7-5 Sodium Chloride Dissolving in Water

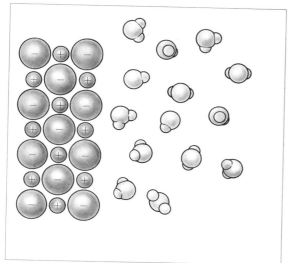
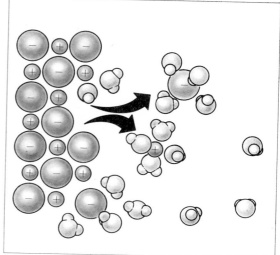

Sodium chloride dissolves readily in water because its ions are attracted to the polar water molecules. When a crystal of NaCl is added to water, the water molecules begin to bombard the surface of the crystal. As a result of the many collisions, eventually a water molecule will be oriented so that its partially negative oxygen is next to a positively charged sodium ion, and the attraction between the two is strong enough to draw the sodium ion out of the crystal and into the water. Similarly water molecules attract the negatively charged chloride ions to the partially positive hydrogen atoms of the solvent. In this way the crystal is broken apart and its ions are dispersed through the water. Each ion becomes surrounded by water molecules, the exact number of which is unknown, in the process called *hydration*. (In solutions having solvents other than water, the more general term *solvation* is used.) We can describe hydration by the following equation:

$$NaCl(s) \xrightarrow{H_2O} Na^+(aq) + Cl^-(aq)$$

solvation: the process in which solvent particles surround solute particles.

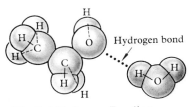

Fig. 7-6 Hydrogen Bonding Between Ethanol and Water Molecules

where NaCl(s) represents solid sodium chloride, and Na^+(aq) and Cl^-(aq) represent the hydrated (aqueous) ions.

Many ionic compounds are soluble in water, as are many molecular compounds. Polar covalent compounds are able to interact favorably with water molecules through polar attractions and, in some cases, hydrogen-bond formation. Ethanol, the liquid commonly called ethyl alcohol, dissolves readily in water by forming hydrogen bonds (see Figure 7-6).

Rates of Dissolution

The rate at which a solute dissolves depends on three factors: surface area, agitation, and temperature.

Surface Area

The surface of the undissolved solute is where solvent comes into contact with solute. For a given mass of solute, the greater the surface area, the more exposure the solute has to the solvent, and the faster the solute will dissolve. Crushing or grinding a solid solute to tiny bits increases its surface area and thereby increases the rate of its dissolution. For liquid solutes, shaking breaks the liquid into small droplets, thus increasing surface area and causing faster dissolution.

Agitation

Agitation by stirring or shaking helps to disperse solute particles through the solution more quickly and thus increases the rate of dissolution.

Temperature

An increase in temperature causes solvent molecules to move faster. Consequently the frequency of bombardment of undissolved solute by solvent molecules is increased, and the dissolving process is more rapid.

Solubility

The maximum amount of a solute that will dissolve in a given quantity of solvent at a particular temperature is called the *solubility* of the solute. Table 7-7 lists solubilities for some common substances in water. As you can see from the table, there is quite a range of solubilities for the various substances. The terms soluble and insoluble are commonly used, even though their meanings are not precise, to describe substances with very high or very low solubilities. For our purposes we will use the word soluble to refer to substances having water solubilities of 1 g/100 mL or greater, and we will reserve the use of insoluble for those substances

Table 7-7 Solubilities of Common Substances in Water at 20°C

Substance	Solubility (g/100 mL)
Barium hydroxide, $Ba(OH)_2$	3.89
Iodine, I_2	0.029
Silver chloride, AgCl	0.00015
Sodium chloride, NaCl	36.0
Sodium hydroxide, NaOH	109
Sugar (sucrose, $C_{12}H_{22}O_{11}$)	203.9

with water solubilities of less than 1 g/100 mL. Many liquids are soluble in other liquids, for example, ethanol in water. We use the term *miscible* to describe mutually soluble liquids.

miscible (MISS-sih-bul)

If we add to a solvent more solute than can be dissolved, dynamic equilibrium sets in between the dissolved particles and the undissolved solute. Thus, if we add 36 grams of NaCl to 100 milliliters of water at 20°C, all of the salt will dissolve. But if we continue adding salt, it will not be possible to exceed the solubility limit of 36 grams per 100 milliliters. However, dynamic equilibrium will exist between the dissolved salt and the undissolved salt, and ions will be exchanged between the crystals and the solution at equal rates (see Figure 7-7). When a solution is in dynamic equilibrium with undissolved solute, the solution is said to be *saturated*. The solubility of a particular solute is determined in the laboratory by making a saturated solution and measuring the amount of dissolved solute.

Fig. 7-7 Dynamic Equilibrium in a Saturated Solution

Solubilities of substances are quite variable depending on the solvent used. There is no good way to predict accurately what the solubility of a given substance will be because of the complex factors involved in the the process of dissolving. However, it is possible to make some useful generalizations. When comparing various solutes, it is frequently true that the more negative the heat change of dissolution (called the *molar heat of solution*, $\Delta H_{solution}$), the greater the solubility (at a given temperature). A negative heat of solution means that heat is released when solute dissolves; thus, a negative heat of solution signifies that dissolving the solute is an exothermic process. A positive heat of solution means that heat is absorbed when solute dissolves, and the dissolution process is endothermic. In these cases, a noticeable cooling of the solution sometimes occurs when the solute dissolves.

molar heat of solution ($\Delta H_{solution}$): the heat change when one mole of solute dissolves in a particular solvent.

As illustrated in Table 7-8, solution formation can be endothermic or exothermic, depending on the particular solute and solvent. Ammonium nitrate, NH_4NO_3, exhibits a large endothermic heat of solution in water, so large that it is used by athletic trainers and hospital workers to make instant cold packs. The solid NH_4NO_3 is placed in a small plastic bag, which in turn is put into a larger plastic bag of water and sealed. Kneading the large bag ruptures the small bag and allows the NH_4NO_3 to dissolve in the water. As the solution forms, the pack gets very cold and

Table 7-8 Molar Heats of Solution and Solubilities at 25°C

Substance	$\Delta H_{solution}$ (kcal/mole)	Solubility (moles/L)
Sodium chloride (NaCl)	0.93	6.18
Sodium iodide (NaI)	−1.8	12.3
Sodium hydroxide (NaOH)	−10.6	38.5
Silver iodide (AgI)	26.71	0.00000016
Silver chloride (AgCl)	15.81	0.000014
Silver fluoride (AgF)	−3.4	14.1
Potassium chloride (KCl)	4.12	4.76
Ammonium chloride (NH_4Cl)	3.53	7.35

can be used in place of a regular ice pack. A large exothermic heat of solution accounts for the use of calcium chloride, $CaCl_2$, in making instant hot packs.

Factors Affecting Solubility

The most important factors which influence solubility are pressure, temperature, and chemical structure.

Pressure

The effect of pressure on solubility is only noticeable when there is a volume change caused by solution formation. Since a gaseous solution can change its volume to match the size of the container, there are not necessarily any volume changes when gases mix to form solutions. Hence, pressure has no effect on the solubilities of gases in one another.

However, when a gas dissolves in a liquid, as when CO_2 dissolves in water, there is a large volume decrease in going from the separate components to the solution. An increase in pressure will therefore cause more of a gas to dissolve in a liquid or a solid (Figure 7-8) and thus increase the solubility of the gas. This principle is applied to the production of carbonated beverages like champagne, beer, and soda pop. These solutions are packaged under a high pressure of CO_2, and when the container is opened, the lowered pressure causes CO_2 to bubble out of solution.

There is very little volume change associated with dissolving liquid and solid solutes in liquid, and thus pressure has a negligible effect on these solubilities.

Temperature

The effect of temperature on solubility depends on the sign of the molar heat of solution for individual solutes. For solutes with positive heats of solution, dissolving requires an input of heat into the solution,

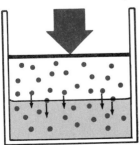

Fig. 7-8 Increasing the applied pressure causes more gas molecules to dissolve in a liquid.

and solubility increases with increasing temperature. On the other hand, if a solution forms exothermically (negative heat of solution), increasing the temperature will decrease the solubility of the solute. The process of dissolution is exothermic for most gases dissolving in water, and increased temperatures usually lower their solubilities. Thus, when soda pop warms up, CO_2 escapes, and the beverage goes "flat."

Chemical Structure

Molecules having strong attractions for water molecules (see Figure 7-9) are said to be *hydrophilic* (water loving). Substances which can form hydrogen bonds with water are especially hydrophilic.

In contrast, there are some substances whose molecules have very little attraction for water molecules. These substances, designated as *hydrophobic* (water hating), are usually nonpolar. If we try to dissolve a hydrophobic substance in water, the attractive forces between water molecules are too strong to be disrupted by the almost nonexistent attractive forces between water and the hydrophobic substance. Consequently, the hydrophobic substance will not dissolve in water but may form isolated clusters of hydrophobic molecules dispersed in the water but not truly dissolved (see Figure 7-10).

Gasoline and oil are examples of hydrophobic substances. They are both composed of mixtures of hydrocarbons derived from petroleum. Since all hydrocarbons are nonpolar and thus hydrophobic, gasoline and oil are not miscible with water to any extent.

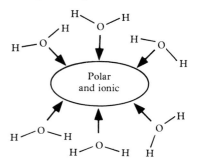

Fig. 7-9 Hydrophilic substances are hydrated by water.

hydrophilic (high-dro-FIL-lic)

hydrophobic (high-dro-FO-bic)

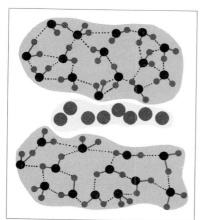

Fig. 7-10 Hydrophobic substances do not dissolve in water but form clusters instead.

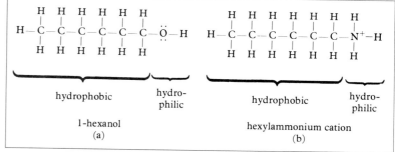

Fig. 7-11 Molecules Having Hydrophobic and Hydrophilic Parts

Some large ions and molecules can have both hydrophilic and hydrophobic parts, as illustrated in Figure 7-11. When such molecules or ions are added to water, the charged or polar parts attract water molecules and become hydrated. The size of the hydrophobic part of the ion or molecule is a factor in the degree of solubility of the substance. Many biological molecules have both hydrophilic and hydrophobic parts, and the tendency for hydrophobic structures to group together in water is an important factor in the behavior of proteins, nucleic acids, biological membranes, and natural detergents.

We can make some fairly reliable guesses about solubility by examining the chemical structures of the solute and solvent. If the solvent is polar, it is likely to dissolve ionic and polar covalent substances. If the solvent is capable of forming hydrogen bonds it will show a definite preference for solutes that are also capable of forming hydrogen bonds. If the solvent is nonpolar, a liquid hydrocarbon, for example, it will probably dissolve nonpolar solutes. A statement often used to summarize these predictions is "like dissolves like."

PREPARATION OF SOLUTIONS

When we stop to think about it, we realize that each of us prepares many solutions every day. Cooking, for example, requires a great deal of solution preparation. Sometimes we add seasonings to a sauce by taste, but in other cases, as when we mix together the basic ingredients of the sauce, we measure carefully. We add chlorine-containing compounds to our swimming pools in measured amounts to prevent growth of bacteria and algae. We dilute liquid fertilizer in water to achieve the appropriate solution of nutrients for our lawns and gardens. In most instances, we make these measurements in order to achieve the correct concentrations of these solutions.

Concentrations

concentration: the amount of solute in a given quantity of solution.

There are several ways of expressing *concentrations* of solutions, and each has its own particular advantages. We will start with the method called molarity.

Molarity

Molarity is concentration expressed as the number of moles of solute per liter of solution. We can formulate the following equation defining molarity, whose symbol is M:

$$M = \frac{\text{moles of solute}}{\text{liters of solution}}$$

This method of expressing concentration is especially useful in solutions containing solutes to be used in chemical reactions, as most compounds react in small whole-number ratios of ions, atoms, or molecules. Thus, a $1M$ (said as 1 molar) solution of NaCl contains 58.5 grams of NaCl per liter of solution. A solution containing 5.85 grams of NaCl per liter has a concentration of $0.1M$, and a solution containing 2.92 grams of NaCl per liter has a concentration of $0.05M$.

EXAMPLE 7.1

How many grams of sodium nitrate, $NaNO_3$, would 1 liter of a $0.75M$ solution contain?

SOLUTION

From the definition of molarity we know that a $0.75M$ solution of $NaNO_3$ contains 0.75 moles of $NaNO_3$ per liter. Thus, we must calculate the weight of 0.75 moles of $NaNO_3$:

$$\text{grams of } NaNO_3 = (0.75 \text{ mole}) \left(\frac{85.0 \text{ g}}{1 \text{ mole}}\right)$$

$$= 64 \text{ g}$$

EXERCISE 7.1 How many grams of glucose, $C_6H_{12}O_6$, would be needed to make one liter of $0.50M$ solution?

The definition of molarity is simply the ratio of moles of solute to volume of solution (in liters). If we know the molarity of a solution, we can use this ratio to calculate the moles of solute, and therefore the mass of solute, in any volume of the solution. For example, how much NaCl is present in 100 mL of a $1.00M$ solution? We can solve this problem as follows. First, write down the definition of molarity:

$$M = \frac{\text{moles of solute}}{\text{liters of solution}}$$

The rearrange the definition to solve for moles of solute:

$$\text{moles of solute} = M \times \text{liters of solution}$$

Now substitute values for molarity and volume:

$$\text{moles of solute} = (1.00 \text{ mole/liter})(0.100 \text{ liter})$$

$$= 0.100 \text{ mole}$$

And finally, we can calculate the mass of NaCl in 0.100 mole by use of a unit factor.

$$\text{grams NaCl} = (0.100 \text{ mole}) \left(\frac{58.5 \text{ g NaCl}}{1 \text{ mole}}\right)$$

$$= 5.85 \text{ g}$$

EXAMPLE 7.2

How many grams of magnesium chloride, $MgCl_2$, are in 500 mL of a $3.00M$ solution of $MgCl_2$?

SOLUTION

From the definition of molarity we can calculate the number of moles of $MgCl_2$ in 500 mL of solution:

$$\text{moles } MgCl_2 = M \times \text{liters of solution}$$

$$= (3.00 \text{ moles/L})(0.500 \text{ L})$$

$$= 1.50 \text{ moles}$$

And then we can calculate the number of grams of $MgCl_2$ in 1.50 moles:

$$\text{g } MgCl_2 = (1.50 \text{ moles}) \left(\frac{95.3 \text{ g}}{1 \text{ mole}}\right)$$

$$= 143 \text{ g}$$

EXERCISE 7.2 How many grams of magnesium phosphate, $Mg_3(PO_4)_2$, are in 250 mL of a 1.50M solution of $Mg_3(PO_4)_2$?

The definition of molarity contains three variables, and we can use that definition, as above, to calculate any one of the three variables if the other two are known. The following example will give you some practice in calculating volume of solution from molarity and moles of solute.

EXAMPLE 7.3

What volume of a 1.75M solution of potassium chlorate, $KClO_3$, would contain 15.0 g of $KClO_3$?

SOLUTION

If we calculate the number of moles of $KClO_3$ in 15.0 g, we can then calculate the volume of the solution which would contain that number of moles:

$$\text{moles of } KClO_3 = (15.0 \text{ g}) \left(\frac{1 \text{ mole}}{122.5 \text{ g}}\right)$$

$$= 0.122 \text{ mole}$$

And,

$$\text{molarity} = \frac{\text{moles of solute}}{\text{liters of solution}}$$

$$\text{liters of solution} = \frac{\text{moles of solute}}{\text{molarity}}$$

$$= \frac{0.122 \text{ mole}}{1.75 \text{ mole/L}}$$

$$= 0.0697 \text{ L}$$

EXERCISE 7.3 What volume of a 3.70M solution of lithium nitrate, $LiNO_3$, would contain 26.2 g of $LiNO_3$?

Percentage Concentrations

In some instances, it is not necessary to know the number of moles of solute in a solution. In these cases, concentration may be expressed as the relative amounts of solute and solvent. One way of expressing relative amounts is by use of *percentage*. For example, a solution of NaCl contains 10 grams per 100 milliliters of solution. In this example, the solute is a solid and the solvent is water, a liquid. Thus we can express percentage as the ratio of grams (weight) of NaCl to the milliliters (volume) of solution. We call this a *weight-to-volume (w/v) percentage concentration*. We can formulate an equation for this kind of concentration as follows:

$$\% \text{ (w/v)} = \frac{\text{g of solute}}{\text{mL of solution}} \times 100$$

Our example, the NaCl solution, thus has a concentration of 10% (w/v). Weight-to-volume percentage concentrations are often used in clinical work.

EXAMPLE 7.4

What is the percent (w/v) concentration of 16.5 grams of glucose dissolved in 150 mL of solution?

SOLUTION

$$\% \text{ (w/v)} = \frac{\text{g of solute}}{\text{mL of solution}} \times 100$$

$$= \frac{16.5 \text{ g}}{150 \text{ mL}} \times 100$$

$$= 1.10$$

EXERCISE 7.4 What is the percent (w/v) concentration of 25.0 g of NaCl dissolved in 200 mL of solution?

In clinical work involving dilute solutions, percentage concentration is often based on milligrams of solid solute per milliliter of solution. This kind of expression is called *milligram percent* (mg %).

$$\text{mg \%} = \frac{\text{mg of solute}}{\text{mL of solution}} \times 100$$

The relationship between mg % and % (w/v) is

$$\text{mg \%} = 1000 \text{ \% (w/v)}$$

EXAMPLE 7.5

What is the concentration in mg % of 100 mg of NaCl dissolved in 500 mL of solution?

SOLUTION

$$\text{mg \%} = \frac{\text{mg of solute}}{\text{mL of solution}} \times 100$$

$$= \frac{100 \text{ mg}}{500 \text{ mL}} \times 100$$

$$= 20.0$$

EXERCISE 7.5 What is the concentration in mg % of 23 mg of KCl dissolved in 125 mL of solution?

When we make solutions containing liquid solutes, it is convenient to measure volume of solute rather than weight. In these cases the percentage concentration is a volume-to-volume ratio (v/v). Thus *volume-to-volume percentage concentration* may be expressed as

$$\% \text{ (v/v)} = \frac{\text{mL of solute}}{\text{mL of solution}} \times 100$$

Volume-to-volume percentage is frequently used to express concentrations of alcoholic beverages.

EXERCISE 7.6 What is the concentration of ethanol in % (v/v) in 80.0 mL of a solution that contains 11.5 mL of ethanol?

The third variation of expressing percentage concentration is that of *weight-to-weight* (w/w). Although this method has limited use, it is valuable for solid solutions.

$$\% \text{ (w/w)} = \frac{\text{g of solute}}{\text{g of solution}} \times 100$$

EXERCISE 7.7 What is the concentration of silver in % (w/w) in 8.36 g of an alloy that contains 85.0 mg of silver?

As is evident from frequent newspaper reports, trace amounts of pollutants in water and air can have significant health implications. The concentrations of very small amounts of such substances are usually measured in parts per million (ppm) or parts per billion (ppb), where the term "parts" refers to any unit of measure (grams, milliliters, moles, etc.). A 1-ppm solution would contain 1 part of solute for every 1 million parts of solution, and a 1-ppb solution would contain 1 part of solute for every 1 billion parts of solution.

Dilutions

reagent (ree-AY-gent): a reactive substance (reactive *agent*) used in the laboratory for chemical analysis.

It is common practice in hospital laboratories to purchase concentrated stock solutions of *reagents* which are then diluted with water or other solvents to achieve desired concentrations. Dilutions are accomplished by adding solvent to concentrated solution, thus lowering the original concentration to a more desirable one. For example, suppose you need a $0.1M$ solution of glucose, $C_6H_{12}O_6$ (MW = 180), and you wish to prepare this from 100 mL of a stock solution of $1.0M$ glucose. You can accomplish this task by adding enough solvent to the $1.0M$ solution to lower its concentration to $0.1M$. The question is, how much solvent should you add? We solve this problem by using a method very similar to that used for gas law problems. Then

$$\text{final concentration} = \text{initial concentration} \times \text{dilution factor}$$

If we rearrange this equation to solve for dilution factor, we get

$$\text{dilution factor} = \frac{\text{final concentration}}{\text{initial concentration}}$$

Thus,

$$\text{dilution factor} = \frac{0.1 \cancel{M}}{1.0 \cancel{M}}$$
$$= 0.1$$

But now we have another question to answer—how do we use the dilution factor? Put very simply, the *dilution factor* has two meanings: it is the factor by which a concentration is reduced, and it is also the ratio of initial volume to final volume. In this case, our original volume of 100 mL must be increased ten times by adding solvent. We can calculate this as follows:

$$\text{dilution factor} = \frac{\text{initial volume}}{\text{final volume}}$$

$$\text{final volume} = \frac{\text{initial volume}}{\text{dilution factor}}$$

$$= \frac{100 \text{ mL}}{0.1}$$

$$= 1000 \text{ mL}$$

By adding enough water to 100 mL of the stock solution to bring its final volume to 1000 mL, we accomplish the necessary dilution. In this case, the dilution would be called a 1:10 (one-to-ten) dilution, since we increased the volume to ten times its original value.

Based on the steps performed above, we can summarize by the following equation:

$$\text{final concentration} = \text{initial concentration} \times \frac{\text{initial volume}}{\text{final volume}}$$

Rearrangement gives

$$\frac{\text{initial concentration}}{\text{final concentration}} = \frac{\text{final volume}}{\text{initial volume}}$$

or

$$\frac{\text{conc}_i}{\text{conc}_f} = \frac{V_f}{V_i}$$

This last equation tells us what we have known intuitively, that the concentration of a solution is lowered by dilution to a larger volume. Another rearrangement gives us

$$\text{conc}_i \times V_i = \text{conc}_f \times V_f$$

This equation of course says the same thing but in a different way. The product of initial concentration and volume is equal to the product of final concentration and volume. Thus, when volume goes up, concentration must go down. All of these equations have exactly the same meaning, and you may use any one you like to solve dilution problems.

EXAMPLE 7.6

What would be the final concentration of a 2.0*M* solution if 100 mL of it is diluted to 500 mL?

SOLUTION

From the preceding equations we know that

$$\text{conc}_\text{f} = \text{conc}_\text{i} \times \frac{V_i}{V_f}$$

Thus,

$$\begin{aligned}\text{conc}_\text{f} &= (2.0 \text{ moles/L}) \left(\frac{100 \cancel{\text{mL}}}{500 \cancel{\text{mL}}}\right)\\ &= 0.40 \text{ mole/L}\\ &= 0.40M\end{aligned}$$

EXERCISE 7.8 What would be the final concentration of a 1.75*M* solution if a 250-mL portion of the solution is diluted to 800 mL?

EXAMPLE 7.7

To what final volume would 50.0 mL of a 2.50*M* solution be diluted to produce a 1.00*M* solution?

SOLUTION

From above we know that

$$V_f = V_i \times \frac{\text{conc}_\text{i}}{\text{conc}_\text{f}}$$

Thus,

$$\begin{aligned}V_f &= (50.0 \text{ mL}) \left(\frac{2.50\cancel{M}}{1.00\cancel{M}}\right)\\ &= 125 \text{ mL}\end{aligned}$$

EXERCISE 7.9 To what final volume would 100 mL of a 2.00*M* solution be diluted to produce a 0.100*M* solution?

PROPERTIES OF SOLUTIONS

Solutions, as homogeneous mixtures, can be distinguished from heterogeneous mixtures on the basis of solution properties described below.

Homogeneity

A solution is uniform throughout. This means that samples taken from anywhere in a solution will have the same amount of solute present and the same physical appearance. Solutions may be colorless, as is a solution of sodium chloride in water, or they may be colored. For example, copper(II) sulfate pentahydrate, $CuSO_4 \cdot 5H_2O$, is a brilliant blue crystalline solid. When it dissolves in water, the solution takes on the blue color of the solute. The solution is clear, as are all solutions, and retains its clarity even when light is shining through it. This clarity is a property of solutions regardless of whether they are colored or colorless.

Solute Particles

Dissolved particles in a solution may be ions, atoms, or molecules. These particles usually have diameters of less than 1 nanometer (nm). Such small particles readily pass through filters, so that other methods must be used to separate the components of a solution. The most common technique used is some form of evaporation. Distillation is used to separate liquid components of a solution according to their boiling points; a distilling apparatus is shown in Figure 7-12. The simplest way to separate solid solutes from solvent is by complete evaporation. This method is used for obtaining salt from seawater. When the liquid solvent has been completely evaporated, any dissolved solids are left as residue.

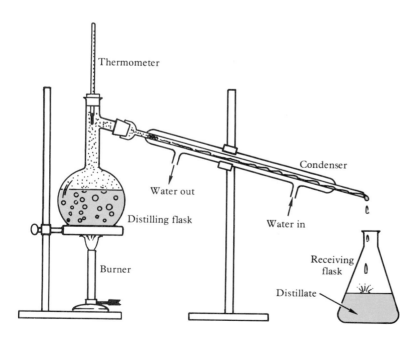

Fig. 7-12 A Simple Laboratory Distillation. Cool water circulating through the jacket of the condenser causes vapor molecules to condense to liquid.

Colligative Properties

There are some properties of solutions which are governed by the number of solute particles in the solution. In these cases, the identity of the solute is immaterial. These properties, called *colligative properties*, are described in the following sections.

colligative (co-LIH-ga-tive)

Vapor-Pressure Lowering

nonvolatile: a descriptive term for substances which do not evaporate readily.

If a nonvolatile solute such as sugar or salt is added to water, the vapor pressure of the solution is lower than the vapor pressure of pure water. This observation is true of any liquid solution containing nonvolatile solute. The result is that solvent evaporates more slowly from solutions than from the pure state. The amount by which the vapor pressure of the liquid solvent is lowered depends only on the number of solute particles present. Thus, when salt is dissolved in water and dissociates into ions, it is the number of ions present which determines the amount of vapor-pressure lowering. On the other hand, sugar is a molecular compound which does not form ions in solution. In this case it is the number of molecules of sugar that determines the amount of vapor-pressure lowering. Thus, a $1M$ solution of NaCl will lower the vapor pressure of the water twice as much as will a $1M$ solution of sugar, because each formula unit of sodium chloride provides two ions, Na^+ and Cl^-.

We can understand the vapor-pressure lowering effect by considering that nonvolatile solute particles are evenly distributed throughout the solution. This means that the surface of the solution contains both solute and solvent particles. However, since solvent particles are present in smaller numbers on the surface of the solution than they would be on the surface of pure solvent, fewer solvent particles are available to break away from the surface of the solution (see Figure 7–13). Thus, the vapor pressure is decreased.

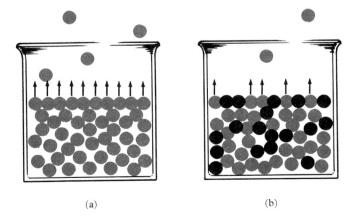

Fig. 7-13 Vapor-Pressure Lowering by a Nonvolatile Solute. (a) Water molecules occupy entire surface in pure liquid. (b) Solute molecules occupy surface positions in a solution, preventing water molecules from escaping easily; thus more energy is required to vaporize water in the solution, and the boiling point is raised.

Boiling-Point Elevation

The vapor pressure of a liquid increases with temperature, and when the vapor pressure reaches atmospheric pressure, the liquid boils. When a nonvolatile solute is present in a liquid solution, the diminished vapor pressure requires a higher temperature to reach atmospheric pressure. Thus, the boiling point of the solution is elevated from that of pure solvent. From this knowledge, we can anticipate that seawater has a higher boiling point than pure water.

Freezing-Point Lowering

In addition to lowering vapor pressure and raising the boiling point, the nonvolatile solute also lowers the freezing point of the solution relative to that of pure solvent. Thus, if salt is poured on icy streets, some of the salt dissolves and causes the ice to melt since the salt solution will still be a liquid at temperatures that would freeze pure water. The freezing-point lowering caused by solutes also explains the addition of antifreeze to the water in automobile radiators.

Osmosis

We have said that the concentration of a solution is uniform throughout, and that is true. But what happens if we add a bit more solute to a small part of a solution? At first the concentration of solute will be higher in the immediate vicinity of addition than elsewhere in the solution. But eventually, the solute particles will diffuse through the bulk of the solution until the concentration is once again uniform throughout. In this diffusion process, solute particles move from a region of higher concentration to one of lower concentration. The difference in concentration that existed temporarily is called a *concentration gradient*. It is known from experimental observation that particles in solution will move from a region of higher concentration to one of lower concentration. This movement in response to a concentration gradient occurs with both solute and solvent particles, and it will continue as long as the concentration gradient exists.

Diffusion along a concentration gradient is an important mode of transport for many biological substances. For example, blood gases diffuse into and out of cells in response to concentration gradients usually expressed as differences in partial pressures. Digested food molecules pass from the intestine (a region of high concentration) into the blood system (a region of low concentration) by the process of diffusion, and they subsequently make their way into cells by the same process.

Most natural membranes and some synthetic ones have structures such that water molecules and certain solute particles can diffuse through them. Such membranes are described as *semipermeable*. Osmotic membranes are a special class of semipermeable membranes which allow only water molecules to pass through. As you might expect, water moves

semipermeable membrane: a membrane which selectively allows only certain solution particles to pass through.

across an osmotic membrane in response to concentration gradient. This movement is called *osmosis*. Consider the situation illustrated in Figure 7-14 where an osmotic membrane separates pure water in compartment A from a sugar solution in compartment B. Water will diffuse through the membrane from compartment A to compartment B in response to its concentration gradient, and the level of solution in compartment B will rise. It is possible to counteract this volume increase by applying pressure to the surface of the sugar solution. The amount of external pressure (in torr) required to prevent the flow of water across the osmotic membrane, thus preventing an increase in volume of the sugar solution, is called the *osmotic pressure* of the sugar solution.

The osmotic pressure of a solution is directly proportional to the concentration of solute particles, as are other colligative properties. Thus, a $2M$ solution will have an osmotic pressure twice as great as a $1M$ solution containing the same solute. When two solutions having the same osmotic pressure are separated by a semipermeable membrane, the two solutions are said to be *isotonic* (literally meaning "same tone"). If one solution is of lower osmotic pressure, it is *hypotonic* ("lower tone")

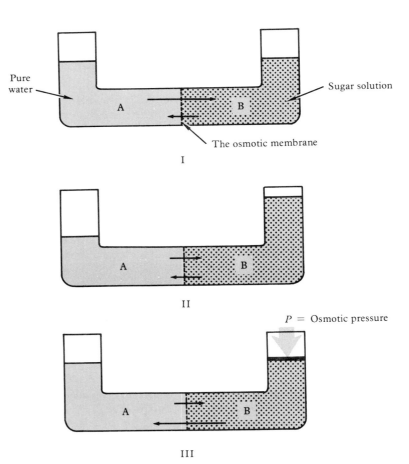

Fig. 7-14 Osmosis. In I, a sugar solution is separated from pure water by an osmotic membrane. In II, water flows through the osmotic membrane to increase the volume of the sugar solution. In III, pressure is applied to prevent the flow of water across the membrane.

relative to the more concentrated solution, and the more concentrated solution is *hypertonic* ("higher tone") relative to the less concentrated solution.

Red blood cells have semipermeable membranes, and the osmotic pressure exerted by the contents of the cell is the same as that of a 0.9% (w/v) solution of NaCl or a 5% (w/v) solution of glucose. When red blood cells are bathed in these isotonic solutions, no noticeable changes take place. For this reason, most intravenous solutions are isotonic with red blood cells.

When red blood cells are placed in a hypotonic solution, such as pure water, the flow of water into the cells causes the membranes to rupture, spilling the cell contents into the solution. This rupturing is called *hemolysis* (shown in Figure 7-15), a term which means red blood cell destruction. Intravenous administration of a hypotonic solution to a patient would have serious consequences because of hemolysis.

hemolysis (hee-MAH-lih-sis)

Placing red blood cells in a hypertonic solution causes water to flow out of the cells, and the cells collapse (Figure 7-15). The collapsing is called *crenation*. Thus, administration of a hypertonic intravenous solution may also have serious consequences unless it is done properly and for a reason. For example, some seriously ill patients must be fed intravenously, and in these cases hypertonic solutions are given in order to hold fluid volume down to a manageable level. The hypertonic solution is injected very slowly into a large artery so that it is quickly diluted by the blood to an isotonic level.

COLLOIDS

If we collect a sample of muddy water and allow it to stand undisturbed, we find that a lot of the mud settles to the bottom of the container in a short while. But there are smaller particles that will remain suspended

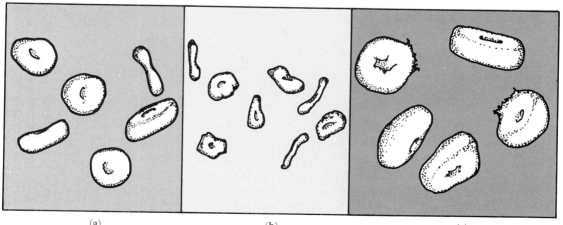

(a) Red cells in an isotonic solution

(b) Red cells in a hypertonic solution

(c) Red cells in a hypotonic solution

188 CH. 7 SOLUTIONS, DISPERSIONS, AND SUSPENSIONS

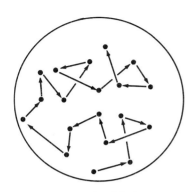

Fig. 7-16 Brownian Motion

Tyndall effect: the visible scattering of light as it passes through a colloid.

for days and perhaps indefinitely. These particles give the water a murky appearance, but we cannot see them individually even when we examine the water under a microscope. Ordinary filters do not remove the particles responsible for the murkiness. This cloudy mixture is an example of a *colloidal dispersion,* or, in a word, a *colloid.* Colloids are homogeneous mixtures of particles larger than most molecules; the particles may be dispersed in a gas, liquid, or solid. The particle diameters range from about 1 nm to 100 nm, and the particles may be very large individual molecules or they may be groups of molecules. Colloidal dispersions are not true solutions, and they appear cloudy or murky. Some everyday examples of colloids are homogenized milk, whipped cream, jellies, and fog.

Colloidal particles move rapidly and randomly through fluids in which they are dispersed. This type of motion, called *Brownian motion* after the Scottish botanist Robert Brown who discovered it in 1827, is a constant but irregular motion of small particles, as illustrated in Figure 7-16. It is a result of bombardment of the dispersed particles by rapidly moving molecules of the fluid. Brownian motion is responsible for preventing colloidal particles from settling.

When bright light passes through colloidal dispersions, as shown in Figure 7-17, the light beam is clearly visible from the side. This is called the *Tyndall effect;* it is sometimes used to distinguish colloidal dispersions from true solutions. True solute particles are not large enough to reflect light, and a light beam passing through a solution is not visible from the side. A familiar example of the Tyndall effect is the shaft of light between a movie projector and the screen caused by light scattered by colloidal dust particles in the air.

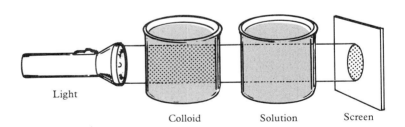

Fig. 7-17 The Tyndall Effect

protoplasm: the viscous, jellylike fluid inside a living cell.

Instead of the terms solute and solvent, which are usually reserved for solutions, we speak of the dispersed substances and the dispersing medium in describing colloids. All three states of matter may function in either role. Table 7-9 gives examples of the various types of colloidal dispersions.

Except for the solid parts of our bodies, such as bones and teeth, most other parts are largely colloidal, with protein and fat dispersed in water to form a gel. Within cells, the gel is called *protoplasm.* The protein

Table 7-9 Types of Colloids

Type	Examples
gas in a gas	none (all are solutions)
gas in a liquid	whipped cream
gas in a solid	marshmallow, soap foam
liquid in a gas	fog
liquid in a liquid	milk, mayonnaise
liquid in a solid	butter
solid in a gas	smoke
solid in a liquid	paint, gelatin
solid in a solid	ruby glass (dispersion of gold in a glass)

molecules are long chains only a few atoms in diameter, but thousands of atoms long. Fat molecules are of smaller size, but they tend to aggregate into colloidal-sized particles. In both cases, the colloidal particles are likely to be hydrogen-bonded to each other and to water, thus forming a semisolid mass.

SUSPENSIONS

Particles having dimensions larger than those of colloidal particles will be visible to the eye and will settle out of a dispersion. When these larger particles are scattered through a liquid, the mixture is called a *suspension*. Suspensions are heterogeneous mixtures and can be separated by filtration. Blood contains dissolved substances (ions and small molecules), dispersed colloids (proteins and fats), and suspended solid matter (red blood cells, white blood cells, and platelets). The heart's pumping action serves to maintain these latter particles in suspension. If a sample of blood is allowed to stand undisturbed in the presence of anticoagulant, the cells and platelets will eventually settle to the bottom. Centrifuges are used to promote fast settling of these substances in the clinical laboratory. Other familiar examples of suspensions are milk of magnesia ($Mg(OH)_2$ suspended in water) and orange juice containing suspended bits of orange pulp. A suspension of barium sulfate, $BaSO_4$, is ingested in preparation for taking gastrointestinal X-rays.

DIALYSIS

The word *dialysis* comes from the Greek language and means "a separation." More specifically, dialysis refers to the process whereby small molecules and ions diffuse through a semipermeable membrane. Many natural membranes are dialysis membranes as are some types of cellophane.

dialysis (dye-AL-lih-sis)

Dialysis is a combination of diffusion and osmosis, since both water molecules and solute particles cross a dialysis membrane. The process is often used in the laboratory to separate colloidal dispersions from solute particles. As illustrated in Figure 7-18, the procedure consists of filling a cellophane bag with colloidal dispersion which also contains dissolved solute. In this case, the colloidal particles are proteins and the solute is NaCl. The cellophane bag is then suspended in pure water, which is gently stirred. Water molecules and solute particles are able to pass through the rather small pores in the bag, and they diffuse through the membrane in response to the concentration gradient. But the protein molecules are too large to pass through the pores in the bag, and they remain inside. By this technique, if we keep replacing the outside solution with pure water, we can reduce the solute concentration inside the dialysis bag to a very low level, thus effectively removing solute particles from the colloid.

TRANSPORT OF BODY FLUIDS AND SOLUTES

The circulation of blood serves to transport oxygen, carbon dioxide, metabolic wastes, nutrients, hormones, enzymes, and water to various

PERSPECTIVE: Hemodialysis

Dialysis can be used to treat patients suffering from kidney disease. One of the functions of the normal kidney is to rid the body of waste products by allowing small molecules and ions to diffuse across kidney membranes into the urine. When kidney function is impaired, artificial kidney machines can be used to accomplish this function. In *hemodialysis*, illustrated in the accompanying figure, blood is pumped from one of the patient's arteries through cellophane tubing. The tubing is immersed in a solution containing ions normally present in blood at their physiological concentrations. The cellophane functions as a dialysis membrane and allows waste solutes to diffuse from the blood into the outside solution. Some of the ions originally in the outside solution may diffuse into the patient's blood if it has lower than normal concentrations of those ions. The blood is pumped back into one of the patient's veins in a continuous process. In this way waste substances are removed and normal ionic balance is restored if necessary. This treatment usually requires several hours, and it may be repeated as often as three times each week. Recent improvements in design have resulted in small, portable hemodialysis units which are more convenient and efficient than earlier models.

Hemodialysis

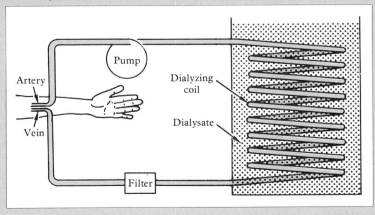

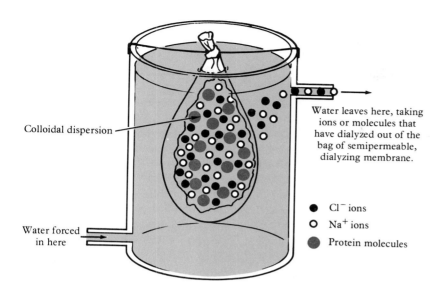

Fig. 7-18 Dialysis can be used to separate solutes from colloidal particles, protein molecules in this case.

parts of the body. However, arteries and veins are enclosed by membranes which are relatively impermeable, and there is probably not much transfer of these substances to or from blood in veins or arteries. The major site of transfer is at the capillary level.

The *capillaries* are a network of very small tubes connected to arteries and veins. Blood from arteries enters one end of the capillary and leaves at the other end to enter veins. The average diameter of a capillary is about 10 micrometers. Because of this extremely small size, the velocity of blood flow in the capillaries is about 1/1000 the flow in a major artery. The relative permeability of capillary walls and the slow flow of blood permits the interchange of substances to take place in response to pressure differences and concentration gradients across the capillary walls.

SUMMARY

Water: With its bent structure and polar nature, water forms an extensive array of hydrogen bonds in the liquid and solid states, causing water to have unusually high values for freezing point, boiling point, heat of vaporization, heat of fusion, and density. Water possesses unique solvent qualities and is able to dissolve a variety of ionic and polar covalent substances. Many compounds form hydrates after being dissolved in water. Hydrates contain a definite number of water molecules (water of hydration) within their crystals. Water provides the medium in which nearly all of the body's chemical reactions take place. It also transports nutrients to cells and takes metabolic wastes away from cells. Water accounts for 60-70% of human body weight; this amount is regulated by balanced intake and output.

Solution Formation: A solution is a homogeneous mixture which forms when a major component (solvent) dissolves one or more minor components (solutes). When a solute dissolves, solute particles are separated from each other and surrounded by solvent particles. The rate of dissolution for a solute depends on the surface area of the solute, the amount of agitation of the system, and the temperature. The solubility of a substance is the maximum amount that will dissolve in a given quantity of solvent at a particular temperature. When the solubility of a substance is exceeded, a saturated solution exists in which undissolved solute and dissolved solute are in dynamic equilibrium. Solution formation may be endothermic or exothermic. In general, the more negative the molar heat of solution is, the greater the solubility will be at a given temperature. The most important factors affecting solubility are pressure, temperature, and chemical structure. Ionic and polar substances are likely to dissolve in polar solvents, whereas nonpolar solutes are more likely to dissolve in nonpolar solvents.

Preparation of Solutions: Solutions are prepared by mixing solute and solvent to achieve the desired concentration of solute. It is common practice to dilute concentrated stock solutions to larger volumes to achieve lower concentrations. Dilution factors can be calculated from the initial concentration of stock solution and the desired final concentration.

Properties of Solutions: Solutions are always clear, but they may be either colored or colorless. Dissolved particles may be ions, atoms, or molecules, but in general they have diameters of less than 1 nm. Colligative properties of solutions—vapor-pressure lowering, boiling-point elevation, freezing-point lowering, and osmotic pressure—depend only on the number of solute particles.

Colloids: Colloids are homogeneous mixtures of particles having diameters usually of 1–100 nm which are dispersed in a gas, liquid, or solid. Colloids are not true solutions, and the dispersions usually appear cloudy or milky.

Suspensions: Particles having diameters greater than 100 nm are visible to the eye. These particles may be temporarily suspended in a gas or liquid, but eventually they will settle out.

Dialysis: Dialysis is a process whereby small molecules and ions diffuse through a semipermeable membrane. It is a combination of diffusion and osmosis. Colloidal dispersions can be separated from solute particles by dialysis.

Transport of Body Fluids and Solutes: The circulation of blood transports oxygen, carbon dioxide, metabolic wastes, nutrients, hormones, enzymes, and water to various parts of the body. Substances enter and leave blood in the capillary beds in response to pressure differences and concentration gradients across capillary walls.

STUDY QUESTIONS AND PROBLEMS

1. Define the following terms:

    (a) solvent
    (b) hydrate
    (c) hormone
    (d) solute
    (e) solution
    (f) dissolution
    (g) hydration
    (h) solubility
    (i) saturated solution
    (j) molar heat of solution
    (k) hydrophilic
    (l) hydrophobic
    (m) concentration
    (n) molarity
    (o) dilution factor
    (p) colligative properties
    (q) osmosis
    (r) osmotic pressure
    (s) isotonic
    (t) hypotonic
    (u) hypertonic
    (v) colloid
    (w) Brownian motion
    (x) Tyndall effect
    (y) protoplasm
    (z) suspension
    (aa) dialysis
    (bb) capillary

2. What types of substances are most likely to dissolve in water? Why?

3. Explain how water intake is balanced with water output in the human body.

4. Give one example each of a gaseous solution, a liquid solution, and a solid solution.

5. Distinguish between a solvent and a solute.

6. Describe the dissolution of NaCl in water.

7. Explain how each of the following factors affects the rate of dissolving:

    (a) surface area
    (b) agitation
    (c) temperature

8. Distinguish between soluble and insoluble substances.

9. What is meant by a "saturated" solution?

10. Describe the dynamic equilibrium that exists in a saturated solution.

11. Explain the significance of the sign of the molar heat of solution for a substance.

12. Predict the relative solubilities of the following compounds in water at 25°C.

Compound	$\Delta H_{solution}$ (kcal/mole)
NaCl	0.93
NaI	−1.8
NaOH	−10.6

13. Predict the relative solubilities of the following compounds in water at 18°C.

Compound	$\Delta H_{solution}$ (kcal/mole)
Na_2SO_4	0.46
$MgSO_4$	20.28
$Al_2(SO_4)_3 \cdot 6H_2O$	56.0

14. Explain the effect of each of the following factors on solubility.

    (a) pressure
    (b) temperature
    (c) chemical structure

15. Explain why soda pop bubbles vigorously when first opened.

16. Distinguish between hydrophilic and hydrophobic substances, and give an example of each.

17. Calculate the number of moles of solute in each of the following:

    (a) 150 mL of 0.10M NaCl
    (b) 200 mL of 0.25M KNO_3
    (c) 600 mL of 1.20M $MgCl_2$
    (d) 400 mL of 0.75M LiBr
    (e) 750 mL of 3.70M KOH
    (f) 450 mL of 0.10M NaOH
    (g) 325 mL of 0.20M $Al_2(SO_4)_3$
    (h) 150 mL of 0.15M Li_2CO_3
    (i) 750 mL of 2.00M NaCl
    (j) 1.25 L of 2.50M $NaNO_3$

18. Calculate the number of grams of solute needed to make each of the following solutions:

    (a) 1.00 L of 0.50M glucose, $C_6H_{12}O_6$
    (b) 800 mL of 0.10M NaCl
    (c) 420 mL of 0.25M Na_3PO_4
    (d) 150 mL of 1.50M $Al_2(SO_4)_3$
    (e) 2.50 L of 0.020M $MgCl_2$
    (f) 750 mL of 2.00M KCl
    (g) 100 mL of 1.25M $NaNO_3$
    (h) 3.75 L of 2.00M NaCl
    (i) 925 mL of 0.15M K_2SO_4
    (j) 225 mL of 0.30M glucose, $C_6H_{12}O_6$

19. Calculate the molarity of each of the following solutions:

    (a) 10 g of NaCl in 1.0 L of solution
    (b) 20 g of KCl in 100 mL of solution
    (c) 15 g of LiBr in 500 mL of solution
    (d) 25 g of $Mg(NO_3)_2$ in 650 mL of solution
    (e) 12.5 g of Na_3PO_4 in 300 mL of solution

20. Calculate the %(w/v) concentration for each of the following solutions:

    (a) 17.0 g of solute in 150 mL of solution
    (b) 24.3 g of solute in 1.00 L of solution
    (c) 35.7 g of solute in 900 mL of solution
    (d) 14.6 g of solute in 500 mL of solution
    (e) 30.6 g of solute in 745 mL of solution
    (f) 9.75 g of solute in 200 mL of solution
    (g) 11.3 g of solute in 425 mL of solution
    (h) 15.7 g of solute in 225 mL of solution
    (i) 37.6 g of solute in 500 mL of solution
    (j) 21.2 g of solute in 100 mL of solution

21. Calculate the number of grams of solute in each of the following solutions:

    (a) 175 mL of 30%(w/v) NaOH
    (b) 900 mL of 10%(w/v) NaCl
    (c) 775 mL of 20%(w/v) $Mg(NO_3)_2$
    (d) 300 mL of 15%(w/v) $CaCl_2$
    (e) 650 mL of 12.5%(w/v) KBr
    (f) 450 mL of 17%(w/v) K_2SO_4
    (g) 50 mL of 1%(w/v) NaCl

22. In most cases, the concentration of ethanol in blood necessary to produce intoxication is 0.3%(w/v). If total blood volume for an adult is 7.5 L, how many grams of ethanol are necessary for intoxication?

23. Hydrogen peroxide solution for bleaching hair is usually at a concentration of 5%(w/v) in water. How many grams of hydrogen peroxide, H_2O_2, are present in 80.0 mL of this solution?

24. Human blood serum contains NaCl at a concentration of 0.15M. What is the percent (w/v) of NaCl in human blood serum?

25. Calculate the mg % concentration for each of the following solutions:

    (a) 1.00 g of solute in 1.50 L of solution
    (b) 17.0 mg of solute in 250 mL of solution
    (c) 12.7 mg of solute in 100 mL of solution
    (d) 15.6 mg of solute in 75.0 mL of solution
    (e) 24.3 mg of solute in 200 mL of solution

26. Calculate the %(v/v) concentration for each of the following solutions:

    (a) 17.0 mL of solute in 150 mL of solution
    (b) 24.7 mL of solute in 250 mL of solution
    (c) 7.80 mL of solute in 125 mL of solution
    (d) 37.4 mL of solute in 750 mL of solution
    (e) 340 mL of solute in 1.50 L of solution

27. Distilled beverages are often described by "proof," defined as twice the percent (v/v) of ethanol in solution at 60°F. How many mL of ethanol, C_2H_5OH, are present in 1.00 L of a "90 proof" beverage?

28. Calculate the %(w/w) concentration for each of the following solutions:

    (a) 17.0 g of solute in 150 g of solution
    (b) 24.8 g of solute in 900 g of solution
    (c) 12.3 g of solute in 49.0 g of solution
    (d) 43.6 g of solute in 275 g of solution
    (e) 2.35 g of solute in 120 g of solution

29. Calculate final volumes for diluting each of the solutions in question 17 to $0.0500M$.

30. Calculate final volumes for diluting each of the following solutions to 2.00%(w/v).

    (a) 10.0 mL of a 7.5%(w/v) solution
    (b) 25.0 mL of a 14.3%(w/v) solution
    (c) 20.0 mL of a 10.0%(w/v) solution
    (d) 5.00 mL of a 17.5%(w/v) solution
    (e) 1.00 mL of a 25.0%(w/v) solution

31. Explain why a nonvolatile solute lowers the vapor pressure of a liquid solvent.

32. Exlain why a nonvolatile solute causes an elevation of the boiling point of a liquid solvent.

33. List the following aqueous solutions in increasing order of their expected boiling points (lowest boiling point first, highest boiling point last).

    (a) $0.05M$ $CaCl_2$
    (b) $0.15M$ NaCl
    (c) $0.10M$ KCl
    (d) $0.05M$ KBr
    (e) $0.15M$ Na_2SO_4

34. What effect does a nonvolatile solute have on the freezing point of a liquid solvent?

35. List the following aqueous solutions in increasing order of their expected freezing points (lowest freezing point first, highest freezing point last).

    (a) $0.05M$ KNO_3
    (b) $0.10M$ NaBr
    (c) $0.15M$ $MgCl_2$
    (d) $0.20M$ $Ca(NO_3)_2$
    (e) $0.10M$ K_3PO_4

36. List the following aqueous solutions in decreasing order of their expected osmotic pressures (highest osmotic pressure first, lowest osmotic pressure last).

    (a) $0.15M$ NaCl
    (b) $0.05M$ KNO_3
    (c) $0.10M$ LiBr
    (d) $0.15M$ $MgCl_2$
    (e) $0.15M$ ethanol (C_2H_5OH)

37. Why does salty soil kill plants? (Clue: Consider the effects of osmosis.)

38. Swelling of tissue caused by excess water in a human is called edema. Why does eating a lot of salty food cause edema? (Clue: consider the effects of osmosis.)

39. A 5%(w/v) NaCl solution is separated from a 10% (w/v) NaCl solution by an osmotic membrane.

    (a) Which is the hypotonic solution?
    (b) Which is the hypertonic solution?
    (c) Into which solution will water flow?

40. In making pickles, cucumbers are soaked in a 16% (w/v) salt solution. Why are the resulting pickles shrunken and shriveled?

41. Explain why solutions are perfectly clear whereas colloidal dispersions appear cloudy.

42. Why is homogenized whole milk more opaque than nonfat milk?

43. Explain how you would distinguish between members of the following pairs:

    (a) a solution and a suspension
    (b) a solution and a colloidal dispersion
    (c) a colloidal dispersion and a suspension

44. Why does exchange of water and solutes occur across the walls of capillaries instead of across the walls of arteries and veins?

CHAPTER 8

CHEMICAL EQUILIBRIUM AND RATES OF REACTIONS

CHEMICAL EQUILIBRIUM
- The Dynamic Equilibrium of Reversible Reactions
- Equilibrium Constants
- Utility of the Equilibrium Constant
- LeChatelier's Principle
- Energy and Chemical Change
- Energy and the Biosphere

RATES OF CHEMICAL REACTIONS
- Energy of Activation
- Factors Affecting Reaction Rates

THE RELATIONSHIP BETWEEN CHEMICAL EQUILIBRIUM AND REACTION RATES

Up until this point we have discussed chemical reactions as if reactants were always converted completely to products. Thus, we have assumed that chemical reactions always go to completion. In this chapter and through the rest of the book you will discover that although many reactions go to completion, most do not. Most reactions never use all of the reactants; instead, they reach a state of dynamic equilibrium in which the amounts of reactants and products do not change.

Energy changes are the basis for the occurrence of chemical reactions. We have observed that fire produces heat energy; fire is the result of combustion, an oxidation-reduction reaction. Just as combustion reactions produce energy, other reactions can require energy in order to proceed. For example, metabolic oxidations produce energy which can then be used for energy-requiring synthetic reactions to produce cellular components. In this chapter you will learn that energy changes have a close relationship to the underlying principles of chemical equilibrium.

Chemical reactions occur at various rates. Some, like the rusting of iron and the formation of petroleum, are notoriously slow. Others, such as the burning of natural gas and the explosion of a firecracker, happen in the blink of an eye. Metabolic reactions must take place rapidly so that our bodies will be immediately responsive to outside stimuli. As illustrated in the topic outline, you will learn about factors which affect rates of reactions and how those rates are related to chemical equilibrium.

metabolic reactions: all of the chemical reactions that go on in an organism.

CHEMICAL EQUILIBRIUM

The Dynamic Equilibrium of Reversible Reactions

Let's begin our study of chemical equilibrium with an isomerization reaction. Glucose is a six-carbon sugar that we obtain in our diets by digestion of table sugar (sucrose), milk sugar (lactose), or starch. (Sugars and starches will be discussed in more detail in chapter 13.) After glucose is absorbed into the blood stream, it is distributed to tissue cells for use. During metabolism, glucose is isomerized to another sugar called fructose. If we allow the isomerization to occur in a test tube, equilibrium will occur,

glucose ($C_6H_{12}O_6$) isomerizes to fructose ($C_6H_{12}O_6$)

and some fructose is produced, but some glucose remains. In fact, at equilibrium the mole ratio of glucose to fructose is 7:3.

Let's examine the isomerization reaction a bit more closely. If we have only glucose to start with, then at the beginning of the reaction glucose produces fructose:

$$\text{glucose} \longrightarrow \text{fructose}$$
$$(C_6H_{12}O_6) \qquad (C_6H_{12}O_6)$$

But if this were the only reaction, then all of the glucose would be converted to fructose. Since this does not happen, it appears that after some fructose is formed, it is converted back to glucose:

$$\text{glucose} \longleftarrow \text{fructose}$$
$$(C_6H_{12}O_6) \qquad (C_6H_{12}O_6)$$

In other words, after some fructose is formed, a reaction begins to occur in the reverse direction. If we were to analyze the components of the reaction in the laboratory, we would find that the amount of glucose decreases through the course of time, and the amount of fructose increases. But as time goes by, the amounts of these two components change less and less until they become constant (see Figure 8-1). At that time the reaction is in a state of dynamic equilibrium. Glucose and fructose molecules are continuously interchanging, but the rate of the forward reaction is equal to the rate of the reverse reaction at equilibrium.

We write the glucose-fructose isomerization equation as

$$\text{glucose} \rightleftharpoons \text{fructose}$$
$$(C_6H_{12}O_6) \qquad (C_6H_{12}O_6)$$

where the two opposing arrows ($\rightleftharpoons$) indicate that the reaction is reversible. *Reversible reactions* are those which proceed in two directions,

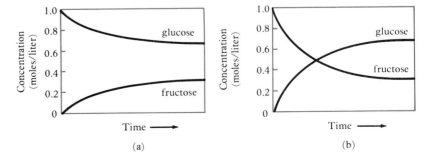

Fig. 8-1 Variation in Concentrations in Formation of the Equilibrium Glucose $\rightleftharpoons$ Fructose. (a) The equilibrium is approached beginning with 1.0 moles/liter of glucose and no fructose. (b) The equilibrium is approached beginning with 1.0 moles/liter of fructose and no glucose.

forward and backward, at the same time. All reversible reactions will achieve a state of dynamic equilibrium if they are left undisturbed for a long enough time.

Equilibrium Constants

When glucose and fructose are in equilibrium, the mole ratio of components in the mixture is 7:3, or 7 moles of glucose for every 3 moles of fructose. This ratio is determined by laboratory analysis; it is independent of the initial amounts of reactants. We would achieve exactly the same equilibrium mixture if we were to start with only fructose, or even with a 1:1 mixture of glucose and fructose. The ratio of components in the equilibrium mixture depends only on the temperature of the mixture. This constant ratio is called the *equilibrium constant* (K_{eq}) for the reaction. If we assume that our reaction consisted of 10 liters of water containing 7 moles of glucose and 3 moles of fructose at equilibrium, we can write the following mathematical expression for the equilibrium constant, where the square brackets denote concentration in moles per liter:

equilibrium constant (K_{eq}): the constant ratio of components in an equilibrium mixture for a reversible reaction at a specific temperature.

$$K_{eq} = \frac{\text{concentration of fructose}}{\text{concentration of glucose}} = \frac{[\text{fructose}]}{[\text{glucose}]}$$

$$= \frac{0.3 \text{ mole/liter}}{0.7 \text{ mole/liter}}$$

$$= 0.43$$

Notice that our expression for the equilibrium constant is of the form

$$K_{eq} = \frac{[\text{product}]}{[\text{reactant}]}$$

This is the standard expression for the equilibrium constant of a reversible reaction involving one reactant and one product, as depicted in the following equation:

$$A \rightleftarrows B$$

$$K_{eq} = \frac{[B]}{[A]}$$

If we choose to write the reaction as

$$B \rightleftarrows A,$$

an equally valid equation, then our equilibrium constant expression becomes

$$K_{eq} = \frac{[A]}{[B]}$$

The glucose-fructose equilibrium is a very simple *system* because it contains only one reactant and only one product. Let us consider a

system: an isolated collection of substances.

slightly more complicated reaction. Hydrogen gas and iodine vapor react according to the following reversible equation:

$$H_2(g) + I_2(g) \rightleftharpoons 2\,HI(g)$$

(We use the notation (g) to emphasize that each component is a gas.) Laboratory analysis of the equilibrium mixture has demonstrated that the expression for the equilibrium constant for the reaction is

$$K_{eq} = \frac{[HI]^2}{[H_2][I_2]}$$

In this case, the concentration of product has an exponent of 2, and the concentrations of the two reactants are multiplied together. If we were to examine more complex reactions, we would find a general pattern for the equilibrium constant expression. Using the following as a generalized reversible reaction,

$$a\,A + b\,B + c\,C + \ldots \rightleftharpoons d\,D + e\,E + f\,F + \ldots$$

then the equilibrium constant can always be expressed as

$$K_{eq} = \frac{[D]^d[E]^e[F]^f\ldots}{[A]^a[B]^b[C]^c\ldots}$$

In these equations, uppercase letters (A, B, C, D, E, F) represent chemical substances, and lowercase letters (a, b, c, d, e, f) are their coefficients in the balanced equation. This general expression for the equilibrium constant can be used for any number of reactants and products if they form a homogeneous mixture. The units of the calculated equilibrium constant will vary, depending on the exponents in the equilibrium expression.

EXAMPLE 8.1

Write the equilibrium constant expression for the reversible formation of nitrogen oxide,

$$N_2(g) + O_2(g) \rightleftharpoons 2\,NO(g)$$

SOLUTION

Referring to our general expression for equilibrium constants,

$$K_{eq} = \frac{[D]^d[E]^e[F]^f}{[A]^a[B]^b[C]^c}$$

In this case our products are 2 NO, and our reactants are N_2 and O_2. Therefore the correct expression is

$$K_{eq} = \frac{[NO]^2}{[N_2][O_2]}$$

EXERCISE 8.1 About 14 million tons of ammonia are used in this country each year as fertilizer. The large-scale production of ammonia is carried out by the Haber process at high temperature and pressure:

$$N_2(g) + 3\,H_2(g) \rightleftharpoons 2\,NH_3(g)$$

Write the equilibrium constant expression for reversible formation of ammonia using the Haber process.

EXAMPLE 8.2

Find the value for the equilibrium constant in the reaction

$$N_2(g) + O_2(g) \rightleftharpoons 2\, NO(g)$$

if the equilibrium concentrations are as follows:

$[N_2] = 0.95$ mole/liter
$[O_2] = 0.95$ mole/liter
$[NO] = 0.10$ mole/liter

SOLUTION

From example 8.1, we found the equilibrium constant expression to be

$$K_{eq} = \frac{[NO]^2}{[N_2][O_2]}$$

Now we must substitute actual concentrations into the expression and carry out the calculations:

$$K_{eq} = \frac{(0.10\text{ mole/L})^2}{(0.95\text{ mole/L})(0.95\text{ mole/L})}$$

$$= 1.1 \times 10^{-2}$$

EXERCISE 8.2 Nitrogen and hydrogen were mixed at 1000 K and allowed to establish equilibrium with ammonia:

$$N_2(g) + 3\,H_2(g) \rightleftharpoons 2\,NH_3(g)$$

Find the value for the equilibrium constant at 1000 K if the equilibrium concentrations were as follows:

$[N_2] = 0.9783$ mole/liter
$[H_2] = 0.935$ mole/liter
$[NH_3] = 0.0434$ mole/liter

Utility of the Equilibrium Constant

The magnitude of the equilibrium constant can be used to estimate the extent of product formation for a given reaction. Since the equilibrium constant is a ratio of product and reactant concentrations with their appropriate exponents, its value provides information about relative amounts of products and reactants existing at equilibrium. In Example 8.2, we calculated the equilibrium constant for the reaction of nitrogen and oxygen to form nitrogen oxide:

$$N_2(g) + O_2(g) \rightleftharpoons 2\,NO(g),\ K_{eq} = 1.1 \times 10^{-2}$$

The very small value for the equilibrium constant indicates that the equilibrium concentrations for products are less than the equilibrium concentrations for reactants. If we use the term *position of equilibrium* to designate the relative amounts of products and reactants at equilibrium, we can say that the position of equilibrium lies on the left, favoring formation of nitrogen and oxygen. Arrows with different lengths are sometimes used to indicate the position of equilibrium:

$$N_2(g) + O_2(g) \rightleftharpoons 2\,NO(g)$$

On the other hand, consider another gaseous reaction, at 1073K:

$$CO(g) + H_2O(g) \rightleftharpoons CO_2(g) + H_2(g),\ K_{eq} = 1.25$$

In this case, the value of the equilibrium constant is greater than 1. This means that the concentrations of products are greater than the concentrations of reactants, when raised to their appropriate powers. In this instance, the position of equilibrium lies on the right, favoring product formation. Our use of arrows with different lengths can be applied to this system also:

$$CO(g) + H_2O(g) \rightleftharpoons CO_2(g) + H_2(g)$$

In general, when the value of an equilibrium constant is less than 1, the equilibrium favors reactants and the position of equilibrium lies on the left. When the value of an equilibrium constant is greater than 1, product formation is favored, and the position of equilibrium lies on the right. When values for equilibrium constants are very large, for example, 10^6, we consider the reaction to go to completion. Likewise, when values are very small, as 10^{-6}, we conclude that the reaction does not occur.

LeChatelier's Principle

LeChatelier (leh-shah-tel-YAY)

In 1888, Henri Louis LeChatelier, a French chemist, formulated what he called the "law of stability of chemical equilibrium." Today we refer to his law as *LeChatelier's principle*. It is summarized as follows: if a chemical system in equilibrium is disturbed, the system will readjust in such a way as to offset the disturbance partially and restore equilibrium.

We can develop an understanding of LeChatelier's principle by remembering that a chemical equilibrium is a dynamic state. The forward and reverse processes are occurring at equal rates, and the system is in balance. But if conditions are altered, thus causing an imbalance of forward and reverse rates, the rates of those processes change to achieve a new state of balance.

Two factors are usually considered to be sources of disturbance for a chemical equilibrium: change in reactant or product concentrations, and change in temperature.

Change in Concentration

Consider the formation of ammonia described in Exercise 8.1:

$$N_2(g) + 3\,H_2(g) \rightleftharpoons 2\,NH_3(g)$$

If we were to change a reactant concentration by adding hydrogen to the equilibrium mixture, the concentrations of nitrogen and ammonia would readjust themselves spontaneously in such a way as to reduce the concentration of hydrogen back toward its equilibrium concentration. This can only occur if more ammonia is formed. Thus, increasing the concentration of hydrogen above the equilibrium value causes formation of more ammonia, and the equilibrium is shifted to the right:

$$N_2(g) + 3H_2(g) \rightleftharpoons 2NH_3(g)$$

increased [H_2] ⟶

And since hydrogen would be used to form the additional ammonia, the concentration of hydrogen would decrease toward its original equilibrium value. Nitrogen would also be used to form the additional ammonia, and thus its concentration would drop.

We can develop another viewpoint for this situation by remembering that the value for the equilibrium constant is always the same at any given temperature.

$$K_{eq} = \frac{[NH_3]^2}{[N_2][H_2]^3}$$

Thus, if the concentration of H_2 is increased beyond its equilibrium concentration, the value of the denominator of the equilibrium expression is raised. In order for K_{eq} to remain constant, the concentration of NH_3 must increase, and the concentration of the other reactant, N_2, must decrease. Hence, more NH_3 is formed at the expense of reactants.

On the basis of our reasoning, we can conclude that for a reaction at equilibrium, increasing the concentration of any or all of the reactants will cause formation of additional product(s) and shift the equilibrium to the right. The reverse situation is also true: for a reaction at equilibrium, increasing the concentration of product(s) will cause an increase in the amount of reactant(s) and shift the equilibrium to the left. The equilibrium constant will not change as long as the temperature is constant, but the equilibrium will shift in the direction indicated below in response to any changes in concentrations:

reactants ⇌ products

increased [reactants] ⟶
⟵ increased [products]

EXAMPLE 8.3

Predict the effect on concentrations of reactants and products and the direction in which each of the following equilibria would shift in response to the indicated changes.

(a) $H_2(g) + I_2(g) \rightleftharpoons 2HI(g)$
increase in concentration of $H_2(g)$

(b) $H_2(g) + CO_2(g) \rightleftharpoons H_2O(g) + CO(g)$
increase in concentration of $H_2O(g)$

(c) $N_2(g) + O_2(g) \rightleftharpoons 2\,NO(g)$
decrease in concentration of $O_2(g)$

SOLUTION

(a) $$K_{eq} = \frac{[HI]^2}{[H_2][I_2]}$$

An increase in concentration of $H_2(g)$ would cause an increase in concentration of $HI(g)$, a corresponding decrease in concentration of $I_2(g)$, and a shift in the equilibrium to the right.

(b) $$K_{eq} = \frac{[H_2O][CO]}{[H_2][CO_2]}$$

An increase in concentration of $H_2O(g)$ would cause an increase in the concentrations of $H_2(g)$ and $CO_2(g)$, a corresponding decrease in concentration of $CO(g)$, and a shift in the equilibrium to the left.

(c) $$K_{eq} = \frac{[NO]^2}{[N_2][O_2]}$$

A decrease in concentration of $O_2(g)$ would cause a decrease in concentration of $NO(g)$, a corresponding increase in concentration of $N_2(g)$, and a shift in the equilibrium to the left.

EXERCISE 8.3 Predict concentration changes and the direction in which each of the following equilibria would shift in response to the indicated changes:

(a) $N_2O_4(g) \rightleftharpoons 2\,NO_2(g)$
increase in concentration of $NO_2(g)$

(b) $2\,CO_2(g) \rightleftharpoons 2\,CO(g) + O_2(g)$
increase in concentration of $CO_2(g)$

(c) $2H_2(g) + O_2(g) \rightleftharpoons 2\,H_2O(g)$
decrease in concentration of $H_2O(g)$

In the case of the glucose-fructose equilibrium,

$$\text{glucose} \rightleftharpoons \text{fructose}$$

the position of equilibrium lies on the left favoring glucose formation:

$$K_{eq} = \frac{[\text{fructose}]}{[\text{glucose}]} = 0.43$$

But certain metabolic processes require the total conversion of glucose to fructose, and indeed total conversion occurs. Application of LeChatelier's principle will help us to understand how that happens. We know that if the system is allowed to establish equilibrium, the concentration of glucose will be greater than that of fructose. But if fructose is removed from the reaction, glucose will form more fructose according to LeChatelier's principle. Indeed, living cells take advantage of LeChatelier's principle by continuously removing fructose by a process in which it serves as a reactant in another reaction. Thus, the glucose-fructose equilibrium is never established, and consequently glucose is completely converted to fructose.

Change in Temperature

Almost every equilibrium constant changes when the temperature is changed. Some chemical reactions liberate heat, while others absorb heat. For example, heat liberation accompanies formation of ammonia,

$$N_2(g) + 3\,H_2(g) \rightleftharpoons 2\,NH_3(g) + \text{heat}$$

while heat absorption is required for decomposition of carbon dioxide:

$$2\,CO_2(g) + \text{heat} \rightleftharpoons 2\,CO(g) + O_2(g)$$

Thus, in these cases we can consider heat energy to be a reactant or a product of chemical reactions. In the case of ammonia formation, increasing the temperature by adding heat can be thought of as an increase in product; the position of equilibrium is shifted to the left, and K_{eq} decreases. The opposite effect occurs in the decomposition of carbon dioxide. In both reactions, the equilibrium constant changes with temperature. This effect of temperature on the position of equilibrium is another manifestation of LeChatelier's principle.

Energy and Chemical Change

Many changes in the world occur spontaneously. These spontaneous changes may be physical or chemical. For example, a stream of water flows downhill spontaneously as a result of the earth's gravitational force. As the population grows, our environment becomes cluttered with waste materials spontaneously. A shiny nail left outdoors turns to rust spontaneously, and dead leaves decay spontaneously. The one feature that all of these spontaneous processes have in common is that energy is released as a result of the change. In these cases, the original state of each system contained more energy than the final state, and the change allowed energy to be released from the system (see Figure 8-2). As you shall see, however, these energy changes are different from any we have yet encountered.

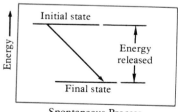

Fig. 8-2 The Energy Change in a Spontaneous Process

Most of the chemical reactions mentioned in this book occur in the presence of constant pressure from the atmosphere. In these cases any heat absorbed or released by a chemical reaction is called the *enthalpy change* (ΔH). For example, burning natural gas (methane) produces heat energy:

$$CH_4(g) + 2\,O_2(g) \longrightarrow CO_2(g) + 2\,H_2O(l) + \text{heat}$$

$$\Delta H = -210.8 \text{ kcal/mole}$$

The heat production per mole of methane burned is the change in enthalpy (ΔH) for the reaction. We designate that heat is released by assigning a negative sign to the enthalpy change. Negative ΔH values mean that the reaction loses heat to its surroundings.

If we wish to cause the reverse reaction to occur, 210.8 kcal of heat would have to be supplied to the reaction:

$$CO_2(g) + 2\ H_2O(l) + \text{heat} \longrightarrow CH_4(g) + O_2(g)$$

$$\Delta H = 210.8\ \text{kcal/mole}$$

We know that heat produced in the first reaction is the same as the heat absorbed by the reverse reaction because energy is neither created nor destroyed in a chemical reaction.

Enthalpy changes for reactions at constant pressure that release heat energy have negative values; these reactions are said to be *exothermic* (heat releasing). *Endothermic* (heat absorbing) reactions are those at constant pressure that require heat energy in order to proceed. Their ΔH values are positive.

Enthalpy changes are measured by the use of a *calorimeter*, illustrated in Figure 8-3. In this device, a vessel containing reactants is immersed in a measured quantity of water. Any heat changes resulting from the reaction will alter the water temperature. The heat absorbed or liberated is reflected by a change in water temperature, and the specific heat of water is used to calculate calories of heat responsible for the change in water temperature.

Some spontaneous changes occur in the absence of heat change. Consider a gas in one side of a container with a movable partition in the middle. If we remove the partition, the gas will flow spontaneously to fill the entire volume of the container (see Figure 8-4). In this case, no

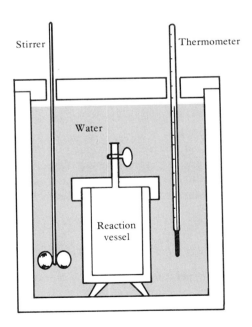

Fig. 8-3 Diagram of a Calorimeter Used for Measuring Heat Changes During a Chemical Reaction

PERSPECTIVE: Chemical Energy from Food

Energy is expended constantly by the human body, and there must be a steady supply of new energy to replace it. The source of that energy is food, whose molecules are rearranged to make other compounds during metabolism. Ultimately the energy is lost from our bodies as heat, corresponding overall to a negative enthalpy change. In order to determine how much food energy a person needs, we must measure the energy which can be furnished by food and the energy expended by the human body.

Nutritionists determine the heat content of food by burning it in a calorimeter. Since the number of calories released is usually quite large, nutritionists report them in units of food Calories (notice the uppercase C), where each Calorie is equivalent to 1 kilocalorie (kcal).

The second task, measuring energy expended by a human, is a bit more difficult. The energy entering the body in the form of food may be transferred into other compounds or it may be converted to heat. The energy of chemical compounds in the body is stored energy, but any heat that is formed is lost to the environment. It is this energy we need to measure, as this lost energy must be replaced. Heat loss from the human body is also measured by a calorimeter, although this kind of calorimeter differs in design from a simple laboratory calorimeter. A person is put into a closed chamber with water circulating in pipes in the ceiling, as shown in the accompanying figure. As heat is released from the person's body, the temperature of the circulating water rises, and the heat gained by the water is equal to the heat lost by the person's body.

As long as the energy intake for a person is equal to the energy lost as heat, the person's weight will remain constant. Weight gain or loss occurs when a person's diet provides more or less energy than is lost as heat to the environment.

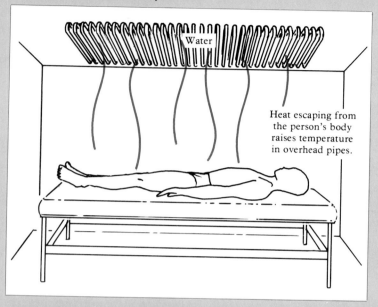

Diagram of a Calorimeter Used for Measuring Heat Loss from the Human Body

Heat escaping from the person's body raises temperature in overhead pipes.

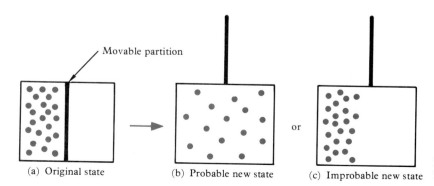

(a) Original state — Movable partition
(b) Probable new state   or   (c) Improbable new state

Fig. 8-4 The Spontaneous Diffusion of a Gas

entropy: a measure of disorder in a system; since a disordered system is more probable than an ordered system, entropy is also a measure of probability.

heat change is associated with diffusion of the gas, and yet it is a spontaneous process. If we analyze our system carefully, we see that the gas molecules have become more disordered as a result of the diffusion. The amount of disorder of the final arrangement of the gas molecules is an important factor in the spontaneous movement.

The technical term used to refer to disorder is *entropy*. Entropy is a measure of disorder in a system, and the less ordered and less structured a system is, the more probable it is. An entropy change is signified as ΔS. In a chemical reaction it is quite possible for the reactants to combine spontaneously to form products even if the products have a higher enthalpy than reactants. The unfavorable increase in enthalpy can be outweighed by a favorable increase in entropy as the reaction proceeds, and the reaction will be spontaneous.

In 1878, the American scientist J. Willard Gibbs proposed a relationship between enthalpy changes and entropy changes for chemical reactions occurring at constant temperature and pressure, as shown below. In words, this equation says that the change in free energy is equal to the change in enthalpy minus the absolute temperature multiplied by the change in entropy:

$$\Delta G = \Delta H - T \Delta S$$

The free energy change (ΔG) is considered to reflect the amount of usable, or free, energy which is made available or required by a process. A chemical reaction having a negative free energy change ($\Delta G < 0$, i.e., ΔG is less than zero) liberates free energy to its environment, and it is a spontaneous reaction. We refer to this kind of reaction as *exergonic*. A reaction that has a positive free energy change ($\Delta G > 0$, i.e., ΔG is greater than zero) is called an *endergonic* reaction. It will not occur spontaneously, but it can be made to occur if there is an available source of free energy, such as an accompanying exergonic reaction. A process in which there is no overall free energy change ($\Delta G = 0$) is in a state of equilibrium.

EXAMPLE 8.4

On the basis of the information regarding free energy changes, classify each of the following reactions as spontaneous, nonspontaneous, or at equilibrium:

(a) $C(s) + O_2(g) \longrightarrow CO_2(g)$
$\Delta G < 0$

(b) $N_2(g) + 3 H_2(g) \longrightarrow 2 NH_3(g)$
$\Delta G = 0$

(c) $2 Al_2O_3(s) \longrightarrow 4 Al(s) + 3 O_2(g)$
$\Delta G > 0$

SOLUTION

(a) Spontaneous reaction because ΔG is negative.

(b) Reaction is at equilibrium because $\Delta G = 0$.

(c) Nonspontaneous reaction because ΔG is positive.

EXERCISE 8.4 On the basis of the indicated free energy changes, classify each of the following reactions as spontaneous, nonspontaneous, or at equilibrium:

(a) $N_2(g) + 3 Cl_2(g) \longrightarrow 2 NCl_3(g)$
$\Delta G > 0$

(b) $N_2(g) + 3 H_2(g) \longrightarrow 2 NH_3(g)$
$\Delta G = 0$

(c) $I(g) + I(g) \longrightarrow I_2(g)$
$\Delta G < 0$

Free energy changes are usually given units of calories. The values may be determined for a reversible chemical reaction by using the following equation for the general reaction

$$A + B \rightleftharpoons C + D$$

$$\Delta G = \Delta G° + 2.303\, RT \log \frac{[C][D]}{[A][B]}$$

where R is the molar gas constant (expressed in cal/mole K) and T is absolute temperature (logarithms are discussed in Appendix A). In this equation $\Delta G°$ is called the *standard free energy change;* it is a reference value which applies to solutions when all components have concentrations of 1 mole/liter at 25°C. Standard free energy changes are determined experimentally and can usually be found in reference literature.

When a reaction is in its equilibrium state, there is no overall change in the system. Because of this, the actual free energy change is zero.

$$\Delta G = 0 \text{ at equilibrium}$$

Also, when a reaction is at equilibrium, the concentrations of all substances are equilibrium concentrations, and

$$K_{eq} = \frac{[C][D]}{[A][B]}$$

Thus,

$$0 = \Delta G° + 2.303\, RT \log K_{eq}$$

$$\Delta G° = -2.303\, RT \log K_{eq}$$

This result indicates that we can calculate $\Delta G°$, the standard free energy change for the reaction, from the equilibrium constant.

We can estimate the position of equilibrium of a reaction from $\Delta G°$ values. When K_{eq} is greater than one ($K_{eq} > 1$), $-2.303\, RT \log K_{eq}$ will have a negative value. Conversely, when K_{eq} is less than 1 ($K_{eq} < 1$), the value for $\Delta G°$ will be positive. Thus, the values for $\Delta G°$ can be used to estimate whether the forward or reverse parts of a reversible reaction will be favored. The following table summarizes these relations:

$\Delta G°$	Equilibrium Constant	Position of Equilibrium
< 0	$K_{eq} > 1$	on the right
> 0	$K_{eq} < 1$	on the left

EXAMPLE 8.5

From the following values for equilibrium constants, give the sign of the standard free energy change and estimate the position of equilibrium:

(a) $K_{eq} = 7.2 \times 10^{-4}$

(b) $K_{eq} = 27.8$ mole/liter

(c) $K_{eq} = 1.07$ liter/mole

SOLUTION

(a) Since $K_{eq} < 1$, $\Delta G°$ is positive and the position of equilibrium lies on the left, favoring reactants.

(b) Since $K_{eq} > 1$, $\Delta G°$ is negative and the position of equilibrium lies on the right, favoring products.

(c) Since $K_{eq} > 1$, $\Delta G°$ is negative and the position of equilibrium lies on the right, favoring products.

EXERCISE 8.5 From the following values for equilibrium constants, give the sign of the standard free energy change and estimate the position of equilibrium:

(a) $K_{eq} = 2.67$ liter/mole

(b) $K_{eq} = 3.79 \times 10^{-3}$ mole/liter

(c) $K_{eq} = 872$

In principle, any process that occurs spontaneously, and thus liberates free energy, can be used for performing useful work. For example, the fall of water in a waterfall is a spontaneous process, and we derive useful work from the process by allowing the falling water to turn a turbine. In another example, the burning of gasoline is a spontaneous process which we use to give motion to automobiles. For processes that are not spontaneous, the free energy change is a measure of the work that must be done to make a process occur. Thus, the extraction of a metal from its ore is a nonspontaneous process, and we must do a certain amount of work to obtain pure metal.

Energy and the Biosphere

The *biosphere* is the part of the earth's crust, waters, and atmosphere where living organisms can subsist. Living organisms require free energy for maintenance, growth, and reproduction. The need for free energy can be traced to two reasons. The first of these is that living organisms depend on nutrients readily available in their environment as raw materials for the synthesis of compounds needed for life processes. The chemical reactions involved in synthesis are for the most part endergonic, and outside sources of free energy are needed to make them occur. The second reason that living organisms require free energy is that they are highly ordered systems and thus low in entropy compared to the disordered world about them. Since the entropy is a very large negative factor in the development of an organism, the contribution to ΔG is a large positive value, and the development process is nonspontaneous. Hence, a second reason exists for the requirement of free energy by living organisms.

The ultimate source of energy for all life is the sun. Animals rely on the ability of plants to convert the sun's energy into usable form. By the process of photosynthesis, plants use solar energy to convert carbon dioxide and water to carbohydrates and oxygen. Animals eat the plants, and the carbohydrates become a usable form of free energy in their diets. Thus the sun's energy is channeled from plants to animals, and ultimately to the entire biosphere.

RATES OF CHEMICAL REACTIONS

The concept of free energy allows us to predict that certain processes will occur spontaneously. But spontaneous processes may occur at such slow rates that they are hardly noticeable. The rusting of iron, a spontaneous reaction, proceeds at an imperceptibly slow rate, as does the tarnishing of silver. Why is it that these spontaneous events are so slow, while others such as those involved in metabolic processes occur at a rapid rate? To answer this question we must consider what is meant by rates of chemical reactions and which factors influence those rates.

Energy of Activation

Free energy changes are derived by comparing only the energy differences between the initial and final states of reactions. In order to understand why some reactions are slow and others are fast, we must examine energy changes along the path that the reactants take as they are transformed into products.

We can compare a *reaction pathway* to a bicycle ride across town. Let's imagine that we wish to ride to a shopping center across town. The most direct route, Path A, has a large hill between our point of origin and the shopping center (see Figure 8-5). Another route, Path B, is longer but there are no major hills along the way. We know that if we take Path A we will have to pedal up the hill, and our rate of travel will be slow. On the other hand, if we take Path B, the ride will take less time because there is not a long uphill portion. Therefore we decide on Path B, and we arrive at the shopping center in less time than we would have used by Path A.

In our analogy, Path A has a high energy barrier which must be crossed to reach our destination, while Path B contains only a negligible

reaction pathway: the hypothetical route whereby reactants are converted to products in a chemical reaction.

Fig. 8-5 A Bicycle Ride Across Town. Path A has a major hill but Path B has no major changes in elevation.

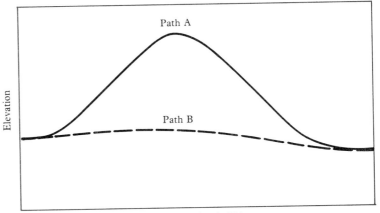

energy barrier. Chemical reactions proceed across similar energy barriers, some of which are high and some of which are low. In the hypothetical reactions A ⟶ B and X ⟶ Y, two reaction diagrams can be drawn (see Figure 8-6). In each case, ΔG is negative, and thus the reactions are spontaneous. However, reaction X ⟶ Y will occur at a faster rate because the energy barrier is lower. In order for reactants to be converted to products in any chemical reaction, the reactant molecules must acquire enough energy to cross the energy barrier in the reaction pathway. The amount of energy needed to cross the barrier is called the *activation energy* (E_a). Reactions with relatively low activation energies will proceed at faster rates than those with relatively high activation energies.

activation energy: the energy maximum in a reaction pathway.

If reactant molecules are to be transformed into products, they must acquire enough energy, the activation energy, to overcome forces that tend to keep them as they are. We might imagine that when the reactant molecules have acquired their activation energy, their structures are in some transition form midway between the structures of unreacted starting compounds and the structures of product molecules. The particular arrangement of atoms, whatever it may be, that possesses the energy of activation is called the *activated complex* or the *transition state*.

Fig. 8-6 Reaction Diagrams. In (a), the reaction A ⟶ B has a relatively high activation energy. In (b), reaction X ⟶ Y has a relatively low activation energy. Both reactions are exergonic.

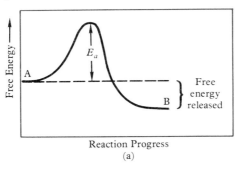
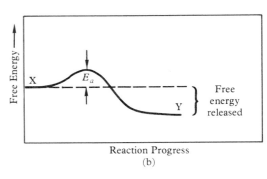

Factors Affecting Reaction Rates

Chemical reactions show a wide range of rates; some take weeks or months to proceed to a measurable extent, while others are over in a fraction of a second. The most important factors that influence rates of chemical reactions are discussed below.

Temperature

An increase in temperature of a substance is an indication that it has gained heat energy. When heat energy has been absorbed by molecules they move faster and their internal vibrations become faster and stronger. In the case of a decomposition reaction, such as

$$CO_2(g) \longrightarrow C(g) + O_2(g)$$

an increase in temperature causes stronger vibrations of atoms within the reactant molecules so that they are more easily broken apart. Thus, a temperature increase provides energy to the reactant, allowing more of its molecules to acquire the activation energy necessary for decomposition. Consequently the reaction is speeded up.

In a chemical system in which two molecules form a product such as

$$H_2(g) + I_2(g) \longrightarrow 2\,HI(g)$$

the reactant molecules must collide with a certain amount of energy to react. Thus, when temperature is increased, the reactant molecules move faster, causing them to collide more frequently and with greater impact. Here, as in the previous example, reaction rate will be increased by higher temperatures. In fact, this statement holds true for all chemical reactions, and the converse is also true: cooling decreases rates of chemical reactions.

Concentration

For those reactions involving two or more reacting compounds, increasing the concentration of one or both brings about more frequent collisions and thus faster reaction rates. In other reactions where there is only one reactant, increasing the concentration of that reactant amounts to providing more molecules to react, and thus there are more of them changing during a given time period. For these reasons, increased concentrations of reactants will increase rates of chemical reactions. For a reaction occurring in the gaseous state, increasing the pressure (confining the reacting molecules to a smaller space) increases the concentration of reactants, and the reaction rate will be increased.

Rates of chemical reactions are experimentally measured quantities; they are usually expressed in units of concentration change per unit time, typically as moles/liter per second or moles/liter per minute. It is possible to determine experimentally how much the rate of a reaction depends on concentration.

Reactions involving only one reactant molecule are called *unimolecular reactions*. This kind of reaction is described by the general equation

$$A \longrightarrow B$$

For unimolecular reactions, the following expression describes concentration dependence:

$$\text{rate} = k[A]$$

Thus, at a constant temperature, the rate of the reaction varies directly with the concentration of reactant. The constant of proportionality, k (called the *rate constant*), is determined by experiment.

rate constant: the constant of proportionality that relates a reaction rate to concentration(s) of reactant(s).

For a reaction involving the collision of two reactants, a *biomolecular reaction*, the rate varies directly with both concentrations:

$$A + B \longrightarrow C$$

$$\text{rate} = k[A][B]$$

In each case, the rate of reaction is related to concentration by a rate constant. Every reaction has its own characteristic rate constant, and only a change in temperature will alter that value.

The Nature of the Reactants

If temperature and concentration were the only factors affecting reaction rates, then all reactions at the same temperature and concentration would have the same rates. Yet experimental evidence demonstrates that this is not the case. Reactions differ in reactivity because each individual reaction has its own characteristic energy of activation and its own rate constant. In general, reactions between oppositely charged ions in solution have low energies of activation, and their rates are fast. In contrast, reactions involving breaking covalent bonds and forming new ones have high activation energies and are usually slower than ionic reactions. Thus, the chemical nature of the reactants has an important effect on the rate of a chemical reaction.

Catalysis

A *catalyst* is a substance that affects the rate of a chemical reaction by its presence in the reaction mixture. The catalyst does not undergo permanent chemical change and may be recovered in its original form after the reaction is over. Most catalysts increase the rate of chemical reactions.

The question of how catalysts work is the subject of much current research activity. There may be many possible answers, but it appears that catalysts speed up reactions by providing a new pathway in which the activation energy is much lower than in the absence of the catalyst, as illustrated by Figure 8-7.

Since most chemical reactions are reversible, catalysts affect both the forward and reverse reaction rates. An example is the glucose-fructose isomerization:

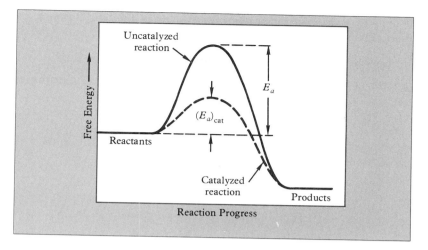

Fig. 8-7 Reaction Diagram of a Catalyzed Reaction Compared to the Reaction Diagram of the Uncatalyzed Reaction

$$\text{glucose} \rightleftharpoons \text{fructose}, K_{eq} = 0.43$$

A solution of glucose in water will eventually reach equilibrium with fructose, but the process might take weeks. If a suitable catalyst is added, both forward and reverse reactions are speeded up, and equilibrium is established in a matter of minutes. But since the forward and reverse reaction rates are enhanced to the same extent by the catalyst, the equilibrium constant is not altered. Thus, a catalyst cannot cause a reaction to occur if the reaction does not take place in the absence of the catalyst.

Each metabolic reaction is catalyzed by a specific protein molecule called an *enzyme*. In their roles as catalysts, enzymes are responsible for causing chemical reactions in plants and animals to proceed at rates fast enough to allow the organism to respond quickly to changes in its environment. Enzymes catalyze reactions involved in digestion, energy storage and release, vision processes, and other metabolic processes. These natural catalysts are often subject to the effect of *inhibitors*, substances that slow down a catalyzed reaction. Inhibitors usually function by inactivating the enzyme or reducing its effectiveness. The action of enzyme inhibitors in providing a control mechanism for regulation of metabolism is a subject we will investigate more thoroughly in later chapters.

enzyme (EN-zime): a protein catalyst that increases the rate of a biochemical reaction.

THE RELATIONSHIP BETWEEN CHEMICAL EQUILIBRIUM AND REACTION RATES

The equilibrium constant for a chemical reaction is usually determined by measuring concentrations of all substances present in the equilibrium mixture. But let us consider the relationship of the equilibrium constant to reaction rates. The reaction between hydrogen and iodine is a good example:

$$H_2(g) + I_2(g) \rightleftharpoons 2\ HI(g)$$

When the reaction is at equilibrium, the rates of the two opposing reactions are equal:

rate of forward reaction = rate of reverse reaction

We can substitute rate expressions into this statement of equality as follows,

$$k_f[H_2][I_2] = k_r[HI][HI] = k_r[HI]^2$$

where k_f is the rate constant for the forward reaction and k_r is the rate constant for the reverse reaction. If we rearrange terms we get

$$\frac{[HI]^2}{[H_2][I_2]} = \frac{k_f}{k_r}$$

And we now recognize that

$$\frac{[HI]^2}{[H_2][I_2]} = K_{eq}$$

With another substitution we get

$$K_{eq} = \frac{k_f}{k_r}$$

As a result of this exercise in algebra, we have proved that the equilibrium constant for this reaction is the ratio of rate constants for the forward and reverse reactions. All equilibrium constants are ratios of rate constants, and this fact allows us to suggest another method for measuring equilibrium constants. If it is possible to measure individual reaction rates and thus calculate rate constants, we can use those values to calculate an equilibrium constant.

The relationship between rates of reaction and chemical equilibrium provides us with a better understanding of LeChatelier's principle. If the equilibrium of a chemical reaction is temporarily disrupted by a change in concentration of one or more components, the system is able to readjust itself because the forward and reverse reaction rates change. Remember that reaction rates are directly proportional to concentration. If concentration changes occur, reaction rates will change in response. However, the rate constants do not change except with temperature, and thus equilibrium constants vary only with temperature.

SUMMARY

Chemical Equilibrium: Reversible reactions, those which proceed in both forward and reverse directions simultaneously, will achieve a state of dynamic equilibrium if left undisturbed. The constant ratio of component concentrations in an equilibrium mixture is the equilibrium constant. When the value of an equilibrium constant is less than 1 ($K_{eq} < 1$), reactants are favored. When the value of an equilibrium constant is greater than 1 ($K_{eq} > 1$), products are favored. LeChatelier's principle states that if a chemical equilibrium is disturbed, the system will readjust in such a way as to offset the disturbance partially and return to equilibrium. A chemical equilibrium will respond to concentration changes by maintaining a constant K_{eq}, but it will respond to a temperature change by an alteration in K_{eq}.

Heat absorbed or released by a chemical reaction is called the enthalpy change (ΔH). Exothermic reactions have negative enthalpy changes ($\Delta H < 0$), whereas endothermic reactions have positive enthalpy changes ($\Delta H > 0$). Entropy is a measure of disorder in a system. The less ordered and structured a system, the more likely it is to exist. Reactions that produce relative disorder have a positive entropy change ($\Delta S > 0$), and reactions producing relative order have a negative entropy change ($\Delta S < 0$). The combination of enthalpy change and entropy change is related to free energy change (ΔG). A release in free energy ($\Delta G < 0$) correlates with a spontaneous (exergonic) chemical reaction. Reactions at equilibrium have no free energy changes ($\Delta G = 0$), and nonspontaneous (endergonic) reactions have positive free energy changes ($\Delta G > 0$). Standard free energy changes ($\Delta G°$) can be calculated from equilibrium constants.

Rates of Chemical Reactions: Each chemical reaction passes through a reaction pathway containing an energy barrier called the activation energy. Reactions with relatively low activation energies proceed at faster rates than those with relatively high activation energies. An increase in temperature speeds up a reaction, as does an increase in the concentration of one or more reactants. Rates of chemical reactions are directly related to concentrations of reactants by a constant of proportionality called the rate constant. The chemical nature of reactants is an important influence on the rate of a chemical reaction. The presence of a catalyst is another factor affecting reaction rates. Most catalysts increase reaction rates and do so by lowering the activation energy. Enzymes are protein molecules which function as catalysts for chemical reactions involved in life processes.

The Relationship Between Chemical Equilibrium and Reaction Rates: Equilibrium constants are ratios of concentrations. Thus, if concentration changes occur in a chemical equilibrium, reaction rates will change in response. However, rate constants do not change except with temperature, and thus equilibrium constants vary only with temperature.

STUDY QUESTIONS AND PROBLEMS

1. Define the following terms:
   (a) reversible reaction
   (b) equilibrium constant
   (c) equilibrium constant expression
   (d) position of equilibrium
   (e) enthalpy change
   (f) calorimeter
   (g) entropy
   (h) free energy change
   (i) exergonic
   (j) endergonic
   (k) reaction pathway
   (l) activation energy
   (m) activated complex
   (n) unimolecular reaction
   (o) rate constant
   (p) bimolecular reaction
   (q) catalyst
   (r) enzyme
   (s) enzyme inhibitor

2. Write the expression for the equilibrium constant for each of the following reactions.
   (a) $4 NH_3(g) + 5 O_2(g) \rightleftharpoons 4 NO(g) + 6 H_2O(g)$
   (b) $2 NO_2(g) \rightleftharpoons N_2O_4(g)$
   (c) $2 SO_3(g) \rightleftharpoons 2 SO_2(g) + O_2(g)$
   (d) $PCl_5(g) \rightleftharpoons PCl_3(g) + Cl_2(g)$
   (e) $H_2(g) + Cl_2(g) \rightleftharpoons 2 HCl(g)$

3. An equilibrium mixture for the reaction
   $$2 H_2S(g) \rightleftharpoons 2 H_2(g) + S_2(g)$$
   had the following concentrations:
   $[H_2S] = 1.0$ mole/liter
   $[H_2] = 0.20$ mole/liter
   $[S_2] = 0.80$ mole/liter
   Calculate the equilibrium constant.

4. Calculate the equilibrium constant for the reaction
   $$H_2(g) + I_2(g) \rightleftharpoons 2 HI(g)$$
   on the basis of the following equilibrium concentrations:
   $[H_2] = 0.9$ mole/liter
   $[I_2] = 0.4$ mole/liter
   $[HI] = 0.6$ mole/liter

5. An equilibrium mixture for the reaction
   $$N_2O_4(g) \rightleftharpoons 2 NO_2(g)$$
   had the following concentrations:
   $[N_2O_4] = 1.40 \times 10^{-3}$ mole/liter
   $[NO_2] = 1.72 \times 10^{-2}$ mole/liter
   Calculate the equilibrium constant.

6. An equilibrium mixture for the reaction
   $$PCl_5(g) \rightleftharpoons PCl_3(g) + Cl_2(g)$$
   had the following concentrations:
   $[PCl_5] = 0.158$ mole/liter
   $[PCl_3] = 0.081$ mole/liter
   $[Cl_2] = 0.081$ mole/liter
   Calculate the equilibrium constant.

7. At 25°C, K_{eq} for the following reaction is 4.0 liter²/mole².

$$3 C_2H_2(g) \rightleftharpoons C_6H_6(g)$$

If the equilibrium concentration of C_2H_2 is 0.70 mole/liter, calculate the concentration of C_6H_6.

8. How is the magnitude of an equilibrium constant used to predict the position of equilibrium for a reversible reaction?

9. Summarize LeChatelier's principle in your own words.

10. Summarize the response of a chemical equilibrium to each of the following:
    (a) an increase in the concentration of a reactant
    (b) an increase in the concentration of a product
    (c) a decrease in the concentration of a reactant
    (d) a decrease in the concentration of a product

11. Predict concentration changes and the direction in which each of the following equilibria would shift in response to the indicated changes:
    (a) glucose ($C_6H_{12}O_6$) $\rightleftharpoons$ fructose ($C_6H_{12}O_6$)
        increase in concentration of glucose
    (b) $N_2(g) + O_2(g) \rightleftharpoons 2 NO(g)$
        decrease in concentration of O_2
    (c) $N_2(g) + 3 H_2(g) \rightleftharpoons 2 NH_3(g)$
        increase in concentration of NH_3
    (d) $CO(g) + H_2O(g) \rightleftharpoons CO_2(g) + H_2(g)$
        decrease in concentration of H_2
    (e) $PCl_5(g) \rightleftharpoons PCl_3(g) + Cl_2(g)$
        decrease in concentration of PCl_5

12. How does increased temperature affect K_{eq} for an exothermic reaction?

13. How does increased temperature affect K_{eq} for an endothermic reaction?

14. A mixture of $H_2(g)$, $O_2(g)$, and $H_2O(g)$ are in equilibrium in a closed container. Assuming that the formation of water is exothermic, predict the effect on the concentration of $H_2O(g)$ of each of the following changes.

$$2 H_2(g) + O_2(g) \rightleftharpoons 2 H_2O(g)$$

    (a) addition of H_2 to the system
    (b) removal of O_2 from the system
    (c) increasing the temperature of the system

15. The following reaction is endothermic:

$$PCl_5(g) \rightleftharpoons PCl_3(g) + Cl_2(g)$$

Predict the effect on the equilibrium of each of the following changes:

    (a) addition of Cl_2 to the system
    (b) increasing the temperature of the system
    (c) removal of PCl_3 from the system

16. Describe how enthalpy changes for chemical reactions are measured in a calorimeter.

17. Distinguish between exothermic reactions and endothermic reactions.

18. Distinguish between exergonic reactions and endergonic reactions.

19. How can free energy changes be used to predict the spontaneity of chemical reactions?

20. On the basis of the indicated free energy changes, classify each of the following reactions as spontaneous, nonspontaneous, or at equilibrium.

    (a) $2 O_3(g) \longrightarrow 3 O_2(g)$
        $\Delta G < 0$
    (b) $2 Cu(s) + SO_2(g) \longrightarrow Cu_2S(s) + O_2(g)$
        $\Delta G > 0$
    (c) glucose $\longrightarrow$ fructose
        $\Delta G = 0$
    (d) $CaO(s) + CO_2(g) \longrightarrow CaCO_3(s)$
        $\Delta G < 0$
    (e) $6 H_2(g) + P_4(g) \longrightarrow 4 PH_3(g)$
        $\Delta G = 0$

21. Oxidation of carbon to carbon dioxide, as illustrated by the following equation, is exergonic at a given temperature:

$$C(s) + O_2(g) \longrightarrow CO_2(g)$$
$$\Delta G = -10 \text{ kcal/mole}$$

Answer the following questions regarding the reaction and give reasons for your answers.

    (a) Is the system at equilibrium?
    (b) Will CO_2 form spontaneously?
    (c) What is ΔG for the reverse reaction?

22. Pretend that you are involved in a research project to investigate the structural features of a large biological molecule. You are able to determine that the following reaction proceeds spontaneously to the right and is endothermic.

    large biological molecule $\longrightarrow$ product

Answer the following questions about the reaction and give reasons for your answers. (Clue: $\Delta G = \Delta H - T\Delta S$)

    (a) Which is larger, ΔH or $T\Delta S$?
    (b) What is the sign of the enthalpy change?
    (c) What is the sign of the free energy change?

23. Consider a growing plant which is actively converting carbon dioxide and water to carbohydrates and oxygen by use of solar energy, and answer the following questions.

    (a) How does the entropy change within the plant?
    (b) How does the free energy change within the plant?

24. How is the sign of the standard free energy change for a reversible reaction related to K_{eq} for the reaction?

25. For the following equilibrium constants, give the sign of the standard free energy change and estimate the position of equilibrium.

    (a) $K_{eq} = 8720$ mole/liter
    (b) $K_{eq} = 6.8 \times 10^{-3}$
    (c) $K_{eq} = 0.72$ liter/mole
    (d) $K_{eq} = 1.63$ liter/mole
    (e) $K_{eq} = 0.28$

26. Summarize how the biosphere is dependent on the sun as the ultimate energy source.

27. How are rates of chemical reactions related to activation energies?

28. How does each of the following factors affect rates of chemical reactions?

    (a) temperature
    (b) concentration
    (c) nature of reactants
    (d) catalysis

29. A rule of thumb is that increasing the temperature 10°C will double the rate of any chemical reaction. How much will a reaction rate increase if the temperature is raised 100°C?

30. For the general reaction

    $$A + B \longrightarrow C + D$$

    what is the effect on the rate of

    (a) doubling the concentration of A?
    (b) doubling the concentration of B?
    (c) doubling the concentration of both A and B?

31. Consider the following reaction:

    $$H_2(g) + Cl_2(g) \longrightarrow 2\ HCl(g)$$
    $$\text{rate} = k[H_2][Cl_2]$$

    What are the units of the rate constant?

32. For the hypothetical reaction

    $$A + B \longrightarrow C,$$

    answer the following questions:

    (a) Write an expression to describe the dependence of the rate of reaction on the concentrations of A and B.
    (b) If the rate of formation of C is 2.5×10^{-5} mole/liter per second, what is the value of the rate constant under the following conditions:

    concentration of A = 0.20 mole/liter
    concentration of B = 0.30 mole/liter

33. The rate of a certain biochemical reaction at body temperature in the absence of enzyme was measured by laboratory experimentation. The rate for the same reaction in the human body when it is catalyzed by an enzyme is 10^5 times faster. Answer the following questions about the reaction, and give reasons for your answers:

    (a) How did the presence of the enzyme affect the activation energy of the reaction?
    (b) How did the presence of the enzyme affect the equilibrium constant for the reaction?
    (c) How did the presence of the enzyme affect the standard free energy change for the reaction?

34. Answer the following questions about a catalyzed reaction, and give reasons for your answers:

    (a) If the catalyst increases the rate of the forward reaction by a factor of 1000, what is the effect on the rate of the reverse reaction?
    (b) If the catalyst increases the rate of the forward reaction by a factor of 500, what is the effect on the equilibrium constant?
    (c) If the catalyst increases the rate of the forward reaction by a factor of 100, what is the effect on the standard free energy change of the reaction?

35. New automobiles are now required by law to have catalytic converters. These devices convert unburned gasoline and carbon monoxide in the automobile exhaust to carbon dioxide, thereby minimizing pollution of the air by the exhaust. Why do you think it is necessary for a catalyst to be present in the converter?

36. How can an equilibrium constant be calculated from rate constants?

37. In terms of reaction rates, explain how a chemical equilibrium responds to a change in concentration of one or more components.

38. In terms of rate constants, explain why a change in temperature causes a change in K_{eq} for a reversible reaction.

CHAPTER 9

ACIDS, BASES, AND SALTS

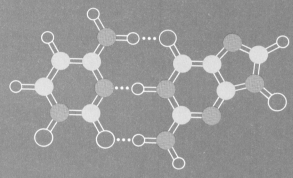

ACIDS AND BASES
- Naming Acids and Bases
- Properties of Acids and Bases
- Common Acids and Bases
- Strong and Weak Acids
- Examples of Strong and Weak Acids
- Strong and Weak Bases
- Normality
- Acid-Base Titrations

SALTS

The idea that some substances are acids and others are bases has its origins buried deeply in the history of chemistry. Acids were probably recognized because of their distinctive sour taste. Bases were noticed at a later date as bitter-tasting substances which react with acids to form salty-tasting solutions. In the modern world, acids and bases are important in industry and in everyday life. Sulfuric acid and sodium hydroxide (a base) are among the chemical products manufactured in largest amount, and a great number of chemical processes involve acids or bases as reactants or catalysts. Almost every biochemical reaction is closely tied to acid-base equilibria in the cell or in the organism as a whole, and the acidity or alkalinity (level of base) of the soil and water are of great importance for plants and animals. In this chapter you will learn about the fundamental properties of acids and bases and how they neutralize each other to form salts.

ACIDS AND BASES

Acids and bases were first defined formally in 1887 by a young Swedish graduate student named Svante Arrhenius. Arrhenius worked mostly with aqueous solutions, and he naturally came to know acids and bases from their behavior in water. His definition of an *acid* is a substance that produces hydrogen ions (H^+) in water. An example of an acid is the aqueous solution of hydrogen chloride. When hydrogen chloride, a gas, dissolves in water, hydrogen ions are produced:

$$HCl(g) \xrightarrow{water} H^+(aq) + Cl^-(aq)$$

Arrhenius (are-REH-nius)

where (aq) means that the ions are hydrated by water molecules. Thus, aqueous solutions of hydrogen chloride came to be known as hydrochloric acid. (Note that hydrogen ions are also referred to as protons.)

Arrhenius defined *bases* as substances that dissolve in water to produce hydroxide ions (OH^-). An example is sodium hydroxide:

$$NaOH(s) \xrightarrow{water} Na^+(aq) + OH^-(aq)$$

We can write generalized equations for Arrhenius definitions as follows, where HA represents any acid, A^- represents the corresponding anion, M represents any metal, and M^+ represents the corresponding cation:

Acid: $\quad HA \xrightarrow{water} H^+(aq) + A^-(aq)$

Base: $\quad MOH \xrightarrow{water} M^+(aq) + OH^-(aq)$

In 1923 two chemists, Johannes Bronsted from Denmark and Thomas Lowry from England, independently arrived at more general definitions of acids and bases that are not restricted to aqueous solutions as the Arrhenius definitions are. According to Bronsted and Lowry, an *acid* is any substance that can donate a proton to another substance, and the acceptor of the proton is a *base*. Thus the reaction between an acid and a base can be illustrated as follows, where hydrogen chloride, HCl(g), is the acid and ammonia, $NH_3(g)$, is the base:

$$HCl(g) + NH_3(g) \longrightarrow NH_4Cl(s)$$

As you can see, the Bronsted-Lowry theory does not require that a base contain hydroxide ions; it simply has to be capable of accepting a proton from an acid donor.

The equation shown below is a generalized representation for Bronsted-Lowry definitions of acids and bases:

$$H\text{-acid} + base \rightleftharpoons H\text{-base}^+ + acid^-$$

Thus we see that an acid-base reaction can be pictured as a reversible substitution reaction in which a proton is transferred from one substance to another. As you learn more about acids and bases, you will see that Bronsted-Lowry theory is perfectly compatible with Arrhenius definitions.

In chapter 7 you learned that when a molecular substance dissolves in water, solute molecules become separated from each other and distributed throughout the solution. However, certain very polar molecules, like hydrogen chloride, dissociate into ions when dissolved in water. As hydrogen chloride dissolves, the attractions between water molecules and the polar ends of the HCl molecules are strong enough to cause the HCl molecules to break apart into ions (Figure 9-1). We can show this process as

$$HCl(g) \xrightarrow{water} H^+(aq) + Cl^-(aq)$$

A more accurate equation for what happens can be written:

$$HCl(g) + H_2O(l) \rightleftharpoons H_3O^+(aq) + Cl^-(aq)$$

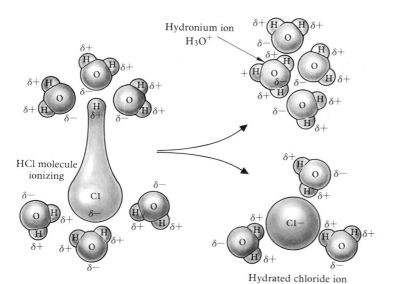

Fig. 9-1 Ionization of the HCl Molecule in Water

Both equations mean that HCl molecules dissociate into hydrogen ions and chloride ions in the aqueous solution. But the second equation is a better description because when dissolution occurs, the hydrogen ions released by the acid become bonded to water molecules to form *hydronium ions*, H_3O^+ (Figure 9-2):

$$H^+ + H_2O \longrightarrow H_3O^+$$

These are formed because the oxygen atoms in water molecules have nonbonded electrons available, and the free protons, with their positive charges, are attracted to the nonbonded electrons. Thus, the covalent bond formed between a proton and a water molecule is composed of electrons donated by an oxygen atom. This kind of covalent bond, in which both bonding electrons are contributed by one atom, is called a *coordinate covalent bond*. Since protons are donated by HCl to water molecules, the process corresponds to Bronsted-Lowry theory, where HCl is the acid and H_2O is the base. Both the hydronium ions and the chloride ions are hydrated by an indeterminate number of water molecules.

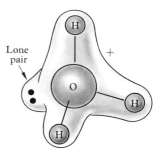

Fig. 9-2 The Hydronium Ion, H_3O^+

According to Bronsted-Lowry theory, for every acid there is a base, and vice versa. In the hydrochloric acid example, HCl is the acid and H_2O is the base. But after the two have reacted by transferring a proton, the products are also an acid and a base. In other words, after H_2O accepts a proton, H_3O^+ is an acid because it is capable of donating a proton to some other substance. And Cl^-, having lost a proton, is capable of accepting one from some other substance. Thus we can write

$$HCl(g) + H_2O(l) \rightleftharpoons H_3O^+(aq) + Cl^-(aq)$$
$$\text{acid}_1 \quad \text{base}_1 \quad \quad \text{acid}_2 \quad \quad \text{base}_2$$

or

$$\text{acid}_1 + \text{base}_1 \rightleftharpoons \text{acid}_2 + \text{base}_2$$

conjugate base: the base formed when an acid loses a proton.

conjugate acid: the acid formed when a base gains a proton.

The acid-base pairs are said to be conjugates of each other. Thus, Cl^- is the *conjugate base* of the acid HCl, and H_3O^+ is the *conjugate acid* of the base H_2O. We say that hydrochloric acid and chloride ion are an acid-base conjugate pair. Most common acids are either neutral compounds or cations, whereas bases are usually neutral compounds or anions.

EXAMPLE 9.1

For each of the following, write the formula for the conjugate acid or base.

Acids	Bases
(a) HBr	(d) H_2O
(b) H_2O	(e) OH^-
(c) HNO_3	(f) ClO^-

SOLUTION

The conjugate base for each acid is the ion or molecule formed when the acid loses a proton. Thus, the conjugate bases are:

(a) $HBr \longrightarrow H^+ + Br^-$
   acid   conjugate base

(b) $H_2O \longrightarrow H^+ + OH^-$
   acid   conjugate base

(c) $HNO_3 \longrightarrow H^+ + NO_3^-$
   acid   conjugate base

The conjugate acid for each base is the ion or molecule formed when the base gains a proton. Thus, the conjugate acids are:

(d) $H_2O + H^+ \longrightarrow H_3O^+$
   base   conjugate acid

(e) $OH^- + H^+ \longrightarrow H_2O$
   base   conjugate acid

(f) $ClO^- + H^+ \longrightarrow HClO$
   base   conjugate acid

EXERCISE 9.1 For each of the following, write the formula for the conjugate acid or base.

Acid	Base
(a) HCO_3^-	(d) ClO_4^-
(b) H_2SO_4	(e) HCO_3^-
(c) HF	(f) NH_3

Naming Acids and Bases

In chapter 3 you were given names of common anions. Most acids can be named easily by a system based on the anions they contain. If the anion name ends in "-ide," the corresponding acid is named as "hydro____ic acid," where the blank space is a portion of the anion name. ("Hydro-" signifies the presence of water.) An example is hydro*chlor*ic acid, based on the name of the chloride anion. If the anion name ends in "-ate" or "-ite," the acids are named as "____ic acid" and "____ous acid," respectively. Table 9-1 gives common acids and the anion names on which the acid names are based. There are a few acids whose names are not covered by this nomenclature, and in those cases, correct names will be furnished as we go along.

Unlike acids, bases are not always easily recognized from their formulas, and they have a variety of structures. Thus, there is no special system for naming bases, and it will be necessary to learn names of individual bases as we encounter them.

Properties of Acids and Bases

The acids that we encounter in daily life are usually dissolved in water, and thus their solutions contain hydronium ions. These impart a sour taste to water, accounting for the tart flavor of vinegar and lemon juice. We identify acids in the laboratory by their ability to change the color of certain dyes called acid-base indicators. One of these is litmus, an indicator obtained from plants. Small strips of paper impregnated with litmus are used; if a drop of solution changes the color of litmus paper

Table 9-1 Common Acids and Their Corresponding Anions

Anion	Name	Acid	Name
BO_3^{3-}	Borate	H_3BO_3	Boric acid
CO_3^{2-}	Carbonate	H_2CO_3	Carbonic acid
CN^-	Cyanide	HCN	Hydrocyanic acid
$CH_3CO_2^-$	Acetate	CH_3CO_2H	Acetic acid
NO_3^-	Nitrate	HNO_3	Nitric acid
NO_2^-	Nitrite	HNO_2	Nitrous acid
SO_4^{2-}	Sulfate	H_2SO_4	Sulfuric acid
SO_3^{2-}	Sulfite	H_2SO_3	Sulfurous acid
PO_4^{3-}	Phosphate	H_3PO_4	Phosphoric acid
F^-	Fluoride	HF	Hydrofluoric acid
ClO_4^-	Perchlorate	$HClO_4$	Perchloric acid
ClO_3^-	Chlorate	$HClO_3$	Chloric acid
ClO_2^-	Chlorite	$HClO_2$	Chlorous acid
ClO^-	Hypochlorite	$HClO$	Hypochlorous acid

from blue to red, the solution is acidic. This test actually corresponds to an acid-base reaction where blue litmus is the base:

$$H_3O^+ + \text{litmus} \rightleftharpoons \text{H-litmus}^+ + H_2O$$
$$\text{acid}_1 \quad \text{base}_1 \quad\quad \text{acid}_2 \quad\quad \text{base}_2$$
$$\quad\quad\quad (\text{blue}) \quad\quad (\text{red})$$

The reaction with litmus represents still another property of acids—they react with bases in what is called a *neutralization* reaction. This reaction gets its name from the fact that as the acid and base react, their characteristic properties disappear, and they thus become neutralized. Acids also react with metal oxides such as Na_2O and MgO,

$$Na_2O(s) + 2\,HCl(aq) \longrightarrow 2\,NaCl(aq) + H_2O(l)$$

and they react with many metals to produce hydrogen gas:

$$Zn(s) + 2\,HCl(aq) \longrightarrow ZnCl_2(aq) + H_2(g)$$

Because of the acid reaction with metals, acids should not be stored in metal containers or allowed to contact metal objects unless a chemical reaction is desired.

As with acids, the more familiar bases are usually dissolved in water. These solutions contain hydroxide ions (OH^-), which give the water solution a bitter taste and a slippery feeling, like soapy water. Bases are identified in the laboratory by their ability to turn red litmus blue, a reversal of the acid reaction with litmus:

$$OH^- + \text{H-litmus}^+ \rightleftharpoons \text{litmus} + H_2O$$
$$\text{base}_1 \quad\quad \text{acid}_1 \quad\quad\quad \text{base}_2 \quad\quad \text{acid}_2$$
$$\quad\quad\quad\quad (\text{red}) \quad\quad\quad (\text{blue})$$

As mentioned earlier, this is an example of an acid-base neutralization reaction.

Concentrated solutions of acids and bases are corrosive to all tissue, and they should be handled cautiously in order to prevent physical contact. If ingested, they can cause severe damage to the gastrointestinal tract.

Common Acids and Bases

Although there are many acids and bases, a few are widely known for their uses in medicine, industry, agriculture, and commercial products. These are described in the following paragraphs. Table 9-2 summarizes the concentrations of commercially available acids and bases.

Hydrochloric Acid

Hydrochloric acid, a colorless liquid, is formed when hydrogen chloride gas dissolves in water.

$$HCl(g) + H_2O(l) \rightleftharpoons H_3O^+(aq) + Cl^-(aq)$$

Table 9-2 Concentrations of Commercially Available Acids and Bases

Substance	Molarity
Acetic acid (99.5%)	17.4
Hydrochloric acid (38%)	12.0
Perchloric acid (72%)	11.6
Hydrofluoric acid (45%)	25.7
Nitric acid (69%)	15.4
Phosphoric acid (85%)	14.7
Sulfuric acid (94%)	17.6
Ammonium hydroxide (27%)	14.3

Hydrochloric acid is usually sold as an aqueous solution containing 38% (w/w) hydrogen chloride, commonly known as concentrated hydrochloric acid. This concentration corresponds to 12.0M, and it represents the maximum solubility in water for hydrogen chloride. Hydrogen chloride vapors or hydrochloric acid can cause serious tissue damage. As little as 0.1% hydrogen chloride by volume in the atmosphere may cause death in a few minutes, and concentrated hydrochloric acid causes burns and inflammation of the skin. For these reasons, we must avoid breathing the vapors and we should be careful not to spill the acid. Because it dissolves many metals readily, hydrochloric acid is used extensively in industrial processing of metals and in concentrating ores.

Sulfuric Acid

Sulfuric acid is formed when sulfur trioxide, a product of burning sulfur, reacts with water:

$$SO_3(g) + H_2O(l) \longrightarrow H_2SO_4(l)$$
$$\text{sulfuric acid}$$

Because of this reaction, sulfur trioxide has an unusually high water solubility, and its dissolution is essentially irreversible. The acid is sold as a 94% (w/w) aqueous solution (17.6M), illustrative of its remarkable water solubility. Sulfuric acid is often called "battery acid" because of its presence in automobile batteries. It is a dense, oily liquid having no color. It has a great affinity for water, and if spilled on the skin, it will draw water out of the cells to cause very painful chemical burns. As with hydrochloric acid, concentrated sulfuric acid should be treated with great caution.

Nitric Acid

Nitric acid, prepared by allowing nitrogen dioxide to react with water,

$$3\ NO_2(g) + H_2O(l) \longrightarrow 2\ HNO_3(aq) + NO(g)$$
$$\text{nitric acid}$$

is a colorless, highly corrosive liquid. It is sold as a 69% (w/w) aqueous solution, corresponding to 15.4M. On standing, nitric acid decomposes to water, nitrogen dioxide, and oxygen, forming a brownish yellow solution due to the presence of brown NO_2. It is used to make fertilizers and explosives.

Phosphoric Acid

Phosphoric acid is another acid formed by the reaction of an oxide of a nonmetal with water. Such acid-forming oxides are called *acid anhydrides*. In this case, the oxide is tetraphosphorus decaoxide, P_4O_{10}, a solid compound. The acid is a syrupy, colorless liquid sold in concentrated form (85% w/w) corresponding to 14.7M.

$$P_4O_{10}(s) + 6\ H_2O(l) \longrightarrow 4\ H_3PO_4(aq)$$
$$\text{phosphoric acid}$$

acid anhydride (an-HIGH-dride): literally, "acid without hydrogen;" a nonmetallic oxide which reacts with water to form an acid.

It is used commercially in the production of fertilizers and detergents, and it serves as a tart, fruitlike flavoring in food products. Concentrated phosphoric acid is an irritant to skin and mucous membranes, but it is not as hazardous as hydrochloric, sulfuric, and nitric acids.

Acetic Acid

Acetic acid is a colorless liquid with a sour smell. Vinegar, a 5% aqueous solution of acetic acid, gets its taste and odor from the presence of acetic acid. The acid is prepared by the oxidation of ethanol:

$$CH_3CH_2OH(l) \xrightarrow{\text{oxidation}} CH_3C{\overset{O}{\underset{OH}{\diagup\!\!\diagdown}}}(l)$$
$$\text{ethanol} \qquad \text{acetic acid}$$

It can be purchased as a 99.5% (w/w) aqueous solution (17.4M). Although it is safe for ingestion in dilute solutions such as vinegar, concentrated acetic acid will damage living tissue.

Carbonic Acid

Carbonic acid forms when the acid anhydride, carbon dioxide, reacts with water. In this case the formation of acid occurs to only a small extent in the reversible reaction:

$$CO_2(g) + H_2O(l) \rightleftharpoons H_2CO_3(aq)$$
$$\text{carbonic acid}$$

Carbonic acid is unstable and does not exist in pure form. When carbon dioxide dissolves in water, 99% stays in solution as hydrated CO_2 molecules, and about 1% reacts to form carbonic acid. Because of the presence of carbonic acid, carbonated beverages have a slightly sour taste. Carbonic acid solutions are so dilute as to be considered relatively safe to handle.

Sodium Hydroxide

Sodium hydroxide, commonly called caustic soda or lye, is a white solid which forms when sodium oxide reacts with water:

$$Na_2O(s) + H_2O(l) \longrightarrow 2\ NaOH(s)$$
$$\text{sodium hydroxide}$$

Since sodium hydroxide is a base, sodium oxide is referred to as a *basic anhydride*. Both compounds are highly injurious to the skin and must be handled with care. Concentrated aqueous solutions of sodium hydroxide can cause blindness if they come into contact with the eyes. Sodium hydroxide is used in textile manufacturing, petroleum refining, soap manufacture, pulp and paper production, food manufacture, and other industrial processes. It is an active ingredient in oven and drain cleaners. A closely related base, potassium hydroxide (KOH), has properties very similar to those of NaOH.

basic anhydride: literally, "base without hydrogen;" a metallic oxide which reacts with water to form a base.

Ammonium Hydroxide

Ammonium hydroxide forms when ammonia dissolves in water:

$$NH_3(g) + H_2O(l) \rightleftharpoons NH_4^+(aq) + OH^-(aq)$$

Ammonia has an extraordinarily high water solubility, and dissolution is accompanied by the formation of ammonium hydroxide. However, despite its high water solubility, very little dissolved ammonia enters the reaction to form ammonium hydroxide. Although aqueous solutions of NH_3 are usually called ammonium hydroxide, the compound NH_4OH has never been isolated. An alternate name for aqueous solutions of NH_3 is "aqua ammonia." Ammonia is sold as a 27% (w/w) aqueous solution (14.3M), and as such it can cause burns on contact with the skin. Ordinary household ammonia, used as a cleaner, is dilute ammonium hydroxide.

Strong and Weak Acids

Acids can be designated as strong or weak depending on the extent of dissociation into ions when dissolved in water. Consider the ionization of an acid, HA, in water:

$$HA(aq) + H_2O(l) \rightleftharpoons H_3O^+(aq) + A^-(aq)$$

The equilibrium expression for this reaction is as follows:

$$K_{eq} = \frac{[H_3O^+][A^-]}{[HA][H_2O]}$$

PERSPECTIVE: Acid Rain

Burning Fossil Fuels Increases Acidity of Rain

In 1959 a Norwegian fisheries inspector connected the decline in Norwegian and Swedish fish populations to the increasing acidity of rain and snow. The source of the so-called acid rain was traced to Europe's heavily industrialized areas, and the situation has worsened to the point that hundreds of Scandinavian lakes no longer support fish. In 1972 acid rain had spread to North America, and fish populations were being wiped out in Canada and the eastern United States. The spread of acid rain continued, and by 1979, biologists had discovered that streams in the Colorado Rocky Mountains were tainted with acids.

The airborne acidity is a result of automobile exhausts and the burning of fossil fuels such as coal, oil, and natural gas. The high temperatures in automobile engines and other combustion processes cause oxidation of atmospheric nitrogen to form nitrogen dioxide (NO_2). When fossil fuels are burned, sulfur compounds in them are oxidized to sulfur dioxide (SO_2) and sulfur trioxide (SO_3). These oxides are acid anhydrides, and they react with moisture in the air to form acids:

$$3\ NO_2(g) + H_2O(l) \longrightarrow 2\ HNO_3(aq) + NO(g)$$
nitric acid

$$SO_2(g) + H_2O(l) \longrightarrow H_2SO_3(aq)$$
sulfurous acid

$$SO_3(g) + H_2O(l) \longrightarrow H_2SO_4(aq)$$
sulfuric acid

As the acidic rain and snow fall, they raise the acid levels of lakes, rivers, and streams, and they cause much damage to agricultural crops, building exteriors, automobile finishes, and human health.

The concentration of water in most aqueous solutions is very nearly the same as for pure water, $55.5M$, and we can consider this value to be a constant. Thus, if we combine the concentration of water with K_{eq}

$$K_{eq}[H_2O] = \frac{[H_3O^+][A^-]}{[HA]}$$

we can derive an expression for a new constant designated as K_a, the *ionization constant* for the acid:

$$K_a = \frac{[H_3O^+][A^-]}{[HA]}$$

(An alternate name is acid dissociation constant.) K_a's have units of molarity, but these are seldom stated.

acid ionization constant, K_a (acid dissociation constant): the ratio of the product of molar concentrations of dissociated ions to the molar concentration of undissociated acid in an acid ionization equilibrium.

Table 9-3 Common Strong Acids

Acid	Formula
Hydrochloric acid	HCl
Hydrobromic acid	HBr
Hydroiodic acid	HI
Nitric acid	HNO_3
Perchloric acid	$HClO_4$
Chromic acid	H_2CrO_4
Sulfuric acid	H_2SO_4

The ionization constant for an acid is similar to an equilibrium constant in that it indicates the position of equilibrium for acid dissociation into ions. Strong acids are those for which the position of equilibrium is far to the right, giving a large value for K_a. The word *strong* is not very precise, but we usually classify *strong acids* as those having ionization constants greater than 1 ($K_a > 1$). Some strong acids are listed in Table 9-3.

Weak acids are those having K_a's of 10^{-1} or less. These acids are only slightly ionized in water, and thus are less harmful to tissue than strong acids. A number of weak acids are listed in Table 9-4.

It should be noted that strong acids may have high concentrations or low concentrations, but they are always strong acids. The same is true for weak acids; regardless of their concentrations, they are still weak acids. Acid strength refers to K_a values, not solution concentrations.

Table 9-5 provides examples of acid-base conjugate pairs. Consider one such pair, hydrofluoric acid and fluoride anion:

$$HF(aq) + H_2O(l) \rightleftharpoons H_3O^+(aq) + F^-(aq), \quad K_a = 6.8 \times 10^{-4}$$
hydrofluoric acid

Hydrofluoric acid is a weak acid as indicated by its K_a. Thus, in the forward reaction, only a few molecules of hydrofluoric acid transfer protons to water, in turn forming only a few fluoride ions. But if we consider the reverse reaction, fluoride ion acts as a proton acceptor in taking the proton from H_3O^+ to form HF. Hence, F^- is a base; in fact, it is the conjugate base of HF. And it must be a stronger base (proton

Table 9-4 Common Weak Acids

Acid	Formula
Acetic acid	CH_3CO_2H
Boric acid	H_3BO_3
Carbonic acid	H_2CO_3
Phosphoric acid	H_3PO_4
Sulfurous acid	H_2SO_3

Table 9-5 Acid-Base Conjugate Pairs

Acid	Formula	Conjugate Base	K_a (25°C)
Acetic acid	CH_3CO_2H	$CH_3CO_2^-$	1.8×10^{-5}
Carbonic acid	H_2CO_3	HCO_3^-	4.3×10^{-7}
Formic acid	HCO_2H	HCO_2^-	1.7×10^{-4}
Hydrocyanic acid	HCN	CN^-	4.9×10^{-10}
Hydrofluoric acid	HF	F^-	6.8×10^{-4}
Nitrous acid	HNO_2	NO_2^-	4.5×10^{-4}
Phosphoric acid	H_3PO_4	$H_2PO_4^-$	7.5×10^{-3}
Sulfurous acid	H_2SO_3	HSO_3^-	1.3×10^{-2}
Base	**Formula**	**Conjugate Acid**	K_b (25°C)
Ammonia	NH_3	NH_4^+	1.8×10^{-5}
Carbonate ion	CO_3^{2-}	HCO_3^-	2.1×10^{-4}
Hypochlorite ion	ClO^-	$HClO$	3.1×10^{-7}

acceptor) than H_2O, because more HF exists in the equilibrium mixture than H_3O^+. We can also say that HF is a weaker acid than water, because most HF molecules are not dissociated in water. According to our reasoning,

HF is a weaker acid than H_2O.

F^- is a stronger base than H_2O.

HF and F^- are an acid-base conjugate pair.

By use of water as our standard for comparison, we can make a more general statement: *The weaker an acid, the stronger its conjugate base will be; and the stronger an acid, the weaker its conjugate base will be.* These relative strengths and weaknesses are indicated in Table 9-5.

EXAMPLE 9.2

In a solution of hydrofluoric acid, the following equilibrium concentrations existed:

$$[HF] = 0.0921 M$$
$$[H_3O^+] = 7.9 \times 10^{-3} M$$
$$[F^-] = 7.9 \times 10^{-3} M$$

Calculate K_a, the ionization constant for hydrofluoric acid.

SOLUTION

Since hydrofluoric acid ionizes as shown below,

$$HF(aq) + H_2O(l) \rightleftharpoons H_3O^+(aq) + F^-(aq)$$

we can write the expression for the ionization constant

$$K_a = \frac{[H_3O^+][F^-]}{[HF]}$$

Substituting the given values for concentrations,

$$K_a = \frac{(7.9 \times 10^{-3})(7.9 \times 10^{-3})}{0.0921}$$

$$= 6.8 \times 10^{-4}$$

EXERCISE 9.2 In a solution of a certain weak acid, HA, the following equilibrium concentrations existed:

$$[HA] = 0.10M$$
$$[H_3O^+] = 2.0 \times 10^{-3}M$$
$$[A^-] = 2.0 \times 10^{-3}M$$

Calculate K_a, the ionization constant for the weak acid.

EXAMPLE 9.3

Hydrofluoric acid has a K_a of 6.8×10^{-4}. If the equilibrium concentration of HF(aq) is 0.20M, what is the concentration of H_3O^+(aq)?

SOLUTION

Hydrofluoric acid ionizes as follows:

$$HF(aq) + H_2O(l) \rightleftharpoons H_3O^+(aq) + F^-(aq)$$

Since HF dissociates to form one H_3O^+ for every F^-, we know that

$$[H_3O^+] = [F^-]$$

Then, substituting into the expression for K_a,

$$K_a = \frac{[H_3O^+][F^-]}{[HF]}$$

$$= \frac{[H_3O^+]^2}{[HF]}$$

and solving for $[H_3O^+]$,

$$[H_3O^+]^2 = K_a[HF]$$
$$[H_3O^+] = \sqrt{K_a[HF]}$$
$$= \sqrt{(6.8 \times 10^{-4})(0.20)}$$
$$= \sqrt{1.36 \times 10^{-4}}$$
$$= 1.2 \times 10^{-2}M$$

EXERCISE 9.3 Hydrocyanic acid has a K_a of 4.9×10^{-10}. If the equilibrium concentration of HCN(aq) is 0.15M, what is the concentration of H_3O^+(aq)?

$$HCN(aq) + H_2O(l) \rightleftharpoons H_3O^+(aq) + CN^-(aq)$$

Examples of Strong and Weak Acids

Hydrochloric acid is a strong acid because of the ease with which water molecules pull protons from the very polar HCl molecule. Hydrochloric acid is secreted by gastric glands in the stomach to aid in digestion of food. Even though such harsh acidic conditions exist in the stomach, mucous secretions protect the stomach lining from harm.

You will notice in Tables 9-3 and 9-4 that several acids are capable of furnishing more than one proton from each formula unit to water. These are called *polyprotic acids*. They react with water to transfer protons in a stepwise manner. Two ionization steps are associated with sulfuric acid, for example:

$$H_2SO_4 \rightleftharpoons H^+ + HSO_4^-, \quad K_a = 1 \times 10^3$$
$$HSO_4^- \rightleftharpoons H^+ + SO_4^{2-}, \quad K_a = 1.2 \times 10^{-2}$$

Note that the first step has an ionization constant corresponding to a strong acid, but the second step has a value characteristic of a weak acid. Thus the second dissociation occurs less readily than the first. Sulfuric acid is considered a strong acid because of its first dissociation. The H_2SO_4 molecule loses a proton to water readily because the sulfur and oxygen atoms draw bonding electrons away from the hydrogen atoms (remember that S and O are quite electronegative), causing the O—H bonds to be highly polarized:

$$\begin{array}{c} O \\ \parallel \\ H-O-S-O-H \\ \parallel \\ O \end{array}$$

sulfuric acid

A proton can be lost from either OH group in sulfuric acid with equal probability. However, after the hydrogen sulfate anion is formed, its negative charge acts to satisfy partially the electronegativities of the sulfur and oxygen atoms, and the remaining O—H bond is only weakly polar:

$$\begin{array}{c} O \\ \parallel \\ H-O-S-O^- \\ \parallel \\ O \end{array}$$

hydrogen sulfate

Thus, HSO_4^- is a weak acid. Another way of looking at the situation is that adding a second negative charge to HSO_4^- requires a large amount of free energy, and thus SO_4^{2-} forms only to a small extent in the equilibrium.

Another common polyprotic acid is phosphoric acid, H_3PO_4. In this case, proton dissociation occurs in three steps:

$$H_3PO_4 \rightleftharpoons H^+ + H_2PO_4^-, \quad K_a = 7.5 \times 10^{-3}$$
$$H_2PO_4^- \rightleftharpoons H^+ + HPO_4^{2-}, \quad K_a = 6.2 \times 10^{-8}$$
$$HPO_4^{2-} \rightleftharpoons H^+ + PO_4^{3-}, \quad K_a = 2.2 \times 10^{-13}$$

The K_a for the first step indicates that H_3PO_4 is a much weaker acid than H_2SO_4. This is because phosphorus is less electronegative than sulfur. Thus the O—H bonds in H_3PO_4 are not as polar as those in H_2SO_4 and other strong *oxyacids*. When the first proton dissociates from H_3PO_4, it can come from any one of the three OH groups with equal probability. The resulting negative charge on dihydrogen phosphate depolarizes the remaining O—H bonds, making this anion still a weaker acid. Similarly, the two negative charges on hydrogen phosphate make it an even weaker acid than dihydrogen phosphate.

oxyacid: an acid in which one or more OH groups and possibly additional oxygen atoms are bonded to a central atom.

```
      O              O              O              O
      ||             ||             ||             ||
H—O—P—O—H      H—O—P—O⁻       ⁻O—P—O⁻        ⁻O—P—O⁻
      |              |              |              |
      O              O              O              O⁻
      |              |              |
      H              H              H
  phosphoric     dihydrogen       hydrogen       phosphate
     acid         phosphate       phosphate
```

Each of the anions formed is a conjugate base and is thus capable of accepting a proton from an acid. As you shall see in chapter 10, two of the anions, $H_2PO_4^-$ and HPO_4^{2-}, play an important role in blood.

Carbonic acid, H_2CO_3, is another example of a polyprotic acid. In this case the central atom, carbon, is less electronegative than sulfur and phosphorus. Thus H_2CO_3 is a weaker acid than H_3PO_4. The bicarbonate anion, HCO_3^-, also has an important function in blood.

$$H_2CO_3 \rightleftharpoons H^+ + HCO_3^-, \quad K_a = 4.3 \times 10^{-7}$$
$$HCO_3^- \rightleftharpoons H^+ + CO_3^{2-}, \quad K_a = 5.6 \times 10^{-11}$$

```
      O              O              O
      ||             ||             ||
H—O—C—O—H      H—O—C—O⁻       ⁻O—C—O⁻
   carbonic       bicarbonate*     carbonate
     acid
```

Acetic acid, another weak acid, is an example of a *carboxylic acid*. These acids are characterized as having the general structure

```
          O
         //
    X—C
         \
          O—H
         └────┘
         carboxyl
          group
```

carboxylic (car-box-SIH-lic) **acid:** an acid containing the carboxyl group,

$$-C\overset{\displaystyle O}{\underset{\displaystyle OH}{\diagup\!\!\!\diagdown}}$$

Most acids present in living organisms are carboxylic acids, and generally speaking, they are weak acids. The reason for their weak acidity lies in their structural features. In acetic acid, for example, the CH_3 group (X in the general structure) is certainly not very electronegative nor is the carbon atom in the carboxyl group:

---

*Even though hydrogen carbonate is the systematic name for this anion, we will use the more popular name bicarbonate.

$$\underset{\text{acetic acid}}{\underset{H}{\overset{H}{\mid}}{H-\underset{|}{\overset{|}{C}}-C\underset{O-H}{\overset{O}{\diagup}}}} \quad \text{or} \quad \begin{array}{c} CH_3CO_2H \\ HC_2H_3O_2 \end{array}$$

However, the double-bonded oxygen atom withdraws electrons strongly enough to allow proton dissociation to proceed to a small extent in water:

$$CH_3-C\underset{O-H}{\overset{O}{\diagup}} \rightleftharpoons H^+ + CH_3-C\underset{O^-}{\overset{O}{\diagup}}, \quad K_a = 1.8 \times 10^{-5}$$

Strong and Weak Bases

Bases can also be classified as strong or weak, depending on their abilities to provide hydroxide ions in aqueous solution. The soluble metal hydroxides NaOH and KOH are ionic solids which dissociate completely in water to form basic solutions:

$$NaOH(s) \xrightarrow{\text{water}} Na^+(aq) + OH^-(aq)$$

$$KOH(s) \xrightarrow{\text{water}} K^+(aq) + OH^-(aq)$$

Thus NaOH and KOH are *strong bases*. Few other metal hydroxides are very soluble in water; for example, $Mg(OH)_2$ is water soluble only to the extent of 0.011 g/L. For this reason, suspensions of $Mg(OH)_2$ (sold as milk of magnesia) are taken to alleviate discomfort caused by high levels of acid in the stomach (a condition called hyperacidity). The $Mg(OH)_2$ slowly dissolves as it neutralizes stomach acid, and thus very few hydroxide ions are present in solution in the digestive tract at any time.

Aside from the metal hydroxides, there are a great many other bases, most of which do not contain hydroxide ions. Ammonia is a common example. It is a base because it reacts with water to form hydroxide ions:

$$NH_3(g) + H_2O(l) \rightleftharpoons NH_4^+(aq) + OH^-(aq)$$

Since the reaction does not proceed very far to the right, ammonia is a *weak base*.

weak base: a base for which K_b is 10^{-1} or less.

For bases like ammonia which do not have hydroxide as part of their structures, we can write the following general equation:

$$B + H_2O(l) \rightleftharpoons BH^+(aq) + OH^-(aq)$$

The expression for the equilibrium constant is

$$K_{eq} = \frac{[BH^+][OH^-]}{[B][H_2O]},$$

and the ionization constant for a base, K_b, is

$$K_b = K_{eq}[H_2O] = \frac{[BH^+][OH^-]}{[B]}$$

(The ionization constant for a base is also called the base dissociation constant.) In these cases, the proton acceptor (B) may be a neutral molecule, such as NH_3, or it may be an anion.

EXAMPLE 9.4

In a solution of ammonium hydroxide, the following equilibrium concentrations existed:

$$NH_3(g) + H_2O(l) \rightleftharpoons NH_4^+(aq) + OH^-(aq)$$

$$[NH_3] = 0.10 M$$
$$[NH_4^+] = 1.35 \times 10^{-3} M$$
$$[OH^-] = 1.35 \times 10^{-3} M$$

Calculate K_b, the ionization constant for ammonium hydroxide.

SOLUTION

The expression for K_b is

$$K_b = \frac{[NH_4^+][OH^-]}{[NH_3]}$$

$$= \frac{(1.35 \times 10^{-3})(1.35 \times 10^{-3})}{0.10}$$

$$= 1.8 \times 10^{-5}$$

EXERCISE 9.4 A weak base, B, reacts with water as follows:

$$B(aq) + H_2O(l) \rightleftharpoons BH^+(aq) + OH^-(aq)$$

From the following equilibrium concentrations, calculate K_b.

$$[B] = 0.40 M$$
$$[BH^+] = 7.7 \times 10^{-8} M$$
$$[OH^-] = 7.7 \times 10^{-8} M$$

BH^+, a protonated base, is actually an acid:

$$BH^+ + H_2O(l) \rightleftharpoons B(aq) + H_3O^+(aq)$$

Thus, an expression for the ionization constant for the acid can be written as follows, and K_a values can be determined:

$$K_a = K_{eq}[H_2O] = \frac{[B][H_3O^+]}{[BH^+]}$$

An example of an anion that is a weak base is the bicarbonate ion, HCO_3^-, which reacts with water:

$$HCO_3^-(aq) + H_2O(l) \rightleftharpoons H_2CO_3(aq) + OH^-(aq)$$

Thus, an aqueous solution of a bicarbonate salt will actually be basic. Sodium bicarbonate, $NaHCO_3$ (known as baking soda), is often an ingredient in antacids. These nonprescription medications are used to combat hyperacidity in the stomach. Hydroxide ions, formed when sodium bicarbonate dissolves in water, neutralize the excess stomach acid. The characteristic belching following ingestion of sodium bicarbonate is caused by formation of carbon dioxide from carbonic acid:

$$H_2CO_3(aq) \rightleftharpoons H_2O(l) + CO_2(g)$$

We will classify weak bases as those having ionization constants of 10^{-1} or less. Some of these are listed in Table 9-6. As you can see, the amines are the most common weak bases. These are compounds in which the hydrogen atoms of the ammonia molecule are replaced by hydrocarbon groups. Amines react with water to produce hydroxide ions as illustrated in Table 9-7.

Note that weak bases have the capability of removing a proton from water to produce hydroxide ion. Each weak base has a structure in which there is an atom with nonbonded electrons. These nonbonded electrons allow the base to form a coordinate covalent bond with a hydrogen ion:

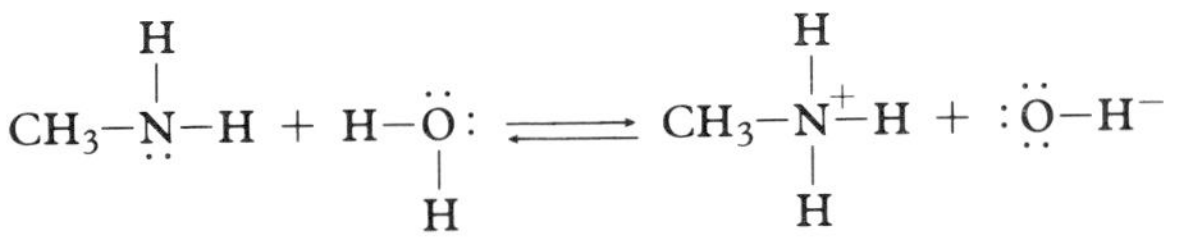

$$CH_3-\underset{H}{\underset{|}{\overset{H}{\overset{|}{N}}}}-H + H-\ddot{O}\!:\!\underset{H}{\underset{|}{{}}} \rightleftharpoons CH_3-\underset{H}{\underset{|}{\overset{H}{\overset{|}{\overset{+}{N}}}}}-H + :\ddot{O}-H^-$$

Normality

Acids react with bases as illustrated below. In this reaction one mole of hydrogen chloride reacts with one mole of ammonia to produce one mole of ammonium chloride:

$$HCl(g) + NH_3(g) \longrightarrow NH_4Cl(s)$$

Similarly, one mole of hydrochloric acid reacts with one mole of sodium hydroxide to produce one mole each of sodium chloride and water:

$$HCl(aq) + NaOH(aq) \longrightarrow NaCl(aq) + H_2O(l)$$

Let's examine the reaction of sodium hydroxide with a polyprotic acid. In this case, one mole of sulfuric acid reacts with two moles of sodium hydroxide:

$$H_2SO_4(aq) + 2\,NaOH(aq) \longrightarrow Na_2SO_4(aq) + 2\,H_2O(l)$$

Table 9-6 Common Weak Bases

Base	Formula	K_b (25°C)
Ammonia	NH_3	1.8×10^{-5}
Carbonate ion	CO_3^-	2.1×10^{-4}
Cyanide ion	CN^-	1.6×10^{-5}
Triethylamine	$(C_2H_5)_3N$	5.2×10^{-4}
Trimethylamine	$(CH_3)_3N$	6.3×10^{-5}

Table 9-7 The Reaction of Ammonia and Amines with Water

$$H-\underset{\cdot\cdot}{\overset{H}{\underset{|}{N}}}-H + H_2O \rightleftharpoons \left[H-\underset{H}{\overset{H}{\underset{|}{N}}}-H\right]^{+} + OH^{-}$$

ammonia

$$CH_3-\underset{\cdot\cdot}{\overset{H}{\underset{|}{N}}}-H + H_2O \rightleftharpoons \left[CH_3-\underset{H}{\overset{H}{\underset{|}{N}}}-H\right]^{+} + OH^{-}$$

methylamine

$$CH_3-\underset{\cdot\cdot}{\overset{H}{\underset{|}{N}}}-CH_3 + H_2O \rightleftharpoons \left[CH_3-\underset{H}{\overset{H}{\underset{|}{N}}}-CH_3\right]^{+} + OH^{-}$$

dimethylamine

$$CH_3-\underset{\cdot\cdot}{\overset{CH_3}{\underset{|}{N}}}-CH_3 + H_2O \rightleftharpoons \left[CH_3-\underset{H}{\overset{CH_3}{\underset{|}{N}}}-CH_3\right]^{+} + OH^{-}$$

trimethylamine

Thus one mole of sulfuric acid has the same neutralizing power as two moles of hydrochloric acid. We can say that one mole of sulfuric acid is equivalent to two moles of hydrochloric acid in neutralizing power. In acid-base neutralization reactions, it is sometimes more useful to think in terms of equivalent amounts instead of molar amounts. An *equivalent* of an acid is the amount that neutralizes one mole of hydroxide ions, and an equivalent of base is the amount which neutralizes one mole of protons. From these definitions, it is possible to calculate the number of equivalents in one mole of an acid or base. For acids having one dissociable proton (monoprotic acids), one mole of acid will neutralize one mole of hydroxide ions, and thus one mole contains one equivalent of acid. But for polyprotic acids, one mole contains more than one equivalent. A mole of a diprotic acid (two dissociable protons) will neutralize two moles of hydroxide ion, and thus one mole of a diprotic acid contains two equivalents of acid. Similarly, a mole of a triprotic acid (three dissociable protons) contains three equivalents of acid.

The number of equivalents in a mole of base is calculated by considering the number of moles of protons which can be neutralized by one mole of base. Thus, one mole of NaOH will neutralize one mole of protons, and one mole of NaOH contains one equivalent of base. For a base like $Ca(OH)_2$, one mole of base neutralizes two moles of H^+, and thus one mole of $Ca(OH)_2$ contains two equivalents of base. These relationships are summarized in Table 9-8.

A solution containing one equivalent of a substance in one liter of solution is called a one-normal ($1N$) solution. *Normality* (N) is simply

PERSPECTIVE: The Chemistry of Antacids

In the United States, several million dollars are spent each year on the purchase of over-the-counter antacids. Although the number of different antacid preparations seems almost infinite, as indicated by numerous advertisements, only a few active ingredients are involved. Most antacids contain just one or two of the following bases: sodium bicarbonate, $NaHCO_3$; calcium carbonate, $CaCO_3$; magnesium carbonate, $MgCO_3$; aluminum hydroxide, $Al(OH)_3$; magnesium hydroxide, $Mg(OH)_2$; and magnesium trisilicate, $Mg_2Si_3O_8$. Table A illustrates the action of these ingredients in neutralizing stomach acid. Table B lists the active ingredients in commercially available antacids.

Antacids are formulated to appeal to a number of personal preferences. For example, effervescent antacids contain sodium bicarbonate and a small amount of carboxylic acid, usually either citric acid or tartaric acid. These ingredients are present as a solid mixture, but when the antacid is added to water, the sodium bicarbonate and acid dissolve, and the reaction between the two forms carbonic acid. Decomposition of carbonic acid to carbon dioxide causes the familiar fizzing.

Plain baking soda (sodium bicarbonate) is often used as an antacid; however, it should be used only occasionally as it has a tendency to make the blood too basic. In addition, baking soda and antacids containing sodium bicarbonate should be avoided by individuals on low-sodium diets.

Table A Reactions of Bases in Antacids

HCO_3^- + H^+ ⟶ H_2CO_3 ⟶ H_2O + CO_2
bicarbonate as $NaHCO_3$ or $KHCO_3$ carbonic acid (unstable)

OH^- + H^+ ⟶ H_2O
hydroxide as $Mg(OH)_2$, $Al(OH)_3$, or $NaAl(OH)_2CO_3$

CO_3^{2-} + $2H^+$ ⟶ H_2CO_3 ⟶ H_2O + CO_2
carbonate as $CaCO_3$, $MgCO_3$, or $NaAl(OH)_2CO_3$ carbonic acid (unstable)

$Si_3O_8^{4-}$ + $4H^+$ ⟶ $3\,SiO_2$ + $2\,H_2O$
trisilicate as $Mg_2Si_3O_8$ silicon dioxide

$C_6H_5O_7^{3-}$ + $3H^+$ ⟶ $H_3C_6H_5O_7$
citrate as $Na_3C_6H_5O_7$ citric acid

$C_4H_4O_6^{2-}$ + $2H^+$ ⟶ $H_2C_4H_4O_6$
tartrate as $Na_2C_4H_4O_6$ tartaric acid

Table B Active Ingredients in Commercial Antacids

Product Name	Active Ingredients
Alka-Seltzer Blue	$NaHCO_3$, citric acid, aspirin
Alka-Seltzer Gold	$NaHCO_3$, citric acid
Bromo-Seltzer	$NaHCO_3$, citric acid, acetaminophen
Milk of Magnesia	$Mg(OH)_2$
Tums	$CaCO_3$
Alka 2	$CaCO_3$
Maalox	$Mg(OH)_2$, $Al(OH)_3$
Mylanta	$Mg(OH)_2$, $Al(OH)_3$, defoaming agent
Di-Gel	$Mg(OH)_2$, $Al(OH)_3$, defoaming agent
Bisodol	$NaHCO_3$, $MgCO_3$
Bisodol (low sodium)	$Mg(OH)_2$, $CaCO_3$
Rolaids	$NaAl(OH)_2CO_3$

Table 9-8 Equivalents of Common Acids and Bases

Acid	Number of Dissociable Protons	Equivalents/Mole
HCl	1	1
CH_3CO_2H	1	1
H_2SO_4	2	2
H_3PO_4	3	3

Base	Moles of Protons Neutralized by One Mole of Base	Equivalents/Mole
NaOH	1	1
KOH	1	1
$Ca(OH)_2$	2	2

concentration expressed as number of equivalents (eq) per liter of solution. Thus a solution having two equivalents per liter is a 2N (two-normal) solution, and 500 mL of a 2N solution contains one equivalent of solute. Normality can be expressed as

$$N = \text{eq/L}$$

EXAMPLE 9.5

What is the normality of a hydrochloric acid solution containing 50.0 g of HCl per liter?

SOLUTION

Since normality is defined as the number of equivalents per liter, we must first find the number of equivalents in one mole of HCl. Since there is only one dissociable proton,

$$1 \text{ eq} = 1 \text{ mole}$$

Then,

$$N = \text{eq/L}$$
$$= \left(\frac{50.0 \text{ g}}{L}\right)\left(\frac{1 \text{ mole}}{36.5 \text{ g}}\right)\left(\frac{1 \text{ eq}}{\text{mole}}\right)$$
$$= 1.37 \text{ eq/L}$$

EXERCISE 9.5 What is the normality of a NaOH solution containing 120 g of NaOH per liter?

EXAMPLE 9.6

How many grams of sulfuric acid, H_2SO_4, are in 1.00 liter of a 1.50N solution?

SOLUTION

We solve this problem by converting the number of equivalents per liter to grams per liter:

$$\frac{g}{L} = \left(\frac{\text{eq}}{L}\right)\left(\frac{g}{\text{eq}}\right) = \left(\frac{\text{eq}}{L}\right)\left(\frac{\text{mole}}{\text{eq}}\right)\left(\frac{g}{\text{mole}}\right)$$

Since sulfuric acid has two dissociable protons, 1 mole contains 2 equivalents:

$$1 \text{ mole} = 2 \text{ eq}$$

Then,

$$\frac{g}{L} = \left(\frac{1.50 \text{ eq}}{L}\right)\left(\frac{1 \text{ mole}}{2 \text{ eq}}\right)\left(\frac{98.1 \text{ g}}{\text{mole}}\right)$$

$$= 73.6 \text{ g/L}$$

Thus, 1.00 liter of solution contains 73.6 g H_2SO_4.

EXERCISE 9.6 How many grams of phosphoric acid, H_3PO_4, are in 500 mL of a 6.00N solution?

For acids and bases present in dilute solutions, it is sometimes useful to state their concentrations in terms of *milliequivalents* per liter, meq/L, where 1 meq = 0.001 eq.

EXAMPLE 9.7

How many meq are in 1.00 mL of a 1.00N solution?

SOLUTION

$$\text{meq/mL} = \left(\frac{1.00 \text{ eq}}{L}\right)\left(\frac{1 \text{ L}}{1000 \text{ mL}}\right)\left(\frac{\text{meq}}{0.001 \text{ eq}}\right)$$

$$= 1.00 \text{ meq/mL}$$

This example illustrates that a 1.00N solution contains 1 equivalent per liter and 1 milliequivalent per milliliter.

EXERCISE 9.7 How many milliequivalents are in 20.0 mL of a 0.100N solution?

For a solution of acid or base whose molarity is known, the normality can easily be calculated. In these cases, the normality is simply the molarity multiplied by the number of equivalents per mole of acid or base.

$$N = \text{moles/L} \times \text{eq/mole}$$

Thus, a 1M solution of HCl would also be a 1N solution, but a 1M solution of H_2SO_4 would be a 2N solution. Similarly, a 1M solution of NaOH would be a 1N solution, but a 1M solution of $Ca(OH)_2$ would be a 2N solution.

EXAMPLE 9.8

What is the normality of a 4.0M sulfuric acid solution?

SOLUTION

Since there are two equivalents of sulfuric acid per mole,

$$N = (4.0 \text{ moles/L})(2 \text{ eq/mole})$$

$$N = 8.0 \text{ eq/L}$$

EXERCISE 9.8 What is the normality of a 0.02M solution of $Ca(OH)_2$?

Acid-Base Titrations

Acid-base neutralization reactions are used to determine the concentration of an aqueous acid or base by the procedure known as *titration*, which is illustrated in Figure 9-3. Titrations are performed as follows:

1. Measure a known volume of the acid and place it in a flask. Add a few drops of *acid-base indicator* (a substance which is itself an acid or a base and which is one color in acid solution and another color in basic solution).

2. Add a solution of base of known concentration (called standard base) to the acid by using a buret. (A buret is a long transparent tube calibrated in units of milliliters from which a solution, the titrant, can be added to another solution.)

3. Add the base slowly to the point where one drop causes a change in the color of the indicator. At this point, the end point, all of the acid in the flask has been neutralized, and there is just the slightest excess of base in the flask. This is why the indicator changes color.

By the end of titration, just enough base has been added to neutralize the acid; this end point is called the *equivalence point*. The number of equivalents of base added are then exactly equal to the number of equivalents of acid originally in the flask. Thus, at the equivalence

titration: a procedure for determining concentration of an acid or base on the basis of equivalents required for neutralization.

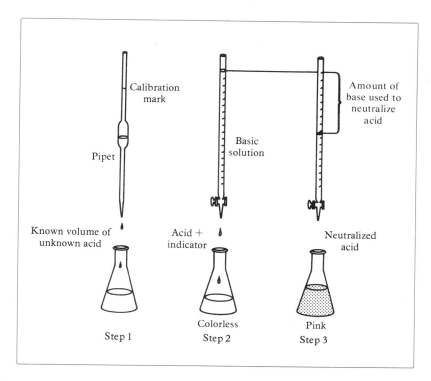

Fig. 9-3 Titration of an Acid with a Base

point, equivalents of base = equivalents of acid. Since we know the normality of the standard base, we can calculate the number of equivalents required to reach the equivalence point:

$$\text{eq of base} = L \times \text{eq/L} = V \times N$$

Since the number of equivalents of base is equal to the number of equivalents of acid at the equivalence point,

$$V_b \times N_b = V_a \times N_a$$

where V_b is the volume of base added, N_b is the normality of the base, V_a is the original volume of acid placed in the flask, and N_a is the unknown normality the acid, the quantity we are seeking. We can solve the equation for N_a, and by substituting known values for V_b, N_b, and V_a, we can calculate N_a:

$$N_a = \frac{V_b \times N_b}{V_a}$$

Concentrations of bases may be determined similarly, except that an acid titrant of known concentration is used. Volume may be expressed in milliliters or in liters, as long as the same units are used for both volumes in the calculation.

EXAMPLE 9.9

A 25.0-mL sample of hydrochloric acid required 37.3 mL of a 0.200N solution of NaOH to reach the end point of a titration. What was the normality of the HCl solution?

SOLUTION

$$V_b \times N_b = V_a \times N_a$$

$$N_a = \frac{V_b \times N_b}{V_a}$$

$$= \frac{(37.3 \text{ mL})(0.200 \text{ eq/L})}{25.0 \text{ mL}}$$

$$= 0.298 \text{ eq/L}$$

EXERCISE 9.9 A 20.0-mL sample of sulfuric acid required 26.2 mL of a 0.100N solution of NaOH to reach the end point of a titration. What was the normality of the H_2SO_4 solution?

EXAMPLE 9.10

You are given a 50.0 mL of a 0.400N solution of sulfuric acid. How many mL of a 0.300N NaOH solution would be required to titrate the sulfuric acid?

SOLUTION

$$V_b \times N_b = V_a \times N_a$$

$$V_b = \frac{V_a \times N_a}{N_b}$$

$$= \frac{(50.0 \text{ mL})(0.400 \text{ eq/L})}{0.300 \text{ eq/L}}$$

$$= 66.7 \text{ mL}$$

EXERCISE 9.10 How many mL of a $0.150N$ NaOH solution will be required to titrate a 25.0-mL solution of $0.250N$ HCl?

SALTS

When acids and bases neutralize each other, one product is always an ionic compound called a *salt*. Hydrogen chloride and ammonia produce the salt called ammonium chloride:

$$HCl(g) + NH_3(g) \longrightarrow NH_4Cl(s)$$

However, if the base contains hydroxide ion, then water is also formed as a product along with the salt, as illustrated by the reaction of hydrochloric acid and sodium hydroxide:

$$HCl(aq) + NaOH(aq) \longrightarrow NaCl(aq) + H_2O(l)$$

Thus, salts are composed of the cations of bases and the anions of acids.

If we rewrite the last equation to show formulas for the dissolved ions, we get a *complete ionic equation*:

$$H^+(aq) + Cl^-(aq) + Na^+(aq) + OH^-(aq) \longrightarrow Na^+(aq) + Cl^-(aq) + H_2O(l)$$

complete ionic equation: an equation which shows all ions and molecules in a reaction.

We have used a single arrow in the equations because neutralization reactions of strong acids and bases go to completion. But notice that sodium and chloride ions did not undergo any actual change. The only chemical change that took place was the combination of H^+ and OH^- to form H_2O. The other ions present that were not involved in any change are called *spectator ions*. If spectator ions are present on both sides of an equation, we can cancel them, and our *net ionic equation* for this neutralization is

net ionic equation: an equation which shows only ions and molecules which undergo change in a reaction.

$$H^+(aq) + OH^-(aq) \longrightarrow H_2O(l)$$

We use net ionic equations when we want to emphasize only the ions which participate in a reaction.

Sodium chloride is very soluble in water. But some salts formed in neutralization reactions are not water soluble. An example is the neutralization of barium hydroxide by sulfuric acid.

$$Ba(OH)_2(aq) + H_2SO_4(aq) \longrightarrow BaSO_4(s) + 2\ H_2O(l)$$

In this case we have no spectator ions, because barium ions react with sulfate ions to form an insoluble material. Thus our net ionic equation is identical to the complete ionic equation:

$$Ba^{2+}(aq) + 2\,OH^-(aq) + 2\,H^+(aq) + SO_4^{2-}(aq) \longrightarrow BaSO_4(s) + 2\,H_2O(l)$$

Salts that are not soluble in water have such strong ionic forces holding them together that water molecules cannot pull many of the ions apart. When ions of these salts find themselves together in the same aqueous solution, they immediately form an insoluble solid called a *precipitate*. We sometimes indicate formation of a precipitate by the use of the symbol ↓ in an equation. Thus, if we mix a barium chloride solution with a sodium sulfate solution, insoluble barium sulfate will form. The complete ionic equation is

$$Ba^{2+}(aq) + 2\,Cl^-(aq) + 2\,Na^+(aq) + SO_4^{2-}(aq) \longrightarrow$$
$$BaSO_4(s)\downarrow + 2\,Cl^-(aq) + 2\,Na^+(aq)$$

The net ionic equation is

$$Ba^{2+}(aq) + SO_4^{2-}(aq) \longrightarrow BaSO_4(s)\downarrow$$

Like all insoluble salts, barium sulfate exists in an equilibrium in water with a very small number of its ions,

$$BaSO_4(s) \rightleftharpoons Ba^{2+}(aq) + SO_4^{2-}(aq)$$

and only trace amounts of the substance can be considered to be dissolved. For this reason, suspensions of barium sulfate can be swallowed. Even though barium ions are quite toxic, the low levels of dissolved ions present in the suspension make it safe for use in taking gastrointestinal X-rays. Because X-rays do not pass through the suspended barium sulfate, a white image of the gastrointestinal tract appears on the photographic film (Figure 9–4).

Fig. 9-4 A Barium Sulfate Gastrointestinal X-ray Taken from the Front. The stomach is the large white image in the upper right and center and the small intestine is to the left of and below the stomach. The spine can be seen on the far left.

It is not possible to make accurate predictions about solubilities of salts, but there are some general guidelines we can use:

1. Salts containing Na^+, K^+, NH_4^+, NO_3^-, and $CH_3CO_2^-$ are usually water soluble.
2. Except for $AgCl$, $PbCl_2$, and Hg_2Cl_2, chloride salts are usually water soluble.
3. Except for $CaSO_4$, $SrSO_4$, $BaSO_4$, Hg_2SO_4, $HgSO_4$, $PbSO_4$, and Ag_2SO_4, sulfate salts are usually water soluble.
4. Most other salts are insoluble in water.

Minerals are naturally occurring ionic substances, usually salts, which are found in the earth's crust. As water flows through the soil it dissolves minerals to the extent allowed by their solubilities. Thus, hard water contains dissolved calcium, magnesium, and iron salts. These metal ions happen to be necessary for proper growth and tissue maintenance in the human body. Other essential dietary minerals are given in Table 9-9. In chapter 23 we will explore the roles played by these minerals in human health.

Table 9-9 Essential Minerals

Major Minerals	Approximate Percentage of Body Weight
Calcium	1.5–2.2
Phosphorus	0.8–1.2
Potassium	0.35
Sulfur	0.25
Sodium	0.15
Chlorine	0.15
Magnesium	0.05
Trace Minerals	
Iron	0.004
Zinc	0.002
Manganese	0.0003
Selenium	0.0003
Copper	0.00015
Iodine	0.00004
Cobalt	trace
Molybdenum	trace
Chromium	trace
Fluorine	trace
Silicon	trace
Vanadium	trace
Nickel	trace
Tin	trace
Arsenic	trace

SUMMARY

Acids and Bases: Under Arrhenius definitions, acids produce protons in water and bases produce hydroxide ions in water. The more general Bronsted-Lowry theory defines acids as proton donors and bases as proton acceptors. Bronsted-Lowry theory also describes acids and bases as conjugate pairs. Names for acids are easily derived from anion names, but there is no special system for naming bases.

In aqueous solutions, acids have a sour taste, turn blue litmus red, and neutralize bases. Bases dissolved in water have a bitter taste and a slippery feeling. They turn red litmus blue and neutralize acids. Six common acids are hydrochloric acid, sulfuric acid, nitric acid, phosphoric acid, acetic acid, and carbonic acid. Three common bases are sodium hydroxide, potassium hydroxide, and ammonium hydroxide. All concentrated acids and bases are extremely damaging to tissue and should be handled with care.

Strong acids and bases are those having ionization constants with values greater than 1. Weak acids and bases have ionization constants of 10^{-1} or less. Examples of strong acids are hydrochloric acid and sulfuric acid. Phosphoric acid is a rather weak acid, and most carboxylic acids are weak acids. Examples of strong bases are sodium hydroxide and potassium hydroxide. Ammonium hydroxide, bicarbonate ion, and amines are weak bases.

Concentrations of acids and bases are often expressed in normality, the number of equivalents per liter of solution. An equivalent of an acid is the amount which neutralizes one mole of hydroxide ions. For bases, an equivalent is the amount that neutralizes one mole of hydrogen ions. Acid-base neutralization reactions are used to determine the concentration of an acid or base dissolved in water by the procedure known as titration. At the end point of a titration, equivalents of acid are equal to equivalents of base.

Salts: Acids neutralize bases to form salts. These are ionic compounds composed of the cations of bases and the anions of acids. Net ionic equations can be written for acid-base neutralization reactions and for precipitation reactions. Not all salts are soluble in water; general guidelines can be used to predict water solubility of salts.

STUDY QUESTIONS AND PROBLEMS

1. Define the following terms:

    (a) Arrhenius acid
    (b) Arrhenius base
    (c) Bronsted-Lowry acid
    (d) Bronsted-Lowry base
    (e) hydronium ion
    (f) conjugate acid
    (g) conjugate base
    (h) litmus
    (i) acid anhydride
    (j) basic anhydride
    (k) strong acid
    (l) weak acid
    (m) polyprotic acid
    (n) oxyacid
    (o) carboxylic acid
    (p) strong base
    (q) weak base
    (r) equivalent
    (s) normality
    (t) titration
    (u) equivalence point
    (v) salt

2. Distinguish between Arrhenius definitions of acids and bases and Bronsted-Lowry definitions of acids and bases.

3. The following substances are strong acids. Write equations for the reaction of each with water, and identify the two conjugate acid-base pairs for each reaction.

    (a) HCl
    (b) HNO_3
    (c) HBr
    (d) $HClO_4$

4. Describe three properties each for acids and bases.

5. The following substances are acid anhydrides; each reacts with one molecule of water to form an acid. Write an equation for the reaction of each acid anhydride with water.

    (a) $SO_3(g)$
    (b) $CO_2(g)$
    (c) $SO_2(g)$

6. The following substances are basic anhydrides; each reacts with one molecule of water to form a solid base. Write an equation for the reaction of each basic anhydride with water.

    (a) $Na_2O(s)$
    (b) $CaO(s)$
    (c) $K_2O(s)$
    (d) $MgO(s)$

7. Distinguish between strong acids and weak acids, and give an example of each.

8. What is the H_3O^+ concentration in a $6.0M$ solution of hydrochloric acid, HCl?

9. In a solution of acetic acid, the following equilibrium concentrations existed:

    $[CH_3CO_2H] = 0.0987M$
    $[H_3O^+] = 1.3 \times 10^{-3}M$
    $[CH_3CO_2^-] = 1.3 \times 10^{-3}M$

    Calculate K_a, the ionization constant, for acetic acid.

10. Hydrogen cyanide, the toxic gas, dissolves in water to form a weak acid solution.

$$HCN(aq) + H_2O(l) \rightleftharpoons H_3O^+(aq) + CN^-(aq)$$

If the following equilibrium concentrations exist,

$$[HCN] = 0.20M$$
$$[H_3O^+] = 9.9 \times 10^{-6}M$$
$$[CN^-] = 9.9 \times 10^{-6}M$$

find the value for K_a, the ionization constant.

11. Hydrofluoric acid is a weak acid because it contains fluorine, the most electronegative element. Fluorine atoms attract bonding electrons and hydrogen atoms in HF so strongly that little dissociation occurs.

$$HF(aq) + H_2O(l) \rightleftharpoons H_3O^+(aq) + F^-(aq), \quad K_a = 6.8 \times 10^{-4}$$

If the equilibrium concentration of HF(aq) is 0.099M, what is the concentration of H_3O^+ at equilibrium?

12. For the weak acid HCO_2H (formic acid), $K_a = 1.7 \times 10^{-4}$. Calculate the H_3O^+ concentration in a solution that contains 0.20M HCO_2H at equilibrium.

$$HCO_2H(aq) + H_2O(l) \rightleftharpoons HCO_2^-(aq) + H_3O^+(aq)$$

13. Fill in the blanks in the following statements:
    (a) A weak acid has a _____ conjugate base.
    (b) A strong acid has a _____ conjugate base.
    (c) A weak base has a _____ conjugate acid.
    (d) A strong base has a _____ conjugate acid.

14. Listed below are several acids. Classify each as monoprotic, diprotic, or triprotic.
    (a) H_2SO_4
    (b) H_3PO_4
    (c) HCl
    (d) H_2CO_3

15. Write equations for the stepwise ionization of sulfuric acid, H_2SO_4. Identify all conjugate acid-base pairs.

16. Write equations for the stepwise ionization of carbonic acid, H_2CO_3. Identify all conjugate acid-base pairs.

17. Write equations for the stepwise ionization of phosphoric acid, H_3PO_4. Identify all conjugate acid-base pairs.

18. Why is phosphoric acid, H_3PO_4, a weaker acid than sulfuric acid, H_2SO_4?

19. Why are most carboxylic acids weak acids?

20. Distinguish between strong bases and weak bases, and give an example of each.

21. What is the OH^- concentration in a 2.0M solution of KOH?

22. In a solution of ammonium hydroxide, the following equilibrium concentrations existed:

$$[NH_3] = 0.148M$$
$$[NH_4^+] = 1.6 \times 10^{-3}M$$
$$[OH^-] = 1.6 \times 10^{-3}M$$

What is the value of K_b, the ionization constant, for the following reaction?

$$NH_3(g) + H_2O(l) \rightleftharpoons NH_4^+(aq) + OH^-(aq)$$

23. The metabolic product urea, which is found in urine, is a weak base:

$$\underset{\text{urea}}{H_2N-\overset{\overset{\displaystyle O}{\|}}{C}-NH_2(aq)} + H_2O(l) \rightleftharpoons$$

$$H_2N-\overset{\overset{\displaystyle O}{\|}}{C}-NH_3^+(aq) + OH^-(aq)$$

A typical urine sample contains the following concentrations:

$$\left[H_2N-\overset{\overset{\displaystyle O}{\|}}{C}-NH_2\right] = 0.40M$$

$$\left[H_2N-\overset{\overset{\displaystyle O}{\|}}{C}-NH_3^+\right] = 7.7 \times 10^{-8}M$$

$$[OH^-] = 7.7 \times 10^{-8}M$$

What is the ionization constant, K_b, for urea?

24. The strong odor of fresh fish comes from amines in the fish. Explain how washing them with lemon juice gets rid of the fish odor.

25. Calculate the normality of a solution containing 392.4 g of sulfuric acid, H_2SO_4, in 2.00 liters of solution.

26. What is the normality of a solution containing 20.0 g of sodium hydroxide, NaOH, in 250 mL of solution?

27. How many grams of phosphoric acid, H_3PO_4, are contained in 5.00 liters of a 2.00N solution?

28. How many grams of potassium hydroxide, KOH, are in 600 mL of a 6.00N solution?

29. How many meq are in 100 mL of a 0.0100N solution?

30. How many meq are in 50.0 mL of a 0.150N solution?

31. What is the normality of a 2.0M solution of phosphoric acid, H_3PO_4?

32. What is the normality of a 1.50M solution of sulfuric acid, H_2SO_4?

33. What is the normality of a 6.00M solution of sodium hydroxide, NaOH?

34. In a titration, 40.0 mL of an acid solution are required to neutralize 20.0 mL of a 1.5N potassium hydroxide. What is the normality of the acid solution?

35. Calculate the volume of 0.400N base necessary to titrate 15.0 mL of 1.00N nitric acid, HNO_3.

36. Potatoes are peeled commercially by soaking them in NaOH solution for a short time and then spraying off the softened peel with a jet of water. If 42.7 mL of 1.00N H_2SO_4 was required to titrate 20.0 mL of NaOH solution, what was the normality of the NaOH?

37. A volume of 50.0 mL of 0.100N NaOH was added to 15.0 mL of 0.400N HCl. Was the resulting solution acidic or basic? Explain your answer.

38. New automobiles come equipped with catalytic converters which reduce NO_2 in exhaust to N_2. How is this beneficial?

39. The catalytic converters mentioned in question 38 have an unfortunate disadvantage in that they oxidize SO_2 to SO_3. Why is this undesirable?

40. Why must hydrocholoric acid and ammonium hydroxide be stored in tightly covered containers?

41. (a) Write a complete ionic equation for the total neutralization of NaOH by H_2SO_4 in water.
    (b) Write a net ionic equation for the reaction described in part (a).

42. (a) Write a complete ionic equation for the total neutralization of $Ca(OH)_2$ by H_2SO_4 in water, keeping in mind that the salt formed in insoluble in water.
    (b) Write a net ionic equation for the reaction described in part (a).

CHAPTER 10

ELECTROLYTES, pH, AND BUFFERS

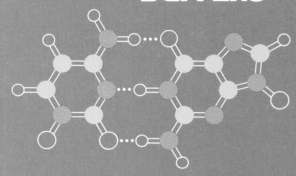

ELECTROLYTES
- Strong Electrolytes, Weak Electrolytes, and Nonelectrolytes
- Concentrations of Electrolytes
- Electrolysis

IONIZATION OF WATER

THE CONCEPT OF pH
- The pH Scale
- Measurement of pH

BUFFERS
- The Henderson-Hasselbalch Equation
- Buffers in Blood
- Acidosis and Alkalosis

Electrolytes are substances that conduct electricity in the melted form or when dissolved in water. These substances can be acids, bases, or salts, as long as they produce ions. Our bodies contain many electrolytes, but the principal ones are K^+, Na^+, and Cl^-. On the average, the concentration of K^+ is 40 times higher inside cells than outside. Just the opposite is true for Na^+ and Cl^-; their concentrations inside cells are on the order of 1/12 and 1/20, respectively, of their extracellular levels. The human body has an intricate set of mechanisms which maintain the balance of electrolytes in correct proportions for normal metabolic processes.

The level of acid in aqueous solution is often measured in units of pH. At one time farmers checked the pH of soil simply by tasting it. If the soil tasted sour, it was acidic, with a pH of less than 7. A sweet taste meant that the soil was alkaline and had a pH greater than 7. Modern farmers may use meters to measure soil pH, and simpler devices are available for measuring pH of swimming pool water and other common substances. The human body is so sensitive to pH changes that it possesses chemical means of responding to potential changes in acid levels. In blood, for example, carbonic acid and bicarbonate ion function as a buffer system to hold the pH relatively constant. In this chapter you will learn how such a buffer system works.

ELECTROLYTES

When we think of electricity, we usually visualize metal wires that carry electric current. In general, metals are good conductors of electricity because their outer electrons are held rather loosely and thus can be made to move from one atom to another. On the other hand, most nonmetals do not conduct electricity because their outer electrons are held tightly or shared by nearby atoms. Molecular compounds are usually not electrical conductors for the same reason. Solid ionic compounds do not conduct electricity because their ions are unable to move from their fixed positions. However, if we melt the ionic compound, or dissolve it in water, it will conduct electricity. This is because the ions are then free to move, and they can carry electrons to charged electrodes (see Figure 10-1). Substances that conduct electricity when they are in the melted state or dissolved in water are called *electrolytes*.

Strong Electrolytes, Weak Electrolytes, and Nonelectrolytes

We can find out whether a substance is an electrolyte by dissolving it in water and measuring the electrical conductivity of the solution. For example, pure water just barely conducts electricity, but tap water is usually a very good conductor because it contains dissolved salts (minerals) that exist as ions in solution. For this reason, we are cautioned not to use electrical appliances in tap water unless they are designed for such use.

Ions in solution can originate from acids, bases, or salts. *Strong electrolytes* are compounds which dissociate completely into ions. All strong acids and strong bases are strong electrolytes, and so are salts that have high water solubility. Aqueous solutions of these substances have a high electrical conductivity. *Weak electrolytes* are solutes for which dissociation into ions is incomplete; aqueous solutions of these show poor electrical conductivity. Weak acids and weak bases are examples of weak electrolytes, as are salts with low water solubility.

Other water-soluble substances exist which produce no ions when dissolved in water. These are called *nonelectrolytes*. They are usually molecular compounds in which the covalent bonds are not polar enough to be separated by water molecules. Most sugars fit this category, as well

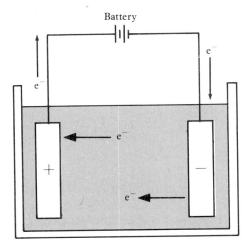

Fig. 10-1 Movement of Electrons in the Flow of Electricity Through a Liquid

as many other biological molecules. Table 10-1 gives examples of strong electrolytes, weak electrolytes, and nonelectrolytes.

Concentrations of Electrolytes

Electrolyte concentrations are commonly expressed in units of molarity or as percentages. However, we can consider electrolyte ions as having been furnished by acids or bases. From this viewpoint, it is understandable that electrolyte concentrations may be expressed in terms of equivalents. In particular, medical laboratory work is often concerned with electrolytes in body fluids. In these cases the ions are present in rather low concentrations, and the use of milliequivalents, meq, is popular, where 1 meq = 0.001 equivalent. An equivalent of an ion is the weight of one mole of the ion divided by the absolute value of the ionic charge (the ion's charge without its sign). Thus, an equivalent of Na^+ is 23.0 g/1, or 23.0 g, and a meq of Na^+ is 0.0230 g. In another example, an equivalent of SO_4^{2-} is 96.1 g/2, or 48.0 g, and a meq of SO_4^{2-} is 0.0480 g. Typical concentrations of the major electrolytes in humans are given in Table 10-2.

Electrolysis

As mentioned earlier, electricity is the flow of electrons. When electricity is conducted through a solution, electrons flow from one metal rod or strip to another. These metal parts are called *electrodes*. They are con-

electrode: a conductor through which an electric current enters or leaves a system.

Table 10-1 Examples of Strong Electrolytes, Weak Electrolytes, and Nonelectrolytes

Substance	Formula	Type of Substance	Classification
Hydrochloric acid	HCl	Strong acid	Strong electrolyte
Sulfuric acid	H_2SO_4	Strong acid	Strong electrolyte
Sodium hydroxide	NaOH	Strong base	Strong electrolyte
Sodium chloride	NaCl	Salt	Strong electrolyte
Calcium nitrate	$CaNO_3$	Salt	Strong electrolyte
Acetic acid	CH_3CO_2H	Weak acid	Weak electrolyte
Formic acid	HCO_2H	Weak acid	Weak electrolyte
Ammonia	NH_3	Weak base	Weak electrolyte
Methylamine	CH_3NH_2	Weak base	Weak electrolyte
Glucose	$C_6H_{12}O_6$	Sugar (molecular solid)	Nonelectrolyte
Sucrose	$C_{12}H_{22}O_{11}$	Sugar (molecular solid)	Nonelectrolyte
Ethanol	C_2H_5OH	Alcohol (molecular liquid)	Nonelectrolyte

Table 10-2 Major Electrolytes in Humans

INTRACELLULAR		EXTRACELLULAR		PLASMA	
Ion	meq/L	Ion	meq/L	Ion	meq/L
K^+	160	K^+	4	K^+	5
Mg^{2+}	35	Mg^{2+}	2	Mg^{2+}	2
Na^+	10	Na^+	145	Na^+	143
		Ca^{2+}	3	Ca^{2+}	5
HPO_4^{2-}	140	HPO_4^{2-}	2	HPO_4^{2-}	2
protein anions	55	protein anions	1	protein anions	16
HCO_3^-	8	HCO_3^-	30	HCO_3^-	25
Cl^-	2	Cl^-	115	Cl^-	105
		SO_4^{2-}	1	SO_4^{2-}	1
		carboxylic acid anions	5	carboxylic acid anions	6

nected to a source of electricity, such as a battery. The electrode from which electrons flow is called the *anode*; because this electrode loses electrons (negative charges), it takes on a positive charge. The other electrode gains electrons and thus becomes negatively charged; it is called the *cathode*. The battery provides the driving force for electron flow, and electric current passes through the solution and then back to the battery. Thus, the system is a closed circuit through which electricity flows, provided there are electrolytes in the solution. Positively charged ions (cations) are attracted to the negatively charged electrode (cathode), and negatively charged ions (anions) are attracted to the positively charged electrode (anode). At the electrodes, anions pass electrons on to the anode, and cations receive electrons from the cathode. The net result is that electrons are carried through the solution by the electrolytes.

If sodium chloride is melted, its ions are free to move about, and it becomes a good conductor of electricity. At the anode, Cl^- loses electrons and forms neutral atoms which quickly unite to produce Cl_2 molecules. At the same time, Na^+ picks up electrons from the cathode to form Na atoms:

$$2\ Cl^- - 2\ e^- \longrightarrow Cl_2$$
$$Na^+ + e^- \longrightarrow Na$$

In this way, molten NaCl is decomposed by electrical energy into its constituent elements:

$$2\ NaCl + energy \longrightarrow 2\ Na + Cl_2$$

As long as anions and cations are present in the melt, current will flow. Thus, when all of the salt has been decomposed, the circuit will be broken, and electricity will cease flowing. The process of decomposing compounds by the use of electricity is called *electrolysis*. Since electrons are transferred in the process, electrolysis is a form of oxidation-reduction in which anions are oxidized and cations are reduced.

electrolysis: decomposition by electricity; *-lysis* means "destruction."

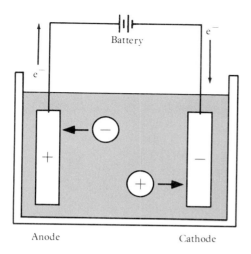

Fig. 10-2 The Movement of Cations Toward the Cathode and Anions Toward the Anode

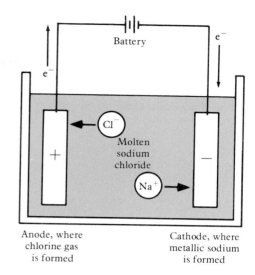

Fig. 10-3 The Electrolysis of Molten Sodium Chloride

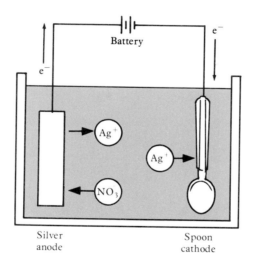

Fig. 10-4 Diagram of an Electroplating Apparatus

Electrolysis is used extensively in extraction and purification of elements from ores or compounds. The electrolysis of molten sodium chloride described above is used for the commercial production of sodium and chlorine. In addition, electrolysis is the basis for electroplating, the process in which one metal is coated with another (see Figure 10-4). Thus, a spoon made of an inexpensive metal can be coated with a thin layer of silver to make it more attractive and resistant to corrosion. The silver-plated spoon is much less expensive than a spoon made entirely of silver. A modern application of electrolysis is the removal of unwanted hair. Small needles are used to transmit electric current to individual hair roots, and the resulting chemical changes kill the hair follicle. Warts and other small skin growths can also be removed this way.

PERSPECTIVE: Charles Martin Hall and the Production of Aluminum

Charles Martin Hall was a chemistry student at Oberlin College in Ohio when one of his professors remarked that a fortune could be made by the person who could discover an inexpensive way to produce aluminum. Hall was so impressed by the professor's remarks that he set out to devise such a process. He worked diligently in the college laboratory, even after graduation, and in 1886 at the age of 21, Charles Martin Hall solved the problem. At about the same time, Paul Heroult, a Frenchman the same age as Hall, made the same discovery. The success of the Hall process is based on the discovery that aluminum ore, Al_2O_3, dissolves in molten cryolite, Na_3AlF_6, to produce a solution that conducts electricity. Thus electrolysis of the solution of aluminum ore in cryolite is used to produce aluminum metal of 99.0–99.9% purity.

After several failures to obtain financial backing, Hall persuaded the Mellon family to establish the Pittsburg Reduction Company in 1888. By 1890, Hall had become vice president of the company, and in 1907 its name was changed to Aluminum Company of America (ALCOA). In the following years Hall did indeed make a fortune for himself, but he remained deeply grateful to his college. He expressed his gratitude by bequeathing more than five million dollars to Oberlin College.

IONIZATION OF WATER

As mentioned above, aqueous solutions of strong electrolytes are good conductors of electricity. But even pure water shows a trace of conductivity, indicative of a very small number of ions present. If we subject pure water to electrolysis, hydrogen is formed at the cathode and oxygen is produced at the anode. These observations are consistent with the following reactions:

$$2\,H^+ + 2\,e^- \longrightarrow H_2$$
$$2\,OH^- - 4\,e^- \longrightarrow O_2 + 2\,H^+$$

Ions are present because water is itself a very weak acid and a very weak base. It is at once a proton donor and a proton acceptor. Thus,

$$H_2O(l) + H_2O(l) \rightleftharpoons H_3O^+(aq) + OH^-(aq)$$

At room temperature only about one out of every 10^8 water molecules is ionized at any one instant.

It is common practice to consider hydronium ions to be hydrated hydrogen ions. With this in mind, we can write a simplified equation for the ionization of water as follows:

$$H_2O(l) \rightleftharpoons H^+(aq) + OH^-(aq)$$

ion-product constant, K_w: the product of molar concentrations of hydrogen ion and hydroxide ion in water.

Since $[H_2O]$ in pure water is a constant, $K_{eq}[H_2O]$ is also a constant, and we call it K_w.

The expression for the ionization constant of the reaction is shown below; for water, this constant has a special name—the *ion-product constant, K_w*.

$$K_{eq} = \frac{[H^+][OH^-]}{[H_2O]}$$

$$K_{eq}[H_2O] = [H^+][OH^-]$$

$$K_w = [H^+][OH^-]$$

From measurements of the extremely low conductivity of water, the value for K_w at 25°C has been determined to be 1×10^{-14}.

As with all other ionization constants, K_w is a true constant. Thus, if $[H^+]$ is increased by adding acid to water, then $[OH^-]$ decreases. On the other hand, if $[OH^-]$ is increased by the addition of base to water, then $[H^+]$ decreases. In the case where $[H^+] = [OH^-]$, the solution is said to be neutral. Pure water is neutral, as is an aqueous solution of equivalent amounts of strong acid and strong base.

THE CONCEPT OF pH

Because the product of $[H^+]$ and $[OH^-]$ in water is constant, the total acid-base balance in any water solution can be expressed by stating the hydrogen ion concentration. A convenient means of expressing hydrogen ion concentration is by the use of *pH*, defined as the negative log of the hydrogen ion concentration:

pH: from the French *pouvoir hydrogene*, meaning "power of hydrogen."

$$pH = -\log [H^+]$$

(If you need a review of logarithms, see Appendix **A**.)

Although the idea of taking a negative logarithm may seem complicated, we can simplify the process by remembering that the logarithm of a number expressed as 1×10^n is simply the value of n. Thus, the log of 1×10^3 is 3, and the log of 1×10^{-3} is -3. Since pH is defined as a negative log, if $[H^+] = 1 \times 10^{-3} M$, then pH is just the value of the exponent with its sign changed, in this instance, 3. Let's see how this works by calculating the pH of pure water:

$$K_w = [H^+][OH^-] = 1 \times 10^{-14}$$

Since

$$[H^+] = [OH^-],$$

then

$$K_w = [H^+]^2 = 1 \times 10^{-14}$$

and

$$[H^+] = 1 \times 10^{-7} M$$
$$[OH^-] = 1 \times 10^{-7} M$$

Now we must take the negative log of $[H^+]$:

$$\begin{aligned} pH &= -\log [H^+] \\ &= -\log (1 \times 10^{-7}) \\ &= -(-7) \\ &= 7 \end{aligned}$$

Thus, pure water has a pH of 7. In fact, all neutral solutions have pH values of 7. If a solution is acidic, the pH will be less than 7, and $[H^+]$

will be greater than $1 \times 10^{-7} M$. A basic solution will have a pH greater than 7, and its $[H^+]$ will be less than $1 \times 10^{-7} M$. Notice that pH values decrease as a solution becomes more and more acidic; in contrast, pH increases as a solution becomes more and more basic.

As an example, consider the acid level in the human stomach. Although somewhat variable, a typical hydrogen ion concentration is $0.01 M$. Let's calculate the pH of this sample of gastric juice:

$$pH = -\log [H^+]$$
$$= -\log 0.01 = -\log (1 \times 10^{-2})$$
$$= -(-2)$$
$$= 2$$

Thus, the pH of gastric juice in this example is 2. For normal stomach activity, the pH of gastric juice is usually in the range of 1–2.

Table 10-3 Concentrations, pH, and pOH

$[H_3O^+]$	pH	$[OH^-]$	pOH
10^0	0	10^{-14}	14
10^{-1}	1	10^{-13}	13
10^{-2}	2	10^{-12}	12
10^{-3}	3	10^{-11}	11
10^{-4}	4	10^{-10}	10
10^{-5}	5	10^{-9}	9
10^{-6}	6	10^{-8}	8
10^{-7}	7	10^{-7}	7
10^{-8}	8	10^{-6}	6
10^{-9}	9	10^{-5}	5
10^{-10}	10	10^{-4}	4
10^{-11}	11	10^{-3}	3
10^{-12}	12	10^{-2}	2
10^{-13}	13	10^{-1}	1
10^{-14}	14	10^{-0}	0

EXAMPLE 10.1

Calculate the pH of a solution in which $[H^+] = 1 \times 10^{-4} M$. Is this solution acidic or basic?

SOLUTION

$$pH = -\log [H^+]$$
$$= -\log (1 \times 10^{-4})$$
$$= -(-4)$$
$$= 4$$

The solution is acidic, because its pH is less than 7.

EXERCISE 10.1 Calculate the pH of a solution in which $[H^+]$ is $1 \times 10^{-9} M$. Is the solution acidic or basic?

EXAMPLE 10.2

Calculate the hydrogen ion concentration of a solution whose pH is 8.

SOLUTION

$$\text{pH} = -\log [\text{H}^+]$$
$$\log [\text{H}^+] = -\text{pH}$$
$$\log [\text{H}^+] = -8$$
$$[\text{H}^+] = 1 \times 10^{-8} M$$

EXERCISE 10.2 Calculate the hydrogen ion concentration of a solution whose pH is 6.0.

Because K_w for water is a constant, the concentration of hydroxide ion is automatically indicated by pH values. We can see how this comes about by the following operation.

$$K_w = [\text{H}^+][\text{OH}^-] = 1 \times 10^{-14}$$
$$\log K_w = \log [\text{H}^+] + \log [\text{OH}^-] = -14$$

Thus the sum of logs of $[\text{H}^+]$ and $[\text{OH}^-]$ in aqueous solution is always -14. Let's go one step further by defining *pOH* as the negative log of $[\text{OH}^-]$. Then,

$$\log \text{H}^+ + \log \text{OH}^- = -14$$
$$-\log \text{H}^+ - \log \text{OH}^- = 14$$
$$\text{pH} + \text{pOH} = 14$$

Now you can see that the sum of pH and pOH is always 14. In the case of our gastric juice calculation, since pH is 2, pOH must be 12, and $[\text{OH}^-] = 1 \times 10^{-12} M$. Thus, when pH is less than 7, $[\text{H}^+]$ must be greater than $1 \times 10^{-7} M$, and that means that there are more hydrogen ions in solution than hydroxide ions. Hence the solution is acidic. Similarly, when the pH is greater than 7, $[\text{H}^+]$ must be less than $1 \times 10^{-7} M$, $[\text{OH}^-]$ must be greater than $1 \times 10^{-7} M$, and the solution is basic.

EXAMPLE 10.3

If the pH of a solution is 3, what is the pOH?

SOLUTION

Since
$$\text{pH} + \text{pOH} = 14$$
then
$$\text{pOH} = 14 - \text{pH}$$
$$= 14 - 3$$
$$= 11$$

EXERCISE 10.3 If the pH of a solution is 11, what is its pOH?

EXAMPLE 10.4

If the pH of a solution is 9, what is $[OH^-]$?

SOLUTION

Since

$$pH + pOH = 14$$

then

$$pOH = 14 - pH$$
$$= 14 - 9$$
$$= 5$$

And then,

$$pOH = -\log [OH^-]$$
$$\log [OH^-] = -pOH$$
$$\log [OH^-] = -5$$
$$[OH^-] = 1 \times 10^{-5} M$$

EXERCISE 10.4 If the pH of a solution is 5, what is $[OH^-]$?

The pH Scale

pH scale: a scale of range 0–14 used for expressing level of acidity.

The *pH scale* has become a popular method for expressing level of acidity because the values are usually small positive numbers in the range of 0–14 for common solutions. pH is measured for all kinds of aqueous solutions—lakes and rivers, gastric juice, blood, rainfall, even shampoos—and the use of small positive numbers simplifies tabulating data and keeping records.

The pH scale is shown in Figure 10-15, along with pH values for

Fig. 10-5 The pH Scale

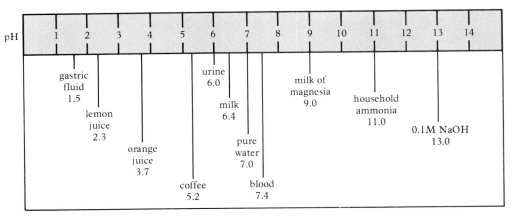

some familiar solutions. A difference of one pH unit anywhere along the scale represents a tenfold change in hydrogen ion concentration. You will notice that pH values for several common solutions are not whole numbers. For example, the pH of milk is 6.4. This means that hydrogen ion concentration is between $1 \times 10^{-6}M$ and $1 \times 10^{-7}M$. If you wish to learn how to convert pH values expressed in decimal numbers to hydrogen ion concentration, refer to Appendix C. However, it is usually sufficient to know just the range of $[H^+]$ corresponding to a decimal pH.

EXAMPLE 10.5

Find the pH of a $0.1M$ solution of hydrochloric acid.

SOLUTION

Since hydrochloric acid is a strong acid, it dissociates completely to produce a solution in which $[H^+] = 0.1M$. Thus,

$$\begin{aligned} pH &= -\log[H^+] \\ &= -\log 0.1 = -\log(1 \times 10^{-1}) \\ &= -(-1) \\ &= 1 \end{aligned}$$

EXERCISE 10.5 Find the pH of a $0.001M$ solution of hydrochloric acid.

EXAMPLE 10.6

Find the pH of a $0.001M$ solution of NaOH.

SOLUTION

Since NaOH is a strong base, it dissociates completely to produce a solution in which $[OH^-] = 0.001M$. Thus,

$$\begin{aligned} [OH^-] &= 0.001M = 1 \times 10^{-3}M \\ pOH &= -\log[OH^-] \\ &= -\log(1 \times 10^{-3}) \\ &= -(-3) \\ &= 3 \end{aligned}$$

$$\begin{aligned} pH + pOH &= 14 \\ pH &= 14 - pOH \\ &= 14 - 3 \\ &= 11 \end{aligned}$$

EXERCISE 10.6 Find the pH of a $0.01M$ solution of NaOH.

Measurement of pH

The pH of a solution can be estimated fairly accurately by use of colored indicators. An indicator is an acid or base, often derived from plant material, which changes color when it is neutralized. Let's examine an indicator acid. The indicator ionizes like any other acid, and the ions are in equilibrium with undissociated indicator in aqueous solution:

$$HInd(aq) \rightleftharpoons H^+(aq) + Ind^-(aq)$$

$$K_{eq} = \frac{[H^+][Ind^-]}{[HInd]}$$

The HInd and Ind$^-$ forms have different colors (a necessity for a compound to be an indicator), and the amount of each existing in a solution is dependent on the pH of the solution, because [H$^+$] affects the extent of ionization. By adding a small amount of indicator to a solution and noting the color, it is possible to tell which of the two forms of indicator is present and thus estimate pH. The many indicators known at present change colors at various pH values between 1 and 14. Some of these are listed in Table 10-4. Paper tape impregnated with various indicators is routinely used for approximate determination of pH.

A second method for measuring pH employs an instrument called a *pH meter*, an electronic device with two electrodes (Figure 10-6). When the electrodes are immersed in an aqueous solution, the hydrogen ions present generate an electrical signal which is registered on the meter. The strength of the signal is directly proportional to [H$^+$], or more properly [H$_3$O$^+$], and the meter is calibrated in units of pH. Thus, pH is read directly from the meter. Measuring pH in this way is not so convenient as using pH paper, but the pH meter is much more precise and can be used in colored solutions which might interfere with the observation of indicator color.

In the past several years, extremely small pH electrodes have been developed, and some of these are so small that they can be inserted directly into living cells. In addition to biological and medical uses, pH meters are routinely used in environmental monitoring and agricultural work.

Table 10-4 Examples of Acid-Base Indicators

Indicator	Color Change (increasing pH)	pH Range
Thymol blue	red to yellow	1.2–2.8
Bromphenol blue	yellow to blue	3.0–4.6
Methyl orange	red to yellow	3.1–4.4
Bromcresol green	yellow to blue	3.8–5.4
Methyl red	red to yellow	4.2–6.2
Litmus	red to blue	4.5–8.3
Bromthymol blue	yellow to blue	6.0–7.6
Phenol red	yellow to red	6.8–8.4
Phenolphthalein	colorless to red	8.3–10.0
Alizarin yellow	yellow to violet	10.1–12.0
1,3,5-Trinitrobenzene	colorless to yellow	12.0–14.0

Fig. 10-6 A pH Meter. The single probe (*right*) is called a combination electrode because it combines both electrodes into a single unit.

BUFFERS

Most body fluids are able to maintain rather constant pH values despite the entry of acidic and basic substances into these fluids. This is because body fluids contain substances which control the acid level so that any pH changes are minimal. Solutions which are able to resist pH change when small amounts of acid or base are added are called *buffer solutions*, or more simply *buffers*.

Buffers consist of a weak acid and a salt of the weak acid, or a weak base and a salt of the weak base. We call the two components a buffer pair; Table 10-5 lists some common buffer pairs. Let's examine one buffer pair in which the weak acid is acetic acid and the salt of the weak acid is sodium acetate. Since acetic acid is a weak acid, it is only slightly dissociated in water:

$$CH_3CO_2H(aq) \rightleftharpoons CH_3CO_2^-(aq) + H^+(aq), K_a = 1.8 \times 10^{-5}$$

Table 10-5 Common Buffer Pairs

Buffer Pair	Formula	Useful pH Range
Formic acid/sodium formate	HCO_2H/HCO_2Na	2.8–4.8
Acetic acid/sodium acetate	CH_3CO_2H/CH_3CO_2Na	3.8–5.8
Sodium dihydrogen phosphate/ disodium hydrogen phosphate	NaH_2PO_4/Na_2HPO_4	6.2–8.2
Boric acid/sodium borate	$H_2B_4O_7/Na_2B_4O_7$	8.0–10.0
Sodium bicarbonate/disodium carbonate	$NaHCO_3/Na_2CO_3$	9.2–11.2

On the other hand, sodium acetate is a strong electrolyte, and it is completely dissociated in water:

$$CH_3CO_2Na \xrightarrow{water} CH_3CO_2^-(aq) + Na^+(aq)$$

Thus acetic acid furnishes the solution with a small number of protons and acetate ions and a relatively large amount of undissociated acid, whereas sodium acetate provides large amounts of both sodium ions and acetate ions. If a small quantity of another acid is added to the solution, the equilibrium adjusts itself by forming more undissociated acetic acid, according to LeChatelier's principle:

$$CH_3CO_2H(aq) \rightleftharpoons CH_3CO_2^-(aq) + H^+(aq)$$
$$\longleftarrow \text{ increase in } [H^+]$$

On the other hand, if a small amount of base is added, it is removed due to neutralization by the hydrogen ions present. Then the equilibrium responds to this loss of hydrogen ions by forming more H^+ from acetic acid:

$$CH_3CO_2H(aq) \rightleftharpoons CH_3CO_2^-(aq) + H^+(aq)$$
$$\text{decrease in } [H^+] \longrightarrow$$

Hence, small amounts of either acid or base can be absorbed by the buffer solution without alteration of pH. The weak acid component of the buffer pair serves as a source of undissociated acid which can provide more H^+ when needed, and the salt of the weak acid functions to provide an abundance of acid anions which can associate with excess H^+ to reduce the level of hydrogen ions.

Another example of a buffer pair is one composed of ammonium hydroxide and ammonium chloride, a weak base and its salt. The salt dissociates completely in aqueous solution into ammonium ions and chloride ions, but because ammonium hydroxide is a weak base, the following equilibrium exists:

$$NH_3(aq) + H_2O(l) \rightleftharpoons NH_4^+(aq) + OH^-(aq), K_b = 1.8 \times 10^{-5}$$

Thus, addition of a small amount of acid causes neutralization of hydroxide ion, and the equilibrium shifts to form more OH^-:

$$NH_3(aq) + H_2O(l) \rightleftharpoons NH_4^+(aq) + OH^-(aq)$$
$$\text{decrease in } [OH^-] \longrightarrow$$

Addition of a small amount of base (hydroxide ions) causes the equilibrium to shift to the left and the excess hydroxide ions are removed:

$$NH_3(aq) + H_2O(l) \rightleftharpoons NH_4^+(aq) + OH^-(aq)$$
$$\longleftarrow \text{ increase in } [OH^-]$$

The Henderson-Hasselbalch Equation

During studies of body fluids, Harvard biochemist Lawrence Henderson developed a very useful equation describing physiological buffers. Henderson's expression, modified by Danish biochemist Karl Hasselbalch, has assumed fundamental importance in the understanding of all buffer systems. The *Henderson-Hasselbalch equation* is remarkably simple in its development, as we will show.

Let's consider the pH of a buffer solution of weak acid HA and its corresponding salt MA, where M could be an alkali metal. The acid dissociation equilibrium is

$$HA(aq) \rightleftharpoons H^+(aq) + A^-(aq)$$

and the ionization constant expression is

$$K_a = \frac{[H^+][A^-]}{[HA]}$$

or

$$[H^+] = K_a \frac{[HA]}{[A^-]}$$

If we take the log of this equation, we get

$$\log [H^+] = \log K_a + \log \frac{[HA]}{[A^-]}$$

If we then multiply by -1,

$$-\log [H^+] = -\log K_a - \log \frac{[HA]}{[A^-]}$$

What we have done up to this point is manipulate our original ionization constant expression to a form where we can substitute pH for $-\log [H^+]$. But if we define a new term pK_a as $-\log K_a$, then our equation becomes

$$pH = pK_a - \log \frac{[HA]}{[A^-]}$$

or

$$pH = pK_a + \log \frac{[A^-]}{[HA]}$$

We now have the Henderson-Hasselbalch equation. Notice that the pH of a buffer solution depends only on the pK_a (remember this is the negative log of K_a) of the weak acid and the ratio of molar concentrations of undissociated acid and its conjugate base. This equation also applies to weak bases and their salts; in those cases, HA corresponds to the protonated base BH^+ (the same as the conjugate acid of the weak

Henderson-Hasselbalch equation: an equation which describes buffer action in terms of pH, pK_a, and $\log([A^-]/[HA])$.

pK_a: by analogy to pH, $pK_a = -\log K_a$

base) and A⁻ corresponds to the unprotonated base, B. Values for pK_a's of common weak acids and weak bases are given in Table 10-6.

An alternate and very practical definition of pK_a can be derived from the Henderson-Hasselbalch equation by setting $pH = pK_a$ and solving for $\log([A^-]/[HA])$:

$$pH = pK_a + \log \frac{[A^-]}{[HA]}$$

$$pH = pH + \log \frac{[A^-]}{[HA]}$$

$$\log \frac{[A^-]}{[HA]} = pH - pH = 0$$

$$\frac{[A^-]}{[HA]} = 1$$

Since $[A^-]/[HA] = 1$ when $pH = pK_a$, then a solution of weak acid at the pH corresponding to the pK_a of the weak acid contains equimolar amounts of anion and undissociated weak acid. In other words, the pK_a is the pH at which a weak acid is half neutralized, and $[A^-] = [HA]$.

EXAMPLE 10.7

What is the pH of a $0.1M$ formic acid (HCO_2H) solution that contains 0.2 moles of sodium formate (HCO_2Na) per liter? Assume that the amount of dissociation of formic acid is negligible but that sodium formate dissociates completely into Na^+ and HCO_2^-. The pK_a of formic acid is 3.75.

SOLUTION

The concentrations of the weak acid and its anion are as follows:

$$[HCO_2H] = 0.1M$$
$$[HCO_2^-] = 0.2M$$

Then, substituting concentrations into the Henderson-Hasselbalch equation,

$$pH = pK_a + \log \frac{[A^-]}{[HA]}$$

$$= 3.75 + \log \frac{0.2M}{0.1M}$$

$$= 3.75 + \log 2.0$$

$$= 3.75 + 0.30$$

$$= 4.05$$

EXERCISE 10.7 What is the pH of a $0.2M$ acetic acid (CH_3CO_2H) solution that contains 0.1 moles of sodium acetate (CH_3CO_2Na) per liter? Assume that the amount of dissociation of acetic acid is negligible but that sodium acetate dissociates completely into Na^+ and $CH_3CO_2^-$. The pK_a of acetic acid is 4.76.

Table 10-6 pK_a Values for Common Weak Acids and Bases (First Dissociation Only)

Weak Acid	Formula	pK_a
Acetic acid	CH_3CO_2H	4.76
Boric acid	H_3BO_3	9.24
Boric acid (tetra-)	$H_2B_4O_7$	9
Carbonic acid	H_2CO_3	6.38
Formic acid	HCO_2H	3.75
Phosphoric acid	H_3PO_4	2.12
Sulfurous acid	H_2SO_3	1.90
Weak Base	**Formula**	pK_a
Ammonia	NH_3	9.24
Carbonate ion	CO_3^{2-}	10.25
Triethylamine	$(C_2H_5)_3N$	10.72
Trimethylamine	$(CH_3)_3N$	9.80

The Henderson-Hasselbalch equation is very useful in the preparation of buffer solutions needed in the clinical laboratory. Since most clinical tests must be performed under simulated physiological conditions, it is possible to prepare buffer solutions having pHs corresponding to those of body fluids. The Henderson-Hasselbalch equation allows us to choose a weak acid and its salt or a weak base and its salt which will provide the desired pH on the basis of the ionization constant for the weak acid or weak base. Then our buffer solution will resist small changes in acid or base level caused by the test itself, and the pH of the test solution will remain constant.

EXAMPLE 10.8

Suppose you wanted to make a buffer solution having a pH of 5.00, and the components are to be sodium acetate and acetic acid. What ratio of molar concentrations of sodium acetate and acetic acid must be used? The pK_a of acetic acid is 4.76. Assume that the amount of free acetate anion contributed by acetic acid is negligible compared to that contributed by sodium acetate.

SOLUTION

We can use the Henderson-Hasselbalch equation to solve this problem:

$$pH = pK_a + \log \frac{[A^-]}{[HA]}$$

$$\log \frac{[A^-]}{[HA]} = pH - pK_a$$

$$\log \frac{[A^-]}{[HA]} = 5.00 - 4.76 = 0.24$$

$$\frac{[A^-]}{[HA]} = 1.7$$

Thus the molar concentration of sodium acetate should be 1.7 times that of acetic acid to provide a pH of 5.00.

EXERCISE 10.8 What ratio of molar concentrations of sodium formate (HCO_2Na) and formic acid (HCO_2H) are needed to make a solution having a pH of 3.50? Assume that the amount of dissociation of formic acid is negligible but that sodium formate dissociates completely into Na^+ and HCO_2^-. The pK_a of formic acid is 3.75.

Buffer concentrations for weak acids and their salts are expressed as molarity of total anion, present both as free anion, A^-, and as anion in undissociated form, HA. Thus, a $0.1M$ acetate buffer, pH 5.00, contains in 1 liter a total of 0.1 moles of acetate, both as $CH_3CO_2^-$ (from the salt) and as CH_3CO_2H.

EXAMPLE 10.9

How many moles each of sodium acetate and acetic acid are needed to make 1 liter of the buffer solution described in example 10.8 at a concentration of $0.100M$ acetate?

SOLUTION

The sum of moles of sodium acetate and moles of acetic acid per liter of solution must be 0.100, since buffer concentration is expressed as moles of anion present both in free form and in undissociated form. Let x = moles HA. Then $1.7 x$ = moles Na^+A^- (see answer to example 10.8), and

$$\text{moles HA} + \text{moles } Na^+A^- = 0.100 \text{ mole}$$

$$x + 1.7 x = 0.100 \text{ mole}$$

$$2.7 x = 0.100 \text{ mole}$$

$$x = \frac{0.100 \text{ mole}}{2.7}$$

$$x = 0.037 \text{ mole}$$

Thus, 0.037 mole of acetic acid is needed for each liter of solution, and

$$\text{moles } Na^+A^- \text{ needed per liter} = 1.7 x$$

$$= 1.7 \, (0.037 \text{ moles})$$

$$= 0.063 \text{ mole}$$

We have calculated that 1 liter of $0.100M$ acetate buffer, pH 5.00, will contain 0.037 mole of acetic acid and 0.063 mole of sodium acetate. The sum of these concentrations is the total acetate concentration:

$$\text{total acetate concentration} = 0.037M + 0.063M$$

$$= 0.100M$$

EXERCISE 10.9 How many moles each of sodium formate (HCO_2Na) and formic acid (HCO_2H) are needed to make one liter of buffer solution, pH 3.75, at a concentration of $0.200M$ formate?

Buffers in Blood

The pH of blood is rather carefully maintained at 7.35–7.45 by three buffer systems. The most important of these is the carbonic acid/bicarbonate buffer pair, as these buffer components are present in highest concentrations. As the cells use oxygen for metabolic processes, carbon dioxide is liberated into the blood. As it dissolves, the following reaction occurs:

$$CO_2(g) + H_2O(l) \rightleftharpoons H_2CO_3(aq)$$

Even though only a little carbonic acid forms, it ionizes as a weak acid,

$$H_2CO_3(aq) \rightleftharpoons H^+(aq) + HCO_3^-(aq), \quad K_a = 4.3 \times 10^{-7}$$

The ionization constant for the second dissociation is so small that for all practical purposes, no further ionization occurs beyond the first step. Thus, protons are liberated, and bicarbonate ions are formed. The combination of carbonic acid and bicarbonate functions just as any other weak acid and its anion would to buffer the solution. As acidic metabolic products enter into the blood, the equilibrium shifts toward the formation of carbonic acid:

$$H_2CO_3(aq) \rightleftharpoons H^+(aq) + HCO_3^-(aq)$$
$$\longleftarrow \text{increase in } [H^+]$$

When basic metabolic products are released into the blood, $[H^+]$ is decreased by neutralization, and more carbonic acid dissociates:

$$H_2CO_3(aq) \rightleftharpoons H^+(aq) + HCO_3^-(aq)$$
$$\text{decrease in } [H^+] \longrightarrow$$

Thus, the carbonic acid/bicarbonate buffer system is extremely effective in balancing blood pH.

A second buffer system, present in much smaller amounts than the bicarbonate system, is composed of dihydrogen phosphate and hydrogen phosphate:

$$H_2PO_4^-(aq) \rightleftharpoons H^+(aq) + HPO_4^{2-}(aq), \quad K_a = 6.2 \times 10^{-8}$$

It responds to entry of acidic and basic substances into the blood much like the bicarbonate system.

The third buffer in blood is a mixture of proteins. These are colloidal-sized molecules which are capable of reacting with protons as bases would. Thus, proteins are able to remove excess hydrogen ions from blood by a type of acid-base neutralization reaction. You will learn more about this reaction in chapter 15.

Acidosis and Alkalosis

In the normal functioning of our bodies, acid-base balance is maintained in the blood by the buffers present. But occasionally ill health or other

PERSPECTIVE: The High Solubility of Carbon Dioxide in Water

Carbon dioxide has a solubility in pure water of 0.145 g per 100 g of water at 25°C. This is unusually high in comparison to other common gases. For example, the solubility of oxygen in pure water at 25°C is 0.004 g per 100 g of water, and the value for nitrogen is 0.002 g per 100 g of water. Thus carbon dioxide is 36 times more soluble in water than is oxygen and 72 times more soluble than nitrogen. The chemical equilibria formed when carbon dioxide dissolves in water help to explain its high solubility:

$$CO_2(g) \xrightleftharpoons{water} CO_2(aq)$$
$$CO_2(aq) + H_2O(l) \rightleftharpoons H_2CO_3(aq)$$
$$H_2CO_3(aq) \rightleftharpoons H^+(aq) + HCO_3^-(aq)$$
$$HCO_3^-(aq) \rightleftharpoons H^+(aq) + CO_3^{2-}(aq)$$

Both ionization equilibria are shifted to the right as pH increases (that is, as [H^+] decreases). An increase in pH not only causes further ionization, but the other equilibria are also shifted to the right to compensate for the drain on H_2CO_3. Thus, blood, with its slightly alkaline pH, has enhanced solubility for CO_2. A rather similar liquid, seawater, also is slightly alkaline and thus has a high solubility for CO_2. This is advantageous because the minute plants (phytoplankton) which grow in the oceans require CO_2 for life-sustaining photosynthesis. Because of its high solubility in seawater, CO_2 is always present in the amounts needed for these tiny plants that constitute the beginning of the aquatic food chain in seawater.

---

unusual conditions create situations which overload the buffering capacity of blood. If the pH of blood falls below 7.35, the condition is characterized as *acidosis*. On the other hand, if blood pH rises above 7.45, a condition of *alkalosis* is said to exist. (The term alkalosis is derived from "alkaline" which means "basic.") If conditions of acidosis or alkalosis are allowed to persist, dire consequences may result.

Respiratory acidosis results from changes in respiration that elevate CO_2 levels in the blood, thus increasing [H_2CO_3]:

$$CO_2(g) + H_2O(l) \rightleftharpoons H_2CO_3(aq)$$
$$H_2CO_3(aq) \rightleftharpoons H^+(aq) + HCO_3^-(aq)$$

respiratory acidosis: an acid condition of the blood brought on by respiratory disorders.

We can understand the consequences of this by referring to the Henderson-Hasselbalch equation:

$$pH = pK_a + \log \frac{[HCO_3^-]}{[H_2CO_3]}$$

Since H_2CO_3 dissociates incompletely, an increase in its concentration causes some increase in [HCO_3^-], but a significant decrease in log ([HCO_3^-]/[H_2CO_3]) occurs. As a result, pK_a + log ([HCO_3^-]/[H_2CO_3]) becomes smaller, and the pH is lowered. Shallow breathing will increase dissolved CO_2 in the blood because CO_2 in the lungs is not expelled in large enough amounts. Thus, there is an increase in the partial pressure of CO_2 and a consequent high blood level of CO_2. Other causes of respiratory acidosis are lung damage (as in emphysema), weak respiratory muscles, and slow breathing (*hypoventilation*).

Acidosis can also be caused by alterations in metabolic or physiological processes. *Metabolic acidosis* can occur as a result of kidney failure, diabetes mellitus, ingestion of acidic substances, loss of bicarbonate (as in prolonged diarrhea), starvation, or dehydration. The body responds to metabolic acidosis by increasing the rate and depth of breathing, thus

metabolic acidosis: an acid condition of the blood brought on by metabolic or physiological disorders (other than respiratory).

lowering CO_2 and H_2CO_3 levels in the lungs and blood. The simplest means of treating metabolic acidosis is by administration of sodium bicarbonate.

Respiratory alkalosis occurs when respiration renders the blood too basic. In this case, CO_2 levels in the blood are too low, and thus $[H_2CO_3]$ is too low. Referring again to the Henderson-Hasselbalch equation,

$$\text{pH} = \text{p}K_a + \log \frac{[HCO_3^-]}{[H_2CO_3]}$$

if $[H_2CO_3]$ decreases, $[HCO_3^-]$ will decrease somewhat, but because of the incomplete dissociation of H_2CO_3, $\log([HCO_3^-]/[H_2CO_3])$ increases significantly. When this happens, $\text{p}K_a + \log([HCO_3^-]/[H_2CO_3])$ becomes larger, and pH is thus raised. Rapid breathing, (*hyperventilation*) as a result of excitement, trauma, or high body temperature can bring on the condition.

respiratory alkalosis: an alkaline condition of the blood brought on by respiratory disorders.

Metabolic alkalosis results from an abnormally high level of HCO_3^- relative to H_2CO_3 in the blood. Referring to the Henderson-Hasselbalch equation, we see that this will cause an increase in pH. The condition can be brought on by excessive loss of stomach acid, as in vomiting (thus causing dissociation of H_2CO_3 to form more H^+ and an excessive amount of HCO_3^- as the blood circulates through the stomach lining), or by diarrhea, ingestion of large amounts of base, or kidney disease. The body adjusts to metabolic alkalosis by decreasing the rate of breathing, thus raising CO_2 and H_2CO_3 levels in the blood. Table 10-7 summarizes both forms of acidosis and alkalosis.

metabolic alkalosis: an alkaline condition of the blood brought on by metabolic or physiological disorders (other than respiratory).

In both types of acidosis and alkalosis, the kidney also plays a role in regaining acid-base balance. In acidosis, the kidney compensates by excreting acidic urine, while basic urine is produced under conditions of alkalosis. These responses by the kidney reflect an attempt to rid the body of excess acid or base.

Table 10-7 Acidosis and Alkalosis

Condition	CO_2	pH	Causes
Respiratory Acidosis	Increases	Decreases	Shallow breathing Hypoventilation Lung damage
Metabolic Acidosis	Decreases	Decreases	Kidney failure Diabetes mellitus Acid ingestion Loss of HCO_3^- Starvation
Respiratory Alkalosis	Decreases	Increases	Hyperventilation High temperature
Metabolic Alkalosis	Increases	Increases	Loss of stomach acid Diarrhea Ingestion of base Kidney disease

SUMMARY

Electrolytes: Electrolytes are substances which conduct electricity when melted or dissolved in water. Strong electrolytes are those which dissociate completely into ions in water, while weak electrolytes partially dissociate into ions. Nonelectrolytes, although water soluble, do not form ions in water. Concentrations of electrolytes in body fluids are often expressed in units of milliequivalents per unit volume.

Electrolysis is the use of electricity to decompose a compound. In the process, cations are attracted to the negative electrode (cathode) where they gain electrons, thus becoming reduced. Anions move to the positive eletrode (anode) where they lose electrons and become oxidized.

Ionization of Water: Water ionizes to a very small extent to produce equal amounts of hydrogen ions and hydroxide ions. The ionization constant for water is called the ion-product constant; it has a value of 1×10^{-14} at 25°C.

The Concept of pH: pH is defined as $-\log [H^+]$. The pH scale is a convenient means of expressing the level of acidity in aqueous solutions in small positive numbers. Values of pH are usually in the range of 0–14 for common solutions. pH can be estimated by use of colored indicators or it can be measured precisely with a pH meter. This instrument has two electrodes which are immersed in the aqueous solution. The electrical signal generated by hydrogen ions registers on the meter, which is calibrated in units of pH.

Buffers: Buffers are solutions which have the ability to resist changes in pH that would be caused by the addition of small amounts of acid or base. Buffers consist of a weak acid and its salt or a weak base and its salt. The Henderson-Hasselbalch equation describes the action of buffers in terms of pH, pK_a, and $\log ([A^-]/[HA])$. According to this relationship, the pH of a buffered solution is dependent on pK_a and $[A^-]/[HA]$.

Blood contains three buffer systems: carbonic acid/bicarbonate, dihydrogen phosphate/hydrogen phosphate, and protein molecules. These buffers maintain blood pH at 7.35–7.45 except in cases of ill health or other unusual conditions.

If the buffering capacity of blood is overloaded, pH variations will occur. Acidosis is characterized as a condition in which blood pH drops below 7.35. In alkalosis, blood pH rises above 7.45. These conditions can be brought on by changes in respiration and/or metabolism. In all cases, kidney function responds in an attempt to compensate for the blood pH variations.

STUDY QUESTIONS AND PROBLEMS

1. Define the following terms:

    (a) electrolyte
    (b) strong electrolyte
    (c) weak electrolyte
    (d) nonelectrolyte
    (e) electrode
    (f) anode
    (g) cathode
    (h) electrolysis
    (i) ion-product constant, K_w
    (j) pH
    (k) pOH
    (l) pH scale
    (m) pH meter
    (n) buffer
    (o) Henderson-Hasselbalch equation
    (p) pK_a
    (q) respiratory acidosis
    (r) metabolic acidosis
    (s) respiratory alkalosis
    (t) metabolic alkalosis

2. (a) Why are most metals good conductors of electricity?
   (b) Why are most nonmetals and molecular compounds nonconductors of electricity?
   (c) Why do ionic compounds conduct electricity when melted or dissolved in water but not when they exist in the solid state?

3. Distinguish between strong electrolytes and weak electrolytes, and give an example of each.

4. Distinguish between weak electrolytes and nonelectrolytes, and give an example of each.

5. Both $Ba(OH)_2$ and H_2SO_4 are strong electrolytes, but when equimolar amounts of these are mixed together in water, the solution shows almost no electrical conductivity. Use a chemical equation to explain this observation.

6. A solution was made by dissolving 5.00×10^{-4} mole of the strong electrolyte $Ca(OH)_2$ to form 100 mL of solution.

    (a) What was the normality of the solution?
    (b) How many meq of Ca^{2+} were in the 100 mL of solution?
    (c) How many meq of OH^- were in the 100 mL of solution?

7. Describe the process of electrolysis.

8. What substances would be formed if each of the following were subjected to electrolysis?
   (a) molten NaI
   (b) molten LiBr
   (c) molten $MgCl_2$
   (d) molten AlF_3

9. When a metal object is to be electroplated, it is always made to be the cathode in the electroplating apparatus. Why?

10. Explain why tap water is usually a good conductor of electricity, but pure water is only a weak conductor.

11. What are the products of electrolysis of water? Explain how these products are formed.

12. If the hydrogen ion concentration of an aqueous solution is $1 \times 10^{-3}M$, what is the hydroxide ion concentration?

13. Calculate the pH for each of the following hydrogen ion concentrations:
    (a) $0.1M$
    (b) $1 \times 10^{-12}M$
    (c) $1 \times 10^{-4}M$
    (d) $1 \times 10^{-9}M$
    (e) $1M$
    (f) $1 \times 10^{-6}M$

14. Calculate the pOH for each of the hydrogen ion concentrations in question 13.

15. The normal pH of urine is 6.0.
    (a) What is the $[H^+]$?
    (b) What is the $[OH^-]$?
    (c) What is the pOH?

16. For a $0.001N$ KOH solution,
    (a) What is the $[OH^-]$?
    (b) What is the $[H^+]$?
    (c) What is the pH?
    (d) What is the pOH?

17. If you were a hospital worker needing 500 mL of HCl, pH 1.0, how would you go about preparing the solution from a stock solution of $12.0M$ HCl?

18. Water that is exposed to the atmosphere dissolves significant amounts of CO_2. Use this information to explain why purified water, after standing exposed to air for a while, has a pH slightly below 7.

19. A sample of blood has a pH of 7.40. Calculate the molarity range for H_3O^+ in blood. What is the molarity range for OH^- in blood?

20. At 37°C (normal body temperature for a human), the ion-product constant for water is 2.42×10^{-14}. What is the concentration of H_3O^+ in pure water at this temperature? What is the concentration of OH^- in pure water at this temperature?

21. Explain how pH can be estimated by use of indicators.

22. Give two advantages of measuring pH with a pH meter instead of pH paper.

23. Explain how a buffer composed of a weak acid and its salt resists change in pH.

24. What is the pH of a $0.20M$ formic acid (HCO_2H) solution that contains 0.15 mole of sodium formate (HCO_2Na) per liter? Assume that the amount of dissociation of formic acid is negligible but that sodium formate dissociates completely into Na^+ and HCO_2^-. The pK_a of formic acid is 3.75.

25. What is the pH of a $0.10M$ acetic acid (CH_3CO_2H) solution that contains 0.05 mole of sodium acetate (CH_3CO_2Na) per liter? Assume that the amount of dissociation of acetic acid is negligible but that sodium acetate dissociates completely into Na^+ and $CH_3CO_2^-$. The pK_a of acetic acid is 4.76.

26. What ratio of molar concentrations of sodium formate (HCO_2Na) and formic acid (HCO_2H) are needed to make a solution having a pH of 4.00? Assume that the amount of dissociation of formic acid is negligible but that sodium formate dissociates completely into Na^+ and HCO_2^-. The pK_a of formic acid is 3.75.

27. Suppose you wanted to make a buffer solution of acetic acid (CH_3CO_2H), pK_a 4.76, and sodium acetate (CH_3CO_2Na), so that $pH = pK_a$. What molar ratio of sodium acetate to acetic acid would you use? Assume that acetic acid does not dissociate but that sodium acetate dissociates completely into Na^+ and $CH_3CO_2^-$.

28. Suppose you want to make an acetic acid/sodium acetate buffer system. You have acetic acid but no sodium acetate. You also have sodium hydroxide. Is it possible to make the buffer? Explain your answer.

29. How many moles each of sodium acetate (CH_3CO_2Na) and acetic acid (CH_3CO_2H) are needed to make 2.00 liters of buffer solution, pH 4.90, at a concentration of $0.100M$ acetate? Assume that acetic acid does not dissociate but that sodium acetate dissolves completely into Na^+ and $CH_3CO_2^-$. The pK_a of acetic acid is 4.76

30. How many moles each of sodium formate (HCO_2Na) and formic acid (HCO_2H) are needed to make 3.50 liters of buffer solution, pH 3.50, at a concentration of $0.200M$ formate? Assume that formic acid does not dissociate but that sodium formate dissolves completely into Na^+ and HCO_2^-. The pK_a of acetic acid is 3.75.

31. What are the three buffer systems in blood? Which is considered the most important? Why?
32. Explain how the carbonic acid/bicarbonate buffer system controls blood pH.
33. In a sample of blood, pH 7.40, calculate the molar concentration ratio of bicarbonate to carbonic acid. The pK_a for the first dissociation of carbonic acid is 6.38.
34. Distinguish between acidosis and alkalosis.
35. Explain how a change in breathing can cause acidosis. What is the body's response to this condition?
36. Explain how a change in breathing can cause alkalosis. What is the body's response to this condition?

CHAPTER 11

AN INTRODUCTION TO ORGANIC CHEMISTRY: THE HYDROCARBONS AND THEIR HALOGEN DERIVATIVES

ORGANIC CHEMISTRY
HYDROCARBONS
- Occurrence and General Uses
- Physical Properties
- Types of Hydrocarbons

ALKANES
- Structure
- Nomenclature
- Physical Properties
- Chemical Properties
- Sources and Uses

ALKENES
- Structure
- Nomenclature
- Physical Properties
- Chemical Properties
- Sources and Uses

ALKYNES
- Structure
- Nomenclature
- Physical Properties
- Chemical Properties
- Sources and Uses

CYCLIC ALIPHATIC HYDROCARBONS

AROMATIC HYDROCARBONS
- Benzene Structure
- Benzene Derivatives
- Polycyclic Aromatic Hydrocarbons

HALOGEN DERIVATIVES OF HYDROCARBONS

Until the nineteenth century it was thought that certain substances could only be produced by living organisms. These substances were thus referred to as "organic" compounds, and it was believed that a mysterious "vital force" present only in living tissue was responsible for their synthesis. However, the "vital force" theory was overthrown in 1828 by a German chemist, Friedrich Wöhler, who accidentally synthesized an organic compound, urea (a constituent of urine), from an inorganic compound, ammonium cyanate:

$$NH_4OCN \xrightarrow{\Delta} H_2N-\underset{\underset{O}{\|}}{C}-NH_2$$

ammonium cyanate   urea

Since that time, organic chemistry has come to mean the study of covalent carbon compounds. Inorganic chemistry, once the study of gases, rocks, and ores, now includes all chemical compounds except those containing covalent carbon. We will begin our study of organic chemistry with the binary compounds called hydrocarbons.

ORGANIC CHEMISTRY

Perhaps the main reason that organic and inorganic chemistry still exist as separate fields is the enormous number of covalent carbon compounds. Of the approximately 3 million known chemical compounds, about 2.7 million (90%) contain covalent carbon. Organic compounds are composed of molecules, and thus they are usually nonelectrolytes with relatively low melting and boiling points. The burning process, in which a substance reacts with oxygen to produce carbon dioxide, water, heat, and light, is for the most part characteristic of organic compounds.

The unique chemical character of carbon accounts for the multitude of organic compounds that exist. Carbon, the first element in Group IVA, has neither a strong tendency to lose electrons nor a strong tendency to gain them. It prefers to share pairs of electrons and thus form covalent bonds. These may be single, double, or triple bonds to other carbon atoms, or single bonds to hydrogen, nitrogen, phosphorous, sulfur, and the halogens. Carbon can also form multiple bonds to nitrogen and oxygen. The unique nature of the carbon atom is such that carbon bonds to itself and to other nonmetallic elements to form a seemingly limitless number of compounds and isomers.

Organic compounds can be classified by distinctive groupings of atoms incorporated into their structures. These atomic groupings, called *functional groups*, are responsible for many of the physical and chemical properties of their corresponding organic compounds. For example, the alcohols contain the hydroxyl group, $-OH$. Because all alcohols have this functional group, we can predict that all of them will participate in reactions characteristic of the hydroxyl group. In chapter 12 you will learn more about properties of the hydroxyl group, but for now it is important to realize that this is only one of many functional groups. By using functional groups as a means of classification, we can simplify the study of organic chemistry. Table 11-1 illustrates this kind of classification. Note that we use the symbol "$R-$" in the general formulas to stand for any hydrocarbon group bonded to the functional group.

EXAMPLE 11.1

With the help of Table 11-1, circle and name each functional group in the following compounds:

(a) $H_3C-CH_2-\overset{\overset{\displaystyle O}{\|}}{C}-CH_3$

(b) H_3C-CH_2-SH

(c) $H_3C-CH_2-C\begin{smallmatrix}\diagup O\\ \diagdown OH\end{smallmatrix}$

(d) H_3C-CH_2-OH

(e) $H_3C-O-CH_2-CH_3$

SOLUTION

(a) $H_3C-CH_2-\overset{\overset{\displaystyle O}{\|}}{C}-CH_3$
   carbonyl group

(b) H_3C-CH_2-SH
   thiol group

(c) $H_3C-CH_2-C\begin{smallmatrix}\diagup O\\ \diagdown OH\end{smallmatrix}$
   carboxyl group

(d) H_3C-CH_2-OH
   hydroxyl group

(e) $H_3C-O-CH_2CH_3$
   ether group

EXERCISE 11.1 With the help of Table 11-1, circle and name each functional group in the following compounds:

(a) $H_3C-CH_2-\overset{\overset{O}{\|}}{C}-CH_2-C\overset{\nearrow O}{\underset{\searrow OH}{}}$

(b) $H_3C-S-S-CH_2-CH_3$

(c) H_3C-CH_2-Cl

(d) $H_3C-\overset{\overset{}{|}}{\underset{H}{N}}-CH_2-CH_3$

(e) $H_3C-\overset{\overset{}{\|}}{\underset{O}{C}}-CH_2-OH$

THE HYDROCARBONS

As you remember from chapter 3, hydrocarbons are binary molecular compounds composed of hydrogen and carbon. In these molecules, carbon atoms form the framework by linking together in chains. Hydrogen atoms are attached to the carbon atoms so that each carbon has four bonds. The carbon atoms may be arranged consecutively, in branches, in rings, or in more complicated structures. Theoretically the number of carbon atoms in a hydrocarbon molecule may be infinitely great, although the largest nonpolymeric hydrocarbon to be synthesized contains 110 carbon atoms.

Occurrence and General Uses

Most hydrocarbons occur naturally as constituents of petroleum and natural gas. In addition, some hydrocarbons are found in trees and other plants. Turpentine, which is obtained from pine trees, contains as its major component a hydrocarbon called α-pinene, $C_{10}H_{16}$. Hydrocarbons are also present in insect waxes such as beeswax. Lycopene, $C_{40}H_{56}$, is the red pigment in tomatoes and watermelon. Certain isomers of lycopene, the carotenes, are the yellow orange color in carrots. Crude plantation rubber is composed almost entirely of a hydrocarbon whose molecules contain approximately one-hundred thousand carbon atoms each. Because they have been so abundant in the past, we have grown dependent on the hydrocarbons as fuel and as raw material in the manufacture of plastics, synthetic rubber, solvents, explosives, lubricants, alcohols, synthetic fabrics, and many other useful products.

Physical Properties

Even though hydrocarbons are quite a diverse group structurally, they have several physical properties in common. Most hydrocarbons are colorless and, as nonpolar molecules, they are insoluble in water but are usually soluble in one another. They may be gases, liquids, or solids at room temperature, depending on molecular weight. Hydrocarbons with fewer than 5 carbon atoms are gases at room temperature, while those with 6-18 carbons are liquids at room temperature. These are less dense

Table 11-1 The Functional Groups of Organic Molecules

Functional Group	Functional Group Name	Class Name of Compounds	General Formula of Compounds	Condensed Formula of Compounds	Examples
$\text{C}=\text{C}$	double bond	alkene	$\underset{(H)}{\overset{H}{R-C}}=\underset{(H)}{\overset{H}{C-R}}$	R–CH=CH–R	$H_2C=CH_2$ ethene
$-C\equiv C-$	triple bond	alkyne	$\underset{(H)}{R-C}\equiv \underset{(H)}{C-R}$	R–C≡C–R	H–C≡C–H ethyne
–OH	hydroxyl	alcohol	R–OH	ROH	CH_3OH methyl alcohol
$-\text{C}-\text{O}-\text{C}-$	ether	ether	R–O–R	ROR	$H_3C-O-CH_3$ dimethyl ether
$\overset{O}{\underset{}{\parallel}}\text{C}$	carbonyl	aldehyde	$\underset{(H)}{R-\overset{O}{\overset{\parallel}{C}}-H}$	RCHO	$H_3C-\overset{O}{\overset{\parallel}{C}}-H$ acetaldehyde
		ketone	$R-\overset{O}{\overset{\parallel}{C}}-R$	RCOR	$H_3C-\overset{O}{\overset{\parallel}{C}}-CH_3$ acetone
$-\overset{O}{\overset{\parallel}{C}}-\text{OH}$	carboxyl	carboxylic acid	$R-\overset{O}{\overset{\parallel}{C}}-OH_{(H)}$	RCOOH	$H_3C-\overset{O}{\overset{\parallel}{C}}-OH$ acetic acid
$-\overset{O}{\overset{\parallel}{C}}-\text{O}-\text{C}-$	ester	ester	$R-\overset{O}{\overset{\parallel}{C}}-O-R_{(H)}$	RCOOR	$H_3C-\overset{O}{\overset{\parallel}{C}}-OCH_2CH_3$ ethyl acetate

Table 11-1 The Functional Groups of Organic Molecules (Continued)

Functional Group	Functional Group Name	Class Name of Compounds	General Formula of Compounds	Condensed Formula of Compounds	Examples
$-\underset{\mid}{N}-$	amino	amine	R—N—H (primary) 　　│ 　　H	RNH_2	H_3C-NH_2 methyl amine
			R—N—R (secondary) 　　│ 　　H	R_2NH	$H_3C-N-CH_3$ 　　│ 　　H dimethyl amine
			R—N—R (tertiary) 　　│ 　　R	R_3N	$H_3C-N-CH_3$ 　　│ 　　CH_3 trimethyl amine
$\underset{NH_2}{\overset{O}{\underset{\|}{C}}}$	amide	amide	$\underset{(H)}{\overset{O}{\underset{\|}{R-C-N}}}\overset{H}{\underset{H}{\diagdown}}$	$RCONH_2$	$H_3C-\overset{O}{\underset{\|}{C}}\diagdown NH_2$ acetamide
$-NO_2$	nitro	nitro	$R-NO_2$	RNO_2	H_3C-NO_2 nitromethane
$-SH$	thiol	thiol	$R-SH$	RSH	H_3C-SH methanethiol
$-S-S-$	disulfide	disulfide	$R-S-S-R$	$RSSR$	$H_3C-S-S-CH_3$ dimethyl disulfide
$-X$ ($-F$, $-Cl$, $-Br$, $-I$)	halo- (fluoro-, chloro-, bromo-, iodo-)	halide	$R-X$	RX	H_3CCl chloromethane

than water, and they float on the surface of water. Hydrocarbons with more than 18 carbon atoms are solids at room temperature.

Types of Hydrocarbons

aliphatic (al-ih-FAAH-tic)

Hydrocarbons can be classified as *aliphatic* or *aromatic*. The term aliphatic comes from the Greek word for "fat," and it was used originally to designate hydrocarbons which could be derived from animal fat. Those hydrocarbon molecules are usually open or branched chains, but the term aliphatic now means all hydrocarbons that are not aromatic. The word aromatic first referred to the pleasant aroma of certain hydrocarbons, but we now use it for hydrocarbons which have closed rings with unusual hybrid bonds. We will come back to aromatic hydrocarbons in this chapter, but for now, let us focus on the group of aliphatic hydrocarbons known as alkanes.

ALKANES

Alkanes are hydrocarbons which contain only single bonds. Each carbon atom is bonded to four other atoms, and because this arrangement corresponds to maximum bonding to other atoms, the alkanes are said to be "saturated." The carbon atoms may be bonded together consecutively, in branched chains, or even in rings.

The open-chain alkanes have the general formula C_nH_{2n+2}, where n represents the number of carbon atoms in the molecule. Table 11-2 lists the first ten linear alkanes with their names and formulas. Notice that as the number of carbon atoms increases from one alkane to the next, the structures differ only by a $-CH_2-$ group.

Structure

The simplest alkane is methane. If we wish to represent the structure of methane in three dimensions, we can use ball-and-stick models, as shown

Table 11-2 The First Ten Linear Alkanes

Number of Carbon Atoms	Structure	Name
1	CH_4	methane
2	CH_3CH_3	ethane
3	$CH_3CH_2CH_3$	propane
4	$CH_3CH_2CH_2CH_3$	butane
5	$CH_3CH_2CH_2CH_2CH_3$	pentane
6	$CH_3CH_2CH_2CH_2CH_2CH_3$	hexane
7	$CH_3CH_2CH_2CH_2CH_2CH_2CH_3$	heptane
8	$CH_3CH_2CH_2CH_2CH_2CH_2CH_2CH_3$	octane
9	$CH_3CH_2CH_2CH_2CH_2CH_2CH_2CH_2CH_3$	nonane
10	$CH_3CH_2CH_2CH_2CH_2CH_2CH_2CH_2CH_2CH_3$	decane

in Figure 11-1. The hydrogen atoms are considered to occupy the corners of a regular tetrahedron, with bond angles between the hydrogens of 109.5°.

Ethane, propane, and butane can also be represented by ball-and-stick models (Figure 11-1). When we examine these models, we find that the bonds to carbon also have angles of 109.5°. This is generally true for a carbon atom with four single bonds.

The alkanes having no branches are called linear alkanes. The use of the term linear is a bit misleading because unbranched alkanes do not actually have a linear structure. Instead, the carbon atoms have a three-dimensional zigzag arrangement. We usually show linear alkanes as having an extended structures as in Figure 11-1. However, groups of atoms are free to rotate about single bonds, and linear alkanes having four or more carbons are quite flexible. They twist and turn continuously in space, and at one moment a linear alkane might have quite a different overall shape than at another moment.

Although models are helpful in visualizing the three-dimensional aspects of alkane structure, we must use formulas to represent alkanes on paper. The *molecular formula* shows the number of atoms present in the molecule, such as

ethane	C_2H_6
propane	C_3H_8
butane	C_4H_{10}

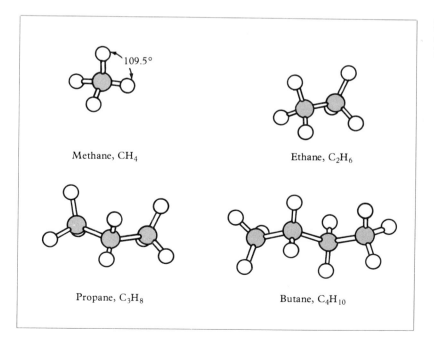

Fig. 11-1 Ball-and-Stick Models of Methane, Ethane, Propane, and Butane

The *condensed structural formula* gives major structural features of a molecule:

ethane	CH_3CH_3
propane	$CH_3CH_2CH_3$
butane	$CH_3CH_2CH_2CH_3$

The *structural formula* provides as many structural features as is possible in two dimensions:

ethane
$$\begin{array}{c} H \, H \\ | | \\ H-C-C-H \\ | | \\ H \, H \end{array}$$

propane
$$\begin{array}{c} H \, H \, H \\ | | | \\ H-C-C-C-H \\ | | | \\ H \, H \, H \end{array}$$

butane
$$\begin{array}{c} H \, H \, H \, H \\ | | | | \\ H-C-C-C-C-H \\ | | | | \\ H \, H \, H \, H \end{array}$$

As was mentioned in chapter 4, alkanes having four or more carbon atoms exist as isomers. More specifically, these are *structural isomers*, compounds having the same molecular formula but different sequences of atoms bonded together. Butane is the smallest alkane to have structural isomers. (Methane, ethane, and propane are such simple molecules that there is only one arrangement each for their atoms.) There are two structural isomers for butane:

$$CH_3-CH_2-CH_2-CH_3 \qquad\qquad CH_3-CH-CH_3$$
$$|$$
$$CH_3$$

linear isomer   branched isomer
boiling point 0°C   boiling point −12°C

butane isomers

Each is an individual compound with its own set of physical properties. Ball-and-stick models of these two isomers are shown in Figure 11-2.

As the number of carbon atoms increases, the number of isomers for an alkane also increases. Thus, there are three isomers of pentane, five of hexane, and nine of heptane, as shown in Table 11-3.

Nomenclature

Through the years, alkanes and other organic compounds have acquired numerous common names. Because of this, a nomenclature system has

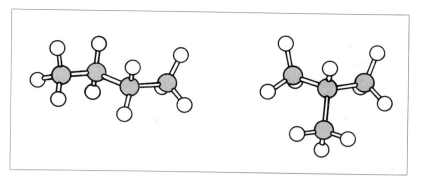

Fig. 11-2 Ball-and-Stick Models of the Two Isomers of Butane (C_4H_{10})

Table 11-3 Isomers of the First Seven Open-Chain Alkanes

Number of Carbon Atoms	Isomeric Structures	Number of Isomers	Boiling Point (°C)
1	CH_4	1	−162
2	CH_3-CH_3	1	−89
3	$CH_3-CH_2-CH_3$	1	−42
4	$CH_3-CH_2-CH_2-CH_3$	2	0
	$CH_3-CH(CH_3)-CH_3$		−12
5	$CH_3-CH_2-CH_2-CH_2-CH_3$	3	36
	$CH_3-CH(CH_3)-CH_2-CH_3$		28
	$CH_3-C(CH_3)_2-CH_3$		10
6	$CH_3-CH_2-CH_2-CH_2-CH_2-CH_3$	5	69
	$CH_3-CH(CH_3)-CH_2-CH_2-CH_3$		60
	$CH_3-CH_2-CH(CH_3)-CH_2-CH_3$		63
	$CH_3-CH(CH_3)-CH(CH_3)-CH_3$		58
	$CH_3-C(CH_3)_2-CH_2-CH_3$		50
7	$CH_3-CH_2-CH_2-CH_2-CH_2-CH_2-CH_3$	9	98
	$CH_3-CH(CH_3)-CH_2-CH_2-CH_2-CH_3$		90
	$CH_3-CH_2-CH(CH_3)-CH_2-CH_2-CH_3$		92

(continued next page)

Table 11-3 Isomers of the First Seven Open-Chain Alkanes (continued)

Number of Carbon Atoms	Isomeric Structures	Number of Isomers	Boiling Point (°C)

7 (cont.)	CH$_3$–CH–CH–CH$_2$–CH$_3$
	    | |
	   CH$_3$ CH$_3$
			90

	CH$_3$–CH–CH$_2$–CH–CH$_3$
	    | |
	   CH$_3$ CH$_3$
			80

	CH$_3$ | CH$_3$–C–CH$_2$–CH$_2$–CH$_3$ | CH$_3$		79
	CH$_3$ | CH$_3$–CH$_2$–C–CH$_2$–CH$_3$ | CH$_3$		86
	CH$_3$ | CH$_3$–C—CH–CH$_3$ | | CH$_3$ CH$_3$		81

	CH$_3$–CH$_2$–CH–CH$_2$–CH$_3$
	           |
	         CH$_2$–CH$_3$
			93

been worked out by the International Union of Pure and Applied Chemistry (IUPAC), which unambiguously assigns one name to each compound. The IUPAC rules for naming alkanes can be summarized:

1. The names of all alkanes end in "-ane."

2. Alkanes are named on the basis of the longest continuous chain of carbon atoms; this "main chain" is considered to be the parent compound of the alkane. Table 11-2 lists the first ten such parent compounds. If there are no branches in the alkane, it is simply named by using Table 11-2.

3. If there are branches in the alkane, a number is assigned to each carbon atom in the longest continuous chain, beginning at the end of the chain nearest a branch. In the example below, the longest continuous chain has seven carbon atoms. Do not be misled by the bend in the main chain.

$$\begin{array}{c} {}^{1}CH_3-{}^{2}CH-{}^{3}CH_2-{}^{4}CH-CH_2-CH_3 \\ | | \\ CH_3 {}^{5}CH_2 \\ | \\ {}^{6}CH_2 \\ | \\ {}^{7}CH_3 \end{array}$$

4. Saturated hydrocarbon substituents, or branches, are called *alkyl groups*, and the name of an alkyl group is derived from the corresponding alkane by changing the "-ane" ending of the alkane to "-yl."

   —CH_3 is "methyl"   (derived from methane)

   —CH_2CH_3 is "ethyl"   (derived from ethane)

   —$CH_2CH_2CH_3$ is "propyl"   (derived from propane)

   and so on.

5. The names of any alkyl groups attached to the main chain are written before the name of the parent compound, in alphabetical order. The Greek prefixes di-, tri-, tetra-, penta-, and so on, are used to indicate two or more identical alkyl group attachments. The position of each substituent on the main chain is designated by the numbers assigned in rule 3. Commas are used to separate numbers, and hyphens are used to separate numbers from words. The name of the compound is written as one word, with the name of the main chain appearing last.

$$CH_3 \!-\! \underset{\underset{CH_3}{|}}{\overset{}{CH}} \!-\! CH_2 \!-\! \underset{\underset{\underset{\underset{^7CH_3}{|}}{^6CH_2}}{\underset{|}{^5CH_2}}}{\overset{}{CH}} \!-\! CH_2 \!-\! CH_3$$
$$\underset{1234}{}$$

4-ethyl-2-methylheptane

$$CH_3 \!-\! \underset{\underset{CH_3}{|}}{\overset{\overset{CH_3}{|}}{C}} \!-\! CH_2 \!-\! CH_2 \!-\! \underset{\underset{\underset{\underset{^8CH_3}{|}}{^7CH_2}}{\underset{|}{^6CH_2}}}{\overset{}{CH}} \!-\! CH_2 \!-\! CH_3$$

2,2-dimethyl-5-ethyloctane

EXAMPLE 11.2

Name the following alkane according to IUPAC rules.

$$CH_3 \!-\! \underset{\underset{CH_3}{|}}{\overset{}{CH}} \!-\! CH_3$$

SOLUTION

First we locate the main chain. In this case, it has three carbon atoms; thus the parent compound is propane.

$$CH_3 \!-\! \underset{\underset{CH_3}{|}}{\overset{}{CH}} \!-\! CH_3$$

Next we number the carbon atoms in the main chain, starting at either end, since each is the same distance from the branch.

$$\underset{1}{CH_3} \!-\! \underset{\underset{CH_3}{|}}{\overset{}{\underset{2}{CH}}} \!-\! \underset{3}{CH_3}$$

There is only one substituent, and its name is methyl; thus, the name of the compound is 2-methylpropane.

EXAMPLE 11.3

Name the following alkane according to IUPAC rules.

$$\begin{array}{c}CH_3 CH_3\\ | |\\ CH_3-CH-CH-CH-CH_3\\ |\\ CH_2\\ |\\ CH_3\end{array}$$

SOLUTION

First we locate the main chain. In this case, the main chain is bent, but the bend is not intended to convey any information about its three-dimensional structure. The main chain has six carbon atoms, and thus its name is hexane.

$$\begin{array}{c}CH_3 CH_3\\ | |\\ CH_3-CH-CH-CH-CH_3\\ |\\ CH_2\\ |\\ CH_3\end{array}$$

Next we number the carbon atoms in the main chain, starting at the end nearest a branch.

$$\begin{array}{c}CH_3 CH_3\\ | |\\ \underset{1}{CH_3}-\underset{2}{CH}-\underset{3}{CH}-\underset{4}{CH}-CH_3\\ |\\ {}^{5}CH_2\\ |\\ {}^{6}CH_3\end{array}$$

The correct IUPAC name is thus 2,3,4-trimethylhexane. Note that all of the following are also 2,3,4-trimethylhexane. They are just drawn differently.

$$\begin{array}{c}CH_3 CH_3\\ | |\\ CH_3-CH-CH-CH-CH_2-CH_3\\ |\\ CH_3\end{array} \qquad \begin{array}{c}CH_3 CH_3 CH_3\\ | | |\\ CH_3-CH-CH-CH\\ |\\ CH_2\\ |\\ CH_3\end{array}$$

$$\begin{array}{c}CH_3 CH_3\\ | |\\ CH-CH-CH-CH_2-CH_3\\ | |\\ CH_3 CH_3\end{array}$$

EXERCISE 11.2 Name the following alkanes according to IUPAC rules.

(a) $\begin{array}{c}CH_3-CH_2-CH-CH_3\\ |\\ CH_2\\ |\\ CH_3\end{array}$

(c) $\begin{array}{c}CH_3-CH-CH_2-CH-CH_2-CH_3\\ | |\\ CH_2 CH_3\\ |\\ CH_3\end{array}$

(b) $\begin{array}{c}CH_3\\ |\\ CH_3-CH_2-C-CH_2-CH_3\\ |\\ CH_3\end{array}$

Physical Properties

The boiling points, melting points, and densities of the first ten alkanes are given in Table 11-4. As you can see, these values increase with molecular weight, a consequence of the fact that as the molecules get larger, London dispersion forces become more powerful. The increased attraction between larger molecules results in higher boiling and melting points. It also increases density by crowding the molecules closer together.

Table 11-3 lists boiling points within sets of isomers. Notice that in each case, the more highly branched isomer has the lowest boiling point. This effect comes about because branching forces the molecules farther apart, and since London dispersion forces operate only over very short distances, the branches prevent maximum attraction between molecules. Thus, in each case, increasing the number of branches lowers the boiling point and melting point.

Chemical Properties

Of all the organic compounds that exist, alkanes are the least reactive. They are resistant to attack by strong acids, strong bases, most oxidizing agents, and most reducing agents. Because of their almost nonexistent reactivity, alkanes are sometimes referred to as *paraffins*, a word derived from Latin meaning "little affinity." The two reactions worthy of mention for alkanes are combustion and halogenation.

Combustion

All alkanes react with oxygen in the process called combustion. Thus all alkanes are flammable. If an alkane is mixed with oxygen, no reaction will take place unless a certain amount of energy is provided. This initial energy requirement is the energy of activation, and once the requirement is met, reaction takes place. In most instances, a small flame or

Table 11-4 Physical Properties of the First Ten Linear Alkanes

Name	Number of Carbon Atoms	Melting Point (°C)	Boiling Point (°C)	Density* (g/mL)
Methane	1	−182	−162	—
Ethane	2	−183	−89	—
Propane	3	−188	−42	—
Butane	4	−138	0	—
Pentane	5	−130	36	0.626
Hexane	6	−95	69	0.659
Heptane	7	−91	98	0.684
Octane	8	−57	126	0.703
Nonane	9	−54	151	0.718
Decane	10	−30	174	0.730

*at 20°C; no data are given for species that are gaseous at this temperature.

spark is enough to furnish the energy of activation. The reaction then proceeds to give carbon dioxide, water, and heat, provided there is adequate oxygen available:

$$2\ CH_3-CH_3 + 7\ O_2 \longrightarrow 4\ CO_2 + 6\ H_2O$$
ethane

If an alkane is burned in a limited supply of oxygen, incomplete combustion occurs, and carbon monoxide is produced:

$$2\ CH_3-CH_3 + 5\ O_2 \longrightarrow 4\ CO + 6\ H_2O$$
ethane

Carbon monoxide is a toxic gas because it combines with hemoglobin and prevents it from transporting oxygen. Thus, breathing carbon monoxide results in an inability of hemoglobin to function normally in delivering oxygen to the tissues. A person inhaling carbon monoxide will become lethargic and sleepy, and may eventually fall into a coma and die if fresh air is not available. Because some CO forms in any combustion, we are cautioned to use properly ventilated space heaters, and we are advised not to use charcoal grills indoors. Even small amounts of carbon monoxide, such as those produced by cigarettes, are detrimental and can produce impaired judgment and reduced muscle control.

Halogenation

halogenation: reaction of any compound with halogen so that one or more halogen atoms are introduced into the compound.

The second important reaction of alkanes is that of *halogenation*. Alkanes undergo substitution reactions with fluorine, chlorine, and bromine to produce halogenated hydrocarbons. The reaction with fluorine is explosive and hard to control, whereas the reactions with chlorine and bromine are much less violent. Iodine is not reactive enough to enter halogenation reactions with alkanes. Since the three halogens react identically, we can illustrate halogenation of methane by using X_2 to represent diatomic halogen molecules. The reaction requires high temperature or ultraviolet light as a source of activation energy, except in the case of fluorination. Note that the reaction proceeds in a series of steps until all of the hydrogen atoms in the alkane have been replaced by halogen:

$$CH_4 + X_2 \longrightarrow CH_3X + HX$$
$$CH_3X + X_2 \longrightarrow CH_2X_2 + HX$$
$$CH_2X_2 + X_2 \longrightarrow CHX_3 + HX$$
$$CHX_3 + X_2 \longrightarrow CX_4 + HX$$

The sum of these four steps is

$$CH_4 + 4\ X_2 \longrightarrow CX_4 + 4\ HX$$

Sources and Uses

Most alkanes are isolated from petroleum. They are considered highly desirable energy sources because of the heat produced in combustion. Methane is the main component of natural gas used for cooking and heating. Propane, because of its higher boiling point, can easily be liquefied by moderate pressures, and it is sold as liquefied propane gas (LPG) for use in camping and in homes not served by natural gas. Butane is also readily liquefied, and it is the familiar fuel of butane cigarette lighters. Inhalation of any of these gases in large amounts can produce sleepiness, unconsciousness, and even death. Since they are colorless and odorless, it is difficult to detect their presence. For this reason, strong-smelling sulfur compounds are added to natural gas so that leaks can be detected.

Alkanes with 5-12 carbon atoms, liquids at room temperature, are blended to make gasoline for automobiles. Those having 12-18 carbon atoms are also liquids; they are used for diesel fuel, jet fuel, and heating oil. Upon contact with the skin, liquid alkanes dissolve body oils and fatty material, thus causing skin irritation. If they enter the lungs, dissolution of fatty material from cell membranes will cause the lungs to lose flexibility, and fluid accumulates as in pneumonia. Thus, anyone who accidentally swallows liquid alkanes should never be made to vomit, as that would risk exposure of the lungs to the fluid.

Alkanes with 19 or more carbon atoms tend to be solids at room temperature and are used as lubricants and greases. They are often present in skin and hair lotions as replacements for natural oils. Petroleum jelly is an example of one formulation used as a skin softener or protective coating. The mixture of solid alkanes called paraffin is used as a wax sealant for homemade jams and jellies.

ALKENES

The *alkenes* are hydrocarbons having one or more double bonds. Because of the double bonds, alkenes do not contain as much hydrogen as the corresponding alkanes, and the alkenes are said to be "unsaturated." As with the alkanes, the carbon atoms may be bonded together consecutively, in branched pairs, or in rings.

The open-chain alkenes have the general formula, C_nH_{2n}, where n represents the number of carbon atoms in the molecule. Table 11-5 lists representative alkenes with their names and formulas. Notice that alkenes also differ from one another by one $-CH_2-$ group as we go from one to the next on the list.

Structure

The simplest alkene is ethene. If we wish to represent the structure of ethene in three dimensions, we can use ball-and-stick models, as shown

Fig. 11-3 The Ethene Molecule, C_2H_4 (a) Ball-and-Stick Model (b) Space-Filling Model

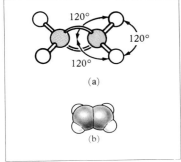

in Figure 11-3. Remember that all of the atoms in ethene lie in the same plane. This is always true of the two atoms connected by a double bond and the four atoms attached to them. In ethene, all of the bond angles are 120°.

Alkenes exist as structural isomers like the alkanes. However, the presence of a double bond allows alkenes to display still another kind of

Table 11-5 Examples of Alkenes

Structure	Molecular Formula	Name
$CH_2=CH_2$	C_2H_4	ethene (ethylene)
$CH_3CH=CH_2$	C_3H_6	propene (propylene)
$CH_3CH_2CH=CH_2$	C_4H_8	1-butene (butylene)
$CH_3CH_2CH_2CH=CH_2$	C_5H_{10}	1-pentene
$CH_3CH_2CH_2CH_2CH=CH_2$	C_6H_{12}	1-hexene
$CH_3-C=CH_2$ $\quad\ \ \vert$ $\quad\ \ CH_3$	C_4H_8	2-methylpropene
$CH_3-CH-CH=CH_2$ $\qquad\ \ \vert$ $\qquad\ \ CH_3$	C_5H_{10}	3-methyl-1-butene
$CH_3CH_2-CH=CH_2$ $\qquad\quad\ \vert$ $\qquad\quad\ CH_3$	C_5H_{10}	2-methyl-1-butene

geometric isomers: isomers of alkenes in which identical substituents on each of the carbons connected by the double bond can be on the same side of the double bond (*cis*) or on the opposite sides of the double bond (*trans*).

isomerism, that of geometric isomerism. *Geometric isomers*, or more specifically *cis-trans* isomers, exist because the double bond between two carbon atoms does not allow rotation of attached groups. Thus, there can be two geometric forms of 2-butene:

$$\begin{array}{cc} CH_3 \quad\quad CH_3 \\ \diagdown C=C \diagup \\ \diagup \quad\quad \diagdown \\ H \quad\quad H \end{array} \qquad \begin{array}{cc} CH_3 \quad\quad H \\ \diagdown C=C \diagup \\ \diagup \quad\quad \diagdown \\ H \quad\quad CH_3 \end{array}$$

cis-2-butene   *trans*-2-butene

In order to distinguish between the two forms of butene, we call the one with alkyl groups on the same side of the double bond *cis*, a word borrowed from Latin meaning "on the near side of." The isomer having alkyl groups on opposite sides of the double bond is called *trans*, a Latin word meaning "across." The isomers *cis*-2-butene and *trans*-2-butene are different compounds, each having its own set of physical properties. Only those alkenes having two different groups on each of the double-bonded carbon atoms can exist as geometric isomers. Thus, alkenes having double bonds at the end of a chain, including ethene and propene, do not have *cis-trans* isomers.

Nomenclature

For reasons lost in history, ethene was once called olefiant gas, and the series of alkenes came to be known as *olefins*. This term is still used as an alternate name for alkenes. In addition, several of the smaller alkenes

PERSPECTIVE: Petroleum Refining

Petroleum, crude oil found in natural underground deposits, is a complicated mixture of alkanes along with smaller amounts of other hydrocarbons and sulfur, nitrogen, and oxygen compounds. Petroleum deposits are the result of spontaneous degradation of plant and animal remains which accumulated through the ages. Converting the complex mixture of chemicals in petroleum to separate, useful products is the process called petroleum refining.

Fractional distillation is the primary refinery process in which crude oil is heated in a distilling tower containing several openings (sidestreams) along the side where vapors are withdrawn. The lower molecular weight molecules come off near the top of the column, and higher molecular weight molecules are taken off from the lower openings. In this way, the petroleum is separated into fractions of molecules grouped by boiling range and thus roughly by molecular weight. The remaining solid residue, called asphalt, is composed of high molecular weight hydrocarbons which were not volatilized under the operating temperatures.

A Petroleum Refinery

have common names ending in "-ylene." For example, the common name for ethene is ethylene. However, because of ambiguities in common names, the following IUPAC rules should be used in naming alkenes:

1. The characteristic name ending for alkenes is "-ene."

2. Alkenes are named on the basis of the longest continuous chain of carbon atoms containing the double bond; this "main chain" is considered the parent compound of the alkene. The main chain is named by selecting the alkane with the same number of carbon atoms in a chain and changing the "-ane" ending of the alkane name to "-ene."

3. The carbon atoms in the main chain are numbered starting at the end nearest the double bond. This gives the first carbon of the double bond the lowest possible number. This number is written in front of the name of the alkene. (Note that numbers are not needed to locate the double bond in ethene and propene.) For example, the alkene below is named 1-butene, not 3-butene or 4-butene.

$$CH_3-CH_2-CH=CH_2$$

4. Groups attached to the main chain are identified by numbers and names individually, as with alkanes.

5. If the alkene contains more than one double bond, the location of each is identified by a number. When there are two double bonds, the name of the parent compound ends in "-diene;" when there are three double bonds, the name of the parent compound ends in "-triene," and so on. In these cases, only the "-ne" ending of the alkane name is changed to "-diene," "-triene," etc.

$$CH_2=CH-CH=CH_2 \qquad CH_3-CH=CH-CH=CH-CH=CH_2$$
1,3-butadiene $\qquad\qquad$ 1,3,5-heptatriene

$$CH_2=\underset{\underset{CH_3}{|}}{C}-CH=CH_2 \qquad CH_3-\underset{\underset{CH_3}{|}}{C}=CH-CH=\underset{\underset{CH_3}{|}}{C}-CH=CH_2$$
2-methyl-1,3-butadiene $\qquad$ 3,6-dimethyl-1,3,5-heptatriene

EXAMPLE 11.4

Name the following alkene according to IUPAC rules.

$$CH_3-CH=CH-CH_2-CH_3$$

SOLUTION

This unbranched alkene has five carbons and one double bond; hence its name must end in "-pentene." We number the carbon atoms starting at the end nearest the double bond,

$$\underset{1}{CH_3}-\underset{2}{CH}=\underset{3}{CH}-\underset{4}{CH_2}-\underset{5}{CH_3}$$

and thus the correct IUPAC name is 2-pentene.

EXAMPLE 11.5

Name the following alkene according to IUPAC rules.

$$\begin{array}{c}CH_3-CH_2-CH_2-CH-CH_3\\|\\CH\\||\\CH_2\end{array}$$

SOLUTION

We select the main chain and number it:

$$\begin{array}{c}\underset{6}{C}H_3-\underset{5}{C}H_2-\underset{4}{C}H_2-\underset{3}{C}H-CH_3\\|\\{}^2CH\\||\\{}^1CH_2\end{array}$$

Thus the correct IUPAC name is 3-methyl-1-hexene.

EXERCISE 11.3 Name the following alkenes according to IUPAC rules.
(a) $CH_3-CH=CH-CH_3$
(b) $\begin{array}{c}CH_3-CH-CH=CH-CH-CH_3\\|\qquad\qquad\qquad|\\CH_3\qquad\qquad\;\;CH_2\\|\\CH_3\end{array}$
(c) $\begin{array}{c}CH_2=CH-CH=CH-CH-CH_3\\|\\CH_3\end{array}$

Physical Properties

The physical properties of alkenes are rather similar to those of alkanes, as illustrated in Table 11-6. The same trend of increasing boiling points with molecular weights is apparent for alkenes. However, it should be noted that after ethene and propene, the alkenes have lower boiling and melting points than the corresponding linear alkanes. This is because the geometry of the double bond does not allow the larger alkenes to

Table 11-6 Physical Properties of Alkenes

Name	Number of Carbon Atoms	Melting Point (°C)	Boiling Point (°C)	Density* (g/mL)
Ethene	2	−169	−104	—
Propene	3	−185	−48	—
1-Butene	4	−185	−6	—
1-Pentene	5	−165	30	0.641
1-Hexene	6	−139	64	0.673
1-Heptene	7	−119	93	0.697
1-Octene	8	−102	121	0.716

*at 20°C; no data are given for species that are gaseous at this temperature.

pack as closely together in the liquid and solid states as linear alkanes do. Hence, most of the alkenes do not have maximum attraction between molecules, and less energy is required to melt them and to vaporize them than for the corresponding linear alkanes. This effect is similar to that observed in branched alkanes.

Chemical Properties

Like all hydrocarbons, alkenes are easily oxidized by oxygen in combustion reactions; however, the double bonds allow alkenes to participate in more chemical reactions than alkanes. Because of this, alkenes are more useful in preparing other organic compounds than as fuels. Since the double bond is responsible for the chemical reactivity of alkenes, it is the functional group by which alkenes are characterized. It serves as a site on the alkene molecule where reactions take place. The two most important reactions of double bonds, and thus of alkenes, are addition and polymerization, a special type of addition reaction.

Addition

Addition reactions are actually combination reactions in which molecules are added to the double bond to produce a saturated compound. An example is the addition of halogen, in which a diatomic halogen molecule reacts with an alkene double bond (since all halogens will participate in this reaction, we will use X_2 to represent the halogen molecule):

$$CH_2=CH-CH_2-CH_3 + X_2 \longrightarrow \underset{\underset{X}{|}}{CH_2}-\underset{\underset{X}{|}}{CH}-CH_2-CH_3$$

1-butene

Addition of bromine is often used as a test for the presence of double bonds. The brownish red color of bromine disappears as the alkene, which is colorless, reacts with bromine to form a dibromoalkane, also colorless.

Hydrogen can also be added to double bonds in alkenes in the process called *hydrogenation*. In this reaction, a platinum or nickel catalyst is necessary, and high temperatures and high pressures of hydrogen are required.

$$CH_3-CH=CH-CH_3 + H_2 \xrightarrow[\Delta]{Pt \text{ or } Ni} CH_3-CH_2-CH_2-CH_3$$

2-butene   butane

Hydrogenation is used to convert double bonds in unsaturated fats to single bonds in the manufacture of margarine and shortening from vegetable oils. This increase in saturation causes a corresponding increase in melting point, so that margarine and shortening have higher melting points than the vegetable oils from which they were made.

A third addition reaction involves reversible combination of hydrogen halide with an alkene. This reaction requires a strong acid as catalyst and leads to products containing one halogen atom. The reaction is illustrated by the following equation, where HX represents a hydrogen halide (HF, HCl, HBr, or HI). Products can be maximized by using a large excess of HX.

$$CH_2=CH_2 + HX \xrightleftharpoons{H^+} \underset{\underset{H}{|}\;\;\underset{X}{|}}{CH_2-CH_2}$$

ethene

In the addition of HX to an alkene like 1-propene, $CH_2=CH-CH_3$, we might expect there to be two products—one in which hydrogen adds to carbon 1 and halogen to carbon 2, and a second with hydrogen added to carbon 2 and halogen added to carbon 1. However, these reactions follow a definite pattern, and just one product is formed from 1-propene and other alkenes having a different number of hydrogens attached to the double-bonded carbons. The added hydrogen bonds to the carbon in the double bond that already had the greater number of hydrogen atoms, and the halogen bonds to the other carbon originally in the double bond:

$$CH_2=CH-CH_3 + HX \xrightleftharpoons{H^+} \underset{\underset{H}{|}\;\;\underset{X}{|}}{CH_2-CH-CH_3}$$

1-propene

This pattern is referred to as *Markovnikov's rule*. If we think of the double-bonded carbon with the greater number of hydrogen atoms as being "rich" in hydrogen, we can remember Markovnikov's rule by the adage "The rich get richer. . . ."

Perhaps the most important addition reaction of alkenes for our purposes is *hydration*, the reversible addition of water to the double bond. This reaction also requires a strong acid catalyst, and a large excess of H_2O promotes maximum yield of product. Hydration reactions are part of the metabolism of fats, and you will be studying those enzyme-catalyzed reactions later. Addition of water to an alkene produces a saturated compound with an —OH group, an alcohol:

$$\underset{\text{2-methyl-1-propene}}{\overset{\overset{CH_3}{|}}{CH_3-C=CH_2}} + HOH \xrightleftharpoons{H^+} \overset{\overset{CH_3}{|}}{\underset{\underset{OH}{|}\;\underset{H}{|}}{CH_3-C-CH_2}}$$

As you can see, hydration follows Markovnikov's rule.

Markovnikov's rule: the statement summarizing the pattern of addition of hydrogen-containing molecules to alkenes; "hydrogen bonds to the carbon atom having the greater number of hydrogens, and the rest of the adding molecule bonds to the other carbon atom formerly in the double bond."

EXAMPLE 11.6

Give structural formulas for the products in each of the following reactions.

(a) $CH_2=CH_2 + Br_2 \longrightarrow$

(b) $CH_3-\underset{\underset{CH_2}{\|}}{C}-CH_3 + H_2 \xrightarrow{Pt}$

(c) $CH_3-\underset{\underset{\underset{CH_3}{|}}{\underset{CH_2}{|}}}{C}=CH_2 + HCl \xrightleftharpoons{H^+}$

(d) $CH_3-CH_2-CH=CH_2 + H_2O \xrightleftharpoons{H^+}$

SOLUTION

(a) 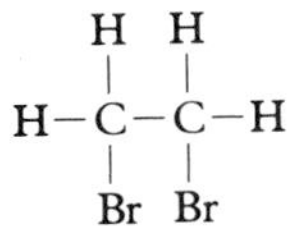 $H-\underset{\underset{Br}{|}}{\overset{\overset{H}{|}}{C}}-\underset{\underset{Br}{|}}{\overset{\overset{H}{|}}{C}}-H$

(b) $H-\underset{\underset{H}{|}}{\overset{\overset{H}{|}}{C}}-\underset{\underset{H-\underset{\underset{H}{|}}{\overset{\overset{H}{|}}{C}}-H}{|}}{\overset{\overset{H}{|}}{C}}-\overset{\overset{H}{|}}{\underset{\underset{H}{|}}{C}}-H$

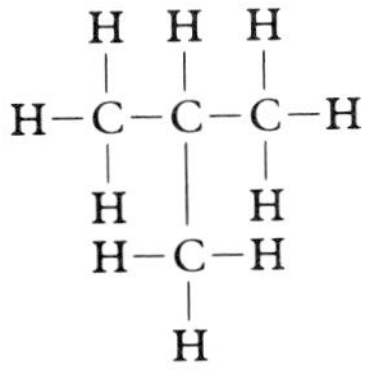

(c) $H-\underset{\underset{H}{|}}{\overset{\overset{H}{|}}{C}}-\underset{\underset{H-\underset{\underset{H-\underset{\underset{H}{|}}{\overset{\overset{H}{|}}{C}}-H}{|}}{\overset{\overset{H}{|}}{C}}-H}{|}}{\overset{\overset{Cl}{|}}{C}}-\overset{\overset{H}{|}}{\underset{\underset{H}{|}}{C}}-H$

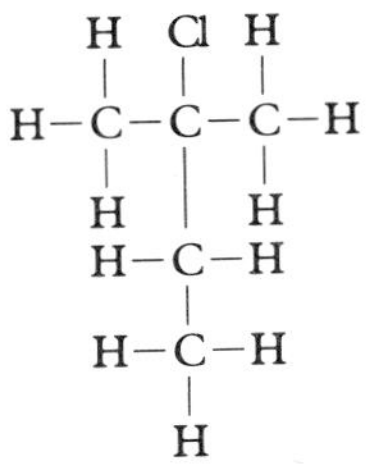

(d) 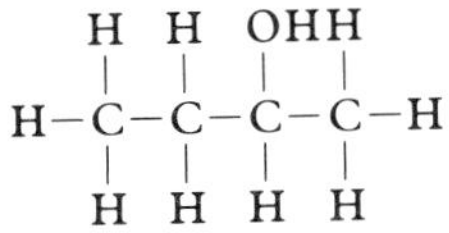 $H-\underset{\underset{H}{|}}{\overset{\overset{H}{|}}{C}}-\underset{\underset{H}{|}}{\overset{\overset{H}{|}}{C}}-\underset{\underset{H}{|}}{\overset{\overset{OH}{|}}{C}}-\overset{\overset{H}{|}}{\underset{\underset{H}{|}}{C}}-H$

EXERCISE 11.4 Give structural formulas for the products in each of the following reactions.

(a) $CH_3-CH=CH_2 + HBr \xrightleftharpoons{H^+}$

(b) $CH_3-CH=CH-CH=CH_2 + 2 H_2 \xrightarrow[\Delta]{Pt}$

(c) $CH_3-CH=\underset{\underset{CH_3}{|}}{C}-CH_3 + H_2O \xrightleftharpoons{H^+}$

(d) $CH_3-CH=CH-CH_3 + Cl_2 \longrightarrow$

Polymerization

polymerization: formation of polymers.

As mentioned in chapter 6, most of the plastics we use are polymers. Alkenes, by virtue of their double bonds, are able to undergo reactions in which many alkene units are joined together to form polymers. The alkene units are called monomers. One of the most common polymers is polyethylene, the material used to make plastic bags, bottles, and toys. The polymerization of ethene (ethylene) requires a special catalyst and results in the formation of a mixture of various chain lengths of repeating ethene units. The polymer product is an alkane, and as such, is chemically inert.

$$n \quad \underset{H}{\overset{H}{>}}C=C\underset{H}{\overset{H}{<}} \xrightarrow{\text{catalyst}} {+}\underset{\underset{H}{|}}{\overset{\overset{H}{|}}{C}}-\underset{\underset{H}{|}}{\overset{\overset{H}{|}}{C}}{+}_n \qquad (n = 100\text{-}1000)$$

ethene polyethylene
(ethylene)

Propene will also undergo polymerization, and the product is this case in polypropylene:

$$n \quad \underset{H}{\overset{H}{>}}C=C\underset{\underset{\underset{H}{|}}{\overset{\overset{H}{|}}{C-H}}}{\overset{H}{<}} \xrightarrow{\text{catalyst}} {+}\underset{\underset{\underset{\underset{H}{|}}{H-C-H}}{\overset{\overset{H}{|}}{|}}}{\overset{\overset{H}{|}}{C}}-\underset{\underset{H}{|}}{\overset{\overset{H}{|}}{C}}{+}_n \qquad (n = 100\text{-}1000)$$

propene polypropylene
(propylene)

This plastic, similar to polyethylene, is used for making many of the same commercial products as well as surgical sutures, tubing, and mesh. Table 11-7 lists polymers made from various alkene derivatives.

Sources and Uses

The low molecular weight alkenes, C_1-C_4, are produced commercially from natural gas or petroleum. Ethene is produced in greatest abundance and is considered to be the most important alkene commercially. Approximately 35% is used to make polyethylene, and 25% is used in the production of antifreeze (ethylene glycol) and related products. Another 20% is converted to ethanol by hydration, and 10% is used in the production of the polymer polystyrene. The remainder of the ethene is consumed in miscellaneous uses. Propene and the butenes are also manufactured in large quantity, and they serve as raw materials for making other chemicals used as solvents or in the manufacture of plastics, detergents, synthetic rubber, food preservatives, and many other products.

ALKYNES

The *alkynes* are hydrocarbons having one or more triple bonds. Because alkynes contain even less hydrogen per carbon atom than alkenes, the alkynes are said to be more unsaturated than alkenes. As with alkanes and alkenes, the carbon atoms may be bonded together consecutively or in branched chains, or they may be joined together to form rings. The open-chain alkynes have the general formula C_nH_{2n-2}. Table 11-8 lists examples of alkynes.

alkynes (al-KINES)

Table 11-7 Examples of Polymers Made from Alkenes

Monomer (Common name used)	Polymer	Uses
$H_2C=CH_2$ Ethylene	$\left[\begin{array}{cccc} H & H & H & H \\ | & | & | & | \\ -C-C-C-C- \\ | & | & | & | \\ H & H & H & H \end{array}\right]$ Polyethylene	plastic film, containers, toys
$H_2C=C{<}^H_{Cl}$ Vinyl chloride	$\left[\begin{array}{cccc} H & H & H & H \\ -C-C-C-C- \\ H & Cl & H & Cl \end{array}\right]$ Polyvinylchloride (PVC)	vinyl film, water pipes, packaging
($H_2C=C{<}^H$ is the vinyl group)		
$H_2C=C{<}^H_{C{\equiv}N}$ Acrylonitrile	$\left[\begin{array}{cccc} H & H & H & H \\ -C-C-C-C- \\ H & CN & H & CN \end{array}\right]$ Polyacrylonitrile	Orlon® and Acrilan® fibers used in clothing
$H_2C=C{<}^H_{C_6H_5}$ Styrene	$\left[\begin{array}{cccc} H & H & H & H \\ -C-C-C-C- \\ H & \phi & H & \phi \end{array}\right]$ Polystyrene	styrofoam, plastic sheeting, insulation
$H_2C=C{<}^{Cl}_{Cl}$ Vinylidene chloride	$\left[\begin{array}{cccc} H & Cl & H & Cl \\ -C-C-C-C- \\ H & Cl & H & Cl \end{array}\right]$ Polyvinylidenechloride	Saran® wrap
$F{>}C=C{<}^F_F$ Tetrafluoroethylene	$\left[\begin{array}{cccc} F & F & F & F \\ -C-C-C-C- \\ F & F & F & F \end{array}\right]$ Polytetrafluoroethylene	Teflon®

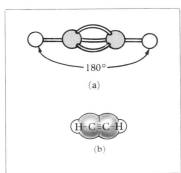

Fig. 11-4 The Ethyne Molecule, C_2H_2 (a) Ball-and-Stick Model (b) Space-Filling Model

Structure

The simplest alkyne is ethyne, commonly known as acetylene. The three-dimensional structure of ethyne is shown in Figure 11-4(a) with ball-and-stick models. In this alkyne all atoms lie in a straight line; this is true for any four atoms surrounding a triple bond. Thus, we say that the triple bond has linear geometry; the bond angles are 180°.

Table 11-8 Examples of Alkynes

Structure	Molecular Formula	Name
H—C≡C—H	C_2H_2	ethyne
CH_3—C≡C—H	C_3H_4	propyne
CH_3CH_2—C≡C—H	C_4H_6	1-butyne
$CH_3CH_2CH_2$—C≡C—H	C_5H_8	1-pentyne
$CH_3CH_2CH_2CH_2$—C≡C—H	C_6H_{10}	1-hexyne
CH_3CH—C≡C—H | CH_3	C_5H_8	3-methyl-1-butyne
CH_3CH_2CH—C≡C—H | CH_3	C_6H_{10}	3-methyl-1-pentyne

Nomenclature

IUPAC names for alkynes are derived in a manner similar to that for alkenes. The name ending for alkynes is "-yne," and they are named on the basis of the longest continuous chain containing the triple bond. The chain is numbered so that the first carbon of the triple bond has the lowest possible number. Examples of alkyne names are given in Table 11-8.

Physical Properties

Alkynes tend to have physical properties similar to those of alkenes and alkanes; however, because of its small, symmetric structure, ethyne (acetylene) has an unusually high melting point, just one degree lower than its boiling point, −84°C. Physical properties for alkynes are given in Table 11-9.

Chemical Properties

As with other hydrocarbons, alkynes are flammable, and combustion of alkynes produces heat. Ethyne (acetylene) is used as fuel for oxyacety-

Table 11-9 Physical Properties of Alkynes

Name	Number of Carbon Atoms	Melting Point (°C)	Boiling Point (°C)	Density* (g/mL)
Ethyne	2	−85	−84	—
Propyne	3	−103	−23	—
1-Butyne	4	−126	9	—
1-Pentyne	5	−106	40	0.691
1-Hexyne	6	−132	71	0.716
1-Heptyne	7	−81	99	0.732
1-Octyne	8	−79	131	0.743

*at 20°C; no data are given for species that are gaseous at this temperature

lene torches to generate temperatures of over 2500°C for welding and cutting metals.

Alkynes undergo addition reactions as do alkenes, but alkynes can add two or four atoms to the triple bond. When conditions are controlled so that only two atoms or groups are added, the product is a *trans*-disubstituted alkene. The following equations illustrate addition reactions for alkynes:

halogenation:

$$H-C\equiv C-H + X_2 \longrightarrow \underset{\underset{X}{|}}{\overset{\overset{X}{|}}{H-C=C-H}} \xrightarrow{X_2} \underset{\underset{X\ X}{|\ |}}{\overset{\overset{X\ X}{|\ |}}{H-C-C-H}}$$

hydrogenation:

$$H-C\equiv C-H + H_2 \longrightarrow \underset{\underset{H}{|}}{\overset{\overset{H}{|}}{H-C=C-H}} \xrightarrow{H_2} \underset{\underset{H\ H}{|\ |}}{\overset{\overset{H\ H}{|\ |}}{H-C-C-H}}$$

addition of HX:

$$H-C\equiv C-CH_3 + HX \underset{}{\overset{H^+}{\rightleftharpoons}} \underset{\underset{H}{|}}{\overset{\overset{X}{|}}{H-C=C-CH_3}} \underset{}{\overset{H^+, HX}{\rightleftharpoons}} \underset{\underset{H\ X}{|\ |}}{\overset{\overset{H\ X}{|\ |}}{H-C-C-CH_3}}$$

hydration:

$$H-C\equiv C-CH_3 + HOH \underset{}{\overset{H^+}{\rightleftharpoons}} \underset{\underset{H}{|}}{\overset{\overset{OH}{|}}{H-C=C-CH_3}} \underset{}{\overset{H^+, H_2O}{\rightleftharpoons}} \underset{\underset{H\ OH}{|\ |}}{\overset{\overset{H\ OH}{|\ |}}{H-C-C-CH_3}}$$

Sources and Uses

Ethyne, the only alkyne produced commercially in large amounts, is derived from soft (bituminous) coal and from high-temperature conversion of methane:

$$2\ CH_4 \xrightarrow{1600°C} H-C\equiv C-H + 3\ H_2$$

About half of the ethyne produced is used in oxyacetylene torches; the remainder serves as raw material for synthesizing other organic chemicals.

CYCLIC ALIPHATIC HYDROCARBONS

Aliphatic hydrocarbons may also exist in closed-ring structures as illustrated in Table 11-10. They are named by adding the prefix "cyclo-" to the name of the corresponding open-chain alkane, alkene, or alkyne.

Table 11-10 Cyclic Aliphatic Hydrocarbons

Structure	Name	Melting Point (°C)	Boiling Point (°C)	Density* (g/mL)
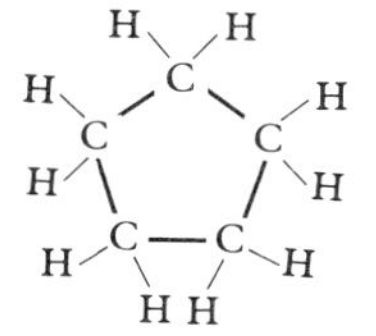 (structure)	cyclopentane	−94	50	0.745
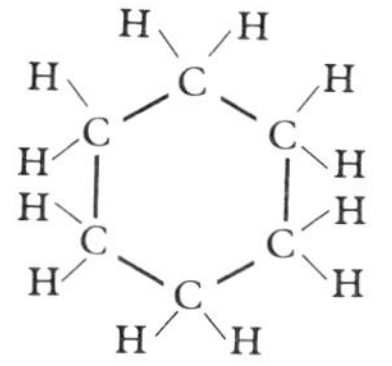 (structure)	cyclohexane	6	81	0.779
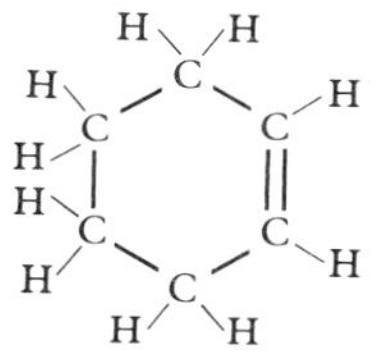 (structure)	cyclohexene	−104	83	0.810

*at 20°C

Cyclic structures are conveniently represented by geometric figures, as shown below, where each line stands for a C—C bond and each corner of the figure stands for a carbon atom with the appropriate number of hydrogen atoms to satisfy carbon's valence of 4.

    cyclopentane    cyclohexane    cyclohexene

Cyclic aliphatic hydrocarbons are colorless compounds obtained from petroleum and by chemical synthesis. These compounds have physical properties similar to those of open-chain aliphatic hydrocarbons, and they find many of the same uses. Cyclopropane was once a popular anesthetic due to its fast action, but because of its flammability, its use has been discontinued as nonflammable replacements have been developed.

AROMATIC HYDROCARBONS

During the mid-nineteenth century, chemists recognized a group of hydrocarbons that have distinctly different chemical properties from the aliphatic hydrocarbons. These mysterious compounds have a high degree

Kekulé (KEH-ku-lay)

of unsaturation, yet they do not participate in addition reactions typical of alkenes and alkynes. Many have pleasant aromas, and some are responsible for the odors of cloves, cinnamon, anise, wintergreen, and other naturally occurring substances. Because of this, the unusual hydrocarbons came to be called *aromatic hydrocarbons.*

It was noticed in 1861 that many aromatic hydrocarbons have formulas suggesting that they were derived from the compound benzene, C_6H_6. Since then the word aromatic has come to mean any compound that is structurally related to benzene. In modern times benzene has come to be regarded as the model for all aromatic hydrocarbons.

Benzene Structure

Friedrich August Kekulé, a German chemist, made the first sound suggestion about the structure of benzene in 1865. He proposed that benzene has two structures which somehow cannot be isolated from each other. Kekulé's structures were

We now recognize that the Kekulé structures are simply two resonance forms since they differ only in the placement of bonding electrons in the ring. Thus, benzene is a hybrid of the two forms, and we often draw its structure as

We have also come to recognize that resonance hybrids such as benzene are unusually stable and therefore much less reactive than would be expected. There are no true double bonds in the benzene ring, and hence benzene does not behave like an alkene chemically. However, the bonds in the ring are not single bonds either; rather, they are more like bonds intermediate between single and double bonds. Thus, even when only a single Kekulé structure is written, we should remember that the resonance-stabilized ring is intended.

Benzene is a flat molecule with all twelve atoms lying in the same plane. It is a colorless liquid, and like other hydrocarbons, it is insoluble in water but soluble in nonpolar liquids.

Benzene Derivatives

Benzene *derivatives* are formed when one or more hydrogens on the ring are replaced by other atoms or groups. When only one substituent exists on the ring, the IUPAC name is just the name of the substituent added as a prefix to the word benzene, as in methylbenzene, hydroxybenzene, and aminobenzene. If two or more substituents appear on the benzene ring, the IUPAC name is derived by numbering the carbon atoms in the ring consecutively so that the substituents have the smallest numbers possible. Examples are 1-ethyl-3-methylbenzene and 1,4-dimethylbenzene.

derivative: a compound made (derived) from another compound.

There are two other methods sometimes used for naming benzene compounds. One of these uses Greek prefixes to designate two identical substituents. The prefix *ortho* (*o-*) is used for the 1,2-disubstituted benzenes; *meta-* (*m-*) is used for the 1,3-disubstituted benzenes, and *para-* (*p-*) refers to 1,4-disubstituted benzenes. The second alternative method relies on common names developed through history for benzene rings having one substituent. The disubstituted compounds are then named as derivatives of the monosubstituted compounds. Examples are *p*-methylaniline and *m*-hydroxytoluene. These and other names of benzene derivatives are given in Table 11–11.

Polycyclic Aromatic Hydrocarbons

The *polycyclic aromatic hydrocarbons* contain two or more aromatic rings fused together. The simplest of these is naphthalene. It is a white solid at room temperature and is often the ingredient responsible for the odor of mothballs. The structures of naphthalene and other examples of polycyclic aromatic hydrocarbons are shown below:

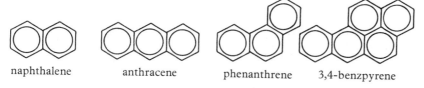

naphthalene   anthracene   phenanthrene   3,4-benzpyrene

Some of those containing four or more fused rings have been found to cause cancer in laboratory animals. For example, 3,4-benzpyrene causes gene mutations and lung cancer in test animals. A product of incomplete combustion of tobacco, coal, and petroleum, it is found in cigarette smoke and automobile exhaust.

HALOGEN DERIVATIVES OF HYDROCARBONS

Halogenated hydrocarbons can be prepared from both aliphatic and aromatic hydrocarbons. Halogen derivatives of aliphatic hydrocarbons are called *alkyl halides*, while those derived from aromatic hydrocarbons are

Table 11-11 Examples of Benzene Derivatives

Structure	Name*
C₆H₅—CH₃	methylbenzene (toluene)
C₆H₅—OH	hydroxybenzene (phenol)
C₆H₅—NH₂	aminobenzene (aniline)
3-H₃C-C₆H₄-CH₂CH₃	1-ethyl-3-methylbenzene (m-ethyltoluene)
H₃C—C₆H₄—CH₃ (1,4)	1,4-dimethylbenzene (p-methyltoluene, p-dimethylbenzene)
C₆H₄Cl₂ (1,2)	1,2-dichlorobenzene (o-dichlorobenzene)
C₆H₄Cl₂ (1,3)	1,3-dichlorobenzene (m-dichlorobenzene)
H₃C—C₆H₄—NH₂ (1,4)	1-amino-4-methylbenzene (p-methylaniline)
HO—C₆H₄—CH₃ (1,3)	1-hydroxy-3-methylbenzene (m-hydroxytoluene)
H₃C—C₆H₂(NO₂)₃ (2,4,6)	1-methyl-2,4,6-trinitrobenzene (2,4,6-trinitrotoluene)
Cl₂C₆H₃—OH (2,4-dichloro-1-hydroxy)	2,4-dichloro-1-hydroxybenzene (2,4-dichlorophenol)

*Common names are given in parentheses.

called *aryl halides*. Examples of common alkyl halides are given in Table 11-12. In the physical sense, halogenated hydrocarbons are similar to alkanes. Most are not soluble in water and are good solvents for other organic compounds such as fats, greases, and oils.

Table 11-12 Examples of Common Alkyl Halides

Compound	Structure	Boiling Point (°C)	Uses
Chloromethane	CH_3Cl	-24	Refrigerant; manufacture of chemical products
Bromomethane	CH_3Br	4	Insecticide and pesticide
Chloroethane	CH_3CH_2Cl	13	Synthesis of tetraethyl lead; local anesthetic
1-Chlorobutane	$CH_3CH_2CH_2CH_2Cl$	79	Chemical synthesis; kills intestinal worms in dogs

An important reaction of alkyl halides is *elimination*. This reaction amounts to reversal of the addition of HX to alkenes catalyzed by strong acid. Hence, an alkyl halide will eliminate HX in the presence of strong base which neutralizes HX as it forms. Elimination reactions by alkyl halides form *trans* alkenes and follow Markovnikov's rule in reverse, so that hydrogen is lost from the carbon adjacent to —C—X which has the fewest hydrogens:

elimination reaction: a reaction in which a hydrocarbon derivative loses the elements of a small molecule to form an alkene.

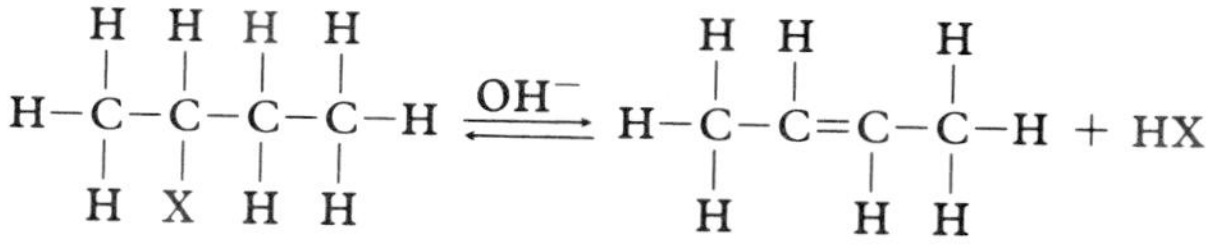

$$\begin{array}{c} H\ H\ H\ H \\ |\ \ |\ \ |\ \ | \\ H-C-C-C-C-H \\ |\ \ |\ \ |\ \ | \\ H\ X\ H\ H \end{array} \xrightleftharpoons{OH^-} \begin{array}{c} H\ H\ \ \ \ \ H \\ |\ \ |\ \ \ \ \ | \\ H-C-C=C-C-H \\ |\ \ \ \ \ \ \ |\ \ | \\ H\ \ \ \ \ \ H\ H \end{array} + HX$$

This statement is called *Saytzeff's rule*. Now we can complete the adage we started earlier: "The rich get richer and the poor get poorer."

Aryl halides, not as reactive as alkyl halides, have rather specialized reactions that we need not consider.

EXAMPLE 11.7

Give structural formulas for the products in each of the following reactions.

(a) 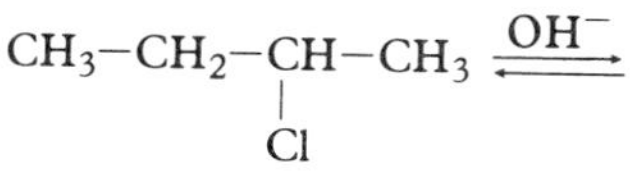 $CH_3-CH_2-\underset{\underset{Cl}{|}}{CH}-CH_3 \xrightleftharpoons{OH^-}$

(b) 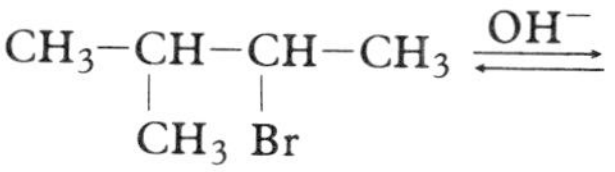 $CH_3-\underset{\underset{CH_3}{|}}{CH}-\underset{\underset{Br}{|}}{CH}-CH_3 \xrightleftharpoons{OH^-}$

(c) 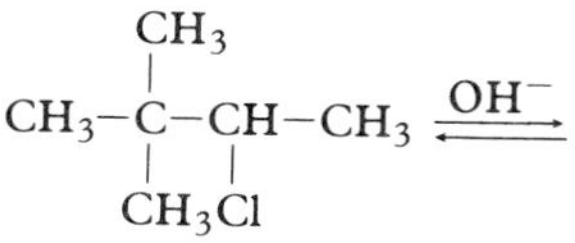 $CH_3-\underset{\underset{CH_3Cl}{|}}{\overset{\overset{CH_3}{|}}{C}}-CH-CH_3 \xrightleftharpoons{OH^-}$

SOLUTION

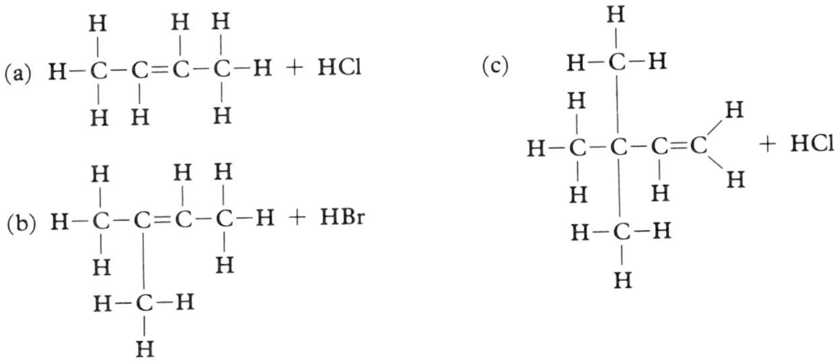

(a) $\text{H}-\underset{\underset{\text{H}}{|}}{\overset{\overset{\text{H}}{|}}{\text{C}}}-\underset{}{\overset{\text{H}}{\text{C}}}=\underset{}{\overset{\text{H}}{\text{C}}}-\underset{\underset{\text{H}}{|}}{\overset{\overset{\text{H}}{|}}{\text{C}}}-\text{H}\;+\;\text{HCl}$

(b) $\text{H}-\underset{\underset{\text{H}}{|}}{\overset{\overset{\text{H}}{|}}{\text{C}}}-\overset{\text{H}}{\text{C}}=\overset{\text{H}}{\text{C}}-\underset{\underset{\underset{\underset{\text{H}}{|}}{\text{H}-\text{C}-\text{H}}}{|}}{\overset{}{\text{C}}}-\text{H}\;+\;\text{HBr}$

(c) $\underset{\underset{\underset{\text{H}}{|}}{\text{H}-\text{C}-\text{H}}}{\overset{\overset{\overset{\text{H}}{|}}{\text{H}-\text{C}-\text{H}}}{\text{H}-\underset{\underset{\text{H}}{|}}{\overset{\overset{\text{H}}{|}}{\text{C}}}-\underset{}{\text{C}}-\overset{\text{H}}{\text{C}}=\text{C}\underset{\text{H}}{\overset{\text{H}}{\diagup\!\!\!\diagdown}}}}\;+\;\text{HCl}$

EXERCISE 11.5 Give structural formulas for the products in each of the following reactions.

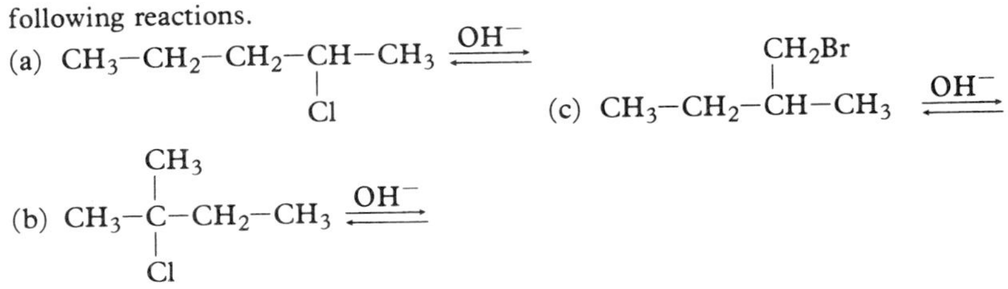

(a) $\text{CH}_3-\text{CH}_2-\text{CH}_2-\underset{\underset{\text{Cl}}{|}}{\text{CH}}-\text{CH}_3 \xrightleftharpoons{\text{OH}^-}$

(b) $\text{CH}_3-\underset{\underset{\text{Cl}}{|}}{\overset{\overset{\text{CH}_3}{|}}{\text{C}}}-\text{CH}_2-\text{CH}_3 \xrightleftharpoons{\text{OH}^-}$

(c) $\text{CH}_3-\text{CH}_2-\underset{\underset{\text{CH}_2\text{Br}}{\;}}{\overset{\overset{\text{CH}_2\text{Br}}{|}}{\text{CH}}}-\text{CH}_3 \xrightleftharpoons{\text{OH}^-}$

Halogenated derivatives of methane and ethane are gases. Because of their relative inertness and ease of liquefication, they have been used for some time as aerosol propellants and as refrigerants in air conditioners and refrigerators. The most well known of these are the "freons," some of whose structures are

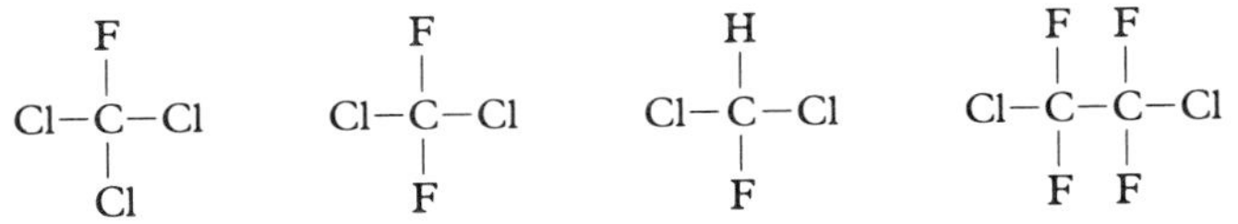

$\underset{\text{Freon-11}}{\text{Cl}-\underset{\underset{\text{Cl}}{|}}{\overset{\overset{\text{F}}{|}}{\text{C}}}-\text{Cl}} \qquad \underset{\text{Freon-12}}{\text{Cl}-\underset{\underset{\text{F}}{|}}{\overset{\overset{\text{F}}{|}}{\text{C}}}-\text{Cl}} \qquad \underset{\text{Freon-21}}{\text{Cl}-\underset{\underset{\text{F}}{|}}{\overset{\overset{\text{H}}{|}}{\text{C}}}-\text{Cl}} \qquad \underset{\text{Freon-114}}{\text{Cl}-\underset{\underset{\text{F}}{|}}{\overset{\overset{\text{F}}{|}}{\text{C}}}-\underset{\underset{\text{F}}{|}}{\overset{\overset{\text{F}}{|}}{\text{C}}}-\text{Cl}}$

(Freon is a trade name of the DuPont Company; the numbers are DuPont codes which refer to the structures.) In more recent times, questions have been raised regarding the possible adverse effect of accumulating concentrations of "freons" in our upper atmosphere. In addition, young people discovered the anesthetic qualities of "freons" and began inhaling aerosol propellants in the late 1960s; several deaths resulted. These factors have caused diminished use of halogenated hydrocarbons in aerosol spray cans. In some instances, alkanes are now being used as propellants in spite of their flammability.

The anesthetic qualities of halogenated alkanes have been known for some time. Trichloromethane (CHCl_3, chloroform) was first used as a general anesthetic in 1847. It is powerful, fast-acting, and nonflammable, but the effective dose is near the lethal dose, and it sometimes causes

PERSPECTIVE: Halogenated Hydrocarbons in the Environment

Large halogenated hydrocarbon molecules, especially the chlorinated ones, are widely used as insecticides. In particular, DDT (dichlorodiphenyltrichloroethane) has been used since World War II.

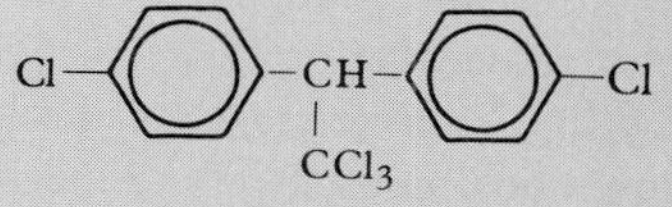

Cl—⌬—CH—⌬—Cl
         |
        CCl$_3$

DDT

(The term "phenyl" refers to the benzene ring when it is a substituent in a compound.) DDT disorganizes the central nervous system, and it indiscriminately kills both insect pests and helpful insects. Since it is nonpolar and nonbiodegradable, it tends to collect in fatty tissues of fish and animals exposed to low levels in the environment. In some instances, DDT has been implicated as a cause of thin shells in certain bird eggs, causing them to break easily; the affected birds are not able to reproduce fast enough to maintain the species.

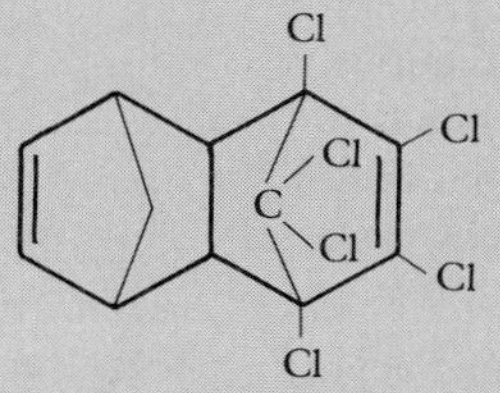

Trace amounts of DDT have been found in human fat, and it has been shown to be carcinogenic in laboratory animals. For these reasons, general use of DDT was banned in 1972, although it may still be used in limited amounts under certain conditions. Other chlorinated hydrocarbon insecticides are shown in the accompanying table.

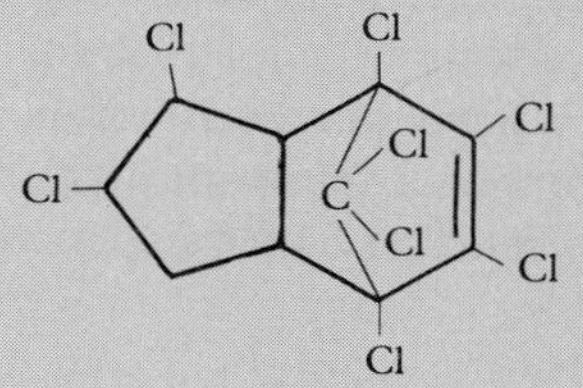

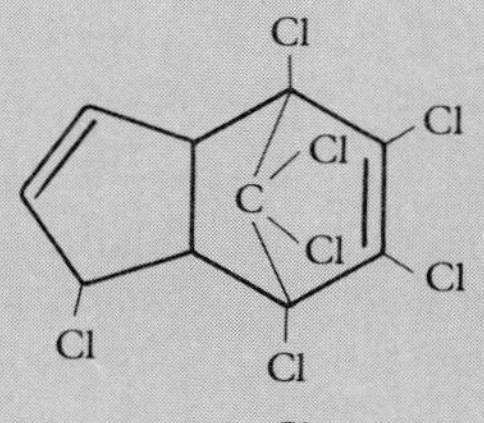

Another group of halogenated hydrocarbons, the PCBs (polychlorinated biphenyls), has received wide use as additives to certain plastics to make them more flexible and as insulating materials in electrical transformers. The PCBs are derived from biphenyl, shown below, by replacement of 1 to 10 of the hydrogens with chlorine atoms:

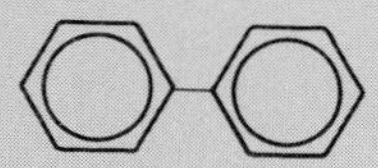

biphenyl

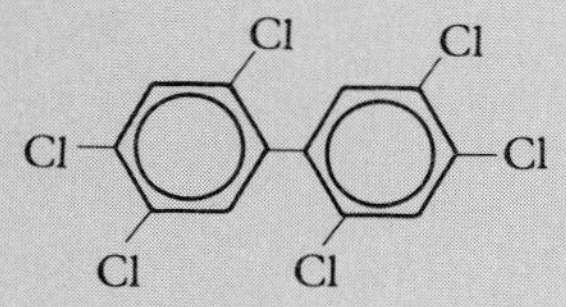

a PCB

The PCBs have made their way into the environment, and since they are not biodegradable, they also collect in fatty tissue of fish and animals. They are known to be toxic to humans, and they are also carcinogenic. Although production of PCBs was discontinued in 1977, they are likely to remain in the environment for many years to come.

Chlorinated Hydrocarbon Insecticides

Insecticide Name	Structure
Lindane	(cyclohexane with 6 Cl and 6 H)
Chlordane	(chlorinated cyclopentane fused to chlorinated bicyclic system)
Heptachlor	(chlorinated cyclopentene fused to chlorinated bicyclic system)
Aldrin	(chlorinated norbornene fused to chlorinated bicyclic system)

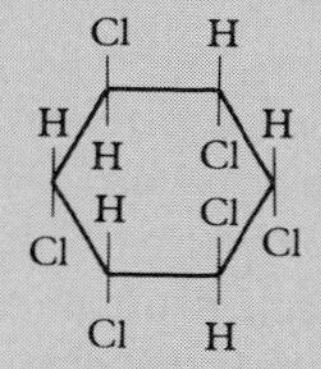

liver damage. Its use has been largely replaced by other substances, notably 2-bromo-2-chloro-1,1,1-trifluoroethane (halothane):

$$\begin{array}{c} FCl \\ || \\ F-C-C-H \\ || \\ FBr \end{array}$$

2-bromo-2-chloro-1,1,1-trifluoroethane
(halothane)

One group of halogenated hydrocarbons includes some rather unreactive polymers. The most well known of these, Teflon, was discovered quite by accident by a DuPont chemist, Roy Plunkett, in 1938. Teflon is now widely used as a nonstick coating for cooking utensils and mechanical parts. A related polymer, Saran, is used as a food wrapper. They are both prepared by polymerization of the corresponding halogenated ethene.

$$\left(\begin{array}{c} FF \\ || \\ C-C \\ || \\ FF \end{array}\right)_n \qquad \left(\begin{array}{c} HCl \\ || \\ C-C \\ || \\ HCl \end{array}\right)_n$$

Teflon   Saran

Distant chemical relatives of Teflon are the group of compounds called perfluorocarbons (PFCs). In these compounds, fluorine has been substituted for hydrogen in low molecular weight hydrocarbons. Almost 20 years ago a Cincinnati physiologist, Leland C. Clark, Jr., discovered that these liquids dissolve oxygen in large enough amounts to support life in laboratory mice that are immersed in the liquids. Since then scientists have found ways of dispersing tiny droplets of the substances in aqueous solutions, and perfluorocarbons are now used effectively as blood substitutes in certain emergencies.

SUMMARY

Organic Chemistry: Organic chemistry is the study of covalent carbon compounds. About 90% of the known chemical compounds are organic. Carbon atoms form single or multiple bonds to other carbon atoms and single bonds to hydrogen, nitrogen, phosphorus, sulfur, and the halogens. Carbon also forms multiple bonds to nitrogen and oxygen. Organic compounds can be classified by distinctive groupings of atoms called functional groups.

The Hydrocarbons: Hydrocarbons are organic compounds in which carbon atoms form the skeleton and hydrogen atoms are attached to the carbon atoms. The hydrocarbons are insoluble in water, less dense than water, and soluble in one another. Hydrocarbons are classified as aliphatic or aromatic.

Alkanes: Alkanes are hydrocarbons containing only single bonds. The open-chain alkanes have the general formula C_nH_{2n+2}. Alkanes have bond angles of 109.5°, and the open-chain alkanes may be linear or branched. Since there is free rotation about single bonds, linear alkanes may exist in the extended conformation as long, zigzag arrangements of atoms, or they may adopt more compact shapes. Alkanes having 4 or more carbon atoms exist as structural isomers. Boiling points, melting points, and densities of alkanes increase with molecular weight because larger molecules have more powerful London dispersion forces than smaller ones. Within sets of isomers, the more highly branched isomer will have the lowest boiling point, as branches prevent maximum intermolecular attractions. The two major reactions of alkanes are combustion and halogenation. Combustion reactions produce water and carbon dioxide if oxygen supplies are abundant, or water and carbon monoxide when oxygen supplies are limited. Halogenation produces a mixture of halogenated alkanes. The lower molecular weight alkanes, C_1-C_4, are used as gaseous fuels for cooking and heating. Liquid alkanes are blended to make gasoline, diesel fuel, jet fuel, and heating oil. Solid alkanes are used as lubricants, greases, and sealants.

Alkenes: Alkenes are hydrocarbons having one or more double bonds. Open-chain alkenes have the general formula C_nH_{2n}. The double bond has a planar structure with bond angles of 120°. Alkenes exist as geometric isomers (*cis-trans* isomers) because double bonds do not allow rotation of attached groups. Alkenes are similar to alkanes in physical properties except that boiling and melting points for alkenes tend to be lower than those for alkanes. Alkenes are easily oxidized by oxygen in combustion reactions, but their most important reactions are addition and polymerization, a special form of addition. Addition reactions include addition of halogen, hydrogenation, addition of hydrogen halide, and hydration. Polymerization requires special catalysts and produces saturated polymers.

Alkynes: Alkynes are hydrocarbons having one or more triple bonds. Open-chain alkynes have the general formula C_nH_{2n-2}. The triple bond has a linear geometry and bond angles of 180°. Alkynes are similar in physical properties to alkanes and alkenes. Alkynes undergo combustion reactions and addition reactions. Under controlled conditions, two atoms or groups may be added to the triple bond, resulting in *trans*-disubstituted alkenes. These may then undergo addition to give saturated products.

Cyclic Aliphatic Hydrocarbons: Aliphatic hydrocarbons may also exist in closed ring structures. These are named by adding "cyclo-" to the beginning of the name of the corresponding open-chain hydrocarbon.

Aromatic Hydrocarbons: Aromatic hydrocarbons are unsaturated cyclic compounds which are structurally related to benzene. Benzene is a hybrid of two resonance forms. As such, benzene, like all aromatic compounds, is unusually stable. The bonds in the benzene ring are intermediate between single and double bonds. Benzene derivatives are formed when one or more hydrogens are replaced by other atoms or groups. Aromatic compounds containing two or more benzene rings fused together are called polycyclic aromatic hydrocarbons.

Halogen Derivatives of Hydrocarbons: Halogenated hydrocarbons are prepared from aliphatic and aromatic hydrocarbons. An important reaction of alkyl halides is elimination, a reversal of addition of HX to alkenes.

STUDY QUESTIONS AND PROBLEMS

1. Define the following terms:
   (a) functional group
   (b) hydrocarbon
   (c) aliphatic hydrocarbon
   (d) aromatic hydrocarbon
   (e) alkane
   (f) structural isomers
   (g) halogenation
   (h) alkene
   (i) geometric isomers
   (j) addition reaction
   (k) hydrogenation
   (l) hydration
   (m) polymerization
   (n) alkyne
   (o) Kekulé structure
   (p) polycyclic aromatic hydrocarbon
   (q) elimination reaction

2. Circle and name each functional group in each of the following compounds.

   (a) vitamin A

   (b) niacin
      (a vitamin)

(continued next page)

(c) thyroxine
   (a hormone)

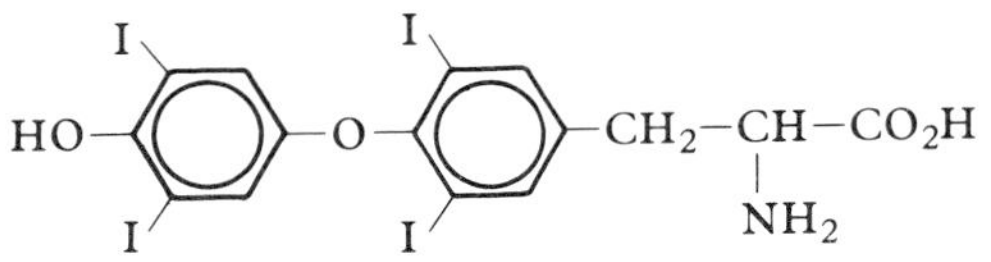

HO—⟨ring with 2 I⟩—O—⟨ring with 2 I⟩—CH$_2$—CH—CO$_2$H
                                            |
                                           NH$_2$

(d) testosterone
    (a hormone)

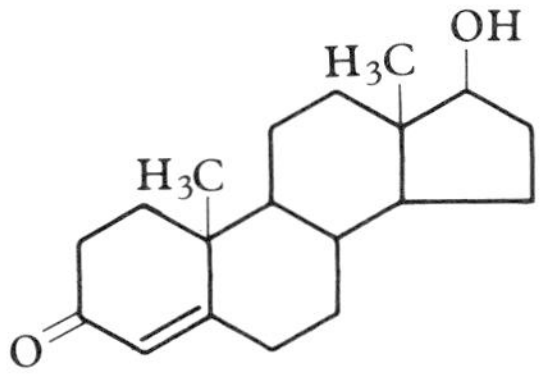

(e) epinephrine
    (a hormone)

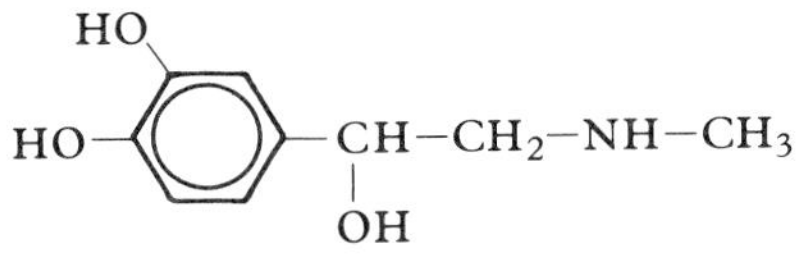

HO—⟨ring, HO⟩—CH—CH$_2$—NH—CH$_3$
              |
              OH

(f) butyl acetate
    (odor of bananas)

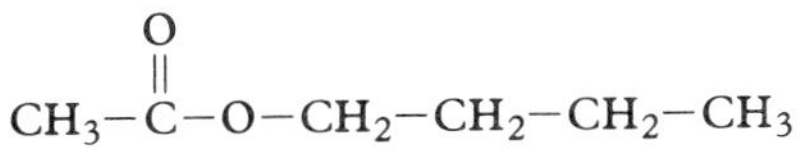

$$CH_3-\overset{O}{\underset{\|}{C}}-O-CH_2-CH_2-CH_2-CH_3$$

3. Summarize the structural features of aliphatic hydrocarbons by completing the following table.

Aliphatic Hydrocarbon	Bond Type	Bond Geometry	Bond Angles
Alkanes	single	tetrahedral	109.5°
Alkenes			
Alkynes			

4. Give IUPAC names for each of the following alkanes:

(a) 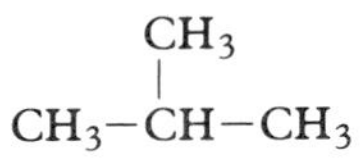 CH$_3$—CH—CH$_3$
         |
         CH$_3$

(b) 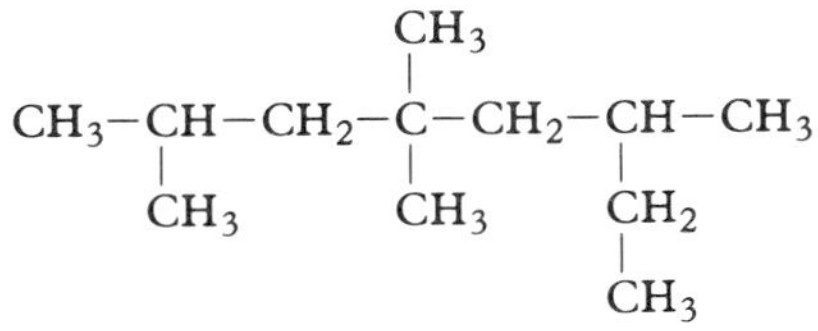 CH$_3$—CH—CH$_2$—C—CH$_2$—CH—CH$_3$
         | | |
         CH$_3$ CH$_3$ CH$_2$
                          |
                          CH$_3$

(c) CH$_3$—CH—CH$_2$—CH$_2$
         | |
         CH$_3$ CH$_3$

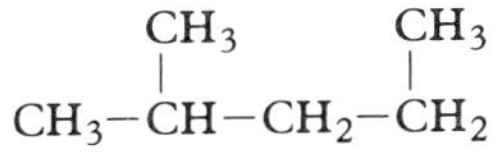

(d) 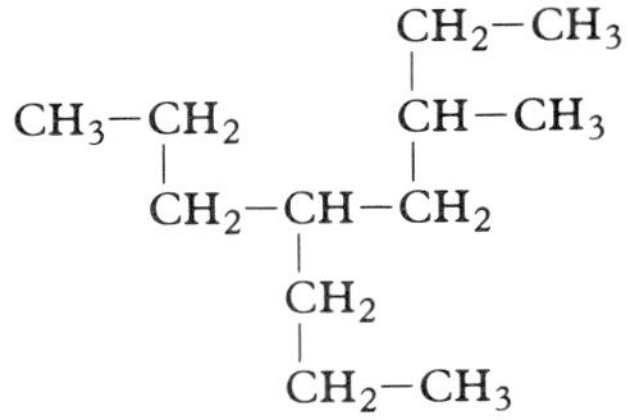 CH$_3$—CH$_2$ CH$_2$—CH$_3$
         | |
         CH$_2$—CH—CH$_2$
                |
                CH$_2$
                |
                CH$_2$—CH$_3$

         CH$_3$ CH$_3$
         | |
(e) 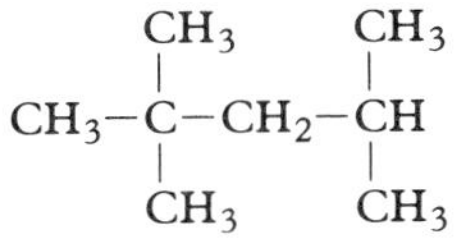 CH$_3$—C—CH$_2$—CH
         | |
         CH$_3$ CH$_3$

5. Draw structures for the three isomers of pentane, C$_5$H$_{12}$, and name them by the IUPAC system.

6. Draw a structure for each of the following alkanes:
   (a) 2-methylpentane
   (b) 2,2-dimethylbutane
   (c) 3,3-dimethyl-5-ethyldecane
   (d) 3,4-diethyl-5-methylnonane
   (e) 2,4-dipropyloctane

7. Why do the boiling and melting points of alkanes increase with molecular weight?

8. Why do branched isomers have lower boiling points than the corresponding linear alkanes?

9. Write balanced equations for the following reactions.
   (a) complete combustion of hexane
   (b) complete combustion of nonane
   (c) incomplete combustion of octane
   (d) complete combustion of octane
   (e) incomplete combustion of butane

10. Why is carbon monoxide toxic?

11. Give a major use for each of the following:
    (a) methane
    (b) propane
    (c) butane
    (d) alkanes with 5–12 carbon atoms
    (e) alkanes with 12–18 carbon atoms
    (f) alkanes with 19 or more carbon atoms

12. How are liquid alkanes harmful to the lungs?

13. Gasolines are blended to achieve desired octane ratings. The octane scale is based on a rating of 100 for 2,2,4-trimethylpentane (an octane isomer) and 0 for heptane. This means that 2,2,4-trimethylpentane burns very smoothly in an automobile engine, and

heptane burns very unevenly and gives rough engine performance. Draw structures for 2,2,4-trimethylpentane and heptane.

14. Give correct IUPAC names to each of the following alkenes:

    (a) CH$_2$=CH
           |
           CH$_3$

    (b) CH$_2$=CH—CH$_2$—CH—CH$_3$
                         |
                         CH$_3$

    (c) CH$_3$—CH=C—CH$_3$
                  |
                  CH$_2$
                  |
                  CH$_3$

    (d) CH$_3$ CH$_2$—CH$_3$
         | |
         CH—CH=CH—CH$_2$

    (e) CH$_3$—CH—CH$_2$—CH=CH—CH$_2$—CH$_3$
             |
             CH$_3$

15. Draw a structure for each of the following alkenes:

    (a) 3-heptene   (d) 2,2-dimethyl-3-nonene
    (b) 2-methyl-2-hexene   (e) 2,4,6-decatriene
    (c) 3-propyl-2-octene

16. Explain why alkenes usually have lower boiling points than alkanes with the same number of carbon atoms.

17. Distinguish between saturated and unsaturated hydrocarbons.

18. Why is it that butene can have geometric isomers but propene cannot?

19. Complete the following equations, using condensed structural formulas for all organic products:

    (a) CH$_3$—CH$_2$—CH=CH$_2$ + HBr $\xrightleftharpoons{H^+}$

    (b) CH$_2$=CH$_2$ + H$_2$O $\xrightleftharpoons{H^+}$

    (c) CH$_3$—C=CH$_2$ + HCl $\xrightleftharpoons{H^+}$
           |
           CH$_3$

    (d) CH$_2$=CH—CH$_3$ + F$_2$ $\longrightarrow$

    (e) n CH$_3$—CH=CH—CH$_3$ $\xrightarrow{catalyst}$

    (f) CH$_3$
         |
         CH$_3$—C=CH—CH$_3$ + H$_2$ $\xrightarrow[\Delta]{Pt}$

20. Describe the use of bromine for detecting unsaturation in a hydrocarbon.

21. Summarize Markovnikov's rule.

22. Name four examples of hydrocarbon polymers, and give a major use for each.

23. Describe four major uses of ethene.

24. Give the common name and major use for ethyne.

25. How does resonance affect the reactivity of benzene?

26. Explain why benzene, an unsaturated hydrocarbon, does not undergo addition reactions.

27. Draw structures for naphthalene and anthracene.

28. Summarize Saytzeff's rule.

29. Draw the structure for the organic product in each of the following reactions:

    (a) CH$_3$—CH—CH—CH$_3$ $\xrightleftharpoons{OH^-}$
             | |
             CH$_3$ Cl

    (b) CH$_3$—CH$_2$—CH—CH$_3$ $\xrightleftharpoons{OH^-}$
                     |
                     CH$_2$Br

    (c) $\xrightleftharpoons{OH^-}$

    (d)   CH$_3$ CH$_3$
                | |
         CH$_3$—CH—C—CH$_3$ $\xrightleftharpoons{OH^-}$
                    |
                    Br

30. Draw the four structural isomers of C$_4$H$_9$Br, and name each one.

31. Draw the eight structural isomers of C$_5$H$_{11}$Cl, and name each one.

32. Give IUPAC names for the following compounds.

    (a) CH$_3$—CH—C≡CH   (d)
             |
             CH$_3$

    (b)

    (c)

(continued next page)

(f) 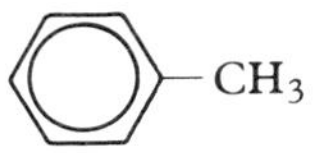 [benzene ring]—CH$_3$

(g) CH$_3$—C≡C—CH$_2$—CH$_3$

(h) 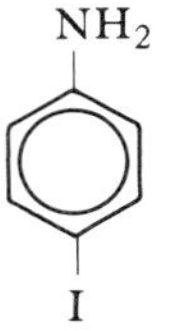 [benzene ring with NH$_2$ on top and I on bottom]

(i) CH$_3$—CH—CH$_2$—CH$_3$
        |
       Br

33. Why are the halogenated hydrocarbons becoming less popular as aerosol propellants?

34. Why is chloroform no longer used as an anesthetic?

35. Calculate the molecular weights of the following compounds:

(a) cyclohexane
(b) cyclopentene
(c) 1,3-butadiene
(d) benzene
(e) naphthalene
(f) Freon-12

36. "Ethyl" or "leaded" gasoline contains tetraethyl lead, a compound that improves the antiknock quality of gasoline. However,

$$\begin{array}{c} CH_2-CH_3 \\ | \\ CH_3-CH_2-Pb-CH_2-CH_3 \\ | \\ CH_2-CH_3 \end{array}$$

this kind of gasoline causes engines to emit toxic lead compounds in the exhaust fumes. Write an equation to illustrate the complete oxidation of tetraethyl lead, and suggest a name for the lead product.

CHAPTER 12

ALCOHOLS, ETHERS, AND RELATED COMPOUNDS

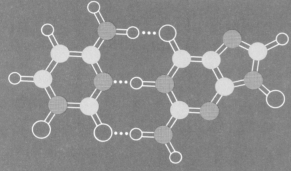

ALCOHOLS
- Structure
- Nomenclature
- Physical Properties
- Chemical Properties
- Sources of Alcohols
- Uses of Alcohols

ETHERS
- Structure
- Nomenclature
- Physical Properties
- Chemical Properties
- Sources and Uses of Ethers

PHENOLS
- Structure and Nomenclature
- Properties
- Sources and Uses of Phenols

THIOLS

The word alcohol is derived from the Arabic *kuhl*, which originally meant "very fine powder." As our ancestors learned to make alcohol by the fermentation of fruits, grains, potatoes, rice, and even cactus, the term was applied to fermented spirits as *alcool*, and finally it became alcohol. Through the development of civilization, alcohols have been an important part of everyday life. Ethers get their group name from the Greek *aither*, which means "clear sky" or "air"; the Greek word suggests the airy, volatile nature of compounds of this class. Ethers, particularly diethyl ether, have been used as anesthetics since the mid-nineteenth century. We shall explore properties and uses of alcohols and ethers in this chapter, along with those of some close relatives, the phenols and thiols.

ALCOHOLS

Alcohols are characterized by the presence of $-OH$, the hydroxyl group; an example is CH_3OH, methanol. We can write a general formula for alcohols as $R-OH$, where R represents any alkyl group. Since water, HOH, also possesses the hydroxyl group, alcohols can be thought of as derivatives of water in which a hydrogen has been replaced with an alkyl group.

Structure

Just as water is a bent molecule, so are alcohols. (Of course,

$$\overset{\ddot{O}}{\underset{H \quad H}{\diagup \diagdown}} \qquad\qquad \overset{\ddot{O}}{\underset{R \quad H}{\diagup \diagdown}}$$

Fig. 12-1 Maximum Hydrogen Bonding in (a) Water and (b) Alcohols

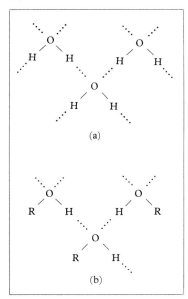

primary, secondary, and tertiary carbon atoms: carbon atoms attached to one, two, and three alkyl groups, respectively.

the R group, the rest of the alcohol molecule, has its own structural features.) However, for the sake of convenience, it is customary to write the —OH group in a straight line, as in

$$CH_3-CH_2-OH$$
ethanol
(ethyl alcohol)

Alcohols are polar molecules, and as in water, the oxygen carries a partial negative charge and the hydrogen carries a partial positive charge:

$$R-\overset{\delta-}{\underset{}{O}}-H^{\delta+}$$

The presence of the —OH group allows alcohols to form hydrogen bonds. In contrast to water, however, alcohols have only one hydrogen attached to oxygen; thus alcohols can form a maximum of only three hydrogen bonds per molecule instead of the maximum of four per molecule that water can form (see Figure 12-1). Liquid and solid alcohols contain hydrogen bonds just as liquid and solid water do, but there are not as many hydrogen bonds for the same number of alcohol molecules.

Because the hydroxyl group can exist anywhere on a hydrocarbon chain, there are three structural *subclasses* of alcohols. Those alcohols in which the hydroxyl group is at the end of a hydrocarbon chain are called *primary (1°) alcohols*. In these, the carbon carrying the —OH group is attached to only one alkyl group.* If, however, the —OH group is attached to a carbon which has two alkyl substituents, as in the interior of a linear hydrocarbon, the alcohol is called a *secondary (2°) alcohol*. If the —OH group is bonded to a carbon atom which is attached to three alkyl groups, the alcohol is a *tertiary (3°) alcohol*. These subclasses are illustrated in Table 12-1. Note that the terms primary, secondary, and tertiary are also used to refer to the carbons which bear the —OH groups. Thus, the primary alcohol ethanol has a primary carbon atom, and the tertiary alcohol 2-methyl-2-propanol has a tertiary carbon atom.

EXAMPLE 12.1

Classify each of the following alcohols as primary, secondary, or tertiary.

(a) $CH_3-CH_2-\underset{\underset{OH}{|}}{CH}-CH_2-CH_3$

(b) $CH_3-CH_2-\underset{\underset{CH_2-OH}{|}}{CH}-CH_3$

(c) $CH_3-\underset{\underset{OH}{|}}{\overset{\overset{CH_3}{|}}{C}}-CH_2-CH_3$

(d) ⬡—OH

(e) ⬠—CH_2—OH

---

*except for methanol, CH_3OH (the simplest alcohol), which is considered to be a primary alcohol.

ALCOHOLS   321

SOLUTION

(a) secondary (c) tertiary   (e) primary
(b) primary   (d) secondary

EXERCISE 12.1 Classify each of the following alcohols as primary, secondary, or tertiary.

(a) 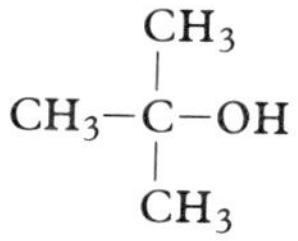 $CH_3-\underset{\underset{CH_3}{|}}{\overset{\overset{CH_3}{|}}{C}}-OH$

(b) $CH_3-CH_2-\underset{\underset{CH_3}{|}}{\overset{\overset{|}{CH-OH}}{CH}}-CH_2-CH_2-CH_3$

(c) $CH_3-CH_2-CH_2-\underset{\underset{CH_2-OH}{|}}{\overset{\overset{CH_3}{|}}{C}}-CH_3$

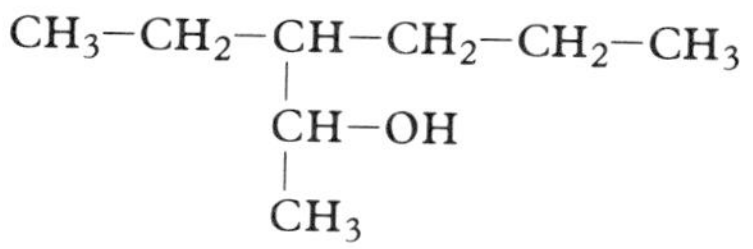

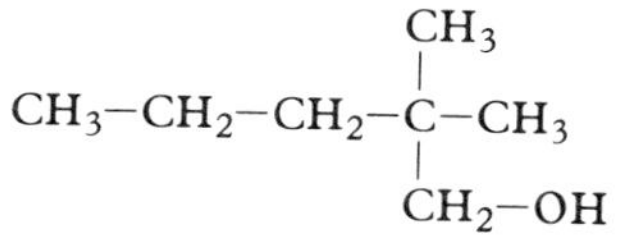

(d) 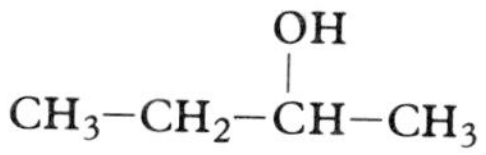 $CH_3-CH_2-\overset{\overset{OH}{|}}{CH}-CH_3$

(e) 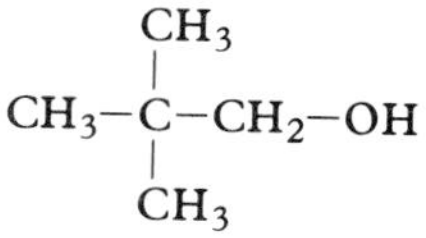 $CH_3-\underset{\underset{CH_3}{|}}{\overset{\overset{CH_3}{|}}{C}}-CH_2-OH$

Nomenclature

Because alcohols are widely known and used by nonchemists, common names are popular, and the lower molecular weight alcohols are often

Table 12-1 Subclasses of Alcohols

Subclass	General Formula	Examples	Name
Primary (1°)	$R-\underset{\underset{H}{\|}}{\overset{\overset{H}{\|}}{C}}-OH$	CH_3OH	methanol
		CH_3CH_2OH	ethanol
		$CH_3CH_2CH_2OH$	1-propanol
Secondary (2°)	$R-\underset{\underset{R}{\|}}{\overset{\overset{H}{\|}}{C}}-OH$	$CH_3-\overset{\overset{OH}{\|}}{CH}-CH_3$	2-propanol
		$CH_3CH_2-\overset{\overset{OH}{\|}}{CH}-CH_3$	2-butanol
		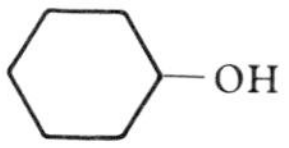 cyclohexyl–OH	cyclohexanol
Tertiary (3°)	$R-\underset{\underset{R}{\|}}{\overset{\overset{R}{\|}}{C}}-OH$	$CH_3-\underset{\underset{CH_3}{\|}}{\overset{\overset{CH_3}{\|}}{C}}-OH$	2-methyl-2-propanol

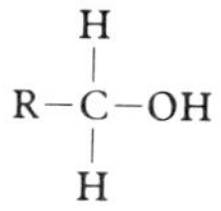

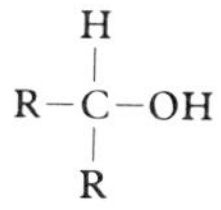

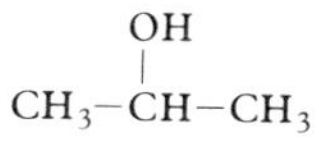

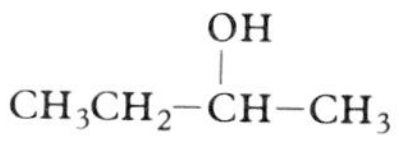

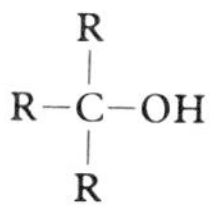

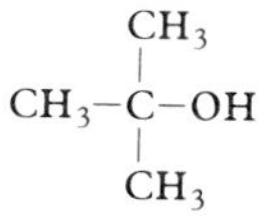

identified by the following names. Alternate common names are given in parentheses.

CH_3-OH CH_3-CH_2-OH $CH_3-CH_2-CH_2-OH$ $CH_3-\underset{\underset{\displaystyle OH}{|}}{CH}-CH_3$

methyl alcohol   ethyl alcohol   propyl alcohol   isopropyl alcohol
(wood alcohol)   (grain alcohol)   (rubbing alcohol)

However, to avoid possible confusion, it is advisable to use IUPAC names for alcohols whenever possible. The rules for naming alcohols are:

1. The characteristic name ending for alcohols is "-ol."

2. Alcohols are named by finding the longest continuous chain of carbon atoms containing the —OH group. This chain is called the main chain.

3. The terminal "-e" of the name of the hydrocarbon corresponding to the main chain is replaced by "-ol" for alcohols containing one —OH group.

4. The main chain is numbered starting from the end nearest the —OH group; however, there is no need to number methanol and ethanol. Examples of IUPAC names are:

CH_3-OH CH_3-CH_2-OH $CH_3-CH_2-CH_2-OH$

methanol   ethanol   1-propanol

$CH_3-\underset{\underset{\displaystyle OH}{|}}{CH}-CH_2-CH_3$ $CH_3-\underset{\underset{\displaystyle OH}{|}}{\overset{\overset{\displaystyle CH_3}{|}}{C}}-CH_3$

2-butanol   2-methyl-2-propanol

5. For aliphatic cyclic alcohols, the ring is numbered by starting with the carbon bearing the —OH group. Numbers can be assigned in either direction. The number 1 does not appear in the name of the alcohol, as illustrated below.

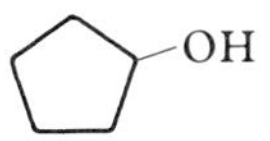

cyclopentanol 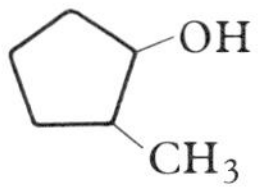   2-methylcyclopentanol 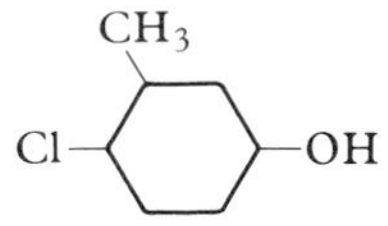   4-chloro-3-methylcyclohexanol

6. If an alcohol contains two —OH groups, it is a "diol"; if it contains three —OH groups, it is a "triol," and so on. In this case, the terminal "-e" of the corresponding hydrocarbon is retained, and the "-diol" or "-triol" is added. Numbers are used to locate each —OH group, as in the examples below.

$\underset{\underset{\displaystyle OH \quad OH}{|\quad\quad |}}{CH_2-CH_2}$ $\underset{\underset{\displaystyle OH \quad\quad OH}{|\quad\quad\quad |}}{CH_2-CH_2-CH_2}$ $\underset{\underset{\displaystyle OH \quad OH \quad OH}{|\quad\quad |\quad\quad |}}{CH_2-CH-CH_2}$

1,2-ethanediol   1,3-propanediol   1,2,3-propanetriol

EXAMPLE 12.2

Give correct IUPAC names for the following alcohols.

(a) CH₃–CH₂–CH₂–CH₂–OH

(b) CH₃–CH(OH)–CH₂–CH₂–CH₃

(c) CH₃–CH(OH)–CH₂–CH(CH₃)–CH₂–CH₃

(d) CH₃–CH₂–C(CH₃)(CH₂CH₃)–OH

(e) CH₃–CH(Br)–CH₂–CH₂–CH(OH)–CH₂–CH₃

(f) cyclopentane with OH groups at 1 and 3 positions

(g) cyclohexane with CH₃, Cl, and OH substituents

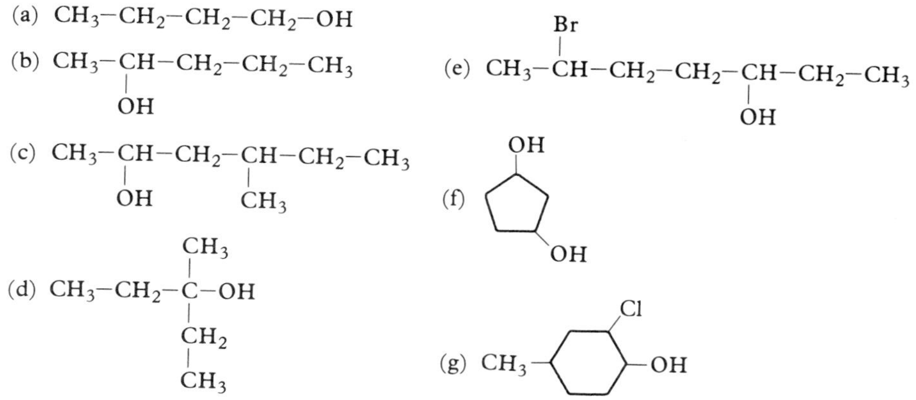

SOLUTION

(a) 1-butanol
(b) 2-pentanol
(c) 4-methyl-2-hexanol
(d) 3-methyl-3-pentanol
(e) 6-bromo-3-heptanol
(f) 1,3-cyclopentanediol
(g) 2-chloro-4-methylcyclohexanol

EXERCISE 12.2 Give the correct IUPAC name for each of the following alcohols:

(a) CH₃–CH(CH₂OH)–CH₂–CH(CH₃)–CH₃ [with CH₃ branch shown and CH₂–OH branch]

(b) cyclohexane–OH

(c) cyclopentane with OH and CH₃ substituents

(d) CH₃–C(CH₃)(CH₃)–OH

(e) CH₃–CH₂–CH(OH)–CH₃

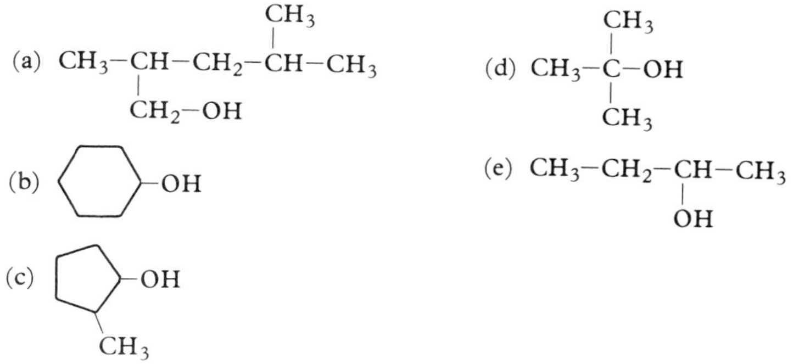

Physical Properties

The boiling points and densities of some alcohols are given in Table 12-2. Alcohols have pleasant odors, and the more common ones are colorless liquids at room temperature. Those with one –OH group are usually less dense than water, while diols and triols are usually more dense than water.

Because alcohols form hydrogen bonds, their boiling points are higher than those for alkanes of similar molecular weight. For example, methanol and ethane have similar molecular weights, but the boiling point of the alcohol is over 150° higher than that of the alkane. But since monohydroxy alcohols cannot form as many hydrogen bonds as water, those

Table 12-2 Boiling Points and Densities of Some Alcohols

Name	Formula	Boiling Point (°C)	Density (g/mL at 20°C)
Methanol	CH_3OH	65	0.792
Ethanol	CH_3CH_2OH	78	0.789
1-Propanol	$CH_3CH_2CH_2OH$	97	0.804
1-Butanol	$CH_3CH_2CH_2CH_2OH$	118	0.810
1-Pentanol	$CH_3CH_2CH_2CH_2CH_2OH$	138	0.818
1,2-Ethanediol	$HOCH_2CH_2OH$	198	1.113
1,3-Propanediol	$HOCH_2CH_2CH_2OH$	214	1.060
1,2,3-Propanetriol	$HOCH_2CHCH_2OH$ $\quad\quad\quad \vert$ $\quad\quad\quad OH$	290	1.261

with fewer than six carbons have lower boiling points than water, even though their molecular weights are higher than that of water. Monohydroxy alcohols with six or more carbons have boiling points higher than water because of increased London dispersion forces. Diols and triols have higher boiling points than monohydroxy alcohols of similar molecular weight due to an increased number of hydrogen bonds.

The hydrogen-bonding ability of alcohols makes the lower molecular weight alcohols very compatible with water and hence water soluble. The hydroxyl group is hydrophilic, but as the hydrophobic hydrocarbon portion of the alcohol structure grows larger, as in butanol, the water solubility of the alcohol decreases (see Table 12-3). Branching increases water solubility because compact alkyl groups are less hydrophobic. Thus, 2-methyl-2-propanol is completely miscible with water, while its isomer 1-butanol is significantly less soluble in water. Increasing the number of —OH groups also makes alcohols more water soluble,

Table 12-3 Water Solubilities of Some Alcohols

Name	Formula	Solubility (g/100 g)
Methanol	CH_3OH	Soluble in all proportions
Ethanol	CH_3CH_2OH	Soluble in all proportions
1-Propanol	$CH_3CH_2CH_2OH$	Soluble in all proportions
1-Butanol	$CH_3CH_2CH_2CH_2OH$	9 (15°C)
2-Methyl-2-propanol	$\quad\quad CH_3$ $\quad\quad\ \vert$ CH_3-C-CH_3 $\quad\quad\ \vert$ $\quad\quad OH$	Soluble in all proportions
1-Pentanol	$CH_3CH_2CH_2CH_2CH_2OH$	3 (22°C)
3-Pentanol	$CH_3CH_2CHCH_2CH_3$ $\quad\quad\quad \vert$ $\quad\quad\quad OH$	5.5 (30°C)
1-Hexanol	$CH_3CH_2CH_2CH_2CH_2CH_2OH$	0.6 (20°C)

and the diols and triols are more soluble in water than the corresponding alcohols having only one hydroxyl group.

As you might expect from the structural similarities of water and alcohols, alcohols are able to dissolve ionic compounds, although not to the extent that water does. Table 12-4 shows the solubility of sodium chloride in water and in various alcohols.

Table 12-4 Solubility of Sodium Chloride in Water and Alcohols

Solvent	Solubility of NaCl, g/100 mL at 25°C
Water	36.2
Methanol	1.4
Ethanol	0.06
1-Propanol	0.01

Chemical Properties

Although alcohols participate in a variety of chemical reactions, three reactions—dehydration, oxidation, and ester formation—are considered to be most important.

Dehydration

As its name implies, *dehydration* of alcohols results in loss of water. The elements of water may be lost from one alcohol molecule at a time to produce an alkene product, or they may be lost from two interacting alcohol molecules. In the second case, the organic product is an ether. Both reactions are catalyzed by strong acid; sulfuric acid is used most often because of its strong affinity for the water produced by the reaction. The structure of the alcohol and the reaction temperature determine whether dehydration produces an alkene or an ether, as discussed below.

Dehydration of Primary Alcohols: If a primary alcohol is mixed with sulfuric acid and heated to moderate temperatures (140°C), an ether is formed from two alcohol molecules.

$$\underset{\underset{\text{OH}}{|}}{CH_3-CH_2} + \underset{\underset{\text{H}}{|}}{O-CH_2-CH_3} \xrightarrow[140°]{H_2SO_4} \underset{\text{diethyl ether}}{CH_3-CH_2-O-CH_2-CH_3} + HOH$$

One alcohol molecule loses its —OH group and the other alcohol loses hydrogen from its —OH group. Combination of OH^- and H^+ leads to formation of water, and the two alcohol fragments join to form an ether. As you can see, this can be considered a double replacement reaction.

In contrast, if the mixture of alcohol and H_2SO_4 is heated to higher temperatures (180°C), then an alkene is produced. In order to describe this reaction, let us designate the carbon atom carrying the $-OH$ group as α, and any adjacent carbon atoms as β-carbons:

$$\overset{\beta}{CH_3} - \overset{\alpha}{CH_2} - OH$$

(In primary alcohols there is only one β-carbon.) At the higher reaction temperature, a single alcohol molecule loses its hydroxyl group and one hydrogen from its β-carbon. These combine to make H_2O, and the electrons left behind from the C—H bond form a second bond between the α- and β-carbons. Thus, an alkene is produced:

$$\underset{\substack{| \\ H}}{\overset{\beta}{CH_2}} - \underset{\substack{| \\ OH}}{\overset{\alpha}{CH_2}} \xrightarrow[180°]{H_2SO_4} CH_2{=}CH_2 + HOH$$

ethanol   ethene

This reaction can be classified as a decomposition reaction. It is frequently referred to as *elimination*, since a molecule of water is eliminated from the alcohol molecule.

Since two different organic products can be formed by dehydration of primary alcohols, we can conclude that there are two separate reaction pathways available. The higher temperature requirement for alkene formation suggests that the elimination pathway has a higher activation energy than the double replacement pathway for primary alcohols. Note that dehydration of alcohols to form alkenes is a reversal of hydration of alkenes to form alcohols.

Dehydration of Secondary and Tertiary Alcohols: In contrast to primary alcohols, secondary and tertiary alcohols are dehydrated by sulfuric acid to form exclusively alkenes. Ethers do not form because the bulky alkyl side chains on the 2° and 3° alcohol molecules prevent them from interacting effectively with each other. This effect is called *steric hindrance*. In addition, secondary and tertiary alcohols form alkenes at lower temperatures than primary alcohols, indicating that these molecules follow reaction pathways having lower activation energies. Probably this is because part of the bulky alcohol molecules is lost as H_2O, and a less strained organic structure is produced.

steric hindrance: the prevention or retardation of a chemical reaction because of the presence of bulky groups; "steric" comes from the Greek *stereo*, which means "solid"; in chemistry, "steric" refers to spatial relationships of atoms.

Examples of dehydration of secondary and tertiary alcohols are shown below. Notice that in each case two alkenes are formed.

$$\underset{\substack{| \\ OH}}{\overset{\beta}{CH_3} - \overset{\alpha}{CH} - \overset{\beta}{CH_2} - CH_3} \xrightarrow[100°C]{H_2SO_4} \begin{cases} CH_3{-}CH{=}CH{-}CH_3 + H_2O \\ \text{2-butene} \\ \text{(major product)} \\ \\ CH_2{=}CH{-}CH_2{-}CH_3 + H_2O \\ \text{1-butene} \\ \text{(minor product)} \end{cases}$$

2-butanol
(2°)

ALCOHOLS 327

$$\underset{\substack{\text{2-methyl-2-butanol} \\ (3°)}}{\overset{\beta}{\text{CH}_3}-\overset{\alpha}{\underset{\underset{\text{OH}}{|}}{\overset{\overset{\text{CH}_3}{|}}{\text{C}}}}-\overset{\beta}{\text{CH}_2}-\text{CH}_3} \xrightarrow[60°C]{H_2SO_4}} \begin{cases} \text{CH}_3-\overset{\overset{\text{CH}_3}{|}}{\text{C}}=\text{CH}-\text{CH}_3 + H_2O \\ \text{2-methyl-2-butene} \\ \text{(major product)} \\ \\ \text{CH}_2=\overset{\overset{\text{CH}_3}{|}}{\text{C}}-\text{CH}_2-\text{CH}_3 + H_2O \\ \text{2-methyl-1-butene} \\ \text{(minor product)} \end{cases}$$

In the case of 2-butanol there are two β-carbons, and both lose hydrogen to form alkenes. However, the alkene having the most alkyl substituents at the double bond is the more stable of the two possibilities, and thus 2-butene forms in greater amount than 1-butene. This relative stability of alkene isomers is the basis for Saytzeff's rule that the β-carbon bearing the fewest hydrogens will lose hydrogen in an elimination reaction. In the case of the tertiary alcohol, 2-methyl-2-butanol, two of the β-carbons have the same number of hydrogen atoms and thus are structurally identical. For this reason, two alkene products are formed instead of the three you might expect from the presence of three β-carbons. You will encounter hydration and dehydration reactions again when you study metabolism. In those cases, enzyme catalysts are able to guide the reactions so that only one product is formed.

EXAMPLE 12.3

Complete the following equations. If more than one organic product is formed in any reaction, indicate major and minor products.

(a) $CH_3-CH_2-CH_2-CH_2-OH \xrightarrow[140°]{H_2SO_4}$

(b) $CH_3-CH_2-\underset{\underset{\text{OH}}{|}}{\text{CH}}-CH_2-CH_3 \xrightarrow[100°]{H_2SO_4}$

(c) $CH_3-\underset{\underset{\text{OH}}{|}}{\overset{\overset{CH_3}{|}}{\text{CH}}}-CH-CH_3 \xrightarrow[100°]{H_2SO_4}$

(d) $CH_3-CH_2-\underset{\underset{\text{OH}}{|}}{\overset{\overset{\overset{CH_3}{|}}{CH_2}}{\text{C}}}-CH_3 \xrightarrow[60°]{H_2SO_4}$

(e) $CH_3-CH_2-CH_2-OH \xrightarrow[180°]{H_2SO_4}$

SOLUTION

(a) $CH_3-CH_2-CH_2-CH_2-O-CH_2-CH_2-CH_2-CH_3 + H_2O$

(b) $CH_3-CH=CH-CH_2-CH_3 + H_2O$ (Although you might think that a second product is formed, $CH_3-CH_2-CH=CH-CH_3$, this is actually the same structure as $CH_3-CH=CH-CH_2-CH_3$.)

(c) $CH_3-\overset{\overset{CH_3}{|}}{C}=CH-CH_3 + H_2O$, $CH_3-\overset{\overset{CH_3}{|}}{CH}-CH=CH_2 + H_2O$
    major product   minor product

(d) 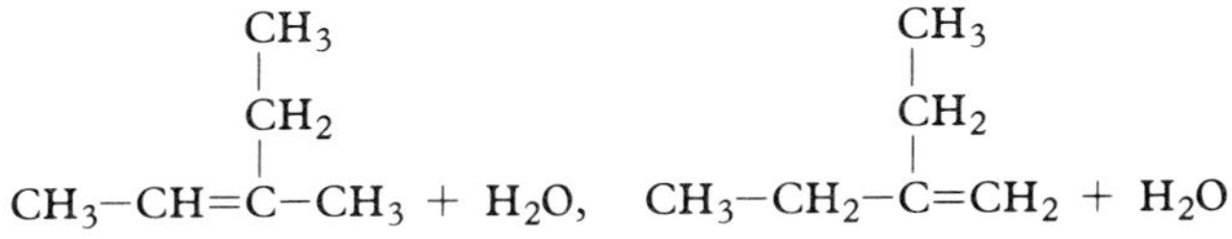 $\underset{\text{major product}}{CH_3-CH=\underset{\underset{CH_3}{|}}{\underset{CH_2}{|}}C-CH_3} + H_2O$, $\underset{\text{minor product}}{CH_3-CH_2-\underset{\underset{CH_3}{|}}{\underset{CH_2}{|}}C=CH_2} + H_2O$

(e) $CH_3-CH=CH_2 + H_2O$

EXERCISE 12.3 Complete the folowing equations. If more than one organic product is formed in any reaction, indicate major and minor products.

(a) 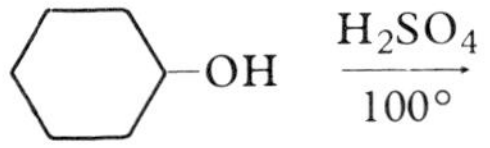 $\text{C}_6\text{H}_{11}-OH \xrightarrow[100°]{H_2SO_4}$

(b) $CH_3-CH_2-CH_2-OH \xrightarrow[140°]{H_2SO_4}$

(c) $CH_3-CH_2-CH_2-CH_2-OH \xrightarrow[180°]{H_2SO_4}$

(d) 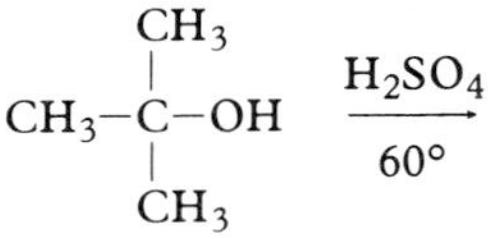 $CH_3-\underset{\underset{CH_3}{|}}{\overset{\overset{CH_3}{|}}{C}}-OH \xrightarrow[60°]{H_2SO_4}$

(e) 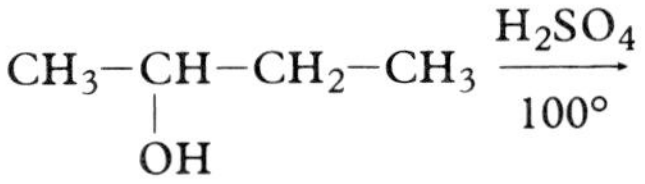 $CH_3-\underset{\underset{OH}{|}}{CH}-CH_2-CH_3 \xrightarrow[100°]{H_2SO_4}$

Oxidation

As with most organic compounds, alcohols are flammable, and combustion produces water and either carbon dioxide or carbon monoxide, as shown below.

$$CH_3-CH_2-OH + 3\,O_2 \longrightarrow 2\,CO_2 + 3\,H_2O$$

$$CH_3-CH_2-OH + 2\,O_2 \longrightarrow 2\,CO + 3\,H_2O$$

However, alcohols can be oxidized in a stepwise fashion by using controlled laboratory conditions. Typical oxidizing agents are a mixture of potassium dichromate ($K_2Cr_2O_7$) and sulfuric acid or a mixture of potassium permanganate ($KMnO_4$) and sulfuric acid. As in the case of alkene formation, the subclasses of alcohols form slightly different products. For example, if methanol is oxidized under controlled conditions, the first product is an *aldehyde* called formaldehyde:

aldehyde (AL-deh-hide): an organic compound having the functional group

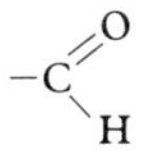

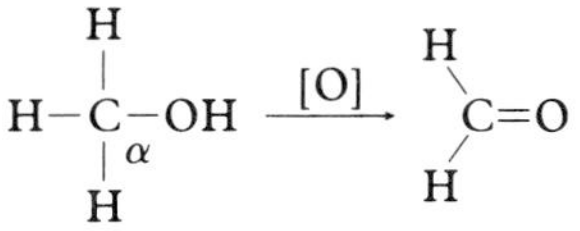

$-\overset{\overset{O}{\parallel}}{C}-H$

$\underset{\text{methanol}}{H-\underset{\underset{H}{|}}{\overset{\overset{H}{|}}{C}}-OH} \xrightarrow{[O]} \underset{\text{formaldehyde}}{\overset{H}{\underset{H}{>}}C=O}$

(where [O] represents an oxidizing agent as described above). Notice that two hydrogens were lost by methanol, one from the $-OH$ group and one from the α-carbon. This loss of hydrogen corresponds to oxidation of methanol. Although we do not show it, the oxidizing agent was reduced. Like methanol, all primary alcohols form aldehydes as initial products when oxidized in the laboratory:

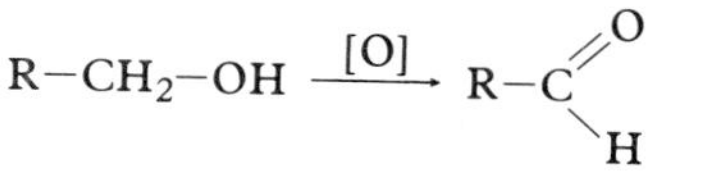

$$R-CH_2-OH \xrightarrow{[O]} R-C\overset{\displaystyle O}{\underset{\displaystyle H}{\diagup\!\!\!\diagdown}}$$

1° alcohol   an aldehyde

However, if the aldehyde product is not removed from the reaction mixture, it too will be oxidized. The product, a *carboxylic acid*, is formed when an oxygen atom is inserted between carbon and hydrogen in the —CHO group:

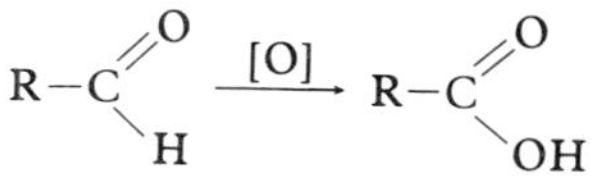

carboxylic (car-box-SIH-lic) **acid:** an organic compound having the carboxyl group,

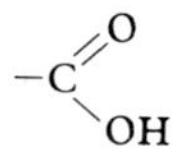

$$-C\overset{\displaystyle O}{\underset{\displaystyle OH}{\diagup\!\!\!\diagdown}}$$

$$R-C\overset{\displaystyle O}{\underset{\displaystyle H}{\diagup\!\!\!\diagdown}} \xrightarrow{[O]} R-C\overset{\displaystyle O}{\underset{\displaystyle OH}{\diagup\!\!\!\diagdown}}$$

an aldehyde   a carboxylic acid

Thus it is possible in the laboratory to oxidize a primary alcohol to either an aldehyde or a carboxylic acid, depending on how one carries out the reaction.

Secondary alcohols are oxidized in a manner similar to that for primary alcohols, except that a *ketone* is formed by loss of hydrogen from the —OH group and from the α-carbon. The following equations illustrate this reaction:

ketone (KEY-tone): an organic compound having the C=O functional group, with alkyl groups attached to the carbon atom.

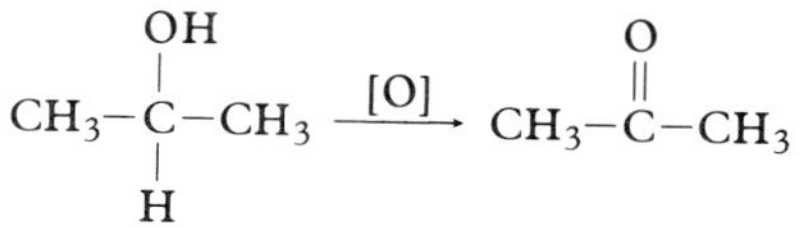

$$\underset{\text{2-propanol}}{CH_3-\underset{\underset{\displaystyle H}{|}}{\overset{\overset{\displaystyle OH}{|}}{C}}-CH_3} \xrightarrow{[O]} \underset{\text{acetone}}{CH_3-\overset{\overset{\displaystyle O}{\|}}{C}-CH_3}$$

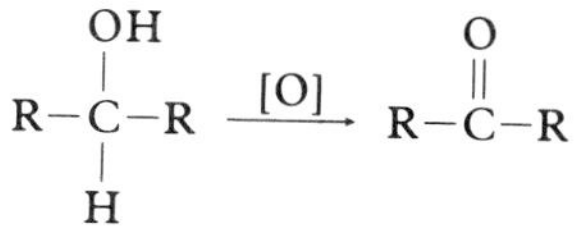

$$\underset{\text{2° alcohol}}{R-\underset{\underset{\displaystyle H}{|}}{\overset{\overset{\displaystyle OH}{|}}{C}}-R} \xrightarrow{[O]} \underset{\text{a ketone}}{R-\overset{\overset{\displaystyle O}{\|}}{C}-R}$$

In this case, the ketone is the only organic product formed, because it has no hydrogens on the carbon bonded to oxygen. Thus ketones do not undergo further oxidation under controlled conditions.

Tertiary alcohols do not undergo controlled oxidation because there are no hydrogens on the α-carbon atoms. Thus, tertiary alcohols can only be oxidized directly to water and carbon dioxide or carbon monoxide. Controlled oxidation of alcohols can be summarized:

1° alcohols are oxidized to aldehydes which are in turn oxidized to carboxylic acids.

2° alcohols are oxidized to ketones.

3° alcohols do not undergo controlled oxidation.

As you will see later, controlled oxidation of alcohols is an important metabolic reaction. In the cell, controlled oxidations are catalyzed by enzymes, so that oxidations are carried out stepwise whenever possible. An example of a metabolic oxidation of an alcohol is the conversion of ethanol first to acetaldehyde and then to acetic acid (this reaction is responsible for vinegar formation in wine):

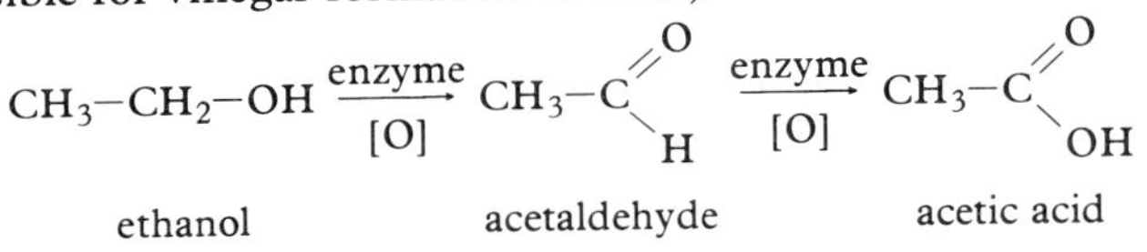

$$CH_3-CH_2-OH \xrightarrow[\text{[O]}]{\text{enzyme}} CH_3-C{\overset{O}{\underset{H}{\lessgtr}}} \xrightarrow[\text{[O]}]{\text{enzyme}} CH_3-C{\overset{O}{\underset{OH}{\lessgtr}}}$$

ethanol · · · · · · · · · acetaldehyde · · · · · · · · · acetic acid

EXAMPLE 12.4

Complete each of the following equations. If there is no reaction, write NR for the answer.

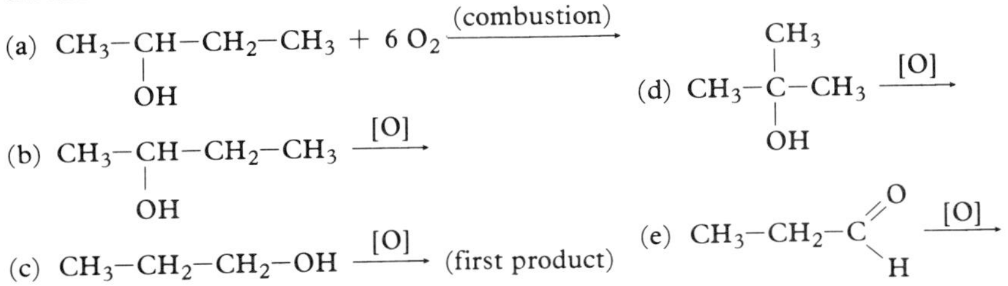

(a) $CH_3-\underset{\underset{OH}{|}}{CH}-CH_2-CH_3 + 6 O_2 \xrightarrow{\text{(combustion)}}$

(b) $CH_3-\underset{\underset{OH}{|}}{CH}-CH_2-CH_3 \xrightarrow{\text{[O]}}$

(c) $CH_3-CH_2-CH_2-OH \xrightarrow{\text{[O]}}$ (first product)

(d) $CH_3-\underset{\underset{OH}{|}}{\overset{\overset{CH_3}{|}}{C}}-CH_3 \xrightarrow{\text{[O]}}$

(e) $CH_3-CH_2-C{\overset{O}{\underset{H}{\lessgtr}}} \xrightarrow{\text{[O]}}$

SOLUTION

(a) $4 CO_2 + 5 H_2O$

(b) $CH_3-\underset{\underset{O}{\|}}{C}-CH_2-CH_3$

(c) $CH_3-CH_2-C{\overset{O}{\underset{H}{\lessgtr}}}$

(d) NR

(e) $CH_3-CH_2-C{\overset{O}{\underset{OH}{\lessgtr}}}$

EXERCISE 12.4 Complete each of the following equations. If there is no reaction, write NR for the answer.

(a) $CH_3-\underset{\underset{CH_3}{|}}{\overset{\overset{CH_3}{|}}{C}}-OH + 6 O_2 \longrightarrow$ (combustion products)

(b) $CH_3-\underset{\underset{CH_3}{|}}{\overset{\overset{CH_3}{|}}{C}}-CH_2-CH_2-OH \xrightarrow{\text{[O]}}$ (first product)

(c) $CH_3-\underset{\underset{\underset{CH_3}{|}}{CH-OH}}{CH}-CH_2-CH_2-CH_3 \xrightarrow{\text{[O]}}$

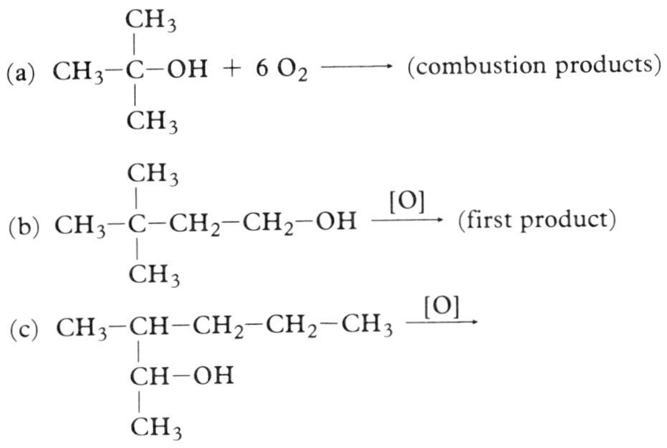

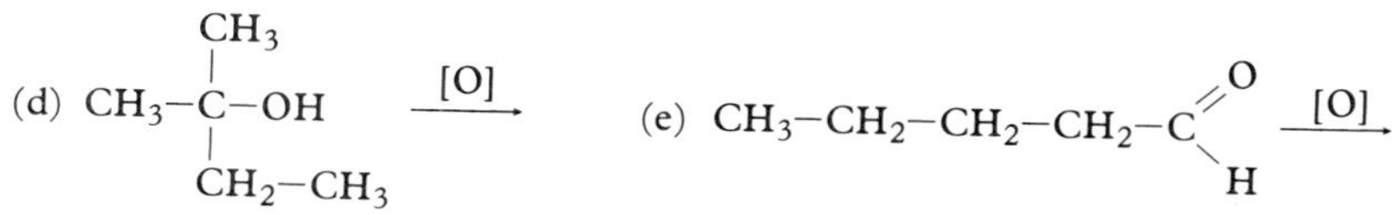

(d) $\underset{\underset{CH_2-CH_3}{|}}{\overset{\overset{CH_3}{|}}{CH_3-C-OH}} \xrightarrow{[O]}$ (e) $CH_3-CH_2-CH_2-CH_2-\overset{\overset{O}{\parallel}}{C}-H \xrightarrow{[O]}$

Ester Formation

Alcohols react with carboxylic acids to form products called *esters*. In this acid-catalyzed reversible reaction, —OH is lost from the carboxylic acid and —H is lost from the alcohol to form water. At the same time, the alcohol and carboxylic acid fragments combine to form an ester:

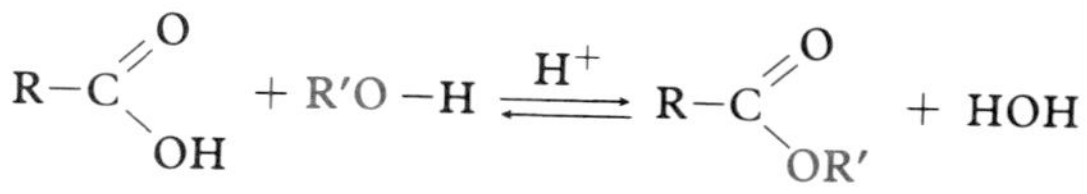

$$R-\underset{OH}{\overset{O}{\overset{\parallel}{C}}} + R'O-H \underset{}{\overset{H^+}{\rightleftharpoons}} R-\underset{OR'}{\overset{O}{\overset{\parallel}{C}}} + HOH$$

carboxylic   alcohol   ester
acid

The equilibrium is shifted toward formation of products by using large concentrations of one or both of the reactants and by removing the water as it forms.

All subclasses of alcohols take part in ester formation. In metabolic processes, ester formation accounts for synthesis of fats. We will discuss esters further in chapter 14.

Sources of Alcohols

Alcohols are found in free form only sparingly in nature, but they can be isolated from volatile oils obtained from flowers, leaves, and stems of plants. The pleasant odors of liquid alcohols make them valuable ingredients for perfumes. The common sugars (for example, glucose) contain hydroxyl groups and therefore can be classified as alcohols, but because other functional groups are present in the sugars, they are not usually thought of as alcohols. Cholesterol also contains a hydroxyl group. This compound and several of its derivatives are found in bile, gallstones, brains, blood, egg yolk, and many animal products.

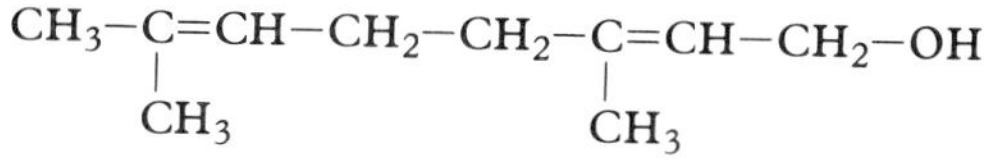

$$CH_3-\underset{\underset{CH_3}{|}}{C}=CH-CH_2-CH_2-\underset{\underset{CH_3}{|}}{C}=CH-CH_2-OH$$

geraniol

(isolated from roses and geraniums)

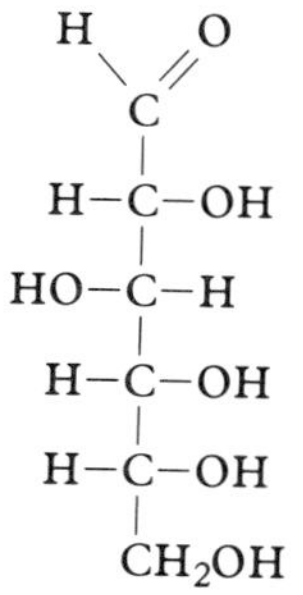

glucose:
$$\begin{array}{c} H\diagdown\!\!\!\!\diagup O \\ C \\ | \\ H-C-OH \\ | \\ HO-C-H \\ | \\ H-C-OH \\ | \\ H-C-OH \\ | \\ CH_2OH \end{array}$$

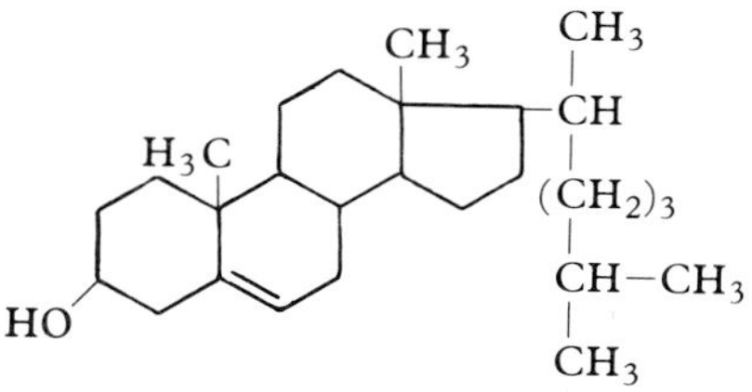

cholesterol

Destructive distillation of wood (the process of heating wood in the absence of air) was a common method in the early part of this century for obtaining methanol, hence the name wood alcohol. However, it is quite difficult to purify this distillate to obtain the pure alcohol, and catalytic hydrogenation of carbon monoxide now accounts for the major industrial production of methanol:

$$CO + 2H_2 \xrightarrow[\Delta]{catalyst} CH_3OH$$

Another means of obtaining alcohols is by fermentation. This process utilizes polysaccharides (polymers of simple sugars) from corn, grain, and vegetable waste or simple sugars from molasses. (If polysaccharides are used, they must first be degraded to simple sugars.) In either case, yeast is added to bring about the fermentation; enzymes present in the yeast convert the sugars to ethanol and carbon dioxide:

$$\underset{\text{glucose}}{C_6H_{12}O_6} \xrightarrow[\text{enzymes}]{\text{yeast}} 2\underset{\text{ethanol}}{CH_3-CH_2-OH} + 2CO_2$$

Fermentation can also produce other alcohols by using bacteria capable of converting sugars to alcohols more complex than ethanol.

In general, the alcohols produced from natural sources are difficult to purify and therefore expensive to produce. For these reasons, producers have had to develop special chemical processes that form alcohols from readily available raw materials under carefully controlled conditions. Some of these processes are:

hydration of alkenes:

$$\underset{}{\overset{}{C}=\overset{}{C}} + H_2O \rightleftharpoons -\underset{H}{\overset{}{C}}-\underset{OH}{\overset{}{C}}-$$

hydrolysis of alkyl halides:

$$R-X + H_2O \xrightarrow{OH^-} R-OH + HX$$

oxidation of hydrocarbons:

$$2R-H + O_2 \xrightarrow[\Delta]{catalyst} 2R-OH$$

reduction of organic compounds:

$$\overset{}{C}=O + H_2 \xrightarrow[\Delta]{catalyst} H-\overset{}{\underset{}{C}}-OH$$

Uses of Alcohols

As judged by the large amounts of alcohols produced in the United States, alcohols are among the most important organic compounds. The largest industrial use of alcohols is in the production of other chemical compounds, mainly because alcohols can readily be converted into a large variety of other compounds by relatively simple reactions. Thus alcohols are the starting materials for synthesizing such things as alkenes, ethers, detergents, plasticizers, lubricants, emollients, and foaming agents. Alcohols are also valuable as solvents. Ethanol is second in importance only to water as a solvent for drugs, lacquers, perfumes, cosmetics, detergents, and plastics.

In addition to the uses mentioned above, several individual alcohols have specialized but well-known uses as described below.

Methanol (CH_3—OH)

This alcohol, commonly called carbinol, wood alcohol, and wood spirit, is used for synthesizing dyes and plastics and is an ingredient in rocket fuels. Methanol is frequently mistaken for ethanol by the uninformed. This is an unfortunate mistake, as methanol is highly toxic, and it causes permanent blindness or death if ingested even in small amounts. Humans metabolize methanol in the liver where it is oxidized to formaldehyde,

$$CH_3\text{—}OH \xrightarrow[\text{[O]}]{\text{liver enzymes}} H\text{—}C\begin{matrix}\diagup O \\ \diagdown H\end{matrix}$$

methanol → formaldehyde

which then circulates in the body to react rapidly with proteins, thereby rendering them nonfunctional. This quality of formaldehyde is used to good advantage in preserving biological tissue specimens and in the embalming process, but its effect on living tissue is disastrous. The antidote for methanol poisoning is feeding large quantities of ethanol, which keeps the liver enzymes occupied so that they cannot convert methanol to formaldehyde.

Ethanol (CH_3—CH_2—OH)

Ethanol is commonly called ethyl alcohol, or just alcohol. It is the intoxicating ingredient of alcoholic beverages, and can cause death if ingested in large quantities. The concentration of ethanol in the blood rises to a peak about 90 minutes after ethanol consumption, and then it gradually decreases until it is normal after 12 to 15 hours. Ethanol is converted first to acetaldehyde and then to acetic acid by liver enzymes:

$$CH_3\text{—}CH_2\text{—}OH \xrightarrow[\text{[O]}]{\text{liver enzymes}} CH_3\text{—}C\begin{matrix}\diagup O \\ \diagdown H\end{matrix} \xrightarrow[\text{[O]}]{\text{liver enzymes}} CH_3\text{—}C\begin{matrix}\diagup O \\ \diagdown OH\end{matrix}$$

ethanol → acetaldehyde → acetic acid

Although the intoxicating effect of ethanol is not entirely understood, it may be a result of certain interactions of acetaldehyde with brain cells.

Ethanol has several other familiar uses, such as the ethanol solution of iodine called "tincture of iodine," used as a topical *germicide*. Ethanol is also used in 70% (v/v) aqueous solutions as an *antiseptic*, and it is gaining popularity as an additive to gasoline in making "gasohol" for automobiles.

germicide: a substance which destroys microorganisms.

antiseptic: a substance used on or in the body to destroy or prevent the growth of infectious microorganisms.

2-Propanol ($CH_3-CHOH-CH_3$)

This alcohol is often called isopropyl alcohol and rubbing alcohol. Most of us have used it as a commercial preparation for rubbing sore muscles. It is used in hospitals for sponge baths and for contracting blood vessels. It is toxic if ingested.

1,2-Ethanediol ($HO-CH_2-CH_2-OH$)

glycol (GLYE-col): a common name for diols.

Commonly called ethylene glycol or just *glycol*, this alcohol is a colorless, oily liquid with a sweet taste and mild odor. It is widely used as antifreeze in automobile water-cooling systems. The diol is toxic if ingested because liver enzymes convert it to oxalic acid, a compound which depresses the central nervous system and destroys kidney cells:

$$HO-CH_2-CH_2-OH \xrightarrow[\text{[O]}]{\text{liver enzymes}} \underset{H}{\overset{O}{\underset{\|}{C}}}-\underset{H}{\overset{O}{\underset{\|}{C}}} \xrightarrow[\text{[O]}]{\text{liver enzymes}} \underset{HO}{\overset{O}{\underset{\|}{C}}}-\underset{OH}{\overset{O}{\underset{\|}{C}}}$$

1,2-ethanediol   glyoxal   oxalic acid
(ethylene glycol)

The vegetable rhubarb owes its tart flavor to the presence of several organic acids, one of them being oxalic acid (in small amounts).

1,2,3-Propanetriol ($CH_2OH-CHOH-CH_2OH$)

This triol is readily recognized by its common names glycerol, glycerin, or glycerine. It is a viscous, sweet-tasting liquid completely miscible with water and frequently found as a moisturizer in hand creams, lotions, and other cosmetics. Glycerol is found in animal fat combined with carboxylic acids in ester compounds called triacylglycerols (or triglycerides). It is a nontoxic material found in the human body because of its involvement in fat synthesis.

ETHERS

Ethers are recognized by the functional group $C-O-C$; an example is dimethyl ether, CH_3-O-CH_3. A general formula for ethers is $R-O-R$, where R represents any aliphatic or aromatic group. The two R groups need not be identical. Like alcohols, ethers can be thought of as derivatives of water; in ethers, both hydrogens of water are replaced by organic groups.

Structure

Like water and alcohols, ethers have bent structures in the vicinity of the oxygen atom:

$$\underset{\text{water}}{\text{H}\overset{\overset{\displaystyle\ddot{\text{O}}}{}}{}\text{H}} \qquad \underset{\text{alcohol}}{\text{R}\overset{\overset{\displaystyle\ddot{\text{O}}}{}}{}\text{H}} \qquad \underset{\text{ether}}{\text{R}\overset{\overset{\displaystyle\ddot{\text{O}}}{}}{}\text{R}}$$

However, it is customary to write the structure for an ether in a straight line, as in R—O—R.

Since ether molecules resemble water and alcohol molecules, we would expect the ether molecules to be polar. In general they are polar, but not as polar as alcohols, which in turn are not as polar as water. The O—R bonds in ethers are not as polar as O—H bonds because the difference in electronegativities between oxygen and carbon in the O—R bond is not as great as the difference in electronegativities between oxygen and hydrogen in the O—H bond. And since ethers have two O—R bonds, we would not expect them to be as polar as alcohols which always have one O—R bond and one O—H bond. The relative dipole moments of water, methanol, and dimethyl ether are

$$\underset{\substack{\text{water}\\1.8\text{ D}}}{\text{H}_2\text{O}} \qquad \underset{\substack{\text{methanol}\\1.7\text{ D}}}{\text{CH}_3-\text{OH}} \qquad \underset{\substack{\text{dimethyl ether}\\1.3\text{ D}}}{\text{CH}_3-\text{O}-\text{CH}_3}$$

Nomenclature

Ethers are named almost always on the basis of their common names; thus we will name them by traditional methods. In the case of ethers in which the two R groups are identical, the prefix "di-" is used with the name of the R group, followed by the word "ether." Thus, CH_3OCH_3 is named dimethyl ether. If the two R groups are different, each is named in alphabetical order, and the word "ether" appears last in the name. In this case, there are three words in the name of the ether. An example is $CH_3OCH_2CH_3$, ethyl methyl ether. Examples of names for several common ethers are given in Table 12-5.

EXAMPLE 12.5

Name the following ethers.

(a) $CH_3CH_2CH_2-O-CH_2CH_2CH_3$ (c) $CH_3CH_2CH_2CH_2CH_2-O-CH_3$
(b) $CH_3CH_2-O-CH_2CH_2CH_2CH_3$

SOLUTION

(a) dipropyl ether   (b) butyl ethyl ether   (c) methyl pentyl ether

Table 12-5 Physical Properties of Some Common Ethers

Name	Structure	Boiling Point (°C)	Density[a] (g/mL)	Water Solubility[a] (g/100 g)
Dimethyl ether	CH_3-O-CH_3	−24	b	7.1
Diethyl ether	$CH_3CH_2-O-CH_2CH_3$	35	0.708	7.5
Ethyl methyl ether	$CH_3CH_2-O-CH_3$	8	0.697	soluble
Methyl phenyl[c] ether	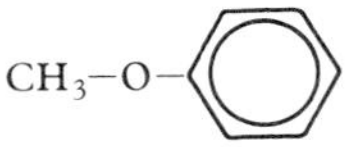 $CH_3-O-C_6H_5$	154	0.990	insoluble

[a] at 20°C
[b] gas at 20°C
[c] The name "phenyl" is used for the benzene ring when it is a substituent in a compound.

EXERCISE 12.5 Name the following ethers.
(a) $CH_3CH_2CH_2CH_2-O-CH_2CH_2CH_2CH_3$
(b) $CH_3-O-CH_2CH_3$
(c) 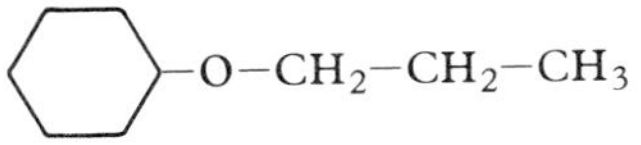 $C_6H_5-O-CH_2-CH_2-CH_3$

Physical Properties

The boiling points and densities of several ethers are given in Table 12-5. The more common ethers are colorless, highly volatile liquids, while dimethyl ether and ethyl methyl ether are gases at room temperature. Like most organic compounds, ethers tend to be less dense than water.

Ether molecules cannot form hydrogen bonds with each other, and thus their boiling points are much lower than those of alcohols. In fact, ether boiling points are rather similar to those for alkanes, which are also unable to form hydrogen bonds among themselves. However, ethers do have the ability to form hydrogen bonds with other molecules. For this reason, the low molecular weight ethers have surprisingly high water solubilities compared to alkanes, as demonstrated in Table 12-5.

The high volatility of ethers presents problems in handling them. Their vapors are usually more dense than air, and thus they tend to flow down the sides of containers along tabletops and floors, where any small spark can ignite them. Ethers should only be handled in areas where there is an extremely good ventilation system and where there are no flames, sparks, or other sources of ignition, such as electrical equipment.

Another hazard in working with ethers stems from the fact that they form *peroxides* when exposed to air for long periods of time:

peroxide: a compound having two oxygen atoms linked together as −O−O−; the word "peroxide" means "excess oxygen."

$$CH_3CH_2OCH_2CH_3 + O_2 \longrightarrow CH_3CH_2OCH(OOH)-CH_3$$
diethyl ether   a peroxide

Peroxides are unstable compounds which can explode without warning. Their formation in ethers can be detected by standard test procedures, and it can be prevented by addition of certain compounds (antioxidants) which prevent the ether from reacting with oxygen in the air.

Chemical Properties

Ethers tend to be rather unreactive compounds, and they behave more like alkanes than like organic compounds containing functional groups. Thus, combustion, illustrated by the following equation for the combustion of diethyl ether, is the only reaction worthy of note.

$$CH_3CH_2OCH_2CH_3 + 6\ O_2 \longrightarrow 4\ CO_2 + 5\ H_2O$$
diethyl ether

Sources and Uses of Ethers

Simple ethers are not found commonly in nature, but ether linkages are present in such natural products as sugars, starches, and cellulose. Ethers are prepared commercially by the dehydration of alcohols (discussed earlier in this chapter) and by specialized chemical reactions.

Because of their inert qualities, ethers as a class of compounds are used as solvents for fats, oils, waxes, perfumes, dyes, gums, and hydrocarbons. Vapors of certain ethers are toxic to insects, and these substances are employed as insecticides and soil fumigants.

Diethyl ether, also called ethyl ether and ether, has been used since 1846 as a general *anesthetic*. It acts on the central nervous system to render the patient insensible. In addition, it is an effective *analgesic*, and it promotes good muscle relaxation. Diethyl ether is toxic to humans, but there is a wide gap between the amount that causes unconsciousness and the lethal dose (one fluid ounce), so it is safe for use as an anesthetic. Because it is a very volatile liquid at room temperature, diethyl ether vapors are administered to a patient in a stream of oxygen. The substance is rapidly absorbed into the body through the lungs because of its high solubility in the fatty material of cell membranes. This same quality makes it a mild skin irritant, and for this reason it should not be swallowed. The compound does not undergo chemical change in the body, but it does have several undesirable features. It has a suffocating and obnoxious odor, it may cause nausea and vomiting after surgery, and it does not always cause immediate loss of consciousness. These factors, combined with its flammable, explosive nature, have caused the use of diethyl ether as a general anesthetic to diminish. For more information on anesthetics, see chapter 24.

anesthetic: a drug that produces loss of the senses of feeling such as pain, heat, cold, and touch.

analgesic: a drug that causes relief from pain.

PHENOLS

The *phenols* are compounds having one or more hydroxyl groups attached to an aromatic ring. The general formula is Ar—OH, where Ar is any aromatic group. Because phenols are somewhat different from their structural relatives, the alcohols, we will examine those differences as well as similarities.

Structure and Nomenclature

As described above, phenols are characterized by the presence of both the aromatic ring and the hydroxyl group. In terms of structure, phenols are very similar to alcohols and can also be considered to be derivatives of water. In the case of phenols, one hydrogen of the water molecule has been replaced by the aromatic ring:

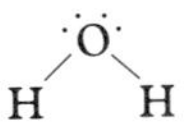

$$\underset{\text{water}}{H-\ddot{O}-H} \qquad \underset{\text{phenol}}{Ar-\ddot{O}-H}$$

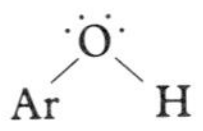

The name phenol is used specifically for the compound having one benzene ring and one hydroxyl group; it is also used as a general name for the entire class of compounds. Although IUPAC rules have been developed for naming phenols, common names have prevailed to the extent that they are used almost exclusively. The following structures represent some of the more well known phenols.

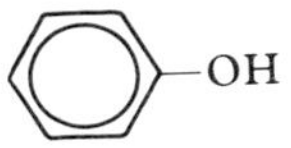

phenol 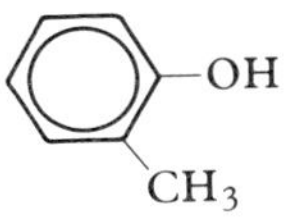   o-cresol 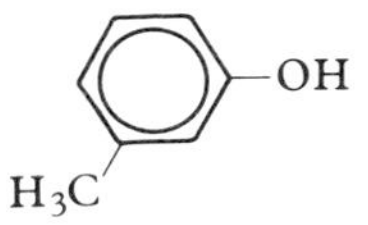   m-cresol    p-cresol

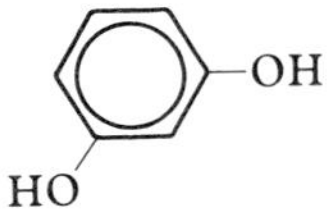

resorcinol 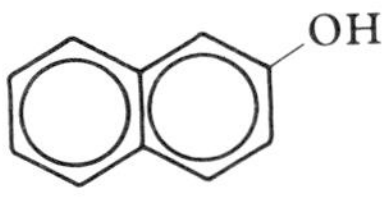   2-naphthol

Properties

Phenols are colorless liquids or white solids at room temperature. The simplest member of the family, phenol, has a sweetish odor, while other phenols are sharp and spicy smelling. Due to their ability to form hydrogen bonds, they are somewhat soluble in water. The physical properties of some common phenols are listed in Table 12-6.

Perhaps the most striking property of phenols is their acidity. Aqueous solutions of phenols contain modest amounts of hydronium ions, demonstrating that phenols dissociate according to the following equation:

$$Ar-\ddot{O}-H + H_2O \rightleftharpoons Ar-\ddot{O}^- + H_3O^+$$

Table 12-6 Physical Properties of Some Phenols

Name	Melting Point (°C)	Boiling Point (°C)	Water Solubility* (g/100g)	K_a (25°C)
Phenol	41	182	8	1.0×10^{-10}
o-Cresol	31	191	2	5.5×10^{-11}
m-Cresol	12	203	0.5	1.0×10^{-10}
p-Cresol	35	202	2	5.5×10^{-11}
2-Naphthol	122	285	trace	2.7×10^{-10}

*at 20°C

Since alcohols are not acidic, what is it about phenols that makes them acids? The answer lies in the presence of the aromatic ring. A number of resonance forms exists for the anion formed when phenol dissociates:

resonance forms for phenoxide ion

We were able to draw these forms by moving electron pairs from one atom to the next, starting with a nonbonded pair on oxygen. As we move pairs of electrons around the benzene ring, the negative charge moves also. The net result is four resonance forms for the phenol anion, called the phenoxide ion. The real structure of the phenoxide ion, an average of the resonance forms, is approximated by

Our approximation of the real structure of phenoxide ion contains partial negative charges on three carbon atoms in the benzene ring as well as on the oxygen atom. You already know that a structure which possesses resonance will be unusually stable. This fact is true for the phenoxide ion, particularly because the negative charge is spread around the whole structure instead of being isolated on just one atom. Whenever a charge can be spread around like this (the technical term is delocalized), the structure is unusually stable. Thus, the benzene ring helps to stabilize phenoxide ion and allows dissociation of the proton from phenol to proceed to a measurable extent. Hence, phenols and their derivatives are weak acids.

In addition to the acidic quality of phenols, they are susceptible to chemical reactions in which hydrogens on the benzene ring are replaced by other groups. Those positions having partial negative charge are particularly reactive to these substitutions. Like alcohols, the phenols also participate in ester formation (discussed earlier in this chapter).

Sources and Uses of Phenols

Phenols may be obtained from the destructive distillation of wood or coal, from petroleum, and from seeds and leaves of plants such as thyme and oregano. However, these natural sources could not possibly satisfy modern demands for phenols, and currently most phenol production is by chemical synthetic procedures with benzene as starting material.

Dilute aqueous solutions of phenols are called carbolic acid. Such solutions have been used for antiseptic purposes since 1865; however, undiluted phenol is highly corrosive to mucous membranes and skin, and it is a nerve poison. The following phenol derivatives have largely replaced the use of phenol itself as a germicide because of their lower toxicities, although phenol is still used as a standard for evaluating activities of germicides.

Hexyl Resorcinol

$$HO-C_6H_3(OH)-CH_2-CH_2-CH_2-CH_2-CH_2-CH_3$$

This phenol derivative is considered to be a better topical antiseptic than phenol; it is also used orally to destroy intestinal worms. Because it is effective against many microorganisms, it is often an ingredient in mouthwashes and throat lozenges.

Thymol

$$HO-C_6H_3(CH(CH_3)_2)-CH_3$$

Thymol's pleasant taste and antiseptic qualities make it a good germicide for use in mouthwashes and toothpastes.

Hexachlorophene

$$\text{(HO)(Cl)_2C_6H_2-CH_2-C_6H_2(Cl)_2(OH)}$$

This compound was once the active ingredient in certain soap preparations used in hospitals. However, in 1972 evidence suggested that it might be implicated in causing brain damage in infants bathed with the soaps. Since that time its use for most purposes has been banned, since it is readily absorbed through the skin.

Picric Acid

[structure: 2,4,6-trinitrophenol]

This phenol derivative is a stronger acid than phenol, and as such, it destroys microorganisms on contact. It is present in burn ointments because of its ability to coagulate proteins in and around the burn area, thus minimizing loss of fluids and risk of infection. Picric acid will detonate if completely dry, and it has been used as an explosive since World War I.

The Cresols

o-cresol   m-cresol   p-cresol

These compounds are used as disinfectants and as wood preservatives. The odor of cresols and phenol used for cleaning purposes is usually readily apparent in hospitals and other medical treatment facilities.

In addition to their uses as germicides, phenol and its derivatives are used in large quantities by chemical and pharmaceutical industries for conversion to many different products such as aspirin, dyes, flavors, fungicides, bacteriocides, weed killers, detergents, tanning agents for leather, nylon, and polymers. The following compounds are examples of some of these products.

aspirin (pain reliever)   Martius yellow (dye)

vanillin (vanilla flavor)   o-phenylphenol (herbicide)

PERSPECTIVE: Joseph Lister, Founder of Modern Surgery

Joseph Lister was a celebrated British surgeon and medical scientist who first used chemicals to prevent surgical infections. His method came to be called antisepsis, and the chemicals used for such purposes are called antiseptics.

Born in Essex, England, in 1827, Lister became interested in medicine early in his childhood. After receiving his medical degree in 1852, he practiced as a physician, and in 1861 he was appointed surgeon to the Glasgow Royal Infirmary, where he began his experiments with antisepsis. He found an effective antiseptic in carbolic acid (phenol), which had already been used to clean foul-smelling sewers and to dress wounds. Lister's use of carbolic acid during and after surgery (see figure) reduced the death rate caused by infection from 45% to 15% at the infirmary.

Despite early skepticism from the medical community, Lister was fortunate to see almost universal acceptance of his principle during his working life. He received much acclaim for his work and was made a baron in 1897. Although he remained gentle, shy, and unassuming thoughout his life, Lister permitted his name to be used on the patented antiseptic product Listerine. He died in 1912.

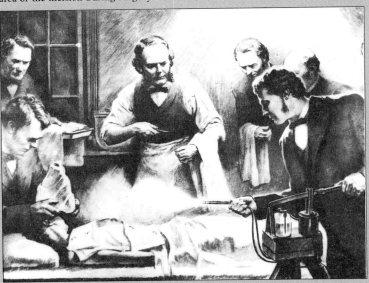

Lister's Antiseptic Method of Surgery. Carbolic acid was sprayed over the area of the incision during surgery.

THIOLS

The sulfur analog of an alcohol is called a *thiol;* an older name for these compounds is mercaptan. The general formula for thiols is R—SH, where the —SH group is called a thiol group or a sulfhydryl group. Examples of thiols and their names are:

CH$_3$—SH	CH$_3$—CH$_2$—SH	CH$_3$—CH$_2$—CH$_2$—SH
methanethiol	ethanethiol	1-propanethiol

Anyone who has ever been exposed to a thiol is sure to remember the most distinctive property of thiols—their odor! The human nose is so sensitive to these foul-smelling compounds that levels of only about 0.02 parts thiol in one billion parts air can be detected. The odor of a skunk's spray is due to the presence of thiols, and small amounts of the compounds are added to natural gas so that leaks can easily be detected by the thiol odor.

Although thiols participate in many chemical reactions, the one of interest to us is the formation of *disulfides*. These are compounds containing the —S—S— linkage. Disulfides are formed when two molecules of thiol react in the presence of a mild oxidizing agent:

$$2\ CH_3-CH_2-SH \xrightleftharpoons[[H]]{[O]} CH_3-CH_2-S-S-CH_2-CH_3$$

ethanethiol   diethyl disulfide

Many proteins contain thiol groups which react together to form disulfide linkages. This structure helps hold protein chains together in proper shapes (Figure 12-2). The location of disulfide linkages determines, for example, whether hair (a protein) is curly or straight.

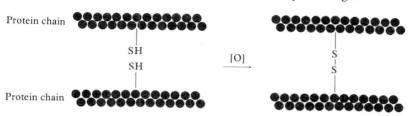

Fig. 12-2 Disulfide Linkages Hold Protein Chains Together

SUMMARY

Alcohols: Alcohols contain the hydroxyl group, —OH. There are three structural subclasses of alcohols: primary, secondary, and tertiary. Alcohols have pleasant odors, and the more common ones are colorless liquids at room temperature. Because of their hydrogen bonds, alcohols have higher boiling points than alkanes of similar molecular weight. Lower molecular weight alcohols are water soluble, but as the hydrophobic hydrocarbon portion of the alcohol structure grows larger, the water solubility of alcohols decreases. Increasing the number of —OH groups makes alcohols more water soluble.

Three important reactions of alcohols are dehydration, oxidation, and ester formation. Dehydration of a primary alcohol at moderate temperatures produces an ether; at higher temperatures an alkene is formed. Dehydration of secondary and tertiary alcohols produces only alkenes. In these cases, more than one alkene can be formed, but the alkene having the most alkyl substituents at the double bond will be favored. Controlled oxidation of primary alcohols produces aldehydes first and then carboxylic acids. In the controlled oxidation of secondary alcohols, only ketones are formed. Tertiary alcohols do not undergo controlled oxidation. Alcohols react with carboxylic acids to form esters.

Ethers: The general formula for an ether is R—O—R, where R represents any aliphatic or aromatic group. Ether molecules are somewhat polar, but less so than alcohol molecules. The more common ethers are colorless, highly volatile liquids. Since they cannot form hydrogen bonds among themselves, ether boiling points are similar to those for alkanes. But because of their ability to form hydrogen bonds with other molecules, the lower molecular weight ethers are more soluble in water than alkanes. The high volatility of ethers requires caution in handling them. Peroxide formation presents another hazard in working with ethers. Ethers are rather unreactive compounds, and combustion is the only reaction worthy of note.

Phenols: Phenols have the general formula Ar—OH, where Ar represents any aromatic group. Phenols are colorless liquids or white solids at room temperature. Because of their ability to form hydrogen bonds, phenols are somewhat soluble in water. Unlike alcohols, phenols are weak acids due to resonance stabilization of the anion formed by proton dissociation.

Thiols: Thiols are sulfur analogs of alcohols; the general formula for thiols is R—SH. The most distinctive physical property of thiols is their bad odor. Thiols form disulfides when two molecules of thiol react in the presence of a mild oxidizing agent. Proteins contain disulfide linkages which help to hold protein chains together in proper shapes.

STUDY QUESTIONS AND PROBLEMS

1. Define the following terms:
   (a) alcohol
   (b) primary alcohol
   (c) secondary alcohol
   (d) tertiary alcohol
   (e) diol
   (f) triol
   (g) dehydration
   (h) ether
   (i) steric hindrance
   (j) peroxide
   (k) anesthetic
   (l) analgesic
   (m) phenol
   (n) thiol
   (o) disulfide

2. Classify each of the following alcohols as primary, secondary, or tertiary.

   (a) 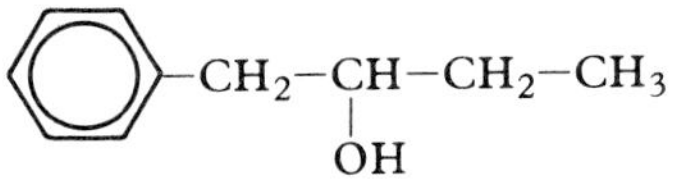 C₆H₅—CH₂—CH(OH)—CH₂—CH₃

   (b) CH₃—CH₂—C(CH₃)₂—OH

   (c) cyclopentanol (OH on ring)

   (d) CH₃—CH(OH)—C(CH₃)₂—CH₃

   (e) CH₂=CH—CH(CH₃)—OH

   (f) CH₃—C(CH₃)₂—CH₂—CH₂—OH

   (g) CH₃—CH₂—C(CH₃)(CH₂CH₂OH)—CH₃

   (h) CH₃—CH(CH₂CH₃)—C(CH₃)₂—OH

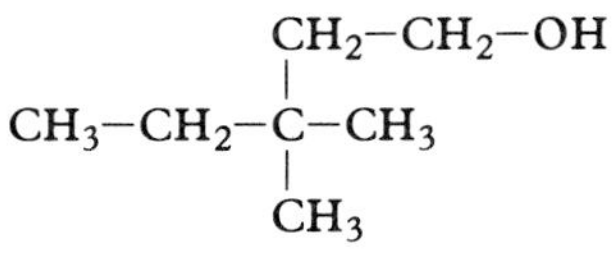

   (i) CH₃—CH₂—CH₂—CH₂—CH₂—OH

   (j) 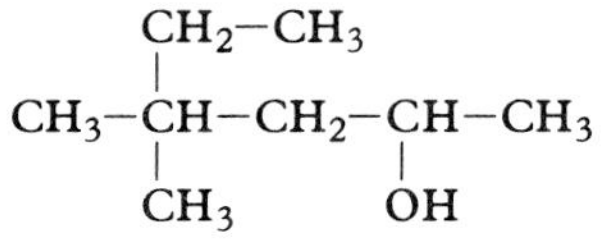 CH₃—CH(CH₃)—CH₂—CH(OH)—CH₂—CH₃

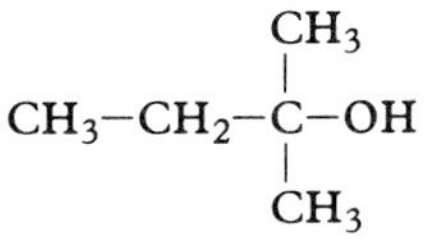

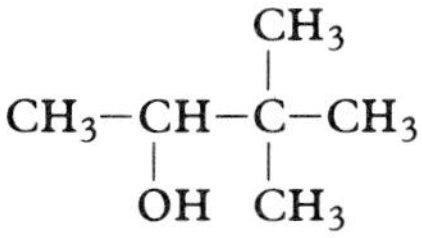

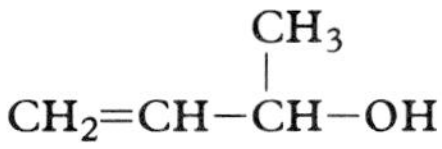

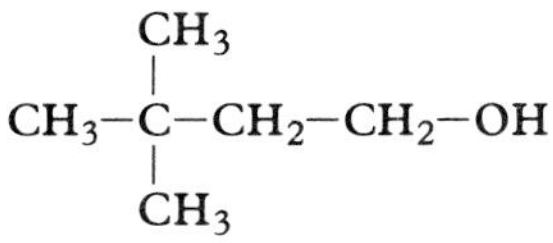

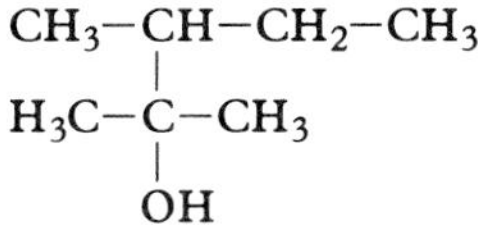

3. Give the IUPAC name for each of the following alcohols.

   (a) 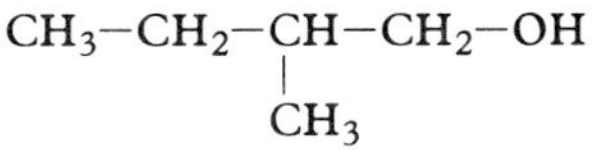 CH₃—CH₂—CH(CH₃)—CH₂—OH

   (b)  1-methylcyclopentanol (cyclopentane ring with CH₃ and OH on same carbon)

   (c) 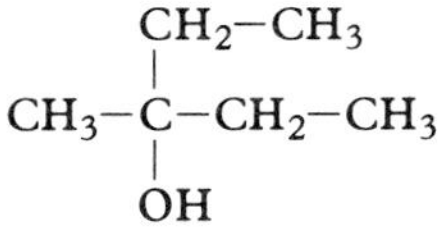 CH₃—C(CH₂CH₃)(OH)—CH₂—CH₃

   (d) HO—CH₂—CH₂—OH

   (e) 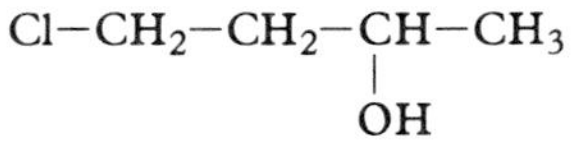 Cl—CH₂—CH₂—CH(OH)—CH₃

   (f) 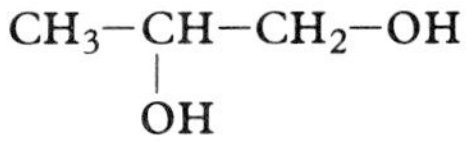 CH₃—CH(OH)—CH₂—OH

   (g) 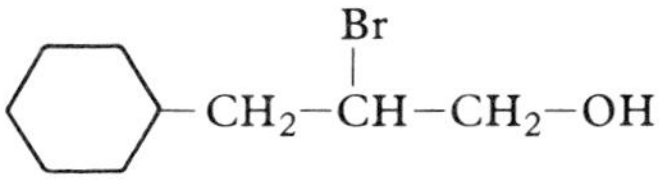 C₆H₁₁—CH₂—CH(Br)—CH₂—OH (cyclohexyl)

   (h) 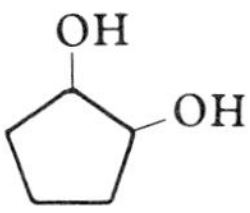 cyclopentane with OH and OH on adjacent carbons

   (i) 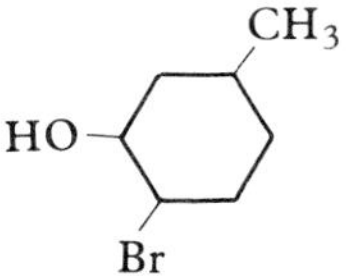 cyclohexane with HO, CH₃, and Br substituents

   (j) 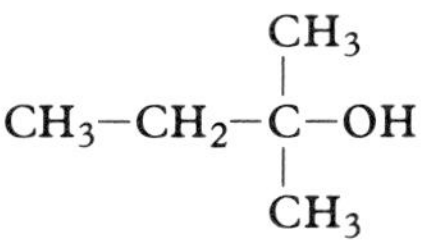 CH₃—CH₂—C(CH₃)₂—OH

4. Draw the structure of each of the following alcohols.
   (a) 2-hexanol
   (b) 2-methyl-1-propanol
   (c) 4-methyl-2-pentanol
   (d) 2,2-dimethyl-1-pentanol
   (e) 3,3-dimethyl-2-butanol
   (f) 3-chloro-1,2-propanediol

5. Give an explanation for each of the following facts.
   (a) Ethanol has a lower boiling point than water but 1,2-ethanediol has a higher boiling point than water.

(b) Ethanol has a higher boiling point than ethane.
(c) 1-Butanol is not as soluble in water as 1-propanol.
(d) 2-Methyl-2-propanol is much more soluble in water than 1-butanol.
(e) 2,3-Hexanediol is much more soluble in water than 1-hexanol.

6. Complete the following equations. If more than one organic product is formed in any reaction, indicate major and minor products. If no reaction takes place, write NR.

(a) $CH_3-\underset{\underset{CH_3}{|}}{CH}-OH \xrightarrow{[O]}$

(b) $CH_3-CH_2-CH_2-OH \xrightarrow[140°]{H_2SO_4}$

(c) ⌬$-\underset{\underset{OH}{|}}{CH}-CH_3 \xrightarrow[100°]{H_2SO_4}$

(d) $CH_3-CH_2-\underset{\underset{CH_3}{|}}{CH}-\underset{\underset{CH_3}{|}}{\overset{\overset{OH}{|}}{C}}-CH_3 \xrightarrow{[O]}$

(e) $CH_3-\underset{\underset{\underset{OH}{|}}{CH_2}}{\overset{|}{CH}}-CH_3 \xrightarrow[180°]{H_2SO_4}$

(f) $CH_3-CH_2-\overset{\overset{O}{\|}}{C}{\diagdown}_H \xrightarrow{[O]}$

(g) $2\ CH_3-CH_2-CH_2-OH + 9\ O_2 \longrightarrow$ (combustion products)

(h) $CH_3-CH_2-OH + CH_3-\overset{\overset{O}{\|}}{C}{\diagdown}_{OH} \xrightleftharpoons{H^+}$

(i) $CH_3-\underset{\underset{OH}{|}}{\overset{\overset{CH_2-CH_3}{|}}{C}}-CH_3 \xrightarrow[60°]{H_2SO_4}$

(j) $CH_3-CH_2-CH_2-CH_2-CH_2-OH \xrightarrow{[O]}$

7. Name the following ethers.
(a) $CH_3-O-CH_2-CH_3$
(b) ⌬$-O-CH_3$
(c) ⬠$-O-CH_2-CH_3$
(d) $CH_3-CH_2-O-CH_2-CH_3$
(e) $CH_3-CH_2-CH_2-O-$⌬
(f) ⌬$-O-$⌬

8. Why are the boiling points of ethers similar to those of alkanes?

9. Explain why diethyl ether is somewhat soluble in water.

10. Describe the hazards of working with ethers.

11. Summarize four disadvantages of diethyl ether as an anesthetic.

12. Explain why phenols are weak acids.

13. Give names and structures for four phenol derivatives used as germicides.

14. Write an equation to demonstrate disulfide formation from a thiol.

15. Name each of the following compounds.
(a) ⌬$-OH$
(b) $CH_3-S-S-CH_3$
(c) $CH_3-CH_2-CH_2-SH$
(d)

16. Classify each of the following as alcohol, ether, aldehyde, ketone, carboxylic acid, ester, phenol, thiol, or disulfide.
(a) ⌬$-O-$⌬
(b) $CH_3-S-S-CH_3$
(c) $H-\overset{\overset{O}{\|}}{C}{\diagdown}_H$
(d) $CH_3-CH_2-\overset{\overset{O}{\|}}{C}{\diagdown}_{OH}$

(e) C₆H₅—CH₂—OH

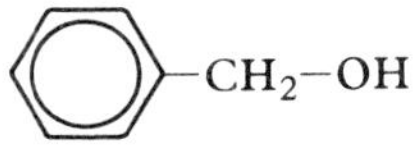

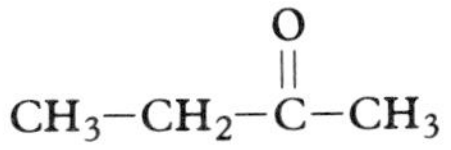

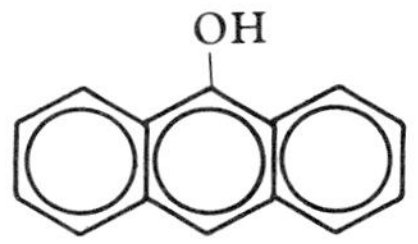

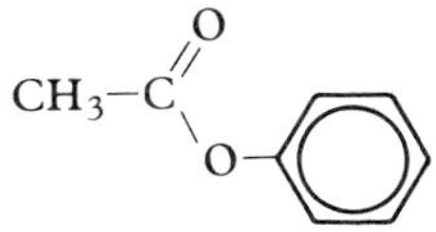

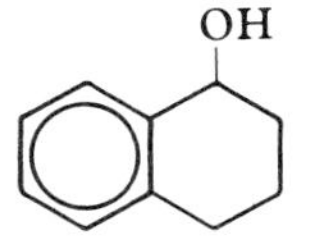

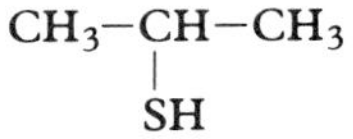

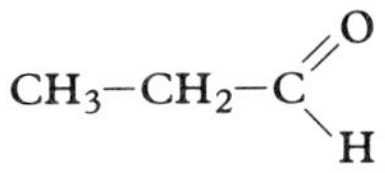

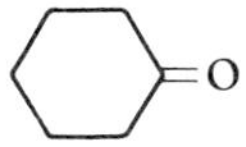

(f) CH₃—CH₂—C(=O)—CH₃

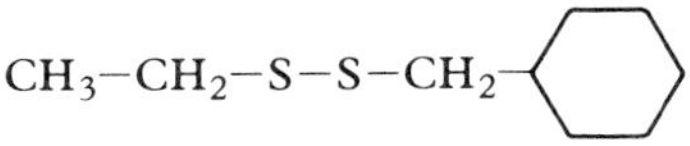

(g) 9-hydroxyanthracene structure

(h) CH₃—C(=O)—O—C₆H₅

(i) 1-hydroxy-tetrahydronaphthalene structure

(j) CH₃—CH(SH)—CH₃

(k) CH₃—CH₂—CHO

(l) cyclohexanone

(m) CH₃—CH₂—S—S—CH₂—C₆H₁₁

(n) cyclopentyl—COOH

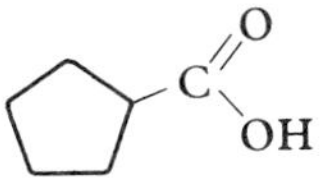

(o) CH₃—O—CH₂—CH₂—CH₃

17. Classify each of the following reactions as hydration, dehydration, or oxidation, and fill in any missing reactants or products.

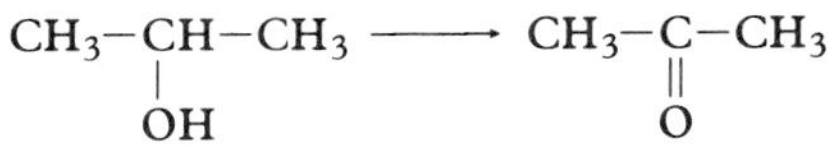

(a) CH₃—CH₂—OH ⟶ CH₃—CHO

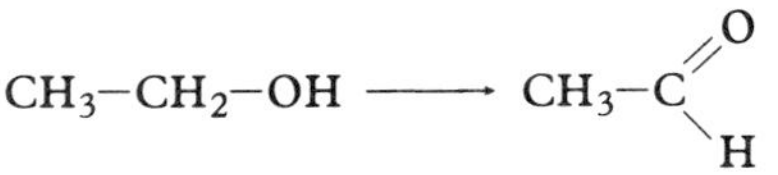

(b) CH₃—CH=CH₂ ⟶ CH₃—CH(OH)—CH₃

(c) CH₃—CH₂—OH ⟶ CH₃—CH₂—O—CH₂—CH₃

(d) CH₃—CHO ⟶ CH₃—COOH

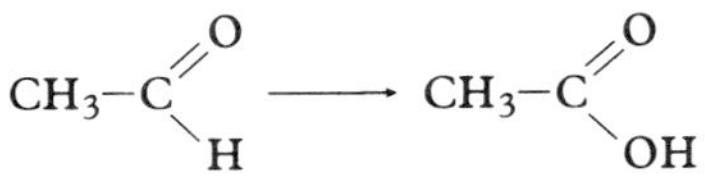

(e) CH₃—CH(OH)—CH₃ ⟶ CH₃—C(=O)—CH₃

(f) CH₃—CH₂—OH ⟶ CH₂=CH₂

(g) CH₃—CHO ⟶ CH₃—CH(OH)₂

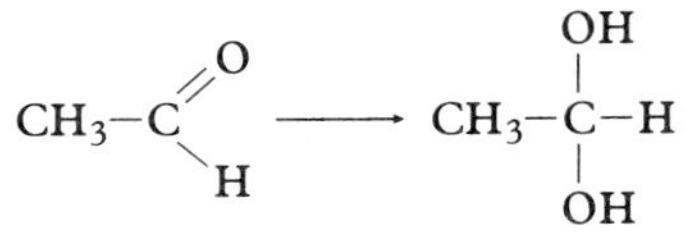

18. The heavy tax on alcoholic beverages in this country is an important source of revenue. In order to prevent consumption of tax-free ethanol intended for purposes other than drinking, small amounts of poisonous substances are added to "denature" the alcohol. A typical denaturant is methanol. Why is this alcohol so much more toxic to humans than ethanol?

19. Give structures and IUPAC names for the compounds having the following common names:
    (a) wood alcohol
    (b) grain alcohol
    (c) rubbing alcohol
    (d) glycerine

20. Why is ethylene glycol toxic to humans?

21. Why are disulfide linkages in proteins important?

22. A seventeenth-century test for whiskey was thought to provide "proof" that the whiskey had not been "watered down." We now use the term "proof" to mean twice the percentage of ethanol in whiskey. What is the percentage of ethanol in 80 proof whiskey?

23. Draw structures for the eight alcohol isomers of $C_5H_{11}OH$; classify each alcohol as primary, secondary, or tertiary.

CHAPTER 13
ALDEHYDES, KETONES, AND CARBOHYDRATES

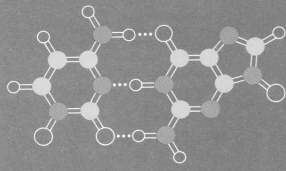

ALDEHYDES AND KETONES
- Structure and Physical Properties
- Nomenclature
- Chemical Properties
- Sources and Uses

CARBOHYDRATES
- Optical Isomerism
- Examples of Monosaccharides
- Disaccharides
- Examples of Disaccharides
- Polysaccharides

Aldehydes and ketones contain the functional group C=O. These compounds are known for their odors and tastes, and they are responsible for aromas and flavors of many processed or cooked foods, such as butter and freshly baked bread.

Many sugars can be classified as aldehydes or ketones, and others are derived from these types of compounds. Sugars belong to the large group of organic compounds known as carbohydrates. These are naturally occurring substances that can be isolated from animals and plants. Carbohydrates can be as simple as glyceraldehyde, $C_3H_6O_3$, or as complex as a glucose polymer of molecular weight 4 million. Carbohydrates are probably the most abundant and widespread organic compounds in nature. Formed in green plants from carbon dioxide and water by the process of photosynthesis, carbohydrates are essential to all living things. They serve as energy sources, structural components, and chemical constituents of genes. The list of topics indicates the scope of our study of aldehydes, ketones, and carbohydrates.

ALDEHYDES AND KETONES

Aldehydes and ketones are the classes of compounds which contain a special grouping of atoms called the *carbonyl group*. This functional group consists of a carbon atom joined to an oxygen atom by a double bond and to other carbon or hydrogen atoms by single bonds:

$$>C=O$$
carbonyl group

Because of their similarities, aldehydes and ketones are discussed together.

Structure and Physical Properties

Aldehydes are distinguished from ketones by the presence of at least one hydrogen atom attached to the carbonyl group. The simplest aldehyde, formaldehyde, has two hydrogen atoms bonded to the carbonyl carbon; all other aldehydes have one hydrogen atom and one alkyl or aromatic group attached to the carbonyl carbon. When two alkyl or aromatic groups are attached to the carbonyl carbon, the compound is a *ketone*.

$$\underset{\text{formaldehyde}}{\overset{O}{\underset{H\quad H}{\|}}\atop C} \qquad \underset{\substack{\text{all other}\\ \text{aldehydes}}}{\overset{O}{\underset{R\quad H}{\|}}\atop C} \qquad \underset{\text{a ketone}}{\overset{O}{\underset{R\quad R}{\|}}\atop C}$$

Ketones may be open-chain or cyclic structures. Condensed structural formulas for aldehydes and ketones are written as follows:

RCHO	RCOR
aldehyde	ketone

The geometry of the carbonyl group resembles that of the double bond in alkenes; bond angles about the double bond are 120°, and the carbonyl group lies in the same plane as its two attached atoms (see Figure 13-1). However, in contrast to carbon-carbon double bonds, a carbon-oxygen double bond is polarized:

$$\delta+\mathrm{C}=\mathrm{O}\,\delta-$$

Fig. 13-1 Geometry of the Carbonyl Group
(a) Formaldehyde (HCHO)
(b) Acetone (CH₃COCH₃)

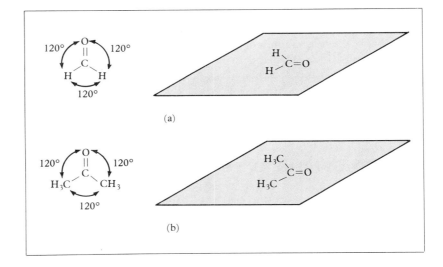

The polarity of the carbonyl group is a result of two factors. One is the difference in electronegativities of carbon and oxygen. However, this difference is relatively small and it alone would not cause much polarity.

The second and more important cause is resonance. The carbonyl group has two resonance forms:

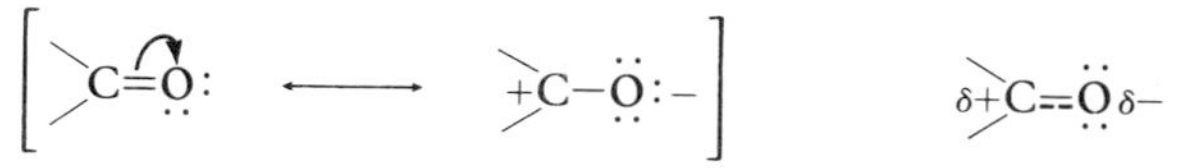

$$\left[\diagup\!\!\!\diagdown\!\!\text{C}=\!\ddot{\ddot{\text{O}}}\!: \longleftrightarrow \diagup\!\!\!\diagdown\!\!\overset{+}{\text{C}}-\ddot{\ddot{\text{O}}}\!:^- \right] \qquad \delta+\!\diagup\!\!\!\diagdown\!\!\text{C}=\!\ddot{\ddot{\text{O}}}\,\delta-$$

The average of these resonance forms (shown on the right) has a greater partial negative charge on oxygen and a greater partial positive charge on carbon than would be predicted from electronegativity values. The polarity of the carbon-oxygen double bond is greater than that of the carbon-oxygen single bond (as in alcohols, for example), because the single bonds do not have resonance stabilization of partial charges.

The unique features of the carbonyl group are the major influence on the physical properties of aldehydes and ketones. Their dipolar attractions cause them to have higher boiling points than nonpolar compounds of similar molecular weight. Because they can participate in hydrogen bond formation with other molecules, aldehydes and ketones of low molecular weight are water soluble. However, since they cannot form hydrogen bonds among themselves, aldehydes and ketones have lower boiling points than the corresponding alcohols. These properties are summarized in Table 13-1. Note that the boiling points increase with molecular weight, but water solubility decreases with molecular weight.

Nomenclature

Common names for aldehydes and ketones were devised long before IUPAC rules existed, and many of those common names are still used almost exclusively. For aldehydes, the common names are derived from those of carboxylic acids having the same number and arrangement of carbon atoms. The "-ic" or "-oic" name ending for the carboxylic acid is replaced by the word "aldehyde," and the common name is written as one word. Common names of aldehydes and corresponding carboxylic acids are given in Table 13-2.

Common names for ketones are similar to those of ethers. They are composed of the names of the groups attached to the carbonyl carbon written in alphabetical order and followed by the word "ketone." If two identical substituents are present, the prefix "di-" is used, and the common name of the ketone is written as two words. If two different substituents are present, the common name of the ketone appears as three words. In addition to this semiorganized system of common names for ketones, many aromatic and cyclic ketones have acquired common names which do not have any obvious derivation. Common names for selected ketones are given in Table 13-1.

Locations of other functional groups in a molecule may be designated by the use of Greek letters. The carbon adjacent to the carbonyl carbon is often referred to as the *alpha* (α) carbon; the next carbon is *beta* (β), the next *gamma* (γ), and the next *delta* (δ). There is hardly ever any need to go beyond *delta*, but if necessary, the Greek alphabet can be continued. In any event, the last letter of the Greek alphabet, *omega* (ω), is

Table 13-1 Physical Properties of Some Aldehydes and Ketones

ALDEHYDES		Molecular Weight	Boiling Point (°C)	Water Solubility* (g/100g)
Common Name	Structure			
Formaldehyde	H–C(=O)–H	30.0	−21	Soluble in all proportions
Acetaldehyde	CH_3–C(=O)–H	44.0	20	Soluble in all proportions
Propionaldehyde	CH_3–CH_2–C(=O)–H	58.0	50	20
Butyraldehyde	CH_3–CH_2–CH_2–C(=O)–H	72.0	76	4
Valeraldehyde	CH_3–CH_2–CH_2–CH_2–C(=O)–H	86.0	103	Very slightly soluble
Benzaldehyde	C_6H_5–C(=O)–H	106	179	0.3

KETONES		Molecular Weight	Boiling Point (°C)	Water Solubility* (g/100g)
Common Name	Structure			
Acetone	CH_3–C(=O)–CH_3	58.0	56	Soluble in all proprotions
Methyl ethyl ketone	CH_3–C(=O)–CH_2–CH_3	72.0	80	37
Diethyl ketone	CH_3–CH_2–C(=O)–CH_2–CH_3	86.0	102	4.7
Cyclohexanone	cyclohexyl=O	98.0	156	2
Acetophenone	CH_3–C(=O)–C_6H_5	120	202	0
Benzophenone	C_6H_5–C(=O)–C_6H_5	182	306	0

*at 20°C

Table 13-2 Common Names of Aldehydes and Corresponding Carboxylic Acids

Aldehyde	Structure	Carboxylic Acid	Structure
Formaldehyde	H–CHO	Formic acid	H–CO$_2$H
Acetaldehyde	CH$_3$–CHO	Acetic acid	CH$_3$–CO$_2$H
Propionaldehyde	CH$_3$–CH$_2$–CHO	Propionic acid	CH$_3$–CH$_2$–CO$_2$H
Butyraldehyde	CH$_3$–CH$_2$–CH$_2$–CHO	Butyric acid	CH$_3$(CH$_2$)$_2$CO$_2$H
Valeraldehyde	CH$_3$–CH$_2$–CH$_2$–CH$_2$–CHO	Valeric acid	CH$_3$(CH$_2$)$_3$CO$_2$H
Benzaldehyde	C$_6$H$_5$–CHO	Benzoic acid	C$_6$H$_5$–CO$_2$H

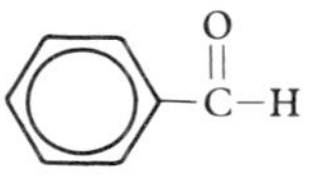

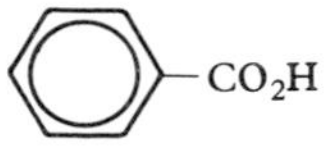

used to designate the last carbon in a chain regardless of chain length. Examples are given below.

$$\underset{\omega}{CH_3}-\underset{\delta}{CH_2}-\underset{\gamma}{CH_2}-\underset{\beta}{CH_2}-\underset{\alpha}{CH_2}-CHO$$

CH$_3$–CH(Br)–CHO

α-bromopropionaldehyde

CH$_3$–CH(Cl)–CH$_2$–CH$_2$–CO–CH$_3$

γ-chlorobutyl methyl ketone

In the IUPAC system, the name of an aldehyde is derived from the name of the parent alkane by changing the alkane name ending "-e" to "-al." No number is needed because the –CHO group is at the end of the carbon chain and thus always contains carbon number 1.

H–CHO   CH$_3$–CHO   CH$_3$–CH$_2$–CHO

methanal   ethanal   propanal

To name ketones by the IUPAC system, it is necessary to change the alkane name ending "-e" to "-one." In this case a number is used to indicate the location of the carbonyl group when necessary.

CH$_3$–CO–CH$_3$ CH$_3$–CO–CH$_2$–CH$_3$ CH$_3$–CO–CH$_2$–CH$_2$–CH$_3$

propanone   butanone   2-pentanone

EXAMPLE 13.1

Give common and IUPAC names for each of the following.

(a) $CH_3-\overset{\overset{O}{\|}}{C}-CH_3$

(b) $CH_3-CH_2-\overset{\overset{O}{\|}}{C}-CH_2-CH_3$

(c) CH_3-CHO

(d) 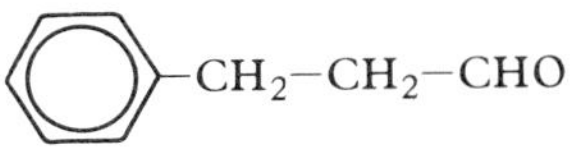 $C_6H_5-CH_2-CH_2-CHO$

(e) $Br-CH_2-CH_2-CHO$

SOLUTION

	Common Name	IUPAC Name
(a)	acetone or dimethyl ketone	propanone
(b)	diethyl ketone	3-pentanone
(c)	acetaldehyde	ethanal
(d)	β-phenylpropionaldehyde	3-phenylpropanal
(e)	β-bromopropionaldehyde	3-bromopropanal

EXERCISE 13.1 Give common and IUPAC names for each of the following.

(a) $CH_3-\overset{\overset{O}{\|}}{C}-CH_2-CH_3$

(b) $H-CHO$

(c) $CH_3-CH_2-CH_2-CH_2-CHO$

(d) 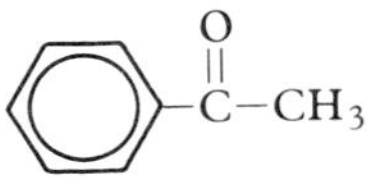 phenyl–$\overset{\overset{O}{\|}}{C}-CH_3$

(e) 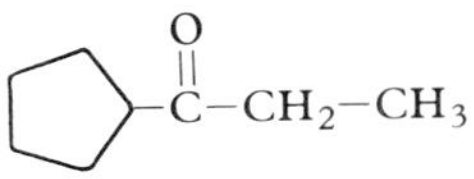 cyclopentyl–$\overset{\overset{O}{\|}}{C}-CH_2-CH_3$

Chemical Properties

Although aldehydes and ketones participate in a number of chemical reactions, the most important ones for our purposes are oxidation and addition.

Oxidation

As you have seen in chapter 11, aldehydes undergo controlled oxidation to form carboxylic acids.

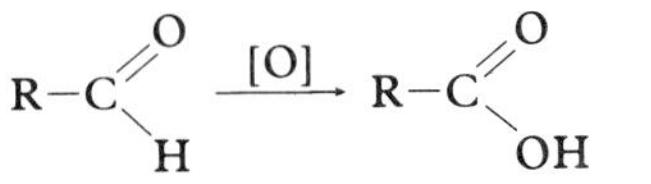

$$R-\overset{\overset{O}{\nearrow}}{C}_{\searrow H} \xrightarrow{[O]} R-\overset{\overset{O}{\nearrow}}{C}_{\searrow OH}$$

aldehyde → carboxylic acid

Ketones do not undergo controlled oxidation, but under vigorous oxidizing conditions, ketones may be broken apart and the fragments oxidized to carboxylic acids. Both aldehydes and ketones can be burned to produce water and either carbon dioxide or carbon monoxide.

Addition

As with alkenes, hydrogen may be incorporated into the structure of aldehydes and ketones by the process of catalytic hydrogenation. This process results in the addition of a hydrogen atom to each atom in the double bond; in this case, the carbonyl group is transformed into an alcohol group. Catalytic hydrogenation of aldehydes produces primary alcohols, and ketones form secondary alcohols. Notice that these *reduction* reactions correspond to the reverse of oxidations of alcohols.

$$\underset{\text{aldehyde}}{R-\overset{\overset{O}{\|}}{C}-H} + H_2 \xrightarrow[\Delta]{\text{Pt or Ni}} \underset{\text{1° alcohol}}{R-\overset{\overset{OH}{|}}{\underset{\underset{H}{|}}{C}}-H}$$

$$\underset{\text{ketone}}{R-\overset{\overset{O}{\|}}{C}-R} + H_2 \xrightarrow[\Delta]{\text{Pt or Ni}} \underset{\text{2° alcohol}}{R-\overset{\overset{OH}{|}}{\underset{\underset{H}{|}}{C}}-R}$$

Because of the polarity of the carbonyl group, aldehydes and ketones are more reactive to additions of polar molecules than are alkenes. In particular, water adds easily to the double bond of the carbonyl group, so that aldehydes and ketones are in equilibrium with their hydrated forms when dissolved in water. In this case, hydrogen atoms from water add to the partially negative carbonyl oxygen, and hydroxyl groups from water add to the partially positive carbonyl carbon. Both acids and bases catalyze the addition of water to the carbonyl group, but the reaction will proceed in the absence of catalyst. Examples are:

$$\underset{\text{acetaldehyde}}{CH_3-\overset{\overset{O}{\|}}{C}-H} + HOH \rightleftharpoons \underset{\text{an organic hydrate}}{CH_3-\overset{\overset{OH}{|}}{\underset{\underset{OH}{|}}{C}}-H}$$

$$\underset{\text{acetone}}{CH_3-\overset{\overset{O}{\|}}{C}-CH_3} + HOH \rightleftharpoons \underset{\text{an organic hydrate}}{CH_3-\overset{\overset{OH}{|}}{\underset{\underset{OH}{|}}{C}}-CH_3}$$

Note that an organic hydrate contains the elements of water incorporated into the structure of the molecule, in contrast to inorganic hydrates, which contain intact water molecules in their crystals.

Most hydrates of aldehydes and ketones are unstable, but there are a few notable exceptions. One is chloral hydrate, a white crystalline compound which is a powerful sedative and sleep inducer. It is the substance known as "knockout drops" used to create the famous "Mickey

Finn" drink. (The combination of chloral hydrate and alcohol is extremely dangerous and can cause death.)

$$\underset{\text{chloral}}{\underset{\text{Cl}_3\text{C}}{\overset{\text{O}}{\overset{\|}{\text{C}}}}\text{H}} + \text{HOH} \rightleftharpoons \underset{\text{chloral hydrate}}{\text{Cl}_3\text{C}-\underset{\overset{|}{\text{OH}}}{\overset{\overset{|}{\text{OH}}}{\text{C}}}-\text{H}}$$

Because alcohols have structures similar to water, they can add to aldehydes and ketones much in the same way as water does. In this case an acid catalyst is necessary. The hydrogen from ROH adds to the carbonyl oxygen, while the OR part of the alcohol adds to the carbonyl carbon:

$$\underset{\text{aldehyde}}{\underset{\text{R}}{\overset{\text{O}}{\overset{\|}{\text{C}}}}\text{H}} + \text{R'OH} \xrightleftharpoons{\text{H}^+} \underset{\text{a hemiacetal}}{\text{R}-\underset{\overset{|}{\text{OR'}}}{\overset{\overset{|}{\text{OH}}}{\text{C}}}-\text{H}}$$

$$\underset{\text{ketone}}{\underset{\text{R}}{\overset{\text{O}}{\overset{\|}{\text{C}}}}\text{R}} + \text{R'OH} \xrightleftharpoons{\text{H}^+} \underset{\text{a hemiketal}}{\text{R}-\underset{\overset{|}{\text{OR'}}}{\overset{\overset{|}{\text{OH}}}{\text{C}}}-\text{R}}$$

The reason for this orientation is that the partially negative oxygen atom in the alcohol is attracted to the partially positive carbon atom in the carbonyl group. As in hydration, the aldehyde or ketone is in equilibrium with the product. Addition of one molecule of alcohol produces what is called a *hemiacetal* if the carbonyl compound was an aldehyde, or a *hemiketal* if the carbonyl compound was a ketone.

hemiacetal (hemi-AS-e-tal)
hemiketal (hemi-KEY-tal)

A second molecule of alcohol can react with the hemiacetal or hemiketal to form an *acetal* or a *ketal*, respectively:

$$\underset{\text{a hemiacetal}}{\text{R}-\underset{\overset{|}{\text{OR'}}}{\overset{\overset{|}{\text{OH}}}{\text{C}}}-\text{H}} + \text{R''OH} \xrightleftharpoons{\text{H}^+} \underset{\text{an acetal}}{\text{R}-\underset{\overset{|}{\text{OR'}}}{\overset{\overset{|}{\text{OR''}}}{\text{C}}}-\text{H}} + \text{HOH}$$

$$\underset{\text{a hemiketal}}{\text{R}-\underset{\overset{|}{\text{OR'}}}{\overset{\overset{|}{\text{OH}}}{\text{C}}}-\text{R}} + \text{R''OH} \xrightleftharpoons{\text{H}^+} \underset{\text{a ketal}}{\text{R}-\underset{\overset{|}{\text{OR'}}}{\overset{\overset{|}{\text{OR''}}}{\text{C}}}-\text{R}} + \text{HOH}$$

ALDEHYDES AND KETONES 357

In this reaction, the —OR part of the alcohol replaces the —OH on the hemiacetal or hemiketal. The —H remaining from the alcohol then combines with —OH to form water. Notice that hemiacetals, hemiketals, acetals, and ketals all have the —O— linkage characteristic of ethers. Complex carbohydrates such as cellulose and starch, which will be discussed later in this chapter, also contain these linkages.

EXAMPLE 13.2

Complete the following equations. If there is no reaction, write NR.

(a) $CH_3-CH_2-CHO \xrightarrow{[O]}$

(b) $CH_3-\underset{\underset{O}{\parallel}}{C}-CH_3 \xrightarrow{[O]}$

(c) $2\ CH_3-CHO + 2\ O_2 \xrightarrow{\text{(combustion)}}$

(d) $CH_3-\underset{\underset{CHO}{|}}{CH}-CH_3 + H_2 \xrightarrow[\Delta]{Pt}$

(e) $CH_3-\underset{\underset{O}{\parallel}}{C}-CH_2-CH_3 + H_2O \rightleftharpoons$

(f) $CH_3-CH_2-CHO + CH_3-OH \xrightarrow{H^+} \rightleftharpoons$

(g) $CH_3-\underset{\underset{O}{\parallel}}{C}-CH_3 + 2\ CH_3-CH_2-OH \xrightarrow{H^+} \rightleftharpoons$

SOLUTION

(a) $CH_3-CH_2-CO_2H$

(b) NR

(c) $4\ CO_2 + 4\ H_2O$

(d) $CH_3-\underset{\underset{CH_2OH}{|}}{CH}-CH_3$

(e) $CH_3-\underset{\underset{OH}{|}}{\overset{\overset{OH}{|}}{C}}-CH_2-CH_3$

(f) $CH_3-CH_2-\underset{\underset{OCH_3}{|}}{\overset{\overset{OH}{|}}{C}}-H$

(g) $CH_3-\underset{\underset{OCH_2CH_3}{|}}{\overset{\overset{OCH_2CH_3}{|}}{C}}-CH_3 + HOH$

EXERCISE 13.2 Complete the following equations. If there is no reaction, write NR.

(a) $H-CHO \xrightarrow{[O]}$

(b) $2\ CH_3-CH_2-CHO + 5\ O_2 \longrightarrow$ (combustion products)

(c) $CH_3-CHO + H_2 \xrightarrow[\Delta]{Pt}$

(d) $CH_3-CHO + CH_3-OH \xrightarrow{H^+} \rightleftharpoons$

(e) $H-CHO + H_2O \rightleftharpoons$

(f) $H-CHO + H_2 \xrightarrow[\Delta]{Pt}$

Sources and Uses

Although aldehydes and ketones are distributed widely in nature, those of commercial importance must be synthesized. Synthesis is usually accomplished by oxidation of the corresponding alcohol, as discussed in chapter 12.

Formaldehyde (HCHO) is one of the more well known aldehydes. It is a gas at room temperature, but it is conveniently supplied as a 37% aqueous solution called formalin. Formaldehyde, the familiar odor of many biology laboratories, is used to preserve biological specimens and to sterilize surgical instruments and gloves. It is the principal ingredient in embalming fluid. Formaldehyde is used in large quantities along with phenol in the industrial production of the heat-resistant plastic called Bakelite (see Figure 13-2), a material used as an electrical insulator, as handles for cooking utensils, and in paints and baked enamel coatings.

Fig. 13-2 Formation of Bakelite from Phenol and Formaldehyde

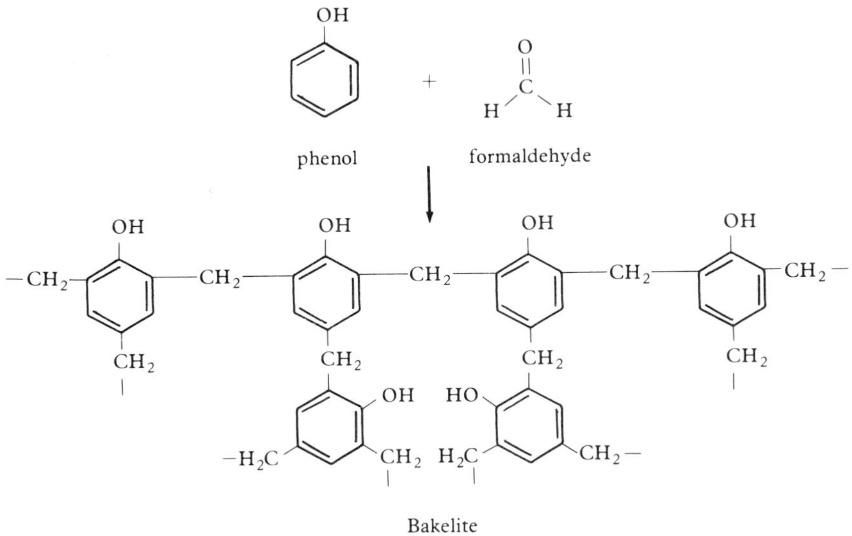

Acetaldehyde (CH_3CHO), a liquid at room temperature, has a sharp, irritating odor like formaldehyde. It is used in the manufacture of acetic acid, synthetic rubber, plastics, perfumes, flavors, and dyes.

Acetone ($CH_3-CO-CH_3$), an industrially important ketone used as a solvent for many organic compounds, is the major component in fingernail polish remover. It is also used in the manufacture of a variety of organic compounds. Acetone is normally produced in the human body in small quantities, but under conditions of starvation or severe diabetes, high levels of acetone can be detected in the odor of the breath.

Many of the higher molecular weight aldehydes and ketones have pleasant odors and tastes. In particular, aldehydes are often used as artificial flavors. Some examples of these compounds are:

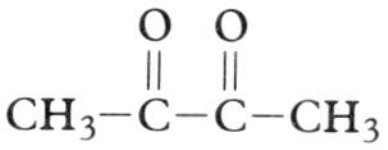

$$CH_3-\underset{\underset{O}{\|}}{C}-\underset{\underset{O}{\|}}{C}-CH_3$$
2,3-butanedione (biacetyl)
(butter flavor)

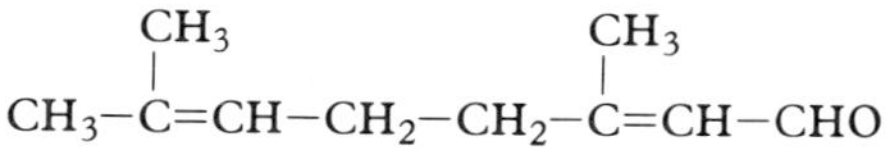

$$CH_3-\underset{\underset{CH_3}{|}}{C}=CH-CH_2-CH_2-\underset{\underset{CH_3}{|}}{C}=CH-CHO$$
citral
(citrus flavor)

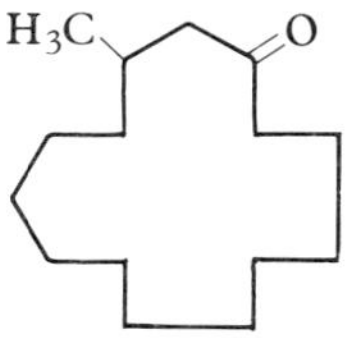

muscone
(musk odor)

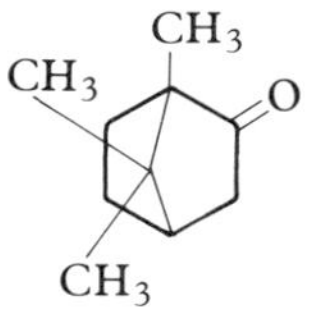

camphor
(incense component)

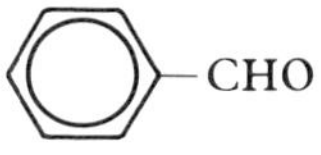

benzaldehyde
(almond flavor)

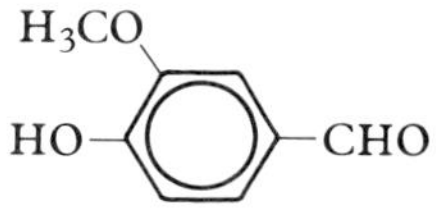

vanillin
(vanilla flavor)

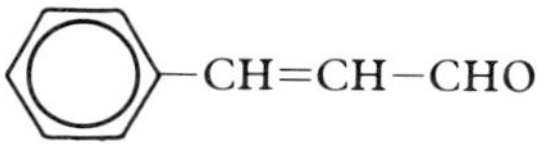

cinnamaldehyde
(cinnamon flavor)

CARBOHYDRATES

The compounds called *carbohydrates* received their name because many of them have the formula $(C \cdot H_2O)_n$, where n is the number of carbon atoms. This formula implies that the compounds are *hydrates* of *carbon*, thus carbohydrates. We now know that carbohydrates are not hydrated carbon compounds; instead, most contain several hydroxyl groups and either an aldehyde or a ketone group. Thus, a more modern description of carbohydrates is that they are polyhydroxy aldehydes, polyhydroxy ketones, or closely related compounds.

Carbohydrates often have a sweet taste, and thus the term *saccharide* is used in reference to them. Carbohydrates which cannot be broken down into simpler molecules by acid hydrolysis (reaction with water in the presence of an acid) are called *monosaccharides*. These compounds are also called simple sugars. Monosaccharides carry common names whose origins are obscure, but they usually end in "-ose." Monosaccharides are subclassified by the number of carbon atoms in their structures.

saccharide (SACK-ah-ride): an organic compound containing a sugar or sugars; derived from the Latin *saccharon*, which means "sugar."

360 CH. 13 ALDEHYDES, KETONES, AND CARBOHYDRATES

Thus, a triose contains three carbon atoms, a tetrose contains four carbon atoms, and so forth. Monosaccharides have the general formula $C_nH_{2n}O_n$.

n	Subclass	Formula	Examples
3	triose	$C_3H_6O_3$	glyceraldehyde, dihydroxyacetone
4	tetrose	$C_4H_8O_4$	erythrose, threose
5	pentose	$C_5H_{10}O_5$	ribose, xylulose
6	hexose	$C_6H_{12}O_6$	glucose, fructose

In addition to subclassification by the number of carbon atoms, monosaccharides can also be categorized as aldehydes or ketones. Those containing aldehyde groups are called *aldoses*, whereas those with ketone groups are called *ketoses*. The aldehyde-ketone designation may be combined with the terminology for the number of carbon atoms, as in ketopentose (a pentose having a ketone group) and aldohexose (a hexose having an aldehyde group).

A carbohydrate that can be hydrolyzed to two monosaccharides units is called a *disaccharide*. These are common in nature, as you shall see. However, carbohydrates having three to ten monosaccharide units are rather uncommon; these are called *oligosaccharides*. Polymers containing hundreds of monosaccharide units make up the class of carbohydrates called *polysaccharides*. These compounds are also common in nature in the form of starch and cellulose. Important examples of the major classes of carbohydrates are discussed later in this chapter.

oligosaccharide (AH-li-go-SACK-ah-ride): *oligo* is the Greek word for "few."

EXAMPLE 13.3

Classify each monosaccharide by combining the aldehyde-ketone designation with terminology indicating the number of carbon atoms.

(a)
```
    CHO
    |
H — C — OH
    |
HO— C — H
    |
H — C — OH
    |
H — C — OH
    |
    CH₂OH
```

(b)
```
    CHO
    |
H — C — OH
    |
H — C — OH
    |
H — C — OH
    |
    CH₂OH
```

(c)
```
    CHO
    |
H — C — OH
    |
    CH₂OH
```

(d)
```
    CH₂OH
    |
    C = O
    |
HO— C — H
    |
H — C — OH
    |
H — C — OH
    |
    CH₂OH
```

SOLUTION

(a) aldohexose
(b) aldopentose
(c) aldotriose
(d) ketohexose

Classify each monosaccharide by combining the aldehyde-ketone designation with terminology indicating the number of carbon atoms.

(a)
```
    CH₂OH
    |
    C=O
    |
  H-C-OH
    |
  H-C-OH
    |
    CH₂OH
```

(b)
```
    CHO
    |
  HO-C-H
    |
  H-C-OH
    |
    CH₂OH
```

(c)
```
    CHO
    |
  HO-C-H
    |
  H-C-OH
    |
  HO-C-H
    |
  H-C-OH
    |
    CH₂OH
```

Optical Isomerism

If we examine the structure of the simplest monosaccharide, glyceraldehyde, we see that there are four different substituents attached to the middle carbon:

$$\begin{array}{c} \text{CHO} \\ | \\ \text{H-C-OH} \\ | \\ \text{CH}_2\text{OH} \end{array}$$

This kind of carbon atom is said to be *asymmetric*. As illustrated in Figure 13-3, ball-and-stick models show that there are two isomers of glyceraldehyde. These are called *stereoisomers*; they have the same structure but a different spatial arrangement of atoms about the asymmetric carbon. We have already had one example of stereoisomers in the *cis-trans* isomers of alkenes that exist because of the carbon-carbon double bond. However, in glyceraldehyde, stereoisomerism is a result of the presence of an asymmetric carbon atom. *Configuration* is the word used to refer to the arrangement of atoms around an asymmetric carbon atom. Thus stereoisomers of glyceraldehyde differ in configuration but

asymmetric: not symmetric.

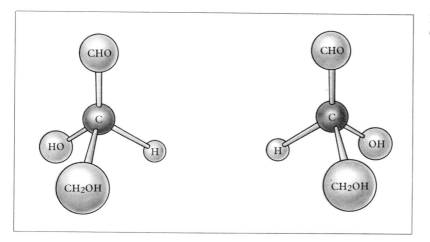

Fig. 13-3 The Two Stereoisomers of Glyceraldehyde

not in structure, while structural isomers of the compound differ in the sequence in which the atoms are bonded together:

$$\underset{\text{glyceraldehyde}}{\begin{array}{c} \text{CHO} \\ | \\ \text{H}-\text{C}-\text{OH} \\ | \\ \text{CH}_2\text{OH} \end{array}} \qquad \underset{\text{dihydroxyacetone}}{\begin{array}{c} \text{CH}_2\text{OH} \\ | \\ \text{C}=\text{O} \\ | \\ \text{CH}_2\text{OH} \end{array}}$$

structural isomers

The two stereoisomers of glyceraldehyde interact differently with light, and for that reason they are called optical isomers. The concept of optical isomerism is developed in the following sections.

Chirality

Notice that the two glyceraldehyde isomers in Figure 13-3 are mirror images of each other, and as such, cannot be *superimposed* on each other. This is similar to the relationship between your two hands. Your right hand is a mirror image of your left hand, and you cannot superimpose one hand on the other any more than you can wear your left glove on

superimpose: to place one object on another so that their parts match exactly.

Fig. 13-4 (a) Chiral Objects: Your two hands are nonsuperimposable mirror images. (b) Achiral Objects: The sphere, cube, and cup can be superimposed on their mirror images.

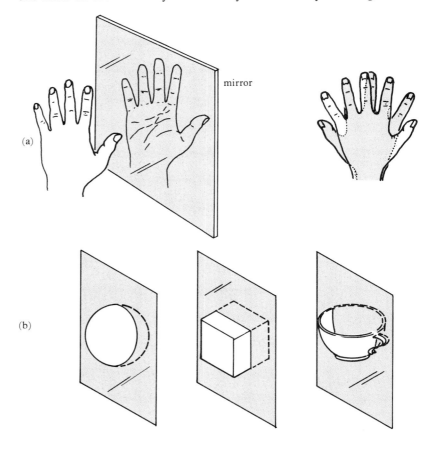

your right hand. Any object that cannot be superimposed on its mirror image is said to be *chiral*. Thus, a hand, a glove, and a shoe are chiral objects; they cannot be superimposed on their mirror images. However, a cup, a sphere, and a cube are *achiral* objects because they can be superimposed on their mirror images. Figure 13-4 illustrates these relationships.

Many molecules in addition to glyceraldehyde are chiral, but fortunately it is not necessary to try to imagine nonsuperimposable mirror images in order to determine whether a molecule is chiral. An organic molecule that contains one asymmetric carbon atom will be chiral and thus will have two nonsuperimposable mirror images. The asymmetric carbon atom responsible for *chirality* is called the *chiral center* or the *center of asymmetry*.

chiral (KYE-ral): a descriptive term used for objects possessing nonsuperimposable mirror images; an organic molecule will be chiral if it contains a carbon atom bonded to four different atoms or groups.

achiral (AY-kye-ral): not chiral.

chirality (kye-RAH-lity): the existence of chiral quality.

Fischer Projections

Since it is not always possible to make models of molecules, it is necessary to be able to draw two-dimensional structures which represent mirror images. Emil Fischer, a German chemist known as the father of carbohydrate chemistry, introduced a method late in the nineteenth century for drawing structures of chiral molecules. His structures, called *Fischer projections*, are illustrated below for glyceraldehyde. Note that in Fischer projections, the carbonyl group is placed at or near the top, and the top carbon is numbered 1.

Fischer projections: a method for drawing structures of monosaccharides in which the carbonyl group is placed at or near the top of the open-chain structure and the spatial arrangement of atoms at the chiral center(s) is indicated by the projection of the attached groups.

$$\begin{array}{cc} ^1CHO & ^1CHO \\ | & | \\ H-^2C-OH & HO-^2C-H \\ | & | \\ ^3CH_2OH & ^3CH_2OH \end{array}$$

mirror images of glyceraldehyde

The configuration at the chiral center is indicated by the projection of the attached groups. Those to the left and right represent groups projected toward the viewer, while groups attached by vertical bonds represent groups projected away from the viewer.

Optical Activity

Molecules which are nonsuperimposable mirror images are called *enantiomers*. In the case of glyceraldehyde, as with other chiral compounds, we would like to differentiate one enantiomer from the other in their names. To do this, we make use of the only physical property that is different for members of a pair of enantiomers. This one difference is the direction in which enantiomers rotate plane-polarized light. Ordinary light moves in the form of a wave which undulates in all directions at right angles to the direction of travel of the light beam. However, if a polarizing filter (the same kind as in polaroid sunglasses) is used, only those waves in a single plane will pass through. This filter behaves as if it has slots which let only one plane of light through, and the light that

enantiomers (ee-NAN-tee-oh-mers): isomeric pairs of chiral compounds whose molecules are mirror images and thus nonsuperimposable.

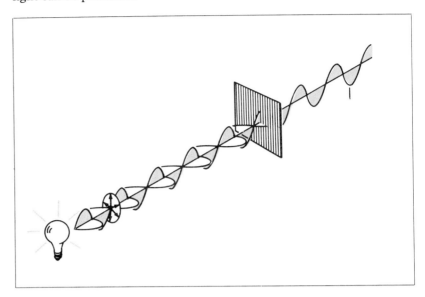

Fig. 13-5 Three-dimensional light waves strike a polarizing filter, but only light waves in a single plane pass through the filter.

If plane-polarized light is passed through a solution containing a single enantiomer, the plane of light is rotated either to the right (clockwise) or to the left (counterclockwise). Rotation of plane-polarized light is called optical rotation, and a compound that rotates plane-polarized light is said to be optically active. Thus, the enantiomers of glyceraldehyde are optically active, and are referred to as *optical isomers*. Members of a pair of enantiomers rotate plane-polarized light to an equal extent but in opposite directions. The amount of rotation is measured in units of degrees by an instrument called a polarimeter, as illustrated in Figure 13-6. The enantiomer which rotates plane-polarized light to the right (clockwise) is said to be *dextrorotatory*. The other enantiomer rotates plane-polarized light the same number of degrees to the left (counterclockwise) and is thus said to be *levorotatory*.

dextrorotatory: term applied to a substance which rotates plane-polarized light in the clockwise (+) direction; the Latin *dextro* means "to the right."

levorotatory: term applied to a substance which rotates plane-polarized light in the counterclockwise (−) direction; the Latin *levo* means "to the left."

The direction of rotation is specified in the name of an enantiomer by (+) for dextrorotatory and (−) for levorotatory. A mixture of exactly equal parts of any pair of enantiomers is called a *racemic mixture*, and is designated by (±) in the name. Thus, racemic glyceraldehyde is named (±)-glyceraldehyde. A racemic mixture does not rotate plane-polarized light because the rotation of each enantiomer is canceled by the equal and opposite rotation of the other:

(+)-glyceraldehyde	+	(−)-glyceraldehyde	=	(±)-glyceraldehyde
+8.7°		−8.7°		0°
(clockwise rotation)		(counterclockwise rotation)		(a racemic mixture) (no rotation)

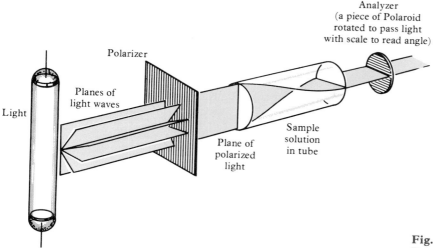

Fig. 13-6 Schematic Diagram of a Polarimeter

We have seen how enantiomers can be named on the basis of optical rotation, but how is this property related to structure? Emil Fischer did not know, and he had no way of determining molecular structure in his day, so he did the only reasonable thing—he guessed! If we draw Fischer projections for the two glyceraldehyde enantiomers, we see that they differ only by the orientation of the secondary hydroxyl group. This hydroxyl can be on the left or the right. Fischer decided to assume that the glyceraldehyde enantiomer with its secondary hydroxyl on the right was the dextrorotatory form, and he placed a D in front of its name:

$$\begin{array}{cc} \text{CHO} & \text{CHO} \\ | & | \\ \text{H}-\text{C}-\text{OH} & \text{HO}-\text{C}-\text{H} \\ | & | \\ \text{CH}_2\text{OH} & \text{CH}_2\text{OH} \\ \text{D-glyceraldehyde} & \text{L-glyceraldehyde} \end{array}$$

He assumed that the other enantiomer was the levorotatory form, and he used a L for that name. Fischer had a 50% chance of being correct that the glyceraldehyde isomer with its number 2 hydroxyl on the right was actually dextrorotatory. As it turned out, he was lucky enough to have guessed right. But it was not until 1951 that a Dutch chemist was able to prove that D-glyceraldehyde rotates plane-polarized light to the right and L-glyceraldehyde rotates plane-polarized light to the left; we thus name the isomers:

$$\begin{array}{cc} \text{CHO} & \text{CHO} \\ | & | \\ \text{H}-\text{C}-\text{OH} & \text{HO}-\text{C}-\text{H} \\ | & | \\ \text{CH}_2\text{OH} & \text{CH}_2\text{OH} \\ \text{D-(+)-glyceraldehyde} & \text{L-(−)-glyceraldehyde} \end{array}$$

Because so much time passed between Fischer's guess and the actual structure determination for glyceraldehyde, chemists adopted his practice of using D and L to designate the position of the $-$OH group next to the terminal alcohol group. However, this designation only gives the configuration of that one asymmetric carbon atom. It says nothing about the actual direction of rotation of plane-polarized light, a property that must be measured experimentally. It is not unusual to find D-monosaccharides which are actually levorotatory, and vice versa (see Table 13-3).

Table 13-3 Specific Optical Rotations of Some Monosaccharides at Anomeric Equilibrium (20°C)

Monosaccharide	Specific Optical Rotation (in degrees)
D-Ribose	−23.7
L-Arabinose	+104.5
D-Xylose	+19.0
D-Glucose	+52.2
D-Galactose	+80.5
D-Fructose	−92.0
D-Mannose	+14.6

Most monosaccharides found in nature belong to the D family, shown in Figure 13-7. In fact, enzymes have the uncanny ability to distinguish between D- and L-isomers, and because they prefer the D-isomers of monosaccharides, humans are unable to metabolize the L-isomers. Yeast can only ferment D-glucose to produce alcohol, and most animals are able to utilize only L-amino acids in the synthesis of proteins.

Examples of Monosaccharides

Ribose and Deoxyribose

These two aldopentoses are components of nucleic acids, the structures which control heredity and protein synthesis. Deoxyribose differs from ribose by not having an oxygen atom on carbon 2. Structures for the two are:

$$
\begin{array}{cc}
^1CHO & ^1CHO \\
| & | \\
H-^2C-OH & H-^2C-H \\
| & | \\
H-^3C-OH & H-^3C-OH \\
| & | \\
H-^4C-OH & H-^4C-OH \\
| & | \\
^5CH_2OH & ^5CH_2OH \\
\text{D-ribose} & \text{D-deoxyribose}
\end{array}
$$

CARBOHYDRATES 367

Fig. 13-7 The D Family of Aldoses. (Members of the D-aldose family can be considered derivatives of D-glyceraldehyde; members of the L-aldoses can be considered derivatives of L-glyceraldehyde and are mirror images of the D-aldoses.)

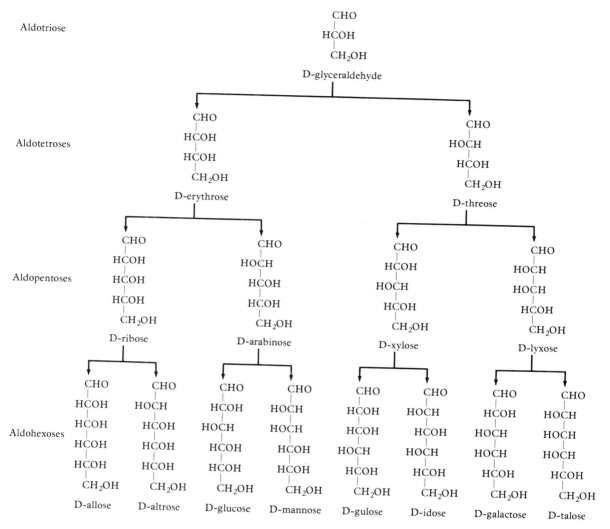

Since the D forms of ribose and deoxyribose are especially important in nature, we will restrict our discussion to them.

The structures shown above are called open-chain structures, and Fischer projections are used to illustrate the sterochemistry of open-chain structures of monosaccharides. However, those having five or more carbon atoms exist mostly in forms other than the open-chain structure. For example, when ribose and deoxyribose are in solution, their flexible chains bend around to form a ring. Ring formation occurs because the hydroxyl group on carbon 4 reacts with the aldehyde group to form a hemiacetal, resulting in a five-membered ring:

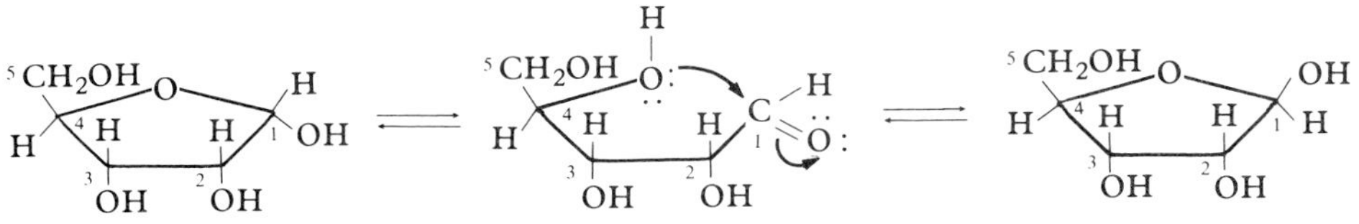

The hemiacetal, and thus the ring, forms easily because the two reactants (the hydroxyl group and the aldehyde group) are in the same molecule and thus rather close together in space.

When the cyclic hemiacetal forms, an hydroxyl group appears on carbon 1. This group can be oriented in either of two directions, above the ring or below it. Because of the variability of orientation of the new hydroxyl group, two isomers called anomers are created. *Anomers* are cyclic stereoisomers which differ from each other only by the the orientation of the newly formed hydroxyl group. Thus the two cyclic forms of D-ribose (and also of D-deoxyribose) are anomers, and carbon 1 is referred to as the anomeric carbon; the cyclic forms of these two monosaccharides are shown in Figure 13–8.

In drawing cyclic structures of monosaccharides, certain rules should be followed so that all of our structures will be consistent. First, we must draw the ring with its oxygen atom to the rear; we must also put the anomeric carbon on the right side of the ring. (The terminal —CH$_2$OH group is always shown above the ring for D-monosaccharides.) Then we imagine that the ring is at right angles to the plane of the page,

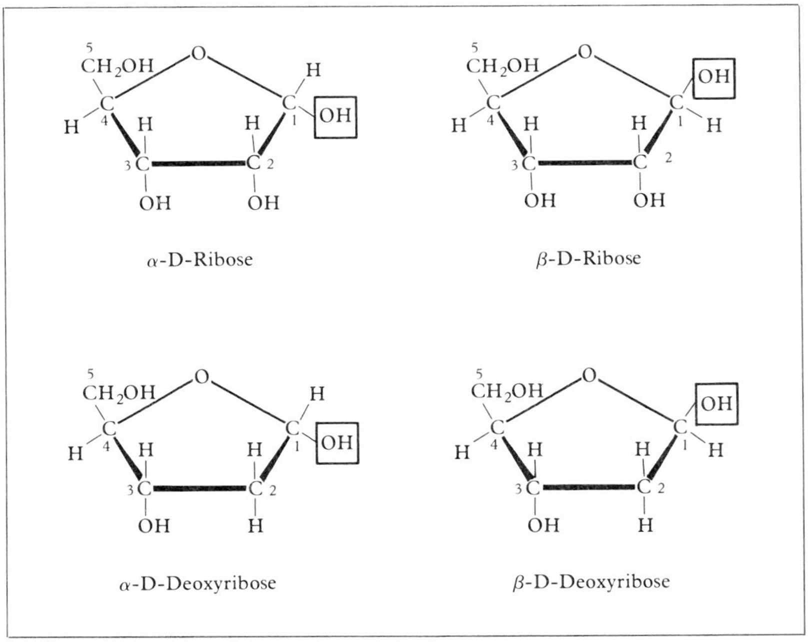

Fig. 13-8 Haworth Structures for D-Ribose and D-Deoxyribose

so that those hydroxyl groups pointing up are projected above the ring. Structures of cyclic hemiacetals drawn in this way are called *Haworth structures*. As you can see, there are two Haworth structures each for D-ribose and D-deoxyribose. In order to distinguish between the anomers, we designate the one in which hydroxyl number 1 is pointed down as *alpha* (α), and the one in which hydroxyl number 1 is pointed up is *beta* (β). Abbreviated Haworth structures are drawn as follows:

Haworth structures: two-dimensional representations of cyclic hemiacetals and hemiketals formed by monosaccharides.

α-D-ribose β-D-ribose

In solution, anomers are interconverted through the open-chain structure. In the case of ribose and deoxyribose, a six-membered ring also forms, but only the five-membered ring is found in biological molecules. Four anomers exist in solution in the relative proportions shown below.

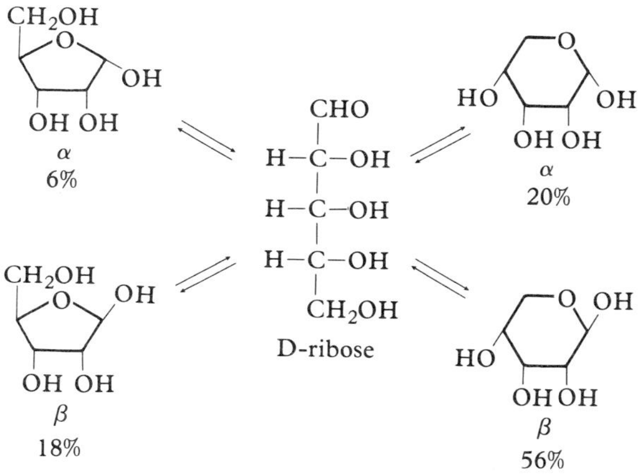

α 6% α 20%

D-ribose

β 18% β 56%

EXAMPLE 13.4

Draw Fischer projections for the open-chain forms of D-ribose, D-deoxyribose, and their enantiomers.

SOLUTION

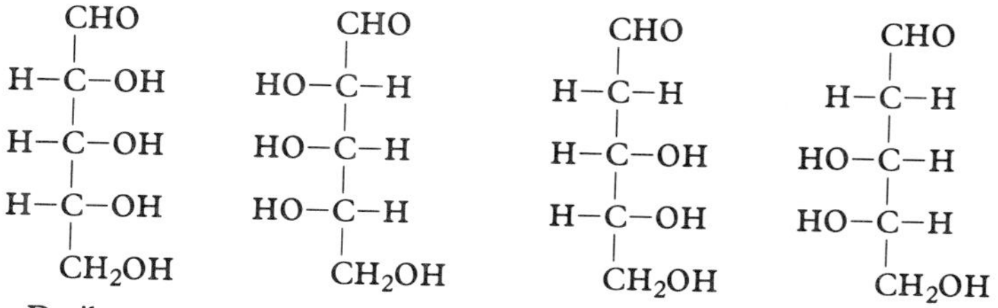

```
   CHO              CHO              CHO              CHO
H—C—OH          HO—C—H           H—C—H            H—C—H
H—C—OH          HO—C—H           H—C—OH           HO—C—H
H—C—OH          HO—C—H           H—C—OH           HO—C—H
  CH₂OH            CH₂OH            CH₂OH            CH₂OH
 D-ribose         L-ribose        D-deoxyribose    L-deoxyribose
               (enantiomer of                    (enantiomer of
                 D-ribose)                        D-deoxyribose)
```

Glucose

Glucose, also called dextrose, is present in greater amounts in nature than any other monosaccharide. In solution it exists as two six-membered cyclic hemiacetals along with the aldohexose open-chain structure:

$$\text{D-glucose} \quad\rightleftharpoons\quad \alpha\text{-D-glucose (36\%)} \quad\rightleftharpoons\quad \text{D-glucose (open-chain) trace} \quad\rightleftharpoons\quad \beta\text{-D-glucose (64\%)}$$

Glucose is found in fruits, and it is the monomer of starch and cellulose, which are polymers. It is sometimes referred to as blood sugar, as it is the primary monosaccharide found in blood. Glucose is absorbed directly by the intestine without need for preliminary digestion. For this reason, it is often administered to patients intravenously as a 5% solution to provide a direct source of energy. The 5% solution is isotonic with blood, and it provides the patient with about 200 kcal of energy per liter. The normal concentration of glucose in blood is approximately 100 mg %. However, larger glucose concentrations may be found in the blood of diabetics; in some instances glucose is found in the urine of these patients when the blood glucose rises above 160 mg %.

It is often necessary for hospital personnel to determine the concentration of glucose in blood to aid in the diagnosis of diabetes and *hypoglycemia*. One test often used employs *o*-toluidine in acetic acid. When a small sample of *blood serum* is mixed with the reagent, the open-chain form of glucose reacts to form a blue green product:

hypoglycemia: condition in which the concentration of glucose in the blood is abnormally low.

blood serum: the liquid that remains after blood clots.

$$\text{glucose} + o\text{-toluidine} \xrightarrow[100°C]{\text{acetic acid}} \text{blue green product} + H_2O$$

The color intensity produced by the reaction is directly proportional to the amount of product formed, which in turn is a measure of glucose present. Although other aldoses will also react with *o*-toluidine, they are normally present in blood in such low concentrations that they do not present serious interferences.

Frequently in testing urine for glucose, it is only necessary to detect the presence of glucose and not the amount, since any glucose in urine may indicate an abnormal condition. In this case, Benedict's or Fehling's test is often used. Both tests employ solutions of copper(II) ion in aqueous base. Benedict's solution contains a small amount of sodium citrate to prevent the precipitation of $Cu(OH)_2$, while Fehling's solution contains sodium tartrate for that purpose. The open-chain form of glucose is oxidized by Cu(II), which in turn is reduced and precipitated as Cu_2O, a reddish brown solid:

$$\begin{array}{c} CHO \\ | \\ H-C-OH \\ | \\ HO-C-H \\ | \\ H-C-OH \\ | \\ H-C-OH \\ | \\ CH_2OH \\ \text{glucose} \end{array} + 2\,Cu^{2+} + 4\,OH^- \longrightarrow \begin{array}{c} CO_2H \\ | \\ H-C-OH \\ | \\ HO-C-H \\ | \\ H-C-OH \\ | \\ H-C-OH \\ | \\ CH_2OH \\ \text{gluconic acid} \end{array} + 2\,H_2O + Cu_2O\downarrow$$

reddish brown precipitate

Because of its ability to reduce Cu(II), glucose is said to be a *reducing sugar*. Any sugar which has a free aldehyde group or a similar functional group is a reducing sugar, and since the open-chain form of glucose contains a free aldehyde group, glucose qualifies in spite of the fact that most glucose in solution exists as cyclic hemiacetal. In fact, all monosaccharides are reducing sugars, but usually none are present in urine to interfere with Benedict's or Fehling's test.

Clinitest tablets are commercially prepared testing materials based on Benedict's test. Other commercial testing materials utilize enzymes to determine whether glucose is present in urine. The advantage in this kind of test is that the enzymes are specific for the presence of glucose only and not for other monosaccharides.

Galactose

Galactose is an aldohexose which closely resembles glucose; the difference in the two hexoses is the orientation of the number 4 hydroxyl group. D-Galactose exists in an open-chain form and in α and β cyclic hemiacetals:

$$\begin{array}{c} {}^1\text{CHO} \\ \mid \\ \text{H}-\overset{2}{\text{C}}\text{OH} \\ \mid \\ \text{HO}-\overset{3}{\text{C}}-\text{H} \\ \mid \\ \text{HO}-\overset{4}{\text{C}}-\text{H} \\ \mid \\ \text{H}-\overset{5}{\text{C}}-\text{OH} \\ \mid \\ {}^6\text{CH}_2\text{OH} \end{array}$$

D-galactose

α-D-galactose ⇌ D-galactose (open-chain) ⇌ β-D-galactose

Galactose is seldom found free in nature, as it is usually combined with glucose to form lactose, a carbohydrate found in milk and often referred to as milk sugar. Lactose is synthesized during pregnancy and lactation, and at these times galactose may appear in the urine to give a false-positive test for glucose.

About one out of every 18,000 infants suffers from the inheritable disease called galactosemia. Characterized by mental retardation and cataracts, the disease is caused by a defective or absent enzyme, a condition which allows abnormal buildup of galactose in the body. The condition can be controlled by removing sources of galactose from the diet of the infant and using a special formula in which sucrose (table sugar) is substituted for lactose. By adulthood, the body develops an alternate metabolic route for disposal of galactose.

EXAMPLE 13.5

Galactose is a reducing sugar because it has a free aldehyde group in its open-chain form that can be oxidized by Cu(II). Write the Fischer structure for the oxidized form of D-galactose produced by reaction with Cu(II).

SOLUTION

Since the aldehyde group of D-galactose can be oxidized by Cu(II) to a carboxylic acid group, the Fischer structure of the oxidized product is

$$\begin{array}{c} \text{CO}_2\text{H} \\ \mid \\ \text{H}-\text{C}-\text{OH} \\ \mid \\ \text{HO}-\text{C}-\text{H} \\ \mid \\ \text{HO}-\text{C}-\text{H} \\ \mid \\ \text{H}-\text{C}-\text{OH} \\ \mid \\ \text{CH}_2\text{OH} \end{array}$$

Fructose

Fructose is a ketohexose present in fruits and honey.

$$\begin{array}{c}
^1CH_2OH \\
| \\
^2C=O \\
| \\
HO-^3C-H \\
| \\
H-^4C-OH \\
| \\
H-^5C-OH \\
| \\
^6CH_2OH
\end{array}$$

D-fructose

It is also found combined with glucose to form sucrose, the sugar obtained from sugar beets and sugar cane and often referred to as table sugar. Fructose is approximately 70% sweeter than table sugar; it is currently receiving attention as a substitute for table sugar in low-calorie diets because less fructose is needed to achieve the same degree of sweetness in foods.

Fructose exists in interchangeable cyclic structures in solution, but because of the presence of the ketone group, fructose forms cyclic hemi-ketals rather than hemiacetals. In this case carbon 2 is the anomeric carbon.

α-D-fructose ⇌ D-fructose (open-chain) ⇌ β-D-fructose

EXAMPLE 13.6

Draw Fischer projections for the open-chain forms of D-glucose, D-fructose, and their enantiomers.

SOLUTION

$$\begin{array}{c}
CHO \\
| \\
H-C-OH \\
| \\
HO-C-H \\
| \\
H-C-OH \\
| \\
H-C-OH \\
| \\
CH_2OH
\end{array}
\quad
\begin{array}{c}
CHO \\
| \\
HO-C-H \\
| \\
H-C-OH \\
| \\
HO-C-H \\
| \\
HO-C-H \\
| \\
CH_2OH
\end{array}
\quad
\begin{array}{c}
CH_2OH \\
| \\
C=O \\
| \\
HO-C-H \\
| \\
H-C-OH \\
| \\
H-C-OH \\
| \\
CH_2OH
\end{array}
\quad
\begin{array}{c}
CH_2OH \\
| \\
C=O \\
| \\
H-C-OH \\
| \\
HO-C-H \\
| \\
HO-C-H \\
| \\
CH_2OH
\end{array}$$

D-glucose   L-glucose (enantiomer of D-glucose)   D-fructose   L-fructose (enantiomer of D-fructose)

EXAMPLE 13.7

Draw abbreviated Haworth structures for the α- and β-anomers of D-ribose, D-deoxyribose, D-glucose, and D-fructose.

SOLUTION

α-D-ribose, β-D-ribose, α-D-deoxyribose, β-D-deoxyribose

α-D-glucose, β-D-glucose, α-D-fructose, β-D-fructose

Disaccharides

Disaccharides are carbohydrates containing two monosaccharides joined by acetal or ketal linkage. Just as open-chain hemiacetals and hemiketals can form acetals and ketals, so can the cyclic structures of monosaccharides. The following equations illustrate acetal and ketal formation:

$$\underset{\text{hemiacetal or hemiketal}}{R-\underset{OR'}{\overset{OH}{C}}-H} + R''-OH \rightleftharpoons \underset{\text{acetal or ketal}}{R-\underset{OR'}{\overset{OR''}{C}}-H} + HOH$$

α-D-glucose (hemiacetal) + β-D-glucose (hemiacetal) ⇌ a disaccharide (acetal) + HOH

Hydrolysis, which can be catalyzed by acid or enzymes, splits disaccharides into the two constituent monosaccharides. The enzyme-catalyzed reactions for the most common disaccharides—maltose, lactose, and sucrose—are illustrated below. Each of these disaccharides is discussed individually in the following paragraphs.

$$\text{maltose} + H_2O \xrightleftharpoons{\text{maltase}} 2 \text{ glucose}$$

$$\text{lactose} + H_2O \xrightleftharpoons{\text{lactase}} \text{glucose} + \text{galactose}$$

$$\text{sucrose} + H_2O \xrightleftharpoons{\text{sucrase}} \text{glucose} + \text{fructose}$$

Examples of Disaccharides

Maltose

Maltose, also called malt sugar, is formed by the action of enzymes in joining two units of D-glucose by an ether linkage to form an acetal:

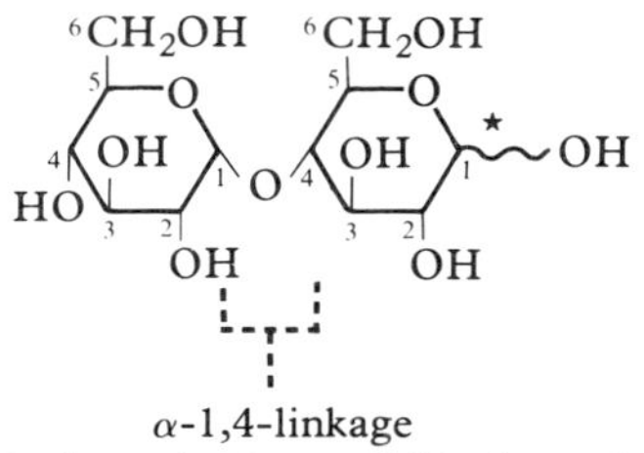

α-1,4-linkage
(α-D-glucose) (α- or β-D-glucose)
Maltose

Note that the number 1 hydroxyl used to form the ether linkage had the α-orientation. Because of this, and because the linkage connects carbon 1 of the first glucose unit to carbon 4 of the second glucose unit, the linkage is said to be α-1,4. The second glucose unit, in which the anomeric hydroxyl group is not involved in an ether linkage, can exist in either the α- or the β-form, or a mixture of the two. Sugar derivatives formed by acetal or ketal linkages are called *glycosides*, and the ether linkage is often referred to as a *glycosidic linkage*. Because the second ring in maltose can open to form the open-chain structure with an aldehyde group, maltose is a reducing sugar. It will react with Benedict's or Fehling's solution to reduce Cu(II).

glycoside (GLYE-co-side)
glycosidic (GLYE-co-SIH-dic)

Maltose is found primarily in germinating grains. Its name is derived from its presence in malt, the substance produced by soaking grain, allowing it to germinate, and then drying the germinated grain.

Lactose

Lactose is the disaccharide in mammalian milk; it is often called milk sugar. The lactose content of milk varies with the species. Human milk

---

*The squiggle is used to indicate that the structure may be α or β or a mixture.

contains about 7% lactose, whereas cow's milk, which is not quite as sweet, contains about 5% lactose.

Lactose is composed of a β-D-galactose unit linked to an α- or β-D-glucose unit. Thus the linkage is β-1,4.

$$\underset{\text{Lactose}}{\underset{(\beta\text{-D-galactose}) \quad (\alpha\text{- or }\beta\text{-D-glucose})}{\underset{\beta\text{-1,4-linkage}}{\text{[structure]}}}}$$

Since the number 1 hydroxyl on the glucose unit is free (and therefore this cyclic structure can form an open-chain structure), lactose is a reducing sugar. It sometimes appears in urine and blood during pregnancy and lactation, and it can be mistaken for glucose in chemical tests at those times.

PERSPECTIVE: Lactose Intolerance

Although milk is the universal food of newborn mammals, some human infants are unable to digest it. They suffer from a condition known as lactose intolerance, caused by insufficient quantities of lactase, the enzyme that catalyzes the hydrolysis of lactose to galactose and glucose. All adult animals except humans also lack the enzyme, and so do most humans after two to four years of age. The main symptom of lactose intolerance is the diarrhea caused by high lactose levels in intestinal juice which draw water out of the tissues by osmosis. Adult tolerance to lactose has been observed only in about 90% of northern Europeans and 80% of two nomadic pastoral tribes in Africa. It is possible that lactose tolerance is a result of an evolutionary pattern that developed about ten thousand years ago when small groups of humans began to milk animals. A chance genetic mutation resulting in lactase synthesis would have conferred an adaptational advantage to those who lived in a milk-drinking society.

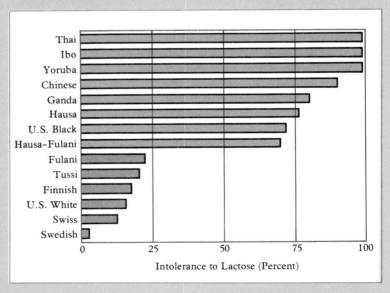

Lactose Intolerance in Various Populations (The blue bars represent populations for which more than 50% of the members exhibit lactose intolerance.)

Sucrose

Probably the most important disaccharide is sucrose, the one we call table sugar. It is found in fruits and honey and is produced commercially from sugar cane and sugar beets. Sucrose is composed of α-D-glucose and β-D-fructose joined in α-1,2-linkage.

Sucrose

Note that in this structure, both anomeric hydroxyl groups were used to form the glycosidic linkage. Because neither ring can open and revert to an aldehyde or ketone structure, sucrose is not a reducing sugar.

Sucrose is the standard of comparison in rating sugars for their sweetness (see Table 13-4). Such ratings are usually done by a panel of tasters, and the scale indicates the relative sweetnesses of various sugars and related substances.

Polysaccharides

Polysaccharides are high molecular weight polymers of monosaccharides. Because of their size, polysaccharides are not water soluble, but their many hydroxyl groups become hydrated individually when exposed to water, and some polysaccharides form thick colloidal dispersions when heated in water. Thus the polysaccharide known as starch can be used as a thickener in sauces, gravies, pie fillings, and other food preparations.

Table 13-4 Relative Sweetness of Some Sugars and Artificial Sweeteners (Based on the reference value of 100 for sucrose)

Sugar or Sweetener	Relative Sweetness
Lactose	16
Galactose	32
Glucose	74
Sucrose	100
Fructose	173
Cyclamate	3,000
Saccharin	45,000

Some important examples of polysaccharides are discussed in the following paragraphs.

Starch

Starch is a mixture of two polymers, amylose and amylopectin. About 10–20% of starch is amylose, a polysaccharide containing 60–300 glucose units joined by α-1,4-linkages to form a long strand which assumes the shape of a helix (Figure 13-9(a)). The rest of starch is amylopectin, which consists of 300–600 glucose units connected by α-1,4-linkages. In contrast to amylose, amylopectin contains short branches of 25–30 glucose units with α-1,4-linkages connected to the main chain by α-1,6-linkages. It exists as a series of small helices connected to a larger helix (Figure 13-9(b)). The helices of amylose and amylopectin form an unusual compound with iodine in which I_2 molecules become trapped inside the starch helices (Figure 13-10) to give a dark blue color. This reaction is often used to test for the presence of starch.

Fig. 13-9 Amylose and Amylopectin

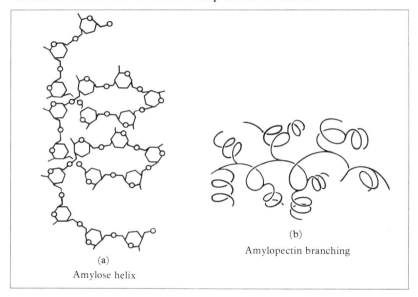

(a) Amylose helix

(b) Amylopectin branching

Starch, which accounts for most of the carbohydrate in our diet, is obtained primarily from grains, potatoes, and vegetables. The species of the plant determines the number of glucose units and their characteristic arrangement. Enzymes in saliva and intestinal juice catalyze starch hydrolyis, degrading it to single glucose units during digestion. These glucose molecules are then absorbed into the blood system and transported to cells throughout the body.

Fig. 13-10 The I_2 Molecule Trapped Inside a Helix of Amylose or Amylopectin

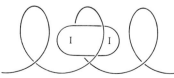

Glycogen

Glycogen is synthesized by animals as a storage form of glucose; an alternate name for glycogen is animal starch. Although it is found throughout the body, it appears in greatest amounts in the liver (1.5–4% glycogen by weight) and muscles (about 1% glycogen by weight). Enzymes in glycogen storage granules catalyze its hydrolysis to glucose when there is a need for energy. The average adult has enough glycogen to meet normal energy needs for about 16 hours. Because of this storage capacity, we need to eat only a few meals each day to satisfy energy requirements. Glycogen structure (Figure 13-11) is similar to that of amylopectin except that branching occurs at shorter intervals on the main strand (about every twelfth glucose unit).

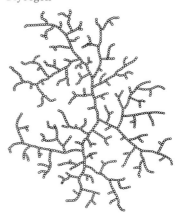

Fig. 13-11 A Simplified Representation of the Structure of Glycogen

Cellulose

Cellulose is the basic structural component of most plants. It imparts rigidity to plants because the linear chains of glucose units in cellulose are frequently aligned within plant cell walls to form fibrous arrangements.

Cellulose is a linear polymer with β-1,4-linkages between glucose units. This structural feature makes cellulose indigestible to humans and other meat-eating animals because their enzymes can only catalyze hydrolysis of α-linkages of polysaccharides. Thus cellulose passes unchanged through the digestive tracts of many animals. It is currently popular to speak of dietary cellulose as fiber, the roughage that provides bulk, stimulates contraction of the intestines, and speeds the passage of food through the digestive system.

As demonstrated by certain trees having life-spans of several thousand years, cellulose possesses remarkable biological stability. Cellulose enzymes (those catalyzing hydrolysis of β-glycosidic linkages of polysaccharides) are generally found only in bacteria and molds. Even termites cannot digest cellulose but instead they must rely on parasitic microorganisms to break it down. Fortunately there are bacteria in the stomachs of sheep and cattle to digest cellulose for these animals and other ruminants.

The chemical characteristics of the polysaccharides we have discussed are summarized in Table 13-5.

Table 13-5 Chemical Characteristics of the Major Polysaccharides

Characteristic	Amylose	Amylopectin	Glycogen	Cellulose
Monomer	D-Glucose	D-Glucose	D-Glucose	D-Glucose
Linkage	α-1,4-	α-1,4- and α-1,6-	α-1,4- and α-1,6-	β-1,4-
Branching (% of units in side chains)	None	4%	9%	None
Molecular weight	4×10^3– 1.5×10^5	5×10^4– 1×10^6	5×10^6	2×10^5– 2×10^6

SUMMARY

Aldehydes and Ketones: Aldehydes and ketones are organic compounds having the carbonyl functional group. Aldehydes have one carbon atom bonded to the carbonyl group, while ketone carbonyl groups are attached to two carbons; their general formulas are RCHO and RCOR, respectively. The carbonyl double bond is more polar than would be expected from electronegativity values because of the effect of resonance. Because of this, aldehydes and ketones have higher boiling points than nonpolar compounds of similar molecular weight. However, they cannot form hydrogen bonds among themselves, and thus they have lower boiling points than the corresponding alcohols. Low molecular weight aldehydes and ketones are water soluble. Common names are used extensively for aldehydes and ketones. For aldehydes, these are derived from common names of carboxylic acids, and common names for ketones are similar to those of ethers. In the IUPAC system, aldehyde names are derived from names of parent alkanes by changing the alkane name ending from "-e" to "-al," and ketone names are derived by changing the parent alkane name ending from "-e" to "-one." The two most important chemical reactions of aldehydes and ketones are controlled oxidation and addition. Addition reactions include hydrogenation, hydration, and formation of hemiacetals and hemiketals (and subsequent formation of acetals and ketals).

Carbohydrates: Carbohydrates are classified as monosaccharides, disaccharides, oligosaccharides, and polysaccharides. Monosaccharides are carbohydrates which cannot be broken down into simpler molecules by acid hydrolysis. These can be categorized by the number of carbon atoms in their structures and by the presence of an aldehyde or ketone group. All monosaccharides belong to either the D or L families. D-Monosaccharides have the —OH group next to the terminal carbon on the right in their Fischer projections. L-Monosaccharides have that —OH group oriented to the left. The D and L designations are not indicative of the direction of rotation of plane-polarized light. Most monosaccharides found in nature belong to the D family. Monosaccharides having five or more carbon atoms can form cyclic hemiacetals or hemiketals. When these ring structures form, pairs of new isomers called anomers are created which differ only by the orientation of the newly formed —OH group. Cyclic structures for monosaccharides exist in equilibrium with open-chain forms in solution. The open-chain form of glucose reacts with o-toluidine and with Cu(II); these reactions are the basis for chemical detection and determination of glucose in blood and urine. Disaccharides are carbohydrates containing two monosaccharides joined by acetal or ketal linkage. Common disaccharides (and their monosaccharide constituents) are maltose (glucose), lactose (glucose and galactose), and sucrose (glucose and fructose). Polysaccharides are high molecular weight polymers of monosaccharides. Examples of polysaccharides are starch, glycogen, and cellulose.

STUDY QUESTIONS AND PROBLEMS

1. Define the following terms:
   - (a) aldehyde
   - (b) ketone
   - (c) hemiacetal
   - (d) hemiketal
   - (e) acetal
   - (f) ketal
   - (g) carbohydrate
   - (h) monosaccharide
   - (i) aldose
   - (j) ketose
   - (k) disaccharide
   - (l) oligosaccharide
   - (m) polysaccharide
   - (n) stereoisomers
   - (o) chiral molecule
   - (p) enantiomers
   - (q) optical isomers
   - (r) racemic mixture
   - (s) anomers
   - (t) reducing sugar
   - (u) glycosidic linkage

2. How do aldehydes differ from ketones?

3. Draw structures for the simplest aldehyde and the simplest ketone. What are their common names and their IUPAC names?

4. Describe the geometry of the carbonyl group.

5. Use structural examples to illustrate how aldehydes and ketones form hydrogen bonds with other molecules.

6. Arrange the following compounds in order of increasing relative boiling points as predicted from their structures.

   (a) $CH_3-\underset{\underset{O}{\|}}{C}-CH_2-CH_3$

   (b) $CH_3-\underset{\underset{CH_3}{|}}{CH}-CH_2-CHO$

   (c) phenyl-$\underset{\underset{O}{\|}}{C}-CH_3$

   (d) $Br-CH_2-CHO$

   (e) cyclopentanone

   (f) 2-chlorobenzaldehyde

(g)
$$\text{C}_6\text{H}_5-\overset{\underset{\displaystyle\|}{O}}{C}-\text{C}_6\text{H}_5$$

(h) $CH_3-CH_2-CH_2-CH_2-CHO$

7. Give a common name and the IUPAC name for each compound in question 6 except (f) and (g).

8. Draw the structure for each of the following compounds.

   (a) formaldehyde
   (b) benzaldehyde
   (c) butyl hexyl ketone
   (d) β-iodopropionaldehyde
   (e) pentanal
   (f) butanone
   (g) 2-phenylethanal
   (h) 3-hydroxyhexanal

9. Complete the following equations. If no reaction occurs, write NR.

   (a) $Cl_3C-CHO + CH_3OH \xrightleftharpoons{H^+}$

   (b) $C_6H_5-CH_2-CHO \xrightarrow{[O]}$

   (c) $CH_3-CH_2-\overset{\underset{\displaystyle\|}{O}}{C}-CH_3 + H_2O \rightleftharpoons$

   (d) $CH_3-\overset{\underset{\displaystyle\|}{O}}{C}-CH_3 + H_2 \xrightarrow[\Delta]{Pt}$

   (e) $2\ CH_3-CH_2-CH_2-CHO + 11\ O_2 \longrightarrow$ (combustion products)

   (f) $CH_3-CHO + 2\ CH_3-CH_2-OH \xrightleftharpoons{H^+}$

   (g) $C_6H_5-\overset{\underset{\displaystyle\|}{O}}{C}-C_6H_5 \xrightarrow{[O]}$

   (h) $H-CHO + H_2 \xrightarrow[\Delta]{Pt}$

10. Give the equation which describes the reaction of one mole of acetaldehyde with each of the following.

    (a) one mole of water
    (b) one mole of methanol
    (c) two moles of methanol

11. Repeat question 10 using acetone instead of acetaldehyde.

12. Undecanal, a sex attractant for certain moths, can be prepared by controlled oxidation of an alcohol. Draw the structure of the alcohol needed to prepare undecanal.

$$CH_3-(CH_2)_9-CHO$$
undecanal

13. Glutaraldehyde, used in hospitals to destroy bacteria, viruses, and spores, reacts with water to form a cyclic structure.

    [structure: tetrahydropyran ring with OH groups]
    OH   O   OH

    The reaction proceeds in two steps. Explain how the cyclic compound forms.

14. Name two common aldehydes and give a major use for each.

15. Draw the structure of acetone and describe a major use for the compound.

16. Classify each of the following monosaccharides by combining the aldehyde-ketone designation with terminology indicating the number of carbon atoms.

   (a) CHO
        |
        HO-C-H
        |
        H-C-OH
        |
        CH$_2$OH

   (b) CH$_2$OH
        |
        C=O
        |
        H-C-OH
        |
        H-C-OH
        |
        CH$_2$OH

   (c) CH$_2$OH
        |
        C=O
        |
        HO-C-H
        |
        H-C-OH
        |
        H-C-OH
        |
        CH$_2$OH

   (d) CHO
        |
        HO-C-H
        |
        H-C-OH
        |
        H-C-OH
        |
        CH$_2$OH

17. Write Fischer projections for the four isomers of the D-aldopentoses.

18. Draw the Fischer projection for D-fructose and label all of the chiral centers with asterisks.

19. Give the direction of rotation of plane-polarized light by each of the following aqueous solutions:

    (a) (+)-ribose
    (b) (−)-glucose
    (c) (−)-galactose
    (d) equal concentrations of (+)-fructose and (−)-fructose in one solution

20. What is meant by D and L families of monosaccharides? Which family is more prevalent in nature?

21. Draw the Fischer projection for D-deoxyribose and then draw ring structures for its two five-membered ring anomers.

22. How are the hexoses glucose and fructose structurally different from each other?
23. How is *o*-toluidine used to measure blood sugar?
24. What are Benedict's and Fehling's tests used for? Write an equation to illustrate your answer.
25. Why is glucose a reducing sugar?
26. Explain why glucose can be injected directly into the bloodstream to serve as an energy source.
27. Use structures to illustrate cyclic hemiacetal formation by D-glucose.
28. How are the hexoses glucose and galactose structurally different from each other?
29. Describe the condition of galactosemia.
30. Why is it that fructose can be used as a low-calorie sweetener?
31. Draw the Fischer projection for D-fructose and then draw ring structures for its two anomers.
32. Mannose differs from glucose only in the configuration of carbon 2. Draw the open-chain structure of D-mannose and its two anomeric ring structures.
33. Give the natural sources for each of the following:
    (a) glucose   (b) fructose   (c) galactose
34. Give the monosaccharide components of each of the following disaccharides:
    (a) maltose   (b) lactose   (c) sucrose
35. Write equations for the hydrolysis of maltose, lactose, and sucrose.
36. Trehalose ($C_{12}H_{22}O_{11}$), a nonreducing sugar found in young mushrooms, gives only D-glucose when hydrolyzed in the presence of acid or maltase.
    (a) Classify trehalose as a monosaccharide, disaccharide, or polysaccharide.
    (b) What kind of glycosidic linkages are present in trehalose?
    (c) How would trehalose react with *o*-toluidine?
37. How is it that maltose and lactose are reducing sugars, but sucrose is not?
38. Describe how starch functions as a thickener in prepared foods.
39. What are the two polymers in starch? How are they structurally different from each other?
40. Describe the biological role of glycogen.
41. Why is cellulose indigestible by humans?
42. What monosaccharides are produced by acid hydrolysis of the following polysaccharides?
    (a) starch   (b) glycogen   (c) cellulose
43. Name 10 commercial products composed mostly of cellulose. Suggest two methods for chemically decomposing these products.

CHAPTER 14

CARBOXYLIC ACIDS, ESTERS, AND LIPIDS

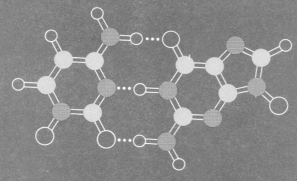

CARBOXYLIC ACIDS
- Structure
- Nomenclature
- Physical Properties
- Chemical Properties
- Examples of Carboxylic Acids

ESTERS
- Nomenclature
- Physical Properties and Uses
- Chemical Properties
- Examples of Esters

LIPIDS
- Classification of Lipids
- Fatty Acids
- Waxes
- Triacylglycerols
- Chemical Properties of Triacylglycerols
- Phosphoglycerides
- Sphingolipids
- Terpenes
- Steroids
- Prostaglandins

Carboxylic acids, the weak organic acids produced by living organisms, have been known for about two thousand years. Greek alchemists realized that a certain kind of substance was responsible for the sour taste of vinegar and the tartness of fruit juices. However, organic chemistry was slow in developing, and by the early eighteenth century, only four carboxylic acids had been identified. Fortunately a Swedish chemist, Carl Wilhelm Scheele, developed a technique for isolating calcium salts of carboxylic acids around 1780, and before his death in 1786, he had identified a host of carboxylic acids.

When carboxylic acids react with alcohols, esters are formed. Esters of low molecular weight carboxylic acids are the substances responsible for the fragrance and flavor of flowers and fruit. Long-chain carboxylic acids react with alcohols to form complex lipids such as fats, oils, and waxes. The word "lipid" originally meant "fat," but its modern usage applies to any greaselike material insoluble in water and produced by a living organism. Lipids are components of all living cells, and they perform a surprising variety of functions in living tissue.

CARBOXYLIC ACIDS

Carboxylic acids are identified by the presence of the carboxyl group,

$$-C\begin{matrix}\nearrow O \\ \searrow OH\end{matrix}$$

The rest of the molecule may be either an alkyl group or an aromatic group, and the formula is written as $RCOOH$ or RCO_2H.

Structure

The carboxyl group is a special form of the carbonyl group in which a hydroxyl group is bonded to the carbonyl carbon. The geometry of the carboxyl group is similar to that of the carbonyl group: bond angles about the carboxyl carbon are 120°, and the atoms attached to the carboxyl carbon are in the same plane as the carbon atom. The carboxyl group is polarized as indicated below.

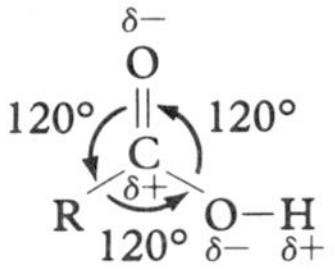

$$\underset{R\ \underset{120°}{}\ \underset{\delta-}{O}-\underset{\delta+}{H}}{\overset{\overset{\delta-}{O}}{\underset{120°}{\overset{\|}{\underset{\delta+}{C}}}}}\ 120°$$

In addition to dipolar attractions between molecules, the carboxylic acids form intermolecular hydrogen bonds. In pure form and in concentrated aqueous solution, carboxylic acid molecules form hydrogen bonds with each other, creating pairs called *dimers*:

dimer (DYE-mer): a pair of chemical units held together by covalent bonds, hydrogen bonds, or other attractive forces.

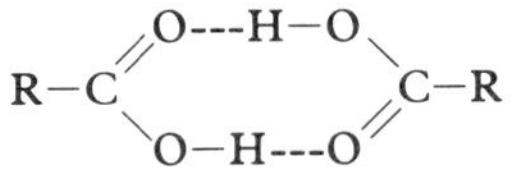

$$R-C\underset{O-H---O}{\overset{O---H-O}{}}C-R$$

a carboxylic acid dimer

Nomenclature

Like aldehydes and ketones, many familiar carboxylic acids were isolated from natural sources before the development of IUPAC nomenclature; these are known by common names which reflect the origin of the carboxylic acid. For example, ants secrete formic acid (from the Latin *formica*, meaning "ant"), acetic acid is present in vinegar (from the Latin *acetum*, meaning "vinegar"), and lactic acid is produced when milk sours (from the Latin *lac* or *lactis*, meaning "milk"). IUPAC names for aliphatic carboxylic acids (Table 14-1) are derived on the

Table 14-1 Examples of Aliphatic Carboxylic Acids

Structure	Common Name	IUPAC Name
H—COOH	Formic acid	Methanoic acid
CH_3—COOH	Acetic acid	Ethanoic acid
CH_3—CH_2—COOH	Propionic acid	Propanoic acid
CH_3—$(CH_2)_2$—COOH	Butyric acid	Butanoic acid
CH_3—$(CH_2)_3$—COOH	Valeric acid	Pentanoic acid
CH_3—$(CH_2)_4$—COOH	Caproic acid	Hexanoic acid
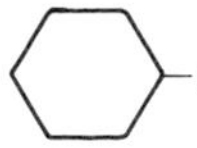 ⬡—COOH	—	Cyclohexanecarboxylic acid

CH_3—CH—COOH
          |
         OH
	Lactic acid	2-Hydroxypropanoic acid

basis of the longest continuous carbon chain containing the carboxyl group. This main chain is named by changing the "-e" name ending of the corresponding alkane to "-oic" and adding the word "acid." The carboxyl carbon is always numbered 1. Aromatic carboxylic acids are almost always identified by common names (Table 14–2). IUPAC names of substituted carboxylic acids are illustrated below, with common names in parentheses.

$HO-CH_2-CH_2-CO_2H$

3-hydroxypropanoic acid
(β-hydroxypropionic acid)

$CH_3-\overset{\overset{\displaystyle CH_3}{|}}{CH}-CH_2-CO_2H$

3-methylbutanoic acid
(β-methylbutyric acid)

$CH_3-\overset{\overset{\displaystyle Br}{|}}{CH}-CH_2-CH_2-CO_2H$

4-bromopentanoic acid
(γ-bromovaleric acid)

$CH_3-CH_2-CH_2-\overset{\overset{\displaystyle CH_3}{|}}{CH}-CH_2-\overset{\overset{\displaystyle Cl}{|}}{CH}-CO_2H$

2-chloro-4-methylheptanoic acid
(α-chloro-γ-methylenanthic acid)

EXAMPLE 14.1

Give either a common name or the IUPAC name for each of the following carboxylic acids.

(a) $Br-\overset{\overset{\displaystyle }{|}}{\underset{\underset{\displaystyle CH_3}{|}}{CH}}-CO_2H$

(b) $CH_3-CH_2-CH_2-CH_2-CH_2-CO_2H$

(c) $CH_3-CH_2-\underset{\underset{\displaystyle OH}{|}}{CH}-CO_2H$

(d) 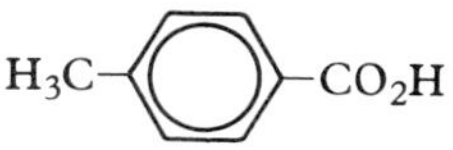 $H_3C-\underset{}{\bigcirc}-CO_2H$

(e) 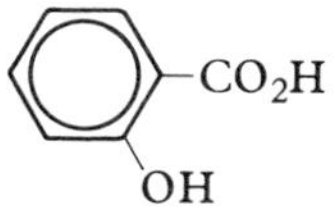 $\underset{\underset{\displaystyle OH}{}}{\bigcirc}-CO_2H$

Table 14-2 Examples of Aromatic Carboxylic Acids

Structure	Common Name	IUPAC Name
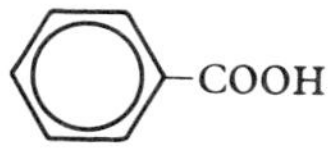 ⌬-COOH	Benzoic acid	Benzoic acid
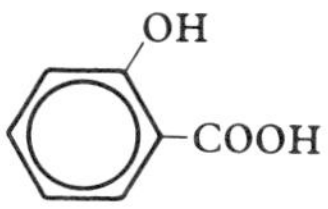 ⌬(OH)-COOH	Salicylic acid	2-Hydroxybenzoic acid
 HOOC-⌬-COOH	Phthalic acid	Phthalic acid

SOLUTION

(a) α-bromopropionic acid (common name) or 2-bromopropanoic acid (IUPAC name)
(b) caproic acid (common name) or hexanoic acid (IUPAC name)
(c) α-hydroxybutyric acid (common name) or 2-hydroxybutanoic acid (IUPAC name)
(d) *p*-methylbenzoic acid (common name)
(e) *o*-hydroxybenzoic acid or salicylic acid (both common names)

EXERCISE 14.1 Give either a common name or the IUPAC name for each of the following carboxylic acids.

(a) $CH_3-CH_2-\overset{Br}{\underset{|}{CH}}-CH_2-CH_2-CO_2H$

(b) $CH_3-CH_2-CH_2-CO_2H$

(c) $CH_3-\overset{}{\underset{\underset{OH}{|}}{CH}}-CO_2H$

(d) 4-chlorobenzoic acid structure (Cl para to CO_2H on benzene ring)

(e) cyclohexane-CO_2H

Physical Properties

Because all of their molecules contain oxygen, it is informative to compare physical properties of alcohols, aldehydes, ketones, and carboxylic acids. Of this group, only alcohols and carboxylic acids can form hydrogen bonds among themselves, and only carboxylic acids can form hydrogen-bonded dimers. Thus, carboxylic acids have the highest boiling points of the group; the relative order of boiling points for simple aliphatic compounds of similar molecular weights is given below.

| hydrocarbons | ethers | aldehydes and ketones | alcohols | carboxylic acids |

increasing boiling points

Since the presence of the carboxyl group allows carboxylic acids to form hydrogen bonds with water,

$$R-C\underset{O-H\cdots}{\overset{O---H}{\Big\langle}}\overset{H}{\underset{}{O}}$$

carboxylic acids are quite water soluble, more so than the corresponding alcohols. However, water solubility of carboxylic acids decreases as the size of the hydrocarbon portion of the structure increases, and carboxylic acids containing more than eight carbon atoms are not soluble in water. Organic compounds may be ordered according to water solubility:

| hydrocarbons | ethers | aldehydes and ketones | alcohols | carboxylic acids |

increasing water solubility

Given that intermolecular forces play important roles in determining both boiling points and water solubility, it is not surprising that the same order is generally observed for these two properties. Some physical properties of a number of carboxylic acids are shown in Table 14-3.

Table 14-3 Physical Properties of Some Carboxylic Acids

Structure	Molecular Weight	Melting Point (°C)	Boiling Point (°C)	Water Solubility* (g/100g)
H—COOH	46.0	8	101	Soluble in all proportions
CH_3—COOH	60.0	16	118	Soluble in all proportions
CH_3—CH_2—COOH	74.0	−21	141	Soluble in all proportions
CH_3—$(CH_2)_2$—COOH	88.0	−6	164	Soluble in all proportions
CH_3—$(CH_2)_3$—COOH	102	−34	186	3
CH_3—$(CH_2)_4$—COOH	116	−4	205	1
⌬—COOH	122	122	250	trace

*at 20°C

Because of their acidic properties, aqueous solutions of carboxylic acids have a characteristic sour taste. Those having one to three carbon atoms have sharp odors resembling vinegar, while carboxylic acids having four to eight carbon atoms have rank, disagreeable odors. In fact, caproic, caprylic, and capric acids received their common names because they typify the odor of goats (from the Latin *caper* or *capri,* meaning "goat"). The odor of butyric acid is characteristic of rancid butter and strong cheese.

Chemical Properties

The most important chemical property of carboxylic acids is their acidity. Since this is the distinguishing feature of the group of compounds, we will start with a discussion of acidity and how carboxylic acids react with bases.

Acidity

It was mentioned in chapter 9 that carboxylic acids are acidic because of the electron-withdrawing effect of the double-bonded oxygen.

$$R-C\overset{\delta-}{\underset{\delta+}{\overset{\displaystyle O}{\diagdown}}}\underset{\delta-\ \delta+}{O-H}$$

A second effect, resonance, also causes carboxylic acids to be acidic. When the proton dissociates, a *carboxylate ion* is formed:

$$R-C\diagup^{O}_{\diagdown OH} \rightleftharpoons R-C\diagup^{O}_{\diagdown O^-} + H^+$$

a carboxylic acid   a carboxylate anion

carboxylate (car-BOX-sih-late)

This anion has two resonance forms,

$$\left[R-C\begin{smallmatrix}\ddot{O}\\ \\\ddot{O}:^-\end{smallmatrix} \longleftrightarrow R-C\begin{smallmatrix}\ddot{O}:^-\\ \\\ddot{O}:\end{smallmatrix} \right]$$

so that resonance stabilization of the anion is an important factor in the acidity of carboxylic acids. A stabilized anion forms more easily than one which is not resonance stabilized, and thus dissociation of the proton will occur. However, the combination of electron withdrawal by the second oxygen and resonance stabilization of the anion results in an oxygen-hydrogen bond that is only weakly polar. Thus carboxylic acids are much weaker acids than the mineral acids (sulfuric acid, nitric acid, and hydrochloric acid). Now that we have studied several kinds of acids, it is helpful to order them according to their relative acid strengths:

$$\text{H}_2\text{O} \quad \text{ArOH} \quad \text{H}_2\text{CO}_3 \quad \text{RCO}_2\text{H} \quad \text{H}_2\text{SO}_4, \text{HCl, and HNO}_3$$

increasing acid strength

Like all acids, carboxylic acids serve as proton donors. For example, when a carboxylic acid dissolves in water, it reversibly donates hydrogen ions to water molecules to form hydronium ions. Since dissociation is slight, only a few hydronium ions are present in the solution.

$$\text{RCO}_2\text{H} + \text{H}_2\text{O} \rightleftharpoons \text{RCO}_2^- + \text{H}_3\text{O}^+$$

Carboxylic acids participate in neutralization reactions to form salts, as shown by the following equations for neutralization by two strong bases, NaOH and KOH, and a weak base, NH_3:

$$\text{RCO}_2\text{H(aq)} + \text{NaOH(aq)} \longrightarrow \text{RCO}_2^-\text{Na}^+(\text{aq}) + \text{H}_2\text{O(l)}$$

$$\text{RCO}_2\text{H(aq)} + \text{KOH(aq)} \longrightarrow \text{RCO}_2^-\text{Na}^+(\text{aq}) + \text{H}_2\text{O(l)}$$

$$\text{RCO}_2\text{H(aq)} + \text{NH}_3(\text{aq}) \longrightarrow \text{RCO}_2^-\text{NH}_4^+(\text{aq})$$

Because of these reactions, carboxylic acids usually dissolve in basic solutions, since their sodium, potassium, and ammonium salts are water soluble. (Formulas for the salts are written with charges in the equations to emphasize that they are ionic.) However, carboxylate ions are conjugate bases of acids, and thus they are proton acceptors. If a strong acid is added to the solution of carboxylate salt, protons will be transferred from the strong acid to the carboxylate ions, and undissociated carboxylic acid will form:

$$\text{H}^+(\text{aq}) + \underbrace{\text{RCO}_2^-(\text{aq}) + \text{Na}^+(\text{aq})}_{\text{soluble salt}} \rightleftharpoons \text{RCO}_2\text{H(aq)} + \text{Na}^+(\text{aq})$$

Thus, water-insoluble carboxylic acids can be solubilized by conversion to sodium, potassium, or ammonium salts, and they can be precipitated from solution by adding strong acid.

Although common names are used most of the time, IUPAC names may be derived for salts of carboxylic acids. In either case, the name of the cation is written first, followed by the name of the carboxylic acid with its "-oic" ending (for common names) or its "-ic" ending (for IUPAC names) changed to "-ate." Some examples are:

$$(CH_3CO_2^-)_2Mg^{2+}$$

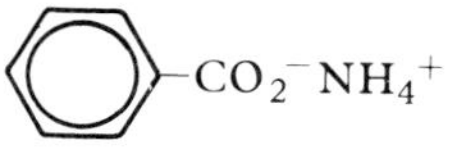

common name: magnesium acetate
IUPAC name: magnesium ethanoate

$$CH_3-\underset{\underset{OH}{|}}{CH}-CO_2^-Na^+$$

sodium lactate
sodium 2-hydroxypropanoate

$$\text{C}_6\text{H}_5-CO_2^-NH_4^+$$

ammonium benzoate
ammonium benzoate

EXAMPLE 14.2

Write a net ionic equation for the neutralization of acetic acid by sodium hydroxide.

SOLUTION

$$CH_3CO_2H(aq) + \cancel{Na^+(aq)} + OH^-(aq) \longrightarrow$$
$$CH_3CO_2^-(aq) + \cancel{Na^+(aq)} + H_2O(l)$$

$$CH_3CO_2H(aq) + OH^-(aq) \longrightarrow CH_3CO_2^-(aq) + H_2O(l)$$

(Note that the formula for acetic acid, a weak electrolyte, is written in the molecular form, as most of it exists in that form in aqueous solution.)

EXERCISE 14.2 Write a net ionic equation for each of the following reactions in water.

(a) $CH_3-CH_2-CO_2H + NaOH$ (c) $CH_3-CH_2-CH_2-CO_2H + NH_3$

(b) 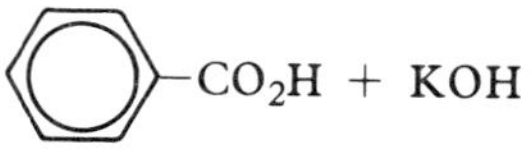 $\text{C}_6\text{H}_5-CO_2H + KOH$

Ester Formation

As mentioned in chapter 12, alcohols react reversibly with carboxylic acids to form *esters*. A strong acid catalyst and heat are required to carry out the reaction in the laboratory.

$$R-C\underset{OH}{\overset{O}{\diagup\!\!\!\diagdown}} + R'OH \underset{\Delta}{\overset{H^+}{\rightleftharpoons}} R-C\underset{OR'}{\overset{O}{\diagup\!\!\!\diagdown}} + H$$

ester: an organic compound having the group

$$-C\underset{OR}{\overset{O}{\diagup\!\!\!\diagdown}}$$

The amount of ester product can be maximized by using excessive amounts of either reactant or by removing one product as it forms, or by a combination of both techniques. Note that the —OH group from the carboxylic acid combines with hydrogen from the alcohol to form water, and the *acyl group*,

$$R-C\underset{}{\overset{O}{\diagup\!\!\!\diagdown}}$$

acyl (ay-SEAL)

combines with the *alkoxy group*, R'O—, from the alcohol to form the ester. Thus the esterification reaction can be thought of as a double replacement reaction. We will come back to esters later in this chapter.

Examples of Carboxylic Acids

Formic Acid

HCO_2H (IUPAC name: methanoic acid)

This is the simplest of the carboxylic acids, and it is also the strongest of this group of weak acids. Formic acid is the substance which causes much of the discomfort of bites and stings by bees and ants, and it was once obtained by distillation of ants. A colorless, fuming liquid with a pungent odor, formic acid is irritating to mucous membranes and produces burns and blisters when it contacts the skin. Its major use is in the processing of textiles and leather.

Acetic Acid

CH_3CO_2H (IUPAC name: ethanoic acid)

Acetic acid, probably the most well known carboxylic acid, is the acid constituent produced in vinegar by fermentation and subsequent oxidation of natural carbohydrates. Because of its relatively high melting point (16°C, 61°F), it is prone to freeze when stored at cool temperatures. Since frozen acetic acid looks icy, the pure liquid is sometimes referred to as "glacial" acetic acid.

Pure acetic acid is a corrosive, colorless liquid at room temperature; it has the odor of vinegar. It occurs naturally in small amounts in body fluids and plant juices as a metabolic intermediate. Industrially, it is produced by oxidation of ethanol or acetaldehyde, and it is used to prepare metal acetate salts (used in certain printing processes), plastics, textiles, and solvents.

Lactic Acid

$$\underset{CH_3CHCO_2H}{\overset{OH}{|}}$$ (IUPAC name: 2-hydroxypropanoic acid)

Lactic acid is present in plant juices, animal blood and muscles, and in the soil. It is formed by bacteria from lactose in milk, and it accounts for the sour taste in sour milk, yogurt, and buttermilk. It is produced in muscles as a product of glucose metabolism during vigorous exercise and it is partly responsible for muscle soreness after hard physical activity.

Lactic acid is a colorless liquid at room temperature; it has a melting point of 18°C (64°F). Manufactured by the fermentation of molasses, starch, or milk whey, it is used in tanning leather and dyeing wool; as a flavoring agent and preservative in processed cheese, salad dressings, pickles, and carbonated beverages; and in numerous chemical processes.

Citric Acid

$$\begin{array}{c} H_2C-CO_2H \\ | \\ HO-C-CO_2H \\ | \\ H_2C-CO_2H \end{array}$$

Citric acid is a colorless (white) crystalline solid at room temperature. Found in practically all plants and in many animal tissues and fluids, it is the characteristic sour taste of citrus fruits. Citric acid is one of a series of compounds involved in the metabolic oxidation of fats, proteins, and carbohydrates to carbon dioxide and water. Because of the prominent role of citric acid in animal metabolism, the primary oxidative pathway is called the *citric acid cycle*. Alternative names for this pathway are the *tricarboxylic acid cycle* (citric acid is a tricarboxylic acid) and the *Krebs cycle* (in honor of the biochemist Hans A. Krebs, who discovered the pathway in 1937).

Citric acid is manufactured by fermentation of cane sugar or molasses, and its major use is as flavoring in candies and soft drinks. Soluble salts of citric acid are widely used as blood anticoagulants and as metal ion scavengers because of the ability of citrate anions to form soluble *complex ions* with metal ions. (Removal of free Ca^{2+} ions from blood in this way prevents blood clot formation.)

complex ion: an ion composed of a metal ion and two or more other ions or groups attached to the metal ion by coordinate covalent bonds.

$$\begin{array}{c} H_2C-CO_2^- \\ | \\ HO-C-CO_2^- \\ | \\ H_2C-CO_2^- \end{array} \rangle Ca^{2+}$$

soluble complex ion formed by citrate and Ca^{2+}

Benzoic Acid

$$\text{C}_6\text{H}_5-CO_2H$$

Benzoic acid, a white crystalline solid at room temperature, is manufactured by oxidation of toluene. The acid and its sodium salt are widely used as food preservatives and in the manufacture of cosmetics, dyes, plastics, and insect repellents. Benzoic acid exists in small amounts in many plants.

Salicylic Acid

(structure: benzene ring with OH and CO_2H groups)

salicyclic (saah-lih-SIH-lic)
antipyretic (an-tie-pie-REH-tic) **agent:** a substance which reduces fever.

Salicylic acid, a white crystalline solid at room temperature, is an *antipyretic* (fever-reducing) *agent* and an analgesic. It was introduced into medicine in the 1870s, but because of its tendency to irritate the stomach lining, some of its ester derivatives (to be discussed later in this chapter) are now preferred for general use. It is still manufactured in large quantity, using phenol as starting material, for use in the preparation of derivatives and for inclusion in various ointments for warts, corns, and calluses. It is also used in the preparation of the sodium salt, which receives some application as an antipyretic and analgesic.

ESTERS

Esters are derivatives of carboxylic acids formed by reaction with alcohols:

$$R-C(=O)O-H + R'OH \xrightleftharpoons[\Delta]{H^+} R-C(=O)O-R' + HOH$$

The general formula is RCO_2R', where R and R' may be aliphatic or aromatic. As explained earlier, formation of esters from carboxylic acids and alcohols is a reversible reaction which requires heat and a strong acid catalyst. In esters, the four atoms closest to the carbonyl double bond all lie in the same plane, as they do in carboxylic acids.

EXAMPLE 14.3

Write equations to illustrate ester formation from

(a) CH_3-CO_2H and CH_3-CH_2-OH

(b) $\text{C}_6\text{H}_5-CO_2H$ and CH_3-OH

(c) $CH_3-CH_2-CH_2-CO_2H$ and C_6H_5-OH

SOLUTION

(a) $CH_3-CO_2H + CH_3-CH_2-OH \xrightleftharpoons[\Delta]{H^+} CH_3-\overset{O}{\overset{\|}{C}}-O-CH_2-CH_3 + H_2O$

(b) $\text{C}_6\text{H}_5-CO_2H + CH_3-OH \xrightleftharpoons[\Delta]{H^+} \text{C}_6\text{H}_5-\overset{O}{\overset{\|}{C}}-O-CH_3 + H_2O$

(c) $CH_3-CH_2-CH_2-CO_2H + \text{C}_6\text{H}_5-OH \xrightleftharpoons[\Delta]{H^+} CH_3-CH_2-CH_2-\overset{O}{\overset{\|}{C}}-O-\text{C}_6\text{H}_5 + H_2O$

EXERCISE 14.3 Write equations to illustrate ester formation from

(a) $CH_3-CH_2-CO_2H$ and CH_3-OH

(b) CH_3-CO_2H and ⬡—OH

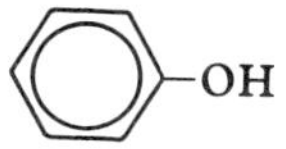

(c) 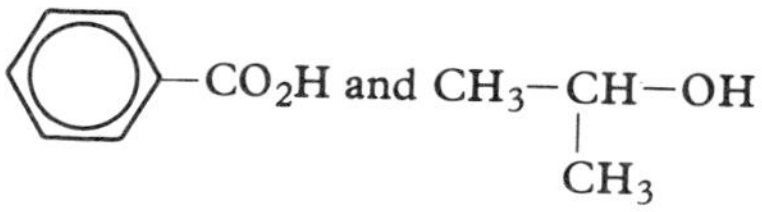 ⬡—CO_2H and $CH_3-CH(CH_3)-OH$

Nomenclature

Esters are named by a trivial system as well as by IUPAC rules, with common names predominating. In both systems, esters are named as derivatives of carboxylic acids, and the "-ic" ending of the acid name is changed to "-ate" for esters. This is similar to the method used for naming carboxylic acid salts. The only difference between common names and IUPAC names for esters is that common names of carboxylic acids are used in one system, and IUPAC names of carboxylic acids are used in the other. The first part of the name of an ester is the name of the alkyl or aromatic group contributed by the alcohol. The carboxylic acid name, with its changed ending, is the second word in the name of an ester. Examples are:

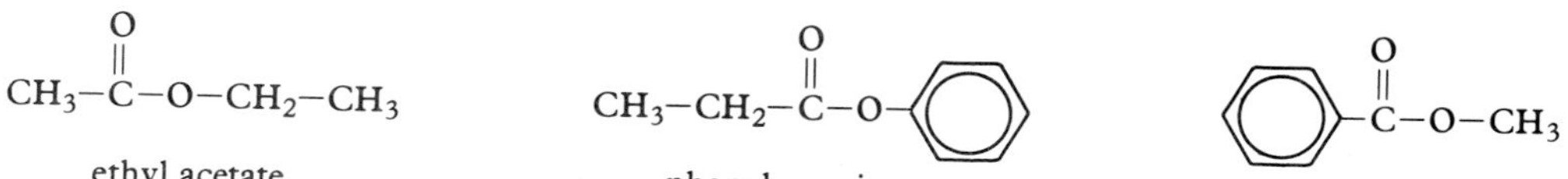

$$CH_3-\overset{O}{\underset{\|}{C}}-O-CH_2-CH_3 \qquad CH_3-CH_2-\overset{O}{\underset{\|}{C}}-O-\text{Ph} \qquad \text{Ph}-\overset{O}{\underset{\|}{C}}-O-CH_3$$

Common name: ethyl acetate   phenyl propionate   methyl benzoate
IUPAC name: ethyl ethanoate   phenyl propanoate   methyl benzoate

EXAMPLE 14.4

Give the common name and the IUPAC name for each of the following esters.

(a) $CH_3-CH_2-\overset{O}{\underset{\|}{C}}-O-CH_3$

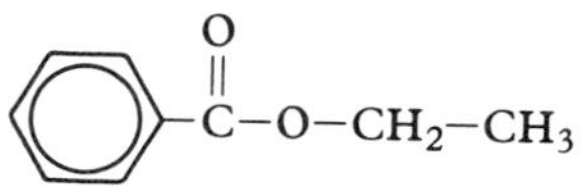

(b) 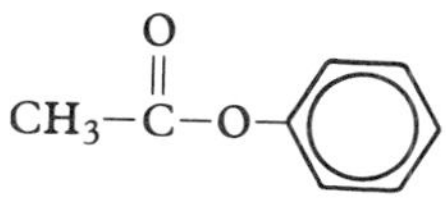 $CH_3-\overset{O}{\underset{\|}{C}}-O-\text{Ph}$

(c) $\text{Ph}-\overset{O}{\underset{\|}{C}}-O-CH_2-CH_3$

SOLUTION

(a) Common name: methyl propionate
     IUPAC name: methyl propanoate

(b) Common name: phenyl acetate
     IUPAC name: phenyl ethanoate

(c) Common name: ethyl benzoate
     IUPAC name: ethyl benzoate

EXERCISE 14.4 Give the common name and the IUPAC name for each of the following esters.

(a) 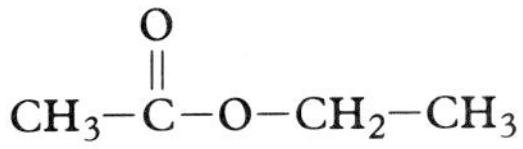 $CH_3-\overset{\overset{O}{\|}}{C}-O-CH_2-CH_3$

(b) $CH_3-CH_2-\overset{\overset{O}{\|}}{C}-O-\text{C}_6\text{H}_5$

(c) $\text{C}_6\text{H}_5-\overset{\overset{O}{\|}}{C}-O-CH_2-CH_2-CH_2-CH_3$

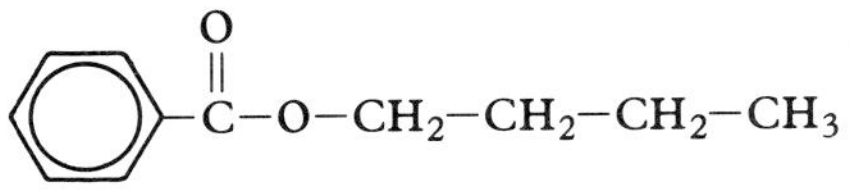

EXAMPLE 14.5

Write a structure for each of the following esters.

(a) propyl acetate
(b) butyl butyrate
(c) ethyl 4-phenylbutanoate
(d) methyl salicylate

SOLUTION

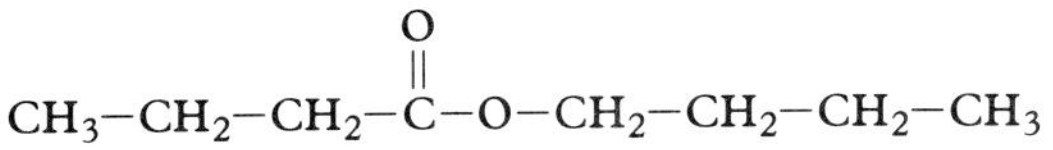

(a) 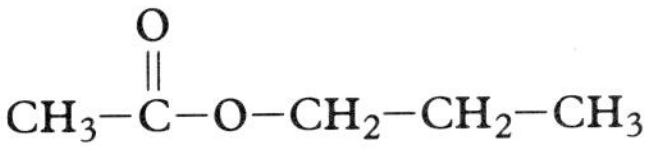 $CH_3-\overset{\overset{O}{\|}}{C}-O-CH_2-CH_2-CH_3$

(b) $CH_3-CH_2-CH_2-\overset{\overset{O}{\|}}{C}-O-CH_2-CH_2-CH_2-CH_3$

(c) $\text{C}_6\text{H}_5-CH_2-CH_2-CH_2-\overset{\overset{O}{\|}}{C}-O-CH_2-CH_3$

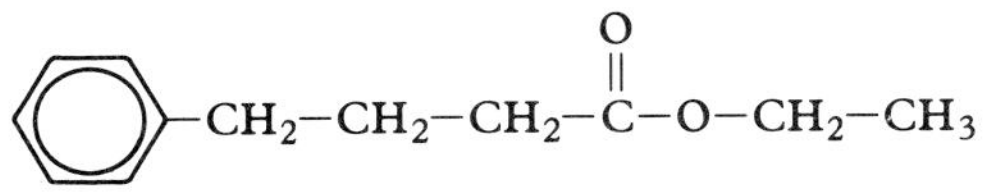

(d) 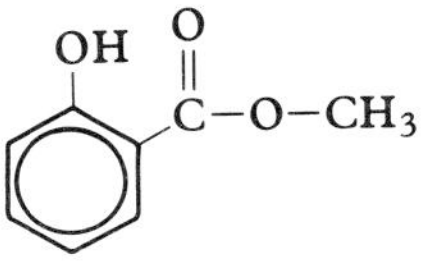 2-hydroxyphenyl-$\overset{\overset{O}{\|}}{C}-O-CH_3$ (methyl salicylate: benzene ring with OH ortho to $-C(=O)-O-CH_3$)

EXERCISE 14.5 Write a structure for each of the following esters.

(a) butyl propionate
(b) propyl acetate
(c) methyl 3-bromopentanoate
(d) ethyl benzoate

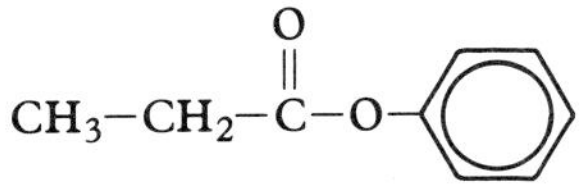

Physical Properties and Uses

Esters of low molecular weight are usually colorless liquids with pleasant odors. Because they cannot form hydrogen bonds among themselves, esters have lower boiling points than carboxylic acids and alcohols of similar molecular weight. Even though the carbonyl group in esters is polar, the polarity is weak, and the attraction for other polar molecules is poor. Thus, we can add esters to our scale of relative boiling points:

hydrocarbons ethers esters aldehydes and ketones alcohols carboxylic acids
⎯⎯⎯⎯⎯⎯⎯⎯⎯⎯⎯⎯⎯⎯⎯⎯⎯⎯⎯⎯⎯⎯⎯⎯⎯⎯⎯⎯⎯⎯⎯⎯⎯⎯⎯⎯⎯⎯⎯→
increasing boiling points

Esters can form hydrogen bonds to other molecules capable of furnishing hydrogen atoms, such as water, and thus they have some degree of water solubility. However, they are not as soluble as carboxylic acids of similar molecular weight.

Esters are responsible for the pleasant odors of fruits and flowers. Accordingly, esters are produced in large quantity by synthetic methods for use as flavoring agents, perfumes, and cosmetics. Some of these are listed in Table 14-4. Esters of low molecular weight are also used as solvents for lacquers, paints, and varnishes; those of high molecular weight are usually solids and can be used as waxes and softeners. Fats and oils are esters of long-chain carboxylic acids and 1, 2, 3-propanetriol (glycerol). The clear plastic called Lucite or Plexiglas is a polymeric ester, polymethyl methacrylate. Another polymeric ester ("polyester"), polyethylene terephthalate, is processed into the fibrous material called Dacron and Fortrel. Structures of some familiar esters are shown in Figure 14-1.

Chemical Properties

As a class of compounds, esters are rather unreactive. They are neutral substances, neither acidic nor basic, and their most important reaction is hydrolysis, a reaction that is the reverse of esterification. Thus, if an ester is heated with water in the presence of strong acid catalyst, it is

Table 14-4 Some Esters Used as Flavoring Agents or in Perfumes

Name	Structure	Odor or Taste
Methyl butyrate	$CH_3-O-\overset{\overset{O}{\|}}{C}-CH_2-CH_2-CH_3$	Apple
Ethyl butyrate	$CH_3-CH_2-O-\overset{\overset{O}{\|}}{C}-CH_2-CH_2-CH_3$	Strawberry
Butyl butyrate	$CH_3-CH_2-CH_2-CH_2-O-\overset{\overset{O}{\|}}{C}-CH_2-CH_2-CH_3$	Pineapple
Amyl acetate	$CH_3-CH_2-CH_2-CH_2-CH_2-O-\overset{\overset{O}{\|}}{C}-CH_3$	Banana
Ethyl formate	$CH_3-CH_2-O-\overset{\overset{O}{\|}}{C}-H$	Rum
Methyl salicylate	(2-hydroxyphenyl)-$\overset{\overset{O}{\|}}{C}-O-CH_3$	Wintergreen
Methyl phenylacetate	$CH_3-O-\overset{\overset{O}{\|}}{C}-CH_2-$(phenyl)	Jasmine

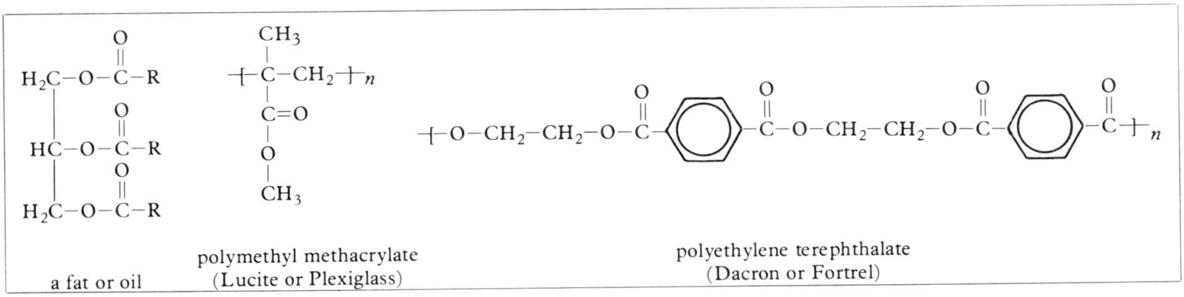

$$\begin{array}{c} O \\ \| \\ H_2C-O-C-R \\ O \\ \| \\ HC-O-C-R \\ O \\ \| \\ H_2C-O-C-R \end{array}$$

a fat or oil

$$\begin{array}{c} CH_3 \\ | \\ +C-CH_2+_n \\ | \\ C=O \\ | \\ O \\ | \\ CH_3 \end{array}$$

polymethyl methacrylate
(Lucite or Plexiglass)

$$+O-CH_2-CH_2-O-\overset{O}{\underset{\|}{C}}-\bigcirc-\overset{O}{\underset{\|}{C}}-O-CH_2-CH_2-O-\overset{O}{\underset{\|}{C}}-\bigcirc-\overset{O}{\underset{\|}{C}}+_n$$

polyethylene terephthalate
(Dacron or Fortrel)

Fig. 14-1 Familiar Esters

reversibly decomposed into the corresponding carboxylic acid and alcohol:

$$R-C\begin{matrix}O\\\diagup\\\diagdown\\OR'\end{matrix} + HOH \underset{\Delta}{\overset{H^+}{\rightleftharpoons}} R-C\begin{matrix}O\\\diagup\\\diagdown\\OH\end{matrix} + R'OH$$

ester   carboxylic acid   alcohol

Since the reaction does not go to completion, use of excessive amounts of water and/or removal of one product as it forms can improve the yield of products.

EXAMPLE 14.6

Write structures for the hydrolysis products of each of the following esters.

(a) $CH_3-\overset{O}{\underset{\|}{C}}-O-CH_2-CH_3$

(b) 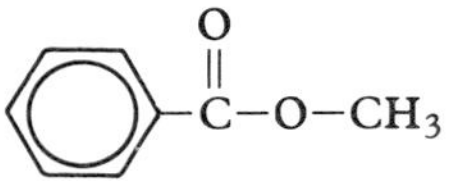 $\bigcirc-\overset{O}{\underset{\|}{C}}-O-CH_3$

(c) $CH_3-\overset{O}{\underset{\|}{C}}-O-\bigcirc$

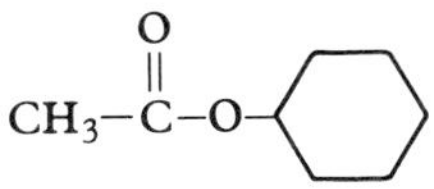

(d) $CH_3-CH_2-\overset{O}{\underset{\|}{C}}-O-CH_2-CH_2-CH_3$

SOLUTION

(a) $CH_3-CO_2H + CH_3-CH_2-OH$

(b) 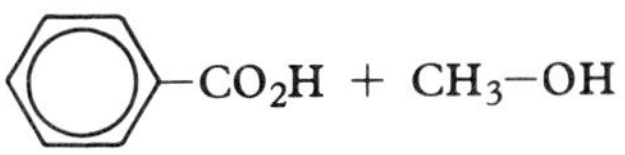 $\bigcirc-CO_2H + CH_3-OH$

(c) $CH_3-CO_2H + \bigcirc-OH$

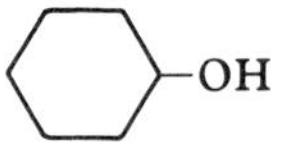

(d) $CH_3-CH_2-CO_2H + CH_3-CH_2-CH_2-OH$

EXERCISE 14.6 Write structures for the hydrolysis products of each of the following esters.

(a) 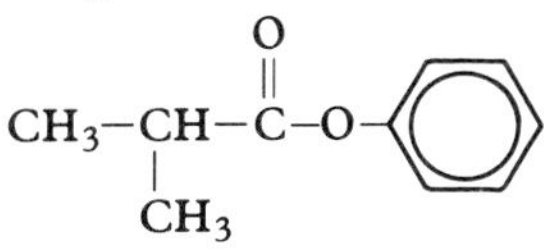 $CH_3-\underset{\underset{CH_3}{|}}{CH}-\overset{O}{\underset{\|}{C}}-O-\bigcirc$

(b) $CH_3-CH_2-CH_2-\overset{O}{\underset{\|}{C}}-O-CH_3$

(c) [structure: benzoyl-O-phenyl, i.e., phenyl benzoate]

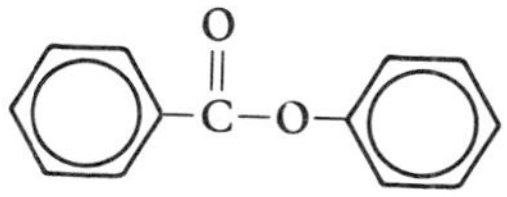

(d) $CH_3-\overset{O}{\underset{\|}{C}}-O-CH_2-$[phenyl]

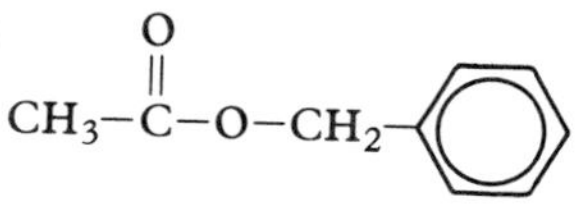

Examples of Esters

Acetylsalicylic Acid (Aspirin)

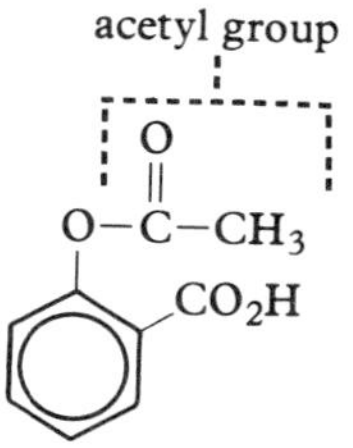

[structure showing benzene ring with O–C(=O)–CH₃ (acetyl group labeled) and CO₂H substituents]

Acetylsalicylic acid is the chemical name of the analgesic commonly called aspirin. It is also effective as an antipyretic agent and an anti-inflammatory agent (reduces inflammation and swelling), and because of these characteristics, it is the medication of choice for relieving symptoms of *rheumatoid arthritis*. Although aspirin is mildly irritating to the stomach lining, it is less so than salicylic acid, because esterification of the phenol group reduces the acidity of the nearby carboxyl group:

rheumatoid (ROOM-ah-toyd) **arthritis:** a chronic disease characterized by painful inflammation of the joints and often accompanied by pronounced deformities.

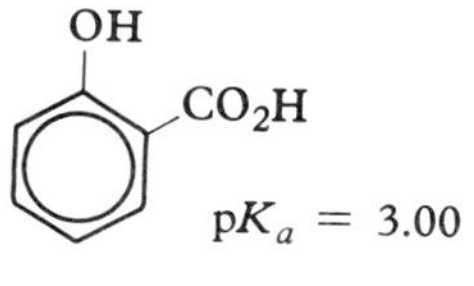

[structure: salicylic acid with OH and CO₂H] $pK_a = 3.00$
salicylic acid

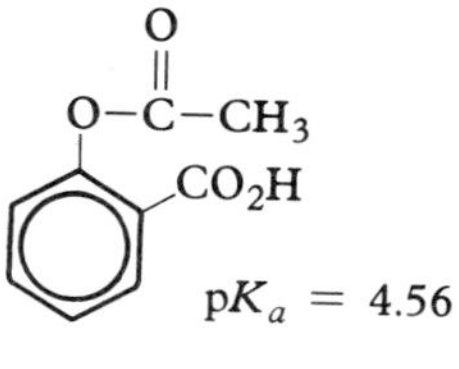

[structure: acetylsalicylic acid with O-C(=O)-CH₃ and CO₂H] $pK_a = 4.56$
acetylsalicylic acid

However, aspirin is hydrolyzed to salicylic acid, the active form of the drug, in the stomach and intestine.

Methyl Salicylate (Oil of Wintergreen)

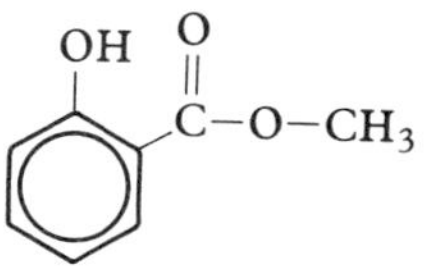

[structure: benzene ring with OH and C(=O)-O-CH₃]

Methyl salicylate is the flavor and odor of wintergreen. It is often an ingredient in various liniments, and when it is rubbed onto the skin, it slowly hydrolyzes to salicylic acid and methyl alcohol. The salicylic acid then behaves as a topical analgesic.

Phenyl Salicylate (Salol)

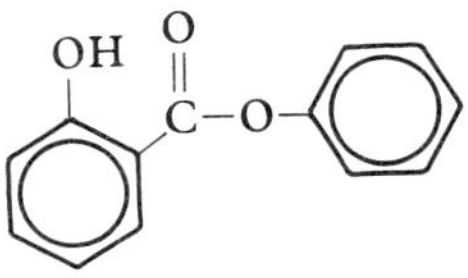

Phenyl salicylate, also known as Salol, is still another salicylic acid ester used as an analgesic. It remains intact in the stomach, but the slightly basic conditions of the intestine cause it to be hydrolyzed. Because of its stability in the stomach, it is sometimes used as a coating for pills intended to dissolve in the small intestine and not in the stomach.

Ascorbic Acid (Vitamin C)

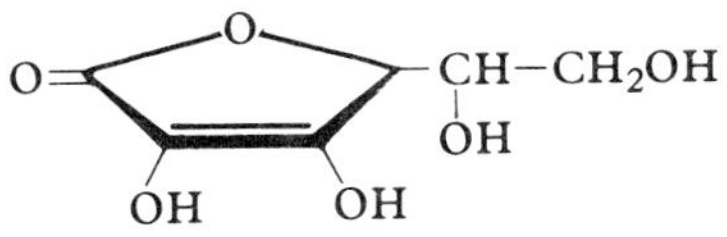

Ascorbic acid is the chemical name of vitamin C, the water-soluble vitamin found in citrus fruits and fresh vegetables. Although most animals can synthesize vitamin C, it is a necessary part of the human diet (as well as that of other primates and the guinea pig). Vitamin C prevents scurvy, a disease characterized by swollen, bleeding gums and skin discoloration caused by hemorrhaging.

The structure of ascorbic acid is interesting in that it is a cyclic ester formed when the carboxyl group in one part of the molecule reacts with the hydroxyl group three carbon atoms away. The resulting five-membered ring undergoes hydrolysis in aqueous solution to the open-chain acid form of the compound, accounting for its acid name:

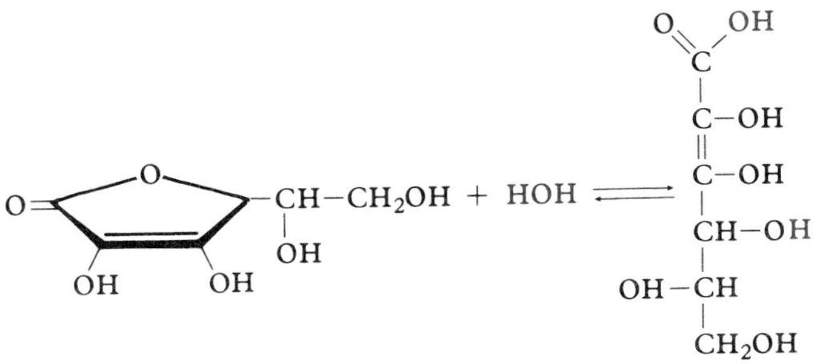

Nitroglycerin (Glyceryl Trinitrate)

Nitroglycerin was first prepared in 1846 by adding glycerol to a mixture of concentrated nitric acid and sulfuric acid:

$$\begin{array}{c}CH_2-OH \\ | \\ CH-OH \\ | \\ CH_2-OH\end{array} + 3\ H-O-N\begin{array}{c}O \\ \diagup \\ \diagdown \\ O\end{array} \xrightarrow{H^+} \begin{array}{c}CH_2-O-NO_2 \\ | \\ CH-O-NO_2 \\ | \\ CH_2-O-NO_2\end{array} + 3\ HOH$$

glycerol   nitric acid   nitroglycerin
(glycerin)

Nitroglycerin is thus an ester of an inorganic acid.

The explosive nature of nitroglycerin was recognized early on to be a result of liberation of heat and gases by decomposition of the compound:

$$4\ \begin{array}{c}CH_2-ONO_2 \\ | \\ CH-ONO_2 \\ | \\ CH_2-ONO_2\end{array} \longrightarrow 6\ N_2 + 12\ CO_2 + 10\ H_2O + O_2$$

nitroglycerin

In fact, detonation of nitroglycerin generates more than twelve hundred times the original volume at room temperature and pressure; in addition, the heat liberated raises the temperature of the gaseous mixture to about 5,000°C. The resulting effect is the instantaneous development of twenty thousand atmospheres of pressure.

Nitroglycerin is extremely sensitive to shock and to rapid heating. It is a colorless, oily liquid at room temperature, but its high freezing point (13°C, 55°F) makes it an even greater hazard, because the solid is more shock-sensitive than the liquid. A nonexplosive formulation of nitroglycerin is one of several drugs used for relief of the chest pain called *angina pectoris*. These drugs act by relaxing the coronary blood vessel, as well as others, so that coronary blood flow increases and blood pressure falls. In high doses, antianginal drugs also depress heart muscle contractions and decrease the heart rate.

angina (an-JINE-ah) **pectoris** (peck-TOH-ris): a medical condition characterized by severe recurring chest pain; caused by restricted blood flow to the heart, usually due to coronary artery disease.

Atropine

atropine (AAH-troe-peen)

Atropine is a poisonous crystalline ester obtained from an herb called belladonna. It is one of several medicinal *alkaloids* which can be isolated

alkaloids: a class of weakly basic organic compounds having rings containing nitrogen atoms; the alkaloids usually have pharmacological activity.

PERSPECTIVE: Alfred Nobel, Inventor of Dynamite

Alfred Nobel, industrialist and inventor, was born in Stockholm, Sweden, in 1833. By the age of 16, Nobel was a competent chemist and fluent in five languages. In 1860, in Sweden, he began manufacturing nitroglycerin to be used in mining, dredging, and construction. However, his factory blew up in 1864 (not an uncommon occurrence in the early days of producing explosives), causing the death of his youngest brother and four other persons. Nobel became characterized as a "mad scientist," and the Swedish government forebade him to rebuild the factory. He then started experimenting on a barge in the middle of a lake to find a safe way of handling nitroglycerin. By chance he discovered that nitroglycerin absorbed into a natural clay called diatomaceous earth (*kieselguhr*, in German) had to be purposely shocked before it would explode, and he marketed this product as dynamite. It is made today by mixing nitroglycerin with wood pulp and sodium or ammonium nitrate.

Alfred Nobel

Nobel went on to develop other explosives, making an immense fortune in the process. Much to his dismay, his explosives were immediately put to use in wars. A lifelong pacifist, Nobel suffered considerable anguish because of these applications, and he sought to make amends through the terms of his will. After Nobel's death in 1895, much of his personal fortune was designated to be used in creating five highly regarded international awards, the Nobel Prizes for Peace, Literature, Physics, Chemistry, and Physiology or Medicine. The Nobel Prize for Economics was established by the National Bank of Sweden in 1968.

---

from this plant. Atropine is used in opthamology to dilate the pupil of the eye for examination of the retina. It also gives symptomatic relief from hay fever and head colds by drying up nasal and eye secretions. The belladonna alkaloids relax certain intestinal spasms and are thus prescribed for some types of intestinal distress. They are also used in sedatives, stimulants, and antispasmodics. However, because of their toxicity and undesirable side effects (decreased production of sweat, mucus, and saliva; increased heart rate), synthetic substitutes are replacing the belladonna alkaloids in many of their uses.

Heroin (Diacetylmorphine)

narcotic: a substance which produces a state of sleep, drowsiness, or unconsciousness.

Heroin is prepared by reacting morphine with acetic anhydride (the anhydride of acetic acid, a more reactive compound than acetic acid) to form the diester (see Figure 14–2). Heroin was introduced into medicine in the latter part of the last century as a *narcotic* analgesic four to eight times more powerful than morphine, but its undesirable side effects (addiction and respiratory depression) outweigh its value as a drug. Heroin is so addictive that it seems that only one or two injections are enough to cause dependence in some people. Because of this problem, any use of heroin is illegal in the United States.

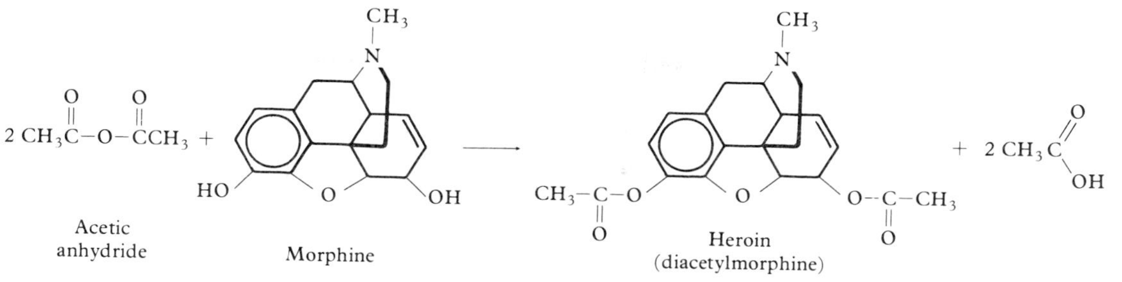

Fig. 14-2 The Preparation of Heroin from Morphine

LIPIDS

The lipids are a hetrogeneous group of organic compounds characterized by a greasy feeling, insolubility in water, and high solubility in such nonpolar solvents as chloroform, diethyl ether, and benzene. Although their structures are quite diverse, all lipid molecules have large hydrocarbon portions, and this feature accounts for their solubility in nonpolar solvents. Lipids have many biological roles, serving as (1) metabolic fuel, (2) highly concentrated energy stores, (3) structural components of membranes, (4) a protective coating, (5) insulation against cold, (6) padding and support for organs, (7) essential factors in enzyme catalysis, and (8) cellular messengers.

Classification of Lipids

There are many ways of classifying this group of compounds; for our purposes, we will separate them into two classes, as shown in Table 14-5. The *complex lipids* contain long-chain carboxylic acids attached to alcohols by ester linkage. These include waxes, triacylglycerols, phosphoglycerides, and sphingolipids. The second class is the *simple lipids*. These do not contain carboxylic acid esters. The simple lipids include terpenes, steroids, and prostaglandins.

Table 14-5 Classification of Lipids

Lipid Type	Alcohol Component
Complex	
Waxes	Nonpolar alcohols of high molecular weight
Triacylglycerols	Glycerol
Phosphoglycerides	Glycerol-3-phosphate
Sphingolipids	Sphingosine or a closely related base
Simple	
Terpenes	
Steroids	
Prostaglandins	

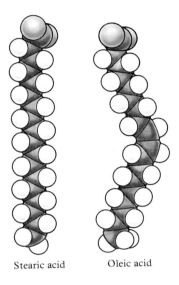

Fig. 14-3 Space-Filling Models of Stearic Acid and Oleic Acid. Both fatty acids have 18 carbon atoms, but the double bond in oleic acid prevents close interaction between oleic acid molecules, causing oleic acid to have a much lower melting point (13°C) than stearic acid (70°C).

Stearic acid   Oleic acid

Fatty Acids

Since carboxylic acids are structural components of the complex lipids, many of which are fats, carboxylic acids which can be isolated by hydrolysis of lipids are called *fatty acids*. These are usually long-chain carboxylic acids having an even number of carbon atoms. Although fatty acids as small as butyric acid are known, most contain 14–22 carbon atoms. Despite their abundance in complex lipids, only traces of fatty acids occur in free (unesterified) form in cells and tissues. Fatty acids may be saturated or unsaturated, with the unsaturated ones predominating in nature. In particular, unsaturated fatty acids are found in the complex lipids of plants and in animals living at low temperatures. The most common saturated fatty acids are palmitic acid (16 carbons) and stearic acid (18 carbons); the most common unsaturated fatty acid is oleic acid (18 carbons). As you would expect, unsaturated fatty acids have lower melting points than the corresponding saturated ones (see Figure 14-3). Common fatty acids and their names and melting points are given in Table 14-6.

Unsaturated fatty acids may have just one double bond, or they may have several. For those having only one double bond, the site of unsaturation is usually located at carbon 9. If more than one double bond exists, the locations usually are at carbon 9 and farther down the chain, each pair of double bonds separated by one $-CH_2-$ group: $=CH-$

Table 14-6 Common Fatty Acids

Structure	Common Name	Melting Point (°C)
Saturated Fatty Acids		
$CH_3(CH_2)_{10}COOH$	Lauric acid	44
$CH_3(CH_2)_{12}COOH$	Myristic acid	54
$CH_3(CH_2)_{14}COOH$	Palmitic acid	63
$CH_3(CH_2)_{16}COOH$	Stearic acid	70
$CH_3(CH_2)_{18}COOH$	Arachidic acid	76
$CH_3(CH_2)_{22}COOH$	Lignoceric acid	86
Unsaturated Fatty Acids		
$CH_3(CH_2)_5CH=CH(CH_2)_7COOH$	Palmitoleic acid	− 0.5
$CH_3(CH_2)_7CH=CH(CH_2)_7COOH$	Oleic acid	13
$CH_3(CH_2)_4CH=CHCH_2CH=CH(CH_2)_7COOH$	Linoleic acid	− 5
$CH_3CH_2CH=CHCH_2CH=CHCH_2CH=CH(CH_2)_7COOH$	Linolenic acid	−11
$CH_3(CH_2)_4(CH=CHCH_2)_3CH=CH(CH_2)_3COOH$	Arachidonic acid	−50

$CH_2-CH=$. These double bonds of nearly all naturally occurring unsaturated fatty acids are in the *cis* configuration. It has been found that mammals can synthesize saturated fatty acids and those with one double bond, but they are incapable of synthesizing certain fatty acids having more than one double bond. For this reason, polyunsaturated fatty acids are necessary in the diet of mammals. These are called essential fatty acids, and for humans, linoleic and linolenic acids have been identified as essential fatty acids, although it is not certain what role these polyunsaturated fatty acids play in the body. It is interesting to note that arachidonic acid, with its four double bonds, is synthesized by humans from linoleic acid, which has two double bonds.

Waxes

Waxes are solid esters of high molecular weight fatty acids and either long-chain alcohols (fatty alcohols) or sterols (sterols will be discussed later in this chapter). These compounds are soft and pliable when warm, but they become hard when cold. Waxes form protective coatings on skin, fur, and feathers of animals, and on the leaves and fruits of plants. Some examples and their natural sources are:

$$CH_3-(CH_2)_{24}-\overset{O}{\underset{\|}{C}}-O-(CH_2)_{27}-CH_3 \qquad CH_3-(CH_2)_{14}-\overset{O}{\underset{\|}{C}}-O-(CH_2)_{15}-CH_3$$
$$\text{beeswax} \qquad\qquad\qquad\qquad \text{whale oil}$$

Bees synthesize wax as a structural support for the honeycomb. The major components of beeswax are palmitic acid esters of fatty alcohols having 26–34 carbon atoms. Another wax, lanolin (also called wool fat) is a coating on wool; it has become a popular ingredient in ointments and creams. One of our very popular natural waxes, whale oil (actually a wax with a low melting point), is becoming scarce because of bans on killing whales. Since this substance is used in cosmetics, candles, transmission oil, and ointments, it will be necessary to find a suitable replacement for it. One promising possibility is the oil from seeds of the jojoba plant, an evergreen shrub which grows in the southwestern United States.

Triacylglycerols

This group of lipids consists of fatty acid esters of glycerol (1,2,3-propanetriol). These were once called neutral fats, but since the group actually includes both fats and oils, we now use the name *triacylglycerols*.

triacylglycerols (try-ace-sil-GLIH-sur-ols): fatty acid esters of glycerol containing three ester groups per molecule.

$$\begin{array}{c} \text{O} \\ \| \\ H_2C-O-C-R \\ | \quad\; \text{O} \\ \quad\; \| \\ HC-O-C-R \\ | \quad\; \text{O} \\ \quad\; \| \\ H_2C-O-C-R \end{array}$$
a triacylglycerol

Those which are solids at room temperature are commonly known as fats, while liquid triacylglycerols are oils. However, these are food oils, and since they are esters, they are chemically different from petroleum oil. Another name which was used for this group of lipids was "triglycerides," but this term is ambiguous, and it is rapidly falling out of use.

It is not necessary that the three fatty acid components in a triacylglycerol be identical, but when they are, the compound is called a simple triacylglycerol, and it is named after the fatty acid component. Examples are given below with their proper names; common names appear in parentheses.

$$\begin{array}{ccc}
\text{H}_2\text{C}-\text{O}-\overset{\text{O}}{\overset{\|}{\text{C}}}-\text{C}_{17}\text{H}_{35} & \text{H}_2\text{C}-\text{O}-\overset{\text{O}}{\overset{\|}{\text{C}}}-\text{C}_{15}\text{H}_{31} & \text{H}_2\text{C}-\text{O}-\overset{\text{O}}{\overset{\|}{\text{C}}}-\text{C}_{17}\text{H}_{33} \\
\text{HC}-\text{O}-\overset{\text{O}}{\overset{\|}{\text{C}}}-\text{C}_{17}\text{H}_{35} & \text{HC}-\text{O}-\overset{\text{O}}{\overset{\|}{\text{C}}}-\text{C}_{15}\text{H}_{31} & \text{HC}-\text{O}-\overset{\text{O}}{\overset{\|}{\text{C}}}-\text{C}_{17}\text{H}_{33} \\
\text{H}_2\text{C}-\text{O}-\overset{\|}{\text{C}}-\text{C}_{17}\text{H}_{35} & \text{H}_2\text{C}-\text{O}-\overset{\|}{\text{C}}-\text{C}_{15}\text{H}_{31} & \text{H}_2\text{C}-\text{O}-\overset{\|}{\text{C}}-\text{C}_{17}\text{H}_{33} \\
\text{tristearoylglycerol} & \text{tripalmitoylglycerol} & \text{trioleoylglycerol} \\
\text{(tristearin)} & \text{(tripalmitin)} & \text{(triolein)}
\end{array}$$

Triacylglycerols found in nature most frequently contain two or more different fatty acid components. These are called mixed triacylglycerols. Since they can have quite complicated names, we will not be concerned with deriving names for mixed triacylglycerols.

The main function of triacylglycerols is to provide a fuel supply for organisms. Deposits of triacylglycerols, called adipose (fat) tissue, are composed of individual fat cells whose primary purpose is to synthesize fat when there is low energy demand by the organism and to degrade the fat when energy is needed. This degradation is accomplished by enzymes called *lipases*, which catalyze the hydrolysis of triacylglycerols to glycerol and fatty acids:

$$\begin{array}{c}
\text{H}_2\text{C}-\text{O}-\overset{\text{O}}{\overset{\|}{\text{C}}}-\text{R} \\
\text{HC}-\text{O}-\overset{\text{O}}{\overset{\|}{\text{C}}}-\text{R} + 3\,\text{HOH} \xrightarrow{\text{lipase}} \begin{array}{c}\text{H}_2\text{C}-\text{OH} \\ \text{HC}-\text{OH} \\ \text{H}_2\text{C}-\text{OH}\end{array} + 3\,\text{R}-\text{C}\overset{\displaystyle\text{O}}{\underset{\displaystyle\text{OH}}{}} \\
\text{H}_2\text{C}-\text{O}-\overset{\|}{\text{C}}-\text{R} \\
\text{a triacylglycerol} \qquad\qquad \text{glycerol} \qquad \text{fatty acids}
\end{array}$$

The fatty acids are then oxidized through a series of metabolic reactions to carbon dioxide, water, and energy.

Triacylglycerols also serve other bodily needs. In particular, deposits of fat under the skin insulate animals against cold. This function is especially important to animals commonly found in cold waters or cold climates, such as whales, walruses, and bears. And triacylglycerols are deposited as fat layers around internal organs in animals to provide structural support or padding for them.

The melting points of triacylglycerols are determined by their fatty acid components. As you might expect, the melting point increases with the size of the fatty acid component. As Table 14-7 indicates, animals are usually the sources of saturated triacylglycerols, while plants provide unsaturated triacylglycerols.

Chemical Properties of Triacylglycerols

The chemical properties of triacylglycerols are typical of esters and alkenes, since these are the two functional groups present.

Hydrolysis

As illustrated in the preceding section, triacylglycerols undergo hydrolysis catalyzed by enzymes. Strong acids will also catalyze hydrolysis of triacylglycerols.

Table 14-7 Major Fats and Oils (Listed in order of domestic production)

Source of Fat or Oil	Iodine Number	Major Fatty Acids
Soybean	120-141	Linoleic acid, Oleic acid
Cottonseed	97-112	Linoleic acid, Palmitic acid
Lard (hogs)	58-68	Oleic acid, Palmitic acid
Butterfat	25-42	Oleic acid, Palmitic acid
Tallow (beef)	35-48	Oleic acid, Palmitic acid
Tallow (lamb)	48-61	Oleic acid, Palmitic acid
Corn	103-128	Linoleic acid, Oleic acid
Peanut	84-100	Oleic acid, Linoleic acid
Safflower	140-150	Linoleic acid
Coconut	8-10	Lauric acid, Myristic acid
Palm	44-58	Palmitic acid, Oleic acid
Olive	80-88	Oleic acid
Chicken fat	64-76	Oleic acid, Palmitic acid
Linseed (flaxseed)	175-202	Linolenic acid

saponification (sah-pon-ih-fih-CAY-shun): the process of soap making.

soap: the sodium or potassium salt of a fatty acid.

micelles (MY-cells): spherical clusters of organic ions held together by hydrophobic forces.

Fig. 14-4 Formation of a Soap Micelle

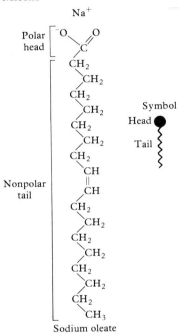

Sodium oleate

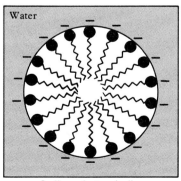

Sodium oleate micelle

Saponification

Reaction with NaOH or KOH occurs readily, and this reaction, called *saponification*, is the traditional method for making *soap*. When a triacylglycerol is heated with concentrated NaOH or KOH, glycerol and salts of the fatty acids are formed in an irreversible reaction:

$$\begin{array}{c}H_2C-O-\overset{O}{\underset{\|}{C}}-R \\ HC-O-\overset{O}{\underset{\|}{C}}-R' \\ H_2C-O-\overset{O}{\underset{\|}{C}}-R''\end{array} + 3\,NaOH(aq) \xrightarrow{\Delta} \begin{array}{c}H_2COH \\ HCOH \\ H_2COH\end{array} + \begin{array}{c}RCO_2^-Na^+ \\ + \\ R'CO_2^-Na^+ \\ + \\ R''CO_2^-Na^+\end{array}$$

triacylglycerol (or 3 KOH) soaps

In traditional soapmaking, animal fat is the source of triacylglycerol, and lye (crude NaOH) or an aqueous extract of wood ashes (which contain minerals that dissolve to produce OH^-) is the source of base.

The sodium and potassium salts of fatty acids are soluble to a certain extent in water, but the long hydrophobic portion of the fatty acid anion limits true solubility somewhat, and at this limit the ions begin to form spherical clusters of 50–100 units. These clusters, called *micelles*, are dispersed as a colloid in water. Micelles are structured so that the hydrophobic "tails" of the fatty acid anions are grouped together at the interior, and Na^+ or K^+ ions are located around the exterior along with water molecules which hydrate the hydrophilic "heads," the charged carboxylate groups (see Figure 14-4). Micelle formation accounts for the cleaning action of soap. Since most dirt and grime on clothing and skin is combined with oils from one source or another, this material is largely hydrophobic, and it dissolves in the hydrophobic interior of the micelle. Thus, the micelles suspend dirt and oil particles, allowing them to be rinsed away with water (Figure 14-5).

Addition

The presence of one or more double bonds in fatty acid components of triacylglycerols confers upon these compounds chemical properties typical of alkenes. In particular, the double bonds are subject to addition reactions such as hydrogenation and halogenation.

Hydrogenation is accomplished commercially by high temperature, high pressures of hydrogen, and use of a catalyst:

$$-CH=CH- + H_2 \xrightarrow[\Delta]{Pt} -CH_2-CH_2-$$

an unsaturated triacylglycerol   a saturated triacylglycerol

This process creates a more highly saturated fat from an unsaturated one, and thus the product triacylglycerol has a higher melting point than the starting triacylglycerol. Hydrogenation is used to convert vegetable oils to margarine and shortening.

Halogenation is often used to determine the extent of unsaturation of a natural fat or oil. Iodine was traditionally used for this purpose; although the reasons for its use are not clear, it was probably because it is less hazardous to handle than chlorine and bromine. However, iodine is less reactive than the other halogens, and its sluggish behavior has led to the use of the curious compounds iodine chloride or iodine bromide, which react more readily than iodine:

$$-CH=CH- \;+\; ICl \quad (or\; IBr) \longrightarrow -\underset{I}{\underset{|}{CH}}-\underset{Cl}{\underset{|}{CH}}-$$

an unsaturated triacylglycerol — iodine chloride — iodine bromide — a halogenated triacylglycerol

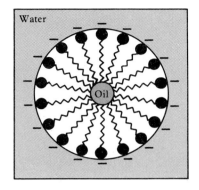

Fig. 14-5 Oil Droplets and Dirt Particles Dissolve in the Interiors of Micelles

The reaction is carried out with a known amount of halogen reactant, and then the amount of halogen left unreacted is measured. The difference in the two values, calculated as if the halogen reactant were I_2, is used to calculate the *iodine number*—the number of grams of iodine that would react with 100 grams of the lipid. A relatively high iodine number indicates a greater degree of unsaturation than a lower iodine number. Thus, so-called polyunsaturated oils have higher iodine numbers than animal fats and oils, which are usually more saturated. Table 14-7 provides iodine numbers for some fats and oils.

Phosphoglycerides

Physphoglycerides, another example of abundant naturally occurring lipids, are characterized by the presence of phosphorus and glycerol in their structures. Phosphorus is present as phosphate diesters, which are similar in structure to phosphoric acid and phosphate monoesters:

$$\underset{\text{phosphoric acid}}{HO-\underset{\underset{OH}{|}}{\overset{\overset{O}{\|}}{P}}-OH} \qquad \underset{\text{a phosphate monoester}}{RO-\underset{\underset{OH}{|}}{\overset{\overset{O}{\|}}{P}}-OH} \qquad \underset{\text{a phosphate diester}}{RO-\underset{\underset{OH}{|}}{\overset{\overset{O}{\|}}{P}}-OR'}$$

Phosphoglycerides are major components of cell membranes. In these compounds, one of the primary hydroxyl groups of glycerol is esterified (joined by an ester linkage) to phosphoric acid, which is in turn esterified to an alcohol. The other hydroxyl groups of glycerol are esterified to fatty acids, resulting in the general form:

$$\begin{array}{c}
\overset{1}{H_2C}-O-\overset{O}{\overset{\|}{C}}-R \\
\overset{2}{HC}-O-\overset{O}{\overset{\|}{C}}-R' \\
\overset{3}{H_2C}-O-\overset{O}{\overset{\|}{P}}-OR'' \\
OH
\end{array}$$

a phosphoglyceride

The most abundant phosphoglycerides in plants and animals are formed when one of the alcohols choline, ethanolamine, or serine,

$$HO-CH_2-CH_2-\overset{+}{N}(CH_3)_3 \qquad HO-CH_2-CH_2-\overset{+}{N}H_3 \qquad \begin{array}{c} HO-CH_2-\overset{+}{C}H-\overset{+}{N}H_3 \\ | \\ CO_2^- \end{array}$$

choline   ethanolamine   serine

is attached to the phosphate group. The resulting phosphogycerides are phosphatidyl choline, phosphatidyl ethanolamine, and phosphatidyl serine:

$$\begin{array}{c}
H_2C-O-\overset{O}{\overset{\|}{C}}-R \\
HC-O-\overset{O}{\overset{\|}{C}}-R' \\
H_2C-O-\overset{O}{\overset{\|}{P}}-O-CH_2-CH_2-\overset{+}{N}(CH_3)_3 \\
OH
\end{array} \qquad
\begin{array}{c}
H_2C-O-\overset{O}{\overset{\|}{C}}-R \\
HC-O-\overset{O}{\overset{\|}{C}}-R' \\
H_2C-O-\overset{O}{\overset{\|}{P}}-O-CH_2-CH_2-\overset{+}{N}H_3 \\
OH
\end{array} \qquad
\begin{array}{c}
H_2C-O-\overset{O}{\overset{\|}{C}}-R \\
HC-O-\overset{O}{\overset{\|}{C}}-R' \\
H_2C-O-\overset{O}{\overset{\|}{P}}-O-CH_2-CH_2-\overset{+}{N}H_3 \\
OH \quad\quad\quad\quad\quad CO_2^-
\end{array}$$

phosphatidyl choline (lecithin)   phosphatidyl ethanolamine   phosphatidyl serine

You may recognize phosphatidyl choline by its common name lecithin, the substance used to give a smooth texture to margarine and chocolate candies.

The phosphate diester group has one —OH group, and it behaves as an acid:

$$RO-\overset{O}{\overset{\|}{P}}-OR \rightleftharpoons RO-\overset{O}{\overset{\|}{P}}-OR + H^+$$
$$\quad\; OH \quad\quad\quad\quad\quad\quad O^-$$

Its pK_a is in the range of 1–2. This means that the acid dissociation constant (10^{-1}–10^{-2}) corresponds to that of a weak acid. However, at physiological pH (approximately 7), the group will be completely ionized. Thus, at neutral pH, each phosphoglyceride is composed of a charged "head" group and a hydrophobic "tail" made up of two long hydrocarbon chains, as shown in Figure 14-6. These structural features cause phosphoglycerides to form micelles in water and thus serve as good *emulsifying agents*. Lecithin is commercially extracted from soybeans for use as an emulsifying agent in food and candies.

emulsifying agent: a substance which promotes suspensions (emulsions) of one liquid in another.

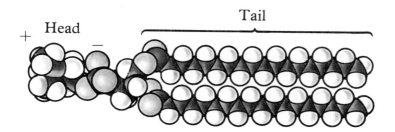

Fig. 14-6 A Space-Filling Model of Phosphatidyl Choline

Sphingolipids

Sphingolipids are complex lipids present in largest amount in the membranes of brain and nerve cells. They contain as their backbone *sphingosine* or a related base.

sphingolipids (SFIN-goe-LIH-pids)

$$CH_3-(CH_2)_{12}-CH=CH-\underset{\underset{\underset{CH_2OH}{|}}{\underset{H-C-NH_2}{|}}}{\overset{H}{\underset{|}{C}}-OH}$$

sphingosine

Structures of the three major types of sphingolipids are:

$$CH_3-(CH_2)_{12}-CH=CH-\underset{\underset{\underset{CH_2-O-\overset{O}{\underset{|}{\overset{||}{P}}}-O-CH_2-CH_2-\overset{+}{N}H_3}{|}}{\underset{|}{H-C-NH-\overset{O}{\overset{||}{C}}-(CH_2)_7-CH=CH-(CH_2)_7-CH_3}}}{\overset{H}{\underset{|}{C}}-OH}$$
$$\underset{O^-}{}$$

a sphingomyelin

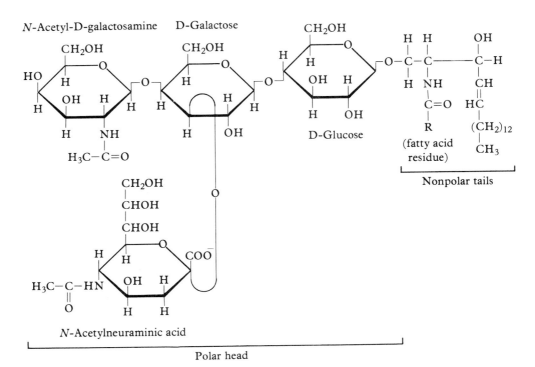

a cerebroside

a ganglioside

Terpenes

Terpenes are nonpolar, hydrocarbonlike molecules synthesized in plants and animals. They are constructed of units of 2-methyl-1,3-butadiene, the hydrocarbon commonly known as isoprene:

$$H_2C=C-CH=CH_2$$
$$\;\;\;\;\;\;\;\;\;|$$
$$\;\;\;\;\;\;\;\;CH_3$$

2-methyl-1,3-butadiene
(isoprene)

Many terpenes have characteristic odors or flavors and are major components of plant oils. For example, geraniol, limonene, menthol, pinene, camphor, and carvone are major components of geranium oil, lemon oil, mint oil, turpentine, camphor oil, and caraway oil, respectively.

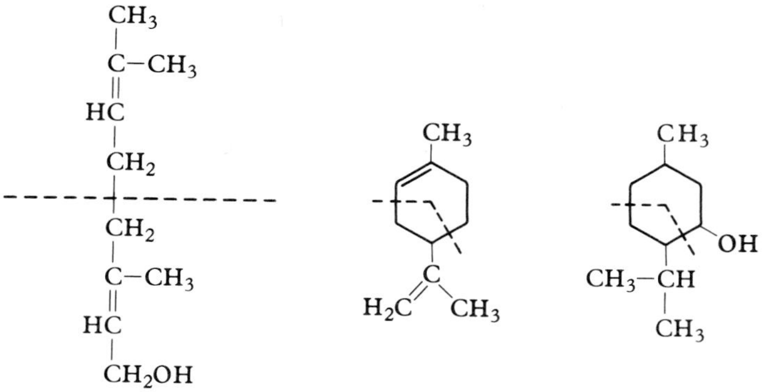

geraniol   limonene   menthol
(dashed lines delineate isoprene units)

Another example of a terpene is squalene (Figure 14-7), an intermediate in the synthesis of cholesterol by animals. The carotenoids are also terpenes; these compounds are responsible for the color in yellow vegetables. The carotenoid in carrots, β-carotene, also shown in Figure 14-7, is converted into vitamin A by enzymatic reactions in the intestinal wall and the liver.

Fig. 14-7 The Structures of Squalene and β-Carotene

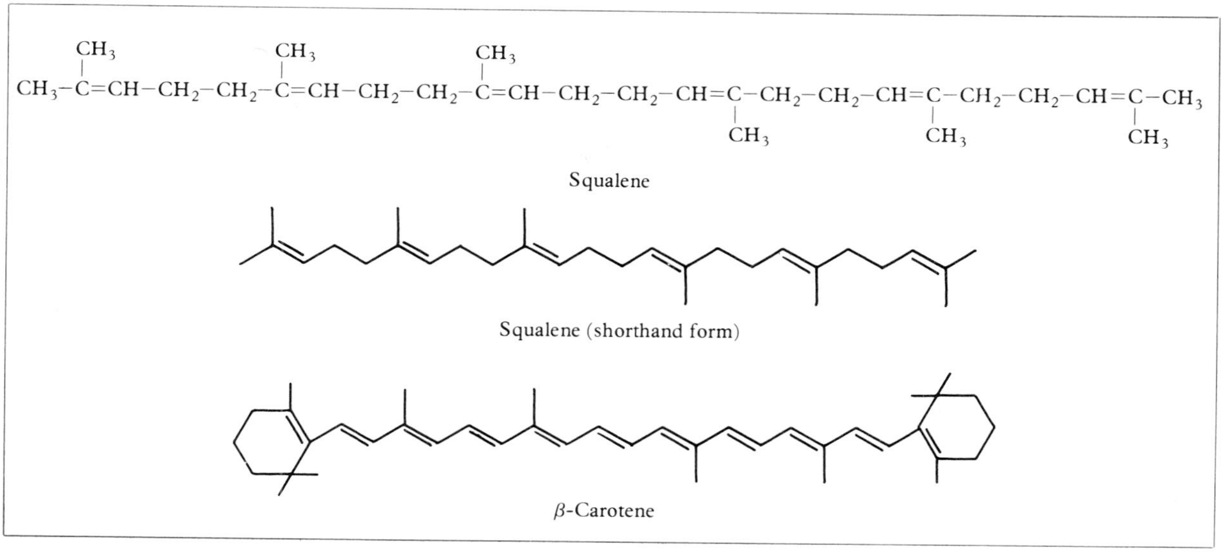

Squalene

Squalene (shorthand form)

β-Carotene

Natural rubber is a polyterpene consisting of long hydrocarbon chains containing hundreds of isoprene units (see Figure 14-8).

Still another important group of terpenes are the three fat-soluble vitamins A, E, and K, whose structures are shown in Figure 14-9. These are designated as fat soluble because they are insoluble in water

Fig. 14-8 The Structure of Natural Rubber

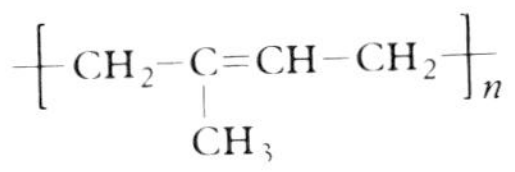

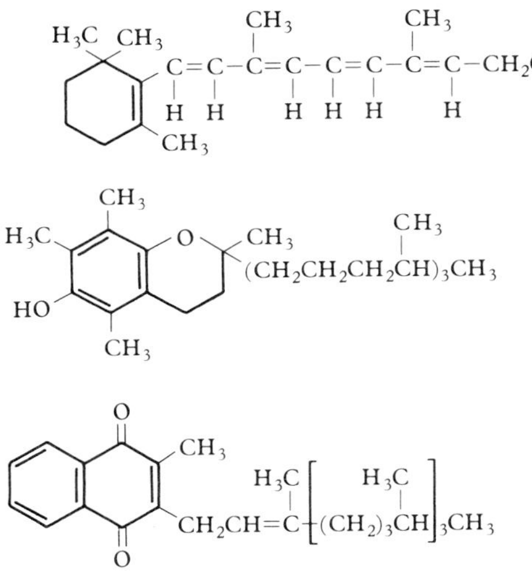

$$\left[CH_2-C=CH-CH_2 \right]_n$$
$$\quad\quad\;\; | $$
$$\quad\quad\; CH_3$$

Fig. 14-9 The Structures of Vitamins A, E, and K

but soluble in nonpolar substances such as fats. Although these characteristics allow them to be classified as lipids, the biological functions of these compounds are so distinctive that they will be considered separately in chapter 23.

Steroids

Steroids have as their basic structure the set of four fused rings shown below.

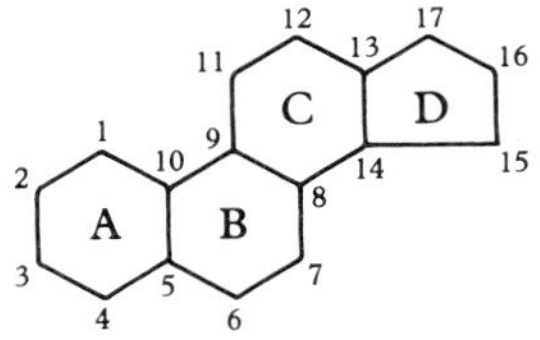

steroid nucleus

Many steroids are known; they usually differ from each other in the number and position of double bonds and in the type, location, and number of substituent functional groups. All steroids originate from the terpene squalene. Figure 14-10 shows the structure of a number of steroids.

The most abundant steroid in animal tissues is cholesterol,

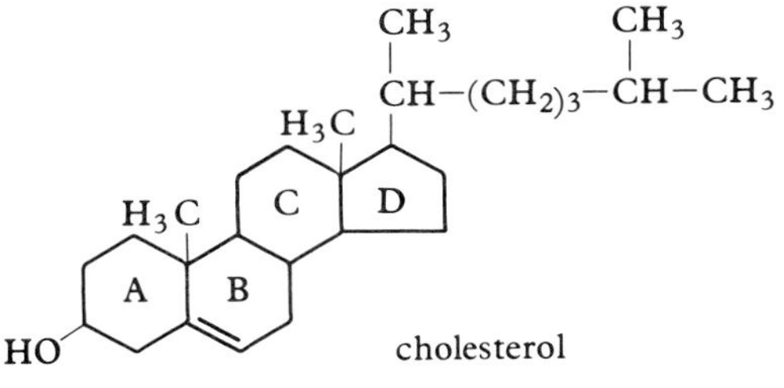

cholesterol

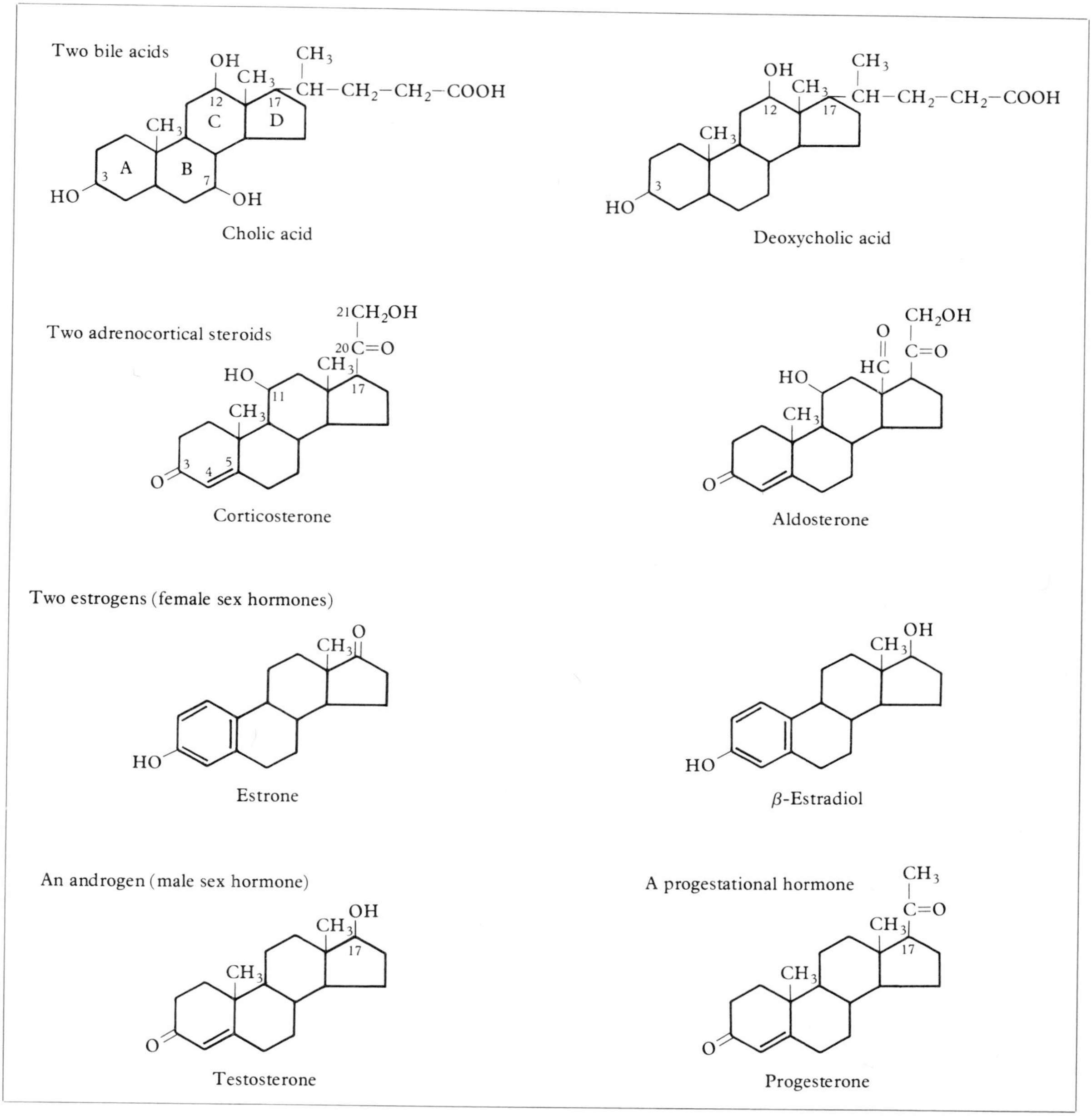

Fig. 14-10 Some Important Steroids

a member of the subgroup of steroids called *sterols* because they have a hydroxyl ("-ol") group at carbon 3 in ring A. Cholesterol appears to be an essential component of cell membranes. It is also an important metabolic intermediate from which many other steroids are synthesized. These include the bile acids, compounds which aid in lipid digestion; the androgens, or male sex hormones; the estrogens, or female sex hormones; the progestational hormone progesterone; and the adrenocortical hormones, secreted by the adrenal glands.

prostaglandins (prah-stah-GLAN-dins): cyclic derivatives of 20-carbon fatty acids found in most tissues and containing a carboxyl group and a cyclopentane ring.

Prostaglandins

This group of compounds was given its name in the 1930s by a Swedish physiologist because of their presence in the prostate gland. It was first thought that prostaglandin was a simple substance secreted by the male genital tract, but more recent research has identified as many as 20 prostaglandins in a variety of tissues.

Because of their low concentrations, prostaglandins were not isolated and purified until recently. All of the natural prostaglandins are cyclic derivatives of the 20-carbon fatty acid prostanoic acid. Several prostaglandins are shown in Figure 14-11; their names are codes for ring substituents and the number of side-chain double bonds.

tissue hormone: a hormone produced by one cell type in a tissue which acts on another cell type in the same tissue.

The prostaglandins are classified as *tissue hormones*, and although they differ from one another in biological activity, all show some capability to induce smooth muscle contraction and to lower blood pressure. From a medical viewpoint, the prostaglandins represent enormous therapeutic possibilities. Some potential uses of them include treatment of blue babies to increase oxygen content in blood, prevention of heart attacks, improvement of poor circulation to the extremities as in Raynaud's disease, protection of the stomach lining against irritation by drugs and perhaps prevention of peptic ulcers, and relieving high blood pressure. The prostaglandin story is still developing, and more uses are being found daily.

Fig. 14-11 Some Prostaglandins and the Parent Compound, Prostanoic Acid

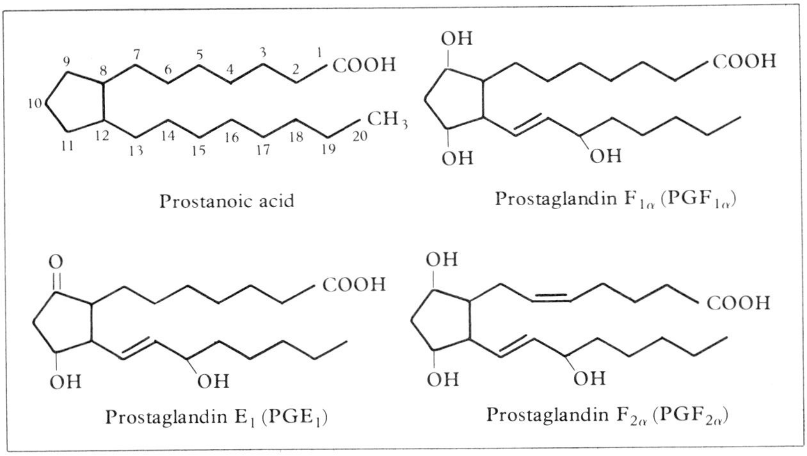

SUMMARY

Carboxylic Acids: Carboxylic acids have formulas written as RCOOH (or RCO_2H). Bond angles about the carboxyl carbon are 120°, and the three atoms attached to the carboxyl carbon are in the same plane with it. The presence of hydrogen-bonded dimers causes carboxylic acids to have higher boiling points than alcohols. They are also more soluble in water than corresponding alcohols, but solubility decreases as the size of the hydrocarbon group increases. The most important chemical property of carboxylic acids is their acidity. They react with bases to form salts and with alcohols to form esters. Esterification requires a strong acid or enzyme as catalyst.

Esters: Esters, carboxylic acid derivatives formed by reaction with alcohols, have the general formula RCO_2R'. Esters of low molecular weight are colorless liquids with pleasant odors. Their boiling points are lower than those of corresponding carboxylic acids and alcohols because they cannot form hydrogen bonds among themselves. However, they can form hydrogen bonds with water, and thus they have some water solubility. Although esters are rather unreactive, they are susceptible to hydrolysis. If a strong acid catalyst is used, the reversible reaction produces carboxylic acid and alcohol.

Lipids: The lipids are characterized by a greasy feeling, insolubility in water, and solubility in nonpolar solvents. Lipids may be classified as complex or simple. Complex lipids contain long-chain carboxylic acids (fatty acids) esterified to alcohols; this class includes waxes, triacylglycerols, phosphoglycerides, and sphingolipids. The simple lipids do not contain fatty acids, and they are not esters. They include terpenes, steroids, and prostaglandins.

Carboxylic acids which can be isolated by hydrolysis of lipids are called fatty acids. Most fatty acids contain an even number of carbon atoms numbering in the range of 14–22. Fatty acids may be saturated or unsaturated, but the unsaturated ones predominate in nature. Unsaturated fatty acids usually have one double bond at carbon 9, with any additional double bonds located farther down the chain. Nearly all of the double bonds are in the *cis* configuration. Essential fatty acids for humans are linoleic and linolenic acids. Waxes are solid esters of high molecular weight fatty acids and either long-chain alcohols or sterols. They form protective coatings on skin, fur, and feathers of animals and on the leaves and fruits of plants. Triacylglycerols are fatty acid esters of glycerol. Solid triacylglycerols are called fats while liquid ones are oils. These lipids provide a fuel supply for cells, insulate animals against cold, and furnish structural support or padding for internal organs. Triacylglycerols undergo hydrolysis catalyzed by strong acid or enzymes (lipases). They can also be saponified to yield glycerol and soaps. Unsaturated triacylglycerols are subject to addition reactions such as hydrogenation and halogenation. The iodine number, determined by reaction with iodine chloride or iodine bromide, indicates the degree of unsaturation of a fat or oil. Phosphoglycerides are formed when the phosphate group of glycerol phosphate is esterified both to glycerol and to an alcohol. The sphingolipids are complex lipids which contain as their backbone sphingosine or a related base.

The terpenes are nonpolar, hydrocarbonlike molecules constructed of units of isoprene. The stereoids have a structural skeleton of four saturated hydrocarbon rings fused together. All steroids are synthesized in their natural sources from squalene. The prostaglandins are a group of fatty acid derivatives classified as tissue hormones. Although they have varied biological activities, all show some capability to induce smooth muscle contraction and to lower blood pressure.

STUDY QUESTIONS AND PROBLEMS

1. Define the following terms:
   - (a) carboxylic acid
   - (b) carboxylate ion
   - (c) ester
   - (d) acyl group
   - (e) alkoxy group
   - (f) complex ion
   - (g) antipyretic agent
   - (h) alkaloid
   - (i) narcotic
   - (j) lipids
   - (k) fatty acid
   - (l) essential fatty acid
   - (m) waxes
   - (n) triacylglycerol
   - (o) lipase
   - (p) saponification
   - (q) soap
   - (r) micelle
   - (s) iodine number
   - (t) phosphoglyceride
   - (u) emulsifying agent
   - (v) sphingolipids
   - (w) terpenes
   - (x) steroid
   - (y) sterol
   - (z) prostaglandin
   - (aa) tissue hormone

2. Give either a common name or the IUPAC name for each of the following carboxylic acids.
   - (a) $CH_3-CH_2-CH_2-CO_2H$
   - (b) $CH_3-\underset{\underset{CH_3}{|}}{\overset{\overset{CH_3}{|}}{C}}-CO_2H$
   - (c) $Br-\text{C}_6\text{H}_4-CO_2H$
   - (d) $\text{C}_6\text{H}_5-CH_2-CO_2H$
   - (e) $Cl-CH_2-CO_2H$
   - (f) $CH_3-\underset{\underset{OH}{|}}{CH}-CO_2H$

3. Draw the structure for each of the following carboxylic acids.
   - (a) 2-methylbutanoic acid
   - (b) *p*-aminobenzoic acid
   - (c) β-chlorobutyric acid
   - (d) octanoic acid
   - (e) citric acid

4. Explain why carboxylic acids have higher boiling points than alcohols of similar molecular weight.

5. Explain why carboxylic acids are more soluble in water than the corresponding alcohols.

418   CH. 14 CARBOXYLIC ACIDS, ESTERS, AND LIPIDS

6. What are the two factors that cause carboxylic acids to be acidic?

7. Why are carboxylic acids weak acids instead of strong acids?

8. Give either a common name or the IUPAC name for each of the following carboxylic acid salts.
   (a) $CH_3-CO_2^- Na^+$
   (b) $CH_3-CH_2-CO_2^- K^+$
   (c) $CH_3-CH-CO_2^- Na^+$
       $\quad\quad\;\;|$
       $\quad\quad\;OH$
   (d) $H_2C-CO_2^- Na^+$
       $\;\;|$
       $HO-C-CO_2^- Na^+$
       $\;\;|$
       $H_2C-CO_2^- Na^+$

9. Complete each of the following equations.
   (a) $CH_3-CH_2-CO_2H(aq) + OH^-(aq) \longrightarrow$
   (b) $CH_3-CO_2H + CH_3-CH_2-OH \underset{\Delta}{\overset{H^+}{\rightleftarrows}}$
   (c) ⌬$-CO_2^- Na^+(aq) + H^+(aq) \longrightarrow$
   (d) $CH_3-CO_2H(aq) + Na_2CO_3(aq) \longrightarrow$
   (e) ⌬$\begin{array}{l}-CO_2H\\-CO_2H\end{array}$ (aq) + NaOH(aq) $\longrightarrow$

10. By a process called extraction, carboxylic acids may be removed from nonpolar solvents. In this process, the solution of carboxylic acid in nonpolar solvent is shaken vigorously with dilute aqueous base. During the shaking, the carboxylic acid passes from the nonpolar solvent into the aqueous base. Explain why the carboxylic acid moves from one solution to the other. (Clue: consider the possibility of salt formation.)

11. An aqueous solution contains acetic acid and sodium acetate, each at a concentration of $1M$. The pH of the solution remains constant at a value of about 4, despite addition of small amounts of strong acids and bases. Explain the basis for this observation, and use chemical equations to illustrate your explanation. (Clue: think about the composition of buffer solutions, chapter 10.)

12. Explain how soluble salts of citric acid function as anticoagulants.

13. Why is salicylic acid no longer widely used as an antipyretic and analgesic?

14. Write equations to illustrate ester formation from
    (a) ⌬$-CO_2H$ and CH_3OH
    (b) CH_3CO_2H and CH_3-CH_2-OH
    (c) CH_3-CO_2H and ⌬$-OH$
    (d) H_2N-⌬$-CO_2H$ and $CH_3-CH-CH_2-OH$
       $\quad\quad\quad\quad\quad\quad\quad\quad\quad\quad\quad\quad\;|$
       $\quad\quad\quad\quad\quad\quad\quad\quad\quad\quad\quad\;CH_3$
    (e) $CH_3-CH=CH-CO_2H$ and ⌬$-OH$

15. Give either a common name or the IUPAC name for each of the following esters.
    (a) ⌬$-\overset{O}{\overset{\|}{C}}-O-CH_2-CH_3$
    (b) $CH_3-CH_2-O-\overset{O}{\overset{\|}{C}}-CH_3$
    (c) ⌬$\begin{array}{l}OH\\\;\;\;\;O\\\;\;\;\;\|\\-C-O-CH_3\end{array}$
    (d) ⌬$-\overset{O}{\overset{\|}{C}}-O-$⌬
    (e) ⌬$\begin{array}{l}O-\overset{O}{\overset{\|}{C}}-CH_3\\-CO_2H\end{array}$

16. Draw the structure of each of the following esters.
    (a) heptyl acetate
    (b) propyl benzoate
    (c) ethyl acetate
    (d) phenyl propionate
    (e) methyl lactate

17. Arrange the following compounds in order of decreasing boiling point.
    CH_4
    ⌬$-CO_2H$
    CH_3-CH_2-OH
    $CH_3-CH_2-CH_2-CH_2-CHO$
    $CH_3-O-CH_2-CH_3$
    ⌬
    $CH_3-CO_2CH_3$
    $CH_3-(CH_2)_8-OH$

18. Arrange the following compounds in order of decreasing water solubility.

   C₆H₅—CO₂CH₂CH₃ $CH_3-CH_2-\overset{\overset{O}{\|}}{C}-CH_3$

   $CH_3-CH_2-CH_2-CH_3$ CH_3-CO_2H

   C₆H₁₁—OH $CH_3-CH_2-O-CH_2-CH_3$

   CH_3-CHO $CH_3-\underset{\underset{OH}{|}}{CH}-CH_2-OH$

19. Write an equation to illustrate hydrolysis of each ester in question 15.

20. Write an equation to illustrate hydrolysis of each ester in question 16.

21. Write an equation to illustrate the hydrolysis of acetylsalicylic acid (aspirin) that occurs in the stomach.

22. Write an equation to illustrate the hydrolysis of methyl salicylate (oil of wintergreen) that occurs when the substance is rubbed on the skin.

23. The chemical name of vitamin C is ascorbic acid. Explain why vitamin C is named as an acid.

24. Distinguish between an organic ester and an inorganic ester. Give an example of each.

25. After a bottle of aspirin tablets has been opened and its contents exposed to moist air for a period of time, it begins to take on the odor of vinegar. What do you think causes this?

26. Summarize structural and physiological differences between morphine and heroin.

27. Give eight biological roles of lipids.

28. What are the two major classes of lipids? Give members of each.

29. Explain why unsaturated fatty acids have lower melting points than saturated fatty acids. How would you expect the extent of unsaturation of the fatty acids to affect melting points of waxes and triacylglycerols?

30. Distinguish between waxes and triacylglycerols, and give an example of each.

31. What is the main difference between fats and (food) oils? Between food oils and petroleum oils?

32. Write a chemical equation for the lipase-catalyzed hydrolysis of tristearin.

33. Write a chemical equation for the saponification of tripalmitin.

34. Give an explanation for the cleaning action of soap.

35. Write a chemical equation to illustrate how hydrogenation converts an oil to shortening.

36. How is the iodine number related to the extent of unsaturation in fats and oils?

37. Arrange the fats and oils in Table 14-7 in order of decreasing unsaturation.

38. Give a specific structural example for each of the following complex lipids:
    (a) a triacylglycerol
    (b) a phosphoglyceride
    (c) a sphingolipid

39. Explain why phosphoglycerides are good emulsifying agents.

40. Give a specific structural example for each of the following simple lipids.
    (a) a terpene
    (b) a steroid
    (c) a sterol
    (d) a prostaglandin

41. Why are some vitamins said to be fat soluble?

42. Give two reasons for the presence of cholesterol in tissues.

43. Suggest possible therapeutic uses of prostaglandins in the future.

CHAPTER 15

ORGANIC NITROGEN COMPOUNDS

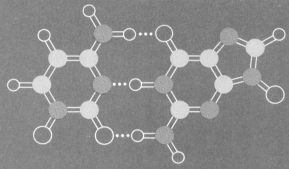

AMINES
- Structure
- Nomenclature
- Physical Properties
- Chemical Properties
- Examples of Biologically Important Amines

AMINO ACIDS
- Structure
- Classification
- Physical Properties
- Chemical Properties
- Unusual Amino Acids

AMIDES
- Structure
- Nomenclature
- Physical Properties
- Chemical Properties
- Examples of Amides

HETEROCYCLIC NITROGEN COMPOUNDS
- Five-Membered Rings
- Six-Membered Rings
- Fused-Ring Nitrogen Heterocyclics

Organic nitrogen compounds include a variety of structures, the most common of which are amines, amino acids, amides, and heterocyclic nitrogen compounds. Amines, considered to be derivatives of ammonia, are abundant in nature and account for the odors of fish and rotting meat. Amino acids, the building blocks of proteins, contain both the amino group and the carboxyl group. Such substances as flesh, hair, horn, and milk solids are composed largely of proteins and thus amino acids. Amides are formed when certain carboxylic acids react with ammonia or amines to form amide bonds. These bonds, also called amide linkages, hold amino acids together in proteins. Finally, heterocyclic nitrogen compounds are structures that have one or more nitrogen atoms incorporated into a ring. Heterocyclic nitrogen structures are found in amino acids, hemoglobin, chlorophyll, certain vitamins, and DNA and RNA.

AMINES

Amines are organic compounds derived from ammonia when one or more hydrogen atoms are replaced by alkyl or aromatic groups:

amines (aah-MEENS)

$$H-\overset{H}{\underset{H}{N:}} \qquad H-\overset{R}{\underset{H}{N:}} \qquad H-\overset{R}{\underset{R}{N:}} \qquad R-\overset{R}{\underset{R}{N:}}$$

ammonia   amines

Thus, amines have the amino group, RNH_2, or one of its substituted forms (R_2NH or R_3N) as their identifying structural feature.

Structure

You will recall that the hydrogen atoms in ammonia are located at the corners of an irregular tetrahedron. The bond angles between hydrogen atoms are slightly less than the normal tetrahedral bond angles of 109.5° because of the extra space taken up by the nonbonded electrons on nitrogen. However, in amines where hydrocarbon groups have replaced all of the hydrogen atoms, bond angles approach the normal values because of steric repulsion between these large groups (see Figure 15-1).

Like alcohols, amines have three subclasses, defined by the number of ammonia hydrogen atoms that have been replaced by hydrocarbon groups. When only one hydrogen atom has been replaced, the formula is RNH_2, and the compound is a primary (1°) amine. If two hydrogen atoms have been replaced, the formula is R_2NH, and we have a secondary (2°) amine. If all three hydrogens have been replaced, as in R_3N, the amine is a tertiary (3°) amine. Thus, tertiary amines have the largest bond angles between substituent groups, and their geometry is very much like that of a carbon atom bonded to four other atoms. The amine subclasses are summarized in Table 15-1.

Fig. 15-1 The Geometry of (a) the Ammonia Molecule and (b) a Tertiary Amine Molecule

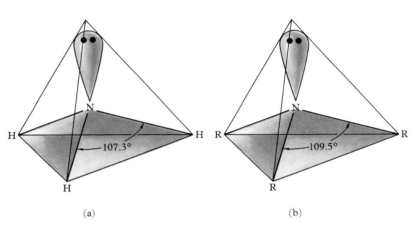

Table 15-1 Subclasses of Amines

Subclass	General Formula	Example
Primary (1°)	$R-NH_2$	$CH_3-N{\begin{smallmatrix}H\\H\end{smallmatrix}}$
Secondary (2°)	R_2-NH	$CH_3-N{\begin{smallmatrix}H\\CH_3\end{smallmatrix}}$
Tertiary (3°)	R_3N	$CH_3-N{\begin{smallmatrix}CH_3\\CH_3\end{smallmatrix}}$

AMINES   423

EXAMPLE 15.1

Classify each of the following amines as primary, secondary, or tertiary.

(a) 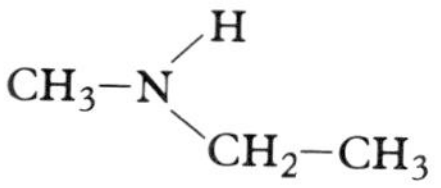 CH$_3$−N(H)(CH$_2$−CH$_3$)

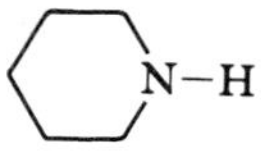

(b) 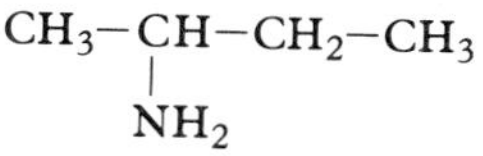 CH$_3$−CH(NH$_2$)−CH$_2$−CH$_3$

(c) (cyclohexyl)N−H

(d) 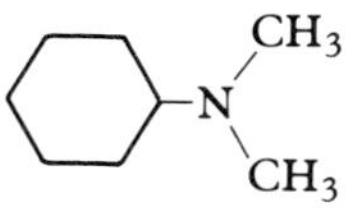 (cyclohexyl)−N(CH$_3$)(CH$_3$)

SOLUTION

(a) secondary
(b) primary
(c) secondary
(d) tertiary

EXERCISE 15.1 Classify each of the following amines as primary, secondary, or tertiary.

(a) 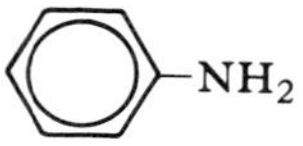 (phenyl)−NH$_2$

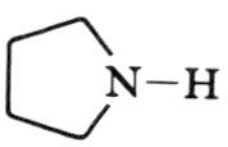

(b) 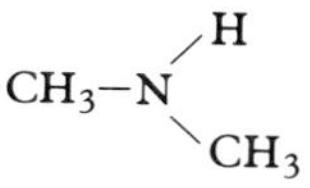 CH$_3$−N(H)(CH$_3$)

(c) (cyclopentyl)N−H

(d) 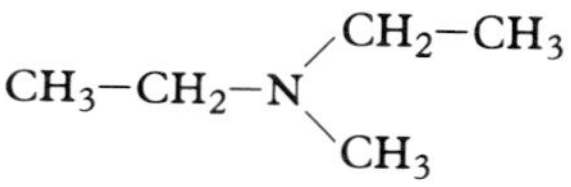 CH$_3$−CH$_2$−N(CH$_2$−CH$_3$)(CH$_3$)

Nomenclature

Common names are used extensively for low molecular weight amines. These are easily derived for aliphatic amines by listing the alkyl groups bonded to nitrogen, either in alphabetical order or in order of increasing size, followed by the suffix "-amine." The name is written as one word, and the prefixes "di-" and "tri-" are used when identical alkyl groups are present. Polyamines are named as di- and triamines. Some examples are:

CH$_3$−NH$_2$
methylamine

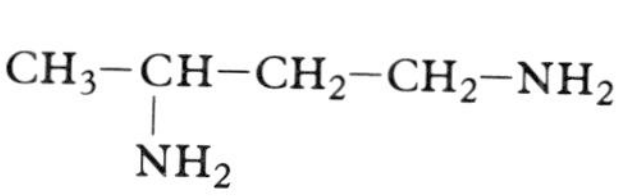

CH$_3$−CH(NH$_2$)−CH$_2$−CH$_2$−NH$_2$
1,3-butanediamine

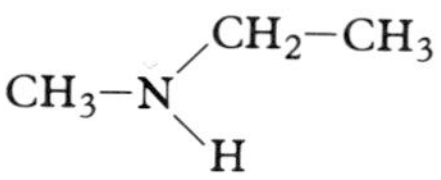

CH$_3$−N(H)(CH$_2$−CH$_3$)
methylethylamine
or ethylmethylamine

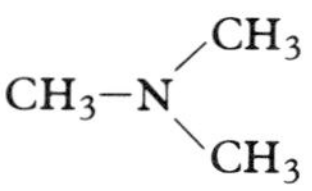

CH$_3$−N(CH$_3$)(CH$_3$)
trimethylamine

Aromatic amines are almost always given common names as derivatives of aniline, the simplest aromatic amine. If alkyl or aromatic groups are attached to the nitrogen atom in aniline, the prefixes "N–" and "N,N–" are used to indicate their attachment to nitrogen.

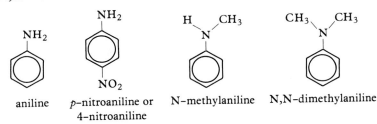

aniline   p-nitroaniline or   N-methylaniline   N,N-dimethylaniline
           4-nitroaniline

Aliphatic amines containing other functional groups are usually named by the IUPAC system. In these cases, the $-NH_2$ group is called the amino group, and the amine is named as a derivative of a parent compound. The following examples illustrate IUPAC names for amines.

$$CH_3-\underset{\underset{NH_2}{|}}{CH}-CH_2-OH \qquad CH_3-CH_2-\underset{\underset{NH-CH_3}{|}}{CH}-CO_2H \qquad CH_3-CH_2-\underset{\underset{N(CH_3)_2}{|}}{CH}-CHO$$

2-amino-1-propanol   2-(N-methylamino)-butanoic acid   2-(N,N-dimethylamino)-butanal

$$H_2N-CH_2-CH_2-NH_2 \qquad H_2N-CH_2-\underset{\underset{NH_2}{|}}{CH}-CH_2-NH_2$$

1,2-diaminoethane   1,2,3-triaminopropane

EXAMPLE 15.2

Give either a common name or the IUPAC name for each of the following compounds.

(a) ⌬—NH_2 (d) $H_2N-CH_2-CH_2-\underset{\underset{NH_2}{|}}{CH}-CH_2-CH_3$

(b) $CH_3-\underset{\underset{NH_2}{|}}{CH}-CO_2H$ (e) CH_3-NH_2

(c) $CH_3-N\overset{H}{\underset{CH_2-CH_3}{\diagdown}}$

SOLUTION

	Common Name	IUPAC Name
(a)	aniline	aniline
(b)	2-aminopropionic acid	2-aminopropanoic acid
(c)	methylethylamine	N-methylaminoethane
(d)	1,3-pentanediamine	1,3-diaminopentane
(e)	methylamine	aminomethane

EXERCISE 15.2 Give either a common name or the IUPAC name for each of the following compounds.

(a) CH$_3$—CH$_2$—N(H)(CH$_2$—CH$_3$)

(b) [benzene ring with OH and NH$_2$ substituents]

(c) H$_2$N—CH$_2$—CO$_2$H

(d) H$_2$N—CH$_2$—CH$_2$—CH$_2$—NH$_2$

(e) CH$_3$—CH(N(CH$_3$)$_2$)—CH$_2$—CH$_2$—CHO

Physical Properties

Although the low molecular weight amines are gases at room temperature, most amines are liquids. The simple amines have distinct odors which resemble ammonia, while the more complex amines have fishy odors. For example, methylamine smells very much like ammonia, and trimethylamine smells like dead fish. In fact, odor of fish is due to the presence of amines. Two of the most offensive-smelling amines, putrescine and cadaverine, occur when fish and other meats decay. These two amines are examples of *ptomaines,* diamino compounds formed during putrefaction of protein-containing substances.

ptomaines (TOE-mains)

	H$_2$N—CH$_2$—CH$_2$—CH$_2$—CH$_2$—NH$_2$	H$_2$N—CH$_2$—(CH$_2$)$_3$—CH$_2$—NH$_2$
Common Name:	putrescine	cadaverine
IUPAC Name:	1,4–diaminobutane	1,5–diaminopentane

Most ptomaines are not toxic and they do not play a role in food poisoning, although "ptomaine poisoning" is an old-fashioned and incorrect term for food poisoning.

Primary and secondary amines form hydrogen bonds among themselves, but since nitrogen is less electronegative than oxygen, hydrogen bonds formed by amines are weaker than those involving oxygen atoms. For this reason, boiling points of primary and secondary amines are usually somewhat lower than those for comparable alcohols. Tertiary amines cannot form hydrogen bonds among themselves, and their boiling points are similar to those of alkanes of corresponding molecular weight.

Because they can form hydrogen bonds with water, as shown in Figure 15-2, amines tend to be soluble in water. This is true for tertiary amines as well as primary and secondary amines, because tertiary amines have nonbonded electrons which can form hydrogen bonds. However, the water solubility of amines containing more than six carbon atoms decreases rapidly, due to the hydrophobic effect of the large hydrocarbon part of the structure. Physical properties of some primary amines are summarized in Table 15-2.

Although they are not appreciably water soluble, aromatic amines are rather soluble in fatty substances, and thus they are absorbed easily through the skin. These amines are toxic and therefore must be handled

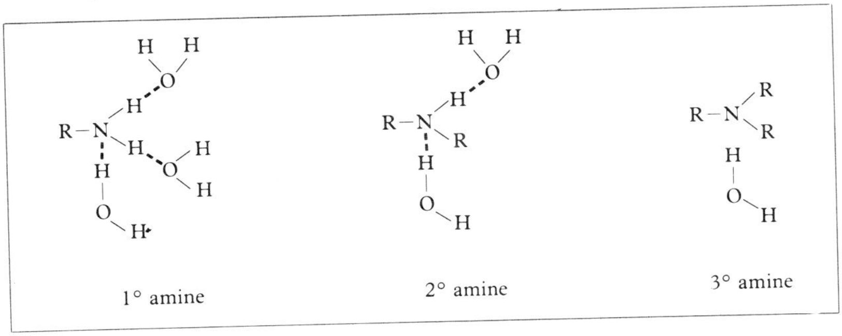

Fig. 15-2 Hydrogen Bonding Between Amine Molecules and Water Molecules

Table 15-2 Physical Properties of Some Amines

Name	Structure	Boiling Point (°C)	Water Solubility* (g/100 g)
Methylamine	CH_3NH_2	−7	Soluble in all proportions
Ethylamine	$CH_3CH_2NH_2$	17	Soluble in all proportions
Propylamine	$CH_3CH_2CH_2NH_2$	49	Soluble in all proportions
Butylamine	$CH_3CH_2CH_2CH_2NH_2$	78	Soluble in all proportions
Pentylamine	$CH_3CH_2CH_2CH_2CH_2NH_2$	104	Soluble
Hexylamine	$CH_3CH_2CH_2CH_2CH_2CH_2NH_2$	130	Slightly soluble
Heptylamine	$CH_3CH_2CH_2CH_2CH_2CH_2CH_2NH_2$	155	Slightly soluble
Aniline	 C₆H₅—NH_2	184	4

*at 20°C

very carefully. At least one aromatic amine, β-naphthylamine, is known to be carcinogenic.

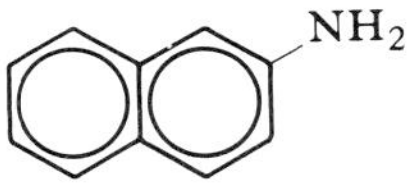

β-naphthylamine

EXAMPLE 15.3

Show how each of the following amines forms hydrogen bonds with water.

(a) $CH_3-N\begin{smallmatrix}H\\H\end{smallmatrix}$

(b) $CH_3-CH_2-N\begin{smallmatrix}H\\CH_3\end{smallmatrix}$

(c) $CH_3-N\begin{smallmatrix}CH_3\\CH_3\end{smallmatrix}$

SOLUTION

(a)
```
        H   H
         \ /
          O
          ⋮
          H
  CH₃—N—H⋯O—H
      |    \
      H    H
      ⋮
      O
      |
      H
```

(b)
```
            H   H
             \ /
              O
              ⋮
              H
  CH₃—CH₂—N—CH₃
          |
          H
          ⋮
          O
          |
          H
```

(c)
```
           CH₃
          /
  CH₃—N
      |   \
      H   CH₃
      ⋮
      O
      |
      H
```

Chemical Properties

Amines undergo many chemical reactions, but we will limit our study to the two chemical properties which are of physiological significance: basicity and amide formation.

Basicity

The single most distinguishing feature of amines is their behavior as bases. In each subclass of amines, the pair of nonbonded electrons on nitrogen can be used to form a bond to a hydrogen ion, as shown in the following equation. Thus amines qualify as bases by being proton acceptors. However, the reaction is incomplete, because amines are only weak bases.

$$R{-}\ddot{N}H_2 + HOH \rightleftharpoons [R{-}NH_3]^+ + OH^-$$

When an amine reacts with an acid such as HCl or H_2SO_4, a salt is formed. This is to be expected, since bases react with acids to form salts.

$$R{-}NH_2(aq) + HCl(aq) \longrightarrow R{-}NH_3^+(aq) + Cl^-(aq)$$

$$R_3N(aq) + H_2SO_4(aq) \longrightarrow R_3NH^+(aq) + HSO_4^-(aq)$$

An amine that is insoluble in water may be made soluble by treatment with acid because the salt will be water soluble. The free amine (free base) may be regenerated from its salt by addition of strong base such as NaOH or KOH.

Another kind of salt is formed by tertiary amines when they react with alkyl halides. In this reaction, a fourth alkyl group becomes covalently bonded to the nitrogen atom, and nitrogen takes on a positive charge.

$$R'-\underset{R''}{\underset{|}{N}}\!\!:\!\overset{R}{\underset{|}{}} + R'''-X \longrightarrow \left[R'-\underset{R''}{\underset{|}{N}}-R'''\overset{R}{\underset{|}{}}\right]^{+} + X^{-} \quad \text{(where X = Cl, Br, or I)}$$

3° amine → quaternary ammonium salt

quaternary ammonium salt: a salt whose cation contains a positively charged nitrogen atom covalently bonded to four alkyl or aromatic groups.

These salts are called *quaternary ammonium salts*. Choline, which we met in chapter 14 as a phosphoglyceride component, is an example of a quaternary ammonium ion.

Amine salts are named as derivatives of ammonium salts. Thus NH_4Cl, or $NH_4^+Cl^-$ to emphasize the ionic nature of the salt, is named ammonium chloride, and $CH_3NH_3^+Cl^-$ is named methylammonium chloride. Other examples are

$(CH_3)_2NH_2^+Cl^-$
dimethylammonium chloride
(salt of a 2° amine)

$CH_3-\underset{CH_2CH_3}{\underset{|}{\overset{CH_3}{\overset{|}{N^+}}}}\!\!-H \quad Br^-$
dimethylethylammonium bromide (salt of a 3° amine)

$(CH_3)_4N^+Cl^-$
tetramethylammonium chloride (a quaternary ammonium salt)

Amine salts have physical properties characteristic of ionic compounds. They are colorless (white) crystalline solids with high melting points, and they are usually water soluble. Since they are solids, they have no odor. Thus the offensive odor of amines can be eliminated by converting the amine to its salt. Fish odor can be controlled by lemon juice or vinegar, both acids, because the fish amines are converted to their salts.

Amide Formation

If ammonia or an amine is reacted with a carboxylic acid at elevated temperatures, the product is an amide:

$$R-N\overset{H}{\underset{H}{}} + R'-C\overset{O}{\underset{OH}{}} \xrightarrow{\Delta} R'-C\overset{O}{\underset{\underset{H}{N}}{}}\!\!R + HOH$$

an amide

The reaction is quite similar to ester formation in that a molecule of water is released, and the remaining fragments of the two reactants join to form the amide. Amides can also be obtained by heating the ammonium or amine salt of a carboxylic acid:

$$R-C(=O)O^-NH_4^+ \xrightarrow{\Delta} R-C(=O)NH_2 + HOH$$

Amides are often synthesized by first converting the carboxylic acid to a more reactive derivative, the *acid chloride*. This compound then reacts readily with ammonia and amines to form amides. No heat is required for this reaction.

acid chloride: a carboxylic acid derivative containing the group

$$-C(=O)Cl$$

$$R-C(=O)Cl + R'-NH_2 \longrightarrow R-C(=O)NH-R' + HCl$$

an acid chloride

In fact, acid chlorides are so reactive that they must be stored in tightly closed containers to prevent reaction with water vapor in the air.

EXAMPLE 15.4

Complete the following equations.

(a) $CH_3NH_2(aq) + HCl(aq) \longrightarrow$

(b) $CH_3CH_2NHCH_3(aq) + H_2SO_4(aq) \longrightarrow$

(c) $(CH_3)_3N + CH_3CH_2I \longrightarrow$

(d) $CH_3CO_2^-NH_4^+ \xrightarrow{\Delta}$

(e) $CH_3CH_2-\overset{O}{\underset{\|}{C}}-Cl + CH_3NH_2 \longrightarrow$

SOLUTION

(a) $CH_3NH_3^+Cl^-(aq)$

(b) $CH_3CH_2\overset{+}{N}H_2CH_3HSO_4^-(aq)$

(c) $(CH_3)_3\overset{+}{N}CH_2CH_3\ I^-$

(d) $CH_3-\overset{O}{\underset{\|}{C}}-NH_2 + H_2O$

(e) $CH_3CH_2-\overset{O}{\underset{\|}{C}}-NHCH_3 + HCl$

EXERCISE 15.3 Complete the following equations.

(a) $CH_3-\underset{NH_2}{\underset{|}{CH}}-CH_3(aq) + H_2SO_4(aq) \longrightarrow$

(b) $CH_3-\underset{CH_2-CH_3}{\underset{|}{N}}-CH_3 + CH_3-Cl \longrightarrow$

(c) $CH_3-NH_2 + CH_3-C(=O)OH \xrightarrow{\Delta}$

(d) 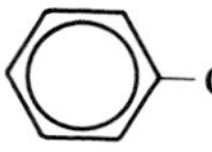 $\text{C}_6\text{H}_5-CO_2^-NH_4^+ \xrightarrow{\Delta}$

Examples of Biologically Important Amines

Epinephrine

epinephrine (eh-pin-EH-frin)

catecholamine (caah-tah-call-aah-MEEN)

Epinephrine, also known as adrenalin, is a hormone synthesized primarily in the adrenal gland and secreted at times of stress or fright. For this reason, it has been called the "fight or flight" hormone. Because of its structural similarity to catechol, epinephrine is often referred to as a *catecholamine.*

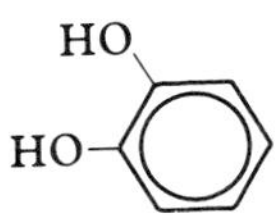

catechol

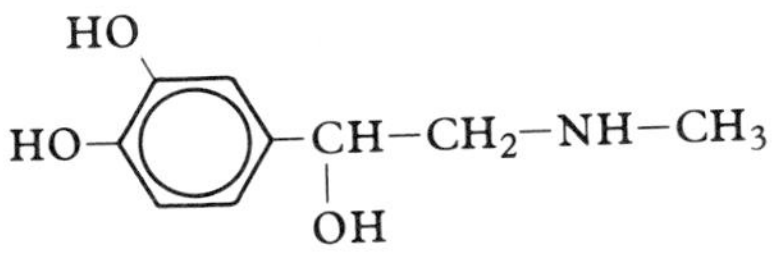

epinephrine

The hormone raises blood pressure by increasing the rate and force of contraction of the heart, and it increases alertness by action on the brain. Thus, it prepares an animal for vigorous activity. It also causes constriction of peripheral blood vessels. Epinephrine is usually administered in the same solution with local anesthetics because it constricts blood vessels in the vicinity of the injection. This allows the anesthetic to remain in the nearby tissue for a longer time, thus prolonging the anesthesia. Epinephrine is also used to stop hemorrhage, to treat asthma attacks, and to combat anaphylactic shock.

norepinephrine (nor-eh-pin-EH-frin)

Norephinephrine

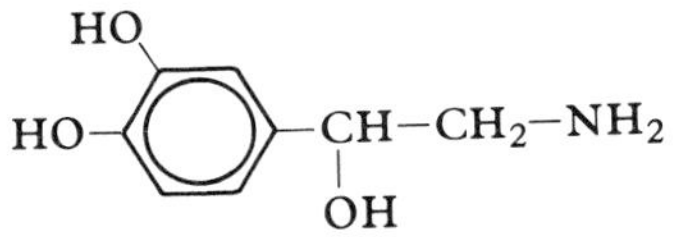

This is a second catecholamine hormone secreted by the adrenal gland. It causes a physiological response similar to that of epinephrine, except that it raises blood pressure by constricting all blood vessels. Norepinephrine is sometimes administered intravenously to combat certain types of shock. Epinephrine and norepinephrine are also liberated at the ends of certain nerve fibers where they serve as neurotransmitters.

The Amphetamines

This series of amines is a set of synthetic drugs which stimulate the central nervous system much in the same way as the naturally occurring catecholamines. The amphetamines include amphetamine itself (also known as Benzedrine), which is a racemic mixture (see Chapter 13) of isomers; dextroamphetamine (Dexedrine), the dextrorotatory enantiomer of amphetamine (can you find the asymmetric carbon?), which is two to four times more active than amphetamine; and methamphetamine (Methedrine).

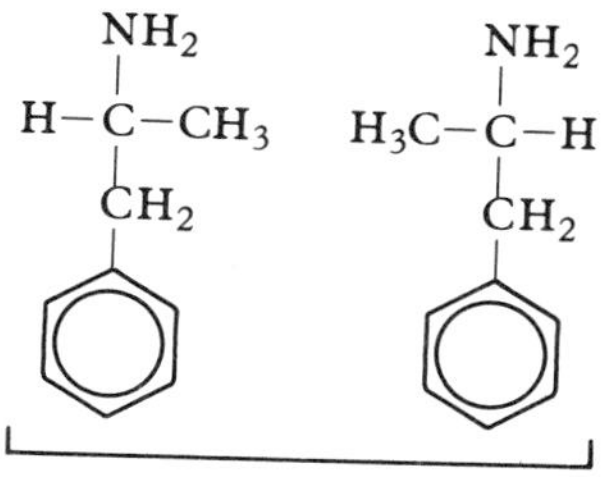

(±)-amphetamine
(Benzedrine)

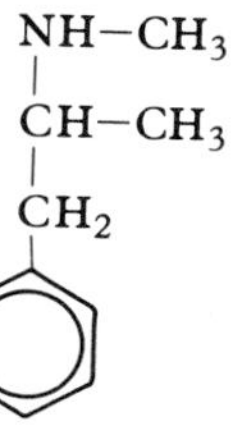

methamphetamine
(Methedrine)

Amphetamines have been abused by indiscriminate use to overcome fatigue or to provide energy, and their use is now limited to prescriptions by physicians.

p-Aminobenzoic Acid (PABA)

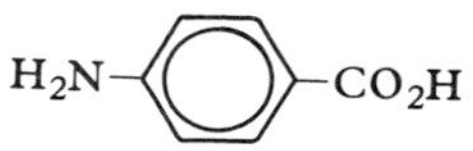

$$H_2N-\text{C}_6H_4-CO_2H$$

This compound is required by several types of microorganisms for the synthesis of folic acid. *Folic acid* is a vitamin for humans, but because intestinal bacteria usually provide enough of the substance to meet human needs, deficiencies of this vitamin are rare. Commercial production of folic acid also relies on *p*-aminobenzoic acid as starting material.

folic (FOE-lic) acid: a vitamin required by most animals: deficiencies result in anemia and diarrhea.

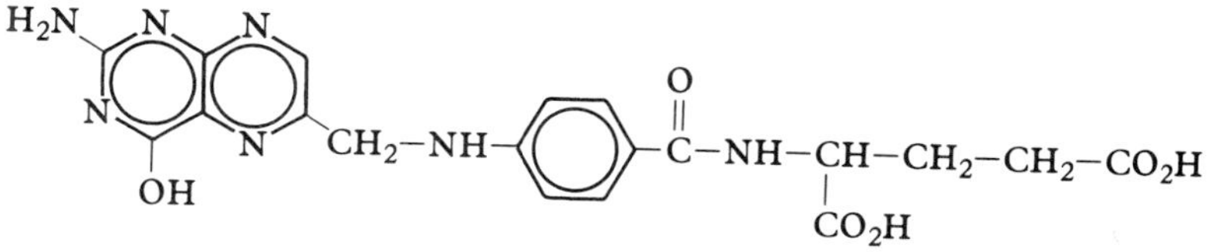

folic acid

Esters of *p*-Aminobenzoic Acid

Certain esters of *p*-aminobenzoic acid act as local anesthetics; some of these are shown along with their trade names. Benzocaine and Butesin are used as ointments, while Novocaine is usually injected. Note that Novocaine is formulated as the hydrochloride salt to improve its solubility in water.

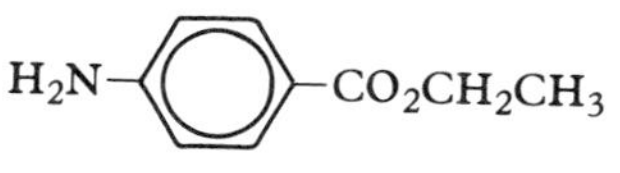

ethyl *p*-aminobenzoate
(Benzocaine)

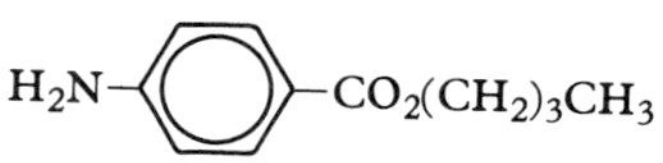

butyl *p*-aminobenzoate
(Butesin)

$$H_2N-\underset{}{\bigcirc}-CO_2(CH_2)_2-\overset{H}{\underset{CH_2-CH_3}{N^+}}-CH_2-CH_3 \quad Cl^-$$

procaine
(Novocaine)

Lidocaine

$$\underset{CH_3}{\overset{CH_3}{\bigcirc}}-NH-\overset{O}{\underset{}{C}}-CH_2-\overset{H}{\underset{CH_2-CH_3}{N^+}}-CH_2-CH_3 \quad Cl^-$$

lidocaine
(Xylocaine)

This is a local anesthetic whose structure is similar to those of *p*-aminobenzoic acid esters. The hydrochloride salt is administered by injection. Lidocaine is frequently used in place of Novocaine because of its prompter, more intense, and longer-lasting anesthesia.

EXERCISE 15.4 Give the names and structures of five biologically important amines, and summarize the biological importance of each.

AMINO ACIDS

Amino acids are organic compounds which contain both an amino group and a carboxyl group. Thus, amino acids have both basic and acidic groups in their structures. Since proteins contain only α-amino acids, these will be the focus of our discussion.

Structure

Amino acids are the building blocks of proteins and are thus considered monomers of protein polymers. The amino acids in proteins are those which have the amino group attached to the same carbon as the carboxyl group. Thus, the amino group is bonded to the α-carbon, and the compounds are called α-amino acids. The general formula is

$$R-\overset{\overset{\alpha}{H}}{\underset{NH_2}{C}}-CO_2H$$

The R group can be hydrogen, a simple alkyl group, a group containing an aromatic ring, or alkyl groups containing various substituents. The R group, often referred to as the side chain, distinguishes one amino acid from another.

Classification

There are 20 amino acids commonly found in proteins. Although systematic names are available, common names are used almost exclusively. Table 15-3 gives the commonly occurring amino acids and their classes: (1) nonpolar (hydrophobic) side chains; (2) neutral (uncharged) polar side chains; (3) basic (positively charged) side chains; and (4) acidic (negatively charged) side chains. In the cases of charged side chains, we are referring to charges which exist at pH 6-7, the pH range ordinarily found within living cells.

Also given in Table 15-3 are three-letter abbreviations for each amino acid and one-letter symbols. These symbols were developed so that long sequences of amino acids could be summarized in a relatively short space on paper.

Amino Acids with Nonpolar (Hydrophobic) Side Chains

This group contains five amino acids with aliphatic side chains: alanine, leucine, isoleucine, valine, and proline. Note that proline has a cyclic saturated amine as its side chain; this structure is called a pyrrolidine ring. In this group there are two amino acids which have aromatic groups in their side chains: phenylalanine, considered to be a derivative of alanine, and tryptophan. And finally, one amino acid, methionine, has in its side chain a sulfur atom linked to two carbon atoms. This linkage is similar to the $-O-$ linkage in ethers. Because of their nonpolar side chains, members of this group of amino acids are not very soluble in water.

pyrrolidine (pih-ROH-lih-deen)

Amino Acids with Neutral (Uncharged) Polar Side Chains

These amino acids are more soluble in water than those with nonpolar side chains. The group includes three hydroxyl-containing structures—serine, threonine, and tyrosine—that can form hydrogen bonds with water. Two members of this group, asparagine and glutamine, contain amides in their side chains, and one member, cysteine, has a thiol group in its side chain. The simplest amino acid, glycine, is included in this group because its side chain (a hydrogen atom) is considered to be more polar than nonpolar.

An amino acid very closely related to cysteine occurs often in proteins. Called cystine, it is the disulfide formed when two units of cysteine are oxidized to create a covalent cross-linkage between them:

$$HO_2C-\underset{NH_2}{\underset{|}{\overset{H}{\overset{|}{C}}}}-CH_2-SH + HS-CH_2-\underset{NH_2}{\underset{|}{\overset{H}{\overset{|}{C}}}}-CO_2H \xrightarrow{[O]} HO_2C-\underset{NH_2}{\underset{|}{\overset{H}{\overset{|}{C}}}}-CH_2-S-S-CH_2-\underset{NH_2}{\underset{|}{\overset{H}{\overset{|}{C}}}}-CO_2H$$

cysteine   cysteine   cystine

Table 15-3 The Amino Acids

Amino Acids with Nonpolar (Hydrophobic) Side Chains

	R Groups	pK'_a α-CO_2H	pK'_a α-$\overset{+}{NH_3}$
Alanine Ala A Mol wt 89	CH_3–C(H)(NH_3^+)–COO^-	2.35	9.69
Valine Val V Mol wt 117	$(CH_3)_2CH$–C(H)(NH_3^+)–COO^-	2.32	9.62
Leucine Leu L Mol wt 131	$(CH_3)_2CH$–CH_2–C(H)(NH_3^+)–COO^-	2.36	9.60
Isoleucine Ile I Mol wt 131	CH_3–CH_2–CH(CH_3)–C(H)(NH_3^+)–COO^-	2.36	9.68
Proline Pro P Mol wt 115	(pyrrolidine ring with COO^-)	1.99	10.60
Phenylalanine Phe P Mol wt 165	C_6H_5–CH_2–C(H)(NH_3^+)–COO^-	1.83	9.13
Tryptophan Trp W Mol wt 204	(indole)–CH_2–C(H)(NH_3^+)–COO^-	2.38	9.39
Methionine Met M Mol wt 149	CH_3–S–CH_2–CH_2–C(H)(NH_3^+)–COO^-	2.28	9.21

Amino Acids with Neutral (Uncharged) Polar Side Chains

	R Groups	pK'_a α-CO_2H	pK'_a α-$\overset{+}{NH_3}$
Glycine Gly G Mol wt 75	H–C(H)(NH_3^+)–COO^-	2.34	9.6
Serine Ser S Mol wt 105	HO–CH_2–C(H)(NH_3^+)–COO^-	2.21	9.15
Threonine Thr T Mol wt 119	CH_3–CH(OH)–C(H)(NH_3^+)–COO^-	2.63	10.43

Table 15-3 The Amino Acids (continued)

Amino Acids with Neutral (Uncharged) Polar Side Chains

	R Groups	pK'_a $\alpha\text{-CO}_2\text{H}$	pK'_a $\alpha\text{-}\overset{+}{\text{NH}}_3$
Cysteine Cys C Mol wt 121	HS–CH$_2$–CH(NH$_3^+$)–COO$^-$	1.71	10.78
Tyrosine Tyr Y Mol wt 181	HO–C$_6$H$_4$–CH$_2$–CH(NH$_3^+$)–COO$^-$	2.20	9.11
Asparagine Asn N Mol wt 132	H$_2$N–CO–CH$_2$–CH(NH$_3^+$)–COO$^-$	2.02	8.8
Glutamine Gln Q Mol wt 146	H$_2$N–CO–CH$_2$–CH$_2$–CH(NH$_3^+$)–COO$^-$	2.17	9.13

Amino Acids with Basic (Positively Charged) Side Chains

	R Groups	pK'_a $\alpha\text{-CO}_2\text{H}$	pK'_a $\alpha\text{-}\overset{+}{\text{NH}}_3$	pK'_a R Group
Lysine Lys K Mol wt 146	H$_3\overset{+}{\text{N}}$–CH$_2$–CH$_2$–CH$_2$–CH$_2$–CH(NH$_3^+$)–COO$^-$	2.18	8.95	10.53
Arginine Arg R Mol wt 174	H$_2$N–C(=$\overset{+}{\text{NH}}_2$)–NH–CH$_2$–CH$_2$–CH$_2$–CH(NH$_3^+$)–COO$^-$	2.17	9.04	12.48
Histidine (at pH 6.0) His H Mol wt 155	imidazolium–CH$_2$–CH(NH$_3^+$)–COO$^-$	1.82	9.17	6.0

Amino Acids with Acidic (Negatively Charged) Side Chains

	R Groups	pK'_a $\alpha\text{-CO}_2\text{H}$	pK'_a $\alpha\text{-}\overset{+}{\text{NH}}_3$	pK'_a R Group
Aspartic Acid Asp D Mol wt 133	$^-$OOC–CH$_2$–CH(NH$_3^+$)–COO$^-$	2.09	9.82	3.86
Glutamic acid Glu E Mol wt 147	$^-$OOC–CH$_2$–CH$_2$–CH(NH$_3^+$)–COO$^-$	2.19	9.67	4.25

Amino Acids with Basic (Positively Charged) Side Chains

Each of these amino acids has a basic nitrogen group in the side chain. In the case of lysine, the basic group is an amino group. Because this group is protonated in aqueous solution at neutral pH, it carries a positive charge at physiological conditions. The basic group in arginine is the sequence

$$H_2N-\underset{\underset{NH_2}{\|}}{C}-NH-,$$

guanidino (gwah-nih-DEE-no)

called the *guanidino group*, which also carries a positive charge at neutral pH (and is then called the guanidinium group). The third member of the group, histidine, contains a very weak base, the imidazole ring, and only about 10% of its side chains are charged at neutral pH.

Amino Acids with Acidic (Negatively Charged) Side Chains

There are only two amino acids in this group: aspartic acid and glutamic acid. Their side chains contain carboxyl groups that are completely ionized at physiological pH.

Physical Properties

Amino acids have two important physical properties: their optical activity and their melting points.

Optical Activity

Except for glycine, each of the common amino acids contains a center of chirality at the α-carbon. This means that these amino acids have a pair of enantiomers that differ by configuration at the α-carbon. For example, alanine has the enantiomers:

$$\underset{\text{D-alanine}}{\overset{CO_2H}{\underset{CH_3}{|}}{H-C-NH_2}} \qquad \underset{\text{L-alanine}}{\overset{CO_2H}{\underset{CH_3}{|}}{H_2N-C-H}}$$

If the amino acid structures are drawn with their carboxyl groups at the top, then the orientation of the α-amino group determines whether the amino acid belongs to the D- or the L-family. This is analogous to the orientation of the secondary hydroxyl group in D- and L-glyceraldehyde. Since optical activity is possessed by all chiral compounds, all amino acids, except for glycine, are optically active. Only L-amino acids are found in proteins. Thus we will be concerned primarily with these forms, and if no designation is given, it should be assumed that the amino acid is of the L-family. Although only L-amino acids are present in proteins, many different D-amino acids are found in living cells in other chemical forms, such as in the cell walls of microorganisms or in certain molecules synthesized by microorganisms.

Melting Points

In their solid forms, amino acids are white, crystalline substances which resemble salts in appearance. Like salts, they have relatively high melting points, so high that amino acids usually decompose before they melt, and they also tend to be more soluble in water than in nonpolar solvents. Studies have shown that amino acids exist as dipolar ions (*zwitterions*) in the solid state if crystallized from aqueous solution at neutral pH:

$$\begin{array}{c} H \\ | \\ R-C-CO_2^- \\ | \\ NH_3^+ \end{array}$$
a dipolar ion

zwitterion (ZVIH-ter-eye-on): a German word meaning an ion with a positive and a negative charge.

Thus, amino acids are positioned in the crystalline lattice by ionic forces. Their ionic nature explains why their properties are more like those of inorganic salts than those of organic molecules.

Chemical Properties

The chemical features of amino acids result from the presence of amino and carboxyl groups. Three important chemical properties will be discussed.

The Amphoteric Nature of Amino Acids

When an amino acid is dissolved in water at neutral pH, it can act as an acid by donating a proton or it can act as a base by accepting a proton:

proton donor: $H_3\overset{+}{N}-\underset{R}{CH}-CO_2^- \rightleftharpoons H^+ + H_2N-\underset{R}{CH}-CO_2^-$

proton acceptor: $H_3\overset{+}{N}-\underset{R}{CH}-CO_2^- + H^+ \rightleftharpoons H_3\overset{+}{N}-\underset{R}{CH}-CO_2H$

Any substance which can behave as both an acid and a base is said to be *amphoteric* (from the Greek *amphi*, meaning "both"), and amphoteric electrolytes such as the amino acids are called *ampholytes* (from *amphoteric* and electro*lyte*).

amphoteric (am-foe-TEH-ric)

Consider what happens if we dissolve an amino acid with an uncharged side chain in water at pH 1 or lower. In this solution, both the carboxyl group and the amino group will be protonated, because the pH of the solution is appreciably below their pK_a's. However, if we add strong base in small increments, as in a titration, the pH of the solution will rise, and the carboxyl group will start to dissociate. At about pH 3, the carboxyl group will be completely ionized, because this pH value is much higher than its pK_a. If we keep adding base, we will reach a point, around pH 10.5, at which the charged amino group loses its extra proton, because the pH will be much higher than its pK_a:

$$\underset{\underset{\text{pH 1}}{}}{\overset{+}{H_3N}-\underset{R}{CH}-CO_2H} \underset{\longleftarrow}{\overset{OH^-}{\longrightarrow}} \underset{\underset{\substack{+H_2O \\ \text{pH 3-8}}}{}}{\overset{+}{H_3N}-\underset{R}{CH}-CO_2^-} \underset{\longleftarrow}{\overset{OH^-}{\longrightarrow}} \underset{\underset{\substack{+H_2O \\ \text{pH 10.5}}}{}}{H_2N-\underset{R}{CH}-CO_2^-}$$

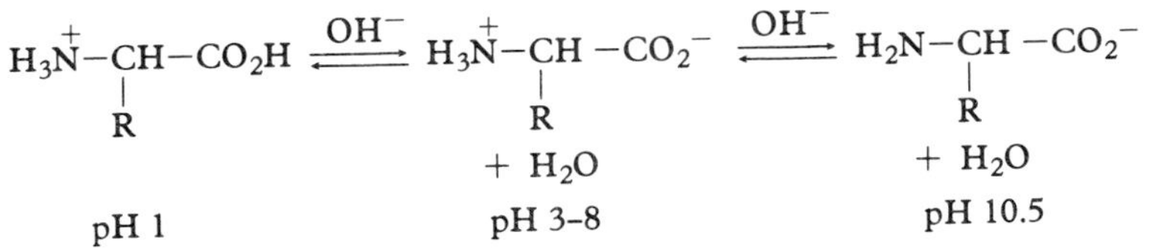

Thus, an amino acid with an uncharged side chain may exist in three different ionic forms in aqueous solution, depending on the pH. In the acid range, the amino acid has a single positive charge; at neutrality, it has a positive charge and a negative charge. At extremely basic pH values, the amino acid will have a single negative charge.

Amino acids with basic or acidic groups in their side chains can exist in additional ionic forms because of protonation of the side-chain groups. In the case of amino acids with basic side chains, such as lysine, the side-chain amino group usually has a pK_a rather close to the pK_a of the α-amino group, as shown in Table 15-3. Those amino acids having carboxyl groups in their side chains, such as aspartic acid, have two pK_a's in the acid pH range. In either type of amino acid, there will be four ionic forms, instead of the usual three, as illustrated here for lysine and aspartic acid:

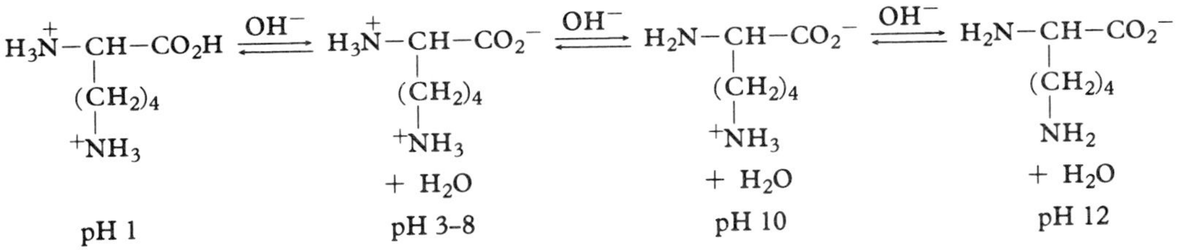

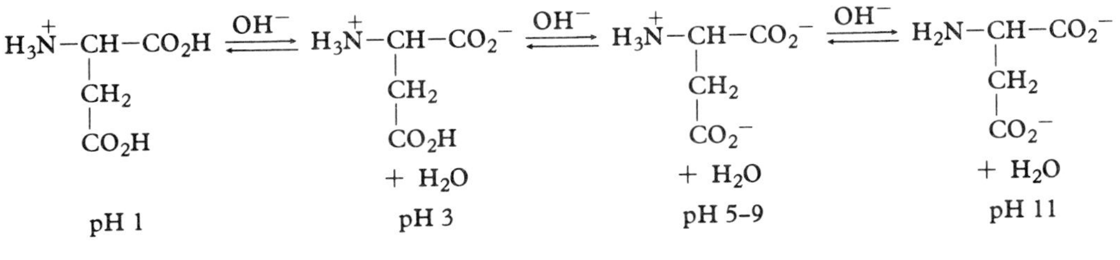

$$\underset{\text{pH 1}}{\overset{+}{H_3N}-\underset{\underset{+NH_3}{(CH_2)_4}}{CH}-CO_2H} \overset{OH^-}{\rightleftarrows} \underset{\substack{+H_2O \\ \text{pH 3-8}}}{\overset{+}{H_3N}-\underset{\underset{+NH_3}{(CH_2)_4}}{CH}-CO_2^-} \overset{OH^-}{\rightleftarrows} \underset{\substack{+H_2O \\ \text{pH 10}}}{H_2N-\underset{\underset{+NH_3}{(CH_2)_4}}{CH}-CO_2^-} \overset{OH^-}{\rightleftarrows} \underset{\substack{+H_2O \\ \text{pH 12}}}{H_2N-\underset{\underset{NH_2}{(CH_2)_4}}{CH}-CO_2^-}$$

$$\underset{\text{pH 1}}{\overset{+}{H_3N}-\underset{\underset{CO_2H}{CH_2}}{CH}-CO_2H} \overset{OH^-}{\rightleftarrows} \underset{\substack{+H_2O \\ \text{pH 3}}}{\overset{+}{H_3N}-\underset{\underset{CO_2H}{CH_2}}{CH}-CO_2^-} \overset{OH^-}{\rightleftarrows} \underset{\substack{+H_2O \\ \text{pH 5-9}}}{\overset{+}{H_3N}-\underset{\underset{CO_2^-}{CH_2}}{CH}-CO_2^-} \overset{OH^-}{\rightleftarrows} \underset{\substack{+H_2O \\ \text{pH 11}}}{H_2N-\underset{\underset{CO_2^-}{CH_2}}{CH}-CO_2^-}$$

When amino acids are linked together in protein molecules, the α-amino groups and α-carboxyl groups are involved in the linkages. In this situation, the side-chain ionizable groups (amino groups and carboxyl groups) are capable of neutralizing small amounts of acid or base. Thus, proteins are able to behave as buffers because of the ionizable groups in the side chains of their constituent amino acids.

EXAMPLE 15.5

Write all possible ionic forms for each of the following amino acids.

(a) CH$_3$—CH—CO$_2^-$
　　　　|
　　　+NH$_3$

(b) HO—CH$_2$—CH—CO$_2^-$
　　　　　　|
　　　　　+NH$_3$

(c) $^-$O$_2$C—CH$_2$—CH$_2$—CH—CO$_2^-$
　　　　　　　　　　|
　　　　　　　　+NH$_3$

SOLUTION

(a) CH$_3$—CH—CO$_2$H , CH$_3$—CH—CO$_2^-$, CH$_3$—CH—CO$_2^-$
　　　　|　　　　　　　　|　　　　　　　　|
　　　+NH$_3$　　　　　+NH$_3$　　　　　NH$_2$

(b) HO—CH$_2$—CH—CO$_2$H , HO—CH$_2$—CH—CO$_2^-$, HO—CH$_2$—CH—CO$_2^-$
　　　　　　|　　　　　　　　　　　|　　　　　　　　　　　|
　　　　　+NH$_3$　　　　　　　　+NH$_3$　　　　　　　　NH$_2$

(c) HO$_2$C—CH$_2$—CH$_2$—CH—CO$_2$H , HO$_2$C—CH$_2$—CH$_2$—CH—CO$_2^-$,
　　　　　　　　　　　　|　　　　　　　　　　　　　　　　　|
　　　　　　　　　　　+NH$_3$　　　　　　　　　　　　　　+NH$_3$

$^-$O$_2$C—CH$_2$—CH$_2$—CH—CO$_2^-$, $^-$O$_2$C—CH$_2$—CH$_2$—CH—CO$_2^-$
　　　　　　　　|　　　　　　　　　　　　　　|
　　　　　　　+NH$_3$　　　　　　　　　　　NH$_2$

EXERCISE 15.5 Write all possible ionic forms for each of the following amino acids.

(a) glycine　　(b) phenylalanine　　(c) arginine

A mixture of amino acids can be separated by their behavior in an electric field. The technique is called *electrophoresis*. If we immerse electrodes in a solution of amino acids and turn on the electricity, those amino acids having net positive charges will move toward the negative electrode (cathode), and those having net negative charges will move toward the positive electrode (anode). In actual practice, a sample of the amino acid mixture is applied to the center of a small strip of paper or other material, and the strip is placed in a container of buffer with immersed electrodes. The net charge on each kind of amino acid determines its rate of movement toward an electrode (see Figure 15-3).

In a group of molecules of any one amino acid, the pH determines the average charge on each molecule. This average charge will more than likely be a fractional value, and for a mixture of amino acids, the average charges at a given pH are all different. Thus, each kind of amino acid will have a different rate of migration. Electrophoresis is a very good method of separating mixtures of amino acids. Even at pH values where molecules of one amino acid may have an average charge of 0 (this pH is called the *isoelectric point*, pI, of the amino acid), separation occurs. At this pH, molecules of this amino acid do not move from the point of application, while all of the others migrate toward their appropriate electrodes.

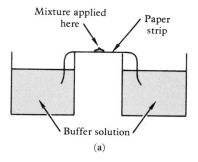
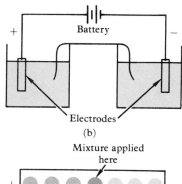
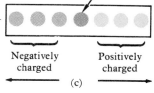

Fig. 15-3 The Process of Electrophoresis. (a) A small amount of the amino acid mixture is applied to a strip of paper saturated with buffer solution. (b) Positive and negative electrodes are connected to create an electric field across the paper strip. (c) As time passes, the amino acids separate as they migrate at different rates to the electrode of opposite charge.

electrophoresis (eh-leck-troe-foe-REE-sis): an electrical method used to separate charged particles in solution.

Reaction with Ninhydrin

ninhydrin (nin-HIGH-drin): an organic compound which reacts with amino acids (and other amines) to produce colored products.

The reaction of amino acids with *ninhydrin* produces a colored compound which can be detected visually. Each molecule of amino acid reacts with two of ninhydrin when heated:

$$R-CH(NH_2)-CO_2H + 2\ \text{ninhydrin} \xrightarrow{\Delta} \text{colored product} + RCHO + CO_2 + 3\ H_2O$$

A purple color is produced by amino acids having a free α-amino group. Proline, which has a substituent on the α-amino group, gives a yellow color. If the paper strip containing the amino acids after electrophoresis is sprayed with a solution of ninhydrin and then heated gently, colored spots will appear where amino acids are present. This pattern of spots can then be compared to the one produced by amino acids whose identities are known, and components in the original mixture can be identified. Unfortunately, the ninhydrin reaction is irreversible, so if there is a need to recover the original amino acids, other means of detection are necessary.

Formation of Amides

Because amino acids contain both an amino group and a carboxyl group, it is possible for them to form amides. However, since amino acids decompose when heated, the conventional methods for making amides from carboxylic acids (discussed earlier in this chapter) cannot be used. Also, it is not possible to form stable acid chlorides from amino acids. But it is possible by means of special "condensing agents" to remove the elements of water from two amino acid molecules and cause them to form an amide. This is usually done in the laboratory.

$$R_1-NH-CH(R_2)-C(=O)OH + H_2N-CH(R_3)-C(=O)-R_4 + \text{condensing agent} \longrightarrow$$

$$R_1-NH-CH(R_2)-C(=O)-NH-CH(R_3)-C(=O)-R_4 + \text{condensing agent} \cdot H_2O$$

Note that the extra amino and carboxyl groups must be reacted ahead of time with certain reagents so that they are blocked; otherwise all free amino and carboxyl groups will react to form a mixture of amide products. The blocking groups, R_1 and R_4, can be removed after the desired amide is produced. It is also necessary to use special techniques to control chain length, since amino acid molecules can form indefinitely long chains. Given all of these problems, chemists can synthesize only certain rather small polyamide chains of specified sequence and length.

Proteins are polymers of amino acids; each protein has its own particular sequence of monomers. Even though we have great difficulty synthesizing even a small protein in the laboratory, enzymes are able to direct polyamide formation so that only the proteins needed by the cell at any given time are formed.

Unusual Amino Acids

Besides the 20 standard amino acids found in most proteins, a few rare ones have been isolated from highly specialized proteins. For example, hydroxyproline is a derivative of proline found in the protein *collagen*. Hydroxylysine, a derivative of lysine, is also found in collagen.

collagen (COH-lah-jen): a fibrous protein found in connective tissue and bones.

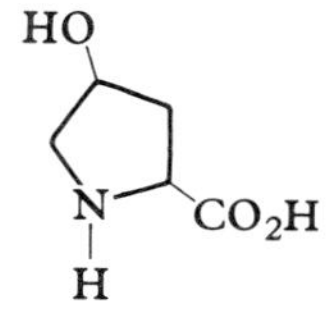

hydroxyproline

$H_2N-CH_2-CH(OH)-CH_2-CH_2-CH(^+NH_3)-CO_2^-$

hydroxylysine

In these cases, the standard amino acid is first incorporated into the protein, after which time it is called an amino acid *residue*. The residue is then modified chemically to produce the unusual amino acid residue, which remains a part of the protein structure.

Other unusual amino acids are known to occur in free or combined form, but never in proteins. Some of these are illustrated here:

$CH_2-CH_2-CO_2^-$
$|$
$^+NH_3$

β-alanine
(a constituent of the vitamin pantothenic acid)

$CH_2-CH_2-CH-CO_2^-$
$|\qquad\qquad|$
$SH\qquad\; ^+NH_3$

homocysteine
(an intermediate in amino acid metabolism)

$CH_2-CH_2-CH-CO_2^-$
$|\qquad\qquad|$
$OH\qquad\; ^+NH_3$

homoserine
(an intermediate in amino acid metabolism)

$H_2N-\underset{\underset{O}{\|}}{C}-NH-(CH_2)_3-CH-CO_2^-$
$\qquad\qquad\qquad\qquad\qquad\;|$
$\qquad\qquad\qquad\qquad\qquad\;^+NH_3$

citrulline
(an intermediate in the biosynthesis of arginine)

$CH_2-CH_2-CH_2-CH-CO_2^-$
$|\qquad\qquad\qquad\quad|$
$^+NH_3\qquad\qquad\;\;^+NH_3$

ornithine
(an intermediate in the biosynthesis of arginine)

$CH_2-CH_2-CH_2-CO_2^-$
$|$
$^+NH_3$

γ-aminobutyric acid
(a neurotransmitter)

Finally, a number of D-amino acids are found in free form in bacteria and insects. For example, D-glutamic acid is found in the cell walls of many bacteria, D-alanine is present in the larvae or pupae of some insects, and D-serine can be isolated from earthworms.

AMIDES

amide (AAH-mid)

Amides are organic compounds derived from carboxylic acids by replacing the —OH group with an amino group to give the amide group,

$$-C\underset{N}{\overset{O}{\diagup}}$$

Structure

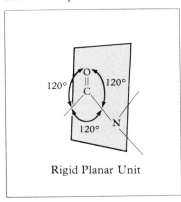

Fig. 15-4 Geometry of the Amide Group

Since amides are nitrogen analogs of esters, the geometry of the amide group is similar to that of the ester group. As illustrated in Figure 15-4, bond angles about the carbonyl carbon are 120°, and the carbon, oxygen, and nitrogen atoms are all in the same plane. The bond between carbon and nitrogen is called an amide bond or an amide linkage. It is not a simple single bond as it appears to be; instead, a resonance form can be drawn for amides in which a double bond appears in this position.

$$R-\overset{\overset{:\ddot{O}:}{\|}}{C}-\ddot{N}\diagup \longleftrightarrow R-\overset{:\ddot{O}:^-}{\underset{}{C}}=\overset{+}{N}\diagup$$

Thus, the carbon-nitrogen bond is somewhat like a double bond, and there is no rotation associated with it. This lack of rotation freezes the amide into a rigid plane; in contrast, an ester has free rotation about the corresponding carbon-oxygen bond. The rigid amide structure is an important factor in the spatial arrangement of protein molecules.

Amides may be formed from ammonia or from amines. Representative structures are

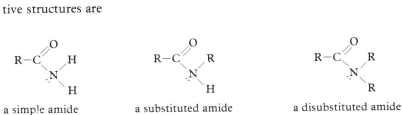

a simple amide   a substituted amide   a disubstituted amide

Nomenclature

Amides are named as derivatives of carboxylic acids, and for simple amides, common names are more popular than IUPAC names. Both types of names are derived by replacing the name ending of the carboxylic acid with "-amide." In the case of common names, it is the "-ic" ending of the common name of the carboxylic acid which is replaced; for IUPAC names, the "-oic" ending of the carboxylic acid IUPAC name is replaced. Examples for simple amides are

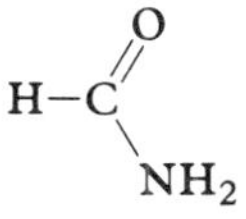

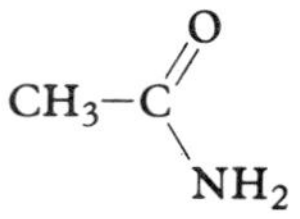

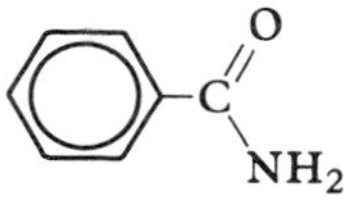

	H–C(=O)NH₂	CH₃–C(=O)NH₂	C₆H₅–C(=O)NH₂
Common Name:	formamide	acetamide	benzamide
IUPAC Name:	methanamide	ethanamide	benzamide

Substituted amides are named much like simple amides, but in these cases it is necessary to use the prefixes "N–" and "N,N–" to indicate the groups bonded to the nitrogen atom. Examples of substituted amides are

	H–C(=O)N(H)CH₃	H–C(=O)N(CH₃)₂	CH₃–C(=O)N(CH₃)(CH₂CH₃)
Common Name:	N-methyl-formamide	N,N-dimethyl-formamide	N-ethyl-N-methyl-acetamide
IUPAC Name:	N-methyl-methanamide	N,N-dimethyl-methanamide	N-ethyl-N-methyl-ethanamide

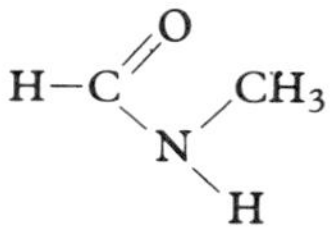

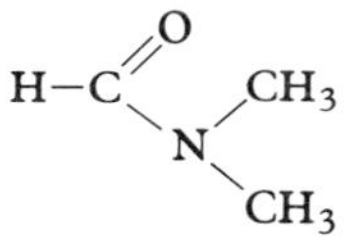

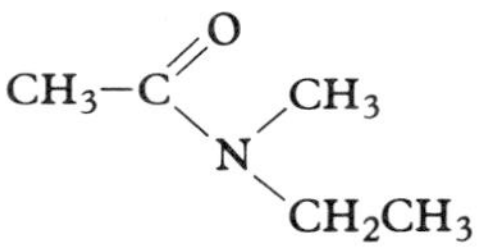

EXAMPLE 15.6

Give either a common name or the IUPAC name for each of the following amides.

(a) 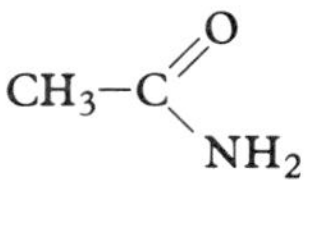 CH₃–C(=O)NH₂

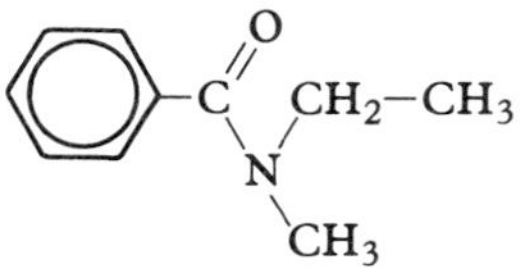

(b) 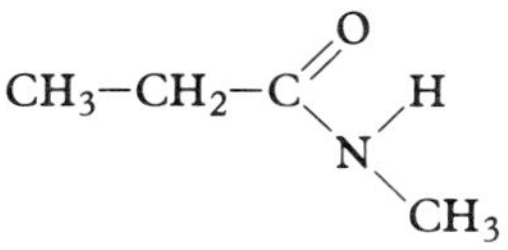 CH₃–CH₂–C(=O)N(H)(CH₃)

(c) C₆H₅–C(=O)N(CH₃)(CH₂–CH₃)

SOLUTION

	Common Name	IUPAC Name
(a)	acetamide	ethanamide
(b)	N-methylpropionamide	N-methylpropanamide
(c)	N-ethyl-N-methylbenzamide	N-ethyl-N-methylbenzamide

EXERCISE 15.6 Give either a common name or the IUPAC name for each of the following amides.

(a) $CH_3-CH_2-CH_2-C\underset{NH_2}{\overset{O}{\lessgtr}}$

(b) $H-C\underset{N}{\overset{O}{\lessgtr}}\underset{CH_2-CH_3}{\overset{H}{}}$

(c) $CH_3-C\underset{N}{\overset{O}{\lessgtr}}\underset{CH_3}{\overset{CH_3}{}}$

Physical Properties

Most amides are solids at room temperature and have little odor. The simplest one, formamide, is a liquid at room temperature.

$$H-C\underset{NH_2}{\overset{O}{\lessgtr}}$$
formamide

It is the only unsubstituted amide which is a liquid; all others are solids. This is because the unsubstituted amides can form the most intermolecular hydrogen bonds; thus their crystals are held together by more hydrogen bonds than those of substituted amides. Substitution of hydrogen atoms by alkyl or aromatic groups reduces the number of intermolecular hydrogen bonds, and thus lowers melting points.

Intermolecular hydrogen bonding in an unsubstituted amide

Disubstituted amides have the lowest melting and boiling points because they cannot form intermolecular hydrogen bonds.

Amides are rather water soluble, especially those having fewer than six carbon atoms. Their water solubility arises from the ability of amides to form hydrogen bonds with water. Even disubstituted amides can do so because of the presence of the carbonyl oxygen. However, the carbon-nitrogen bond in amides is not polar enough for nitrogen to participate in hydrogen bonding.

Hydrogen bonding between water and an unsubstituted amide

EXAMPLE 15.7

Show how each of the following amides forms hydrogen bonds with water.

(a) $CH_3-\overset{\overset{O}{\|}}{C}-NH_2$ (b) $CH_3-\overset{\overset{O}{\|}}{C}-NH-CH_3$ (c) $CH_3-\overset{\overset{O}{\|}}{C}-N\overset{CH_3}{\underset{CH_3}{\diagdown}}$

SOLUTION

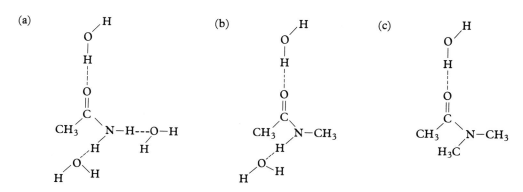

EXERCISE 15.7 Show how each of the following amides forms hydrogen bonds with water.

(a) $H-\overset{\overset{O}{\|}}{C}-NH_2$ (b) $C_6H_5-\overset{\overset{O}{\|}}{C}-N\overset{H}{\underset{CH_3}{\diagdown}}$ (c) $H-\overset{\overset{O}{\|}}{C}-N\overset{CH_3}{\underset{CH_3}{\diagdown}}$

Chemical Properties

Like esters, amides are rather inert, and the only chemical properties of interest are their neutrality and their susceptibility to hydrolysis.

Neutrality

Although they are formed from carboxylic acids and amines, amides are neither acidic nor basic in aqueous solution. The carbonyl group bonded to nitrogen destroys the basicity of the original amine, because the electron-withdrawing effect of the carbonyl group makes the non-bonded pair of electrons on nitrogen unavailable for bond formation to a proton:

$$R-C\overset{O}{\underset{N:}{\diagup}}$$

This is also the reason why the nitrogen atom in amides does not participate in hydrogen bonding.

Hydrolysis

Amides are very resistant to reaction with water, and hydrolysis occurs only in the presence of a catalyst. In the laboratory, strong acids and strong bases serve this purpose.

$$R-\overset{O}{\underset{\|}{C}}-NH_2 + H_2O \underset{}{\overset{H^+ \text{ or } OH^-}{\rightleftharpoons}} R-\overset{O}{\underset{\|}{C}}-O^- + NH_4^+$$

Since proteins are polymers of amino acids joined by amide linkages, they can be hydrolyzed to produce the constituent amino acids. Hydrochloric acid is often used as a catalyst for protein hydrolysis in the laboratory, but in an animal, enzymes called proteases (or proteolytic enzymes) serve as natural catalysts.

a protein

$$\{NH-\underset{R_1}{CH}-\overset{O}{\underset{\|}{C}}-NH-\underset{R_2}{CH}-\overset{O}{\underset{\|}{C}}-NH-\underset{R_3}{CH}-\overset{O}{\underset{\|}{C}}-NH-\underset{R_4}{CH}-\overset{O}{\underset{\|}{C}}-NH-\underset{R_5}{CH}-\overset{O}{\underset{\|}{C}}\} + H_2O \xrightleftharpoons{\text{protease}}$$

$$H_3\overset{+}{N}-\underset{R_1}{CH}-CO_2^- + H_3\overset{+}{N}-\underset{R_2}{CH}-CO_2^- + H_3\overset{+}{N}-\underset{R_3}{CH}-CO_2^- + H_3\overset{+}{N}-\underset{R_4}{CH}-CO_2^- + H_3\overset{+}{N}-\underset{R_5}{CH}-CO_2^- + \text{etc.}$$

Several proteases exist in the digestive tract of animals for catalyzing hydrolysis of protein in the diet. Hydrolysis in the stomach and the intestine degrades protein molecules into amino acids. These are then absorbed into the bloodstream and delivered to cells where they are utilized.

EXAMPLE 15.8

Give the hydrolysis products for each of the following amides.

(a) 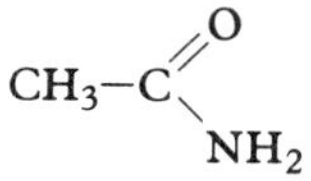 $CH_3-C\underset{NH_2}{\overset{O}{\Vert}}$ (b) $H-C\underset{\underset{CH_3}{N}}{\overset{O}{\Vert}}H$ (c) $CH_3-C\underset{\underset{CH_3}{N}}{\overset{O}{\Vert}}CH_3$

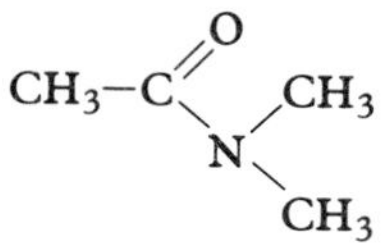

SOLUTION

(a) $CH_3-CO_2^- + NH_4^+$ (c) $CH_3-CO_2^- + (CH_3)_2NH_2^+$

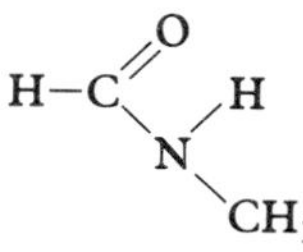

(b) $H-CO_2^- + CH_3-NH_3^+$

EXERCISE 15.8 Give the hydrolysis products for each of the following amides.

(a) 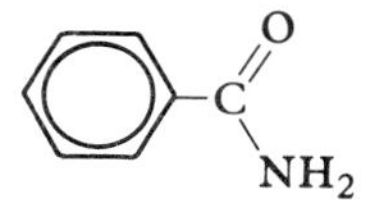 $C_6H_5-C(=O)-NH_2$

(b) $CH_3-C(=O)-N(H)(CH_2-CH_3)$

(c) $C_6H_5-C(=O)-N(CH_3)(CH_2-CH_3)$

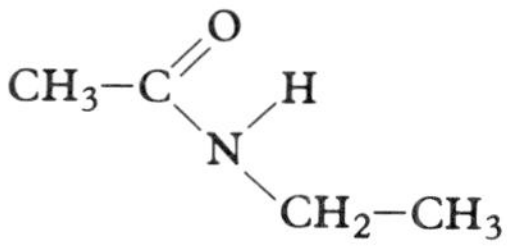

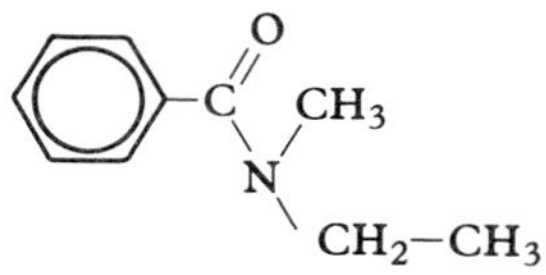

Examples of Amides

Amides may be simple compounds such as urea, a substance found in urine, or they may be such complicated structures as the polymers of nylon and proteins. Some common amides are

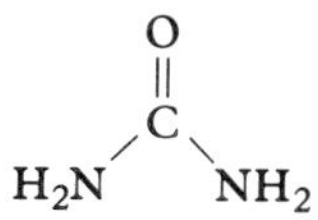

urea (excreted in the urine)

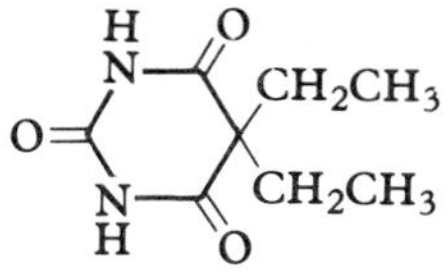

Barbital (a barbiturate)

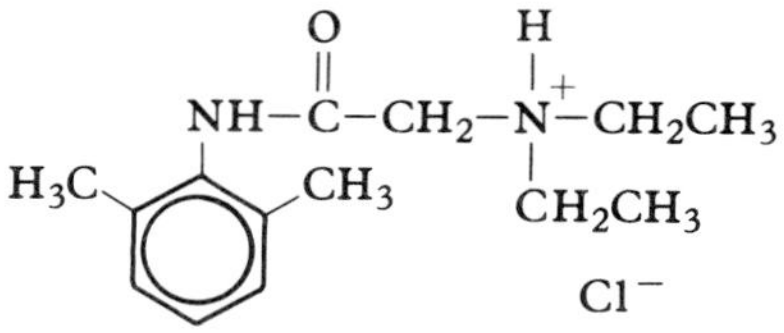

lidocaine (Xylocaine, a local anesthetic)

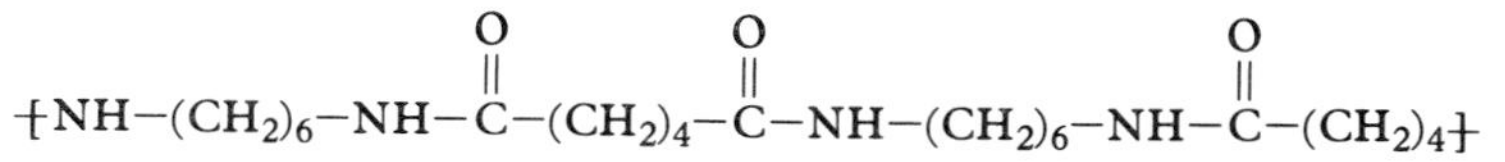

$\{NH-(CH_2)_6-NH-\overset{O}{\underset{\Vert}{C}}-(CH_2)_4-\overset{O}{\underset{\Vert}{C}}-NH-(CH_2)_6-NH-\overset{O}{\underset{\Vert}{C}}-(CH_2)_4\}$

Nylon (a polymeric synthetic fiber)

PERSPECTIVE: Nylon, A Synthetic Polyamide

Nylon, an example of a human attempt to mimic nature, is a polyamide whose structure is remarkably similar to that of proteins. It was first developed in the 1930s by a DuPont research team. One type of nylon is prepared from 1,6-hexanediamine and adipic acid, as illustrated below. This type of nylon is designated Nylon 66 because each monomer contains 6 carbon atoms.

Nylon can be drawn, cast, or extruded through spinnerets from a melt or solution to form fibers, filaments, bristles, or sheets, and it can be formed into molded products. By varying the structures of the diamine and the dicarboxylic acid starting materials, products can be made that are hard and tough or soft and rubbery.

$$H_2N-(CH_2)_6-NH_2 + HO_2C-(CH_2)_4-CO_2H \longrightarrow \{NH-(CH_2)_6-NH-\overset{O}{\underset{\|}{C}}-(CH_2)_4-\overset{O}{\underset{\|}{C}}\}$$

1,6-hexanediamine   adipic acid   Nylon 66

HETEROCYCLIC NITROGEN COMPOUNDS

Cyclic compounds whose rings contain one or more atoms of elements other than carbon are called *heterocyclic compounds*. Many nitrogen heterocyclics occur in living cells, and for this reason, we will briefly consider the most common examples of those.

Five-Membered Rings

pyrrole (pih-ROLE)
imidazole (ih-MIH-dah-zole)

This group of nitrogen heterocyclics includes *pyrrolidine, pyrrole,* and *imidazole*. The structures for these compounds are given here along with examples of biomolecules in which they occur.

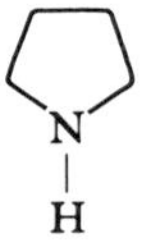

pyrrolidine

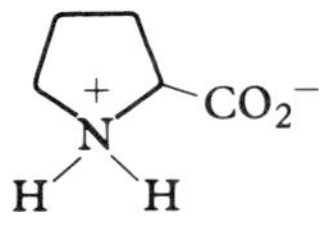

proline
(an amino acid)

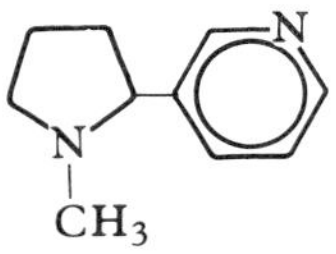

nicotine

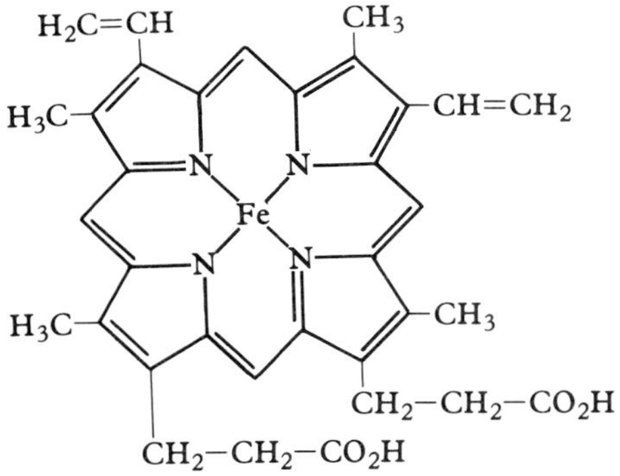

pyrrole

the heme group of hemoglobin

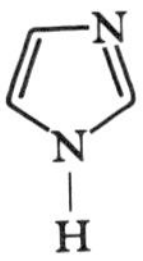

imidazole

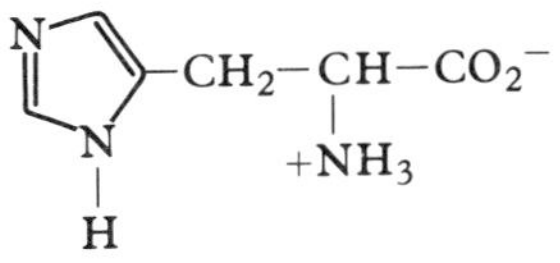

histidine
(an amino acid)

Six-Membered Rings

These nitrogen heterocyclics include *pyridine* and *pyrimidine*, whose structures are shown along with those of biomolecules containing pyridine and pyrimidine.

pyridine (PIH-rih-deen)
pyrimidine (pih-RIH-mih-deen)

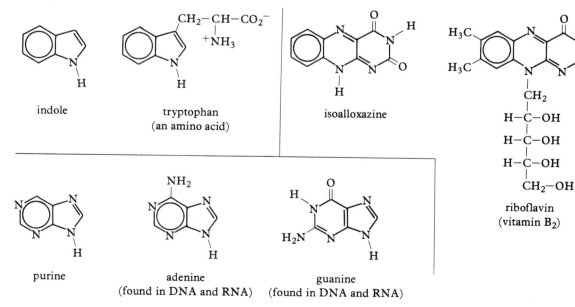

Fused-Ring Nitrogen Heterocyclics

Examples of heterocyclic nitrogen compounds containing fused rings are *indole*, *purine*, and *isoalloxazine*. These ring systems are also found in many naturally occurring compounds, some of which are illustrated here.

purine (PURE-reen)
isoalloxazine (eye-so-al-LOCKS-ah-zeen)

SUMMARY

Amines: Amines are derived from ammonia by replacing one or more hydrogen atoms with alkyl or aromatic groups. Amines have tetrahedral geometry with bond angles approaching 109.5° in those with three substituents. Amines are divided into three subclasses—primary, secondary, and tertiary—according to the number of substituents bonded to nitrogen. Most amines are liquids; the simple ones have odors resembling ammonia, while the more complex amines have fishy odors. Primary and secondary amines can form intermolecular hydrogen bonds, but nitrogen makes weaker hydrogen bonds than oxygen, and these amines usually have lower boiling points than comparable alcohols. The boiling points of tertiary amines resemble those of alkanes of similar molecular weight. All subclasses of amines containing six or fewer carbons are soluble in water, but the water solubility of amines with more than six carbon atoms drops off because of the large hydrocarbon part of their structures. Amines have two important chemical properties: basicity and formation of amides. Amines are bases because the nitrogen atom has nonbonded electron pairs that can accept protons. Amines react with acids to form salts that are usually water soluble. Ammonia and amines react with carboxylic acids at elevated temperatures to produce amides. Amides are also obtained by heating the ammonium or amine salt of a carboxylic acid. Amides are often synthesized by first converting the carboxylic acid to a more reactive derivative, such as an acid chloride.

Amino Acids: Amino acids, organic compounds containing both an amino acid and a carboxyl group, are the building blocks of proteins. There are 20 amino acids commonly found in proteins; all of these are α-amino acids. They can be classified by their side chains as nonpolar (hydrophobic), neutral (uncharged) polar, basic (positively charged), and acidic (negatively charged). Only L-amino acids are found in proteins. Amino acids have high melting points because they exist as dipolar ions (zwitterions) in the solid state. Amino acids have three important chemical features: their amphoteric nature, their ability to react with ninhydrin, and their reaction under special conditions to form amides and polyamides. Amino acids are amphoteric because they can act as both acids and bases in aqueous solution. Those containing ionizable side chains behave as buffers when incorporated into proteins. Because the average charge on a group of amino acid molecules depends on the pH of the solution, a mixture of amino acids can be separated by electrophoresis. Besides the 20 standard amino acids found in most proteins, a few rare ones have been isolated from highly specialized proteins such as collagen. β-Amino acids and D-amino acids are found in free or combined form in nature, but never in proteins.

Amides: Amides are derived from carboxylic acids by replacing the —OH of the carboxyl group with an amino group. Amides are nitrogen analogs of esters and thus have bond angles of 120° about the carbonyl carbon. The carbon-nitrogen bond possesses double-bond character due to resonance, and there is no rotation associated with it. This lack of rotation freezes the amide into a rigid plane, an important factor in protein structure. Most amides are solids at room temperature and thus have no odor. Unsubstituted amides have the highest melting points because of their large number of intermolecular hydrogen bonds. Substitution of hydrogen atoms by other groups reduces the number of intermolecular hydrogen bonds and thus lowers melting points. Amides are rather water soluble, especially those with fewer than six carbon atoms. They are rather inert compounds, and hydrolysis occurs only in the presence of strong acids, strong bases, or enzymes.

Heterocyclic Nitrogen Compounds: Cyclic compounds which have other atoms in the rings in addition to carbon are called heterocyclic compounds. Heterocyclic nitrogen compounds, common in living cells, can be classified as those containing five-membered rings, six-membered rings, and fused rings. Five-membered nitrogen heterocyclics include pyrrolidine, pyrrole, and imidazole, while examples of six-membered nitrogen heterocyclics are pyridine and pyrimidine. Common fused-ring nitrogen heterocyclic compounds are indole, purine, and isoalloxazine.

STUDY QUESTIONS AND PROBLEMS

1. Define the following terms:
   - (a) amine
   - (b) ptomaine
   - (c) quaternary ammonium salt
   - (d) acid chloride
   - (e) epinephrine
   - (f) catecholamine
   - (g) norepinephrine
   - (h) amphetamines
   - (i) folic acid
   - (j) amino acid
   - (k) guanidino group
   - (l) zwitterion
   - (m) amphoteric substance
   - (n) ampholyte
   - (o) electrophoresis
   - (p) isoelectric point

(q) ninhydrin (t) protease
(r) collagen (u) heterocyclic compounds
(s) amide

2. Classify each of the following amines as primary, secondary, or tertiary.

(a) N—CH₃

(b) CH₃—CH₂—CH—CH₂—NH₂
              |
              CH₃

(c) ⌬—NH₂

(d)
       H
       |
       N
      / \
     /___\

(e) CH₃—CH₂—CH₂—NH—CH₃

(f) ⌬N

(g) CH₃—N⟨CH₃ / CH₃

(h) CH₃—NH—C(CH₃)—CH₃ (with CH₃ above and below the C)

3. Give an acceptable name for each of the following amines.

(a) CH₃—NH—CH₂—CH₃
(b) CH₃—CH—NH₂
         |
         CH₃
(c) CH₃—CH—CH₂—CH—CO₂H
         | |
         NH₂   Cl
(d) ⌬—NH—⌬
(e) H₂N—CH₂—CO₂H
(f) Br—⌬—NH₂

4. Draw the correct structure for each of the following amines.
(a) dimethylethylamine
(b) 2-aminoethanol
(c) β-aminobutyric acid
(d) N-methylaminoacetic acid
(e) m-chloroaniline
(f) 2-amino-4-methylheptane

5. What is wrong with the term "ptomaine poisoning"?

6. Explain why boiling points of amines are lower than those for corresponding alcohols.

7. Explain why boiling points of tertiary amines are much lower than those of corresponding primary and secondary amines.

8. Explain why all classes of low molecular weight amines are soluble in water.

9. Write an equation to illustrate the weak basicity of amines.

10. When diethylamine is dissolved in water, the solution turns red litmus blue. Give an explanation to account for this observation.

11. Why is lemon juice often used as a seasoning to eliminate fish odor?

12. Complete the following equations.

(a) CH₃—CH—CO₂H + CH₃—NH₂ $\xrightarrow{\Delta}$
        |
        CH₃

(b) ⌬—NH₂ + HCl ⟶

(c) ⌬—CO₂⁻NH₄⁺ $\xrightarrow{\Delta}$

(d) ⌬—CH₂—NH—CH₃ + HBr ⟶

(e) CH₃—CO₂⁻ $\overset{+}{N}H_3CH_3$ $\xrightarrow{\Delta}$

(f) CH₃—CH₂—$\overset{O}{\overset{\|}{C}}$—Cl + ⌬—NH₂ ⟶

13. Give a structural example of a quaternary ammonium salt. How is this kind of amine salt different from others?

14. How are the two hormones epinephrine and norepinephrine alike? How are they different?

15. Why are local anesthetics often formulated as amine salts?

16. Why is epinephrine administered in the same solution with local anesthetics?

17. Classify the 20 common α-amino acids by their side chains. Include complete structures and names.
18. Arrange the 4 classes of amino acids in order of decreasing water solubility.
19. Explain why amino acids have physical properties like those of inorganic salts.
20. Show how the amino acid glycine behaves as both an acid and a base.
21. Write all possible ionic forms for each of the following amino acids.
    (a) valine
    (b) tyrosine
    (c) aspartic acid
    (d) arginine
22. Estimate the overall charge at pH 7 on each of the amino acids as +, −, or 0.
    (a) isoleucine
    (b) methionine
    (c) glutamine
    (d) glutamic acid
    (e) lysine
23. Explain how ninhydrin is used as a visual detector for amino acids.
24. Why is a special condensing agent necessary in the laboratory synthesis of amides from amino acids?
25. Explain why it is difficult to synthesize polyamide chains of specific sequence and length from amino acids in the laboratory.
26. What do proteins and nylon have in common?
27. How does an α-amino acid differ from a β-amino acid?
28. Give five examples of unusual amino acids not included in the 20 standard amino acids. What are the sources of your examples?
29. Explain why there is no rotation about an amide bond.
30. Give either a common name or the IUPAC name for each of the following amides.

(a) $H-C\overset{O}{\underset{}{\|}}-N(CH_3)_2$

(b) $CH_3-CH_2-\overset{O}{\underset{\|}{C}}-NH_2$

(c) $C_6H_5-\overset{O}{\underset{\|}{C}}-NH-CH_3$

(d) $C_6H_5-\overset{O}{\underset{\|}{C}}-NH_2$

(e) $CH_3-\overset{O}{\underset{\|}{C}}-N(CH_2-CH_3)(CH_2-CH_2-CH_3)$

31. Explain why disubstituted amides have lower melting points than unsubstituted and monosubstituted amides.
32. Show how each of the following forms hydrogen bonds with water.

    (a) $CH_3-CH_2-\overset{O}{\underset{\|}{C}}-NH-CH_3$

    (b)

    (c) $CH_3-\overset{O}{\underset{\|}{C}}-N(CH_2-CH_3)_2$

    (d) CH_3-NH_2

    (e) piperidine

33. Why are amides neutral compounds?
34. Write an equation, using a specific reactant, to illustrate hydrolysis of an amide.
35. Give the hydrolysis products for each amide in question 30.
36. Give a structural example for each of the following heterocyclic nitrogen compounds.
    (a) pyrrolidine   (e) pyrimidine
    (b) pyrrole   (f) indole
    (c) imidazole   (g) purine
    (d) pyridine   (h) isoalloxazine

CHAPTER 16
PROTEINS

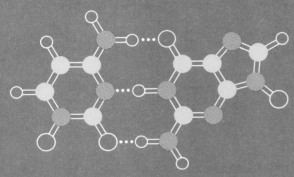

FUNCTIONS OF PROTEINS
PROTEIN STRUCTURE
- Primary Structure
- Secondary Structure
- Tertiary Structure
- Quaternary Structure

CLASSES OF PROTEINS
PROPERTIES OF PROTEINS
SEPARATION OF PROTEIN MIXTURES
- Liquid Chromatography
- Electrophoresis

Proteins are probably the most important organic material in living cells; they are fundamental to cell structure and function. Thousands of proteins are known, and hundreds have been isolated in pure crystalline form. All contain carbon, hydrogen, nitrogen, and oxygen, and most contain sulfur. Other elements such as phosphorus, iron, zinc, and copper are found in some proteins.

Proteins are one of the three major classes of nutrients for animals; the other two are fats and carbohydrates. Proteins are necessary dietary components because they are the source of amino acids needed by animals to synthesize their own proteins. Because plants contain relatively small amounts of proteins, animals that eat only plant materials (ruminants, for example) must eat very large quantities. Other animals, such as humans, have traditionally obtained proteins mainly from animals and animal products—meat, milk, and eggs. In modern times, however, beans and peas are being used more and more as inexpensive sources of dietary protein.

In this chapter and others to follow, we will immerse ourselves completely in the world of biological chemistry—the study of substances found in living organisms and the changes they undergo during the life process. You will find that this branch of chemistry draws heavily on concepts of inorganic and organic chemistry developed earlier.

FUNCTIONS OF PROTEINS

The importance of proteins stems from the fact that they play crucial roles in all biological processes. The various functions of proteins are discussed below and summarized in Table 16-1.

Catalysis

Enzymes represent the largest category of proteins. Several thousand enzymes are known; each one is responsible for catalyzing a chemical reaction in a living organism. Enzyme-catalyzed reactions range from simple processes, such as hydration of carbon dioxide to produce carbonic acid, to such complex systems as the duplication of hereditary material in the nucleus of a cell.

Transport

Another important function of proteins is the transport of small molecules and ions. For example, serum albumin is a protein which carries fatty acids between fat (adipose) tissue and other tissues or organs. Other transport proteins are hemoglobin, which carries oxygen from the lungs to other body tissues, and transferrin, a carrier of iron in blood plasma.

Storage

Some proteins are known for their roles in storing small molecules or ions. Ovalbumin, for example, is a storage form of amino acids to be used as nutrients by chick embryos in eggs. Casein in milk and gliadin in wheat seeds also store amino acids for future use by young animals and plants, respectively. Ferritin, a liver protein, attaches to iron ions to form a storage complex in humans and other animals.

Motion

Proteins are the major components of muscle. In particular, actin and myosin are long filament proteins which slide along each other during muscle contraction. Other forms of motion involving proteins are the movement of chromosomes during cell division and the swimming action of sperm as they seek out eggs in the fertilization process.

Structural Support

Collagen, a fiberlike protein, is responsible for the mechanical strength of skin and bone. Other examples of structural proteins are α-keratin, found in hair, horn, skin, nails, and feathers, and elastin, found in yellow elastic tissue.

Protection

The most important protective proteins are the *antibodies*. These are highly specific proteins that combine with viruses, bacteria, and other foreign substances that have made their way into the blood or tissues. In this way antibodies neutralize foreign material and defend animals from outside invaders. Other protective proteins are those involved in blood clotting, such as thrombin and fibrinogen. These proteins protect animals by helping to prevent loss of blood.

Chemical Messengers

The *hormones* are chemical messengers produced in endocrine glands to act on distant parts of the body. Although hormones have a variety of structures, many of them are proteins. Examples of these are growth hormone, which is responsible for the rate of growth of young animals, and thyrotropin, a protein which stimulates activity of the thyroid gland.

hormones: internally secreted compounds that affect the functions of target organs or tissues when transported to them by the body fluids.

Transmission of Nerve Impulses

Certain proteins serve as receptors of small molecules at junctions between nerve cells. In this way, they participate in the transmission of nerve impulses. One such receptor protein is rhodopsin, a red protein involved in the vision process in the rod cells of the retina.

Toxins

Toxins are substances which are poisonous to plants or animals. Many natural toxins are proteins; these include diphtheria toxin and the toxin of the bacterium *Clostridium botulinum*, responsible for the food poisoning called botulism.

Physiological Regulation

Many enzymes are involved in the regulation of metabolism. These enzymes are sensitive to inhibitors or activators of specific metabolic reactions. Other proteins serve roles in controlling gene expression, so that only specific segments of a gene are active at any one time. Still other regulatory proteins play a role in guiding cell differentiation, the developmental process in which cells in one kind of tissue take on different characteristics than those in another kind of tissue.

PROTEIN STRUCTURE

Since the structures of proteins are immensely complicated, we will approach the subject in a series of steps, each successive step corresponding to a higher level of structural complexity.

Primary Structure

The *primary structure* of a protein is simply the order in which the amino acids are linked together in the protein. The amide linkages that hold the amino acids together are called *peptide bonds*. If only a few

Table 16-1 Biological Functions of Proteins

Protein Function and Examples	Occurrence or Role
Catalysis	
Lactate dehydrogenase	Oxidizes lactic acid
Cytochrome c	Transfers electrons
DNA polymerase	Replicates and repairs DNA
Transport	
Hemoglobin	Transports O_2 in blood
Myoglobin	Transports O_2 in muscle cells
Serum albumin	Transports fatty acids in blood
Transferrin	Transports iron in blood
Ceruloplasmin	Transports copper in blood
Storage	
Ovalbumin	Egg-white protein
Casein	A milk protein
Ferritin	Stores iron in the spleen and liver
Gliadin	Stores amino acids in wheat
Zein	Stores amino acids in corn
Motion	
Myosin	Thick filaments in muscle fiber
Actin	Thin filaments in muscle fiber
Dynein	Movement of cilia and flagella
Structure	
Viral-coat proteins	Sheath around nucleic acid of viruses
Glycoproteins	Cell coats and walls
α-Keratin	Skin, hair, feathers, nails, and hoofs
β-Keratin	Silk of cocoons and spider webs
Collagen	Fibrous connective tissue
Elastin	Elastic connective tissue
Protection	
Antibodies	Form complexes with foreign proteins
Complement	Complexes with some antigen-antibody systems
Fibrinogen	Involved in blood clotting
Thrombin	Involved in blood clotting
Chemical Messengers (Hormones)	
Insulin	Regulates glucose metabolism
Adrenocorticotrophic hormone	Regulates corticosteroid synthesis
Growth hormone	Stimulates growth of bone
Nerve Impulse Transmission	
Rhodopsin	Involved in vision
Acetylcholine receptor protein	Impulse transmission in nerve cells
Toxins	
Clostridium botulinum toxin	Causes bacterial food poisoning
Diphtheria toxin	A bacterial toxin
Snake venoms	Enzymes that hydrolyze phosphoglycerides
Physiological Regulation	
Repressor proteins	Regulate gene expression in bacteria
Nerve growth factor	Guides formulation of neural networks in animals

amino acids are joined together, as in the following diagram, the structure is called a *peptide*. Often the prefixes "di-", "tri-", "tetra-", and so on, are used to indicate the number of amino acids in a peptide.

$$\underset{\text{Gly}}{H_3\overset{+}{N}-\underset{H}{\overset{H}{C}}-\overset{O}{\overset{\|}{C}}}-\underset{\text{Ala}}{\underset{CH_3}{\overset{H}{N}-\underset{H}{\overset{H}{C}}-\overset{O}{\overset{\|}{C}}}}-\underset{\text{Ser}}{\underset{\underset{OH}{CH_2}}{\overset{H}{N}-\underset{H}{\overset{H}{C}}-\overset{O}{\overset{\|}{C}}}}-\underset{\text{Phe}}{\underset{CH_2-C_6H_5}{\overset{H}{N}-\underset{H}{\overset{H}{C}}-\overset{O}{\overset{\|}{C}}}O^-$$

a tetrapeptide

Another term for an amino acid in a polymeric chain is *residue*. Thus, a dipeptide contains two amino acid residues, a tripeptide contains three residues, and so forth. The term *polypeptide* refers to peptides containing 20 or more residues. Proteins usually have at least 50 residues, but sometimes the number may be as high as 2,500.

Peptide structures are written with the *terminal amino group* to the left, the *terminal carboxyl group* to the right, and the amino acid side chains extending from the main chain of peptide linkages. They may also be written in shorthand form by using three-letter abbreviations or one-letter symbols for the amino acid residues, as in the following example. (Please note that some polypeptides are circular and thus do not have terminal amino and carboxyl groups.)

terminal amino group (N-terminus): the free α-amino group at one end of a peptide or protein chain; this amino group is protonated at physiological pH.

terminal carboxyl group (C-terminus): the free α-carboxyl group at the opposite end from the terminal amino group of a peptide or protein chain; this carboxyl group is ionized at physiological pH.

a pentapeptide

Tyr-Gln-Arg-Cys-Val

Y - Q - R - C - V

In addition to the peptide bonds in proteins, another type of covalent bond exists. This is the disulfide bond which forms between two cysteine residues. The cystine disulfide bond can cross-link two separate polypeptide chains or it can join cysteine residues in the same chain, as in oxytocin:

```
            Ile—Gln
         Tyr    Asn
          |      |
    Cys—S—S—Cys—Pro—Leu—Gly—NH₂*
```

$$*(Gly-NH_2 = -\underset{H}{\overset{H}{C}}-\underset{NH_2}{\overset{O}{C}})$$

Oxytocin
(a peptide hormone)

amino acid composition: the relative proportions of amino acids in a protein.

Proteins can be hydrolyzed completely to their constituent amino acids by use of strong acid or enzyme catalysts. Then the amino acids are identified and their quantities measured in order to determine the *amino acid composition* of a protein; examples of the amino acid composition of several proteins are shown in Table 16-2. In addition, biochemists have devised techniques for finding the order in which the residues are linked together. Thus, the primary structures for many proteins are known now, and more are being determined each day.

Table 16-2 Amino Acid Composition of Some Proteins (Number of Residues per Molecule of Protein)

Constituent	Insulin (Bovine)	Cytochrome c (Horse)	Hemoglobin (Human)	Myoglobin (Human)	β-Keratin (Silk)
Alanine	3	6	72	12	45
Arginine	1	2	12	2	3
Asparagine	3	5	20	3	1
Aspartic acid	0	3	30	8	4
Cysteine	0	2	6	1	2
Cystine	6	0	0	0	0
Glutamic acid	4	9	24	14	4
Glutamine	3	3	8	7	0
Glycine	4	12	40	15	74
Histidine	2	3	38	9	2
Isoleucine	1	6	0	8	1
Leucine	6	6	72	17	2
Lysine	1	19	44	20	5
Methionine	0	2	6	3	0
Phenylalanine	3	4	30	7	4
Proline	1	4	28	5	7
Serine	3	0	32	7	17
Threonine	1	10	32	4	2
Tryptophan	0	1	6	2	1
Tyrosine	4	4	12	2	23
Valine	5	3	62	7	9
Total	51	104	574	153	206

Many peptides are found in nature, often with unusual features. For example, glutathione contains a peptide bond involving the γ-carboxyl group of glutamic acid. The muscle dipeptide carnosine contains a β-amino acid, and some peptide antibiotics, tyrocidin A, for example, contain D-amino acids. These peptides, along with examples of peptide hormones, are shown in Figure 16–2.

Fig. 16-1 Some Peptides of Nonprotein Origin. (The arrows, where used, indicate the direction of peptide bonds, going from the α-carboxyl group of one residue to the α-amino group of the next.)

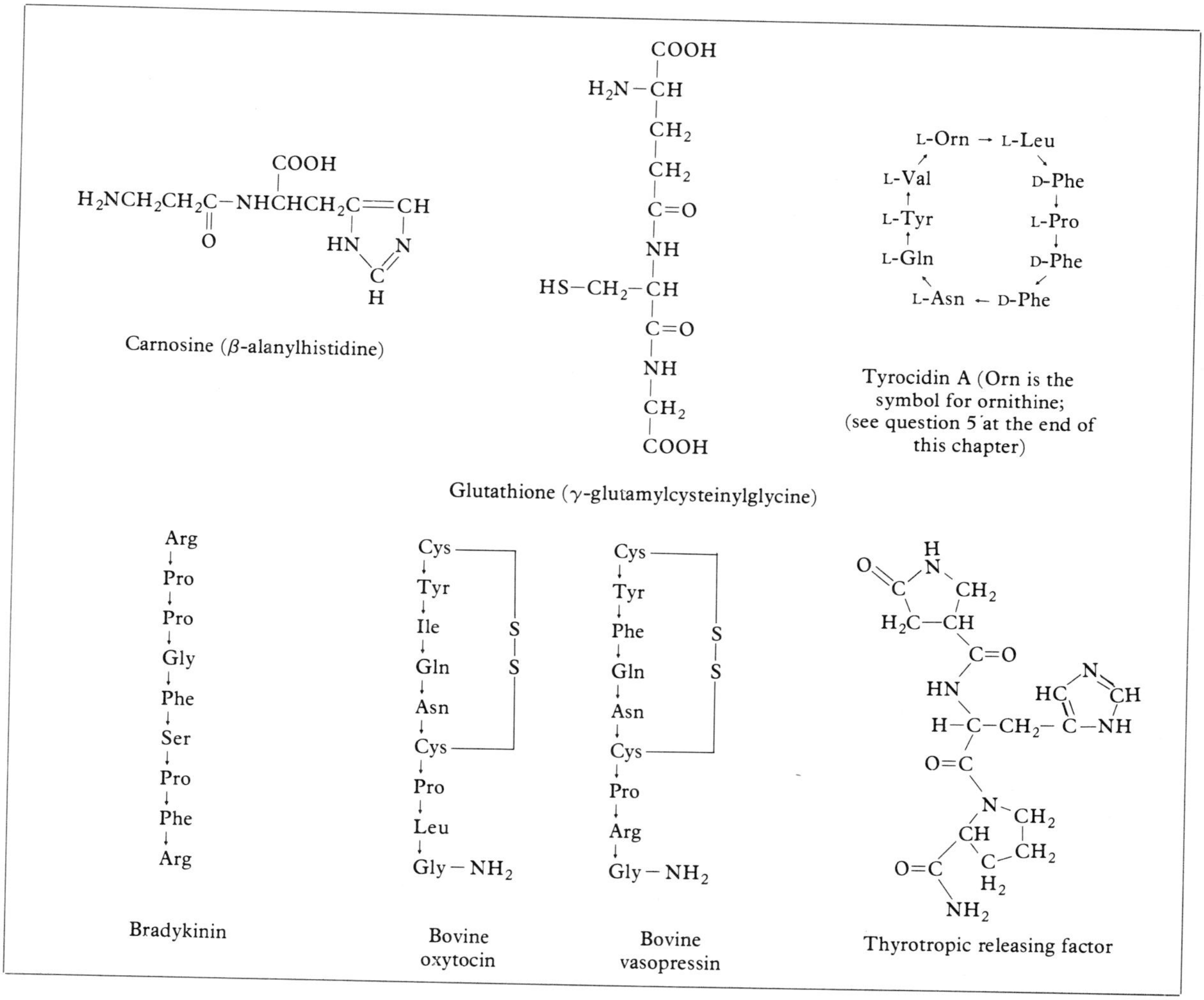
Carnosine (β-alanylhistidine)
Glutathione (γ-glutamylcysteinylglycine)
L-Orn → L-Leu → D-Phe → L-Pro → D-Phe → D-Phe → L-Asn → L-Gln → L-Tyr → L-Val
Tyrocidin A (Orn is the symbol for ornithine; (see question 5 at the end of this chapter)
Arg → Pro → Pro → Gly → Phe → Ser → Pro → Phe → Arg
Bradykinin
Cys → Tyr → Ile → Gln → Asn → Cys → Pro → Leu → Gly−NH2
S–S
Bovine oxytocin
Cys → Tyr → Phe → Gln → Asn → Cys → Pro → Arg → Gly−NH2
S–S
Bovine vasopressin
Thyrotropic releasing factor

EXAMPLE 16.1

Histidine and alanine can combine to form two dipeptide isomers. Draw structures for these isomers.

SOLUTION

The two amino acids form dipeptides by joining carboxyl and amino groups in peptide bonds. Thus, the α-carboxyl group of histidine can attach to the α-amino group of alanine, or the α-carboxyl group of alanine can attach to the α-amino group of histidine:

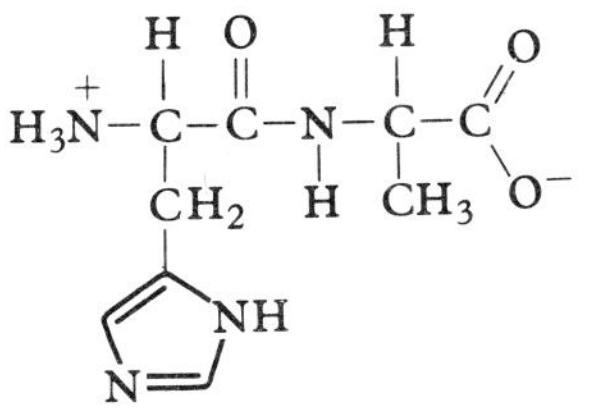

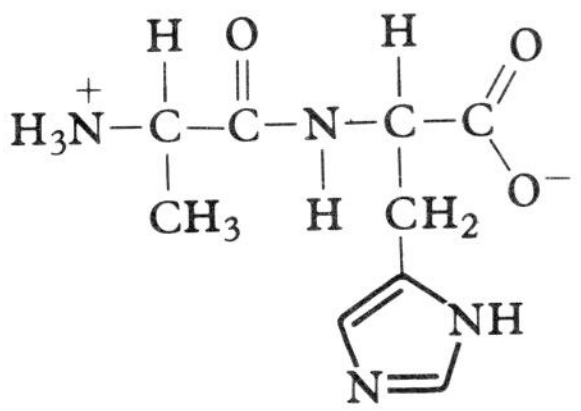

EXERCISE 16.1 Phenylalanine and glycine can combine to form two dipeptide isomers. Draw structures for these isomers.

EXAMPLE 16.2

Draw the structure of the following tripeptide, and name the amino acid residues.

Asn-Val-Thr

SOLUTION

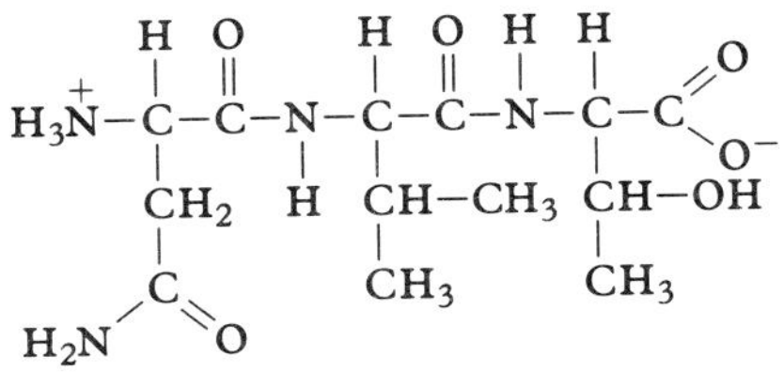

asparagine, valine, threonine

PERSPECTIVE: Endorphins, The Morphine from Within

Morphine and other opiates have been used for centuries to relieve pain. However, it has long puzzled scientists that the human brain should be so sensitive to these poppy seed extracts. One theory that developed was that morphine might mimic molecules that are normally present in animals. This theory received support in 1975 when two pentapeptides with opiatelike activity were isolated from pig brains. These peptides were called enkephalins because of their presence and activity in the brain (*en* means "in" and *kepha* refers to "brain").

Tyr-Gly-Gly-Phe-Met
methionine enkephalin

Tyr-Gly-Gly-Phe-Leu
leucine enkephalin

A year later, polypeptides with similar activity were isolated from the pituitary gland. These polypeptides and the enkephalins have the same amino acids in the first four positions of their primary structures. Mole for mole, the polypeptides have the same remarkable capacity as morphine for relieving pain, and they were named endorphins, a word which means "morphine from within." For example, β-endorphin induces profound analgesia in the whole body of animals which lasts for several hours, accompanied by lowered body temperature and narcosis.

Tyr-Gly-Gly-Phe-Leu-Met-Thr-Ser-Glu-Lys-Ser-Gln-Thr-Pro-Leu-Val-Thr-Leu-Phe-Lys-Asn-Ala-Ile-Val-Lys-Asn-Ala-His-Lys-Lys-Gly-Gln
β-endorphin

The behavior caused by endorphins suggests that they may normally play a role in regulating emotional responses. Research is currently under way to find out whether they may be beneficial in treating mental and emotional disorders.

EXERCISE 16.2 Draw the structure of the following tetrapeptide, and name the amino acid residues.

Glu-Lys-Tyr-Pro

Secondary Structure

The next level of structural complexity in proteins is the *secondary structure*. At this level we are concerned with the spatial arrangement of the peptide chain, the backbone of the protein. There are three types of orderly secondary structure in proteins. We will discuss them in order of their discovery.

The α-Helix

In the 1930s, Linus Pauling and Robert Corey began a study of amino acid and peptide structures. One of the things they found was that the *peptide group* is rigid and planar. This was the first indication that amide groups in general are planar (see chapter 15). Pauling and Corey were able to determine the total geometry of the peptide group, as illustrated in Figure 16-2. Note that the double bond between carbon and oxygen is shorter than the single bond between the α-carbon and the carbonyl carbon. (It is generally true that double bonds are shorter than single bonds.) Notice also that the bond between the carbonyl carbon and nitrogen is of intermediate length, corresponding to its partial double bond character. Since there is no rotation about this bond, the peptide group is frozen into a plane.

In 1951, Pauling and Corey suggested that proteins could exist in highly organized structures. They called one of these the *α-helix*. In this type of secondary structure, the protein takes on a rodlike shape with the tightly coiled peptide backbone on the inside and the side chains extended outward.

peptide group: the atoms involved in a peptide bond,

$$\begin{matrix} & O & \\ & \| & \\ -C & - & N- \\ & & | \\ & & H \end{matrix}$$

Fig. 16-2 (a) The Rigid Planar Unit of the Peptide Group (Bond distances are shown in angstroms.) (b) The bonds linking peptide groups have a large degree of rotational freedom.

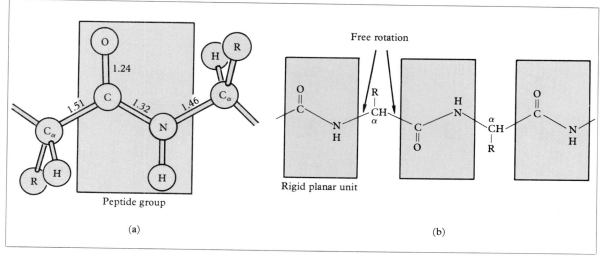

α-helix: the ordered helical structure of some proteins; the first type of organized secondary protein structure to be proposed.

The α-helix is stabilized by hydrogen bonds between carbonyl oxygens and amide hydrogens in adjacent turns of the helix:

$$\text{C}=\text{O}-----\text{H}-\text{N}$$

The carbonyl group of each amino acid residue is hydrogen bonded to the amide group of the amino acid four residues ahead in the chain, so that all of the carbonyl and amide groups in the peptide backbone are hydrogen bonded:

$$-\underset{\underset{H}{|}}{\overset{\overset{H}{|}}{N}}-\underset{\underset{R_1}{|}}{\overset{\overset{H}{|}}{C}}-\underset{\overset{\parallel}{O}}{C}-\underset{\underset{R_2}{|}}{\overset{\overset{H}{|}}{N}}-\underset{\underset{|}{|}}{\overset{\overset{H}{|}}{C}}-\underset{\overset{\parallel}{O}}{C}-\underset{\underset{H}{|}}{\overset{\overset{H}{|}}{N}}-\underset{\underset{R_3}{|}}{\overset{\overset{H}{|}}{C}}-\underset{\overset{\parallel}{O}}{C}-\underset{\underset{|}{|}}{\overset{\overset{H}{|}}{N}}-\underset{\underset{R_4}{|}}{\overset{\overset{H}{|}}{C}}-\underset{\overset{\parallel}{O}}{C}-\underset{\underset{H}{|}}{\overset{\overset{H}{|}}{N}}-\underset{\underset{R_5}{|}}{\overset{\overset{H}{|}}{C}}-\underset{\overset{\parallel}{O}}{C}-$$

Notice that these hydrogen bonds run lengthwise through the helix (see Figure 16-3).

Six years after Pauling and Corey proposed the α-helix, its presence in proteins was detected. Since that time it has been found that the amount of α-helical content is quite variable in proteins. In some it is the major structural theme, while in others there may be little or no α-helix. In the proteins α-keratin (found in hair), myosin and tropomy-

Fig. 16-3 The α-Helix (a) The Entire Helix (b) Cross-sectional View of the α-Helix (c) Schematic Diagram of the α-Helix, Emphasizing Hydrogen Bonds and Amino Acid Side Chains.

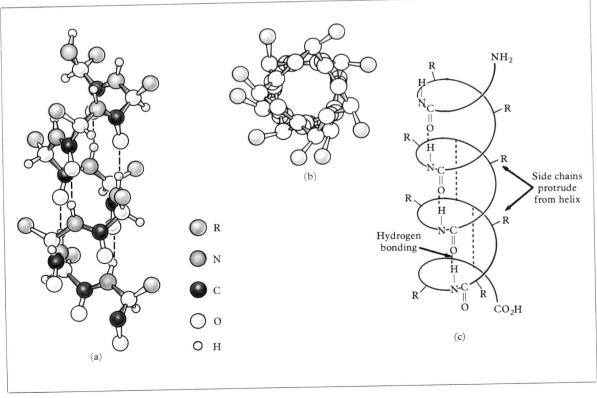

osin (found in muscle), epidermin (found in skin), and fibrin (found in blood clots), two or more helices interact to form a cable (Figure 16-4). The cables make up bundles of fibers which lend strength to the tissue in which they are found.

The β-Pleated Sheet

A second type of secondary structure was proposed by Pauling and Corey to account for the behavior of α-keratin when heated by steam. Under these conditions, fibers of α-keratin can be stretched to almost double their original length. In this stretched form, α-keratin fibers take on a structural resemblance to β-keratin, a similar protein found in spider web fibers, silk, and the scales, claws, and beaks of reptiles and birds. Pauling and Corey thought that perhaps the energy from the steam ruptures the hydrogen bonds that normally stabilize the α-helix. They suggested that the subsequent stretching forces the α helix into a more extended, zigzag conformation, which they designated as β. They went on to propose that polypeptide chains in the β-conformation arrange themselves in pleated sheets, which are cross-linked by hydrogen bonds. Hence the name *β-pleated sheet* was used. In this secondary structure, the hydrogen bonds run at right angles to the length of the sheet, joining different polypeptide chains together. The side chains extend above and below the pleated sheet (see Figure 16-5).

Fig. 16-4 A Cable Formed by Supercoiling of Two α-Helices

Two α-helical polypeptide chains

β-pleated sheet: the ordered pleated sheet structure of some proteins; the second type of organized secondary protein structure to be proposed.

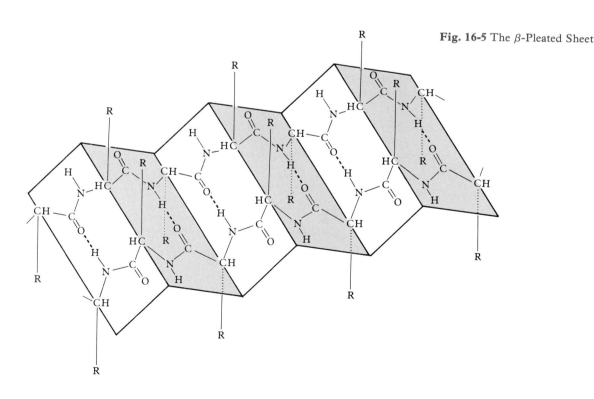

Fig. 16-5 The β-Pleated Sheet

Fig. 16-6 Collagen Fibers Obtained from Human Skin

Fig. 16-7 The Triple Helix of Collagen. (Only α-carbon atoms are shown.)

The high proportion of relatively small side chains, especially glycine and alanine (see Table 16-2), in β-keratin accounts for its β-pleated sheet structure. If the side chains in a protein are large and bulky, or if they have like charges, the pleated sheet structure cannot exist. This is because the bulk and/or the charge repulsions prevent the side chains from coming close enough together to form the β-pleated sheet. The high content of bulky and highly charged side chains of α-keratin causes it to shrink back into an α-helix when cooled.

The Collagen Triple Helix

Collagen is a fiberlike protein which comprises up to 30% of total body protein in animals. The most abundant protein in mammals, it is found primarily in connective tissues such as skin, bone, tendon, cartilage, and teeth. When magnified by the electron microscope (Figure 16-6), collagen fibers show a cross-striation pattern. If it is boiled in water, collagen is converted to gelatin, a mixture of polypeptides with little structural organization.

Collagen has a secondary structure in which the polypeptide chains are arranged in a *triple helix* (Figure 16-7) held together by hydrogen bonds. Proline residues occur at frequent intervals in the primary structure, and these side chains project outwardly from the triple helix. In contrast, glycine residues, which occur in every third position along the chains, occupy interior positions within the triple helix. Thus, the small glycine residues allow the chains to intertwine. No other proteins seem to have the ability to form triple helices.

The triple-stranded cables of collagen are composed of recurring subunits having knobby-shaped "heads." The subunits are arranged head-to-tail in a staggered array (see Figure 16-8) to form the dark cross-bands shown in Figure 16-6.

EXAMPLE 16.3

Indicate the type of secondary structure (α-helix, β-pleated sheet, triple helix) to which each of the following features belongs.

(a) hydrogen bonds between different polypeptide chains
(b) cross-striations
(c) hydrogen bonds within the same polypeptide chain

SOLUTION

(a) β-pleated sheet
(b) triple helix
(c) α-helix

EXERCISE 16.3 Indicate the type of secondary structure to which each of the following features belongs.
(a) glycine residues in every third position
(b) a high proportion of negatively charged amino acid side chains
(c) a high proportion of small amino acid residues

Tertiary Structure

Tertiary structure, the next higher level of complexity in protein structure, refers to the bending and folding of protein helices and pleated sheets. This bending and folding may seem rather disorganized, but nevertheless it results in a favored arrangement for a given protein. The protein *myoglobin* will be used to illustrate features of tertiary structure.

Myoglobin is a relatively small, red protein containing a single polypeptide chain of 153 amino acids of known sequence (see Figure 16-9). It contains a heme group, just as hemoglobin does, which functions as a site for attachment of an oxygen molecule. The biological role of myoglobin is to store oxygen and aid its diffusion through cells in the *skeletal muscle* of animals. It is found in especially high concentrations in deep-sea diving animals such as the whale, seal, and walrus, which must store large quantities of oxygen to last while they are underwater.

Myoglobin was the first protein whose structure was analyzed in detail; the work was reported by the British biochemist John Kendrew in 1960. As mentioned, myoglobin contains a *heme* group. This is a planar organic unit made up of four pyrrole groups with a Fe^{2+} ion held in the center of the heme molecule by four coordinate covalent bonds

myoglobin (MY-oh-glow-bin)

skeletal muscle: the muscle tissue responsible for movement of an entire organism or parts of its skeleton.

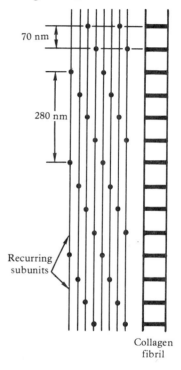

Fig. 16-8 The Staggered Alignment of Subunits in a Collagen Fiber

Val-Leu-Ser-Glu-Gly-Glu-Trp-Gln-Leu-Val-	10
Leu-His-Val-Trp-Ala-Lys-Val-Glu-Ala-Asp-	20
Val-Ala-Gly-His-Gly-Gln-Asp-Ile-Leu-Ile-	30
Arg-Leu-Phe-Lys-Ser-His-Pro-Glu-Thr-Leu-	40
Glu-Lys-Phe-Asp-Arg-Phe-Lys-His-Leu-Lys-	50
Thr-Glu-Ala-Glu-Met-Lys-Ala-Ser-Glu-Asp-	60
Leu-Lys-Lys-His-Gly-Val-Thr-Val-Leu-Thr	70
Ala-Leu-Gly-Ala-Ile-Leu-Lys-Lys-Lys-Gly-	80
His-His-Glu-Ala-Glu-Leu-Lys-Pro-Leu-Ala-	90
Gln-Ser-His-Ala-Thr-Lys-His-Lys-Ile-Pro-	100
Ile-Lys-Tyr-Leu-Glu-Phe-Ile-Ser-Glu-Ala-	110
Ile-Ile-His-Val-Leu-His-Ser-Arg-His-Pro-	120
Gly-Asn-Phe-Gly-Ala-Asp-Ala-Gln-Gly-Ala-	130
Met-Asn-Lys-Ala-Leu-Glu-Leu-Phe-Arg-Lys-	140
Asp-Ile-Ala-Ala-Lys-Tyr-Lys-Glu-Leu-Gly-	150
Tyr-Gln-Gly	153

Fig. 16-9 The Amino Acid Sequence of Sperm Whale Myoglobin

Fig. 16-10 The Structure of Heme

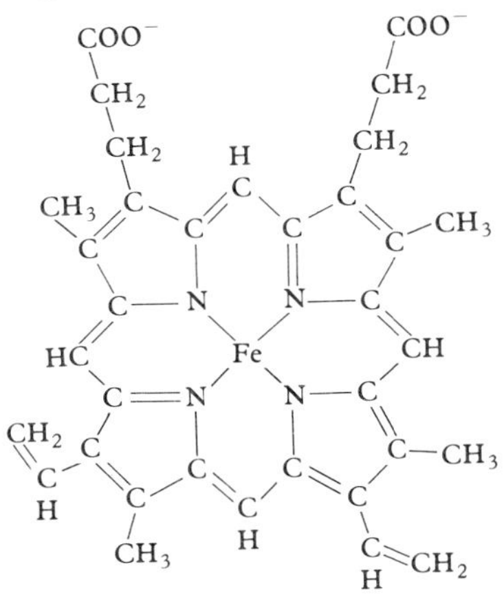

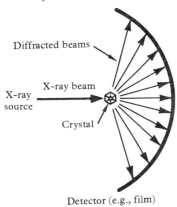

Heme
(Fe-protoporphyrin IX)

(see Figure 16-10). Heme is an example of a *prosthetic* group, a nonpolypeptide chemical unit attached to a protein molecule. Certain proteins require the presence of prosthetic groups for biological activity; a protein without its characteristic prosthetic group is called an *apoprotein*.

Kendrew used *X-ray diffraction* (Figure 16-11) to study the structure of myoglobin. In this technique, a narrow beam of X-rays strikes the protein crystal, and part of the beam is scattered by the atoms in the protein crystal into a variety of directions. The scattered X-rays can then be analyzed to yield information about the positions of atoms in the protein molecule. As a result of his analyses, Kendrew found the following features about myoglobin structure:

1. Myoglobin is a very compact molecule, and there is very little empty space inside.

2. About 75% of the myoglobin molecule is folded into the α-helical structure (see Figure 16-12).

3. There is a bend in the α-helix at each proline residue, suggesting that pyrrolidine rings do not fit well into a straight section of an α-helix.

4. The interior of the myoglobin molecule contains mostly nonpolar amino acid residues; the only polar residues inside the molecule are two histidines. The exterior of the molecule contains both polar and nonpolar residues.

Fig. 16-11 The X-Ray Diffraction Technique

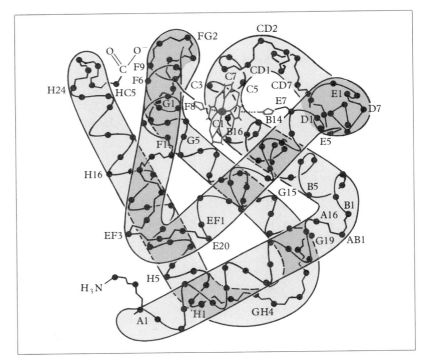

Fig. 16-12 The Structure of Sperm Whale Myoglobin. (The letters *A, B, C, D, E, F, G,* and *H* refer to the eight major helical regions; *CD, EF, FG, GH,* and *HC* refer to the five nonhelical segments between helical regions.)

5. The heme group is held in a crevice in the myoglobin molecule by hydrophobic forces, with the ionized propionate side chains of heme located on the surface of myoglobin. One of the interior histidines, the proximal histidine, is bonded to the Fe^{2+} ion by a coordinate covalent bond. The second interior histidine, the distal histidine, is located on the opposite side of the heme plane, too far away to be bonded to iron. The Fe^{2+} ion is located slightly out of the plane of the heme unit. (See Figure 16–13).

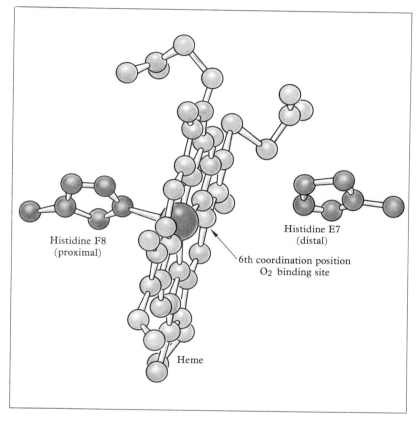

Fig. 16-13 The Oxygen-Binding Site in Myoglobin

Fig. 16-14 Bond Formation to Oxygen Draws Fe^{2+} into the Plane of Heme

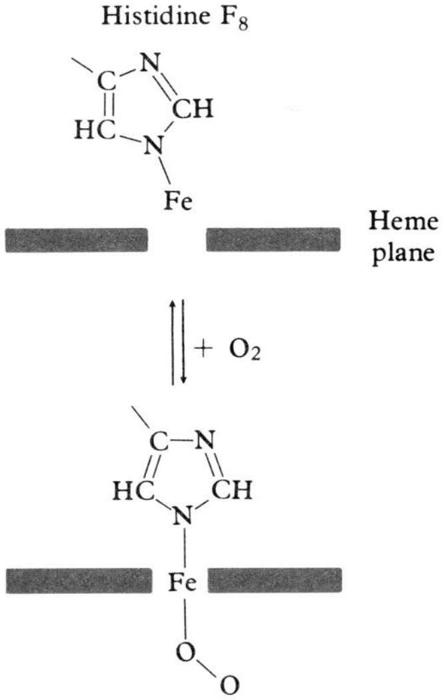

Knowledge of myoglobin structure has led to an explanation of how it attaches to oxygen. Oxygen molecules are delivered to muscle tissue by hemoglobin. As hemoglobin releases oxygen, each molecule diffuses into a myoglobin molecule where it forms a coordinate covalent bond with Fe^{2+} on the side of the heme ring near the distal histidine. This bond draws Fe^{2+} into the plane of the heme unit, as shown in Figure 16–14. Thus, in oxygenated myoglobin (oxymyoglobin), the oxygen molecule occupies the sixth bonding (coordination) position of Fe^{2+}. When myoglobin releases oxygen to be used by the cell, this sixth position becomes vacant.

Experiments have demonstrated that the protein part of myoglobin is necessary to prevent the oxidation of Fe^{2+} to Fe^{3+}, which would result in reduction of the oxygen molecule. If oxygen were to be reduced by heme, it would no longer be available for more important oxidation reactions. In addition, myoglobin containing Fe^{3+} (called ferrimyoglobin) is not functional as an oxygen carrier.

If the primary structure of myoglobin did not have the proper sequence of polar and nonpolar residues, the molecule would not fold into its biologically active structure. It has become an accepted principle of biochemistry that the amino acid sequence (primary structure) of a protein specifies its *conformation*. Although we might expect a number of conformations to be possible, each protein usually has just one that is preferred under normal physiological conditions of temperature and pH. This predominate conformation, called the *native conformation* of the protein, is the one which possesses normal biological activity. It is usually stable enough to be isolated and studied.

conformation: the shape of a flexible molecule.

Studies of myoglobin and other proteins have demonstrated that tertiary structure is an overall effect of many forces within each protein molecule. Hydrophobic forces among nonpolar amino acid side chains seem to be the most important of these, as most proteins contain 30–50% nonpolar residues. As is the case with myoglobin, most of the nonpolar residues in a protein are located in the interior of the molecule, away from the polar aqueous surroundings. In addition, hydrogen bonds play a strong role in stabilizing tertiary structure, with residues such as tyrosine and serine forming hydrogen bonds with other residues and with water. Other noncovalent interactions are ionic attractions between opposite charges, such as those found on lysine and aspartic acid side chains and the terminal amino and carboxyl groups,

ionic attractions: also called salt bonds.

$$\text{Protein}-NH_3^+----^-O-\overset{\overset{\displaystyle O}{\|}}{C}-\text{Protein}$$

and attractions among polar side chains. The only covalent linkages involved in tertiary structure are disulfide bonds formed between pairs of cysteine residues.

EXAMPLE 16.4

How will the side chains of each of the following pairs of amino acid residues interact if they are close together in the tertiary structure of a protein?

(a) cysteine and cysteine
(b) tyrosine and aspartic acid
(c) lysine and glutamic acid
(d) phenylalanine and leucine
(e) serine and asparagine

SOLUTION

(a) A disulfide bond will form.

$$R-S-S-R$$

(b) A hydrogen bond will form.

$$\text{R}-\!\!\bigcirc\!\!-\text{O}-\text{H}\cdots\text{O}-\overset{\displaystyle\text{O}}{\underset{\|}{\text{C}}}-\text{R}$$

(c) An ionic attraction will exist.

$$\text{R}-\text{NH}_3^+ \cdots \cdots \text{O}-\overset{\displaystyle}{\underset{\|}{\text{C}}}-\text{R}$$
$$\qquad\qquad\qquad\quad\text{O}$$

(d) A hydrophobic attraction will exist.

$$\text{R}-\!\!\bigcirc\!\! \qquad \text{CH}_3-\underset{\underset{\text{CH}_3}{|}}{\text{CH}}-\text{CH}_2-\text{R}$$

(e) A hydrogen bond will form.

$$\text{R}-\text{O}-\text{H}\cdots\text{O}=\underset{\text{H}_2\text{N}}{\text{C}}-\text{R}$$

EXERCISE 16.4 How will the side chains of each of the following pairs of amino acid residues interact if they are close together in the tertiary structure of a protein?
(a) arginine and glutamic acid
(b) isoleucine and tryptophan
(c) serine and aspartic acid
(d) valine and phenylalanine

Quaternary Structure

Many proteins contain two or more polypeptide chains held together by forces such as ionic attractions, disulfide bonds, hydrogen bonds, and hydrophobic forces. The polypeptide chains are called subunits; each subunit has its own primary, secondary, and tertiary structure. The arrangement of subunits to form the larger protein is called the *quaternary structure* of the protein. Collagen, mentioned earlier, is such a protein, as is hemoglobin. The following discussion will focus on hemoglobin as an illustration of quaternary protein structure.

Hemoglobin is the red protein found primarily in red blood cells. It has a molecular weight of 64,500, and each hemoglobin molecule contains four heme units. Hemoglobin transports oxygen from the lungs through the blood to the tissues.

Max Perutz, an Austrian graduate student studying in England, began structural studies of hemoglobin in 1936. At the time, he knew that determining the structure of hemoglobin by X-ray diffraction would be a difficult problem, but he had no idea that it would take him 23 years to finish the work. Perutz and Kendrew were joint recipients of the Nobel Prize in chemistry in 1962 for their work on the structure of heme proteins. Perutz found the following features about hemoglobin structure:

1. The hemoglobin molecule is nearly spherical, containing four subunits packed together rather tightly by hydrophobic forces (Figure 16–15).

2. The four subunits are actually two identical pairs, designated as α and β subunits. Each α-chain is in contact with both β-chains, and each β-chain is in contact with both α-chains. Very little contact exists between one α-chain and the other or between one β-chain and the other.

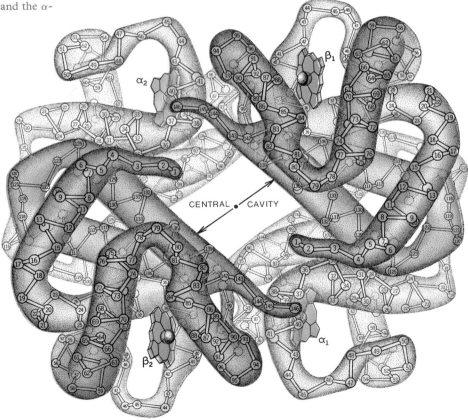

Fig. 16-15 Top View of Horse Deoxyhemoglobin. (The β-chains are shown in dark color and the α-chains in light color.)

3. The tertiary structures of myoglobin and each of the hemoglobin subunits are quite similar (see Figure 16-16), although there are many differences in primary structure of the three polypeptide chains.

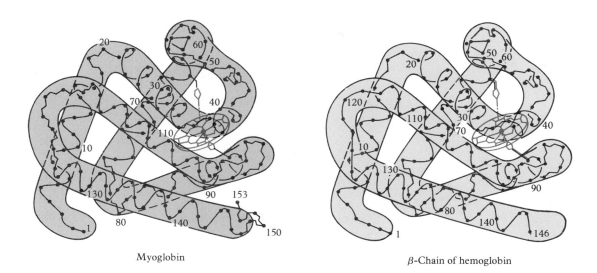

Fig. 16-16 Comparison of the Tertiary Structures of Myoglobin and the β-Chain of Hemoglobin

4. The heme groups are located in crevices near the exterior of the molecule, one in each subunit. Thus the four oxygen-binding sites are rather far apart in space. As with myoglobin, there are proximal and distal histidine residues on opposite sides of each heme unit.

5. Oxygenated hemoglobin (oxyhemoglobin) has a much more compact structure than deoxygenated hemoglobin (deoxyhemoglobin). Deoxyhemoglobin has a strained conformation held in place by a series of eight ionic attractions between oppositely charged groups.

Perutz's structural studies opened the door to understanding how hemoglobin binds oxygen. When deoxyhemoglobin is exposed to oxygen, the protein attaches one oxygen molecule at a time to each of the four heme units (Figure 16-17). When the first oxygen molecule binds, it draws the Fe^{2+} into the plane of the heme unit of that subunit. This rather subtle movement triggers a series of conformational changes in that subunit, the first of which is inward movement of the proximal histidine. The conformational changes in one subunit affect the contacts it makes with the other two subunits. And as the first subunit changes conformation, this change is conveyed to the other subunits, so that their conformations also change. Although we do not understand why, these conformational changes make it easier for the other subunits to bind oxygen. Thus, binding the first oxygen molecule at one subunit affects the other subunits so that their affinities for oxygen are increased.

$$Hb + O_2 \rightleftharpoons HbO_2 \qquad K_1 = \frac{[HbO_2]}{[Hb][O_2]}$$

$$HbO_2 + O_2 \rightleftharpoons Hb(O_2)_2 \qquad K_2 = \frac{[Hb(O_2)_2]}{[HbO_2][O_2]}$$

$$Hb(O_2)_2 + O_2 \rightleftharpoons Hb(O_2)_3 \qquad K_3 = \frac{[Hb(O_2)_3]}{[Hb(O_2)_2][O_2]}$$

$$Hb(O_2)_3 + O_2 \rightleftharpoons Hb(O_2)_4 \qquad K_4 = \frac{[Hb(O_2)_4]}{[Hb(O_2)_3][O_2]}$$

$$K_4 > K_3 > K_2 > K_1$$

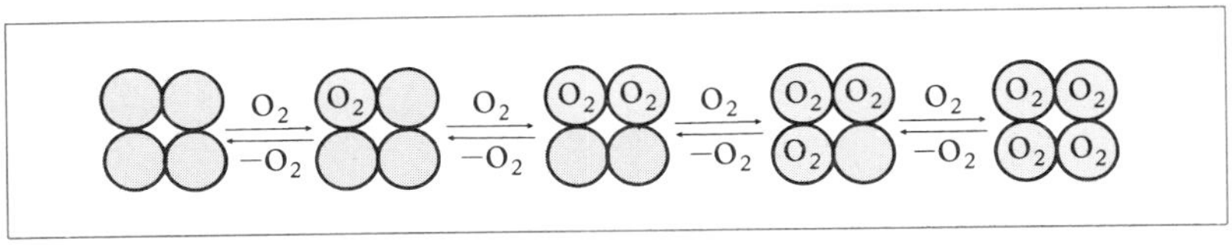

Fig. 16-17 Schematic Diagram of the Binding of Oxygen to Hemoglobin

allosteric (aah-low-STEH-ric): from the Greek words *allo stereo*, meaning "other space."

The term applied to interactions between spatially different sites on a protein molecule is *allosteric interactions*. These interactions often occur in proteins which contain subunits, and they are thought to play critical roles in regulation of metabolism and other molecular events in living organisms.

The communication among the heme groups in the hemoglobin molecule is also referred to as a *cooperative effect*. The subunits seem to cooperate with each other, so that one heme facilitates the binding of oxygen at another heme. This cooperation works in reverse also. The unloading of oxygen from one subunit makes it easier for other subunits to lose their oxygen molecules. The cooperative binding of oxygen by hemoglobin makes the protein a very efficient oxygen transporter. Hemoglobin is much more sensitive to changes in the partial pressure of oxygen than it would be if the oxygen-binding sites operated independently of each other.

EXERCISE 16.5 Name the attractive forces involved in each level of protein structure.

CLASSES OF PROTEINS

Because proteins have such diverse structures and functions, it has not been possible to devise a good classification system for them. However, two broad classes of proteins can be described on the basis of their conformations. These are the globular proteins and the fibrous proteins.

The *globular proteins* are spherical, or nearly so, with hemoglobin as an example. Because of their compact interiors, globular proteins are soluble in water. Other examples of globular proteins are the albumins, the globulins, and most enzymes.

Fibrous proteins are long rod-shaped or stringlike molecules which can intertwine with each other to form strong fibers. Since most fibrous proteins are not soluble in water, they are usually found as structural proteins of connective tissue, elastic tissue, hair, and skin. Examples are collagen, elastin, and keratin.

Many other descriptive terms are used to refer to proteins. Those proteins that have prosthetic groups are sometimes called *conjugated proteins*, and they can be referred to in terms of their prosthetic groups. For example, myoglobin, hemoglobin, and the cytochromes are *hemoproteins*—that is, they contain the heme prosthetic group. Proteins containing lipids, carbohydrates, or metal ions are called *lipoproteins, glycoproteins,* and *metalloproteins,* respectively. Thus, myoglobin, hemoglobin, and the cytochromes are also metalloproteins.

lipoprotein (LYE-po-protein)
glycoprotein (GLYE-co-protein)

Proteins having the same function are found widely distributed in nature. For example, hemoglobin is present in a variety of species, although each type of hemoglobin has a somewhat different primary structure. Nevertheless, all of these hemoglobins are oxygen carriers with tetrameric subunit structure, the same general conformation, and the same biological function.

PROPERTIES OF PROTEINS

In their pure states, proteins are fluffy solids, many of which have been crystallized. Although most proteins are colorless, some are colored by the presence of prosthetic groups. For example, the hemoproteins are red, and ceruloplasmin, a copper-transporting protein in blood, is blue due to its copper content. Those proteins containing a prosthetic group derived from the vitamin riboflavin are yellow. Proteins that are said to be water soluble do not truly dissolve but instead form colloidal dispersions because of their large sizes. Thus aqueous solutions of proteins usually appear opalescent due to the Tyndall effect.

Protein molecules are too large to pass through cell membranes, and thus they are trapped inside normal cells. However, if cells are damaged by disease or trauma, the protein contents leak out. Thus, persistent excessive amounts of proteins in the urine are indicative of kidney damage which has allowed protein to escape from kidney cells. A routine urine analysis usually includes a test for protein. Similarly, a heart attack can be confirmed by the presence of certain enzymes in the blood which ordinarily are confined to cells in heart tissue.

Most proteins are biologically active only over a narrow temperature and pH range, typically 0–40°C and a pH of 6–8. If soluble (globular) proteins are exposed to extreme temperatures or pH values, even for only a short time, they may undergo a physical change known as *denaturation*. In the strictest sense, denaturation means a change from the native state to some other state. Usually denaturation of a protein is the result of unfolding of tertiary structure, although secondary and/or

denaturation: any physical change which alters the native state of a protein to render it biologically inactive.

PERSPECTIVE: The Glycoproteins, a Diverse Group of Biomolecules

The glycoproteins are a group of biological molecules with enormously diverse roles. For example, they serve as enzymes, hormones, blood-clotting factors, carriers of vitamin B_{12}, and nutritive components of milk and egg whites. These molecules are composed of a protein segment covalently linked to polysaccharide branches. Although the specific role of a glycoprotein may be known, the reason for the attachment of sugars to the protein is not always clear.

One type of glycoprotein in which the role of the sugar portion is understood is the mucins. These are responsible for the high viscosity of the various mucus secretions in our bodies. Many mucins contain polysaccharides which carry a negatively charged carboxyl group. Repulsions among adjacent negatively charged polysaccharide chains force the peptide backbone of the glycoprotein to assume a linear shape rather than a more compact conformation. Consequently, the long, rod-shaped molecules encounter much resistance in moving past one another, causing high viscosity in their solutions. This is a significant property for body fluids which act as lubricants and protective coatings. In addition, the highly hydrophilic nature of the polysaccharide branches aids in keeping the mucins dissolved, and the negative charges keep them from grouping together to form fibers.

Attachment of a Polysaccharide Branch to the Polypeptide Backbone of a Glycoprotein

quaternary structure may also be affected. Temperatures above those of a normal living organism can cause intramolecular vibrations to intensify to the point that hydrogen bonds or other noncovalent stabilizing forces are disrupted. When this happens, the folded native structure is no longer stabilized, and the protein takes on a random, disorganized conformation. The denatured protein then becomes nonfunctional, and it may precipitate, as many of its hydrophobic residues are likely to be exposed to the aqueous surroundings. High concentrations of H^+ or OH^- will also disrupt hydrogen bonds and cause denaturation. In some cases, if the source of denaturation is removed, the protein will resume its native conformation. Thus, if the protein molecule is not damaged beyond repair, denaturation may be reversible.

Heavy metal ions such as Hg^{2+}, Ag^+, and Pb^{2+} form bonds with thiol groups and with the side-chain carboxyl groups of aspartic and glutamic acids in proteins. This latter reaction causes the protein to precipitate out as an insoluble metal salt. Solutions of Hg^{2+} (mercurochrome and merthiolate) are commonly used as antiseptics because they denature proteins in bacteria. Mercury, silver, and lead salts are also poisonous to humans for the same reason. Large doses of raw egg white are often given to victims of heavy metal poisoning. In these cases, the high albumin content in egg white precipitates with the heavy metal ions, and the metal is temporarily localized and kept out of circulation. Subsequent medical attention is necessary for total removal of the metal ions.

In addition to those already mentioned, a variety of substances and conditions will denature proteins. The most common of these are listed in Table 16-3.

SEPARATION OF PROTEIN MIXTURES

Liquid Chromatography

In order to study proteins, it is necessary to separate them from mixtures obtained from biological specimens. A common method of separation is *liquid chromatography*. In this technique, a glass or metal column is filled with a finely divided water-insoluble material that has an affinity for the proteins. Two examples of packing material are chemically

chromatography (crow-mah-TOH-grah-fee)

Table 16-3 Substances and Conditions that Denature Proteins

Substance or Condition	Effect on Proteins
Heat and ultraviolet light	Disrupt hydrogen bonds and ionic attractions by making molecules vibrate too violently; produce coagulation, as in cooking an egg.
Urea ($H_2N-\overset{\overset{O}{\|\|}}{C}-NH_2$) and guanidine hydrochloride ($H_2N-\overset{\overset{NH_2^+}{\|\|}}{C}-NH_2\ Cl^-$)	Disrupt hydrogen bonds in proteins by forming new ones to the protein.
Organic solvents (ethanol and others miscible with water)	Disrupt hydrogen bonds in proteins and probably form new ones with the proteins.
Strong acids or bases	Disrupt hydrogen bonds and ionic attractions; prolonged exposure results in hydrolysis of protein.
Detergents	Disrupt hydrogen bonds and ionic attractions.
Heavy metal ions	Form bonds to thiol groups and precipitate proteins as insoluble heavy metal salts.
Foaming	Protein solutions are foamy, and when shaken or whipped, the protein molecules in the foam film are stretched from their normal secondary and tertiary structures.

ion-exchange chromatography: a form of liquid chromatography that utilizes charged packing material; separations result from different charges on protein molecules.

Fig. 16-18 Ion-Exchange Chromatography

modified cellulose, in which there are positively charged amino groups or negatively charged carboxyl groups, and uncharged polymeric material with predetermined pore sizes.

When a solution of a mixture of proteins is poured through a column of charged packing material, each different protein will be attracted to the packing with a different strength, depending on the net charge per molecule of protein. Thus, the mixture will be separated on the column, because each different protein will move through the column at its own rate. The proteins can be eluted with a salt solution and collected separately as they emerge from the column. This kind of liquid chromatography is called *ion-exchange chromatography*; it is illustrated in Figure 16-18.

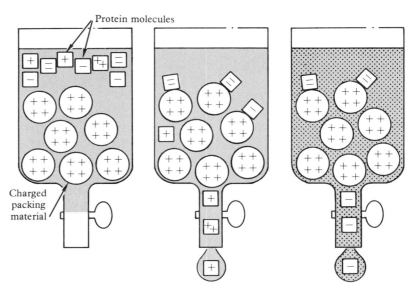

Proteins with same net charge as beads pass through; those with opposite charge stick.

Weak salt solution displaces weakly held protein.

gel filtration (gel permeation) chromatography: a form of liquid chromatography that utilizes a packing material composed of tiny polymeric beads with definite pore sizes; separations are based on the sizes (and thus the molecular weights) of molecules. (When the hydrated beads are packed together in a column, they have a gelatinous appearance, hence the name "gel.")

The second kind of liquid chromatography, called *gel filtration chromatography* or *gel permeation chromatography*, separates proteins by molecular weight (see Figure 16-19). In this case, the column is packed with tiny beads of a polymeric material which have small but definite pore sizes. As the mixture of protein molecules flows through the column, the smaller protein molecules pass through the pores into the beads and then out again. Since the larger molecules are not able to get inside the beads, they pass around them and make their way down the column. Thus, the smaller molecules follow a longer path through the column than the larger ones, which emerge first from the column. In this way, larger protein molecules pass through the column faster than smaller ones, with each type of protein moving at its own rate. The separated proteins can be collected as they flow from the bottom of the column.

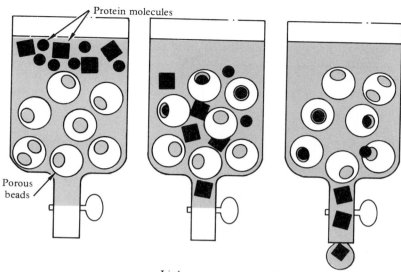

Fig. 16-19 Gel Filtration Chromatography

Little ones penetrate pores in the beads.

Little ones are retarded in their passage, since they keep blundering into pores.

There are several other kinds of liquid chromatography, but they operate on the same general principles described above. Liquid chromatography is usually the preferred separation method for large quantities of proteins.

Electrophoresis

Small quantities of protein mixtures can be separated by electrophoresis (see chapter 15), and the *homogeneity* of a protein sample can be examined by this technique. One of the most famous examples of this method is the separation of sickle-cell hemoglobin (HbS) from normal adult hemoglobin (HbA) performed by Linus Pauling in 1949. Pauling found that HbS migrates faster toward the negative electrode than HbA. The two forms of hemoglobin are otherwise very similar to each other. Pauling's initial separation led others to examine hemoglobin structures more carefully, and Vernon Ingram subsequently demonstrated a small difference in primary structure between HbS and HbA. In particular, Ingram found that in each β-chain of HbS there is a valine residue at position 6 instead of the usual glutamate:

homogeneity (hoe-moe-gih-NAY-ih-tee): purity

```
HbA    Val-His-Leu-Thr-Pro-Glu-Glu-Lys-
HbS    Val-His-Leu-Thr-Pro-Val-Glu-Lys-
  β     1   2   3   4   5   6   7   8
```

Thus HbS, with two fewer negatively charged side chains than HbA, has a greater net positive charge, and it moves toward the negative electrode at a faster rate.

anemia: a deficiency of hemoglobin; from the Greek words *an emia*, meaning "without blood."

Fig. 16-20 (a) Normal Red Blood Cells (b) Deoxygenated Red Blood Cells from a Sickle-Cell Anemia Patient

The altered primary structure of HbS is related to its abnormal behavior in the body. When HbS unloads its oxygen, the solubility of HbS becomes much lower, and deoxy HbS precipitates inside red blood cells. This causes the cells to become deformed and take on their characteristic half-moon (sickle) shape (Figure 16-20). The irreversible deformation leads to premature destruction of blood cells, accounting for the symptoms of *anemia*.

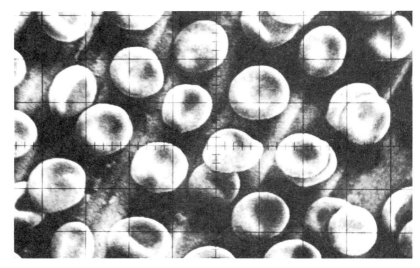

(a)

(b)

Sickle-cell anemia has been characterized by Pauling as a "molecular disease," a disease which can be explained on a molecular basis. HbS, with its two valines substituted for glutamates, has abnormal hydrophobic patches on the exterior of each β-chain (Figure 16-21). These hydrophobic patches are attracted to other hydrophobic patches normally existing on the surface of deoxyhemoglobin. Thus, deoxy HbS molecules form long insoluble chains held together by hydrophobic forces, as shown in Figure 16-21. When these fibers precipitate, they distort the shape of red blood cells and create an anemic condition.

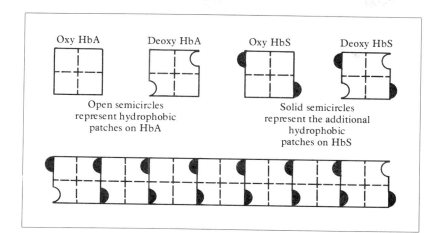

Fig. 16-21 Interaction of Complementary Hydrophobic Patches on Deoxy HbS Molecules to Form Long, Insoluble Fibers

SUMMARY

Functions of Proteins: Proteins, the most important class of biomolecules, play crucial roles in biological processes. Proteins are involved in catalysis, transport and storage of small molecules and ions, the motion process, structural support, protection systems, delivering chemical messages, transmitting nerve impulses, natural toxicity, and metabolic regulations.

Protein Structure: The structure of proteins is described by four levels of complexity. Primary structure is the sequence of amino acids in a protein. Secondary structure is the arrangement of the peptide backbone. The three types of organized secondary structure are the α-helix, the β-pleated sheet, and the collagen triple helix. Tertiary structure is the bending and folding in space of protein helices and pleated sheets. Myoglobin was the first protein for which the tertiary structure was determined. For proteins containing subunits quaternary structure is the term applied to the arrangement of those subunits. An example of a protein possessing quaternary structure is hemoglobin. It has four subunits—a pair of α-chains and a pair of β-chains. Each subunit has a heme group attached which is responsible for binding one molecule of oxygen. Hemoglobin is an allosteric protein; its four oxygen-binding sites communicate by conformational changes so that binding of the first oxygen molecule facilitates binding of subsequent oxygen molecules.

Classes of Proteins: Proteins are often divided into two classes on the basis of their conformations: globular and fibrous. In addition, conjugated proteins may be referred to in terms of their prosthetic groups, such as hemoproteins, lipoproteins, glycoproteins, and metalloproteins.

Properties of Proteins: Proteins are colorless solids for the most part, although some are colored by the presence of prosthetic groups. Those proteins said to be water soluble actually form colloidal dispersions in water. Because of their large sizes, proteins cannot pass through normal cell membranes. Extreme temperatures and/or pH values tend to denature proteins, as do heavy metal ions.

Separation of Protein Mixtures: Mixtures of proteins must be separated for careful study of individual components. Two types of liquid chromatography are often used for large-scale separations: ion-exchange chromatography and gel filtration (gel permeation) chromatography. Small amounts of protein mixtures can be separated by electrophoresis. Because of a small difference in primary structure in their β-chains, sickle-cell hemoglobin and normal adult hemoglobin can be separated by electrophoresis.

STUDY QUESTIONS AND PROBLEMS

1. Define the following terms:
   (a) protein
   (b) enzyme
   (c) antibody
   (d) primary structure
   (e) hormone
   (f) peptide bond
   (g) peptide
   (h) polypeptide
   (i) amino acid composition
   (j) secondary structure
   (k) α-helix
   (l) β-pleated sheet
   (m) collagen triple helix
   (n) tertiary structure
   (o) myoglobin
   (p) heme
   (q) prosthetic group
   (r) apoprotein
   (s) hemoglobin
   (t) allosteric interactions
   (u) globular proteins
   (v) fibrous proteins
   (w) denaturation
   (x) liquid chromatography

2. Briefly discuss ten functions of proteins.

3. What is the main difference between polypeptides and proteins?

4. Draw structures for peptides which could be formed by each group of amino acids.
   (a) glycine and serine (2 isomers)
   (b) alanine, leucine, and lysine (6 isomers)
   (c) arginine and asparagine (2 isomers)
   (d) glutamine, proline, and phenylalanine (6 isomers)

5. The primary structure of Gramicidin S, a peptide antibiotic, is as follows. Draw the complete structure for Gramicidin S. (The arrows indicate the direction of peptide bonds, going from the α-carboxyl group of one residue to the α-amino group of the next.)

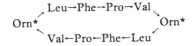

*Orn stands for the amino acid ornithine, whose structure is given below.

$$\overset{+}{H_3N}-CH-CO_2^-$$
$$|$$
$$(CH_2)_3$$
$$|$$
$$^+NH_3$$

6. Summarize major features of the three types of organized secondary structure found in proteins.

7. Wool is easily stretched, but silk is quite resistant to stretching. Explain these observations on the basis of the major protein present in each type of fiber.

8. Explain why α-keratin can be stretched when heated by steam.

9. "Permanent" hair curling is obtained by treating hair with one agent which reduces the disulfide bonds in α-keratin and then another agent which reforms them after arranging the desired curls. Explain how this process curls hair. Could the same process be used for straightening hair?

10. The collagen triple helix is characterized by extensive sequences of $(Gly-X-Pro)_n$ and $(Gly-X-Hypro)_n$ in which X is any amino acid and Hypro is hydroxyproline, a hydroxylated derivative of proline.
    (a) Suggests a reason for glycine being present every third residue.
    (b) What are the principal forces holding the triple helix together?

11. Explain how collagen and gelatin are related.

12. Indicate the type of secondary structure to which each of the following features belongs.
    (a) a high proportion of positively charged amino acid side chains
    (b) a high proportion of proline residues
    (c) hydrogen bonding between every four residues
    (d) hydrogen bonds at right angles to the length of the polypeptide chain
    (e) a high proportion of bulky side chains
    (f) a rodlike shape

13. Which of the following amino acid residues are likely to be found in the interior of a globular protein?
    (a) Pro   (e) Met   (i) Leu
    (b) Gly   (f) Phe   (j) Glu
    (c) Asp   (g) Cys
    (d) Arg   (h) Gln

14. Illustrate how the side chains of each of the following pairs of amino acids would interact if they were close together in the tertiary structure of a protein.
    (a) serine and glutamic acid
    (b) arginine and aspartic acid
    (c) leucine and isoleucine
    (d) lysine and aspartic acid
    (e) phenylalanine and valine

15. Differentiate between tertiary structure and quaternary structure of proteins.

16. Compare myoglobin and hemoglobin on the basis of the following features:
    (a) primary structure
    (b) secondary structure
    (c) tertiary structure
    (d) quaternary structure
    (e) biological function
    (f) allosteric interactions
    (g) heme content

17. Although myoglobin and each hemoglobin subunit have similar tertiary structures, their primary structures are rather different. In particular, a number of polar residues in myoglobin are replaced by nonpolar residues in hemoglobin subunits. Do you think these replacements have anything to do with the quaternary structure of hemoglobin? Explain your answer.

18. Describe the cooperative effect observed in hemoglobin.

19. Distinguish between globular proteins and fibrous proteins.

20. Give the main distinguishing feature of each of the following types of proteins.
    (a) lipoproteins
    (b) glycoproteins
    (c) metalloproteins

21. Why do some proteins have color?

22. List six ways in which proteins are denatured, and explain the basis for each.

23. "Cooking an egg results in denaturation of the egg protein." Explain the meaning of this statement.

24. Explain how raw egg whites are used in treating heavy metal poisoning.

25. Raw fish can be "cooked" by marinating with citrus juice. Explain how this happens.

26. Explain the operating principles of ion-exchange chromatography.

27. Explain the operating principles of gel filtration chromatography.

28. Suppose that you had a mixture of myoglobin and hemoglobin. Explain how gel filtration chromatography could be used to separate the two proteins.

29. Ninhydrin was described as a detecting agent for amino acids in chapter 15. Do you think it can be used for proteins? Explain your answer.

30. Explain the basis for the electrophoretic separation of sickle-cell hemoglobin (HbS) from normal adult hemoglobin (HbA).

31. How does the structure of sickle-cell hemoglobin (HbS) account for its abnormal physiological behavior?

CHAPTER 17
ENZYMES AND COENZYMES

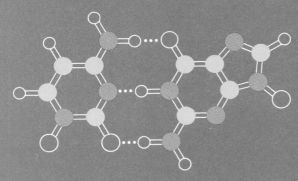

NAMES AND CLASSES OF ENZYMES
PROPERTIES AND FUNCTIONS OF ENZYMES
ENZYME CATALYSIS
- Rates and Equilibria
- Enzyme Specificity
- Theories of Enzyme Action

FACTORS AFFECTING ENZYME ACTIVITY
- Cofactor Concentration
- Substrate Concentration
- Enzyme Concentration
- Temperature
- The Effect of pH

COENZYMES
- The Pyridine Coenzymes
- The Flavin Coenzymes
- Coenzyme A

REGULATION OF ENZYME ACTIVITY
- Activation of Zymogens
- Covalent Modification
- Effect of Inhibitors

USES OF ENZYMES

Enzymes are the specialized proteins which catalyze biological reactions. If these remarkable catalysts did not exist, probably none of the chemical reactions in a living cell would occur fast enough to support life. The secret of the power of enzymes is their efficiency and specificity. All living tissues contain enzymes, and depending on the cell type, as many as three thousand different enzymes may be found in a single cell.

Enzyme-catalyzed reactions are responsible for a number of processes related to food. The leavening of bread, the brewing of beer, the formation of vinegar, and the tenderizing of meat all rely on enzyme activity. Because they are involved in fermentation, enzymes were known as "ferments" for a long time, but the name enzyme (taken from the Greek words *en zume*, meaning "in yeast") came into use in 1878. The first crystallization of an enzyme was accomplished by the American biochemist J. B. Sumner in 1926. Sumner's work demonstrated that enzymes are proteins, but other biochemists were reluctant to accept his findings. It was not until 1946 that he was honored with a Nobel Prize. Today, thousands of enzymes have been studied, and all are known to be proteins.

An interesting feature of many enzymes is their requirement of cofactors for proper function. Although some cofactors are metal ions, a great many are small organic molecules called coenzymes. As we shall see, coenzymes are chemically derived from vitamins.

NAMES AND CLASSES OF ENZYMES

The first name for a specific enzyme was proposed in 1833 by two chemists working in a Paris sugar factory. The factor they isolated from malt extract was responsible for the degradation of starch. Since this factor split starch into simpler components, the workers called the substance diastase, a term derived from the Greek word *diastasis,* meaning "a separation." Since then, the ending "-ase" has been used in naming most enzymes.

A systematic classification for enzymes has been devised in which each enzyme is assigned a recommended name and a systematic name. The *recommended name,* usually short and convenient, is derived by adding the suffix "-ase" to the name of the substrate. (The *substrate* is the substance which undergoes chemical change under the catalytic influence of the enzyme.) The *systematic name* of an enzyme, usually long and cumbersome, unambiguously identifies the enzyme by the reaction it catalyzes. Systematic names are derived from six classes of enzymes, which will be discussed in the following sections. Although it is important to be familiar with these classes, we will use only recommended names for enzymes.

Oxido-reductases

These enzymes catalyze oxidation-reduction reactions involving the simultaneous transfer of electrons and hydrogen to or from a substrate. An example of an oxido-reductase is *alcohol dehydrogenase,* which catalyzes the reversible oxidation of ethanol to produce acetaldehyde:

$$CH_3-\underset{H}{\overset{H}{\underset{|}{\overset{|}{C}}}}-OH \rightleftharpoons CH_3-C\overset{O}{\underset{H}{\diagdown}} + H:^- + H^+$$

Although the enzyme's recommended name is derived from the forward reaction, it also catalyzes the reverse reaction that produces ethanol in yeast fermentation. The name *dehydrogenase* is used for a number of important oxido-reductases.

Transferases

Members of this class of enzymes catalyze the transfer of groups of atoms from one compound to another. Although there are many types of transferases, the two most important are the phosphotransferases and the transaminases. The following equations illustrate the types of reactions these enzymes catalyze.

Phosphotransferase reaction

$$R-O-\underset{O^-}{\overset{O}{\underset{|}{\overset{||}{P}}}}-O^- + H-O-R' \rightleftharpoons R-O-H + {}^-O-\underset{{}^-O}{\overset{O}{\underset{|}{\overset{||}{P}}}}-O-R'$$

Transaminase reaction

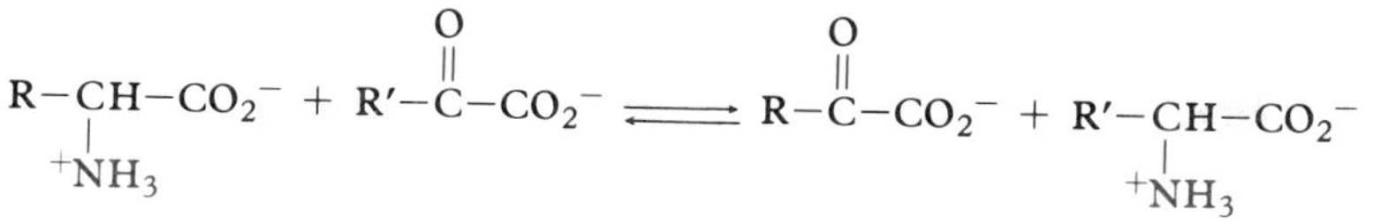

$$R-\underset{^+NH_3}{\underset{|}{CH}}-CO_2^- + R'-\overset{O}{\overset{\|}{C}}-CO_2^- \rightleftharpoons R-\overset{O}{\overset{\|}{C}}-CO_2^- + R'-\underset{^+NH_3}{\underset{|}{CH}}-CO_2^-$$

A number of phosphotransferase reactions are important to carbohydrate metabolism. Transaminations are responsible for shifting amino groups from amino acids to other compounds, thus allowing mobility of nitrogen among biological molecules.

Hydrolases

Enzymes in this class catalyze the hydrolysis of esters, acetals, and amides. Many of these enzymes are involved in the digestive process in animals. Some examples follow.

Lipase reaction (hydrolysis of triacylglycerols)

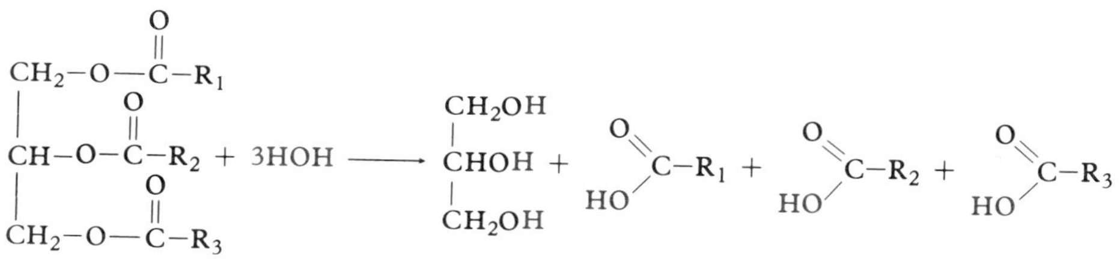

$$\begin{array}{l} CH_2-O-\overset{O}{\overset{\|}{C}}-R_1 \\ | \quad\quad O \\ CH-O-\overset{\|}{C}-R_2 + 3HOH \\ | \quad\quad O \\ CH_2-O-\overset{\|}{C}-R_3 \end{array} \longrightarrow \begin{array}{l} CH_2OH \\ | \\ CHOH \\ | \\ CH_2OH \end{array} + \underset{HO}{\overset{O}{\nwarrow}}C-R_1 + \underset{HO}{\overset{O}{\nwarrow}}C-R_2 + \underset{HO}{\overset{O}{\nwarrow}}C-R_3$$

Phosphatase reaction (hydrolysis of phosphate esters)

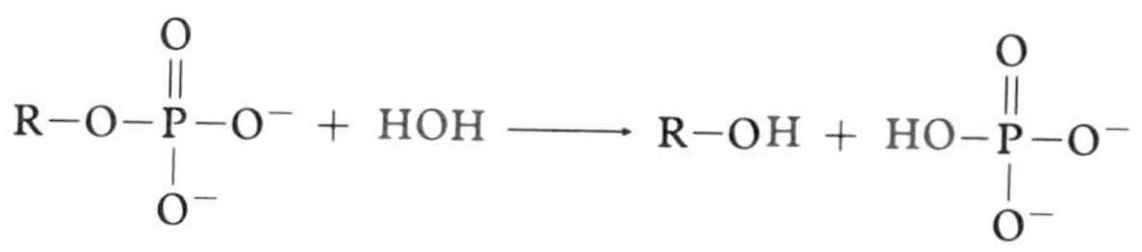

$$R-O-\underset{O^-}{\overset{O}{\overset{\|}{P}}}-O^- + HOH \longrightarrow R-OH + HO-\underset{O^-}{\overset{O}{\overset{\|}{P}}}-O^-$$

Carbohydrate hydrolase reaction (hydrolysis of carbohydrate)

$$\text{starch} + H_2O \longrightarrow \text{oligosaccharides}$$
$$\text{oligosaccharides} + H_2O \longrightarrow \text{monosaccharides}$$

Protease reaction (hydrolysis of proteins and peptides)

$$\text{protein} + H_2O \longrightarrow \text{peptides}$$
$$\text{peptides} + H_2O \longrightarrow \text{amino acids}$$

Lyases

Lyases catalyze both the addition of water, ammonia, and or carbon dioxide to compounds and also the removal of these substances to create double bonds. Decarboxylation reactions are responsible for production of carbon dioxide in cells. Examples of reactions catalyzed by lyases are:

lyase (LYE-ase)

Hydrase/dehydrase reaction (addition and removal of water)

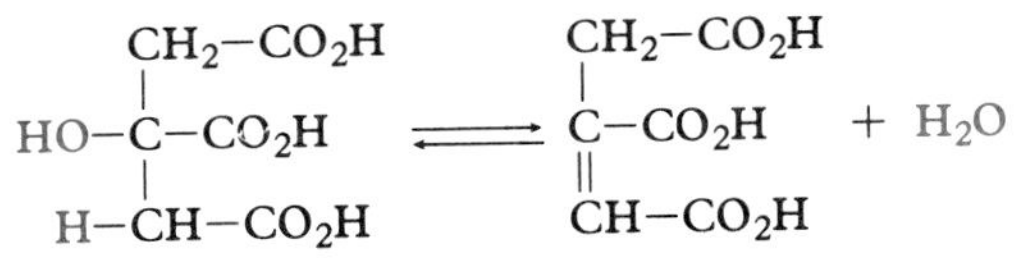

$$\begin{array}{c} CH_2-CO_2H \\ | \\ HO-C-CO_2H \\ | \\ H-CH-CO_2H \end{array} \rightleftharpoons \begin{array}{c} CH_2-CO_2H \\ | \\ C-CO_2H \\ || \\ CH-CO_2H \end{array} + H_2O$$

citric acid aconitic acid

Deaminase reaction (addition and removal of ammonia)

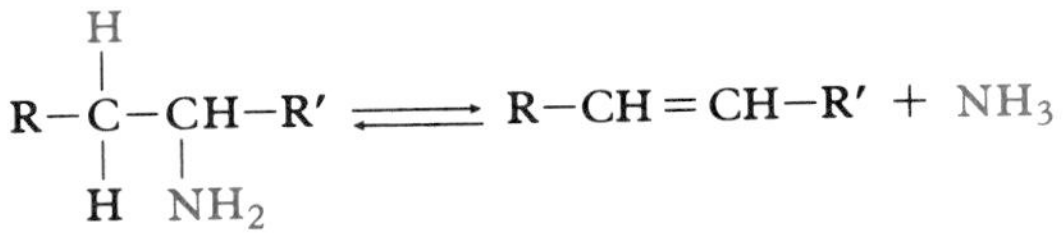

$$\begin{array}{c} H \\ | \\ R-C-CH-R' \\ | \quad | \\ H \quad NH_2 \end{array} \rightleftharpoons R-CH=CH-R' + NH_3$$

Carboxylase/decarboxylase reaction (addition and removal of carbon dioxide)

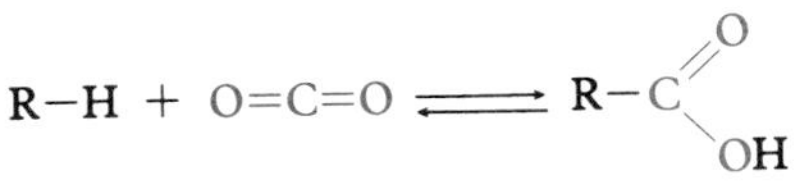

$$R-H + O=C=O \rightleftharpoons R-C\begin{array}{c} O \\ \diagup \\ \diagdown \\ OH \end{array}$$

Isomerases

isomerase (eye-SAH-meh-race)

These enzymes, which catalyze isomerization reactions, exist in several subclasses, illustrated by the following reactions. Isomerization reactions are especially important in carbohydrate and fat metabolism.

Cis-trans isomerase reaction

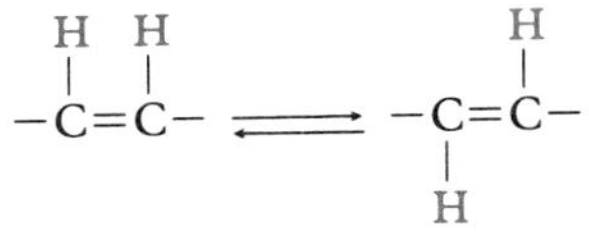

$$\begin{array}{c} H \quad H \\ | \quad | \\ -C=C- \end{array} \rightleftharpoons \begin{array}{c} H \\ | \\ -C=C- \\ | \\ H \end{array}$$

cis   trans

Aldose-ketose isomerase reaction

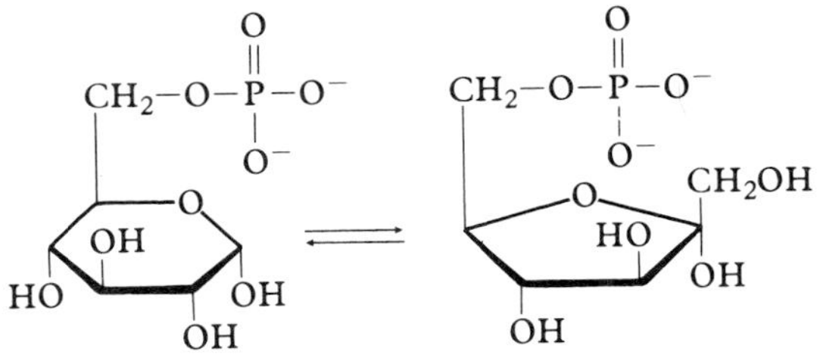

glucose-6-phosphate   fructose-6-phosphate

Mutase reaction

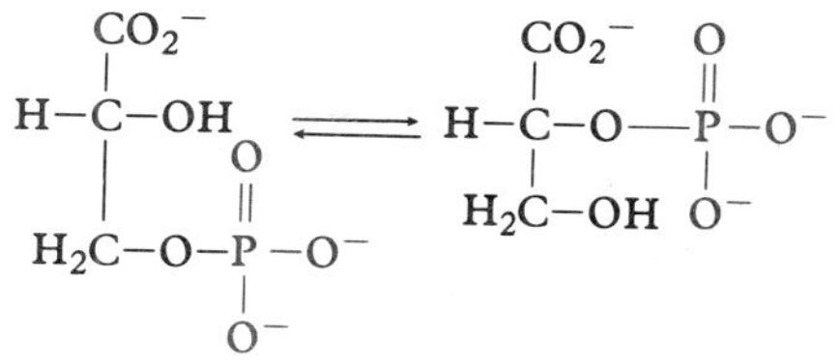

$$\begin{array}{c} CO_2^- \\ | \\ H-C-OH \\ | \\ H_2C-O-\overset{O}{\underset{|}{\overset{\|}{P}}}-O^- \\ O^- \end{array} \rightleftharpoons \begin{array}{c} CO_2^- \\ | \\ H-C-O-\overset{O}{\underset{|}{\overset{\|}{P}}}-O^- \\ | \\ H_2C-OH \quad O^- \end{array}$$

3-phosphoglycerate   2-phosphoglycerate

Ligases

ligase (LYE-gase)

These enzymes catalyze the formation of covalent bonds, usually by joining two molecules to form a larger one. Because ligases catalyze synthesis reactions, they are also called *synthetases*. These reactions are not spontaneous, and a source of energy is needed to make them occur. In metabolic processes, the compound called adenosine triphosphate (ATP) releases energy when hydrolyzed (hydrolysis of ATP also produces adenosine diphosphate (ADP)). Ligases catalyze simultaneous covalent bond formation and hydrolysis of ATP for the energy needed to form the covalent bond. Ligases are involved in the synthesis of lipids, carbohydrates, proteins, and nucleic acids. Typical reactions are

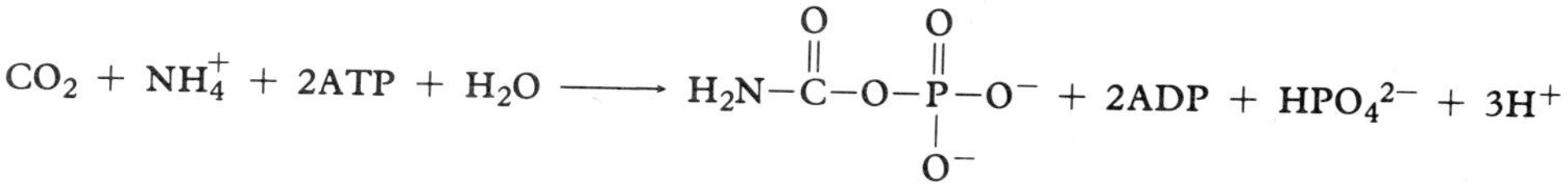

$$CO_2 + NH_4^+ + 2ATP + H_2O \longrightarrow H_2N-\overset{O}{\overset{\|}{C}}-O-\overset{O}{\underset{\underset{O^-}{|}}{\overset{\|}{P}}}-O^- + 2ADP + HPO_4^{2-} + 3H^+$$

carbamoyl phosphate

$$\begin{array}{c} H \\ | \\ H_3\overset{+}{N}-C-CO_2^- \\ | \\ CH_2 \\ | \\ CH_2 \\ | \\ CO_2H \end{array} + NH_4^+ + ATP \rightleftharpoons \begin{array}{c} H \\ | \\ H_3\overset{+}{N}-C-CO_2^- \\ | \\ CH_2 \\ | \\ CH_2 \\ | \\ \underset{O}{\overset{}{C}}\diagdown NH_2 \end{array} + ADP + HPO_4^{2-} + 2H^+$$

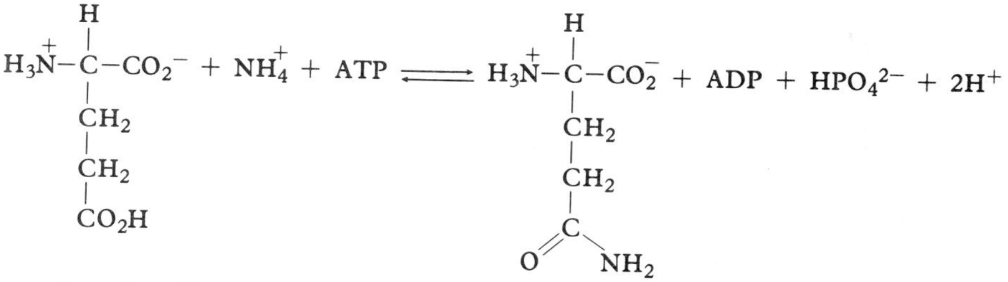

glutamic acid   glutamine

EXAMPLE 17.1

Classify the enzyme for each of the following reactions according to the six major types. Suggest a recommended name for each enzyme.

(a) sucrose + $H_2O \longrightarrow$ glucose + fructose

(b) $\begin{array}{c} \overset{O}{\underset{}{\overset{\diagdown}{C}}}\overset{O^-}{\diagup} \\ | \\ C=O \\ | \\ CH_3 \end{array} + H^+ \longrightarrow \begin{array}{c} \overset{O}{\underset{}{\overset{\diagdown}{C}}}\overset{H}{\diagup} \\ | \\ CH_3 \end{array} + CO_2$

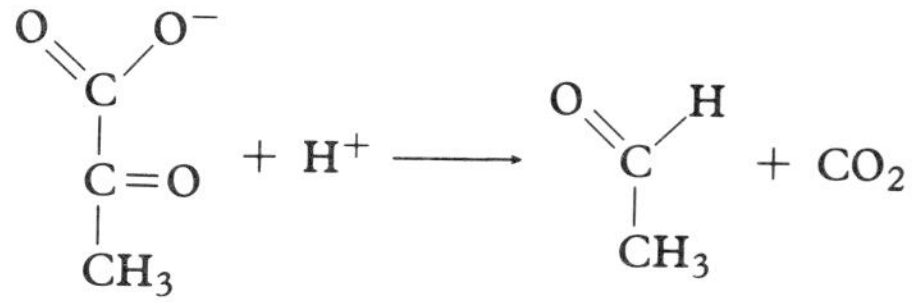

    pyruvate   acetaldehyde
(anion of pyruvic acid)

(c) protein + H₂O ⟶ peptides

(d) 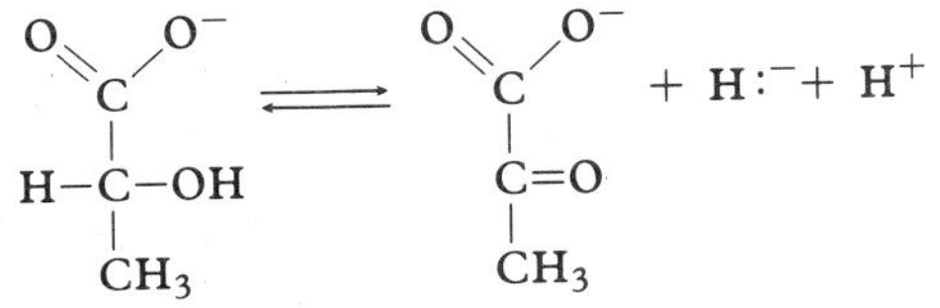
$$\underset{\text{lactate}}{\underset{\text{(anion of lactic acid)}}{\begin{array}{c} O\diagup\diagdown O^- \\ C \\ | \\ H-C-OH \\ | \\ CH_3 \end{array}}} \rightleftharpoons \underset{\text{pyruvate}}{\begin{array}{c} O\diagup\diagdown O^- \\ C \\ | \\ C=O \\ | \\ CH_3 \end{array}} + H:^- + H^+$$

SOLUTION

(a) Class: hydrolase   Recommended name: sucrase
(b) Class: lyase   Recommended name: pyruvate decarboxylase
(c) Class: hydrolase   Recommended name: protease
(d) Class: oxido-reductase   Recommended name: lactate dehydrogenase

EXERCISE 17.1 Classify the enzyme for each of the following reactions, and suggest a recommended name for each enzyme.

(a) $^-O_2C-CH_2-CH_2-\underset{\underset{NH_3^+}{|}}{CH}-CO_2^- + {}^-O_2C-CH_2-\overset{\overset{O}{\|}}{C}-CO_2^- \rightleftharpoons$

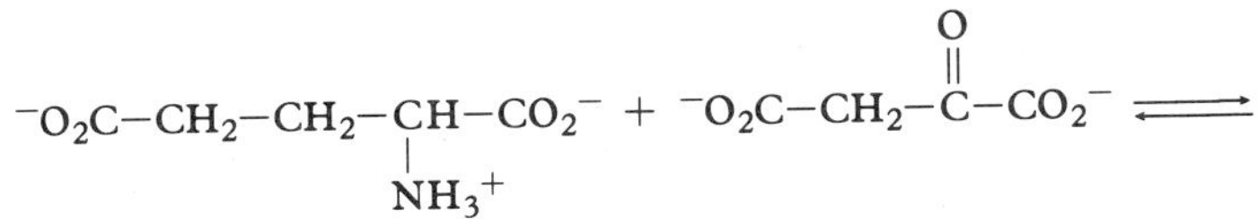

        glutamate   oxaloacetate

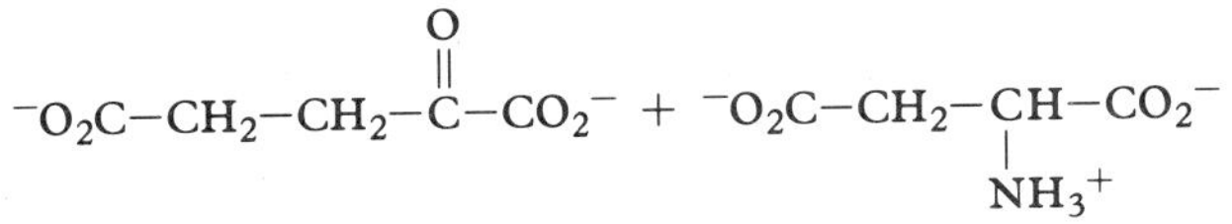

$^-O_2C-CH_2-CH_2-\overset{\overset{O}{\|}}{C}-CO_2^- + {}^-O_2C-CH_2-\underset{\underset{NH_3^+}{|}}{CH}-CO_2^-$

     α-ketoglutarate   aspartate

(b) $^-O_2C-CH_2-\underset{\underset{OH}{|}}{CH}-CO_2^- \rightleftharpoons {}^-O_2C-CH_2-\overset{\overset{O}{\|}}{C}-CO_2^- + H:^- + H^+$

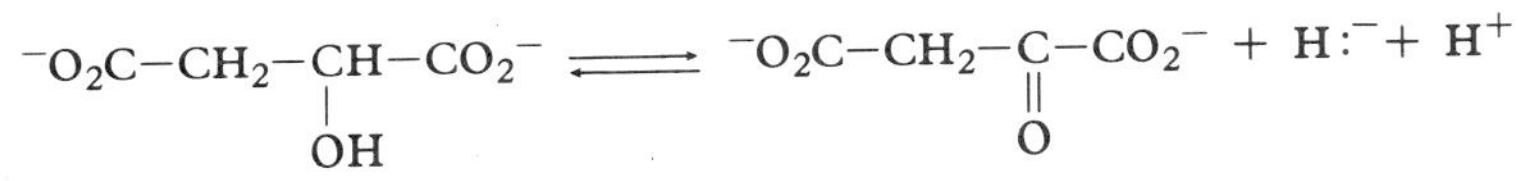

        malate   oxaloacetate

(c) $H_3\overset{+}{N}-\underset{\underset{CH_2Ph}{|}}{CH}-\overset{\overset{O}{\|}}{C}-NH-\underset{\underset{CH_3}{|}}{CH}-CO_2^- + H_2O \rightleftharpoons H_3\overset{+}{N}-\underset{\underset{CH_2Ph}{|}}{CH}-CO_2^- + H_3\overset{+}{N}-\underset{\underset{CH_3}{|}}{CH}-CO_2^-$

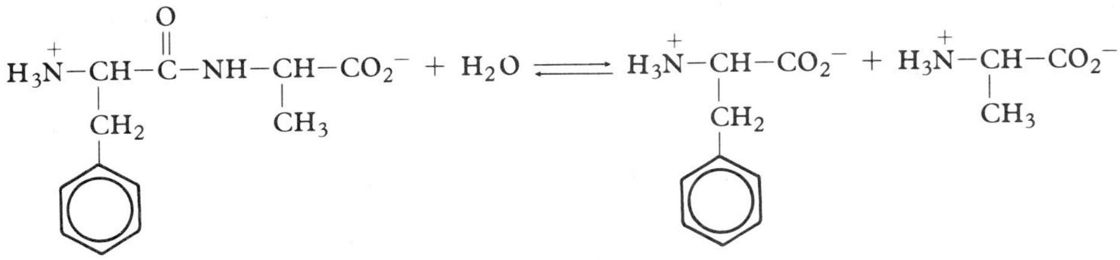

(d) $CH_3-\underset{\underset{NH_2}{|}}{CH}-CO_2H \rightleftharpoons CH_3-\underset{\underset{NH_2}{|}}{CH_2} + CO_2$

        alanine   ethylamine

PROPERTIES AND FUNCTIONS OF ENZYMES

Since the time when the first enzyme was crystallized and studied, all enzymes have proven to be proteins. Some enzymes consist of a single polypeptide chain, but most are composed of two or more subunits. As many as 60 subunits may be incorporated into a functional enzyme, and in many cases, the subunits are structurally different. As proteins, enzymes are sensitive to denaturing conditions, and this sensitivity accounts for the effects of poisons. Cyanide ion, for example, attaches to enzymes needed for proper use of oxygen, rendering them inactive. Heavy metal ions bond to thiol and carboxyl groups in enzymes to cause denaturation, and nerve gases and insecticides produce chemically modified enzymes which lack catalytic activity.

Virtually all of the biochemical reactions responsible for life are catalyzed by enzymes. An important function of enzymes is to take part in breaking down the large nutrient molecules of proteins, fats, and carbohydrates into smaller molecules which can be absorbed into the bloodstream. This process, known as digestion, occurs in the mouth, stomach, and intestines of animals. A second function of enzymes is to catalyze the formation of large, complex molecules from small, simple ones in response to the needs of the cell. Other roles of enzymes are catalysis in the storage and release of energy, in cell reproduction, in respiration, and in the processes involved in visual perception.

Each enzyme catalyzes one type of chemical reaction. Thus, a particular substrate is recognized by an enzyme which then catalyzes the conversion of substrate to product. Some enzymes are catalytically active as they are, but others require one or more nonprotein components, called *cofactors*, for activity. A cofactor may be a metal ion, such as Zn^{2+}, Mg^{2+}, Fe^{2+}, or Fe^{3+}, or it may be an organic molecule called a coenzyme. Some enzymes require both kinds of cofactors. The catalytically active enzyme-cofactor complex is called a *holoenzyme;* if the cofactor is removed, a catalytically inactive protein, the *apoenzyme*, is produced. If the coenzyme is bound tightly to the enzyme molecule, the organic molecule is called a *prosthetic group*.

ENZYME CATALYSIS

A catalyst is usually defined as a substance which speeds up a chemical reaction but is not itself consumed by the reaction. This definition applies to enzymes as well as other catalysts, but enzymes bring about rate enhancements that are much greater than those caused by any man-made catalyst. The following sections examine the major features of enzyme catalysis.

Rates and Equilibria

As discussed in chapter 8, almost all chemical reactions are reversible, and they will achieve equilibrium if left undisturbed. When enzymes speed up chemical reactions, activation energies are lowered (Figure 17-1), and those reactions achieve equilibrium more rapidly than if the enzymes were absent. Thus, enzymes (and other catalysts) accelerate the approach to equilibrium, but they do no alter the relative amounts of compounds present at equilibrium.

Enzyme Specificity

enzyme specificity: the ability of an enzyme to choose both the type of reaction to be catalyzed and the particular substrate to be reacted.

Unlike other catalysts, enzymes choose both the type of reaction to be catalyzed and the particular substrate which will be involved. For example, strong acids show no preference in catalyzing hydrolysis of all amides, esters, and acetals. On the other hand, proteases catalyze the preferential hydrolysis of amide bonds:

$$R-\underset{\underset{H}{|}}{\overset{\overset{O}{\|}}{C}}-N-R' + HOH \rightleftharpoons R-\overset{\overset{O}{\|}}{C}-OH + R'-N\overset{H}{\underset{H}{\diagdown}}$$

amide hydrolysis

They catalyze hydrolysis of esters only poorly, and they have no effect on the rates of acetal hydrolysis. Furthermore, each protease shows a marked preference for a certain type of amide bond. *Trypsin*, a digestive protease, prefers protein and peptide substrates having lysine or arginine residues on the carboxyl side of the amide bond that is hydrolyzed. A closely related digestive protease, *chymotrypsin*, has a specificity for aromatic residues on the carboxyl side of the amide bond. Still another protease, *thrombin* (involved in blood clotting), is even more particular. It catalyzes hydrolysis of proteins and peptides having an arginine residue on the carboxyl side of the amide bond to be hydrolyzed.

trypsin (TRIP-sin)

chymotrypsin (KYE-moe-trip-sin)

thrombin (THRAHM-bin)

specificity of trypsin	specificity of chymotrypsin	specificity of thrombin
Lys or Arg	Phe, Tyr, or Trp	Arg

Thus, these proteases have the ability to distinguish one substrate from another in their catalytic actions. All enzymes have a specificity for substrate and reaction type, and they complement each other in their roles as biological catalysts.

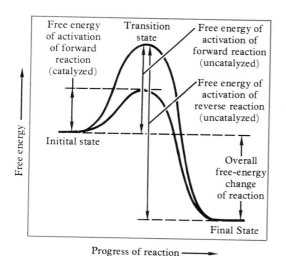

Fig. 17-1 Energy Diagram for a Chemical Reaction, Uncatalyzed and Catalyzed

Theories of Enzyme Action

Before catalysis occurs, an enzyme binds its substrate at a location on the enzyme called the *active site*. Binding occurs when the substrate, usually a small molecule, is attracted to the active site. The binding forces may be hydrophobic attractions, hydrogen bonds, and/or electrical attractions. When the substrate is bound to enzyme, the complex is called the *enzyme-substrate (ES) complex,* or the *Michaelis complex,* after the German biochemist who first suggested its existence. Once this complex is formed, conversion of substrate (S) to product (P) takes place:

Michaelis (mih-KAY-lis)

$$E + S \rightleftharpoons ES \longrightarrow E + P$$

The chemical transformation occurs at the active site, usually aided by functional groups that are part of amino acid residues present in the enzyme. After chemical conversion has occurred, product is liberated from the active site, and the enzyme is free for another round of catalysis.

The substrate specificity of an enzyme exists because the enzyme has the ability to recognize and then bind only one of a multitude of molecules present in a living cell. Two theories have been proposed to explain how recognition and subsequent catalysis occur.

The Lock-and-Key Theory

In addition to his work with carbohydrates (see chapter 13), Emil Fischer also studied proteins, and in particular, enzymes. It was Fischer who suggested in 1890 that enzymes might bind only certain substrates because of the geometry of the active site. He used the metaphor of a

Fig. 17-2 The Lock-and-Key Theory of Enzyme Action

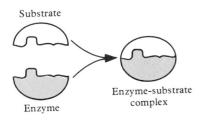

subtilisin (suh-tih-LYE-sin)

Fig. 17-3 The Induced-Fit Hypothesis of Enzyme Action

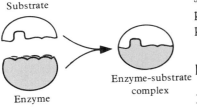

carboxypeptidase (car-BOX-sih-PEP-tih-dase): a protease that catalyzes hydrolysis of proteins and peptides to release the C-terminal amino acid.

lock and its key in postulating that a substrate has a shape that complements the geometry of the active site much the way that a key fits into a lock (see Figure 17-2). It was thought that after the substrate is bound to the enzyme, conformational changes cause certain bonds in the substrate to become strained. The strain results in bond rearrangement to form product. This model of enzyme action suggests that the enzyme and its substrate are somewhat rigid structures initially.

The Induced-Fit Hypothesis

Although Fischer's lock-and-key theory explains the action of enzymes having very narrow substrate specificities, it does not apply to enzymes with broad substrate specificities. For example, *subtilisin*, a protease found in certain bacteria, appears to accept any protein or peptide as substrate, and thus its substrate specificity is very broad. Daniel E. Koshland, Jr., a biochemist at the University of California, Berkeley, has proposed a new explanation of enzyme action to account for broad substrate specificity. Koshland's idea is that perhaps the essential functional groups at the active site of an enzyme are not positioned properly until a substrate molecule binds to the enzyme. The binding of substrate is thought to cause (induce) a conformational change in the enzyme in which certain functional groups are moved into better positions for catalysis (see Figure 17-3). Probably this active conformation of the enzyme is unstable in the absence of substrate. Thus, a range of substrates could induce the enzyme to change its conformation to a more active one, and the degree of activity would depend on how well a particular substrate induces conformational change. The induced-fit hypothesis may well explain some features of enzyme catalysis, but probably there is no single explanation that can completely uncover all facets of this remarkable phenomenon.

FACTORS AFFECTING ENZYME ACTIVITY

Five factors affect the activity of enzymes: cofactor concentration, substrate concentration, enzyme concentration, temperature, and pH. Each of these is discussed below.

Cofactor Concentration

As mentioned earlier, some enzymes require cofactors for catalytic activity. For example, *carboxypeptidase* is a digestive protease that contains Zn^{2+} at its active site. The carbonyl oxygen of the peptide bond to be hydrolyzed is electrically attracted to the zinc ion. This arrangement probably imposes strain on the substrate, making it more susceptible to hydrolysis. A proposed mechanism for the reaction is shown in Figure 17-4. In this example, the amount of Zn^{2+} available for incorporation into carboxypeptidase is a factor in determining the amount of active enzyme which can exist.

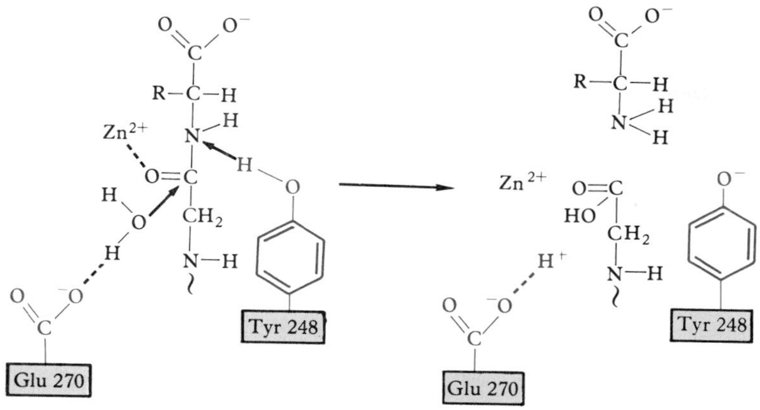

Fig. 17-4 A Proposed Catalytic Mechanism for Carboxypeptidase (Glu 270 and Tyr 248 are amino acid residues at the active site of carboxypeptidase which participate in catalysis.)

Substrate Concentration

In an enzyme-catalyzed reaction, enzyme and substrate are an equilibrium with Michaelis complex that heavily favors formation of ES:

$$E + S \rightleftharpoons ES$$
$$\text{increased [S]} \longrightarrow$$

However, since formation of ES is incomplete, the concentration of free substrate, S, is an important influence on the amount of ES which can be formed. An increase in substrate concentration favors formation of more ES complex, and the more ES that is formed, the faster the catalyzed reaction will be. Figure 17-5 illustrates this relationship. Low substrate concentrations produce slower reaction rates than high substrate concentrations. In theory, an infinitely high substrate concentration would convert all of the free enzyme to ES, and the reaction rate would reach a maximum. This theoretical maximum rate is a constant called the *maximal velocity*, V_{max}. When all of the enzyme has been converted to ES, the enzyme is said to be saturated with substrate. Although it is not possible to achieve infinitely high substrate concentrations in practice (and thus it is not possible to convert all of E to ES), substrate concentrations 10–100 times higher than enzyme concentration produce a reaction rate very close to V_{max}. Under such circumstances, the reaction is said to be proceeding under saturating conditions.

Enzyme Concentration

When enzyme concentration is increased, the result is also an increase in concentration of ES. Thus, the more enzyme molecules there are to catalyze a reaction, the more ES will be formed, and the faster the reaction rate will be. Under saturating conditions, the reaction rate will be directly proportional to enzyme concentration, as illustrated in Figure 17-6.

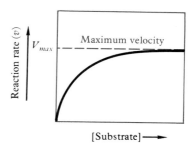

Fig. 17-5 The Effect of Substrate Concentration on the Rate of an Enzyme-Catalyzed Reaction

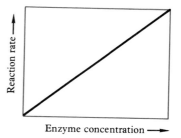

Fig. 17-6 The Effect of Enzyme Concentration on an Enzyme-Catalyzed Reaction Rate Under Saturating Levels of Substrate

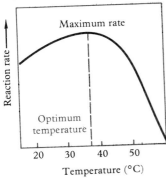

Fig. 17-7 The Effect of Temperature on the Rate of an Enzyme-Catalyzed Reaction

Temperature

As with all chemical reactions, the rates of enzyme-catalyzed reactions increase with temperature. However, since enzymes are proteins, there is a temperature limit at which the enzyme becomes vulnerable to denaturation. Thus, every enzyme-catalyzed reaction has an optimal temperature, and above or below that value, the reaction rate will be slower. This effect is illustrated in Figure 17-7. Although enzymes differ in their heat stabilities, many are inactivated at temperatures above 60°C.

The effect of temperature on enzyme activity has many practical applications. Heating food to be canned or otherwise preserved destroys not only bacteria but also degradative enzymes. In another application, degradative enzymes from *thermophilic* bacteria (bacteria that live at elevated temperatures) were once put into detergents. The high temperature of hot wash water promoted maximal enzyme activity for the degradation of food stains. Unfortunately, the manufacture of enzyme detergents caused lung and skin irritation in workers exposed to the dust for prolonged periods, and most production was stopped by 1971.

The Effect of pH

As was the case with temperature, every enzyme has a pH optimum for its activity (Figure 17-8). The reasons for this behavior are complex, but in general, most enzymes are denatured by pH extremes. However, some enzymes have adapted to unusual pH values. *Pepsin,* the protease that exists in the highly acid environment of the stomach, functions best in the low pH range, near pH 2, and its activity drops off on either side of this value.

pepsin (PEP-sin): a protease having broad specificity but showing some preference for aromatic residues, methionine, and leucine on the carboxyl side of the amide bond to be hydrolyzed.

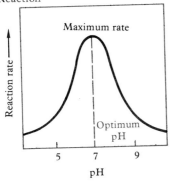

Fig. 17-8 A Hypothetical pH Profile of an Enzyme-Catalyzed Reaction

EXAMPLE 17.2

Indicate how each of the following affects enzyme activity.

(a) increase in cofactor concentration
(b) decrease in substrate concentration
(c) increase in enzyme concentration
(d) increase in temperature
(e) decrease in pH

SOLUTION

(a) increase in enzyme activity
(b) decrease in enzyme activity
(c) increase in enzyme activity
(d) increase in enzyme activity up to point of denaturation; then decrease in enzyme activity
(e) variable effect, depending on the pH optimum for each particular enzyme; extreme pH usually denatures enzymes

EXERCISE 17.2 Indicate how each of the following affects enzyme activity.
(a) decrease in cofactor concentration
(b) increase in substrate concentration
(c) decrease in enzyme concentration
(d) decrease in temperature
(e) increase in pH

COENZYMES

A coenzyme serves as a type of substrate for certain enzymes that use two substrates. A coenzyme may, for example, accept hydrogen from one compound and later donate hydrogen to another compound. Other coenzymes may aid in the transfer of chemical groups from one compound to another. In this way, coenzymes act as shuttle systems in the exchange of chemical substances among various metabolic pathways.

An interesting feature of coenzymes is that they are closely related to vitamins. In fact, each water-soluble vitamin (with the possible exception of vitamin C) is required for the formation of a coenzyme. With this in mind, it is not hard to understand why vitamins are essential to life. We will discuss several examples of coenzymes in the following sections. Others are listed in Table 17–1.

Table 17-1 The Water-Soluble Vitamins and Their Coenzyme Forms

Vitamin	Coenzyme Form	Function
Niacin	Nicotinamide adenine dinucleotide (NAD^+)	Hydrogen transfer
	Nicotinamide adenine dinucleotide phosphate ($NADP^+$)	Hydrogen transfer
Riboflavin	Flavin mononucleotide (FMN)	Hydrogen transfer
	Flavin adenine dinucleotide (FAD)	Hydrogen transfer
Pantothenic acid	Coenzyme A (CoA)	Acyl group carrier and acyl group transfer
Thiamin (B_1)	Thiamin pyrophosphate (TPP)	Aldehyde group transfer
Pyridoxal, pyridoxamine, pyridoxine (B_6 group)	Pyridoxal phosphate	Amino group transfer
Biotin	Biocytin	Carboxyl group removal or transfer
Folic acid	Tetrahydrofolic acid	One-carbon group transfer
Vitamin B_{12}	Coenzyme B_{12}	Shift of hydrogen atoms between adjacent carbon atoms; methyl group transfer
Lipoic acid	Lipoamide	Acyl group transfer
Ascorbic acid (C)	—	Role uncertain, but may serve as a cofactor in hydroxylation reactions

The Pyridine Coezymes

Members of this group of coenzymes are derived from the vitamin niacin, which has two structures (niacin is one of several vitamins which exist in two or more biologically active forms):

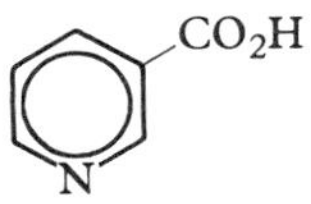

nicotinic acid 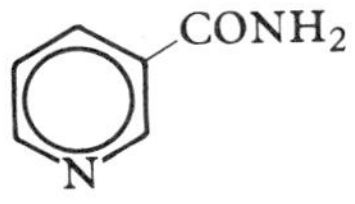   nicotinamide

The two coenzymes in this group contain nicotinamide as their essential component. They are named as *nucleotides* and are shown in Figure 17-9. A nucleotide is a compound that produces a pentose, a heterocyclic nitrogen compound, and phosphoric acid on hydrolysis. Thus, nicotinamide adenine dinucleotide and nicotinamide adenine dinucleotide phosphate qualify as nucleotides because they yield these compounds on hydrolysis. These two coenzymes are also referred to as pyridine coenzymes and pyridine nucleotides, since nicotinamide is a derivative of pyridine.

The pyridine coenzymes function in a large number of reactions catalyzed by dehydrogenases. The coenzymes transfer electrons and hydrogen to and from substrate molecules. The abbrevations for their reduced forms are NADH and NADPH, respectively. The oxidized forms have a positive change, and their names are abbreviated as NAD^+ and

Fig. 17-9 The Pyridine Coenzymes, Nicotinamide Adenine Dinucleotide (NAD^+) and Nicotinamide Adenine Dinucleotide Phosphate ($NADP^+$)

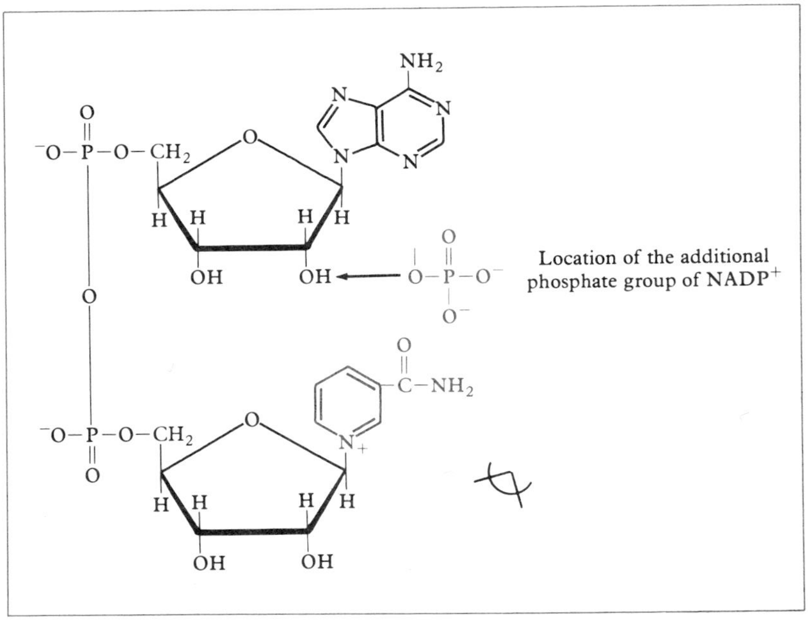

Location of the additional phosphate group of $NADP^+$

NADP$^+$ to emphasize their charged nature. The oxidized forms of the coenzymes accept hydrogen and electrons in the form of a hydride ion, H:$^-$, in position 4 on the nicotinamide ring. The rest of the coenzyme structure is not involved in chemical reactions.

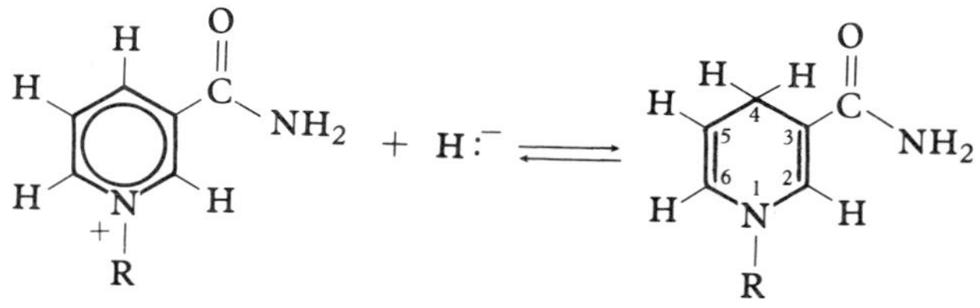

NAD$^+$ or NADP$^+$
(oxidized)

NADH or NADPH
(reduced)

The Flavin Coenzymes

This group of yellow compounds is derived from the vitamin riboflavin, also known as vitamin B$_2$. Riboflavin contains the sugar alcohol called ribotol and the isoalloxazine ring sometimes called flavin.

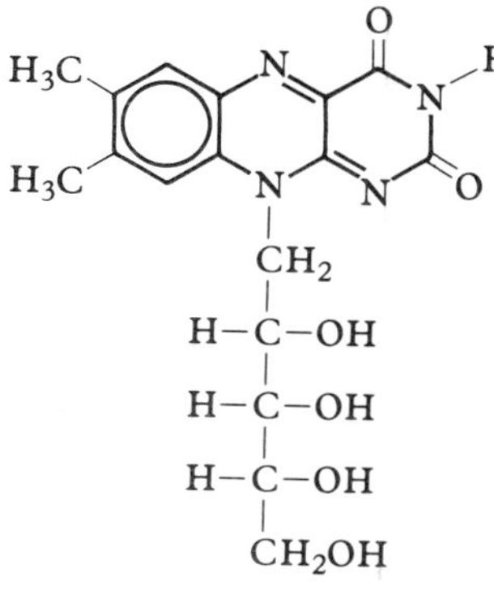

riboflavin
(vitamin B$_2$)

The two coenzymes in this group are flavin monoucleotide (FMN) and flavin adenine dinucleotide (FAD) (Figure 17-10). They function as coenzymes for a group of oxido-reductases called flavoenzymes. In this case, the coenzymes are so tightly attached that they are referred to as prosthetic groups. (Flavin coenzymes are usually attached to their enzymes by hydrophobic forces, but in a few instances a covalent bond is formed.) As with the pyridine coenzymes, flavin coenzymes serve as

Fig. 17-10 The Flavin Coenzymes, Flavin Mononucleotide (FMN) and Flavin Adenine Dinucleotide (FAD)

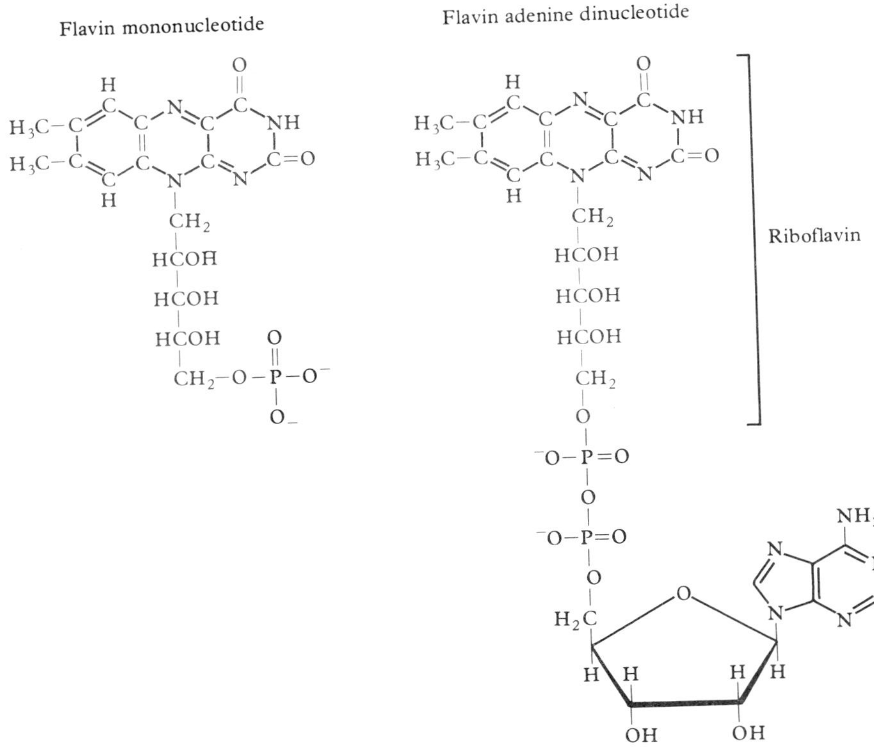

hydrogen acceptors and donors, with the flavin ring reduced as illustrated below. The rest of the coenzyme structure is not involved in chemical reaction.

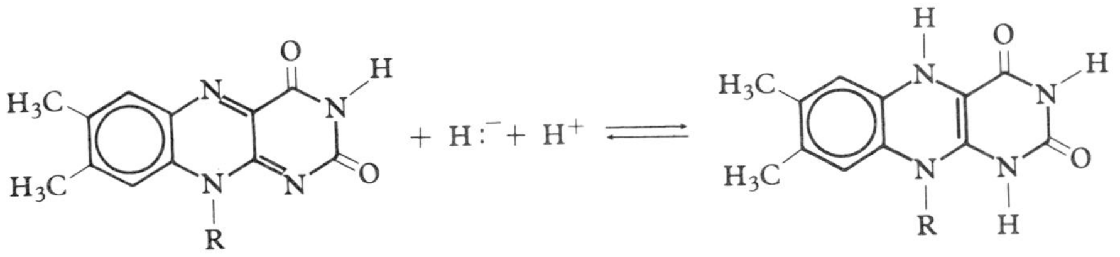

FMN or FAD (oxidized) + H:⁻ + H⁺ ⇌ FMNH$_2$ or FADH$_2$ (reduced)

Coenzyme A

This coenzyme, a central figure in metabolism, is derived from the vitamin pantothenic acid. Coenzyme A (CoA) has a purine ring structure (adenine), a ribose unit, and a β-mercaptoethylamine unit added to pantothenic acid, as shown in Figure 17-11. The coenzyme was given the letter A as part of its name to signify its participation in the transfer

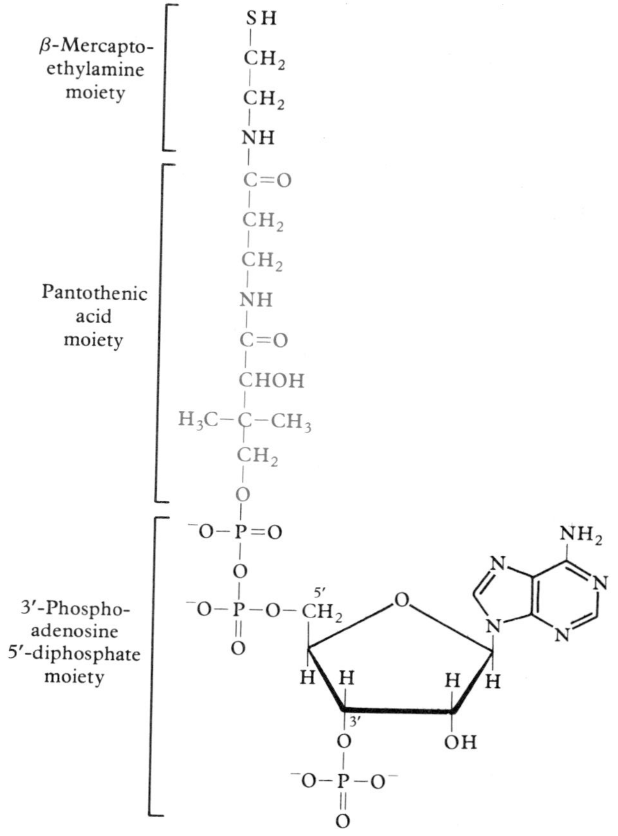

Fig. 17-11 Coenzyme A (CoA)

β-Mercapto-ethylamine moiety:
SH–CH$_2$–CH$_2$–NH–

Pantothenic acid moiety:
C(=O)–CH$_2$–CH$_2$–NH–C(=O)–CHOH–C(CH$_3$)$_2$–CH$_2$–O–

3'-Phosphoadenosine 5'-diphosphate moiety: –O–P(=O)(O⁻)–O–P(=O)(O⁻)–O–CH$_2$–(ribose, 5')–adenine, with 3'-O–P(=O)(O⁻)–O⁻

of acetyl units. We now know that it is a more general acyl group carrier in fatty acid oxidations, fatty acid synthesis, and other biological reactions. Coenzyme A accepts acyl groups at its thiol group to form thiol esters. After it has become acylated, the coenzyme is called acyl coenzyme A, or if the acyl group is acetyl, acetyl coenzyme A (acetyl CoA).

$$H^+ + R-\overset{O}{\underset{\|}{C}}-O^- + HS-CoA \rightleftharpoons R-\overset{O}{\underset{\|}{C}}-S-CoA + H_2O$$

anion of carboxylic acid → acyl CoA

EXAMPLE 17.3

Give the name of the vitamin from which each of the following coenzymes is derived.

(a) coenzyme A
(b) NAD$^+$
(c) FMN
(d) pyridoxal phosphate
(e) thiamin pyrophosphate

SOLUTION

(a) pantothenic acid
(b) niacin
(c) riboflavin (vitamin B_2)
(d) vitamin B_6
(e) thiamin (vitamin B_1)

EXERCISE 17.3 Give the name of the vitamin from which each of the following coenzymes is derived.
(a) tetrahydrofolic acid
(b) lipoamide
(c) FAD
(d) NADH
(e) coenzyme B_{12}

REGULATION OF ENZYME ACTIVITY

Enzymes work together in a very complicated orchestration to conduct all of the biochemical reactions in a living organism. In order for an organism to make the best use of its enzymes, control mechanisms must exist. For example, lowered levels of substrate and cofactor molecules decrease enzyme activity, and temperature and pH are two other means of control. However, many enzymes are susceptible to more complex regulations, as discussed in the following sections.

Activation of Zymogens

zymogen (ZYE-moe-jen): an inactive precursor of an enzyme; many zymogen names end in "-ogen," but some have instead the prefix "pro-," as in procarboxypeptidase.

Many enzymes, particularly the proteases, are synthesized in the cell in an inactive form called a *zymogen*. These proteins are actually the precursors of active enzymes. They usually contain extra amino acid residues

---

PERSPECTIVE: Pernicious Anemia—A Vitamin Deficiency?

Pernicious anemia is a severe, slow-developing disease characterized by a lack of red blood cells. It is caused by a deficiency of vitamin B_{12}, occasionally from dietary lack (the vitamin is found in eggs, milk, fish, and meat, and strict vegetarians may not get enough in their diets) but usually from underproduction of a glycoprotein called intrinsic factor. This glycoprotein, produced by the stomach lining, is essential for normal absorption of vitamin B_{12} from the intestine. Lack of vitamin B_{12} leads to impaired DNA synthesis and red blood cell production becomes abnormal. Pernicious anemia may develop after stomach ulcer, cancer, or removal of the stomach. However, the most common cause is a lowered production of intrinsic factor by older people. Reasons for this problem are not known, but injection of vitamin B_{12} produces quick improvement in the anemia. Since, in most cases, the vitamin is not actually missing from the diet, we might characterize pernicious anemia as a defect in protein synthesis rather than a true vitamin deficiency caused by poor diet.

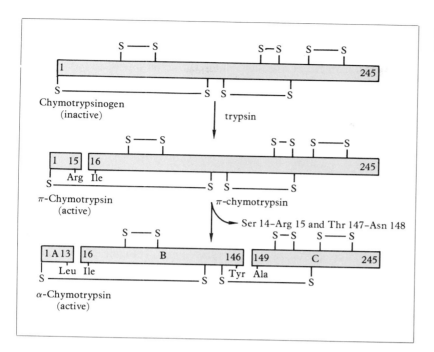

Fig. 17-12 Activation of Chymotrypsinogen

which are excised in an activation process to produce active enzymes. For example, the digestive enzyme chymotrypsin is synthesized in the pancreas as *chymotrypsinogen*, a zymogen containing a single chain of 245 amino acid residues and five disulfide bonds. When chymotrypsinogen is secreted into the intestine, it encounters trypsin, another digestive enzyme, which activates the zymogen by catalyzing hydrolysis of a peptide bond. The subsequent loss of two dipeptides from the protein structure leaves three polypeptide chains connected by disulfide bonds. Apparently this action uncovers the active site of chymotrypsin and allows it to function as a catalyst. The process of chymotrypsinogen activation is shown in Figure 17-12.

Synthesis of proteases as zymogens protects the cells in which they are made from destruction by the proteases. Other examples of zymogens are given in Table 17-2.

chymotrypsinogen (KYE-moe-trip-SIH-no-jen)

Table 17-2 Examples of Zymogens

Zymogen	Active Enzyme
Chymotrypsinogen	Chymotrypsin
Pepsinogen	Pepsin
Procarboxypeptidase	Carboxypeptidase
Proelastase	Elastase
Prothrombin	Thrombin
Trypsinogen	Trypsin

Covalent Modification

Another system that exists for the regulation of enzyme activity is the covalent attachment of a small group to the active site of an enzyme. An example of this is the control of the rate of glycogen breakdown to form glucose. Glycogen reacts with hydrogen phosphate ion, HPO_4^{2-} (often called *inorganic phosphate*, P_i) in a reaction catalyzed by glycogen phosphorylase.

$$\text{glycogen} + P_i \rightleftharpoons \text{glucose-1-phosphate} + \text{glycogen}$$
$$(n \text{ glucose residues}) \qquad\qquad (n-1 \text{ glucose residues})$$

This enzyme is only active when a serine residue at its active site has phosphate attached. Thus, when there is a need for glucose formation from glycogen, inactive glycogen phosphorylase is converted to active, phosphorylated glycogen phosphorylase, and glycogen degradation is promoted.

$$\boxed{\text{Phosphorylase b}} + \text{ATP} \rightleftharpoons \boxed{\text{Phosphorylase a}} + \text{ADP} + H^+$$
$$\underset{\text{Ser—OH}}{} \qquad\qquad\qquad \underset{\text{Ser—O—P(=O)(O}^-\text{)—O}^-}{}$$

inactive   active

Effect of Inhibitors

Enzymes are also influenced by certain molecules which bind to them but which do not undergo chemical change. Since these substances inhibit normal enzyme action when bound, they are called *inhibitors*.

Some inhibitors bind to the active site, preventing the binding of substrate molecules. These inhibitors are called *competitive inhibitors*, since they compete with substrate for binding at the active site of an enzyme. An example of a competitive inhibitor is malonic acid. This compound closely resembles succinic acid, the natural substrate for succinate dehydrogenase:

$$\begin{array}{cc} CO_2H & CO_2H \\ | & | \\ CH_2 & CH_2 \\ | & | \\ CH_2 & CO_2H \\ | & \\ CO_2H & \\ \text{succinic acid} & \text{malonic acid} \\ \text{(substrate)} & \text{(inhibitor)} \end{array}$$

The reaction catalyzed by this enzyme is one of many steps involved in overall oxidation of nutrients in the cell. If malonic acid is present, the oxidation of succinic acid will be inhibited.

Another kind of inhibition, *noncompetitive inhibition,* occurs when the inhibitor binds to the enzyme at a location other than the active site. The inhibitor does not compete with substrate for binding, but the bound inhibitor reduces the efficiency of the enzyme, probably by causing a conformational change. In noncompetitive inhibition, inhibitor and substrate bind to the enzyme simultaneously.

These two types of inhibition can be summarized by the following equations:

$$\text{Competitive inhibition:} \quad E + S \rightleftharpoons ES$$
$$E + I \rightleftharpoons EI$$
$$\text{Noncompetitive inhibition:} \quad E + S \rightleftharpoons ES$$
$$E + I \rightleftharpoons EI$$
$$ES + I \rightleftharpoons ESI$$

Biological molecules are synthesized in a series of steps, each requiring a different enzyme. In many instances, the first enzyme in the sequence is sensitive to inhibition by the end product of the sequence. This type of enzyme regulation is called *feedback inhibition.*

$$A \longrightarrow B \longrightarrow C \longrightarrow D \longrightarrow \underset{\text{end product}}{Z}$$
$$\longleftarrow \text{inhibition of first step} \longrightarrow$$

It is a situation where the accumulation of a high level of end product is a signal to the cell that no more of that material is needed for the time being. In cases where the enzyme is composed of several subunits, the end product binds to a specific site called the *regulatory site.* These enzymes are called *allosteric enzymes;* when the level of end product drops sufficiently, an allosteric enzyme regains its activity.

The types of inhibition discussed so far have involved noncovalent binding of inhibitors. But there are inhibitors that bind to an enzyme and react with important functional groups to render the enzyme permanently inactive. In these reactions, the inhibitor becomes attached to the enzyme by a covalent bond. Heavy metal ions inactivate, or "poison," enzymes by forming covalent bonds with essential thiol groups in the enzyme. Another type of poison, cyanide ion, covalently attaches to a critical enzyme called cytochrome oxidase to cause death. Nerve gases act as poisons by reacting with the essential serine residue at the active site of acetylcholinesterase, an enzyme involved in the transmission of nerve impulses.

regulatory site: a site other than the active site where an inhibitor can bind to an enzyme and thus regulate its activity.

EXAMPLE 17.4

Explain how each of the following processes is involved in the regulation of enzyme activity.

(a) zymogen activation
(b) covalent modification of the active site of an enzyme
(c) feedback inhibition

SOLUTION

(a) An inactive form of an enzyme is converted to an active form when the active enzyme is needed. In this way, cells are protected from protease activity.
(b) Some enzymes are active only when a small chemical group is covalently attached to the active site. When there is a need for enzyme activity, the chemical group is added to the active site.
(c) Feedback inhibition causes loss of activity for an enzyme at the first of a series of steps when the end product accumulates at high levels.

USES OF ENZYMES

Because of their powerful catalysis under very mild conditions of temperature and pH, many practical uses of enzymes have been developed. The addition of enzymes to detergents has already been mentioned. A current popular use of enzymes is in the formulation of meat-tenderizing products. These substances contain the proteases papain (from papaya) or bromelain (from pineapples), which help break down tough, fibrous proteins in meat to provide a more tender consistency.

The medical field has also made use of enzymes in various ways. It is possible to diagnose the occurrence of a heart attack by measuring the levels of three enzymes which are released into the blood by a damaged heart (Figure 17-13). The rates of appearance of creatine phosphokinase (CPK), serum glutamate-oxaloacetate transaminase (SGOT) (also called aspartate transferase, AST), and lactate dehydrogenase (LDH) confirm that a heart attack has occurred, and the extent of rise in enzyme levels indicates the amount of heart damage. Other serum enzyme levels may be elevated in certain forms of cancer and in diseases of the liver, pancreas, lungs, and bone.

PERSPECTIVE: Recent Advances in the Use of Enzymes in Medicine

The field of enzyme research has grown at an enormous rate over the past thirty years, and new uses for enzymes are being developed daily. The discovery in the 1950s that the enzyme urokinase (found in human urine) is able to dissolve blood clots has led to clinical studies proving that urokinase is extremely effective in treating massive internal blood clots. The enzyme is now being marketed for such use. Other enzymes isolated from bacteria have been patented as drugs in the treatment of the itchy rash caused by poison ivy, poison oak, and poison sumac. These enzymes act right on the skin to catalyze the rapid hydrolysis of the plants' toxic irritant (urushiol) to nontoxic products. In experimental work, enzymes that destroy the amino acid phenylalanine have been covalently attached to artificial blood vessel bypasses in monkeys, resulting in destruction of phenylalanine in the monkeys' blood. Researchers foresee a similar technique being used in the treatment of phenylketonuria in children, a condition that can cause mental retardation if the excessively high levels of phenylalanine in the blood are not removed. These are only a few of the new and expanding uses of enzymes in medicine, and now with the use of genetic engineering to produce enzymes in large quantities, the possibilities seem limitless.

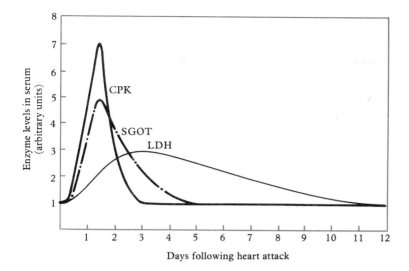

Fig. 17-13 Enzyme Changes Following a Heart Attack

Enzymes are also used in hospital laboratories as *reagents*. The following examples illustrate clinical determinations using enzymes.

Urea (BUN, *Blood Urea Nitrogen*)

The enzyme urease is used to determine urea levels in blood by converting the urea to a more easily measured substance, ammonia.

$$\underset{\text{urea}}{H_2N-\overset{\overset{\displaystyle O}{\|}}{C}-NH_2} + H_2O \xrightarrow{\text{urease}} 2\ NH_3 + CO_2$$

Glucose

Glucose levels in blood and urine are measured by use of the enzyme glucose oxidase, which catalyzes oxidation of glucose to gluconic acid and hydrogen peroxide, a more easily measured substance. Commercially available paper test strips are impregnated with glucose oxidase and other reagents so that urine specimens can be tested easily and quickly for glucose.

$$\underset{\text{glucose}}{\begin{array}{c}H\diagdown_{\displaystyle C}\diagup^{\displaystyle O}\\|\\H-C-OH\\|\\HO-C-H\\|\\H-C-OH\\|\\H-C-OH\\|\\CH_2OH\end{array}} + O_2 + H_2O \xrightarrow{\text{glucose oxidase}} \underset{\text{gluconic acid}}{\begin{array}{c}HO\diagdown_{\displaystyle C}\diagup^{\displaystyle O}\\|\\H-C-OH\\|\\HO-C-H\\|\\H-C-OH\\|\\H-C-OH\\|\\CH_2OH\end{array}} + H_2O_2$$

Alcohol (Ethanol)

Blood and urine samples may be tested for alcohol levels by use of alcohol dehydrogenase. The enzyme, along with its coenzyme, NAD^+, converts ethanol to acetaldehyde and NADH:

$$CH_3-CH_2-OH + NAD^+ \xrightarrow{\text{alcohol dehydrogenase}} CH_3-C{\overset{O}{\underset{H}{\diagdown}}} + NADH + H^+$$

The NADH is easily measured by its light absorbence at 340 nm wavelength.

SUMMARY

Names and Classes of Enzymes: Enzymes are assigned a recommended name, based on names of their substrates, and a systematic name derived from the following classes: oxido-reductases; transferases; hydrolases; lyases; isomerases; and ligases.

Properties and Functions of Enzymes: Of the thousands of known enzymes, several hundred have been crystallized, and all are known to be proteins. Some enzymes consist of a single polypeptide chain, but most are composed of two or more subunits. Because they are proteins, enzymes are sensitive to denaturing conditions. The important catalytic functions of enzymes are: digestion; synthesis of large, complex molecules; storage and release of energy; cell reproduction; respiration; and the processes involved in visual perception. Some enzymes require cofactors for activity. Cofactors may be metal ions or organic molecules called coenzymes.

Enzyme Catalysis: Enzymes cause rate enhancements that are much greater than any man-made catalyst. Like other catalysts, enzymes lower activation energies, and their reactions achieve equilibrium sooner than if enzymes were absent. Enzymes and other catalysts do not alter concentrations at equilibrium. Enzymes are specific for both the type of reaction catalyzed and the particular substrate. Two theories have been proposed for enzyme specificity and catalysis. The lock-and-key theory suggests that enzymes and their substrates initially are rigid molecules, with the substrate fitting the active site like a key fits a lock. The induced-fit hypothesis recognizes that enzyme and substrate molecules are flexible, and that the substrate may induce conformational changes in the enzyme to make it active as a catalyst.

Factors Affecting Enzyme Activity: Enzyme activity increases with increasing concentrations of cofactor, substrate, or enzyme. However, a given amount of enzyme may become saturated with substrate; then the rate of the catalyzed reaction is at a maximum, V_{max}. Enzyme activity will increase with temperature until the denaturation temperature is reached. Every enzyme has a pH optimum for activity; most enzymes are denatured by extremes in pH.

Coenzymes: Coenzymes, which serve as a type of substrate for enzymes that use two substrates, act as shuttle systems exchanging chemical substances among various metabolic pathways. Coenzymes are closely related to vitamins, and each water-soluble vitamin (with the possible exception of vitamin C) is required for the formation of a coenzyme. Important coenzymes (and corresponding vitamins) are the pyridine coenzymes (niacin), the flavin coenzymes (riboflavin, also known as vitamin B_2), and coenzyme A (pantothenic acid).

Regulation of Enzyme Activity: In order for an organism to make the best use of its enzymes, control mechanisms must exist. The most obvious of these are cofactor concentration, substrate concentration, temperature, and pH. In addition, proteases are synthesized as inactive zymogens, which are subsequently activated when needed by excision of peptide fragments. Another system of regulation is the covalent addition of a small group to the active site of an enzyme to promote activity. Enzymes are also influenced by competitive and noncompetitive inhibitors. Allosteric enzymes are subject to feedback inhibition caused by accumulated end product from a series of reactions. In some instances, an inhibitor can become covalently attached to the enzyme, thus rendering it permanently inactive.

Uses of Enzymes: Society has found many practical uses for enzymes, as in enzyme detergents, meat tenderizers, medical diagnosis, and clinical reagents.

STUDY QUESTIONS AND PROBLEMS

1. Define the following terms:
   - (a) substrate
   - (b) oxido-reductase
   - (c) dehydrogenase
   - (d) transferase
   - (e) hydrolase
   - (f) lyase
   - (g) isomerase
   - (h) ligase
   - (i) cofactor
   - (j) coenzyme
   - (k) holoenzyme
   - (l) apoenyzme
   - (m) trypsin
   - (n) chymotrypsin
   - (o) thrombin
   - (p) active site
   - (q) Michaelis complex
   - (r) lock-and-key theory
   - (s) induced-fit hypothesis
   - (t) carboxypeptidase
   - (u) maximal velocity
   - (v) pepsin
   - (w) pyridine coenzyme
   - (x) nucleotide
   - (y) flavin coenzyme
   - (z) coenzyme A
   - (aa) zymogen
   - (bb) enzyme inhibitor
   - (cc) competitive inhibition
   - (dd) noncompetitive inhibition
   - (ee) feedback inhibition
   - (ff) regulatory site
   - (gg) allosteric enzyme

2. Name the six classes of enzymes and give an example of a reaction for each class.

3. Classify the enzyme for each of the following reactions, and suggest a recommended name for each enzyme.

   (a) succinate + FAD ⇌ fumarate + FADH$_2$

   (b) glucose-6-phosphate (pyranose) ⇌ fructose-6-phosphate (furanose)

   (c) malate + NAD$^+$ ⇌ oxaloacetate + NADH + H$^+$

   (d) ribulose-5-phosphate ⇌ ribose-5-phosphate

   (e) lactose + H$_2$O ⇌ glucose + galactose

4. Describe six functions of enzymes.

5. Distinguish among cofactor, coenzyme, and prosthetic group.

6. In which ways do enzymes exhibit specificity?

7. How are the following enzymes alike? How are they different?

   trypsin
   chymotrypsin
   thrombin

8. Explain what is meant by the following equation. Be sure to define each symbol.

   $$E + S \rightleftharpoons ES \longrightarrow E + P$$

9. Summarize the two major theories of enzyme action.

10. What is the role of Zn^{2+} at the active site of carboxypeptidase?

11. Suggest how the following enzymes might recognize their specific substrates.

    trypsin
    chymotrypsin
    carboxypeptidase

12. Explain how each of the following factors affects enzyme activity.

    (a) cofactor concentration
    (b) substrate concentration
    (c) enzyme concentration
    (d) temperature
    (e) pH

13. Give differences and similarities that exist between enzymes and ordinary chemical catalysts.

14. The enzyme pepsin catalyzes hydrolysis of dietary protein in the stomach. Pepsin functions well in the acid environment of the stomach but becomes inactive in the intestine. Suggest a reason for this behavior. (Clue: The pH of intestinal juice is slightly basic.)

15. Explain why enzymes are needed by cells in only small quantities but coenzymes are needed in larger amounts.
16. Use equations to illustrate how NAD^+ and FAD participate in oxidation-reduction reactions.
17. Give the role of each of the following coenzymes and name the vitamin from which it is derived.
    (a) NAD^+
    (b) FAD
    (c) thiamin pyrophosphate
    (d) biocytin
    (e) CoA
    (f) $NADP^+$
    (g) tetrahydrofolic acid
    (h) FMN
    (i) pyridoxal phosphate
    (j) coenzyme B_{12}
    (k) ascorbic acid
18. Describe how most zymogens are activated. Why is it advantageous in some cases for cells to synthesize zymogens instead of active enzymes?
19. An example of regulation of enzyme activity by covalent modification is the phosphorylation of glycogen phosphorylase. How does covalent modification affect this enzyme?
20. Describe the two major types of reversible enzyme inhibitors.
21. Explain how feedback inhibition regulates metabolic activity.
22. Explain why paint containing lead is poisonous.
23. Describe two commercial uses of enzymes.
24. Summarize how the levels of certain enzymes in the blood aid in diagnosis of a heart attack.
25. Describe three clinical determinations in which enzymes are used as reagents.

CHAPTER 18

BODY FLUIDS AND HORMONES

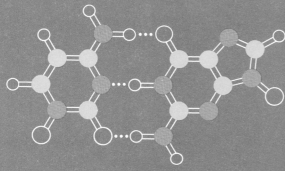

BODY FLUIDS IN DIGESTION
- Saliva
- Gastric Juice
- Pancreatic Juice
- Intestinal Juice
- Bile
- The Large Intestine

BLOOD
- Composition of Blood
- Blood Plasma
- Red Blood Cells
- Hemoglobin
- Blood Clot Formation

URINE
- Diuretics
- Hormones Affecting Urine Formation
- The Role of the Kidneys in Acid-Base Balance
- The Role of the Kidneys in Electrolyte Balance
- The Role of the Kidneys in Regulation of Blood Pressure

HORMONES
- Classes of Hormones
- Prohormones
- The Molecular Basis of Hormone Action

The solutions within our bodies are called body fluids, and they account for about two thirds of total body weight. Most (about 70%) of the total body fluid is found inside cells and is called *intracellular fluid*. The other 30% of total body fluid, *extracellular fluid*, is divided into several types. The fluid found in the tissue spaces between cells, *interstitial fluid*, constitutes about 20% of total body fluid, and the liquid part of blood (plasma) makes up about 7% of total body fluid. The remaining extracellular fluid is mostly digestive juices and cerebrospinal fluid. The distribution of the various body fluids is illustrated in Figure 18-1.

interstitial (in-ter-STIH-shul) **fluid:** fluid found in the tissue spaces (interstices) between cells.

Although there are similarities in composition, there are several distinct differences between intra- and extracellular fluid. For example, the protein content of intracellular fluid is approximately four times the amount present in blood plasma, which in turn contains roughly ten times the amount of protein found in interstitial fluid. A second difference is the ionic content of body fluids. Potassium, magnesium, phosphate, and sulfate are the principal ions in intracellular fluids, whereas sodium, chloride, and bicarbonate ions predominate in extracellular fluids. A third difference is in the concentrations of nonelectrolytes such as glucose and urea, which are different in the various body fluid compartments. These differences exist because of varying membrane permeabilities and transport systems and also because of the metabolic activities of individual tissues. In this chapter we will focus on the principal extracellular fluids and their contents, including hormones.

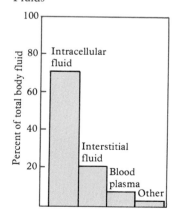

Fig. 18-1 Distribution of Body Fluids

BODY FLUIDS IN DIGESTION

Digestion is the process in which food macromolecules are degraded to smaller molecules capable of being absorbed into the circulatory system. Digestion occurs in the *gastrointestinal tract*, a flexible muscular tube about 26 feet long stretching from the mouth to the anus. Food enters the mouth, where it is ground up by chewing, and proceeds down the esophagus to the stomach. Since the stomach serves as a kind of storage tank in addition to participating in the digestive process, we need to eat only three meals a day instead of a series of smaller, more frequent meals. After storage in the stomach, the semifluid food, now called *chyme*, is moved bit by bit into the small intestine, where most digestion and absorption occur. What is left moves as a result of muscular contractions through the large intestine, where it encounters bacteria which further degrade the food for their own purposes. All that is left after this is undigestible waste, which is excreted from the body as fecal material. Figure 18-2 outlines the important parts of the gastrointestinal tract. At various points along this system, fluid secretions are involved in digestion. These fluids and their functions are discussed in the following sections.

chyme (KIME)

Saliva

Saliva is the secretion of the salivary glands, located on the inner sides of the mouth. Various proteins are present in saliva, including those called *mucins*. These are glycoproteins containing 30–45% carbohydrate. The mucins form viscous solutions which serve as lubricants and protective coatings of the mouth, esophagus, and stomach.

mucins (MU-sins)

Saliva also contains *salivary amylase* (sometimes called *ptyalin*), an enzyme which catalyzes hydrolysis of α-1,4-glycosidic linkages in dietary polysaccharides. Since it has no effect on α-1,6-glycosidic linkages, salivary amylase produces maltose and a mixture of *dextrins*. Amylase requires chloride ions for activity, and fortunately these ions are present in saliva, along with others such as potassium and phosphate. Due to the action of amylase, polysaccharide digestion begins in the mouth. It is possible to detect the slightly sweet taste of maltose in the mouth if starchy food is held there for a while. However, digestion of polysaccharides in the mouth is far from complete, and it occurs to a greater extent in the small intestine.

amylase (AIM-ah-lace)
ptyalin (TIE-ah-lin)

dextrins: a heterogeneous group of polymers of D-glucose intermediate in size between starch and maltose.

Gastric Juice

Swallowing carries food down the esophagus to the stomach where it encounters *gastric juice*. This fluid secreted by the stomach wall contains water (99%), hydrochloric acid, mucins, and enzymes. The secretion of the hydrochloric acid component is stimulated by a group of four hormones called *gastrins*. These are produced in the stomach by the thought, taste, or smell of food, or by the presence of food in the stomach.

Acidity

The pH of gastric juice is dependent on the hydrochloric acid concentration, which may be altered by the presence of food. At the instant of secretion, the hydrochloric acid has a concentration of about $0.17M$ (pH

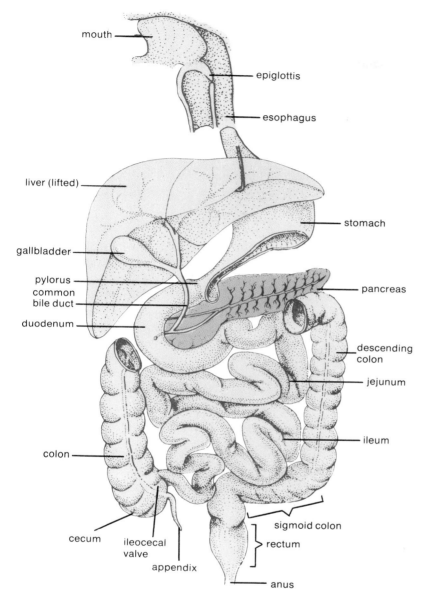

Fig. 18-2 The Gastrointestinal Tract

0.87), but subsequent dilution and mixing with stomach contents usually results in a pH in the range of 1–2. The stomach secretes two mucins which coat the wall and prevent damage from acid or enzymes. Gastric ulcers occur when normal mechanisms for protecting the stomach wall fail. In these cases, secretion of mucins may be inadequate, or the stomach acid may not be buffered well enough by the stomach contents.

Any salivary amylase which might have been swallowed, as well as other proteins, is denatured by the low pH of the stomach, and polysaccharide digestion stops. However, it resumes in the small intestine, as will be discussed later.

Enzymes

Pepsin is the principal protease in gastric juice. This enzyme is synthesized and secreted as the zymogen *pepsinogen*, which has a molecular weight of 40,400. When pepsinogen enters the stomach, it is initially activated by hydrolysis catalyzed by the hydrochloric acid present. In the activation, 42 amino acid residues are removed from the N-terminus as a mixture of peptides. Subsequent activation of pepsinogen is catalyzed by pepsin itself.

$$\underset{\text{MW 40,400}}{\text{pepsinogen}} \xrightarrow{\text{H}^+ \text{ or pepsin}} \underset{\text{MW 32,700}}{\text{pepsin}} + \text{peptide mixture}$$

The enzyme catalyzes hydrolysis of proteins to a mixture of polypeptides. At its pH optimum of 2, pepsin shows a preference for peptide bonds involving the carboxyl group of aromatic amino acids, but it also accepts as substrates portions of proteins having other amino acids such as leucine, methionine, aspartic acid, and glutamic acid. Traces of several other proteases have been found in gastric juice, but they all seem similar to pepsin in action.

Pancreatic Juice

duodenum (doo-ah-DEE-num): the upper part of the small intestine.

secretin (see-KREE-tin)

cholecystokinin (koe-lih-sis-toh-KYE-nin)

When the acidic chyme enters the *duodenum,* two polypeptide hormones, *secretin* and *cholecystokinin,* are released by the duodenum. These hormones are carried by the bloodstream to the pancreas where secretin stimulates the flow of *pancreatic juice* (a basic solution rich in bicarbonate ions), and cholecystokinin stimulates the release of enzymes and zymogens. Thus, the pancreatic juice which enters the duodenum contains trypsinogen, chymotrypsinogen, proelastase, procarboxypeptidase, prophospholipase, lipase, pancreatic amylase, deoxyribonuclease, and ribonuclease. After activation in the small intestine, the zymogens yield, respectively, trypsin, chymotrypsin, elastase, carboxypeptidase, and phospholipase. The catalytic functions of the enzymes in pancreatic juice are summarized in Table 18-1. In addition, the basic fluid (pH about 8) neutralizes the acid chyme, so that the contents of the small intestine are at neutral or slightly basic pH.

Intestinal Juice

enterocrinin (ehn-teh-row-CRYE-nin)

When chyme enters the duodenum, a third hormone is secreted, this one by the small intestine itself. It is a polypeptide called *enterocrinin,* and it stimulates the small intestine to put forth a fluid rich in additional digestive enzymes such as enteropeptidase, aminopeptidases, dipepti-

Table 18-1 Enzymes in Pancreatic Juice

Enzyme	Catalytic Function	Specificity
Amylase*	Hydrolysis of polysaccharides to monosaccharides, disaccharides, and oligosaccharides	α-1,4-Glycosidic linkages
Carboxypeptidase*	Hydrolysis of proteins to polypeptides and peptides	Peptide bonds of C-termini
Chymotrypsin*	Hydrolysis of proteins to polypeptides and peptides	Aromatic residues on the carboxyl side of peptide bonds
Deoxyribonuclease	Hydrolysis of DNA to oligonucleotides	Phosphodiester bonds in DNA
Elastase*	Hydrolysis of proteins to polypeptides and peptides	Neutral aliphatic residues on the carboxyl side of peptide bonds
Lipase	Hydrolysis of triacylglycerols to fatty acids and mono- and diacylglycerols	Ester bonds at the 1 and 3 positions of triacylglycerols
Phospholipase*	Hydrolysis of phosphoglycerides	Ester bonds
Ribonuclease	Hydrolysis of RNA to mono- and oligonucleotides	Phosphodiester bonds in RNA
Trypsin*	Hydrolysis of proteins to polypeptides and peptides	Lysine and arginine residues on the carboxyl side of peptide bonds

*Secreted as zymogens

dases, maltase, sucrase, lactase, lipase, and others (see Table 18-2). It is the function of these enzymes to insure that digestion is complete by catalyzing hydrolysis of any remaining peptides, disaccharides, and lipids.

Bile

Another fluid needed for digestion in the small intestine is the substance called *bile*. The liver continuously produces bile, which is stored in the gall bladder until needed in the intestine. The gall bladder releases bile when cholecystokinin is released by the pancreas. Bile is a thick yellow green fluid of slightly basic pH. In addition to water, bile contains many inorganic salts, which probably help to buffer the fluid, and mucins, which account for its high viscosity. However, the most important components of bile are the bile salts and the bile pigments.

Bile Salts

The principal bile salts are the Na^+ and K^+ salts of *glycocholic* and *taurocholic acids*, which are synthesized in the liver by joining glycine and taurine to cholic acid through amide linkages.

glycocholic (glye-coe-COE-lic)
taurocholic (tah-row-COE—lic)

Table 18-2 Enzymes in Intestinal Juice

Enzyme	Catalytic Function	Specificity
Aminopeptidases	Hydrolysis of protein to polypeptides and peptides	Peptide bonds of N-termini
Dipeptidases	Hydrolysis of dipeptides to amino acids	Peptide bonds in dipeptides
Enteropeptidase	Activation of trypsinogen	A specific lysine-isoleucine bond in trypsinogen
Lactase	Hydrolysis of lactose to glucose and galactose	The glycosidic linkage in lactose
Lipase	Hydrolysis of triacylglycerols to fatty acids and mono- and diacylglycerols	Ester bonds at the 1 and 3 positions in triacylglycerols
Maltase	Hydrolysis of maltose to glucose	The glycosidic linkage in maltose
Nucleases	Hydrolysis of DNA and RNA	Variable
Nucleotidases	Hydrolysis of nucleotides to nucleosides and HPO_4^{2-}	Phosphomonoester bonds in nucleotides
Phosphatase	Hydrolysis of phosphate esters	Phosphomonoester bonds
Sucrase	Hydrolysis of sucrose to glucose and fructose	The glycosidic linkage in sucrose

$$H_2N-CH_2-CO_2H$$
glycine

$$H_2N-CH_2-CH_2-SO_3H$$
taurine

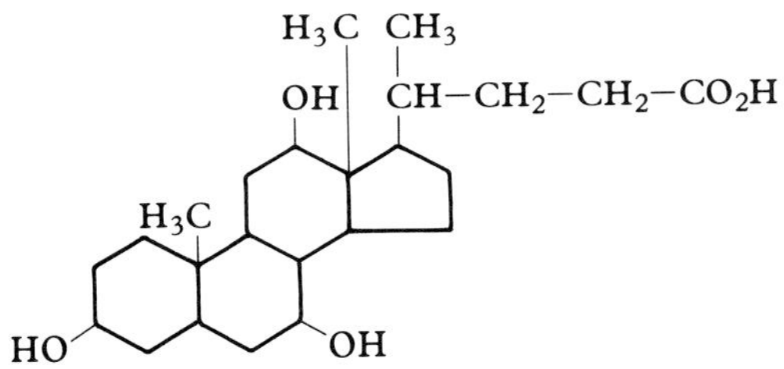

cholic acid

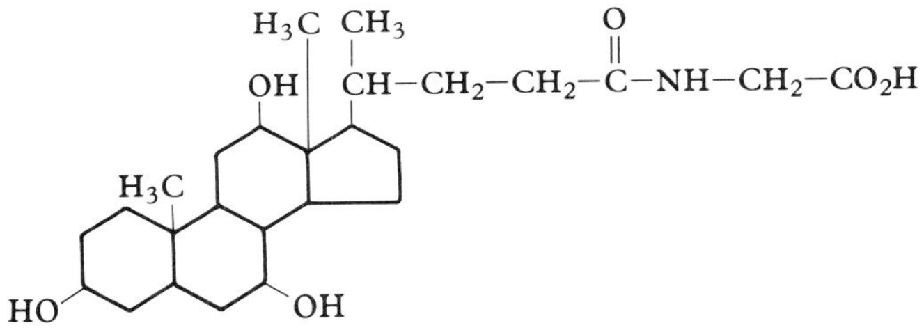

glycocholic acid

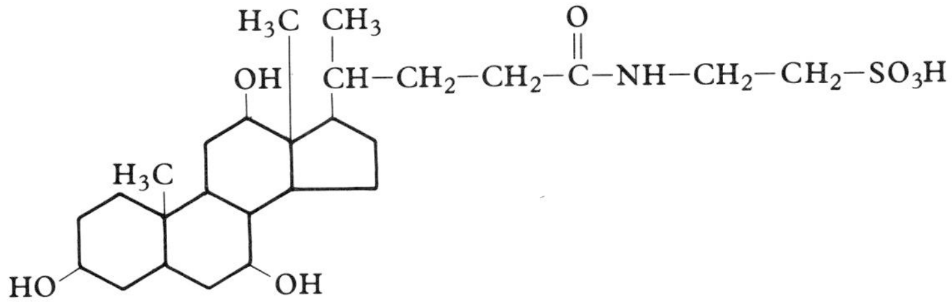

taurocholic acid

(Remember that cholic acid, a steroid, is synthesized from cholesterol.)

The bile salts make it possible for pancreatic lipase to function efficiently. Hydrolysis of lipids occurs only at the interface between the lipid droplets and the water in which they are suspended. The bile salts increase the surface area of the lipid droplets by causing formation of a finely dispersed emulsion. This action aids pancreatic lipase by allowing greater contact between the suspended lipids and the surrounding water.

A second function of the bile salts is to keep cholesterol in solution. Excess cholesterol is excreted in bile. If cholesterol levels are too high in bile, or if levels of bile salts are too low, precipitation of cholesterol and entrapped bile pigment will occur, and gallstones are formed. The most common ones contain about 80% cholesterol, but occasionally gallstones are found that contain 90–98% cholesterol.

The third function of bile salts is to aid in the intestinal absorption of fat-soluble vitamins. This is especially important for vitamin K, needed for blood clotting. Patients who suffer from blocked bile ducts must take bile salts by mouth to insure proper absorption of fat-soluble vitamins.

Bile Pigments

The bile pigments *bilirubin* and *biliverdin* give bile its color. These pigments are derived from the heme of hemoglobin from worn-out red blood cells and from other heme proteins that have undergone natural degradation. The liver removes these pigments from blood and releases them in bile.

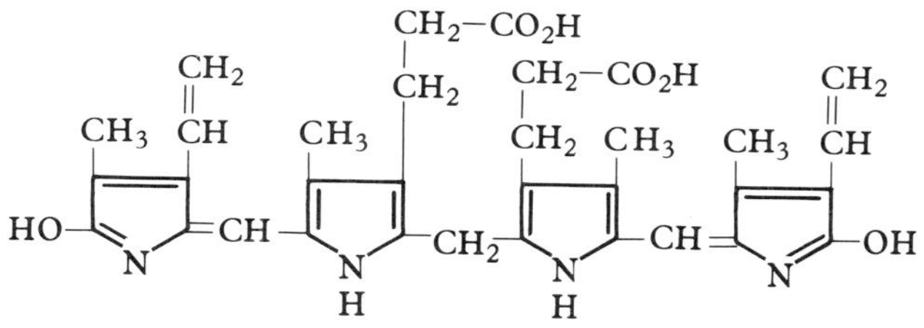

bilirubin (red)

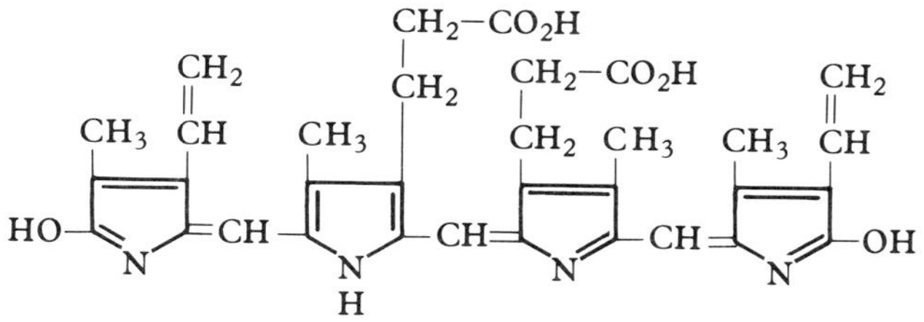

biliverdin (green)

Occasionally a medical condition called *jaundice* exists, in which excessive amounts of bile pigments, primarily bilirubin, accumulate in the blood and make their way into the skin. This gives a yellow color to skin and to the whites of the eyes. Jaundice can be caused by excessive destruction of red blood cells, blocked bile ducts, or liver damage caused by toxins, poisons, heart failure, or disease. Jaundice is especially prevalent in premature babies, usually because their liver enzymes have not

jaundice (JAWN-dis)

fully developed. If the condition is allowed to persist, brain damage can occur due to bilirubin deposits in the brain. Fortunately, the condition is easily treated by exposing the infants to fluorescent light. This treatment degrades bilirubin to less toxic products that can be either metabolized or excreted by the infant.

The Large Intestine

As the chyme moves through the small intestine, digestion produces monosaccharides, amino acids, fatty acids, and acylglycerols. These molecules are absorbed through the walls of the small intestine into the general blood circulation. Digestion in the small intestine is so efficient that the food that reaches the large intestine is not usable by the body. However, this residue is acted on by bacteria living in the large intestine to produce vitamins B and K, as well as some amino acids. These substances are absorbed along with water by the walls of the large intestine, and the resulting fecal material is excreted as waste.

EXAMPLE 18.1

Give the locations for digestion of each of the following:

(a) fats (b) carbohydrates (c) proteins

SOLUTION

(a) small intestine
(b) mouth, small intestine
(c) stomach, small intestine

EXERCISE 18.1 Give the principal constituents dissolved in each of the following body fluids:

(a) saliva
(b) gastric juice
(c) pancreatic juice
(d) intestinal juice
(e) bile

BLOOD

The average adult human body contains about 8% blood, roughly 85 mL per kilogram of body weight. Thus, a person weighing 150 pounds (68.1 kilograms) has about 5.8 liters of blood. Loss of 1 pint of blood (about 0.5 liter) usually does not have serious consequences, and many of us donate a pint of blood occasionally with only slight discomfort. Blood has many functions that are important to life:

1. transportation of food and digestion products
2. transportation of gases between the lungs and the tissues
3. transportation of metabolic waste products
4. transportation of hormones, enzymes, and vitamins
5. participation in the body's immune system

6. participation in maintenance of acid-base, electrolyte, and water balance in the body
7. heat regulation of the body by its ability to shift water to or from the surface of the body
8. protection against excessive blood loss by clot formation

Blood gets its red color from hemoglobin in the red blood cells. Arteries, which carry blood away from the lungs, contain oxyhemoglobin, which has a bright crimson color. Thus, arterial blood has a more vivid color than venous blood, which contains the dark red deoxyhemoglobin. Blood is more viscous than water because of the many cells and proteins present. The pH of blood has a normal range of 7.35 to 7.45, maintained principally by the bicarbonate buffer system (see chapter 10). Since blood is simultaneously a solution, a colloidal dispersion, and a suspension, its specific gravity is noticeably higher than that of water, usually in the range of 1.035 to 1.075.

Composition of Blood

Blood is composed of a liquid portion and the suspended *formed elements*. The liquid portion, called *plasma*, constitutes 55–60% (v/v) of whole blood. The formed elements are the red blood cells (*erythrocytes*), the white blood cells (*leukocytes*), and the platelets. The composition of blood is shown in Table 18-3.

erythrocyte (eh-RIH-throw-site): from the Greek *erythro* ("red") and *cyto* ("cell").

leukocyte (LU-koe-site): from the Greek *leuko* ("white").

When blood is drawn from a person, it clots within a few minutes unless certain procedures are followed. During the time the clot is forming, the blood becomes rather jellylike. Then a clear, straw-colored fluid separates from the solid mass; this fluid is *serum*. The formed elements are trapped within the clot, and serum is the fluid left afterwards. Serum contains neither formed elements nor fibrinogen, one of the clotting factors, whereas plasma contains all dissolved substances but no formed elements.

Table 18-3 Composition of Normal Whole Blood (Human)

Substance	% (w/w)
Total solids	19–23
Water	77–81
Hemoglobin	
Adult males	15.8
Adult females	13.8
Children	12.0
Total nitrogen	3.5
	% (v/v)
Carbon dioxide (venous)	50–60
Carbon dioxide (arterial)	45–55
Oxygen capacity	16–24
Oxygen content (venous)	10–18
Oxygen content (arterial)	15–23

Addition of anticoagulants to blood as it flows from a vessel prevents clot formation. These anticoagulants are often used when blood is to be analyzed in the laboratory, because it is inconvenient to work with clotted blood. Typical anticoagulants either precipitate calcium ions as insoluble compounds or form soluble but un-ionized calcium compounds. Since calcium ions are required in the process, removal of them prevents blood clot formation.

Blood Plasma

Blood plasma, a light straw-colored fluid, can be obtained from whole blood by centrifuging in the presence of anticoagulants to produce sedimentation of the formed elements.

Proteins in Plasma

Proteins comprise most of the dissolved matter in plasma. The major types of proteins are the *albumins* and the *globulins*. The albumins account for 80% of the colloidal osmotic pressure (oncotic pressure) of blood, and they also participate in the transport of fatty acids. The globulins are composed of many different proteins, all of which have important roles. In addition, the globulins are responsible for 20% of the colloidal osmotic pressure of blood. The major plasma proteins are listed in Table 18-4.

If normal human blood plasma is subjected to electrophoresis at pH 8.6, various protein fractions are separated, as indicated in Figure 18-3. The general types of globulins are named by their positions in the electrophoretic pattern as *alpha*-1 (α_1), *alpha*-2 (α_2), *beta*-1 (β_1), *beta*-2 (β_2), and *gamma* (γ). It is in the γ-globulin fraction that the *antibodies* are found. An antibody is a specialized protein (also known as an *immunoglobulin*) that combines with foreign substances to render them inactive. At least 25–30 different immunoglobulins have been isolated from human γ-globulins. These immunoglobulins have a wide range of molecular weights (150,000 to 1,000,000), and invariably they are glycoproteins.

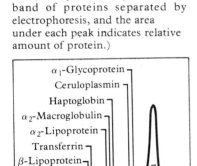

Fig. 18-3 Electrophoretic Pattern of Normal Human Serum at pH 8.6. (Each peak corresponds to a band of proteins separated by electrophoresis, and the area under each peak indicates relative amount of protein.)

Table 18-4 Major Proteins in Human Plasma

Component	Concentration (g/dL)	Percentage of Total Protein
Total Protein	6–8	100
Albumins	4–5	50–60
Globulins	2–3	40–50
α_1	0.09–0.22	4.2–7.2
α_2	0.14–0.36	6.8–12
β	0.19–0.45	9.3–15
γ	0.26–0.69	13–23
Fibrinogen	0.3	7

Other Organic Constituents

In addition to proteins, plasma contains such dissolved materials as amino acids, various peptides, glucose, lipids, lactic acid, citric acid, other organic acids, ketone bodies, nitrogenous waste products, pigments, and small amounts of various enzymes, vitamins, and hormones. In general, the amounts of glucose, amino acids, and lipids increase after meals until they reach a high level, and then they decrease as these substances are taken into the tissue cells. Lactic acid, a product of carbohydrate metabolism,

$$CH_3-\underset{\underset{\text{lactic acid}}{|}}{\overset{OH}{\underset{|}{CH}}}-CO_2H$$

is normally present in small amounts and increases with exercise and under certain pathological conditions. The *ketone bodies*—acetone, acetoacetic acid, and β-hydroxybutyric acid—are derived from fatty acids, and their levels become elevated with increased fat metabolism.

$$\underset{\text{acetone}}{CH_3-\overset{O}{\overset{\|}{C}}-CH_3} \qquad \underset{\text{acetoacetic acid}}{CH_3-\overset{O}{\overset{\|}{C}}-CH_2-CO_2H} \qquad \underset{\text{β-hydroxybutyric acid}}{CH_3-\overset{OH}{\underset{|}{CH}}-CH_2-CO_2H}$$

Inorganic Constituents

The inorganic salts dissolved in plasma provide the ions Na^+, K^+, Ca^{2+}, Mg^{2+}, Cl^-, HCO_3^-, SO_4^{2-}, and HPO_4^{2-}. Concentrations of these ions are shown in Table 18-5. In plasma and other extracellular

Table 18-5 Electrolytes in Human Plasma

	Concentration in Plasma (meq/L)
Cations	
Na^+	143
K^+	5
Ca^{2+}	5
Mg^{2+}	2
Total	155
Anions	
Cl^-	105
HCO_3^-	25
Protein anions	16
Carboxylic acid anions	6
HPO_4^{2-}	2
SO_4^{2-}	1
Total	155

fluids, the concentration of Na^+ is higher than that of K^+, which is concentrated in cells in the blood and tissues. Bicarbonate and phosphate are important as buffers, and the shifting of Cl^- in and out of the red blood cells plays an important role in electrolyte balance in the body.

Red Blood Cells

Red blood cells (erythrocytes) are formed in the bone marrow. They contain about 60% water, less water than is found in the cells of most tissues. Most of the dissolved matter is the red protein hemoglobin, but there are small amounts of other proteins and the lipids cholesterol, phosphatidyl choline, and phosphatidyl ethanolamine. On the average, there are about 5 million red blood cells per cubic millimeter of blood.

The red blood cells do not contain nuclei and thus are not capable of reproduction. However, the many enzymes present allow red blood cells to carry out metabolic reactions to produce energy from glucose. The average lifespan for red blood cells in humans is 120 days. After this, the cells are destroyed in the spleen, liver, and bone marrow. In the liver, heme from hemoglobin is degraded to bilirubin, which is excreted in the bile (as discussed earlier in this chapter), and iron and globin (the protein part of hemoglobin) are conserved for use in synthesis of new red blood cells.

Hemoglobin

The structure of hemoglobin was discussed in chapter 16; you will recall that the molecule is composed of four subunits arranged as two pairs of polypeptide chains. Each subunit has a heme group which binds an oxygen molecule.

Types of Hemoglobin

Normal human hemoglobin exists in several types, with subunits made up of different kinds of polypeptide chains. These are designated as α, β, γ, δ, and ϵ. Most human hemoglobins contain two α-chains and two other chains, usually β-, γ-, or δ-. Adult hemoglobin, HbA, contains two α- and two β-subunits. Approximately 90% of the hemoglobin in a normal adult is of this type ($\alpha_2\beta_2$). Human fetal hemoglobin, HbF, contains two α-subunits and two γ-subunits ($\alpha_2\gamma_2$). A minor component (about 2.5%) of normal adult hemoglobin is HbA_2; it contains two δ-chains in addition to two α-chains ($\alpha_2\delta_2$). And finally, human embryonic hemoglobin, HbE, contains two ϵ-chains ($\alpha_2\epsilon_2$). The amino acid sequences of the various hemoglobin subunits are given in Table 18-6.

Hemoglobin is synthesized in the developing red blood cells in the bone marrow. Globin is made from amino acids, and the heme units are added to the newly synthesized globin. (About 14% of the amino acids in the diet of a human are used for globin synthesis.)

Table 18-6 Amino Acid Sequences of the α-, β-, γ-, and δ-Chains of Human Hemoglobin. The complete sequences of the α- and β-chains are shown; in the γ- and δ-chains, only residues that differ from those in the β-chain are shown. (The sequence of the ε-chain, which appears in human embryos of less than 12 weeks postconceptual age, has not been established.)

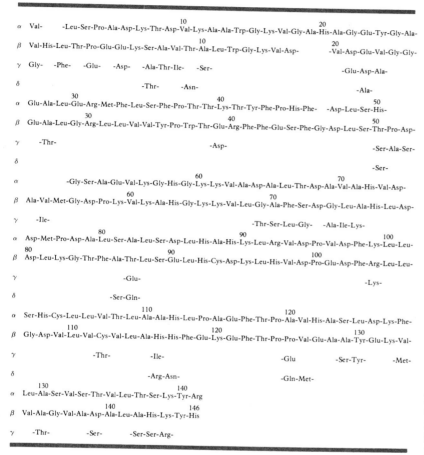

```
                         10                                  20
α  Val-    -Leu-Ser-Pro-Ala-Asp-Lys-Thr-Asp-Val-Lys-Ala-Ala-Trp-Gly-Lys-Val-Gly-Ala-His-Ala-Gly-Glu-Tyr-Gly-Ala-
                          10                                 20
β  Val-His-Leu-Thr-Pro-Glu-Glu-Lys-Ser-Ala-Val-Thr-Ala-Leu-Trp-Gly-Lys-Val-Asp-    -Val-Asp-Glu-Val-Gly-Gly-
γ  Gly-    -Phe-    -Glu-    -Asp-    -Ala-Thr-Ile-    -Ser-                        -Glu-Asp-Ala-
δ                              -Thr-         -Asn-                                  -Ala-
               30                              40                           50
α  Glu-Ala-Leu-Glu-Arg-Met-Phe-Leu-Ser-Phe-Pro-Thr-Thr-Lys-Thr-Tyr-Phe-Pro-His-Phe-    -Asp-Leu-Ser-His-
               30                              40                        50
β  Glu-Ala-Leu-Gly-Arg-Leu-Leu-Val-Val-Tyr-Pro-Trp-Thr-Glu-Arg-Phe-Phe-Glu-Ser-Phe-Gly-Asp-Leu-Ser-Thr-Pro-Asp-
γ        -Thr-                                  -Asp-                                           -Ser-Ala-Ser-
δ                                                                                               -Ser-
                                        60                              70
α       -Gly-Ser-Ala-Gln-Val-Lys-Gly-His-Gly-Lys-Lys-Val-Ala-Asp-Ala-Leu-Thr-Asn-Ala-Val-Ala-His-Val-Asp-
                                   60                                  70
β  Ala-Val-Met-Gly-Asp-Pro-Lys-Val-Lys-Ala-His-Gly-Lys-Lys-Val-Leu-Gly-Ala-Phe-Ser-Asp-Gly-Leu-Ala-His-Leu-Asp-
γ                -Ile-                                       -Thr-Ser-Leu-Gly-    -Ala-Ile-Lys-
                                     80                              90                              100
α  Asp-Met-Pro-Asn-Ala-Leu-Ser-Ala-Leu-Ser-Asp-Leu-His-Ala-His-Lys-Leu-Arg-Val-Asp-Pro-Val-Asn-Phe-Lys-Leu-Leu-
                         80                              90                           100
β  Asp-Leu-Lys-Gly-Thr-Phe-Ala-Thr-Leu-Ser-Glu-Leu-His-Cys-Asp-Lys-Leu-His-Val-Asp-Pro-Glu-Asn-Phe-Arg-Leu-Leu-
γ                                    -Glu-                                                     -Lys-
δ                      -Ser-Gln-
                          110                             120
α  Ser-His-Cys-Leu-Leu-Val-Thr-Leu-Ala-Ala-His-Leu-Pro-Ala-Glu-Phe-Thr-Pro-Ala-Val-His-Ala-Ser-Leu-Asp-Lys-Phe-
                       110                             120                              130
β  Gly-Asn-Val-Leu-Val-Cys-Val-Leu-Ala-His-His-Phe-Gly-Lys-Glu-Phe-Thr-Pro-Pro-Val-Gln-Ala-Ala-Tyr-Gln-Lys-Val-
γ             -Thr-              -Ile-                    -Glu-         -Ser-Tyr-         -Met-
δ                         -Arg-Asn-                       -Gln-Met-
       130                          140
α  Leu-Ala-Ser-Val-Ser-Thr-Val-Leu-Thr-Ser-Lys-Tyr-Arg
                                140                           146
β  Val-Ala-Gly-Val-Ala-Asp-Ala-Leu-Ala-His-Lys-Tyr-His
γ       -Thr-       -Ser-       -Ser-Ser-Arg-
```

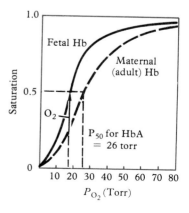

Fig. 18-4 Oxygen Dissociation Curves for Adult Hemoglobin (HbA) and Fetal Hemoglobin (HbF). (The arrow represents transfer of oxygen from maternal to fetal hemoglobin.)

Oxygen Transport by Hemoglobin

The binding of oxygen to hemoglobin in red blood cells is described by the *oxygen dissociation curve* shown in Figure 18-4. This is a plot of the fraction of hemoglobin that is saturated with oxygen as a function of the partial pressure of oxygen. Oxygen affinity is often expressed as a P_{50} value, the partial pressure of oxygen at which 50% of the hemoglobin is oxygenated. The P_{50} value for adult hemoglobin is 26 torr. In other words, an oxygen partial pressure of 26 torr will lead to 50% saturation of adult hemoglobin. The S-shape of the curve is typical of a binding reaction for an allosteric protein, and it reflects the cooperative effect observed in oxygen binding by hemoglobin.

Hemoglobin functions in oxygen transport by obtaining oxygen from air in the lungs and delivering it to body tissues. In the lungs, where P_{O_2} is about 100 torr, hemoglobin is almost fully oxygenated. However, when oxyhemoglobin reaches the tissue capillaries, P_{O_2} is only about 20–40 torr. This lowered P_{O_2} causes hemoglobin to release oxygen to the tissues:

$$\text{Hb} + 4\,O_2 \underset{\text{low } P_{O_2} \text{ (tissues)}}{\overset{\text{high } P_{O_2} \text{ (lungs)}}{\rightleftharpoons}} \text{Hb}(O_2)_4$$

In this way, tissue cells are supplied with oxygen to be used for energy-producing metabolic reactions.

When oxyhemoglobin releases oxygen, the conformational changes cause six proton-binding groups to become exposed to the surroundings. These groups then take up protons from the aqueous environment. Increased concentrations of H^+ (lowered pH) decrease the oxygen affinity of hemoglobin, and conversely, increased concentrations of oxygen lower the affinity for H^+. Since metabolic reactions in the tissues produce acids, the release of oxygen to tissue cells is enhanced. An added benefit is that hemoglobin carries protons away from the tissues.

$$\text{Hb}(O_2)_4 + H^+ \underset{O_2 \text{ in lungs}}{\overset{H^+ \text{ in tissues}}{\rightleftharpoons}} H^+\text{—Hb} + 4\,O_2$$

Carbon monoxide (CO) binds to heme in hemoglobin about 210 times more strongly than oxygen. If sufficient quantities of CO are inhaled, O_2 will be displaced from hemoglobin, and the blood will not have enough oxyhemoglobin to carry out normal respiration. In such cases, asphyxiation may occur. Because the gas is odorless and colorless, and the victim is likely to lose consciousness within minutes, carbon monoxide poisoning is often fatal.

Regulation of Oxygen Affinity in Red Blood Cells

The oxygen affinity of hemoglobin inside the red blood cell is much lower than that of hemoglobin free in solution. In 1967 it was found that the anion of 2,3-diphosphoglyceric acid interacts with hemoglobin to decrease its affinity for oxygen. Diphosphoglycerate (DPG), a normal chemical intermediate in glucose metabolism, is usually present in red blood cells.

2,3-diphosphoglycerate (DPG)

$$\begin{array}{c} CO_2^- \quad O \\ | \quad\quad\; \| \\ H-C-O-P-O^- \\ |\quad\quad\quad\; | \\ H-C-H \;\; O^- \\ | \\ O \\ | \\ O=P-O^- \\ | \\ O^- \end{array}$$

It regulates oxygen affinity by binding in the central cavity of deoxyhemoglobin (Figure 16-15) in such a way that six positively charged residues on the β-chains interact with the negatively charged DPG. In this way DPG stabilizes the structure of deoxyhemoglobin by forming a bridge between the two β-subunits. DPG must be expelled before oxygen can bind to hemoglobin,

$$\text{DPG} \cdot \text{Hb} + O_2 \rightleftharpoons \text{Hb}(O_2) + \text{DPG}$$

and thus more energy is required to convert deoxyhemoglobin to oxyhemoglobin in the presence of DPG. The increased energy requirement for oxygen binding results in a lower oxygen affinity. The ultimate result of DPG regulation is that the lowered oxygen affinity enables hemoglobin to release a greater proportion of its oxygen to the tissues.

Another facet of DPG regulation of oxygen affinity involves fetal hemoglobin, HbF. This hemoglobin has a lower P_{50} value than HbA (see Figure 18-4). The basis for the higher oxygen affinity of HbF is that it binds DPG less strongly than HbA. The weaker binding occurs because a histidine residue in each β-chain of HbA is replaced by a serine residue in HbF. Thus the number of positively charged residues available in HbF to interact with DPG is decreased by 2. This situation is beneficial to the fetus because HbF is oxygenated at the expense of HbA, and the transfer of oxygen across the placental membrane is optimized.

Transport of Carbon Dioxide by Hemoglobin

In addition to its role in oxygen transport, hemoglobin also participates in transport of carbon dioxide. When there is plenty of oxygen available to the tissues, metabolism produces about one molecule of CO_2 for each O_2 molecule consumed. Most of the CO_2 in blood is transported as HCO_3^-, but hemoglobin accounts for about 3-4% of the total. The basis for the carbon dioxide transport by hemoglobin is the reaction between CO_2 and the amino terminus of each of the subunits of hemoglobin:

$$\underset{\text{hemoglobin}}{\text{Hb}-\overset{\overset{\displaystyle H}{|}}{N}-H} + CO_2 \rightleftharpoons \underset{\text{carbaminohemoglobin}}{\text{Hb}-\overset{\overset{\displaystyle H}{|}}{N}-CO_2H}$$

It has been found that deoxyhemoglobin combines with more CO_2 than oxyhemoglobin. This is probably because the amino termini of deoxyhemoglobin are more likely to exist in the unprotonated form required for the reaction. Thus, the presence of high levels of CO_2 and H^+ in metabolically active tissues promotes release of O_2 from oxyhemoglobin. Conversely, the high concentration of O_2 in the lungs causes unloading of H^+ and CO_2 from deoxyhemoglobin. This exchange of O_2 for H^+ and CO_2 and its reverse are known as the *Bohr effect* (Figure 18-5). It was discovered by Christian Bohr in 1904.

Fig. 18-5 The Bohr Effect: The Exchange of O_2 for H^+ and CO_2

Lungs: $Hb(O_2)_4 + H^+ + CO_2$

Tissues: $H^+ - Hb - CO_2 + 4\,O_2$

EXAMPLE 18.2

How does DPG affect oxygen-binding by hemoglobin?

SOLUTION

DPG stabilizes the structure of deoxyhemoglobin by electrostatic binding. This imposes a higher energy requirement on oxygen binding and increases the P_{50} for hemoglobin.

EXERCISE 18.2 Summarize the Bohr effect.

Blood Clot Formation

Hemostasis is the name given to the natural system that prevents blood loss. The process of hemostasis is composed of three mechanisms. The first of these, vasoconstriction, occurs when a small blood vessel is injured. The vessel becomes narrower after injury as a response to stem the flow of blood. The second mechanism is the formation of a platelet clump (thrombus) at the site of damage. While the first and second events are occurring, the blood that has already escaped begins to form a clot, the third mechanism in hemostasis.

A blood clot is formed by a cascade of events involving more than ten different proteins (see Table 18-7). Furthermore, there are two different ways of triggering clot formation, the *intrinsic pathway* and the *extrinsic pathway*. In the intrinsic pathway, clotting occurs when blood encounters an abnormal or foreign surface. In this case, the abnormal surface causes reactions by components already present in the blood, triggering formation of a clot. Clotting can also be caused by addition of substances not normally present in the blood, such as certain tissue

Table 18-7 Factors Involved in Human Blood Clotting

Factor	Pathway	Function of Active Form
Hageman factor (XII)	Intrinsic	Activates XI
Kallikrein	Intrinsic	Required for activation of XI
Kinin	Intrinsic	Required for activation of XI
Plasma thromboplastin antecedent (PTA) (XI)	Intrinsic	Activates IX
Christmas factor (IX)	Intrinsic	Activates X
Antihemolytic factor (VIII)	Intrinsic	Required for activation of X
Tissue factor	Extrinsic	Activates X
Proconvertin (VII)	Extrinsic	Required for activation of X
Stuart factor (X)	Common	Activates prothrombin
Accelerin (V)	Common	Stimulates activation of prothrombin
Prothrombin	Common	Activates fibrinogen
Fibrinogen	Common	FORMS FIBRIN CLOT
Fibrin-stabilizing factor (XIII)	Common	Stabilizes clot by crosslinking fibrin

components that enter the blood as a result of trauma; this initiation of blood clotting is called the extrinsic pathway. The two pathways merge into a common pathway that ultimately results in formation of a clot, and proper hemostasis requires the interplay of both the intrinsic and extrinsic pathways, as illustrated in Figure 18-6.

Formation of Clot Fibers (The Common Pathway)

The ultimate reaction in clotting is the conversion of the soluble protein *fibrinogen,* dispersed in plasma as a colloid, to the insoluble protein called *fibrin.* Fortunately this reaction does not normally occur in circulating blood. However, it seems that both the intrinsic and the extrinsic pathways result in the formation of proteases that activate a zymogen designated as Factor X. When this zymogen is activated, a protease (Factor X_a) is formed that is responsible for activating a subsequent step of the clotting sequence.

Factor X_a catalyzes conversion of the zymogen prothrombin to the protease thrombin. Vitamin K is required for synthesis of prothrombin in the liver, and thus vitamin K deficiencies result in slow blood clot formation. Thrombin requires Ca^{2+} for activity, and it has a specificity for certain arginine-glycine bonds in fibrinogen. Cleavage at these sites causes the formation of fibrin molecules which spontaneously associate to make long, insoluble fibers.

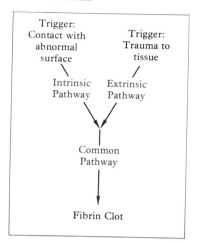

Fig. 18-6 The Pathways of Blood Clot Formation

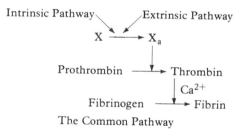

The Common Pathway

The fibrin threads trap formed elements and plasma as they interlace to form the clot. This mass then undergoes contraction, which pulls the edges of the wound together, and serum is squeezed out.

The Intrinsic Pathway

In this pathway, a series of three zymogen activations results in the eventual production of Factor X_a. These events are summarized below, where the subscript *a* designates active enzymes. (See Table 18-7 for identification of the factors.)

abnormal or foreign surface
$$XII \rightarrow XII_a$$
$$XI \rightarrow XI_a$$
$$IX \rightarrow IX_a$$
$$\downarrow VIII$$
$$X \rightarrow X_a$$

The Intrinsic Pathway

The Extrinsic Pathway

The extrinsic pathway also results in activation of Factor X_a, but only two steps are involved. Injury to the blood vessel releases a lipoprotein called tissue factor (also known as thromboplastin). This substance then forms a complex with a protein called Factor VII, and the complex catalyzes the activation of Factor X. These reactions are summarized as follows, and the factors for this pathway are listed in Table 18-7.

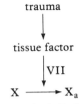

The Extrinsic Pathway

For some people, clotting occurs at an abnormally slow rate. This condition, known as *hemophilia*, is hereditary and usually occurs in males. Hemophilia is most often caused by a deficiency or reduction in activity of Factor VIII (called the antihemolytic factor), but deficiencies in other clotting factors can also result in prolonged or excessive bleeding. Transfusion of blood from a normal individual to a hemophiliac can be effective treatment of the disease, as this allows normal clotting to occur.

EXAMPLE 18.3

What circumstances trigger blood clot formation (a) by the intrinsic pathway, and (b) by the extrinsic pathway?

SOLUTION

(a) Clotting occurs by the intrinsic pathway when blood encounters an abnormal or foreign surface.
(b) Clotting occurs by the extrinsic pathway when substances not normally present in blood make their way into blood.

EXERCISE 18.3 What is the role of Factor X_a in blood clot formation?

URINE

Urine is the solution of metabolic waste products and other substances, often toxic, excreted by the body. It is formed in the kidneys by filtration of blood plasma. The fluid found within the filtering units of the kidneys (see Figure 18-7) is similar to blood plasma, but it does not contain formed elements or proteins. As this fluid passes along through the kidney, water, amino acids, glucose, and other useful components are reabsorbed into the blood. The remaining concentrated solution is urine. It contains water, urea (the end product of amino acid metabolism), inorganic salts, creatinine, ammonium ions, uric acid, and colored

PERSPECTIVE: Hemophilia and the Fall of the Russian Monarchy

Tsar Nicholas II (1868–1918), the last Russian monarch, was a timid, vacillating man who was ill-suited for ruling the vast Russian empire. By 1904, when his son Alexis was born, Nicholas' position as tsar was on shaky ground. Alexis was the first male heir born to a reigning Russian tsar since the seventeenth century, and his birth gave hope for the future to the ruling family. Unfortunately, Alexis developed symptoms of hemophilia, transmitted to him from Queen Victoria of England through his mother, the Empress Alexandra, who was Victoria's granddaughter.

During the first year of his life, Alexis' hemorrhages became severe, and Alexandra sought the services of Rasputin, a mysterious holy man known for his healing powers. In 1905, when he demonstrated an ability to ease the suffering of young Alexis, Rasputin was welcomed into the royal family as a close and trusted friend. Because of the empress' faith in his healing abilities, Rasputin held a position of great power in the Russian court for many years.

By 1911, Rasputin had abused his position by engaging in scandal and treachery. The empress managed to protect him until 1916, when he drowned after a group of extreme conservatives shot him and threw him into a river. The next two years brought a series of misfortunes to Russia, and in 1917, Tsar Nicholas was forced to abdicate. In July, 1918,

Rasputin

he and his entire family, including the delicate fourteen-year-old Alexis, were killed by Bolshevik forces.

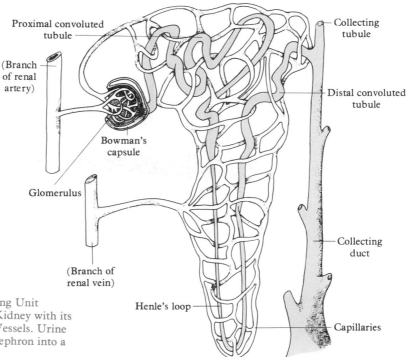

Fig. 18-7 A Filtering Unit (Nephron) of the Kidney with its Associated Blood Vessels. Urine passes from each nephron into a collecting duct.

products of hemoglobin decomposition. The composition of a sample of human urine is shown in Table 18-8.

Table 18-8 Major Components of Human Urine and a Comparison of Urine and Plasma Concentrations. (The volume and composition of urine vary widely; these data are for an average 24-hour specimen of total volume 1,200 mL.)

Component	Grams	Approximate Urine/Plasma Concentration Ratio
Glucose	0.02	0.02
Amino acids	0.5	1.0
Ammonia	0.8	100
Urea	25	70
Creatinine	1.5	70
Uric acid	0.7	20
H^+	pH 5-8	Up to 300
Na^+	3.0	1.0
K^+	1.7	15
Ca^{2+}	0.2	5
Mg^{2+}	0.15	2
Cl^-	6.3	1.5
HPO_4^{2-}	1.2 (P)	25
SO_4^{2-}	1.4 (S)	50
HCO_3^-	0-3	0-2

Diuretics

diuretics (dye-you-REH-tics)

diuresis (dye-you-REE-sis): increased discharge of urine.

edema (ih-DEE-ma): the condition in which fluid collects in tissue spaces between cells and causes swelling.

Diuretics are substances that cause an increase in the quantity of urine produced. They work either by increasing the filtration through the kidneys or by decreasing the reabsorption by the kidneys. Salty and spicy foods tend to cause *diuresis*, as do caffeine and alcohol. Certain diuretic drugs are administered for the relief or prevention of *edema*, for speeding up the excretion of ingested poisons, or for dilution of urine to prevent precipitation of drugs in the kidneys.

Hormones Affecting Urine Composition

The composition of urine is under the control of two hormones. *Vasopressin* (also known as the antidiuretic hormone, ADH), a nonapeptide synthesized in the pituitary, is released if the blood becomes hypertonic by as much as 2%.

$$H_3\overset{+}{N}-Cys-Tyr-Phe-Gln-Asn-Cys-Pro-Arg-Gly-\overset{O}{\overset{\|}{C}}-NH_2$$
$$\phantom{H_3\overset{+}{N}-Cys}\underline{S-S}$$

vasopressin

This hormone acts on the kidneys to cause them to retain more water; this in turn causes formation of less concentrated blood. On the other hand, if blood becomes hypotonic by 2%, vasopressin is not released by the pituitary, and the kidneys remove more water from blood to make urine. This action increases the concentration of dissolved substances in the blood. In *diabetes insipidus*, production of vasopressin is deficient or absent, and urine output is greatly increased.

Aldosterone is a steroid hormone synthesized in the adrenal gland; it is secreted if the sodium level in blood drops too low. The hormone acts on the kidneys by causing them to retain Na^+; this action also requires some retention of water. If the sodium level in blood rises, aldosterone secretion stops, and excess Na^+ ions are excreted by the kidneys.

diabetes insipidus: a disease in which there is a persistent, abnormally large quantity of urine discharged; not to be confused with diabetes mellitus, the disease in which the body's ability to use sugar is impaired and glucose appears in the urine.

aldosterone (al-DOH-steh-rone)

aldosterone

The Role of the Kidneys in Acid-Base Balance

In addition to the buffers in blood which help to maintain acid-base balance, the kidneys serve as a back-up system by excreting excess acid or bicarbonate. Normal urine is slightly acidic, with an average pH of 6, but urine pH may drop as low as 4 at times of acidosis. Under conditions of alkalosis, the pH of urine may be 8 or higher.

The Role of the Kidneys in Electrolyte Balance

The ability of the kidneys to regulate acid-base balance is related to their role in electrolyte balance. If there is a need for the body to retain cations such as Na^+, K^+, Ca^{2+}, or Mg^{2+}, then NH_4^+ or H^+ can be excreted in their place. Aldosterone action on the kidneys also influences Na^+ excretion, as described earlier. In addition, the amount of HCO_3^- in urine can be adjusted by the kidneys, depending on HCO_3^- levels in blood.

The Role of the Kidneys in Regulation of Blood Pressure

If the blood pressure should drop, as in hemorrhaging, for example, the kidneys secrete an enzyme called *renin* into the blood. This enzyme brings about formation of *angiotensin II*, a polypeptide hormone consisting of nine amino acid residues. This hormone is the most powerful vasoconstrictor known, and thus its action on blood vessels increases blood pressure. When the blood pressure is normal, renin is retained by the kidneys. A secondary effect of angiotensin II is to trigger the release of aldosterone, thus increasing blood volume and blood pressure.

EXAMPLE 18.4

How do the kidneys help in maintaining (a) acid-base balance, (b) electrolyte balance, and (c) blood pressure?

SOLUTION

(a) The kidneys help to maintain acid-base balance by excreting excess acid or bicarbonate.
(b) The kidneys help to maintain electrolyte balance by retaining essential cations (Na^+, K^+, Ca^{2+}, Mg^{2+}) when needed and excreting NH_4^+ or H^+ in their place. Also, the kidneys adjust the amount of HCO_3^- in the urine in response to HCO_3^- levels in the blood.
(c) The kidneys help to maintain blood pressure by secretion of renin when the blood pressure falls. This enzyme activates the hormone angiotensin II, which acts to constrict blood vessels and thus increase blood pressure.

HORMONES

The endocrine system is composed of many widely scattered tissue masses called *endocrine glands* (see Figure 18-8). Some endocrine tissues

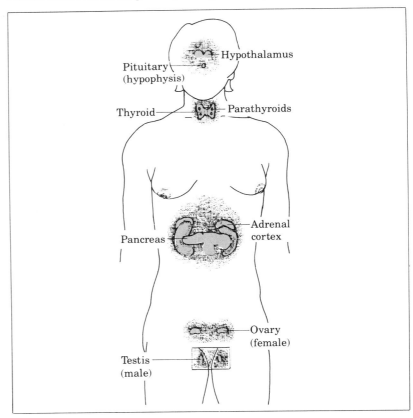

Fig. 18-8 The Major Organs of the Endocrine System

are a part of other organs. For example, the pancreas, a gland that secretes digestive enzymes into the small intestine, also contains endocrine tissues which produce the hormone insulin. Endocrine glands are ductless and secrete their products—hormones—directly into the bloodstream, which transports them throughout the body.

The endocrine system and the nervous system are closely related. Special cells in the hypothalamus, a part of the brain, secrete polypeptide hormones called *regulatory hormones,* which act on the pituitary to stimulate or inhibit synthesis and release of other hormones. Most of the hormones secreted by the pituitary are *trophic hormones*. These are polypeptides which act on the other endocrine glands, causing them to release still other hormones. The pituitary also secretes two hormones (oxytocin and vasopressin) which act directly on target organs. The organization of the endocrine network is diagrammed in Figure 18-9.

Hormones were defined in the early 1900s as chemical transmitters produced and released by a specialized body part and transported by the blood to act on another body part. This definition is still used, but since the prostaglandins (see chapter 14) were discovered, we have had to realize that tissue hormones can be produced by one cell type in a tissue to act on another cell type in the same tissue. Hormones exert potent biological influence and are found in extremely small quantities in the body, often at concentrations of 10^{-7} to 10^{-10} M in the blood. Despite

Fig. 18-9 Organization of the Endocrine Network Controlled by the Hypothalamus.

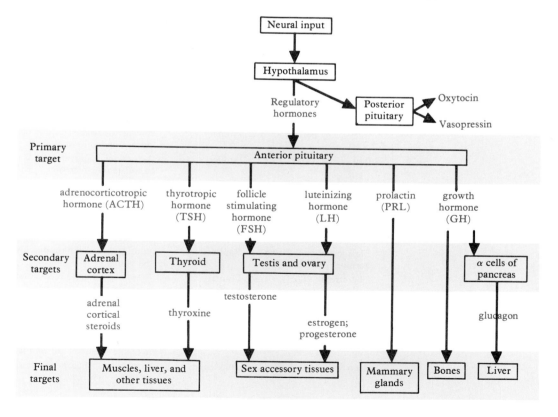

Table 18-9 Major Mammalian Hormones

Hormone	Origin	Action
Polypeptide regulatory hormones		
Adrenocorticotropin regulatory hormone (CRH)	Hypothalamus	Stimulates release of adrenocorticotropin (ACTH) from pituitary
Follicle-stimulating regulatory hormone (FSH-RH)	Hypothalamus	Stimulates release of follicle-stimulating hormone (FSH) from pituitary
Growth-hormone regulatory hormone (GH-RH)	Hypothalamus	Stimulates release of growth hormone from pituitary, insulin and glucagon from pancreas, and secretin from duodenum
Growth-hormone inhibitory hormone (GH-RIH)	Hypothalamus	Inhibits release of growth hormone from pituitary
Luteinizing regulatory hormone (LH-RH)	Hypothalamus	Stimulates release of luteinizing hormone from pituitary
Melanocyte regulatory hormone (MSH-RH)	Hypothalamus	Stimulates release of melanocyte-stimulatory hormone from pituitary
Melanocyte inhibitory hormone (MSH-RIH)	Hypothalamus	Inhibits release of melanocyte-stimulatory hormone from pituitary
Prolactin regulatory hormone (PRH)	Hypothalamus	Stimulates release of prolactin from the pituitary
Prolactin inhibitory hormone (PRIH)	Hypothalamus	Inhibits release of prolactin from the pituitary
Thyrotropin regulatory hormone (TRH)	Hypothalamus	Stimulates release of thyrotropin from the pituitary
Polypeptide hormones		
Adrenocorticotropin (ACTH)	Pituitary	Stimulates release of steroid hormones from adrenal gland
Thyroid-stimulating hormone (TSH)	Pituitary	Stimulates thyroid to release thyroid hormones
Follicle-stimulating hormone (FSH)	Pituitary	Acts on the ovaries to stimulate development of follicles
Luteinizing hormone (LH)	Pituitary	Acts on the ovaries to promote corpus luteum formation
Prolactin (LTH)	Pituitary	Acts on the mammary glands to stimulate milk secretion
Growth hormone (GH, somatotropin)	Pituitary	Causes growth of bone and muscle
Vasopressin	Pituitary	Acts on the kidneys to increase water retention and blood pressure
Oxytocin	Pituitary	Acts on the uterus to induce labor
Melanocyte-stimulating hormone (MSH)	Pituitary	Stimulates melanin production in the skin
Thyrotropin (TSH)	Pituitary	Acts on the thyroid to stimulate release of thyroid hormones
Insulin	Pancreas	Stimulates uptake and utilization of sugar by many tissues; stimulates protein synthesis
Glucagon	Pancreas	Stimulates breakdown of glycogen in liver and release of fat from fat tissue
Secretin	Duodenum	Acts on the pancreas to stimulate release of pancreatic juice

Table 18-9 Major Mammalian Hormones (continued)

Hormone	Origin	Action
Cholecystokinin	Duodenum	Acts on the pancreas to stimulate release of digestive enzymes
Gastrins	Stomach	Act on the stomach to stimulate release of HCl
Enterocrinin	Small intestine	Acts on the small intestine to stimulate release of digestive enzymes
Angiotensin	Blood	Acts on the adrenal gland to stimulate release of aldosterone
Steroid hormones		
Aldosterone	Adrenal gland	Acts on the kidneys to cause increased retention of Na^+ and water, thus increasing blood pressure
Glucocorticoids	Adrenal gland	Stimulate conversion of proteins to carbohydrates; reduce inflammatory responses; stimulate antibody production
Testosterone	Testes	Maintains secondary sexual characteristics of the male
Estrogen	Ovaries	Stimulates thickening of uterine walls during menstrual cycle; maintains secondary sexual characteristics of the female
Progesterone	Corpus luteum	Prepares the uterus for ovum implantation and maintains pregnancy
Amine hormones		
Thyroxine	Thyroid gland	Stimulates general body metabolism
Triiodothyronine	Thyroid gland	Stimulates general body metabolism
Epinephrine	Adrenal gland	Stimulates glucose release by breakdown of glycogen in muscle and liver; increases heart rate
Norepinephrine	Adrenal gland	Similar to epinephrine but has a strong effect on constriction of blood vessels; also a neurotransmitter

their low levels, most mammalian hormones have been purified and studied; their origins and actions are summarized in Table 18-9.

Hormones have very specific functions, as indicated in Table 18-9. Several different hormones from different endocrine glands often interact to bring about changes in body functions; an example is the regulation of a woman's menstrual cycle. Malfunctions of endocrine glands and imbalances of hormones cause a variety of health problems, including goiter, diabetes, and sterility. Several specific hormones and their actions have already been described, and others will be discussed in later chapters in the context of their metabolic influences.

Classes of Hormones

Most hormones can be classified as polypeptides, steroids, or amines (as shown in Table 18-9). The polypeptide hormones, a diverse group,

Fig. 18-10 Examples of the Three Major Classes of Hormones

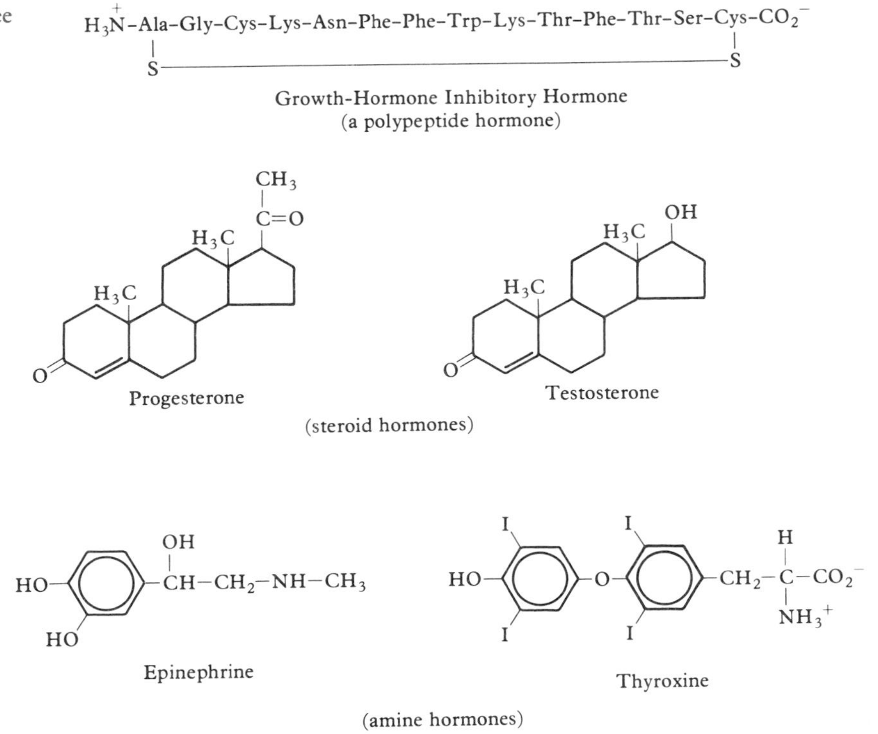

$$\overset{+}{H_3N}\text{-Ala-Gly-Cys-Lys-Asn-Phe-Phe-Trp-Lys-Thr-Phe-Thr-Ser-Cys-}CO_2^-$$
$$\qquad\qquad\ \ |\underline{\qquad\qquad\qquad\qquad\qquad\qquad\qquad\qquad}|$$
$$\qquad\qquad\ \ S\qquad\qquad\qquad\qquad\qquad\qquad\qquad\qquad\ \ S$$

Growth-Hormone Inhibitory Hormone
(a polypeptide hormone)

Progesterone   Testosterone

(steroid hormones)

Epinephrine   Thyroxine

(amine hormones)

include the regulatory hormones produced by the hypothalamus. The steroid hormones, all synthesized from cholesterol, include the female sex hormones (estrogens), male sex hormones (androgens), and the steroid hormones produced by the adrenal gland. The amine hormones are small molecules, several of which are synthesized from the amino acid tyrosine. The structures of the members of the three hormone classes are quite different; examples are shown in Figure 18-10.

Prohormones

Just as some enzymes are synthesized in inactive zymogen forms, many polypeptide hormones are synthesized as inactive prohormones (also called hormonogens). Some of these are listed in Table 18-10. Like zymogens, prohormones are inactive until activated extracellularly by peptidases.

Table 18-10 Some Prohormones

Prohormone	Source
Proinsulin	Pancreas
Proparathyroid hormone	Parathyroid
Angiotensinogen	Liver
Progastrin	Stomach

The Molecular Basis of Hormone Action

Two theories have been proposed to explain the action of hormones. The first of these was suggested by Earl Sutherland, a pharmacologist and physiologist who received the Nobel Prize in 1971. Sutherland's theory is referred to as the "two-messenger hypothesis" because it involves a second messenger in addition to the hormone (the first messenger). This model, which applies to polypeptide and amine hormones (except for the thyroid hormones), was originally proposed as a result of studies on the action of epinephrine.

Epinephrine, often referred to as the "fight or flight" hormone, is secreted by the adrenal gland at times of stress or excitement. Its action results in increased glycogen degradation to produce energy. In the two-messenger hypothesis, epinephrine is secreted into the blood and it then binds to specific receptor sites on the outer surface of membranes of muscle cells (the target organ). Apparently this binding induces conformational changes in membrane components which then stimulate production of cyclic adenosine monophosphate (cyclic AMP or cAMP) within the cells (see Figure 18-11). Thus, cAMP is the "second messenger"; it conveys the influence of epinephrine even though the hormone does not enter the cells. High levels of cAMP activate glycogen phosphorylase, the enzyme needed for glycogen degradation (see chapter 17), and deactivate the enzyme needed for synthesis of glycogen from glucose. In both cases, the enzyme activities are regulated by a complex series of events leading to covalent modification.

A second model has been proposed for hormone action; this one explains the action of steroid hormones. (Even though they are amine hormones, the thyroid hormones seem to have actions similar to those of steroid hormones.) This model, proposed by Elwood V. Jenson in the early 1970s, suggests that steroid hormones enter their target cells and make their way to the cell nucleus. There they bind to specific sites on chromosomes. The result of this interaction is increased RNA and protein synthesis and changes in membrane structure and function. Although far from complete, these proposed models have brought about a better understanding of the molecular basis of hormone action.

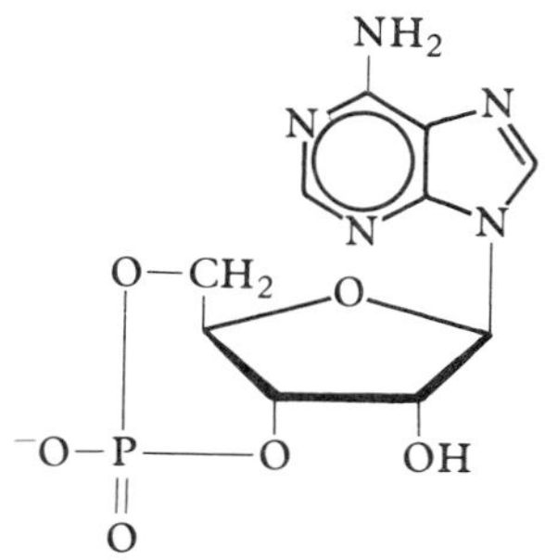

Fig. 18-11 Cyclic Adenosine Monophosphate

SUMMARY

Body Fluids in Digestion: Digestion occurs in the gastrointestinal tract in the mouth, stomach, and small intestine. Saliva, a body fluid secreted by the salivary glands in the mouth, contains mucins, amylase, and inorganic ions. Carbohydrate digestion begins in the mouth where amylase catalyzes hydrolysis of α-1,4-glycosidic linkages in polysaccharides to produce maltose and a mixture of dextrins. In the stomach, food encounters gastric juice, which contains hydrochloric acid, mucins, and enzymes. Conditions in the stomach activate pepsinogen to pepsin, a protease which catalyzes partial hydrolysis of proteins.

When chyme enters the duodenum, hormones stimulate the flow of pancreatic juice containing bicarbonate ions, zymogens, and enzymes. The small intestine also secretes fluid containing additional enzymes which insure that digestion is complete. Another fluid needed for digestion in the small intestine is bile, which contains inorganic salts, mucins, bile salts, and bile pigments. The bile salts help to emulsify lipids, making them more susceptible to hydrolysis catalyzed by pancreatic lipase. Bile salts also solubilize cholesterol and aid in the intestinal absorption of fat-soluble vitamins. The small molecules produced by digestion are absorbed through the walls of the small intestine. The residue is acted on by bacteria in the large intestine to produce vitamins B and K, as well as some amino acids. These are absorbed, and any remaining solid material is excreted as waste.

Blood: Blood is composed of a liquid portion (plasma) and suspended formed elements. If blood is allowed to clot, the clear fluid called serum separates. Serum does not contain formed elements or fibrinogen. Blood plasma contains many proteins, usually classified as albumins and globulins. Both types of proteins are responsible for the colloidal osmotic pressure of blood. The antibodies (immunoglobulins) are found in the γ-globulin fraction of blood proteins. Plasma also contains various dissolved organic substances and inorganic salts. In plasma and other extracellular fluids, the concentration of Na^+ is higher than that of K^+; the reverse is true of intracellular fluids.

The red blood cells, formed in bone marrow, contain hemoglobin as their principal protein, but there are small amounts of other proteins and lipids. There are several types of normal hemoglobin, with subunits made up of different kinds of polypeptide chains. The oxygen affinity of hemoglobin inside red blood cells is regulated by DPG, which binds to deoxyhemoglobin and stabilizes its structure. Hemoglobin also participates in the transport of CO_2 and H^+.

A blood clot is formed by a cascade of events involving more than ten different proteins. Blood clot formation can proceed through the intrinsic pathway or the extrinsic pathway. Both merge into a common pathway which ultimately converts the soluble protein fibrinogen to the insoluble material called fibrin.

Urine: Urine is the solution of metabolic waste products and other substances excreted by the body. It is formed by the kidneys, and it usually contains water, urea, inorganic salts, creatinine, ammonium ions, uric acid, pigments, and other substances. Urine formation is under control of the hormones vasopressin and aldosterone. The kidneys have the ability to alter the pH of urine to help maintain acid-base balance in the body. The kidneys also participate in the regulation of electrolyte balance and blood pressure.

Hormones: Hormones are chemical transmitters produced and released into the blood to act at a target site. They are classified as polypeptides, steroids, and amines. The "two-messenger hypothesis," involving cyclic AMP, has been proposed to explain the action of polypeptide and amine hormones. The model for steroid hormones suggests that these hormones act at the chromosome level, in contrast to polypeptide and amine hormones, which bring about their effects by binding to the outside of cell membranes.

STUDY QUESTIONS AND PROBLEMS

1. Define the following terms:

    (a) intracellular fluid
    (b) interstitial fluid
    (c) extracellular fluid
    (d) digestion
    (e) chyme
    (f) saliva
    (g) mucins
    (h) amylase
    (i) gastric juice
    (j) gastrins
    (k) pepsin
    (l) secretin
    (m) cholecystokinin
    (n) pancreatic juice
    (o) enterocrinin
    (p) intestinal juice
    (q) bile
    (r) bile salts
    (s) bile pigments
    (t) jaundice
    (u) blood plasma
    (v) formed elements
    (w) blood serum
    (x) immunoglobulin
    (y) ketone bodies
    (z) DPG
    (aa) carbaminohemoglobin
    (bb) fibrin
    (cc) thrombin
    (dd) urine
    (ee) diuretic
    (ff) vasopressin
    (gg) aldosterone
    (hh) renin
    (ii) angiotensin II
    (jj) hormone
    (kk) epinephrine
    (ll) cyclic AMP

2. Why is chewing beneficial to digestion?
3. Describe the role of saliva in digestion.
4. Describe the role of pepsin in digestion.
5. Summarize the hormonal controls exerted on the release of pancreatic juice.
6. What is the role of pancreatic juice in digestion?
7. What do you think would happen if the zymogens in pancreatic juice were activated before they left the pancreas?

8. Explain how the secretion of intestinal juice is under hormone control.
9. What is the role of intestinal juice in digestion?
10. Describe three functions of bile salts.
11. How are cholesterol levels in bile related to gallstone formation?
12. What is the source of bile pigments?
13. Explain how jaundice might result from hepatitis.
14. Draw representative structures for the end products of digestion for each of the following:
    (a) fats
    (b) carbohydrates
    (c) proteins
15. Give the principal function for each of the following constituents of body fluids:
    (a) Cl^- in saliva
    (b) pepsin
    (c) aminopeptidases
    (d) hemoglobin
    (e) DPG
    (f) HCO_3^-
    (g) fibrinogen
    (h) epinephrine
16. Describe the eight major functions of blood.
17. Why is arterial blood a brighter color than venous blood?
18. How does blood plasma differ from blood serum?
19. What is the principal role of the immunoglobulins? Where are they found?
20. What are the ketone bodies? Draw their structures.
21. From information given in this chapter, calculate the number of red blood cells in one mL of blood.
22. Myoglobin, the protein that stores oxygen in muscle tissue, has a P_{50} of 1 torr. How is this value related to the fact that hemoglobin releases O_2 to myoglobin in tissues?
23. Explain why fetal hemoglobin has a higher oxygen affinity than adult hemoglobin. How is this beneficial to the fetus?
24. What is meant by the Bohr effect?
25. How does a drop in blood pH affect oxygen transport?
26. How does a drop in blood pH affect CO_2 transport?
27. Trace the reactions of the common pathway for blood clot formation.
28. How are the intrinsic and extrinsic pathways of blood clot formation different from each other? How are they similar?
29. Why is vitamin K needed for blood clot formation?
30. Why is Ca^{2+} needed for blood clot formation?
31. Heparin, a polysaccharide derivative, is a natural anticoagulant of blood found in high concentration in the liver and the walls of arteries. It is said to be an antithrombin agent because it prevents formation of thrombin. How does this action prevent blood clot formation?
32. Warfarin, an ingredient in rat poison that promotes internal hemorrhaging, inhibits the transport of vitamin K to its site of action in the liver. How is this inhibition related to internal hemorrhaging?
33. Explain how a blood transfusion would be effective treatment for a hemophiliac.
34. Write representative reactions, indicating specificity, for the action of each of the following enzymes.
    (a) pepsin
    (b) trypsin
    (c) carboxypeptidase
    (d) thrombin
    (e) sucrase
    (f) pancreatic lipase
    (g) dipeptidase
    (h) lactase
35. How do diuretics work? For what reasons might diuretic drugs be administered?
36. Describe the hormonal controls which govern urine formation.
37. Alcohol in blood is known to suppress vasopressin secretion. What does this information have to do with the diuretic action of alcohol?
38. How does the body respond to a drop in Na^+ levels in blood?
39. What is the average pH of urine? Why is urine sometimes more acidic and sometimes more basic than this average pH?
40. How does the body respond to a drop in blood pressure?
41. How do the kidneys serve to regulate electrolyte balance?
42. What are the three classes of hormones?
43. Why is epinephrine called the "fight or flight" hormone?
44. What is meant by the "two-messenger hypothesis"?
45. How are steroid hormones thought to work?
46. Give the principal effect of each of the following:
    (a) gastrins
    (b) secretin
    (c) cholecystokinin
    (d) enterocrinin
    (e) vasopressin
    (f) aldosterone
    (g) angiotensin II

CHAPTER 19

AN INTRODUCTION TO METABOLISM: BIOENERGETICS AND THE CITRIC ACID CYCLE

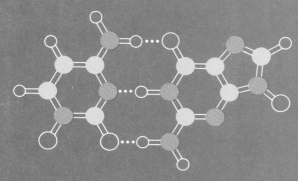

THE CELL AND AN OVERVIEW OF METABOLISM
- Features of Eucaryotic Cells
- Overview of Metabolism

MEMBRANES
- Membrane Lipids
- Lipid Bilayers
- Membrane Permeability
- Membrane Proteins
- Membrane Fluidity
- The Fluid Mosaic Model
- Membrane Transport

BIOENERGETICS
- The Central Figure—ATP
- The Pathway of Electron Transport
- Oxidative Phosphorylation

THE CITRIC ACID CYCLE
- Reactions of the Citric Acid Cycle
- Regulation of the Citric Acid Cycle
- Summary and Major Features of the Citric Acid Cycle
- Energetics of the Citric Acid Cycle

Most biochemical reactions take place within small, living cells that are usually only a few microns long. Cells have a surprising degree of internal organization, and the biochemical specialization in the various cellular compartments gives rise to distinct metabolic roles. The membranes that enclose these compartments have many important functions in addition to the obvious one of separating one compartment from another.

An important component of cells is the tiny unit called the mitochondrion. It is in this compartment that aerobic oxidation occurs, leading to the production of ATP (adenosine triphosphate), the principal currency of free energy in biological systems. For this reason, mitochondria are referred to as the energy factories of the cell. The central metabolic pathway, the citric acid cycle, is also located in the mitochondria. This pathway completes the oxidation of fats, carbohydrates, and amino acids begun in other pathways. In doing so, the citric acid cycle generates reduced coenzymes needed for ATP production.

THE CELL AND AN OVERVIEW OF METABOLISM

A biological cell is a functional unit of biological activity; it is surrounded by a membrane and capable of reproduction. The size of a cell is quite variable, ranging from about 1 micron (μ, 10^{-6} meter) up to about 100 microns. Table 19-1 compares the sizes of cells to those of viruses and molecules. The human body probably contains as many as 10^{14} cells of countless shapes and sizes; each plays an important role in sustaining life.

There are two basic types of cells: *procaryotic* and *eucaryotic*. The main difference is that a eucaryotic cell has a nucleus surrounded by a membrane, while the procaryotic cell has nuclear material concentrated

procaryotic (pro-KAAH-ree-AH-tic): from the Greek words *pro* ("before") and *karyo* ("nucleus").

eucaryotic (you-KAAH-ree-AH-tic): from the Greek words *eu* ("good") and *karyo* ("nucleus").

Table 19-1 Relative Sizes of Cells, Viruses, and Molecules (Approximate values)

Cell or Particle	Size in Microns (10^{-6} meter)
Human Ova	100
Human Red Blood Cells	10
Bacteria	1
Viruses	0.1
Protein Molecules	0.01
Amino Acid Molecules	0.001

in a small section but does not have a nuclear membrane. Examples of procaryotic organisms are bacteria and certain algae. Since animal cells belong to the eucaryotic class, we will focus on the principal features of this type of cell in the following discussion. A eucaryotic cell is illustrated in Figure 19-1.

Features of Eucaryotic Cells

Cell Membrane

Animal cells are enclosed in a *membrane* composed mostly of lipids and proteins. The membrane insures that the cell contents do not spill out into the surroundings. It is also responsible for the selective passage of small molecules, ions, and water into and out of the cell. We will return to the topic of membranes later in this chapter.

Fig. 19-1 Diagram of a Eucaryotic Cell

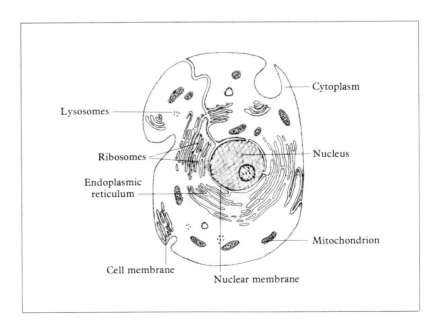

Nucleus

The *nucleus* is a distinctive feature of animal cells, as illustrated in Figure 19-1. Except in the case of very specialized cells, such as red blood cells and platelets, the nucleus is an essential part of eucaryotic cells. It is enclosed by a thin membrane which allows for exchange of substances with the surrounding cytoplasm. Chromosomes, the complex fibers composed of DNA (deoxyribonucleic acid) and proteins are present in the nucleus. It is here that DNA is duplicated in preparation for cell division, and the genetic code is transcribed from DNA into RNA in preparation for protein synthesis.

Mitochondria

Mitochondria are the distinctive, cigar-shaped *organelles* found in all cells except red blood cells and bacteria. Each mitochondrion is surrounded by an inner and an outer membrane. The inner membrane folds in and out through the mitochondrial *matrix* (Figure 19-2). This space is especially rich in enzymes, and it is here that oxidation of derivatives of carbohydrates, fats, and amino acids produces CO_2 and H_2O. Part of the energy released from these oxidations is used to synthesize ATP (adenosine triphosphate), a molecule which then can deliver its energy to locations throughout the cell. However, much of the energy liberated by oxidation is released to the surroundings as heat to maintain the temperature of the organism.

mitochondria (mite-oh-KON-dree-ah): plural of mitochondrion.

organelle: a specialized part of a cell.

mitochondrial matrix: the interior space enclosed by the inner mitochondrial membrane.

Endoplasmic Reticulum

The *cytoplasm* of most eucaryotic cells contains a system of flattened, interconnected channels. This system is called the *endoplasmic reticulum;* it serves as a network of canals for the transport of various substances to and from the nucleus. Some portions of the endoplasmic reticulum have tiny granules attached to their outer surfaces. These granules are rich in RNA, ribonucleic acid, and are called *ribosomes*. Ribosomes present in

cytoplasm (SITE-oh-plas-im): the colorless, semifluid substance inside a cell, exclusive of cellular organelles.

endoplasmic (in-doe-PLAAHS-mic) **reticulum** (reh-TICK-you-lum)

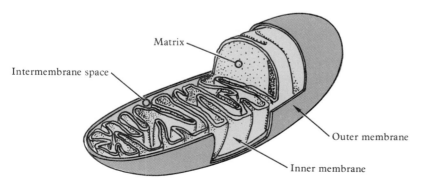

Fig. 19-2 Diagram of a Mitochondrion

this rough endoplasmic reticulum are the sites of protein synthesis in the cell. The portions of endoplasmic reticulum without ribosomes are referred to as smooth endoplasmic reticulum.

Lysosomes

lysosome (LYE-so-zome)

These organelles are small vesicles which contain hydrolytic enzymes necessary for degradation of various substances. The membrane enclosing lysosomes is very fragile, and when a cell dies, the lysosome membrane ruptures to release enzymes which digest the cell parts. Before the death of a cell, lysosomes digest foreign substances and cell parts needing replacement.

EXAMPLE 19.1

List the five major parts of a eucaryotic cell, and give their major functions.

SOLUTION

The five major parts of a eucaryotic cell and their functions are as follows.

cell membrane: encloses the cell and allows selective passage of small molecules, ions, and water.

nucleus: responsible for duplication of DNA and transcription of the genetic code from DNA into RNA in preparation for protein synthesis.

mitochondria: oxidation of derivatives of carbohydrates, fats, and amino acids, and synthesis of ATP.

endoplasmic reticulum: transport of substances to and from the nucleus, and protein synthesis (rough endoplasmic reticulum).

lysosomes: degradation of foreign substances and cell parts.

Overview of Metabolism

catabolism (caah-TAAH-bo-lis-im)
anabolism (an-AAH-bo-lis-im)

Metabolism is defined as the network of chemical reactions responsible for extraction of energy from foods and synthesis of needed cellular materials. Extraction of energy occurs when substances are degraded in the cell; this breakdown of molecules is called *catabolism*. The second part of metabolism, synthesis of cellular materials, is called *anabolism*. Both processes are essential to an organism, but we will focus our attention on the decomposition of cell foods and the energy released in this process.

Three stages are involved in the extraction of energy from food, as illustrated in Figure 19-3. In the first stage, large molecules in food are broken down into smaller units. This stage corresponds to digestion: proteins are hydrolyzed to amino acids, polysaccharides are hydrolyzed to simple sugars, and fats are hydrolyzed to fatty acids, glycerol, monoacylglycerols, and diacylglycerols. In the second stage, these small molecules are degraded to even simpler units, the most common being the acetyl unit of acetyl CoA:

$$CH_3-C\overset{O}{\diagdown}$$

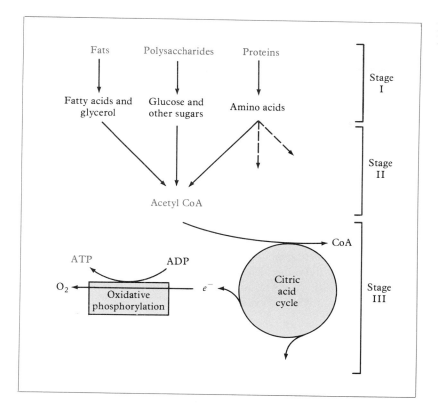

Fig. 19-3 The Extraction of Energy from Food

Some energy is generated in this second stage, but much more is produced from the complete oxidation of the acetyl units of acetyl CoA. This occurs in the third stage and involves the action of the final common pathways in the oxidation of fuel molecules: the citric acid cycle and oxidative phosphorylation. Acetyl CoA delivers acetyl units to the citric acid cycle, where they are completely oxidized to CO_2. For each acetyl unit that is oxidized, four pairs of electrons are transferred to NAD^+ and FAD to produce NADH and $FADH_2$. As these reduced coenzymes pass electrons on to other receivers, and ultimately to O_2 to form H_2O, ATP is generated by the process called oxidative phosphorylation. It is in this third stage that most of the ATP is generated by the degradation of food molecules.

EXAMPLE 19.2

Summarize the three stages of energy extraction from food.

SOLUTION

The three stages are as follows.

Stage 1: digestion of large food molecules to smaller units.
Stage 2: degradation of small units into even simpler units, usually acetyl units of acetyl CoA.
Stage 3: complete oxidation of acetyl units to CO_2 (in the citric acid cycle) and H_2O, and generation of ATP by oxidative phosphorylation.

MEMBRANES

As mentioned earlier, animal cells are enclosed in membranes which protect the contents of the cells. In addition, internal membranes in eucaryotic cells surround organelles such as mitochondria, the nucleus, and lysosomes. Although their structures and functions are quite diverse, these membranes have a number of common features. They usually exist as thin sheets, about six to ten nanometers (nm, 10^{-9} meter) thick, and they are composed mainly of lipids and proteins. The lipids spontaneously associate to form double layers, and the proteins are found on the surfaces and inside the double layers of lipid.

Membrane Lipids

The three major membrane lipids are phospholipids, glycolipids, and cholesterol. Typical phospholipids found in membranes are phosphatidyl serine, phosphatidyl ethanolamine, phosphatidyl choline, and sphingomyelin. To refresh your memory, their structures are shown here (R represents a long hydrocarbon chain, usually C_{13}-C_{21}):

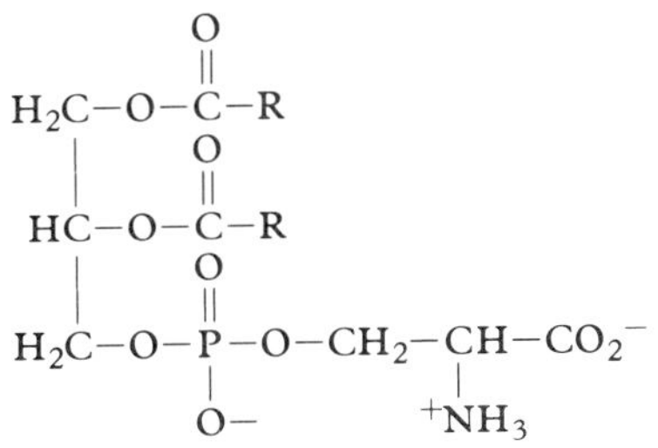

$$\begin{array}{c}
\text{H}_2\text{C}-\text{O}-\overset{\overset{\text{O}}{\|}}{\text{C}}-\text{R} \\
| \\
\text{HC}-\text{O}-\overset{\overset{\text{O}}{\|}}{\text{C}}-\text{R} \\
| \\
\text{H}_2\text{C}-\text{O}-\overset{|}{\underset{\text{O}^-}{\text{P}}}-\text{O}-\text{CH}_2-\underset{^+\text{NH}_3}{\text{CH}}-\text{CO}_2^-
\end{array}$$
phosphatidyl serine

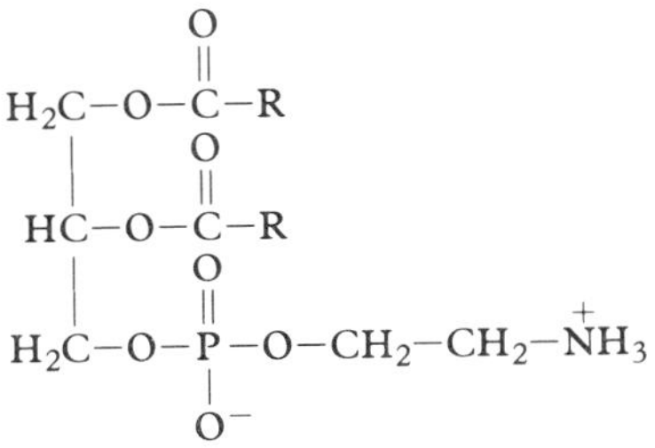

$$\begin{array}{c}
\text{H}_2\text{C}-\text{O}-\overset{\overset{\text{O}}{\|}}{\text{C}}-\text{R} \\
| \\
\text{HC}-\text{O}-\overset{\overset{\text{O}}{\|}}{\text{C}}-\text{R} \\
| \\
\text{H}_2\text{C}-\text{O}-\overset{|}{\underset{\text{O}^-}{\text{P}}}-\text{O}-\text{CH}_2-\text{CH}_2-\overset{+}{\text{N}}\text{H}_3
\end{array}$$
phosphatidyl ethanolamine

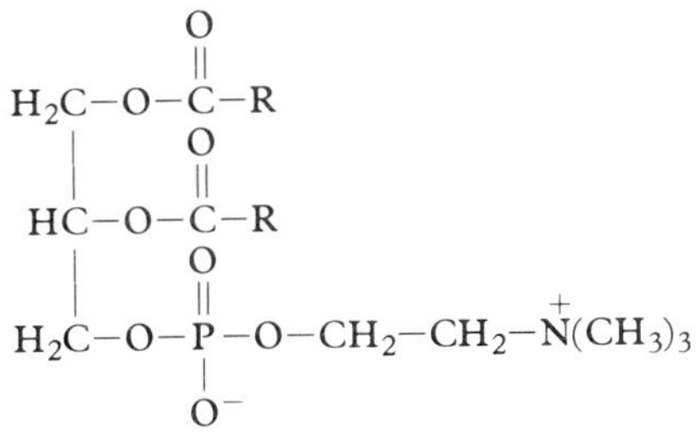

$$\begin{array}{c}
\text{H}_2\text{C}-\text{O}-\overset{\overset{\text{O}}{\|}}{\text{C}}-\text{R} \\
| \\
\text{HC}-\text{O}-\overset{\overset{\text{O}}{\|}}{\text{C}}-\text{R} \\
| \\
\text{H}_2\text{C}-\text{O}-\overset{|}{\underset{\text{O}^-}{\text{P}}}-\text{O}-\text{CH}_2-\text{CH}_2-\overset{+}{\text{N}}(\text{CH}_3)_3
\end{array}$$
phosphatidyl choline

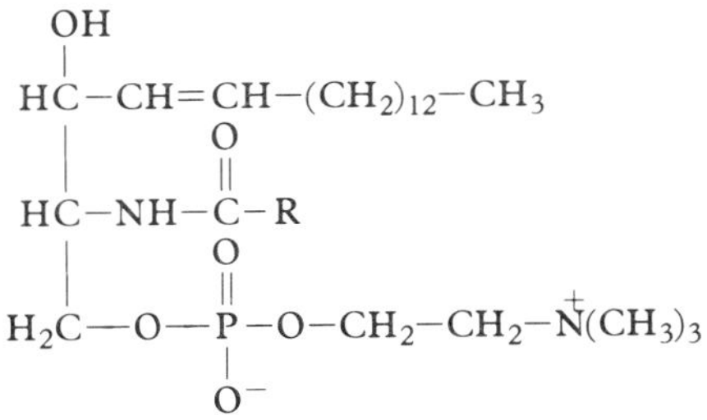

$$\begin{array}{c}
\text{OH} \\
| \\
\text{HC}-\text{CH}=\text{CH}-(\text{CH}_2)_{12}-\text{CH}_3 \\
| \\
\text{HC}-\text{NH}-\overset{\overset{\text{O}}{\|}}{\text{C}}-\text{R} \\
| \\
\text{H}_2\text{C}-\text{O}-\overset{|}{\underset{\text{O}^-}{\text{P}}}-\text{O}-\text{CH}_2-\text{CH}_2-\overset{+}{\text{N}}(\text{CH}_3)_3
\end{array}$$
sphingomyelin

Glycolipids present in biological membranes include the cerebrosides and the gangliosides; their structures are shown in Figures 19-4 and 19-5.

Lipid Bilayers

Phospholipids such as phosphatidyl choline and sphingomyelin have a shape that is almost rectangular, as illustrated by the space-filling models in Figure 19-6. As you can see, the hydrocarbon chains are roughly

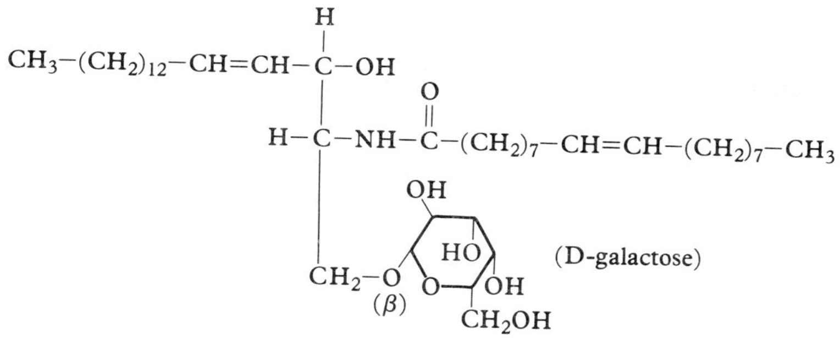

Fig. 19-4 A Cerebroside

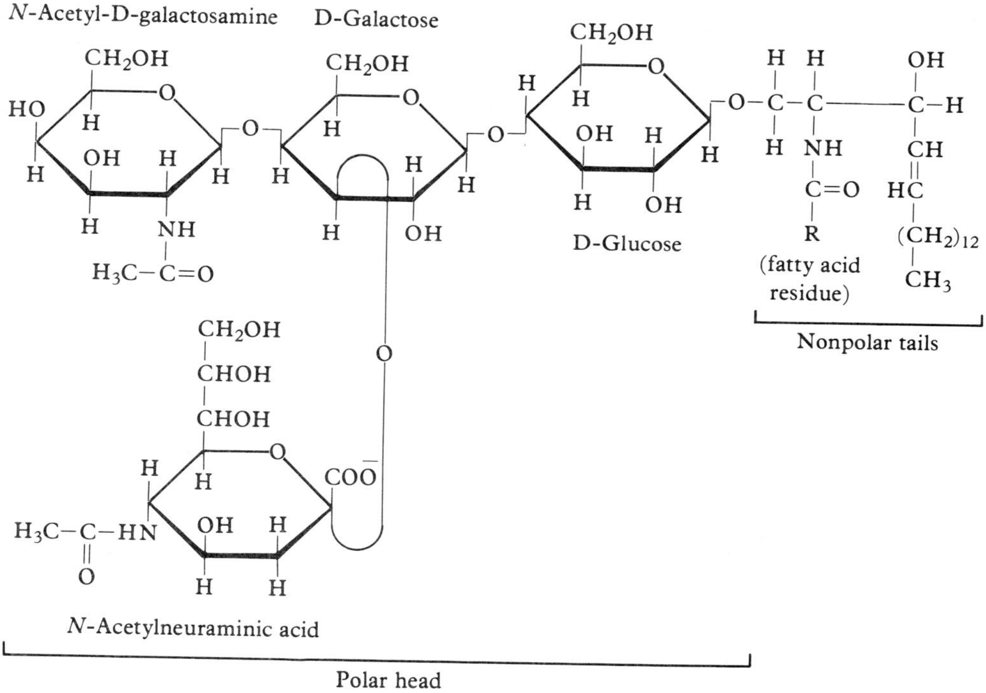

Fig. 19-5 A Ganglioside

parallel to each other, and the polar parts of the molecules extend in the opposite direction. We can approximate the structures of these molecules by using wavy lines for the hydrocarbon portions and circles for the polar head groups. (Figure 19-7).

Phospholipids form micelles in aqueous solution when dissolved in moderate amounts. But when large amounts of phospholipids are added to water, the lipids have a pronounced tendency to form extensive sheets composed of two layers of lipid molecules. (Both of these forms are shown in Figure 19-8.) In this way, their hydrophobic tails interact effectively, and their hydrophilic heads receive maximum exposure to the polar surroundings. Glycolipids also form bimolecular sheets readily in an aqueous environment. In both cases, the formation of the bilayer is

Fig. 19-6 Space-Filling Models of Two Phospholipids
(a) Phosphatidyl choline
(b) Sphingomyelin

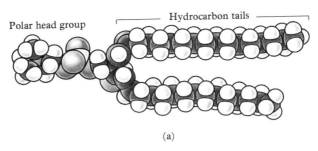

(a)

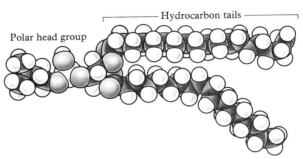

Fig. 19-7 Simplified Representation of a Phospholipid Molecule

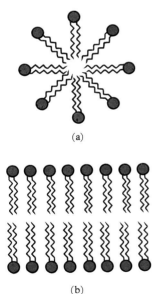

Fig. 19-8 A Micelle (a) and a Bilayer Sheet (b) Formed From Phospholipid Molecules

spontaneous, and the attractive forces cause the bilayers to close on themselves to form spherical shells. Thus, a cell membrane is a continuously closed lipid bilayer which surrounds the cell contents.

Membrane Permeability

Lipid bilayer membranes have a low permeability for ions and polar molecules. In fact, the more soluble a molecule is in nonpolar solvents, the more likely it is to be able to cross a lipid bilayer membrane. This fact suggests that molecules cross membranes by first dissolving in them and then diffusing across. A dramatic exception to this theory is water, which readily crosses biological membranes, perhaps aided by membrane proteins.

Membrane Proteins

The amount of protein present in a biological membrane depends on the function of the particular membrane. For example, the membrane called myelin is an insulator for certain nerve fibers. Lipids function well in this capacity, and the lipid content of myelin is about 80%; the remaining 20% is protein. In contrast, most cell membranes allow certain ions and molecules to pass into and out of cells. The protein content of these membranes is likely to reach as high as 50% because of the participation by the proteins in transport processes. In even greater contrast are mitochondrial membranes, which play an important role in energy conversions within cells. These membranes contain a number of enzymes, and they have the highest protein content, usually about 75%.

Some proteins can be extracted from membranes by treatment with sodium chloride solutions. Presumably these proteins are attached to the surface of membranes by electrical attractions and hydrogen bonds. They are called *peripheral proteins* (see Figure 19-9). Their purpose is probably to stabilize the overall structure of the membrane. Other membrane proteins are bound much more strongly and can only be dissociated by using detergents or organic solvents. These are the *integral proteins,* also shown in Figure 19-9. It is thought that integral proteins interact positively with membrane lipids through hydrophobic forces. These proteins are probably involved in membrane transport processes and enzymatic reactions within membranes.

Fig. 19-9 Types of Membrane Proteins. Protein a is a peripheral protein, while proteins b, c, and d are integral proteins.

Glycoproteins are often found in membranes, usually as integral proteins. In many of these structures, one or more carbohydrate residues are linked to proteins by attachment to serine or asparagine residues, as illustrated in Figure 19-10. It has been found that the carbohydrate groups of glycoproteins and glycolipids are located on the outer surface of the membrane. These carbohydrate groups seem to have two functions. One is that they may help to orient glycoproteins in membranes (see Figure 19-11). The sugar groups are attracted to the aqueous environment around the cell, and their protein portions are thus anchored within the membrane. A second function for membrane carbohydrates is involvement in intercellular recognition. This process is necessary for the grouping of cells to form tissue and for recognition of foreign cells by the immune system.

Membrane Fluidity

Biological membranes are rather fluid structures. They have constant internal motion, and membrane proteins can diffuse a distance of several microns in just one minute. Since phospholipids are smaller molecules than proteins, they diffuse even faster through membranes, perhaps several microns in a second. Many bacterial cells are only a few microns long, and a membrane lipid can probably travel from one end of

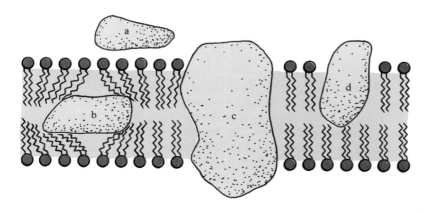

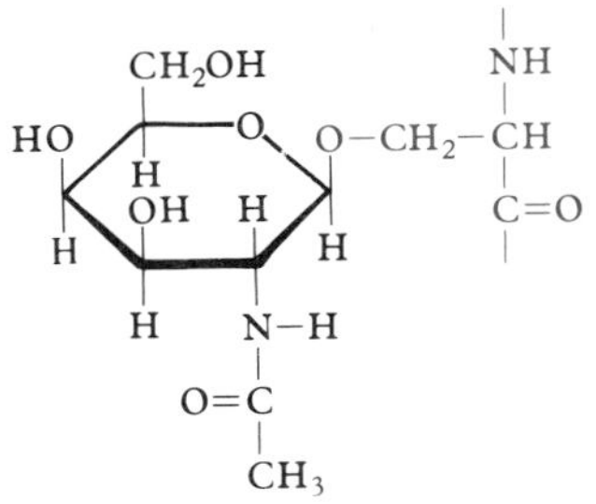

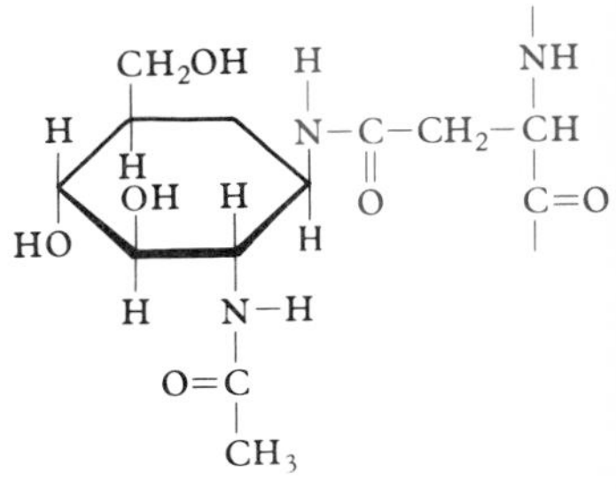

Fig. 19-10 Many glycoproteins have carbohydrate groups attached to (a) serine or (b) asparagine. (The amino acid residues are shown in color.)

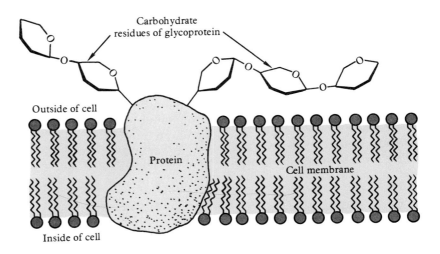

Fig. 19-11 A Membrane Glycoprotein

a bacterial cell to the other in a second. Studies have shown that the viscosity of a membrane is about the same as that of olive oil.

The fatty acid composition of membrane lipids largely determines the degree of membrane fluidity. The less orderly the packing of the lipids in the membrane, the more fluid the membrane will be. Thus, saturated fatty acid components, with their preferred linear conformation, will pack together well, producing a membrane of relatively low fluidity. On the other hand, the *cis* double bonds normally found in unsaturated fatty acids produce bends in the hydrocarbon chains, causing looser packing. Thus, unsaturated fatty acids give greater fluidity to membranes than saturated fatty acids (Figure 19-12). Similarly, short-chain fatty acids increase membrane fluidity because they do not pack together as strongly as long-chain fatty acids. A third influence on membrane fluidity is the amount of cholesterol present. Because cholesterol molecules disrupt the orderly packing of fatty acid chains, the more cholesterol present, the more fluid the membrane will be.

Cells are able to control the composition of their membranes so that an optimal viscosity is maintained for a given temperature. If, for example, there is a temperature decrease in the environment of an organism, it will synthesize new lipids for its membranes to increase the fluidity, so that the membranes will function properly at the lower temperature.

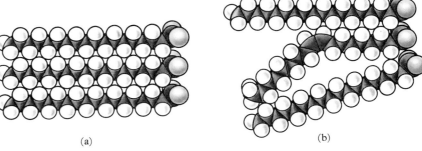

Fig. 19-12 The Packing of Fatty Acid Chains (a) The packing of three molecules of stearic acid (C_{18}, saturated) is very tight and orderly. (b) The *cis* double bond in a molecule of oleic acid (C_{18}, one double bond) between two molecules of stearic acid causes loose, disorderly packing.

The Fluid Mosaic Model

The current model for the structure of membranes (Figure 19-13) was proposed by S. Jonathan Singer and Garth Nicolson in 1972. The *fluid mosaic model*, as it is called, suggests that membranes are solutions of globular proteins and lipids. The lipid bilayer has two roles: it is both a permeability barrier and a solvent for integral proteins. The proteins are distributed through the membrane in a mosaic pattern, and certain membrane lipids interact specifically with particular membrane proteins. However, some membrane proteins are free to move laterally through the membrane, as are membrane lipids. In contrast, diffusion from one surface of the membrane to the other (transverse diffusion) occurs very slowly with lipids and probably does not occur at all with proteins.

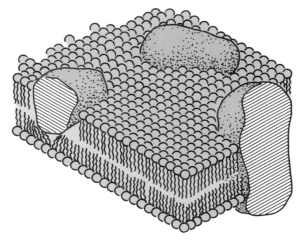

Fig. 19-13 The Fluid Mosaic Model of a Biological Membrane

EXAMPLE 19.3

What is meant by each of the following terms?

(a) lipid bilayer
(b) peripheral protein
(c) integral protein
(d) membrane fluidity

SOLUTION

(a) A lipid bilayer is the double-layered sheet formed spontaneously in an aqueous environment by phospholipids and glycolipids.
(b) A peripheral protein is a protein loosely attached to the surface of a membrane by electrical attractions and/or hydrogen bonds.
(c) An integral protein is a protein firmly embedded in a membrane, held by hydrophobic forces.
(d) Membrane fluidity is the constant internal flowing motion of membranes.

Membrane Transport

As was mentioned earlier, membranes are impermeable to most ions and molecules. This is advantageous because the ions and molecules necessary for cell function are retained inside the cell. But cells must obtain

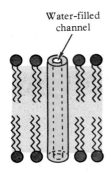

Fig. 19-14 A Water-Filled Channel Formed by Membrane Proteins

nutrients from the outside, and they must also secrete their waste materials. Thus it is necessary for cell membranes to have ways of transporting materials into and out of cells and cellular compartments.

A few solutes are able to cross membranes by simple diffusion. This involves spontaneous movement across the membrane from a region of high concentration to one of lower concentration. However, this process is rare in biological membranes. Instead, transport is usually aided by membrane proteins and is thus called *facilitated transport*. In this process, a membrane protein binds certain ions or small molecules and carries them across the membrane. Facilitated transport systems can be saturated with the substance being transported just as enzymes can be saturated with substrate. It is thought that the transport proteins exist as aggregates of two or more molecules that span the thickness of the membrane. A water-filled channel (Figure 19-14) probably stretches through this aggregate, and a binding site specific for the transported substance is thought to exist on a polypeptide region within the channel.

There are two types of facilitated transport. *Facilitated diffusion* involves movement of solute as in diffusion, from a region of high concentration to a region of low concentration (Figure 19-15(a)). In this case, the binding site on the carrier is probably accessible from both sides of the membrane. Once the substance to be transported is bound, a small conformational change may occur, which moves the substance from one

Fig. 19-15 Facilitated Diffusion (a) and Active Transport (b)

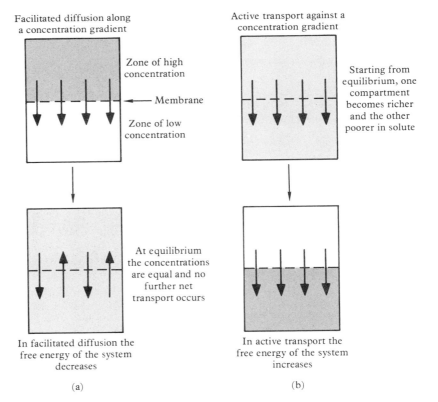

side of the membrane to the other. Entry of glucose into red blood cells and muscle cells occurs by facilitated diffusion.

The second type of facilitated transport, *active transport*, has more complex features. Movement of solute occurs in a nonspontaneous direction, from a region of low concentration to one of high concentration (Figure 19-15 (b)). Active transport requires free energy input because it is nonspontaneous, and in most cases, the source of the energy is hydrolysis of ATP (adenosine triphosphate). Active transport systems exist for calcium ions, amino acids, certain sugars, and for sodium and potassium ions. This latter transport system is referred to as the *"sodium pump"* (see Figure 19-16). Na^+ is pumped from a low intracellular concentration of less than $0.01M$ to a high extracellular concentration of about $0.14\ M$. Coupled to this extrusion of Na^+ is the pumping of K^+ from an extracellular concentration of $0.004M$ to an intracellular concentration of about $0.11M$. The system involves an enzyme, Na^+/K^+-ATPase, which is an integral protein in cell membranes. This enzyme catalyzes hydrolysis of ATP to provide energy for the transport process. ATPase is inhibited by a group of compounds called the *cardiac glycosides*, of which *ouabain* is an example.

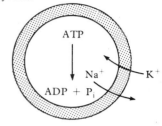

Fig. 19-16 The "Sodium Pump". Free energy from ATP hydrolysis is used to transport Na^+ and K^+ against their concentration gradients; hence, the "sodium pump" is an active transport system.

ouabain (WAH-bane)

ouabain

The cardiac glycosides may be administered as drugs to alter the excitability of heart muscle. Such treatment regulates heart rhythm, which is dependent on the Na^+/K^+ balance across the membrane.

BIOENERGETICS

Bioenergetics is the study of the production and use of energy by living organisms. When food is catabolized, energy is released. Part of this energy is the heat used to maintain body temperature. The rest of the energy is used primarily to synthesize molecules whose formation is nonspontaneous, thereby storing the energy until it is needed. These high-energy compounds can later be hydrolyzed by enzyme catalysts to deliver their energy when and where it is needed.

PERSPECTIVE: Obesity—An Enzyme Deficiency?

It is estimated that 80 million Americans are overweight, with about half of these technically classified as obese. For these, accumulated fat makes them weigh 20% or more than the normal range for their age, sex, and height. Researchers are currently investigating causes for obesity, long thought to be a simple matter of overeating, and they have found some surprising information. Certain obese mice get fat, even when they do not eat too much, because the enzyme ATPase is deficient in their tissues. This enzyme catalyzes ATP hydrolysis to provide energy for the "sodium pump"; in doing so, calories are consumed and heat is generated. Reduced activity of the enzyme probably means that the animals expend less energy in maintaining the Na^+/K^+ balance and thus accumulate extra weight, even on a moderate diet.

Harvard scientist Jeffrey Flier recently compared red blood cells from obese people to those of people of normal weight. He found that nearly all of the overweight people had considerably lower levels of ATPase. Other investigations have demonstrated that obese people tend to have slightly low body temperatures, supporting the ATPase theory. Even when Flier's obese subjects dieted and lost weight, their ATPase levels stayed low. It has been suggested that perhaps low ATPase levels developed over countless generations, enabling people to store up calories when food was plentiful in order to use them later during times of famine.

The Central Figure—ATP

Although there are many high-energy compounds produced in cells, one is the most abundant and the most versatile. This principal energy-storage molecule is *adenosine triphosphate, ATP*. This molecule has three main structural components: adenine, ribose, and three phosphate groups joined together.

adenosine triphosphate (ATP)

pyro- (PYE-roe): Greek prefix meaning "fire" or "heat."

ATP is a nucleotide, just as NAD^+, FAD, and FMN are, because it yields a cyclic nitrogen base, a pentose, and phosphoric acid upon total hydrolysis. The linkages between the phosphate units are called *anhydride linkages*, because similar linkages occur in the anhydride of phosphoric acid. An alternate name for the linkages is *pyro*phosphate linkages; this term results from the fact that pyrophosphoric acid, a type of acid anhydride, can be formed by heating phosphoric acid:

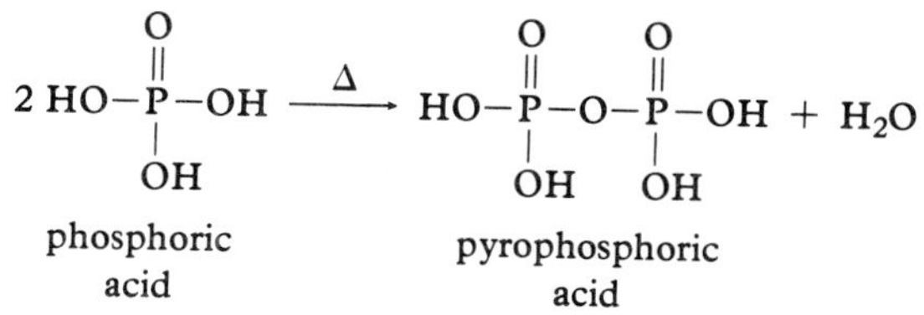

$$2\ \underset{\underset{\text{phosphoric acid}}{}}{\text{HO}-\overset{\overset{O}{\|}}{\underset{\underset{\text{OH}}{|}}{P}}-\text{OH}} \xrightarrow{\Delta} \underset{\underset{\text{pyrophosphoric acid}}{}}{\text{HO}-\overset{\overset{O}{\|}}{\underset{\underset{\text{OH}}{|}}{P}}-\text{O}-\overset{\overset{O}{\|}}{\underset{\underset{\text{OH}}{|}}{P}}-\text{OH}} + \text{H}_2\text{O}$$

The oxygen atoms in ATP's triphosphate group attract electrons away from the phosphorus atoms, and partial charges exist on these atoms:

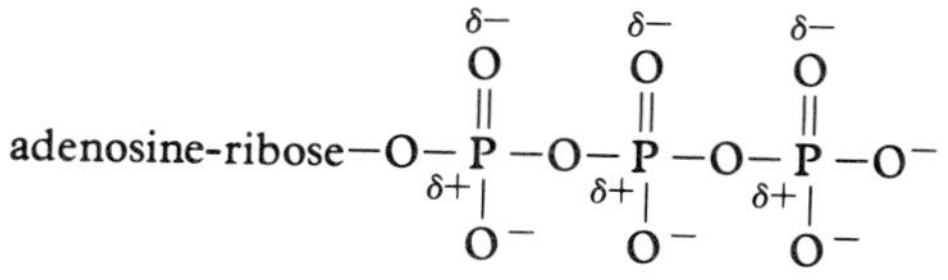

$$\text{adenosine-ribose}-\text{O}-\overset{\overset{\delta-}{\overset{O}{\|}}}{\underset{\underset{\text{O}^-}{\underset{\delta+}{|}}}{P}}-\text{O}-\overset{\overset{\delta-}{\overset{O}{\|}}}{\underset{\underset{\text{O}^-}{\underset{\delta+}{|}}}{P}}-\text{O}-\overset{\overset{\delta-}{\overset{O}{\|}}}{\underset{\underset{\text{O}^-}{\underset{\delta+}{|}}}{P}}-\text{O}^-$$

The electrical repulsions between like charges within the triphosphate group make ATP a rather unstable molecule. It is especially susceptible to hydrolysis at the anhydride linkages, and cleavage at these points liberates free energy (ΔG). Under the standard conditions of 25°C, 1 atmosphere of pressure, and $1M$ concentration, the energy change is referred to as the standard free energy change, $\Delta G°$. When a standard free energy change ($\Delta G°$) is corrected for physiological pH (pH 7), a prime (') is used in the symbol for standard free energy change ($\Delta G°'$).

Hydrolysis of ATP to adenosine diphosphate (ADP) is the principal energy-releasing reaction for ATP. Under physiological conditions, both ATP and ADP form a complex ion with Mg^{2+}, and the metal ion is a cofactor for most enzymatic reactions involving the two compounds.

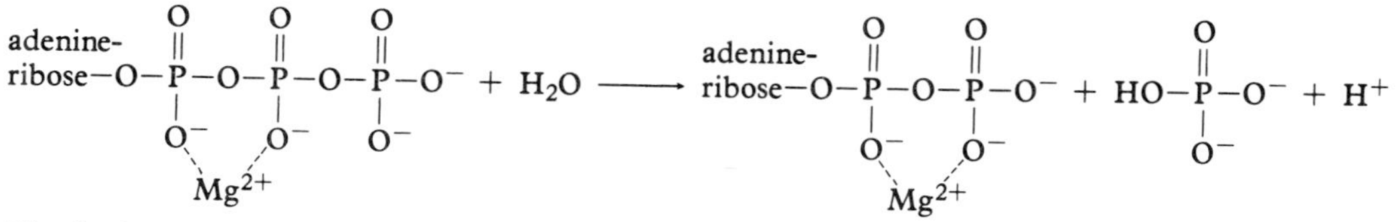

adenine-ribose—O—P(=O)(O⁻)—O—P(=O)(O⁻)—O—P(=O)(O⁻) ··· Mg^{2+} + H₂O ⟶ adenine-ribose—O—P(=O)(O⁻)—O—P(=O)(O⁻) ··· Mg^{2+} + HO—P(=O)(O⁻)—O⁻ + H⁺

The hydrogen phosphate ion is often referred to simply as phosphate, or inorganic phosphate, P_i. The reaction, which can be abbreviated as

$$\text{ATP} + \text{H}_2\text{O} \longrightarrow \text{ADP} + P_i + \text{H}^+$$
$$\Delta G°' = -7.3 \text{ kcal/mole}$$

has a free energy change of -7.3 kcal per mole of ATP hydrolyzed.

ATP also undergoes hydrolysis to yield adenosine monophosphate (AMP) and pyrophosphate (PP_i), as shown below.

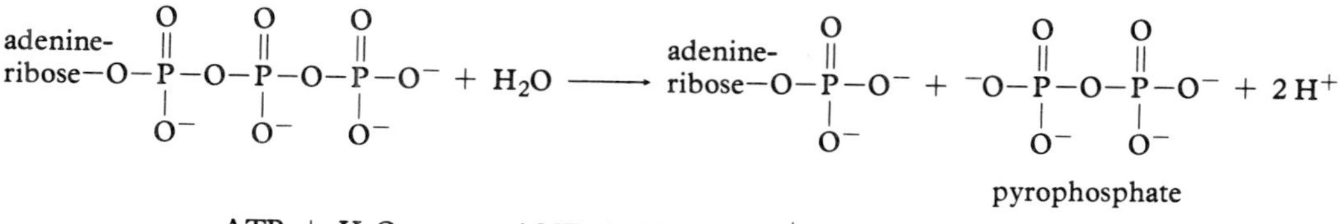

adenine-ribose—O—P(=O)(O⁻)—O—P(=O)(O⁻)—O—P(=O)(O⁻)—O⁻ + H₂O ⟶ adenine-ribose—O—P(=O)(O⁻)—O⁻ + ⁻O—P(=O)(O⁻)—O—P(=O)(O⁻)—O⁻ + 2 H⁺

pyrophosphate

$$\text{ATP} + \text{H}_2\text{O} \longrightarrow \text{AMP} + PP_i + 2\,\text{H}^+$$
$$\Delta G°' = -8.0 \text{ kcal/mole}$$

This reaction occurs frequently in metabolism and is usually followed by immediate hydrolysis of pyrophosphate to release even more free energy:

$$^-O-\overset{\overset{O}{\|}}{\underset{\underset{O^-}{|}}{P}}-O-\overset{\overset{O}{\|}}{\underset{\underset{O^-}{|}}{P}}-O^- + H_2O \longrightarrow 2\ HO-\overset{\overset{O}{\|}}{\underset{\underset{O^-}{|}}{P}}-O^-$$

$$\Delta G^{\circ\prime} = -6.5 \text{ kcal/mole}$$

ADP is also a high energy compound, as its hydrolysis releases a significant quantity of free energy. However, this reaction is seldom used by biological systems.

$$ADP + H_2O \longrightarrow AMP + P_i + H^+$$
$$\Delta G^{\circ\prime} = -7.3 \text{ kcal/mole}$$

These hydrolysis reactions and their free energy changes are summarized in Table 19-2.

Table 19-2 Hydrolyses of Some High-Energy Phosphate Compounds

Reaction	$\Delta G^{\circ\prime}$, kcal/mole
$ATP + H_2O \longrightarrow ADP + P_i + H^+$	-7.3
$ATP + H_2O \longrightarrow AMP + PP_i + 2\ H^+$	-8.0
$PP_i + H_2O \longrightarrow 2\ P_i$	-6.5
$ADP + H_2O \longrightarrow AMP + P_i + H^+$	-7.3

The formation of ATP from ADP and P_i, called *phosphorylation*, requires a free energy input of exactly the same amount as that released by hydrolysis. We can consider ATP and ADP, the two most common forms of adenine nucleotides, to be in equilibrium with each other in a cell. When ATP levels fall due to energy comsumption, ADP and P_i are used to synthesize more ATP, a process requiring oxidation of nutrients.

$$\begin{array}{c} ATP + H_2O \\ \rightleftarrows \\ ADP + P_i \end{array}$$

When ATP levels are high, oxidation of nutrients is slowed, and energy from excess ATP is used to synthesize other storage molecules such as triacylglycerols and glycogen. In this way, the cell regulates ATP levels so that there will be enough ATP present to provide energy for chemical synthesis, muscle contraction, nerve transmission, active transport, and other cellular needs.

The Pathway of Electron Transport

Oxidation of food materials occurs by sequences of chemical reactions organized in metabolic pathways. Each pathway consists of many steps, each of which is only a small alteration in chemical structure. However, these oxidations liberate free energy which can be used to synthesize ATP. In essence, the products of digestion of the three major types of food are degraded to acetyl units attached to coenzyme A by thioester linkage:

$$CH_3-\overset{\overset{O}{\|}}{C}-S-CoA$$
acetyl CoA

As indicated in Figure 19-3, units of acetyl CoA feed into the citric acid cycle. This metabolic pathway produces carbon dioxide, $FADH_2$, NADH, and H^+ by a series of redox reactions which also regenerate oxaloacetate, a key intermediate in the cycle. However, it is the NADH and $FADH_2$ which ultimately lead to the reduction of O_2 to form H_2O, and the free energy liberated in this process is used to synthesize ATP.

The *electron transport system* (ETS) is a sequence of reactions, each catalyzed by its own particular enzyme. Since it is this system that is responsible directly for oxygen consumption, it is also called the *respiratory chain*. Its chemical product is H_2O. At least seven different steps are involved in electron transport. The process occurs within the inner membrane of mitochondria, and all of the necessary enzymes are embedded in this membrane. The inner mitochondrial membrane is impermeable to many small molecules and ions, and specific protein carriers transport needed molecules into and out of the mitochondrial matrix. It appears that electrons are transferred as hydride ions from NADH and $FADH_2$ to acceptors, which then pass electrons on through a series of spontaneous reactions, as indicated in Figure 19-17. Each step in the electron transport chain regenerates the oxidized form of the electron carrier: for example, NADH is oxidized to NAD^+ by a loss of a hydride ion:

$$NADH \rightleftharpoons NAD^+ + H:^-$$

The hydride ion acceptor for NADH is *flavin mononucleotide* (FMN), a prosthetic group on the enzyme called *NADH dehydrogenase*. After FMN is reduced to $FMNH_2$, it passes hydride ions to an acceptor called *coenzyme Q* (CoQ), and $FMNH_2$ becomes reoxidized, as did NADH. Coenzyme Q represents a branch point in the system, as it can also accept hydride ions from $FADH_2$, through the reaction shown in Figure 19-18. From CoQ, electrons are transferred through a series of

electron transport system (ETS): the series of spontaneous reactions in which electrons are transferred from reduced coenzymes to oxygen through a series of acceptors.

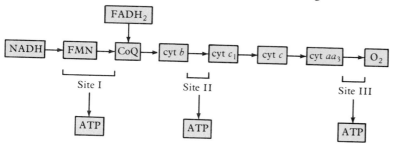

Fig. 19-17 The Electron Transport System

Fig. 19-18 The Reduction of Coenzyme Q

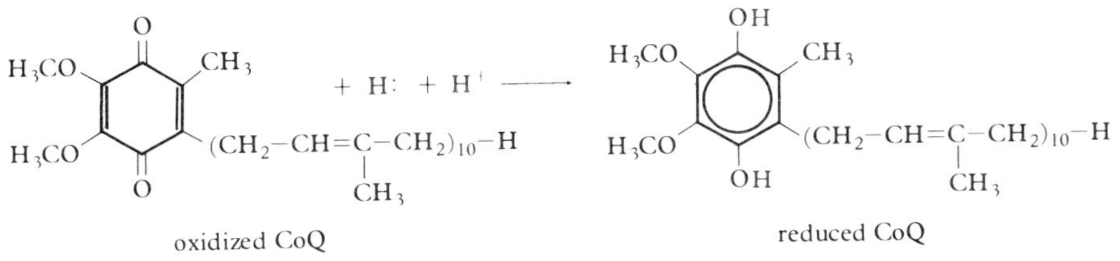

oxidized CoQ + H:⁻ + H⁺ ⟶ reduced CoQ

heme proteins called the *cytochromes*. It appears that CoQ passes along single electrons rather than hydride ions, and as the cytochromes accept and pass on electrons, their iron ions change between the +2 and the +3 oxidation states.

Cytochromes a and a_3 are the last members of the respiratory chain. They exist as a complex called *cytochrome oxidase*, represented by cyt a a_3; although the exact details are not known, this complex somehow causes four electrons to reduce a molecule of O_2 to form H_2O.

$$O_2 + 4e^- + 4H^+ \longrightarrow 2 H_2O$$

Electron transport through the respiratory chain is spontaneous, and if any one step is blocked, the entire process stops, since the reverse reactions are nonspontaneous. Certain drugs and poisons have the ability to block electron transport. For example, rotenone, an insecticide and fish poison, blocks the transfer of hydride ions from $FMNH_2$ to CoQ. The barbiturate amytal has the same effect. Cyanide ions ($C \equiv N^-$) are extremely toxic because they form coordinate covalent bonds to Fe^{3+} in the cytochrome oxidase complex, preventing the final step in the electron transport system.

Oxidative Phosphorylation

oxidative phosphorylation: synthesis of ATP from ADP and P_i utilizing free energy released by the electron transport system.

In the spontaneous passage of electrons through the electron transport chain, free energy is liberated at each step of the way, as illustrated in Figure 19-19. For three of these steps, enough free energy is released to allow the synthesis of one molecule of ATP. This process of *oxidative phosphorylation* is noted in Figures 19-17 and 19-19. Although much research has been done on the subject, it still is not clear exactly how the liberated free energy is channeled into ATP synthesis. It does appear, however, that members of the electron transport system are firmly anchored in the inner mitochondrial membrane, and that the entire set of participants is arranged in the proper order for efficient electron flow. Note that the free energy changes in the system result in production of 3 ATP molecules for each NADH oxidized and 2 ATP molecules for each $FADH_2$ oxidized.

The free energy change in direct oxidation of NADH by O_2 is -52.0 kcal/mole:

$$NADH + H^+ + \tfrac{1}{2} O_2 \longrightarrow NAD^+ + H_2O$$
$$\Delta G^{\circ\prime} = -52.0 \text{ kcal/mole}$$

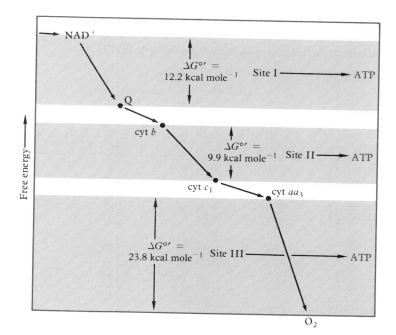

Fig. 19-19 Free Energy Changes in Electron Transport

Thus, of the total free energy released by the electron transport system, starting with NADH, 21.9 kcal are stored in the synthesis of 3 moles of ATP (3 moles × 7.3 kcal/mole):

$$ADP + P_i \longrightarrow ATP$$
$$\Delta G^{\circ\prime} = +7.3 \text{ kcal/mole}$$

This means that ATP synthesis conserves about 42% (21.9 kcal/52.0 kcal = 0.42) of the free energy released when NADH is oxidized. The rest of the free energy is dissipated to the surroundings, primarily as heat used to maintain body temperature.

EXAMPLE 19.4

Calculate the standard free energy change for the overall process given below.

$$ATP + H_2O \longrightarrow AMP + PP_i + 2H^+ \qquad \Delta G^{\circ\prime} = -8.0 \text{ kcal/mole}$$
$$PP_i + H_2O \longrightarrow 2P_i \qquad \Delta G^{\circ\prime} = -6.5 \text{ kcal/mole}$$

$$ATP + 2H_2O \longrightarrow AMP + 2P_i + 2H^+$$

SOLUTION

Since the two individual reactions can be added to get the overall reaction, the standard free energy changes for the individual steps can also be added to get the overall free energy change. Thus, for the overall process,

$$\Delta G^{\circ\prime} = (-8.0 \text{ kcal/mole}) + (-6.5 \text{ kcal/mole})$$
$$= -14.5 \text{ kcal/mole}$$

EXAMPLE 19.5

If the standard free energy change for oxidation of $FADH_2$ is -40.5 kcal/mole,

$$FADH_2 + \tfrac{1}{2}O_2 \longrightarrow FAD + H_2O$$
$$\Delta G^{o\prime} = -40.5 \text{ kcal/mole}$$

calculate the percentage of free energy conserved by the electron transport system starting with $FADH_2$.

SOLUTION

When one mole of $FADH_2$ is oxidized by the electron transport system, 2 moles of ATP are synthesized. Thus, of the total free energy released (40.5 kcal), the amount conserved by ATP synthesis is 14.6 kcal (2 mole $\times$ 7.3 kcal/mole). The percentage of free energy conserved is

$$\% \text{ conserved} = \frac{14.6 \text{ kcal}}{40.5 \text{ kcal}} \times 100$$
$$= 36.0 \%$$

THE CITRIC ACID CYCLE

As was mentioned earlier, the *citric acid cycle* (CAC) is the principal metabolic pathway for generating NADH and $FADH_2$. It is called the citric acid cycle because one of the key intermediates is citric acid. However, it is also called the *tricarboxylic acid* (TCA) *cycle* and the *Krebs cycle* in honor of Sir Hans A. Krebs, who deduced its sequence in 1937. The citric acid cycle is the central metabolic pathway; it receives acetyl units (in the form of acetyl CoA) that have been formed by the degradation of fatty acids, carbohydrates, and some amino acids. Thus we can say that the fuel for the citric acid cycle is the acetyl group.

Reactions of the Citric Acid Cycle

The enzymes catalyzing the various steps of the citric acid cycle are located in the mitochondrial matrix. The intermediates are also found in the matrix, and they are present only in small amounts. Thus, this metabolic cycle is similar to a merry-go-round which picks up an acetyl unit and converts it to carbon dioxide, which is released at two different points. The cycle is summarized in Figure 19-20, and Figure 19-21 is a more detailed presentation.

The reactions and their corresponding enzymes are summarized in Table 19-3. Significant features of each reaction are discussed in the following paragraphs, which are numbered to correspond to Figure 19-21 and Table 19-3.

Reaction 1. Formation of Citrate

As acetyl CoA enters the cycle, the acetyl group is transferred to oxaloacetate to produce citrate, the anion of citric acid:

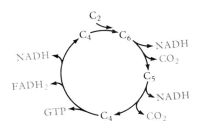

Fig. 19-20 An Overview of the Citric Acid Cycle

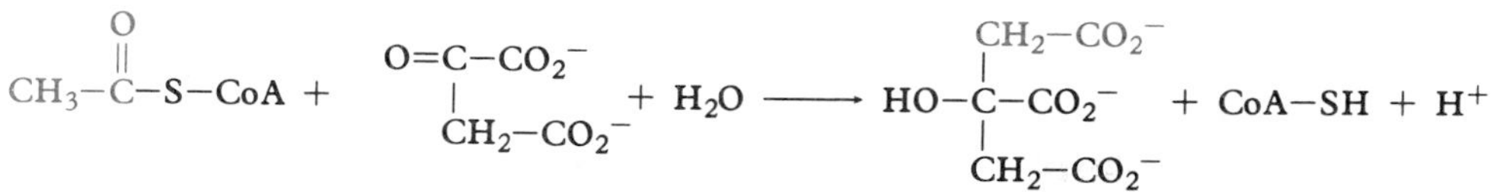

THE CITRIC ACID CYCLE 563

$$CH_3-\overset{O}{\underset{\|}{C}}-S-CoA + \underset{\text{oxaloacetate}}{\begin{array}{c}O=C-CO_2^-\\|\\CH_2-CO_2^-\end{array}} + H_2O \longrightarrow \underset{\text{citrate}}{\begin{array}{c}CH_2-CO_2^-\\|\\HO-C-CO_2^-\\|\\CH_2-CO_2^-\end{array}} + CoA-SH + H^+$$

acetyl CoA

Thus, a 2-carbon unit combines with a 4-carbon compound to form the 6-carbon citrate ion. Water is also involved in this reaction; it is used to

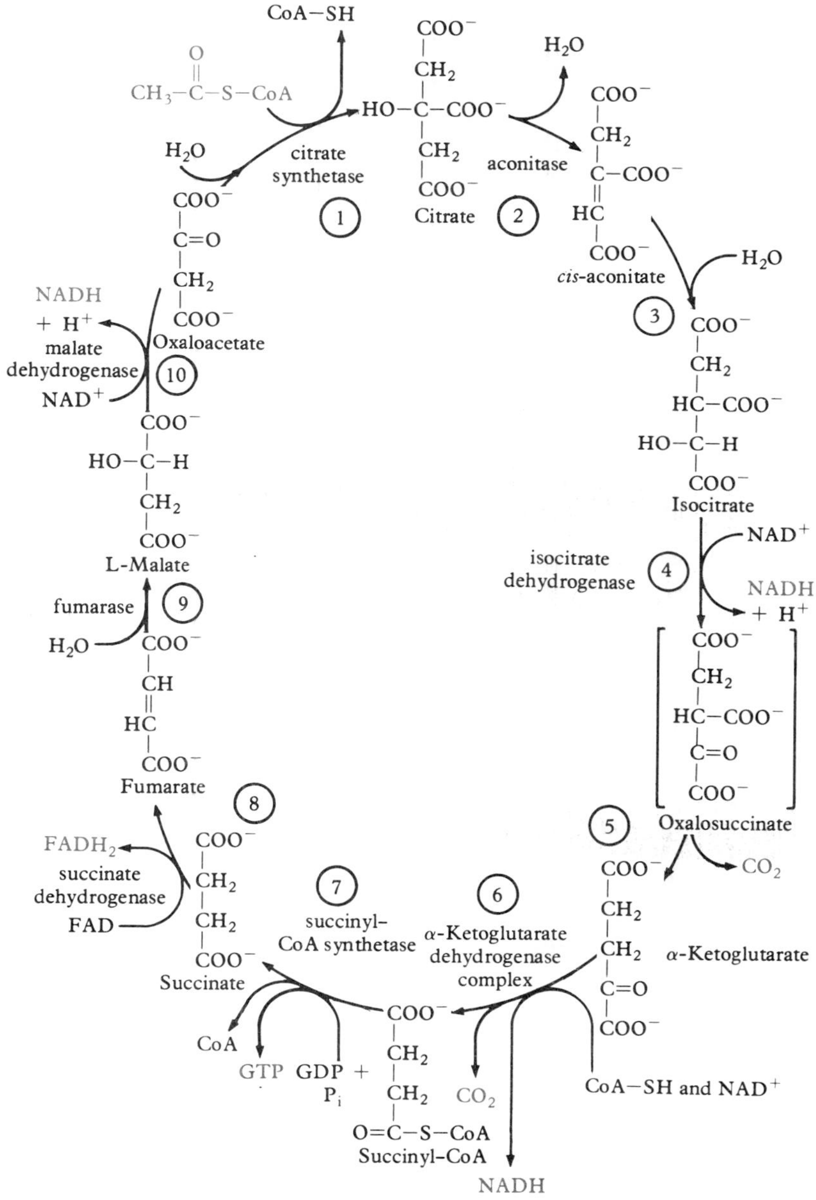

Fig. 19-21 The Citric Acid Cycle

Table 19-3 Reactions of the Citric Acid Cycle

Step	Reaction	Enzyme	Cofactor
1	Acetyl CoA + oxaloacetate + $H_2O \longrightarrow$ citrate + CoA + H^+	Citrate synthetase	CoA
2	Citrate $\rightleftharpoons$ cis-aconitate + H_2O	Aconitase	Fe^{2+}
3	cis-Aconitate + $H_2O \rightleftharpoons$ isocitrate	Aconitase	Fe^{2+}
4	Isocitrate + $NAD^+ \rightleftharpoons$ oxalosuccinate + NADH + H^+	Isocitrate dehydrogenase	NAD^+
5	Oxalosuccinate + $H^+ \rightleftharpoons$ α-ketoglutarate + CO_2	Isocitrate dehydrogenase	None
6	α-Ketoglutarate + CoA + $NAD^+ \longrightarrow$ succinyl CoA + CO_2 + NADH	α-Ketoglutarate dehydrogenase complex	CoA NAD^+ Mg^{2+} TPP Lipoic acid FAD
7	Succinyl CoA + GDP + Pi $\rightleftharpoons$ succinate + GTP + CoA	Succinyl CoA synthetase	None
8	Succinate + FAD $\rightleftharpoons$ fumarate + $FADH_2$	Succinate dehydrogenase	FAD
9	Fumarate + $H_2O \rightleftharpoons$ L-malate	Fumarase	None
10	L-Malate + $NAD^+ \rightleftharpoons$ oxaloacetate + NADH + H^+	Malate dehydrogenase	NAD^+

hydrolyze acetyl CoA to release the acetyl unit. The enzyme is citrate synthetase.

Reaction 2. Formation of cis-Aconitate

In the second reaction of the cycle, citrate is dehydrated to form cis-aconitate, catalyzed by aconitase. The reaction is reversible.

$$\begin{array}{c} CH_2-CO_2^- \\ | \\ HO-C-CO_2^- \\ | \\ H-CH-CO_2^- \\ \text{citrate} \end{array} \rightleftharpoons \begin{array}{c} CH_2-CO_2^- \\ | \\ C-CO_2^- \\ || \\ H-C-CO_2^- \\ \text{cis-aconitate} \end{array} + HOH$$

Reaction 3. Formation of Isocitrate

In this step, cis-aconitate is reversibly hydrated to form the citrate isomer called isocitrate. The enzyme for this reaction is also aconitase.

$$\begin{array}{c} CH_2-CO_2^- \\ | \\ C-CO_2^- \\ || \\ H-C-CO_2^- \\ \text{cis-aconitate} \end{array} + HOH \rightleftharpoons \begin{array}{c} CH_2-CO_2^- \\ | \\ H-C-CO_2^- \\ | \\ HO-CH-CO_2^- \\ \text{isocitrate} \end{array}$$

Thus, the real function of aconitase is to catalyze the isomerization of citrate to isocitrate, and *cis*-aconitate is an intermediate in the interconversion of the two isomers. Note that the enzyme is able to direct hydration of *cis*-aconitate in violation of Markovnikof's rule (see chapter 11).

Reaction 4. Formation of Oxalosuccinate

The next reaction of the cycle is the reversible oxidation of isocitrate to produce oxalosuccinate (the oxalo group is $-CO-CO_2^-$):

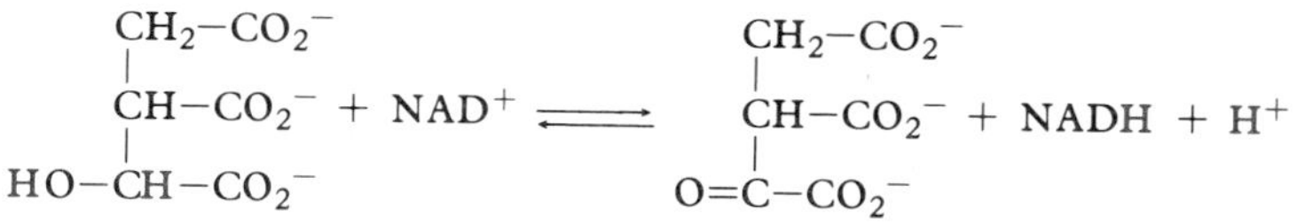

$$\begin{array}{c} CH_2-CO_2^- \\ | \\ CH-CO_2^- \\ | \\ HO-CH-CO_2^- \end{array} + NAD^+ \rightleftharpoons \begin{array}{c} CH_2-CO_2^- \\ | \\ CH-CO_2^- \\ | \\ O=C-CO_2^- \end{array} + NADH + H^+$$

isocitrate → oxalosuccinate

In this reaction, two hydrogens are removed from isocitrate; one is transferred to NAD^+ as a hydride ion, and the other is released as H^+. Since hydrogen is removed from isocitrate, the enzyme is called isocitrate dehydrogenase. This reaction is the first in the cycle to generate NADH.

Reaction 5. Formation of α-Ketoglutarate

Oxalosuccinate formed in the preceding reaction remains bound to isocitrate dehydrogenase, and the same enzyme catalyzes an immediate decarboxylation to form α-ketoglutarate:

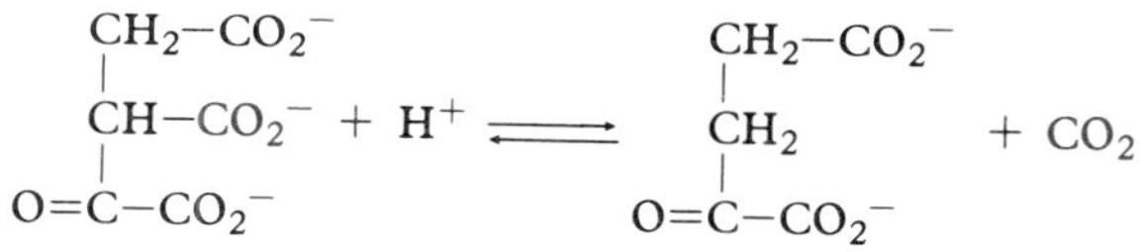

$$\begin{array}{c} CH_2-CO_2^- \\ | \\ CH-CO_2^- \\ | \\ O=C-CO_2^- \end{array} + H^+ \rightleftharpoons \begin{array}{c} CH_2-CO_2^- \\ | \\ CH_2 \\ | \\ O=C-CO_2^- \end{array} + CO_2$$

oxalosuccinate → α-ketoglutarate

This reaction results in release of the first molecule of CO_2 in the cycle. Note that the CO_2 comes from the central carbon atom in oxalosuccinate.

The existence of α-ketoglutarate as a citric acid cycle intermediate carries important consequences. One is that 5 amino acids (arginine, glutamate, glutamine, histidine, and proline) can be degraded to α-ketoglutarate, and thus their carbon atoms can enter the citric acid cycle to be utilized for energy production. In addition, α-ketoglutarate can be used to synthesize glutamate.

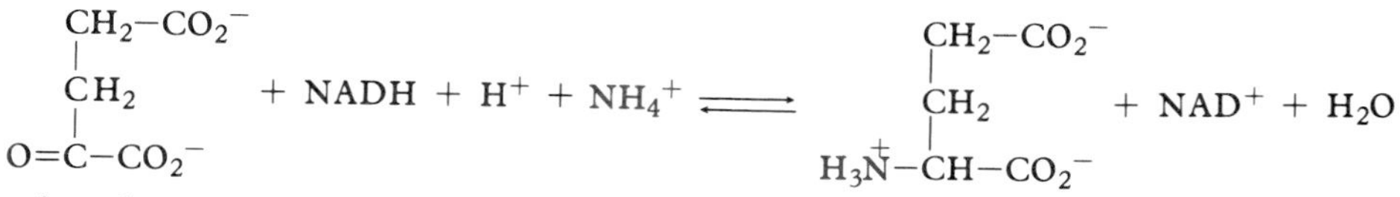

$$\begin{array}{c} CH_2-CO_2^- \\ | \\ CH_2 \\ | \\ O=C-CO_2^- \end{array} + NADH + H^+ + NH_4^+ \rightleftharpoons \begin{array}{c} CH_2-CO_2^- \\ | \\ CH_2 \\ | \\ H_3\overset{+}{N}-CH-CO_2^- \end{array} + NAD^+ + H_2O$$

α-ketoglutarate → glutamate

This reaction is one of the few which convert inorganic nitrogen (NH_4^+) to organic nitrogen (the amino group in an organic compound), and it is an entry point for nitrogen into biological systems.

Reaction 6. Formation of Succinyl CoA

The conversion of α-ketoglutarate to succinyl CoA is summarized by the equation

$$\underset{\alpha\text{-ketoglutarate}}{\begin{array}{c}CH_2-CO_2^-\\|\\CH_2\\|\\O=C-CO_2^-\end{array}} + CoA-SH + NAD^+ \longrightarrow \underset{\text{succinyl CoA}}{\begin{array}{c}CH_2-CO_2^-\\|\\CH_2\\|\\O=C-S-CoA\end{array}} + CO_2 + NADH$$

It is actually much more complicated than this, and it occurs in a series of steps catalyzed by the α-ketoglutarate dehydrogenase complex. This aggregate contains three separate enzymes and requires four cofactors in addition to CoA and NAD^+. This reaction accounts for production of the second and final CO_2 molecule by the cycle. Also, the second molecule of NADH in the citric acid cycle is produced in this step. This is the only irreversible step in the citric acid cycle. Succinyl CoA represents another entry point for carbon atoms derived from amino acids. The carbon skeletons of isoleucine, methionine, threonine, and valine can be converted to succinyl CoA and thus be oxidized in the citric acid cycle.

Reaction 7. Formation of Succinate

In this step, catalyzed by succinyl CoA synthetase, succinyl CoA is cleaved, and the free energy released is used to synthesize a molecule of guanosine triphosphate (GTP), a high-energy compound that is structurally similar to ATP.

$$\underset{\text{succinyl CoA}}{\begin{array}{c}CH_2-CO_2^-\\|\\CH_2\\|\\O=C-S-CoA\end{array}} + GDP + P_i \rightleftharpoons \underset{\text{succinate}}{\begin{array}{c}CH_2-CO_2^-\\|\\CH_2-CO_2^-\end{array}} + GTP + CoA-SH$$

This synthesis is referred to as *substrate-level phosphorylation*, because a high-energy phosphate compound is formed with energy released by a substrate reaction rather than the electron transport system. The GTP produced here can readily transfer its terminal phosphate to ADP to produce an ATP molecule and GDP.

Reaction 8. Formation of Fumarate

In this reaction, succinate is oxidized to fumarate, a *cis* alkene with two carboxyl groups. Note that two hydrogen atoms are transferred from succinate to coenzyme FAD. This reaction is the only step in the citric acid cycle that generates $FADH_2$. The enzyme is succinate dehydrogenase.

$$\underset{\text{succinate}}{\begin{array}{c}\text{H–CH–CO}_2^-\\|\\{}^-\text{O}_2\text{C–HC–H}\end{array}} + \text{FAD} \rightleftharpoons \underset{\text{fumarate}}{\begin{array}{c}\text{H–C–CO}_2^-\\\|\\{}^-\text{O}_2\text{C–C–H}\end{array}} + \text{FADH}_2$$

Fumarate represents a third entry point for carbon atoms from amino acids, since aspartate, phenylalanine, and tyrosine can be degraded to fumarate.

Reaction 9. Formation of L-Malate

This reaction, catalyzed by fumarase, involves the reversible hydration of the double bond of fumarate to form L-malate.

$$\underset{\text{fumarate}}{\begin{array}{c}\text{H–C–CO}_2^-\\\|\\{}^-\text{O}_2\text{C–C–H}\end{array}} + \text{HOH} \rightleftharpoons \underset{\text{L-malate}}{\begin{array}{c}\text{HO–CH–CO}_2^-\\|\\\text{H–CH–CO}_2^-\end{array}}$$

Reaction 10. Regeneration of Oxaloacetate

In this last reaction of the citric acid cycle, oxaloacetate is regenerated by oxidation of L-malate.

$$\underset{\text{L-malate}}{\begin{array}{c}\text{HO–CH–CO}_2^-\\|\\\text{CH}_2\text{–CO}_2^-\end{array}} + \text{NAD}^+ \rightleftharpoons \underset{\text{oxaloacetate}}{\begin{array}{c}\text{O=C–CO}_2^-\\|\\\text{CH}_2\text{–CO}_2^-\end{array}} + \text{NADH} + \text{H}^+$$

The reaction is catalyzed by malate dehydrogenase, and NAD^+ is the required coenzyme. This step generates the third and final molecule of NADH in the citric acid cycle, and it provides oxaloacetate to begin a new round of the cycle. Oxaloacetate represents the fourth entry point in the cycle for carbon atoms from amino acids; asparagine and aspartate can be degraded to oxaloacetate.

Figure 19-22 illustrates the points where carbon atoms from amino acids can enter the citric acid cycle.

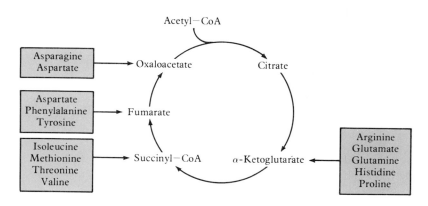

Fig. 19-22 Points of Direct Entry into the Citric Acid Cycle for Carbon Atoms from Amino Acids

Regulation of the Citric Acid Cycle

The rate at which the citric acid cycle runs can be precisely adjusted to meet cellular needs. This adjustment is possible because of the sensitivity of three steps in the cycle to the energy needs of the cell. The synthesis of citrate from oxaloacetate and acetyl CoA (reaction 1) is an important control point in the cycle. The catalyst for this reaction, citrate synthetase, is an allosteric enzyme which is inhibited by ATP and NADH and activated by ADP. When the cell has large amounts of stored energy (ATP), there are also usually high levels of NADH. Thus, the presence of large amounts of either of these compounds signals the citric acid cycle to slow down production of NADH and $FADH_2$ by inhibiting citrate synthetase. On the other hand, when ATP levels are low, ADP levels are usually high. Thus, large amounts of ADP stimulate the citric acid cycle to produce NADH and $FADH_2$ to be used by the electron transport system for ATP synthesis. This stimulation occurs by allosteric activation of citrate synthetase.

A second controlling enzyme is isocitrate dehydrogenase (reaction 4). This allosteric enzyme is activated by ADP and inhibited by NADH in very much the same way as citrate synthetase. The third control point (reaction 6) is catalyzed by the α-ketoglutarate dehydrogenase complex. This group of enzymes is inhibited by succinyl CoA and NADH, products of the reaction it catalyzes, and also by ATP. Thus, healthy energy conditions of the cell exert their effects on this enzyme system to slow down the citric acid cycle. These effects are reversed by high levels of CoA and NAD^+, usually signs that the cell is in need of energy. These control points are indicated in Figure 19-23.

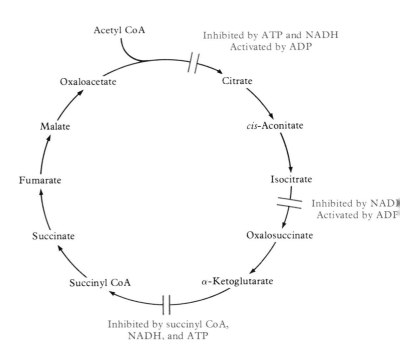

Fig. 19-23 Control Points in the Citric Acid Cycle

Summary and Major Features of the Citric Acid Cycle

The citric acid cycle will continue to run as long as acetyl CoA molecules are available and oxaloacetate is regenerated. The individual reactions of the cycle can be added to give a summary equation, as shown below. Overall, the citric acid cycle forms 3 molecules of NADH, 1 of $FADH_2$, and 1 of GTP for each acetyl unit that enters.

$$AcCoA + 3\,NAD^+ + FAD + GDP + P_i + 2\,H_2O \longrightarrow$$
$$2\,CO_2 + CoA-SH + 3\,NADH + 2\,H^+ + FADH_2 + GTP$$

We can list the major features of the citric acid cycle as follows:

1. Two carbon atoms enter the cycle as an acetyl unit. Two carbon atoms leave the cycle as two molecules of CO_2. However, these carbon atoms correspond to ones entering in the previous cycle; there is a one-cycle delay between entry of two carbon atoms and their release as CO_2.
2. Four redox reactions produce 3 NADH and 1 $FADH_2$.
3. One high-energy phosphate compound (GTP) is generated. Its energy content may be considered equivalent to that of one ATP, and the two compounds are interconvertible.
4. Two water molecules are consumed, one each in the synthesis of citrate and the hydration of fumarate.

Energetics of the Citric Acid Cycle

If we refer to the earlier discussion of electron transport, we can see that each NADH generated by the citric acid cycle is ultimately responsible for the production of 3 ATP molecules. In addition, the $FADH_2$ generated is responsible for production of 2 ATP molecules. If we consider one GTP to be equivalent to one ATP, we can calculate the total ATP production:

3 NADH	ultimately produce	9 ATP
1 $FADH_2$	ultimately produces	2 ATP
1 GTP	is equivalent to	1 ATP
	Total ATP produced:	12 ATP

Thus, one round of the citric acid cycle results in production of 12 high-energy phosphate molecules, or 12 ATP. Note that only one of these was formed directly in the citric acid cycle; the others (11) are a result of electron transport through the respiratory chain. Note also that molecular oxygen does not participate directly in the citric acid cycle. However, the cycle can only operate if NAD^+ and FAD are continuously regenerated from NADH and $FADH_2$, and this occurs only by proper function of the electron transport chain. Since this system requires molecular oxygen, the citric acid cycle can only function in an aerobic environment.

You can see from the preceding discussion that oxidation of foods proceeds in a very complex manner. First, digested food molecules are degraded to acetyl units attached to CoA. These units are then oxidized to CO_2 in the citric acid cycle. The reduced coenzymes generated by the citric acid cycle pass their electrons through the electron system ultimately to reduce O_2 and form H_2O. The net result is that organic molecules are oxidized, forming CO_2 and H_2O. The energy liberated by the oxidations is used primarily to warm the organism and to synthesize ATP.

EXAMPLE 19.6

At which points in the citric acid cycle are reduced coenzymes produced?

SOLUTION

NADH is produced with the formation of oxalosuccinate (reaction 4), the formation of succinyl CoA (reaction 6), and the formation of oxaloacetate (reaction 10). $FADH_2$ is produced with the formation of fumarate (reaction 8).

EXAMPLE 19.7

At which points in the citric acid cycle are CO_2 molecules produced?

SOLUTION

One CO_2 molecule is produced by decarboxylation of oxalosucinate (reaction 5), and a second is released by decarboxylation of α-ketoglutarate (reaction 6).

SUMMARY

The Cell and An Overview of Metabolism: A biological cell is a functional unit of biological activity surrounded by a membrane and capable of reproduction. The two basic cell types differ in that a eucaryotic cell has a nuclear membrane while a procaryotic cell does not. The principal components of eucaryotic cells are the membrane, nucleus, mitochondria, endoplasmic reticulum, and lysosomes.

Metabolism is the network of chemical reactions responsible for the extraction of energy from foods and synthesis of cellular materials. Catabolism, the breakdown of molecules, produces energy needed for anabolism, the synthesis of biological molecules. The three stages involved in energy extraction from food are digestion, degradation of small molecules to acetyl CoA, and oxidation of acetyl CoA in the citric acid cycle with utilization of reduced coenzymes to provide free energy for ATP synthesis.

Membranes: An animal cell contains an external membrane as well as internal membranes surrounding mitochondria, the nucleus, and lysosomes. Membranes are composed mainly of lipids and proteins, the major lipids being phospholipids, glycolipids, and cholesterol. In an aqueous environment, large amounts of phospholipids and glycolipids spontaneously form bimolecular sheets called lipid bilayers. Lipid bilayer membranes have a low permeability for ions and polar molecules. The protein content of biological membranes varies from 20% to 75%. Membrane proteins consist of peripheral and integral proteins. Membrane proteins are involved in transport, catalysis, structure stabilization, and, in the case of glycoproteins, intercellular recognition.

Biological membranes have constant internal motion, with membrane proteins and lipids diffusing throughout the membranes. Unsaturated and short-chain fatty acids increase membrane fluidity as does the presence of cholesterol. The fluid mosaic model describes a membrane as both a permeability barrier and a solvent for integral proteins, which are distributed through the membrane in a mosaic pattern. A few solutes are able to cross membranes by simple diffusion, but most transport is facilitated by membrane proteins. Facilitated transport is classified into two processes: facilitated diffusion and active transport. The latter is nonspontaneous and requires free energy from ATP hydrolysis. The active transport system for Na^+ and K^+, called the "sodium pump," involves participation of the enzyme ATPase.

Bioenergetics: Bioenergetics is the study of the production and use of energy by living organisms. The principal energy-storage molecule in cells is ATP. It is susceptible to hydrolysis in a number of different ways, all of which liberate free energy. Cells regulate ATP levels so that there will be enough present to meet the energy needs of the cells. Oxidation of food materials produces reduced coenzymes which pass electrons along to acceptors in a series of events called the electron transport system. This network of enzymes and substrates is inside the inner mitochondrial membrane. The free energy released by the electron transport system is used to synthesize ATP from ADP and P_i by the process of oxidative phosphorylation.

The Citric Acid Cycle: The citric acid cycle, the central metabolic pathway, is responsible for generating NADH and $FADH_2$. Acetyl units attached to coenzyme A enter the cycle and are oxidized to CO_2. The citric acid cycle has three control points; they are regulated by the enzymes citrate synthetase, isocitrate dehydrogenase, and the α-ketoglurate dehydrogenase complex. These enzymes are especially sensitive to energy levels in the cell. The reduced coenzymes generated in the citric acid cycle pass electrons into the electron transport system to produce ATP and H_2O. The citric acid cycle is responsible ultimately for the production of 12 ATP per turn of the cycle.

STUDY QUESTIONS AND PROBLEMS

1. Define the following terms:

   (a) biological cell
   (b) procaryotic cell
   (c) eucaryotic cell
   (d) cell membrane
   (e) cell nucleus
   (f) mitochondria
   (g) endoplasmic reticulum
   (h) lysosome
   (i) lipid bilayer
   (j) peripheral protein
   (k) integral protein
   (l) membrane fluidity
   (m) fluid mosaic model
   (n) facilitated transport
   (o) facilitated diffusion
   (p) active transport
   (q) "sodium pump"
   (r) cardiac glycosides
   (s) bioenergetics
   (t) high-energy compound
   (u) ATP
   (v) electron transport system
   (w) cytochrome oxidase
   (x) oxidative phosphorylation
   (y) citric acid cycle

2. What is the major difference between procaryotic and eucaryotic cells?

3. Describe two functions of a cell membrane.

4. Give the major function of each of the following:

   (a) cell nucleus
   (b) mitochondria
   (c) endopolasmic reticulum
   (d) lysosomes

5. Describe three common features of biological membranes.

6. Draw representative structures for three types of membrane lipids.

7. What forces are responsible for holding lipid bilayers together? What would an organic solvent such as hexane do to a lipid bilayer?

8. In light of your knowledge of membrane structure and function, explain why inhaling hydrocarbon vapors produces lung damage.

9. Why does the amount of protein present in a biological membrane vary with the type of cell?

10. Why is it that salt solutions can dislodge peripheral proteins from membranes?

11. Suggest how detergents and organic solvents cause dissociation of integral proteins from membranes.

12. Describe four functions of proteins in cell membranes.

13. Explain how lipid structure affects membrane fluidity.

14. Summarize the major features of the fluid mosaic model.

15. Suggest reasons for the rapid lateral diffusion and the very slow transverse diffusion of membrane lipids.

16. Differentiate between facilitated diffusion and active transport.

17. What is the function of the ATPase in the "sodium pump"?

18. Why is ATP called the "principal currency of free energy in biological systems"?

19. Why is ATP an unstable molecule?

20. Draw the structure of ATP and use equations to illustrate its principal energy-releasing reactions in biological systems.

21. How do cells regulate their ATP levels? Why is this necessary?

22. What is the significance of acetyl CoA in metabolism?

23. In which metabolic pathway are NADH and FADH$_2$ generated? Where in the cell is this pathway located?
24. How are NADH and FADH$_2$ reoxidized? Where in the cell does this occur?
25. Name, in order, the participants of the electron transport system. At which steps is ATP synthesized?
26. How does iron behave differently in the cytochromes than in hemoglobin?
27. Explain how H$_2$O is produced by the electron transport system.
28. Why is cyanide such a toxic substance?
29. Why does oxidation of FADH$_2$ by the electron transport system produce fewer ATP molecules than oxidation of NADH?
30. Explain the meaning of the following statement: "Oxidation of NADH by the electron transport system results in 42% energy conservation."
31. Describe the citric acid cycle by identifying the following:
    (a) the fuel needed by the cycle
    (b) the oxidation products of the cycle and where they are formed
    (c) the intermediates in the cycle
    (d) the reduction products of the cycle and where they are formed
    (e) the types of reactions in the cycle
    (f) energy production by the cycle
32. Give the entry points in the citric acid cycle for carbon atoms derived from amino acids. Which amino acids provide the carbon atoms?
33. Malonate, $^-O_2C-CH_2-CO_2^-$, is an extremely effective inhibitor of the citric acid cycle. Which enzyme do you think malonate inhibits? Explain your reasoning.
34. How many ATP molecules can be formed by oxidation of six acetyl units in the citric acid cycle?
35. How is α-ketoglutarate important to nitrogen metabolism?
36. Describe how the citric acid cycle is responsive to cellular energy needs.
37. The citric acid cycle does not utilize oxygen, yet it cannot function without oxygen. Why?

CHAPTER 20
METABOLISM OF CARBOHYDRATES

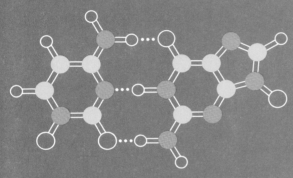

GLUCOSE—THE CENTRAL FIGURE
- Blood Sugar

GLYCOGEN METABOLISM
- Glycogen Synthesis
- Glycogen Breakdown
- Inborn Errors of Glycogen Metabolism

GLYCOLYSIS
- Reactions
- Energy Yield

FATES OF PYRUVATE
- Reduction to Lactate
- Reduction to Ethanol
- Conversion to Acetyl CoA

COMPLETE OXIDATION OF GLUCOSE

GLUCONEOGENESIS

THE PENTOSE PHOSPHATE PATHWAY

REGULATION OF CARBOHYDRATE METABOLISM

INBORN ERRORS OF CARBOHYDRATE UTILIZATION AND GLYCOLYSIS

All living organisms derive their energy ultimately from the sun. Some, such as green plants and a few microorganisms, absorb the sun's energy directly and use it to synthesize carbohydrates from carbon dioxide and water. Thus, plants serve as a source of carbohydrates to be used as food by animals. After digestion, carbohydrates are metabolized principally by the pathway called glycolysis. This pathway converts glucose into pyruvate, with production of a small amount of ATP, and under aerobic conditions, pyruvate is a source of acetyl units for the citric acid cycle. If oxygen supplies are limited, as in actively exercising muscles, pyruvate is converted to lactate. In some microorganisms (for example, yeast) that live in anaerobic environments, pyruvate is converted to ethanol.

Lactate generated in muscles can be transported by the blood to the liver where it can be used to synthesize new glucose. At times of high energy levels (high ATP levels) in liver and muscle cells, glucose is stored away in polymer form as glycogen, which in turn can be degraded when its glucose units are needed for oxidation to produce energy.

GLUCOSE—THE CENTRAL FIGURE

The principal carbohydrates that we obtain in our diets are starch, cellulose, and sucrose. In addition, we receive small amounts of lactose from dairy products, and fructose and glucose from honey, fruits, and vegetables. As mentioned in chapter 13, humans cannot digest cellulose; however, its fibers absorb water and cause fecal matter to be moist and soft. The other complex carbohydrates are hydrolyzed to glucose, fructose, and galactose, and these simple sugars pass through the membranes in the small intestine to the bloodstream. When fructose and galactose reach the liver, they are converted by appropriate enzymes to glucose or other compounds that are metabolized by the same pathway as glucose. Thus, glucose is the focal point of carbohydrate metabolism, as illustrated by Figure 20-1.

Fig. 20-1 Glucose is the Central Figure of Carbohydrate Metabolism.

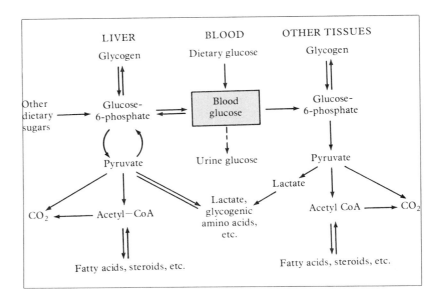

Blood Sugar

Glucose is by far the most plentiful sugar in blood, and the term *blood sugar level* is usually considered to refer to glucose. The blood sugar level of a normal adult, measured 8–12 hours after a meal, is in the range of 60–100 mg/100 mL. (In clinical reports, the value is often expressed in units of milligrams per deciliter, mg/dL.) About an hour after a meal containing sugars or carbohydrates, the blood sugar level reaches a maximum of approximately 140–160 mg/100 mL and then decreases back to normal after 2–2½ hours.

The ability of the body to maintain a relatively constant blood sugar level can be measured by a *glucose tolerance test*. A typical curve is shown in Figure 20-2. In this test, a person ingests 100 g of glucose in a short period of time, and the blood sugar level is monitored for the next several hours. When the glucose concentration is above the fasting level (60–100 mg/100 mL), the condition is referred to as *hyperglycemia*. If the level exceeds approximately 160–180 mg/100 mL, the sugar is not completely reabsorbed by the kidneys, and glucose is excreted in the urine; a condition of *glucosuria* is said to exist. On the other hand, if blood sugar levels fall below the normal values, a condition of *hypoglycemia* exists. Since glucose is the only nutrient normally used by the brain for energy, severe hypoglycemia can cause such symptoms as convulsions and shock. Even mild hypoglycemia can be responsible for dizziness and fainting.

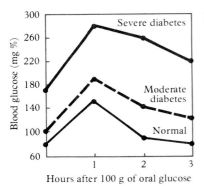

Fig. 20-2 Typical Response of Three Patients in a Glucose Tolerance Test

hyperglycemia: an abnormally high level of blood sugar.

glucosuria: excretion of glucose in the urine.

hypoglycemia: an abnormally low level of blood sugar.

Although diet contributes largely to blood glucose level, most of the digestible carbohydrate is absorbed within hours after a meal. Thus, other sources of glucose must be available for constant supply to the blood between meals. The major back-up source of glucose is glycogen stored in the liver and muscle tissues. Blood glucose is utilized by the cells to produce energy or it can be converted to glycogen for storage if a surplus exists.

GLYCOGEN METABOLISM

Glucose enters most cells by a facilitated diffusion process, but it is able to enter liver cells by simple diffusion. Once inside the cells, glucose is converted to glucose-6-phosphate by the enzyme *hexokinase*:

hexokinase: an enzyme which catalyzes phosphate transfer from ATP to hexoses; in general, a "kinase" catalyzes transfer of phosphate to ADP or from ATP.

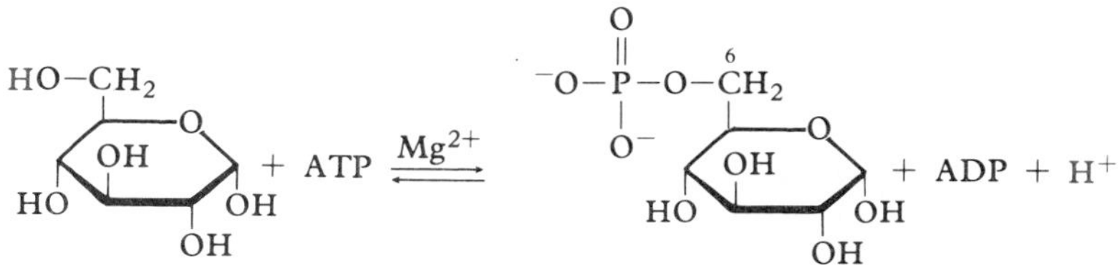

$$\text{glucose} + \text{ATP} \xrightleftharpoons{\text{Mg}^{2+}} \text{glucose-6-phosphate} + \text{ADP} + \text{H}^+$$

This form of glucose and all other phosphorylated intermediates are unable to cross cell membranes; thus, glucose-6-phosphate is locked into the cell, where it can be oxidized or stored in the form of glycogen.

Glycogen Synthesis

Synthesis of glycogen from glucose is called *glycogenesis*. All cells are capable of glycogenesis, but it is an especially important function of liver and muscle cells. In essence, glucose units are attached to a growing glycogen polymer, and hydrolysis of a high-energy nucleotide, UTP (uridine triphosphate), provides the energy for this anabolic process.

glycogenesis (glye-co-JEH-nih-sis)

In the first step, glucose-6-phosphate is converted to an isomer, glucose-1-phosphate, in a reaction catalyzed by phosphoglucomutase:

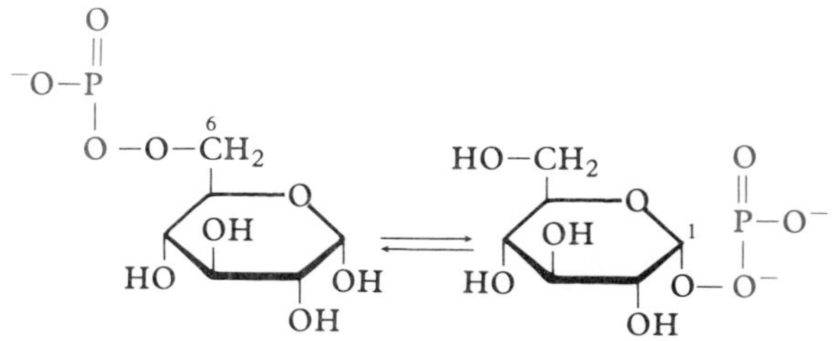

$$\text{glucose-6-phosphate} \rightleftharpoons \text{glucose-1-phosphate}$$

In the second step, glucose-1-phosphate is converted to a high-energy intermediate, UDP-glucose, and energy contained in UTP is transferred to this intermediate. The enzyme is UDP-glucose pyrophosphorylase, and the PP_i produced is rapidly hydrolyzed to drive the reaction to completion.

$$\text{glucose-1-phosphate} + \text{UTP} \rightleftharpoons \text{UDP-glucose} + PP_i$$
$$PP_i + H_2O \longrightarrow 2\,P_i$$
$$\overline{\text{glucose-1-phosphate} + \text{UTP} + H_2O \longrightarrow \text{UDP-glucose} + 2\,P_i}$$

In the third step of glycogen synthesis, UDP-glucose reacts with a growing glycogen polymer to add a unit of glucose:

$$\text{UDP-glucose} + (\text{glucose})_n \longrightarrow (\text{glucose})_{n+1} + \text{UDP}$$
$$\qquad\qquad\qquad\quad \text{glycogen} \qquad\quad \text{glycogen with an additional glucose unit}$$

The enzyme for this step is glycogen synthetase. Thus, glycogen synthesis requires a preformed glycogen molecule, which is formed by a different synthetase. In subsequent reactions, UDP is converted back to UTP,

$$\text{UDP} + \text{ATP} \longrightarrow \text{UTP} + \text{ADP}$$

and branches are added to the glycogen chain.

Glycogen synthesis greatly increases the amount of glucose that is quickly available between meals and during muscle activity. The glucose in the body fluids of a typical 70-kg male has an energy content of only 40 kcal, but the total body glycogen has an energy potential of more than 600 kcal.

Glycogen Breakdown

Glycogen is found in the cytoplasm of cells in the form of granules with diameters ranging from 10 to 40 nm (one nanometer = 10^{-9} meters).

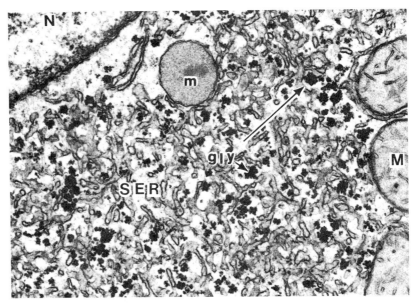

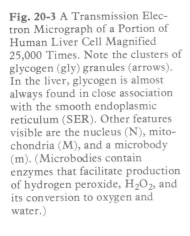

Fig. 20-3 A Transmission Electron Micrograph of a Portion of Human Liver Cell Magnified 25,000 Times. Note the clusters of glycogen (gly) granules (arrows). In the liver, glycogen is almost always found in close association with the smooth endoplasmic reticulum (SER). Other features visible are the nucleus (N), mitochondria (M), and a microbody (m). (Microbodies contain enzymes that facilitate production of hydrogen peroxide, H_2O_2, and its conversion to oxygen and water.)

(Glycogen granules are illustrated in Figure 20-3.) This size range reflects the fact that glycogen molecules have a range of molecular weights. The granules contain enzymes necessary for synthesis and degradation of glycogen and even some enzymes to regulate these processes.

Degradation of glycogen to glucose is called *glycogenolysis*. Although glycogen is stored mostly in muscle and liver tissue, glycogenolysis can only occur in the liver, kidney, and intestine, because one essential enzyme is absent from other tissues. The first step in glycogen breakdown is the cleavage of α-1,4-linkages, catalyzed by glycogen phosphorylase. In this reaction, glucose units are released from the glycogen chain as glucose-1-phosphate:

glycogenolysis (glye-co-jeh-NOH-lih-sis)

$$(\text{glucose})_n + P_i \longrightarrow (\text{glucose})_{n-1} + \text{glucose-1-phosphate}$$

glycogen　　　　　　　　glycogen with one fewer glucose units

In the second reaction, phosphoglucomutase catalyzes isomerization of glucose-1-phosphate to glucose-6-phosphate:

$$\text{glucose-1-phosphate} \rightleftharpoons \text{glucose-6-phosphate}$$

The final step in glycogen breakdown is the hydrolysis of glucose-6-phosphate to produce free glucose and inorganic phosphate. The enzyme for this reaction, glucose-6-phosphatase, is the one found only in liver, kidney, and intestinal cells.

$$\text{glucose-6-phosphate} + H_2O \longrightarrow \text{glucose} + P_i$$

Although muscle and brain cells do not contain glucose-6-phosphatase and thus cannot form free glucose from glycogen, they can produce glucose-6-phosphate from glycogen, and this form of glucose can then be used to produce energy. In this way, glucose-6-phosphate is retained by muscle and brain cells, which need large amounts of fuel for proper function. On the other hand, an important function of the liver is to maintain a relatively constant level of blood glucose. Thus, it has the capacity to degrade glycogen all the way to glucose, which is released into the blood during muscular activity and in between meals. This glucose is taken up primarily by the brain and skeletal muscle.

It is important to note that glycogenesis and glycogenolysis are not simple reversals of each other. Except for the phosphoglucomutase reaction, different enzymes and reactions are used in each process. These differences illustrate an important principle of biochemistry: biosynthetic and degradative pathways are almost always separate from each other. This arrangement allows for much greater flexibility, because each pathway has its own energy requirements and may be controlled separately from the other pathway.

EXAMPLE 20.1

Glycogen degradation is not a simple reversal of glycogen synthesis, even though both occur in glycogen granules. How are the two pathways kept separate?

SOLUTION

In glycogen synthesis, the first step (isomerization of glucose-6-phosphate to glucose-1-phosphate) is catalyzed by phosphoglucomutase, an enzyme also involved in glycogen degradation. However, the other steps involve different enzymes and different types of reactions in the two pathways.

Inborn Errors of Glycogen Metabolism

Inborn errors of metabolism are diseases in which genetic defects cause deficiencies in particular proteins. These deficiencies may be a result of diminished synthesis of the protein or of synthesis of the protein in a nonfunctional form.

Several defects of glycogen metabolism have been studied extensively. In one of these, glycogen storage disease I (von Gierke's disease), glucose-6-phosphatase activity is deficient. This makes the liver unable to hydrolyze glucose-6-phosphate to provide glucose to the blood. Hypoglycemia is typical of the disease, and since an abnormally large part of glucose-6-phosphate is channeled into glycogen synthesis, increased levels of glycogen are observed. In another glycogen storage disease, Pompe's disease, certain enzymes needed for glycogen degradation are deficient. This disease is not understood very well, but infants with the disease usually die within a year.

Von Gierke's disease and Pompe's disease exert their effects primarily on glycogen metabolism in the liver. However, other defects are known to affect glycogen metabolism in muscle. One of these is McArdle's disease. In this case, glycogen phosphorylase activity in muscle is lacking, causing a buildup of glycogen in muscles. This impairs muscle function and causes muscle cramps during exercise. Fortunately, the painful consequences of this disease can be avoided by limiting strenuous activity.

GLYCOLYSIS

glycolysis (glye-CAH-lih-sis): the metabolic pathway in which glucose is degraded to pyruvate, the anion of pyruvic acid.

The metabolic pathway known as *glycolysis* was studied for about fifty years before all of its reactions were discovered. By 1940 the major features of the pathway were known, thanks to the efforts of scientists like Gustav Embden, Otto Meyerhof, Carl Neuberg, Jacob Parnus, Otto Warburg, Gerty Cori, and Carl Cori. The pathway is also referred to as the Embden-Meyerhof pathway in honor of two of these researchers. It is the first stage in the extraction of energy from glucose.

All of the enzymes needed for glycolysis are present in the cytoplasm of the cell. Under aerobic conditions, glucose is converted to pyruvate, which then enters the mitochondria where it is completely oxidized by the citric acid cycle. However, if the supply of oxygen is limited, as in

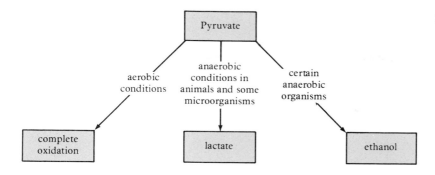

Fig. 20-4 The Fates of Pyruvate

actively exercising muscle, pyruvate is converted to lactate. Some anaerobic organisms, such as yeast, convert pyruvate to ethanol. Figure 20-4 shows the various pathways pyruvate can follow.

Reactions

Glucose is converted to pyruvate in a sequence of 10 reactions as listed in Table 20-1 and illustrated in Figure 20-5. Significant features of each reaction are discussed in the following paragraphs.

Table 20-1 The Reactions of Glycolysis

Step	Reaction	Enzyme	Cofactor
1	Glucose + ATP ⟶ glucose-6-phosphate + ADP + H^+	Hexokinase	Mg^{2+}
2	Glucose-6-phosphate ⇌ fructose-6-phosphate	Phosphoglucoisomerase	None
3	Fructose-6-phosphate + ATP ⟶ fructose-1,6-diphosphate + ADP + H^+	Phosphofructokinase	Mg^{2+}
4	Fructose-1,6-diphosphate ⇌ dihydroxyacetone phosphate + glyceraldehyde-3-phosphate	Aldolase	None
5	Dihydroxyacetone phosphate ⇌ glyceraldehyde-3-phosphate	Triose phosphate isomerase	None
6	Glyceraldehyde-3-phosphate + NAD^+ + P_i ⇌ 1,3-diphosphoglyceric acid + NADH + H^+	Glyceraldehyde-3-phosphate dehydrogenase	NAD^+
7	1,3-Diphosphoglyceric acid + ADP ⇌ 3-phosphoglycerate + ATP	Phosphoglycerate kinase	Mg^{2+}
8	3-Phosphoglycerate ⇌ 2-phosphoglycerate	Phosphoglyceromutase	None
9	2-Phosphoglycerate ⇌ phosphoenolpyruvate + H_2O	Enolase	Mg^{2+}
10	Phosphoenolpyruvate + ADP + H^+ ⟶ pyruvate + ATP	Pyruvate kinase	Mg^{2+}

Fig. 20-5 The Pathway of Glycolysis

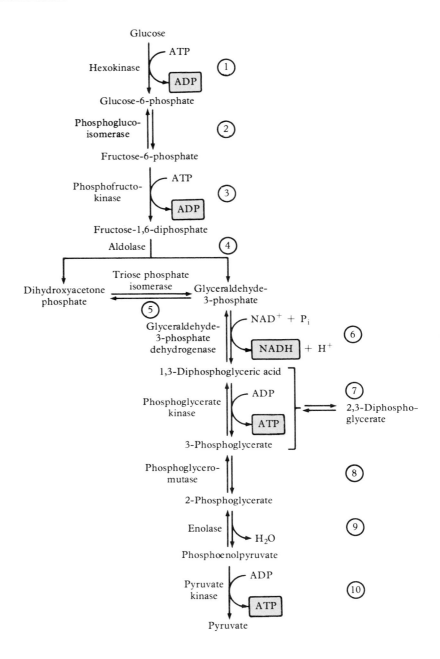

Reaction 1. Formation of Glucose-6-Phosphate

This reaction, mentioned earlier under the topic of glycogen metabolism, involves transfer of a phosphate group from ATP to glucose. Catalyzed by hexokinase, it is an irreversible reaction and requires Mg^{2+} as a cofactor.

[glucose + ATP →(Mg²⁺) glucose-6-phosphate + ADP + H⁺]

Reaction 2. Formation of Fructose-6-Phosphate

In the second step of glycolysis, glucose-6-phosphate is reversibly isomerized to fructose-6-phosphate under the influence of phosphoglucoisomerase (an alternate name is glucose phosphate isomerase):

[glucose-6-phosphate ⇌ fructose-6-phosphate]

Reaction 3. Formation of Fructose-1,6-Diphosphate

Next, another phosphate group is irreversibly transferred from ATP to form fructose-1,6-diphosphate. The reaction is catalyzed by phosphofructokinase, and once again Mg^{2+} is a necessary cofactor.

[fructose-6-phosphate + ATP →(Mg²⁺) fructose-1,6-diphosphate + ADP + H⁺]

Reaction 4. Formation of Triose Phosphates

The reactions up to this point have had as their goal the formation of a six-carbon compound that can be easily cleaved into two three-carbon fragments. Fructose-1,6-diphosphate is just such a compound; it undergoes reversible cleavage, catalyzed by aldolase, to form two triose phosphates. The open-chain structure of fructose-1,6-diphosphate is shown for clarity in illustrating the cleavage:

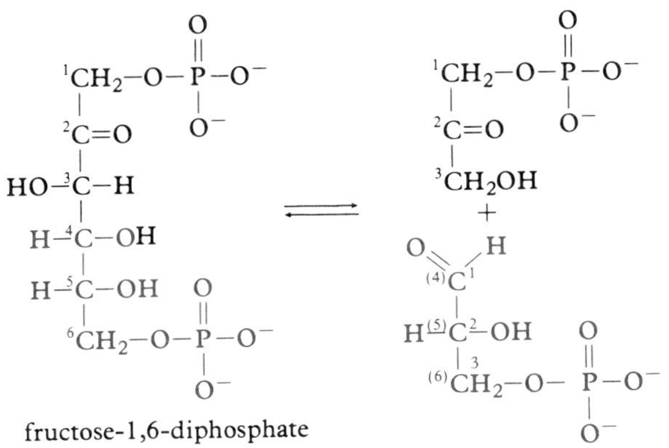

fructose-1,6-diphosphate

dihydroxyacetone phosphate

glyceraldehyde-3-phosphate

Reaction 5. Interconversion of Triose Phosphates

Dihydroxyacetone phosphate is not used directly by the glycolytic pathway, but its isomer, glyceraldehyde-3-phosphate, continues in the pathway. Reversible interconversion of the two isomers is catalyzed by triose phosphate isomerase. Keep in mind that one molecule of glucose thus produces two molecules of glyceraldehyde-3-phosphate due to the cleavage in reaction 4 and the following isomerization:

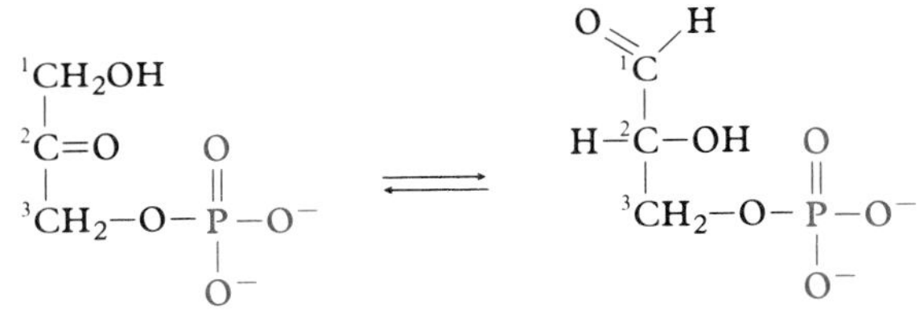

dihydroxyacetone phosphate

glyceraldehyde-3-phosphate

Reaction 6. Formation of 1,3-Diphosphoglyceric Acid

In this reaction, glyceraldehyde-3-phosphate is oxidized at carbon 1, and phosphate is added to make a high-energy carboxylic acid–phosphoric acid anhydride. The reversible reaction is catalyzed by glyceraldehyde-3-phosphate dehydrogenase, utilizing NAD^+ as coenzyme.

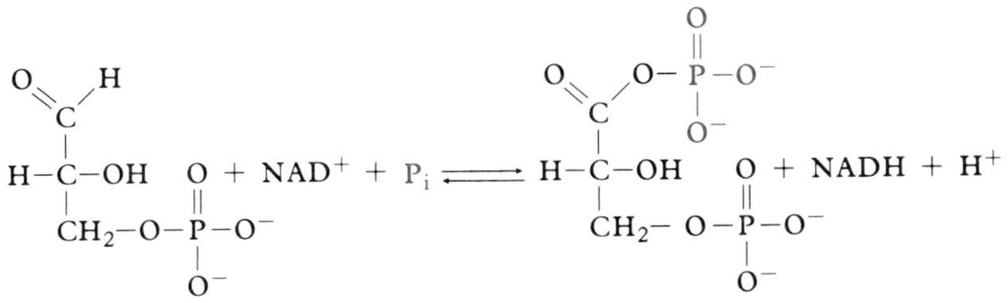

glyceraldehyde-3-phosphate

1,3-diphosphoglyceric acid

Reaction 7. Formation of 3-Phosphoglycerate

At this point, energy released from 1,3-diphosphoglyceric acid is used to synthesize ATP, an example of substrate-level phosphorylation. This reversible reaction, catalyzed by phosphoglycerate kinase, requires Mg^{2+} as cofactor and represents the first synthesis of ATP in glycolysis.

$$\text{1,3-diphosphoglyceric acid} + \text{ADP} \underset{}{\overset{Mg^{2+}}{\rightleftharpoons}} \text{3-phosphoglycerate} + \text{ATP}$$

In addition to their roles in glycolysis, 1,3-diphosphoglyceric acid and 3-phosphoglycerate interact in an offshoot of the pathway to produce 2,3-diphosphoglycerate, the regulator of oxygen-hemoglobin binding. The enzyme for this reaction is diphosphoglyceromutase.

$$\text{1,3-diphosphoglyceric acid} + \text{3-phosphoglycerate} \rightleftharpoons \text{3-phosphoglycerate} + \text{2,3-diphosphoglycerate} + H^+$$

Reaction 8. Formation of 2-Phosphoglycerate

At this point, 3-phosphoglycerate reversibly isomerizes to 2-phosphoglycerate, catalyzed by phosphoglyceromutase:

$$\text{3-phosphoglycerate} \rightleftharpoons \text{2-phosphoglycerate}$$

Reaction 9. Formation of Phosphoenolpyruvate

Next, 2-phosphoglycerate is reversibly dehydrated to form the high-energy compound phosphoenolpyruvate, catalyzed by enolase:

$$\text{2-phosphoglycerate} \xrightleftharpoons{Mg^{2+}} \text{phosphoenolpyruvate} + HOH$$

Reaction 10. Formation of Pyruvate

In this last step of glycolysis, a second molecule of ATP is formed when phosphate from phosphoenolpyruvate is irreversibly transferred to ADP, catalyzed by pyruvate kinase. The enolpyruvate product rapidly rearranges to its more stable isomer, pyruvate.

$$\text{phosphoenolpyruvate} + ADP + H^+ \xrightarrow{Mg^{2+}} \text{enolpyruvate} + ATP$$

$$\text{enolpyruvate} \rightleftharpoons \text{pyruvate}$$

Energy Yield

We can add together all of the reactions leading to pyruvate if we multiply equations 6–10 by two to reflect the fact that each glucose molecule produces two three-carbon fragments. The summary equation is

glucose + 2 P_i + 2 ADP + 2 NAD^+ ⟶
  2 pyruvate + 2 ATP + 2 NADH + 2 H^+ + 2 H_2O

This equation indicates a net gain of two moles of ATP for every mole of glucose converted to pyruvate. A summary of the steps involving ATP is given in Table 20-2.

Table 20-2 Glycolytic Reactions Involving ATP

Step	Reaction
1	Glucose + ATP ⟶ glucose-6-phosphate + ADP + H$^+$
3	Fructose-6-phosphate + ATP ⟶ fructose-1,6-diphosphate + ADP + H$^+$
7	1,3-Diphosphoglyceric acid + ADP ⇌ 3-phosphoglycerate + ATP
10	Phosphoenolpyruvate + ADP + H$^+$ ⟶ pyruvate + ATP

EXAMPLE 20.2

Most of the reactions in glycolysis are reversible, but three are not. Which ones are these?

SOLUTION

The three irreversible reactions of glycolysis are

Reaction 1: Formation of glucose-6-phosphate from glucose
Reaction 3: Formation of fructose-1,6-diphosphate from fructose-6-phosphate
Reaction 10: Formation of pyruvate from phosphoenolpyruvate

FATES OF PYRUVATE

Reduction to Lactate

When muscles are exercised vigorously, the blood cannot deliver oxygen to them fast enough, and oxygen supplies become limited. Under anaerobic conditions, the NADH produced in step 6 of glycolysis is used by animals and some microorganisms to reduce pyruvate to lactate. This reaction is catalyzed by lactate dehydrogenase.

$$CH_3-\overset{O}{\overset{\|}{C}}-CO_2^- + NADH + H^+ \rightleftharpoons CH_3-\overset{OH}{\overset{|}{C}H}-CO_2^- + NAD^+$$
pyruvate   lactate

We can summarize the conversion of glucose to lactate by adding the equation given above (multiplied by 2) to the series of equations for glycolysis, giving

$$\text{glucose} + 2\,P_i + 2\,ADP \longrightarrow 2\,\text{lactate} + 2\,ATP + 2\,H_2O$$

Note that NAD$^+$ and NADH do not appear in this equation. In fact, pyruvate is reduced to lactate so that the NAD$^+$ produced can be used to sustain oxidation of glyceraldehyde-3-phosphate (step 6 in glycolysis). In this way, the glycolytic pathway can continue to function in the absence of oxygen, and ATP is produced. However, there is no net oxidation-reduction in the conversion of glucose to lactate.

When we exercise vigorously, the oxygen supply to our muscles rapidly becomes depleted. In this situation, anaerobic glycolysis is the primary means of producing ATP. It is the accumulation of lactic acid (lactate) that results in the muscle soreness caused by strenuous physical activity. The lactate formed in muscle tissue eventually makes its way to the liver where it is oxidized to pyruvate either for entry into the citric acid cycle or for conversion back to glucose. However, prolonged anaerobic glycolysis and its buildup of lactic acid (lactate) causes a mild form of metabolic acidosis. This state then triggers an increase in the rate and depth of breathing. Since there is a limit to how fast and how deeply we can breathe, vigorous physical exertion must be confined to brief periods for all but well-trained athletes.

Conversion of glucose directly to lactate has a standard free energy change of -47.0 kcal/mole.

$$\text{glucose} \longrightarrow 2 \text{ lactate}$$
$$\Delta G^{o\prime} = -47.0 \text{ kcal/mole}$$

Thus, on a molar basis, 47.0 kcal are released by glycolysis and subsequent lactate formation. Since 14.6 kcal are used to synthesize 2 moles of ATP, the efficiency of energy conservation is 31% (14.6 kcal/47.0 kcal = 0.31). However, if glycogen is the source of glucose for glycolysis, then the net synthesis of ATP is 3 moles, because no ATP is expended in forming glucose-6-phosphate by this route. Thus, 21.9 kcal are used to synthesize 3 moles of ATP. In this case, direct conversion of glycogen to lactose liberates 44.0 kcal, giving an efficiency of energy conservation of 50% (21.9 kcal/44.0 kcal = 0.50) for lactate production from glycogen.

Anaerobic glycolysis releases some of the free energy content of glucose, but that amount is only a small fraction of the total energy available from the complete oxidation of glucose by the citric acid cycle and the electron transport system. The anaerobic production of lactate releases 47.0 kcal, while total oxidation of glucose produces 686.0 kcal of free energy. Thus, anaerobic glycolysis releases only about 7% of the free energy available from glucose (47.0 kcal/686.0 kcal = 0.069).

Anaerobic: glucose $\longrightarrow$ 2 lactate $\qquad \Delta G^{o\prime} = -47.0$ kcal/mole

Aerobic: glucose $+ 6 O_2 \longrightarrow 6 CO_2 + 6 H_2O$ $\Delta G^{o\prime} = -686.0$ kcal/mole

Reduction to Ethanol

Yeast and several other microorganisms living in oxygen-poor environments are able to form ethanol from pyruvate in a two-step reduction process. The first step, decarboxylation of pyruvate, is catalyzed by pyruvate decarboxylase. Thiamin pyrophosphate (TPP), the coenzyme derived from the vitamin thiamin, functions as coenzyme in this reaction.

$$\underset{\text{pyruvate}}{CH_3-C\begin{smallmatrix}\diagup O\\ \diagdown CO_2^-\end{smallmatrix}} + H^+ \xrightarrow{\text{TPP}} \underset{\text{acetaldehyde}}{CH_3-C\begin{smallmatrix}\diagup O\\ \diagdown H\end{smallmatrix}} + CO_2$$

The second step is the reduction of acetaldehyde to ethanol, catalyzed by alcohol dehydrogenase with NADH as coenzyme.

$$CH_3-C{\overset{O}{\underset{H}{\diagup\hspace{-0.5em}\diagdown}}} + NADH + H^+ \rightleftharpoons CH_3-\overset{H}{\underset{H}{C}}-OH + NAD^+$$

acetaldehyde   ethanol

The overall conversion of glucose to ethanol is an example of *fermentation*, the anaerobic metabolism of glucose by microorganisms. We can write a summary equation for this fermentation by adding the two reactions given above (multiplied by 2) to those of the glycolytic pathway; the result is

$$\text{glucose} + 2\,P_i + 2\,ADP + 2H^+ \longrightarrow 2\,\text{ethanol} + 2\,CO_2 + 2\,ATP + 2H_2O$$

This conversion of glucose to ethanol is the basis for the production of alcoholic beverages. As was the case with lactate formation, fermentation does not involve any net oxidation-reduction.

EXAMPLE 20.3

Reduction of pyruvate to lactate (or to ethanol, as in yeast) allows glycolysis to continue in the absence of oxygen. How is this accomplished?

SOLUTION

Reduction of pyruvate regenerates NAD^+ needed for glycolysis.

Conversion to Acetyl CoA

The mitochondrial membrane is permeable to pyruvate formed by glycolysis in the cytoplasm, and under conditions of plentiful oxygen supply, pyruvate is converted inside the mitochondria to acetyl CoA by the multiple-enzyme system called the pyruvate dehydrogenase complex.

$$CH_3-C{\overset{O}{\underset{CO_2^-}{\diagup\hspace{-0.5em}\diagdown}}} + CoA-SH + NAD^+ \longrightarrow CH_3-C{\overset{O}{\underset{S-CoA}{\diagup\hspace{-0.5em}\diagdown}}} + NADH + CO_2$$

pyruvate   acetyl CoA

This system contains three separate enzymes and is similar to the α-ketoglutarate dehydrogenase complex (see chapter 19). In addition to CoA and NAD^+, four other cofactors are required: Mg^{2+}, thiamin pyrophosphate (TPP, derived from the vitamin thiamin), the vitamin lipoic acid, and FAD. The acetyl CoA formed in this reaction enters the citric acid cycle for complete oxidation to CO_2. The NAD^+ needed for the pyruvate dehydrogenase complex and for the oxidation of glyceraldehyde-3-phosphate (step 6 in glycolysis) is regenerated when NADH transfers its hydride ion to the electron transport system.

EXAMPLE 20.4

What is the purpose of converting pyruvate to acetyl CoA?

SOLUTION

This conversion allows for complete oxidation of pyruvate via the pyruvate dehydrogenase complex reaction and the citric acid cycle; thus maximum energy yield is obtained via the electron transport system and oxidative phosphorylation.

COMPLETE OXIDATION OF GLUCOSE

We are almost at the point where we can calculate the ATP yield from total oxidation of glucose. However, one complicating factor remains, and that is the fact that mitochondrial membranes are impermeable to NADH and NAD^+. How then does NADH produced in the cytoplasm by glycolysis enter the mitochondria for oxidation by the electron transport system? The answer is that NADH does not cross the membrane, but its electrons, as a hydride ion, are carried across by glycerol-3-phosphate formed by reduction of dihydroxyacetone phosphate in the cytoplasm. Mitochondrial membranes are permeable to glycerol-3-phosphate, and once inside the mitochondria, the compound is oxidized back to dihydroxyacetone phosphate, which then moves back into the cytoplasm. This system, called the glycerol phosphate shuttle, is illustrated in Figure 20-6. The reactions are catalyzed by different glycerol-3-phosphate dehydrogenases, and the mitochondrial enzyme uses FAD

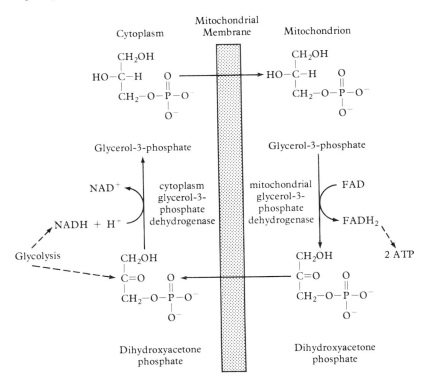

Fig. 20-6 The Glycerol Phosphate Shuttle

as coenzyme. The net result is that cytoplasmic NADH indirectly passes electrons to mitochondrial FAD to form $FADH_2$. Remember that oxidation of $FADH_2$ by the electron transport system produces only two ATP instead of the three ATP formed when NADH is oxidized. Since the concentration of NADH inside the mitochondria is higher than in the cytoplasm, this shuttle system transports electrons against an NADH concentration gradient at the cost of one ATP per pair of electrons.

We can now calculate the number of moles of ATP formed when a mole of glucose is completely oxidized. The summary equation is:

$$\text{glucose} + 36\ ADP + 36\ P_i + 36\ H^+ + 6\ O_2 \longrightarrow 6\ CO_2 + 36\ ATP + 42\ H_2O$$

Of the 36 moles of ATP generated, the vast majority (32 moles) are formed as a result of oxidative phosphorylation, as outlined in Table 20-3. The efficiency of energy conservation from glucose oxidation is comparable to previous calculations. Since $\Delta G^{\circ\prime}$ for total oxidation of glucose is 686.0 kcal/mole, we can calculate the energy conservation on the basis of 36 moles of ATP as 262.8 kcal (36 moles of ATP × 7.3 kcal/mole = 262.8 kcal). This gives an efficiency of 38.3% (262.8 kcal/686.0 kcal = 0.383). This compares favorably to the range of 20–40% efficiency for gasoline and diesel engines.

Table 20-3 ATP Yield From the Complete Oxidation of Glucose

Reaction Sequence	ATP Yield per Glucose
Glycolysis: glucose into pyruvate (in the cytoplasm)	
Phosphorylation of glucose	− 1
Phosphorylation of fructose-6-phosphate	− 1
Dephosphorylation of 2 molecules of 1,3-DPG	+ 2
Dephosphorylation of 2 molecules of phosphoenolpyruvate	+ 2
2 NADH are formed in the oxidation of 2 molecules of glyceraldehyde-3-phosphate	
Conversion of pyruvate into acetyl CoA (inside mitochondria)	
2 NADH are formed	
Citric acid cycle (inside mitochondria)	
2 molecules of guanosine triphosphate are formed from 2 molecules of succinyl CoA	+ 2
6 NADH are formed in the oxidation of 2 molecules each of isocitrate, α-ketoglutarate, and malate	
2 $FADH_2$ are formed in the oxidation of 2 molecules of succinate	
Oxidative phosphorylation (inside mitochondria)	
2 NADH formed in glycolysis; each yields 2 ATP (assuming transport of NADH by the glycerol phosphate shuttle)	+ 4
2 NADH formed in the oxidative decarboxylation of pyruvate; each yields 3 ATP	+ 6
2 $FADH_2$ formed in the citric acid cycle; each yields 2 ATP	+ 4
6 NADH formed in the citric acid cycle; each yields 3 ATP	+18
Net Yield Per Glucose	+36

GLUCONEOGENESIS

gluconeogenesis (glu-co-nee-oh-JEH-nih-sis): literally, the synthesis of new glucose.

The synthesis of glucose from noncarbohydrate precursors is called *gluconeogenesis*. In this pathway, glucose is synthesized from pyruvate. The two principal sites of gluconeogenesis are the liver and the kidneys, with the liver accounting for about 90% of the process. Very little gluconeogenesis occurs in the brain, skeletal muscle, or heart, but these organs have a high demand for glucose. Hence, gluconeogenesis in the liver and kidneys helps to maintain the blood glucose level so that those tissues needing glucose can extract it from blood to meet their metabolic demands.

As might be expected, gluconeogenesis is a separate pathway from glycolysis. Referring to Figure 20-5, there are three irreversible steps in glycolysis; these are steps 1, 3, and 10. In gluconeogenesis, these steps are bypassed by the following new reactions:

(1) pyruvate + CO_2 + ATP + H_2O $\xrightleftharpoons{\text{pyruvate carboxylase}}$ oxaloacetate + ADP + P_i + 2 H^+

oxaloacetate + GTP $\xrightleftharpoons{\text{phosphoenolpyruvate carboxykinase}}$ phosphoenolpyruvate + GDP + CO_2

(2) fructose-1,6-diphosphate + H_2O $\xrightarrow{\text{fructose-1,6-diphosphatase}}$ fructose-6-phosphate + P_i

(3) glucose-6-phosphate + H_2O $\xrightarrow{\text{glucose-6-phosphatase}}$ glucose + P_i

The enzymatic differences between glycolysis and gluconeogenesis are summarized in Table 20-4.

The summary equation for gluconeogenesis is:

$$2 \text{ pyruvate} + 4 \text{ ATP} + 2 \text{ GTP} + 2 \text{ NADH} + 6 \text{ H}_2\text{O} \longrightarrow$$
$$\text{glucose} + 4 \text{ ADP} + 2 \text{ GDP} + 6 P_i + 2 \text{ NAD}^+ + 2\text{H}^+$$
$$\Delta G^{o\prime} = -9 \text{ kcal/mole}$$

In contrast to the release of free energy by gluconeogenesis, a direct reversal of glycolysis would require an input of 20 kcal/mole. Six high-energy phosphate compounds are used to synthesize glucose from pyruvate, but only two ATP are generated when pyruvate is formed from glucose. Thus, it costs four high-energy phosphate compounds to make

Table 20-4 Enzymatic Differences between Glycolysis and Gluconeogenesis

Glycolysis	Gluconeogenesis
Hexokinase	Glucose-6-phosphatase
Phosphofructokinase	Fructose-1,6-diphosphatase
Pyruvate kinase	Pyruvate carboxylase
	Phosphoenolpyruvate carboxykinase

glucose from pyruvate. Because of this cost, our bodies make new glucose at the expense of other energy-requiring processes or when energy levels in cells are high.

Most cell membranes are permeable to lactate, and when lactate levels build up in muscle tissue, the compound diffuses out into the blood. It is carried to the liver where it is readily converted back to pyruvate by lactate dehydrogenase, the enzyme that catalyzes lactate formation in muscle.

$$CH_3-\underset{lactate}{CH(OH)}-CO_2^- + NAD^+ \rightleftharpoons CH_3-\underset{pyruvate}{C(=O)}-CO_2^- + NADH + H^+$$

Pyruvate is then converted into glucose by the gluconeogenesis pathway, and glucose enters the blood. In this way, the liver furnishes glucose to muscle, which derives ATP from the glycolytic pathway. This cyclic conversion is called the *Cori cycle* and is illustrated in Figure 20-7.

Cori cycle: named for the 1947 Nobel Prize winners, Carl and Gerti Cori.

THE PENTOSE PHOSPHATE PATHWAY

Glycolysis, the citric acid cycle, and oxidative phosphorylation are concerned primarily with the generation of ATP. There is, however, another important metabolic pathway which appears to be designed for a rather different purpose. The *pentose phosphate pathway* is a complicated sys-

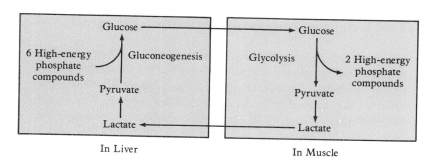

Fig. 20-7 The Cori Cycle

PERSPECTIVE: Calories from Alcohol

Ethanol is the intoxicating ingredient in beer, wine, and whiskey. This alcohol is metabolized primarily in the liver, where it is first oxidized to acetaldehyde by the enzyme alcohol dehydrogenase.

$$CH_3-CH_2-OH + NAD^+ \rightleftharpoons$$

$$CH_3-C\overset{O}{\underset{H}{\diagdown}} + NADH + H^+$$

Acetaldehyde then is attached to coenzyme A to form acetyl CoA, which is readily accepted as fuel by the citric acid cycle. GTP and the reduced coenzymes from the citric acid cycle plus NADH from the alcohol dehydrogenase reaction ultimately result in synthesis of 14 moles of ATP per mole of ethanol. The total energy available from ethanol is 7 kcal/gram, or 322 kcal/mole.

Since ethanol contributes acetyl units for the citric acid cycle, pyruvate from glycolysis is shunted toward lactate formation. The buildup of pyruvate and lactate is accompanied by an increase in H^+, and sometimes acidosis sets in during ethanol metabolism. Also, the increased conversion of pyruvate to lactate slows the rate of gluconeogenesis and often results in hypoglycemia. Because of these effects, drinking as much as 20 ounces (about 600 mL) of 90-proof whiskey rapidly could be fatal for the average person.

---

tem in which carbon atoms are interchanged among 3-, 4-, 5-, 6-, and 7-carbon carbohydrates (see Figure 20-8). Because of its complexities, we will not examine the pathway in detail, but it has three principal functions:

1. Interchange of carbon atoms among various carbohydrates.

2. Production of NADPH, the principal reduced coenzyme used in biosynthetic pathways.

3. Production of ribose needed for synthesis of nucleotides and nucleic acids (DNA and RNA).

REGULATION OF CARBOHYDRATE METABOLISM

It is important that metabolic pathways be responsive to cellular conditions so that energy is not wasted in producing unneeded materials. Regulation of carbohydrate metabolism is accomplished by two principal types of control, enzymatic and hormonal, which are summarized in Table 20-5. There are two major features of enzymatic control:

1. In glycolysis, the rate of glucose utilization is controlled by the level of activity of phosphofructokinase at step 3. This allosteric enzyme is stimulated by ADP and AMP, and it is inhibited by ATP and citrate, the citric acid cycle intermediate. Thus, when the energy level of the cell is low, ADP and AMP are high, and glycolysis is stimulated. However, when the energy level of the cell is high, ATP and citrate are high, and glycolysis is inhibited. Glucose would then be used to make glycogen for storage.

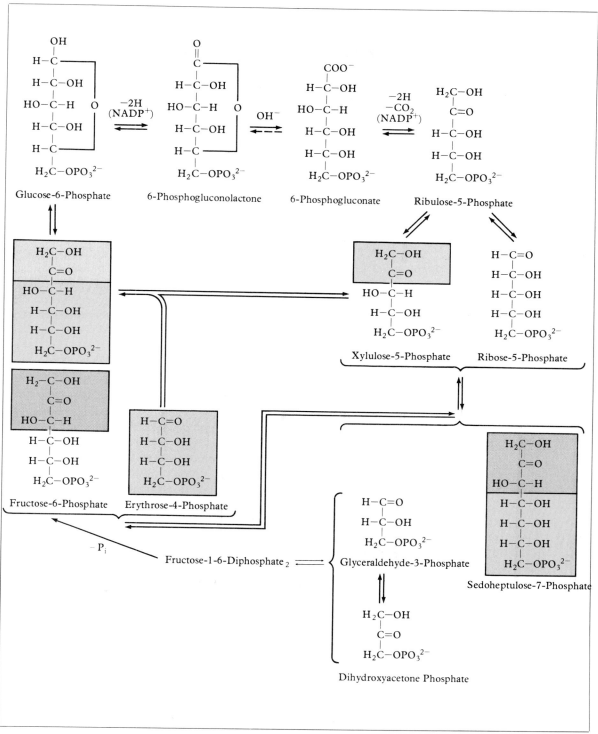

Fig. 20-8 The Pentose Phosphate Pathway

Table 20-5 Regulation of Carbohydrate Metabolism

Enzymatic Control Points	Action
Phosphofructokinase (glycolysis)	Activated by ADP and AMP Inhibited by ATP and citrate
Pyruvate carboxylase (gluconeogenesis)	Activated by acetyl CoA

Hormonal Control	Action
Epinephrine	Activates glycogen phosphorylase (glycogenolysis) in muscle
	Inactivates glycogen synthetase (glycogenesis) in muscle
Glucagon	Activates glycogen phosphorylase (glycogenolysis) in the liver
	Inactivates glycogen synthetase (glycogenesis) in the liver
Insulin	Facilitates entry of glucose, some other sugars, and amino acids into muscle, liver, and fat cells; increases the rate of synthesis of glycogen, fatty acids, and proteins; stimulates glycolysis; inhibits glycogenolysis and gluconeogenesis

2. In gluconeogenesis, pyruvate carboxylase is activated allosterically by acetyl CoA. Since the product from this reaction, oxaloacetate, is involved in both the citric acid cycle and gluconeogenesis, a high level of acetyl CoA is an indication of a need for increased production of oxaloacetate. At that point, if there is a surplus of ATP, oxaloacetate will be used primarily for gluconeogenesis (and subsequently for glycogen synthesis). But if ATP levels are low, oxaloacetate will be used mostly in the citric acid cycle to bring about increased production of ATP.

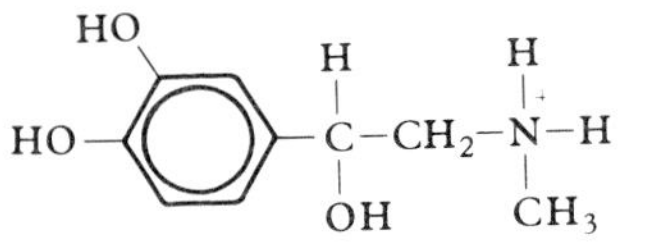

The hormones involved in regulating carbohydrate metabolism are *epinephrine, glucagon,* and *insulin.* Figure 20-9 shows the structures of epinephrine and glucagon. Epinephrine, the catecholamine synthesized by the adrenal gland, stimulates glycogen breakdown in muscles and, to a small extent, in the liver. However, the liver is more responsive to glucagon, a polypeptide hormone liberated by the pancreas. Glucagon increases the blood sugar level by enhancing the degradation of glycogen in the liver.

The mode of action of epinephrine and glucagon is typical of the "cyclic AMP" hormones, which bind to specific protein receptors present only in the cell membranes of their target tissues. Their binding stimulates an enzyme, *adenyl cyclase,* which catalyzes the formation of cyclic AMP (cAMP) from ATP:

Epinephrine

$H_3\overset{+}{N}$-His-Ser-Glu-Gly-Thr-Phe-Thr-
-Ser-Asp-Tyr-Ser-Lys-Tyr-Leu-Asp-
-Ser-Arg-Arg-Ala-Gln-Asp-Phe-Val-
-Gln-Trp-Leu-Met-Asn-Thr-CO_2^-

Glucagon

Fig. 20-9 The Structure of Epinephrine and the Amino Acid Sequence of Glucagon

$$\text{ATP} \longrightarrow \text{cyclic AMP (cAMP)} + PP_i + H^+$$

Through a complex set of reactions, increased levels of cyclic AMP, the "second messenger," affect the activities of a number of enzymes. In the case of epinephrine and glucagon, the increased cyclic AMP indirectly activates glycogen phosphorylase, the enzyme which initiates glycogenolysis. Thus, epinephrine stimulates glycogen breakdown in muscle, and glucagon has a similar effect on the liver. At the same time, the increased cyclic AMP indirectly inactivates glycogen synthetase, the enzyme responsible for lengthening chains of glycogen. In this way, the two hormones accelerate glycogen degradation and retard glycogen synthesis. The net result is that additional glucose is made available as a source of quick energy at times of stress or fright.

Insulin is a polypeptide hormone synthesized as an inactive prohormone in the pancreas. The prohormone is activated by enzyme-catalyzed hydrolysis just before release by the pancreas (see Figure 20-10). The mode of action for insulin is not well understood, but it binds to protein receptors in the cell membranes of muscle, liver, and fat tissue. This binding facilitates entry of glucose, some other sugars, and amino acids into those cells. Insulin thus serves to lower the level of glucose in blood, and in fact, its release by the pancreas is triggered by high blood glucose levels. In addition to its effect on transport, insulin increases the rate of synthesis of glycogen, fatty acids, and proteins, and it stimulates glycolysis. It also inhibits glycogenolysis and gluconeogenesis. The basis for these effects is not known, but it is possible that it may function *via* the cyclic AMP intermediate.

Diabetes mellitus is characterized by the inability of the body to regulate the blood glucose level. It is usually caused by an insufficient supply of insulin. This condition, thought to be hereditary, results in increased blood glucose levels but decreased glucose within the cells. In essence, glucose is usually plentiful in blood because of dietary supplies, but the transport system responsible for getting glucose into cells does not function. Thus the condition is a situation of starvation in the midst of plenty. Only the brain remains unaffected because its supply of glucose is not dependent on insulin. Diabetes mellitus is usually treated by strict control of the diet and administration of drugs to stimulate the production of insulin, or by daily injection of carefully measured doses of insulin itself.

diabetes mellitus (MEH-lih-tus): a condition of abnormal carbohydrate metabolism resulting from an insufficiency of insulin production and/or an inability of the body to respond to insulin.

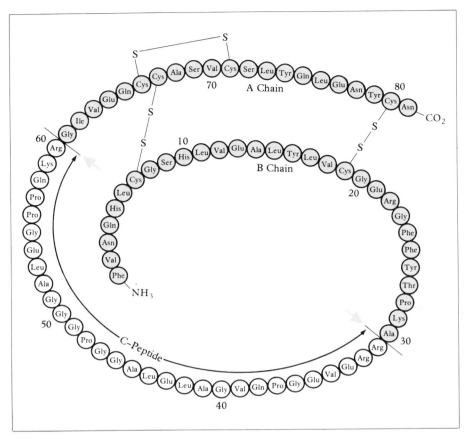

Fig. 20-10 The Structure of Bovine Proinsulin. The arrows indicate the points of cleavage in the conversion of proinsulin to insulin, represented by the colored circles.

INBORN ERRORS OF CARBOHYDRATE UTILIZATION AND GLYCOLYSIS

Two separate defects in sugar utilization produce the diseases known as *galactosemia* and *fructosemia*. The first of these, galactosemia, is characterized by abnormally high levels of galactose in the blood. Victims are unable to utilize dietary galactose because an enzyme needed to convert galactose to glucose is deficient. The resulting buildup of galactose can cause liver damage, cataracts, and mental retardation unless detected within the first year of life. Because of the severe consequences of galactosemia, the blood and urine of newborn infants are routinely tested for galactose levels. After detection, the disease is treated by reducing the amount of galactose in the diet. In fructosemia, there are high levels of fructose in the blood. When persons with fructosemia eat fruit, the effects are nausea, vomiting, sweating, tremors, confusion, convulsions, and coma. The condition is caused by a deficiency of an enzyme that normally allows fructose to enter the glycolytic pathway. The resulting high levels of fructose inhibit enzymes involved in glycogenolysis, and hypoglycemia is produced. The condition is treated by removing fruits and other sources of fructose from the diet.

Another example of an inborn error involving sugar utilization is *hexokinase deficiency* in red blood cells. Lack of this enzyme, which catalyzes the conversion of glucose to glucose-6-phosphate, causes insufficient production of glucose-6-phosphate, and glycolysis is impaired. Not only does this defect cause premature destruction of red blood cells (hemolytic anemia) but it also reduces the amount of 2,3-diphosphoglycerate produced. Thus, victims have hemoglobin with an unusually high affinity for oxygen. In addition to the problems of anemia, oxygen transport to the tissues is further reduced because of tighter binding by hemoglobin.

A defect in glycolysis is caused by *pyruvate kinase deficiency* in red blood cells. In this metabolic error, deficient pyruvate kinase, which catalyzes step 10 in glycolysis, causes reduced conversion of phosphoenolpyruvate to pyruvate. The glycolytic intermediates formed before this step accumulate; one of these is 2,3-diphosphoglycerate, and its accumulation leads to a lower affinity for oxygen by hemoglobin. Pyruvate kinase deficiency in red blood cells causes a severe anemia that can result in death during the first few years of life unless treated by transfusions and removal of the spleen to reduce destruction of red blood cells. The glycolytic pathway is extremely important in normal maintenance of red blood cells, and impaired glycolysis usually results in severe anemia.

SUMMARY

Glucose—The Central Figure: Dietary carbohydrates are digested to simple sugars. In the liver, fructose and galactose are converted to glucose or other compounds that can be metabolized by the same pathway as glucose. The ability of the body to maintain a constant blood sugar level is measured by the glucose tolerance test.

Glycogen Metabolism: After glucose enters cells, it is converted to glucose-6-phosphate, a compound which may be used to synthesize glycogen by the process of glycogenesis. When energy is needed by the body, glycogen can be degraded by the process of glycogenolysis to regenerate glucose-6-phosphate and, in the liver, kidney, and intestine, free glucose. Glycogenesis and glycogenolysis utilize different enzymes and reactions and thus are not simple reversals of each other. Inborn errors of glycogen metabolism result in increased glycogen storage in the liver and muscles.

Glycolysis: Glycolysis, the metabolic pathway that converts glucose to pyruvate, consists of a sequence of ten reactions. For every mole of glucose that enters the pathway, two moles each of pyruvate, ATP, and NADH are produced.

Fates of Pyruvate: Under anaerobic conditions, pyruvate is reduced to lactate, allowing glycolysis to continue in the absence of oxygen. Yeast and certain other microorganisms live in oxygen-poor environments by converting pyruvate to ethanol. Under aerobic conditions, pyruvate is converted to acetyl CoA, which then enters the citric acid cycle.

Complete Oxidation of Glucose: Complete oxidation of one mole of glucose *via* glycolysis and the citric acid cycle results in production of 36 moles of ATP instead of the expected 38 because 2 moles of ATP are sacrificed by the glycerol phosphate shuttle. The efficiency of energy conservation of complete oxidation of glucose is 38.3%.

Gluconeogenesis: In gluconeogenesis, glucose is synthesized from pyruvate. This system is separate from glycolysis since it uses different reactions to bypass the three irreversible steps in glycolysis. The net cost of synthesizing one mole of glucose by gluconeogenesis is four moles of ATP.

The Pentose Phosphate Pathway: The pentose phosphate pathway is a system in which carbon atoms are interchanged among 3-, 4-, 5-, 6-, and 7-carbon carbohydrates. Its major importance is the interchange of carbon atoms and the production of NADPH and ribose.

Regulation of Carbohydrate Metabolism: Carbohydrate metabolism is regulated by enzymatic and hormonal mechanisms. The two principal enzymatic control points are the phosphofructokinase step in glycolysis and the pyruvate carboxylase step in gluconeogenesis. The hormones epinephrine and glucagon stimulate glycogenolysis in muscle and liver, respectively, by indirect activation of glycogen phosphorylase. They also inactivate glycogen synthetase. Another hormone, insulin, exerts its effects by facilitating entry of glucose, some other sugars, and amino acids into muscle and fat cells. Insufficient supplies of insulin result in the disease called diabetes mellitus.

Inborn Errors of Carbohydrate Metabolism: Galactosemia and fructosemia result from deficient enzyme levels, causing lack of utilization of galactose and fructose. Hexokinase and pyruvate kinase deficiencies lead to impaired glycolysis and hemolytic anemia.

STUDY QUESTIONS AND PROBLEMS

1. Define the following terms:
   - (a) blood sugar
   - (b) glucose tolerance test
   - (c) hyperglycemia
   - (d) glucosuria
   - (e) hypoglycemia
   - (f) hexokinase
   - (g) glycogenesis
   - (h) glycogenolysis
   - (i) inborn error of metabolism
   - (j) glycolysis
   - (k) triose phosphate
   - (l) pyruvate
   - (m) lactate
   - (n) fermentation
   - (o) gluconeogenesis
   - (p) Cori cycle
   - (q) pentose phosphate pathway
   - (r) epinephrine
   - (s) glucagon
   - (t) adenyl cyclase
   - (u) insulin
   - (v) galactosemia

2. Explain why glucose is said to be the central figure of carbohydrate metabolism.

3. What are the principal dietary sources of the following hexoses?
   - (a) glucose
   - (b) fructose
   - (c) galactose

4. What does the glucose tolerance test measure?

5. Summarize how your blood sugar level changes when you treat yourself to an ice cream sundae. How does this response compare to that after eating a meal rich in starch? Explain your answer.

6. Why is stored glycogen important to health?

7. How is pyrophosphate hydrolysis important to glycogenesis?

8. Why are biosynthetic and degradative pathways usually separate systems?

9. Summarize the metabolic differences between glycogenesis and glycogenolysis.

10. Why is glycogenolysis restricted to the liver, kidney, and intestine?

11. What is the metabolic explanation for the hypoglycemia and increased glycogen levels in glycogen storage disease I (von Gierke's disease)?

12. Write equations to illustrate the following reactions in glycolysis:
    - (a) ATP utilization
    - (b) ATP production
    - (c) NADH production

13. In which glycolytic reactions is Mg^{2+} required? What is its function?

14. Give the coenzyme utilized in glycolysis and its vitamin source.

15. The total amount of energy available from a mole of glucose is 686 kcal. What per cent of this is conserved in ATP production by glycolysis?

16. Fructose can be converted to a glycolytic intermediate in one step. Suggest how this might be accomplished.

17. What are the three major fates of pyruvate?
18. Explain how lactate formation allows continuation of glycolysis under anaerobic conditions.
19. Write a summary equation for conversion of glucose to lactate with accompanying ATP production.
20. Explain how fermentation allows yeast to survive with limited oxygen supplies.
21. How is thiamin involved in fermentation?
22. In addition to Mg^{2+}, the pyruvate dehydrogenase complex requires five cofactors. What are these cofactors, and what are their vitamin sources?
23. Why does utilization of glycogen as a source of glucose for glycolysis conserve more energy than direct entry of free glucose into the glycolytic pathway?
24. How is the reducing power of cytoplasmic NADH transferred to mitochondria?
25. A total of 36 moles of ATP can be produced by total metabolic oxidation of a mole of glucose. Indicate the moles of ATP produced by:
    (a) glycolysis
    (b) the citric acid cycle
    (c) the electron transport system
26. Summarize the differences between glycolysis and gluconeogenesis.
27. What is the net energy cost of gluconeogenesis?
28. What is the Cori cycle, and how is it beneficial to muscles?
29. What are the three major functions of the pentose phosphate pathway?
30. What is the primary control point in glycolysis? How is it responsive to cellular needs?
31. How does the cell regulate gluconeogenesis?
32. How are the actions of epinephrine and glucagon similar? How are they different?
33. Why are epinephrine and glucagon called "cyclic AMP" hormones?
34. Why is the activity of adenyl cyclase important to carbohydrate metabolism?
35. What is the major effect of insulin? How does this effect influence carbohydrate metabolism?
36. How do the structures of glucose and galactose differ? How might galactose be converted to glucose for entry into glycolysis?
37. How is regulation of oxygen-hemoglobin binding related to carbohydrate metabolism?
38. How does hexokinase deficiency increase the oxygen affinity of hemoglobin?
39. What effect would a deficiency of each of the following in red blood cells have on the oxygen affinity of hemoglobin?
    (a) enolase
    (b) phosphoglyceromutase
    (c) aldolase
40. Drinking methanol can be fatal, as it is converted to formaldehyde in the following reaction catalyzed by alcohol dehydrogenase:

$$CH_3-OH + NAD^- \rightleftharpoons H-C\!\!\underset{H}{\overset{O}{\diagup\!\!\diagdown}} + NADH + H^+$$

One treatment for methanol poisoning is administration of large quantities of ethanol. Can you explain why this treatment is effective? (Clue: Metabolism of both alcohols utilizes the same enzyme, alcohol dehydrogenase.)

CHAPTER 21

METABOLISM OF LIPIDS AND AMINO ACIDS

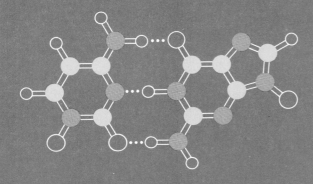

METABOLISM OF LIPIDS
- Absorption of Fats
- Blood Lipids
- Mobilization of Fatty Acids
- Fatty Acid Oxidation
- Energetics of Fatty Acid Oxidation
- Ketone Bodies
- Fatty Acid Synthesis
- Synthesis of Triacylglycerols and Other Lipids
- Cholesterol Metabolism
- Relationships Between Carbohydrate and Fatty Acid Metabolism
- Inborn Errors of Lipid Metabolism

METABOLISM OF AMINO ACIDS
- The Amino Acid Pool
- Amino Acid Degradation
- The Urea Cycle
- Fates of the Carbon Atoms from Amino Acids
- Amino Acid Synthesis
- Inborn Errors of Amino Acid Metabolism

METABOLIC INTERRELATIONSHIPS

Fatty acids and amino acids can be oxidized to provide energy to cells to supplement that derived from glucose. Thus, fatty acids and amino acids constitute additional energy sources. Fatty acids in the form of triacylglycerols represent a very compact storage form of energy, and about 9 kcal are available from one gram of fat. In contrast, carbohydrates and amino acids from proteins provide about 4 kcal per gram. In this chapter you will see that fatty acid and amino acid metabolism are related to carbohydrate metabolism in an elegant array of interchanges, an intricate network that allows derivation of energy and atoms from food molecules for use in cellular processes.

METABOLISM OF LIPIDS

Lipid metabolism generally includes digestion and absorption of lipids, oxidation of fatty acids, and synthesis of lipids. Lipid digestion was discussed briefly in chapter 18, but there are some special aspects of triacylglycerol absorption that are pertinent to the later topics in metabolism. Thus, we will begin with a discussion of absorption of fats.

Absorption of Fats

Digestion of triacylglycerols is accomplished primarily by the action of pancreatic lipase in the small intestine. However, triacylglycerol hydrolysis is incomplete, and a mixture of glycerol, fatty acids, monoacylglycerols, and diacylglycerols is produced.

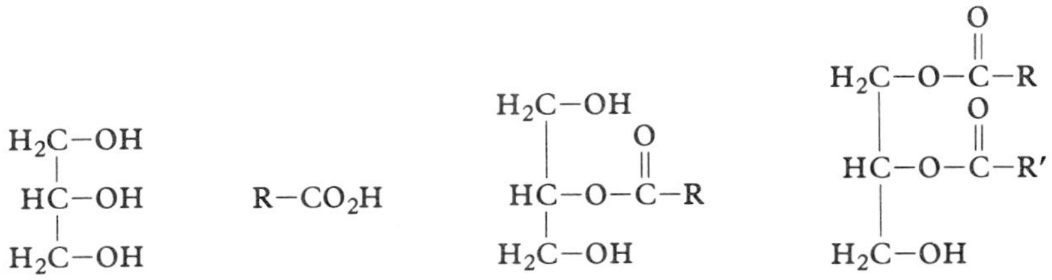

$$\begin{array}{cccc}
\text{H}_2\text{C}-\text{OH} & & \text{H}_2\text{C}-\text{OH} & \text{H}_2\text{C}-\text{O}-\overset{\overset{\text{O}}{\|}}{\text{C}}-\text{R} \\
| & & | & | \\
\text{HC}-\text{OH} & \text{R}-\text{CO}_2\text{H} & \text{HC}-\text{O}-\overset{\overset{\text{O}}{\|}}{\text{C}}-\text{R} & \text{HC}-\text{O}-\overset{\overset{\text{O}}{\|}}{\text{C}}-\text{R}' \\
| & & | & | \\
\text{H}_2\text{C}-\text{OH} & & \text{H}_2\text{C}-\text{OH} & \text{H}_2\text{C}-\text{OH} \\
\text{glycerol} & \text{fatty acid} & \text{monoacylglycerol} & \text{diacylglycerol}
\end{array}$$

Glycerol and short-chain fatty acids (C_{10} and shorter) are absorbed across the wall of the small intestine directly into the blood system. Long-chain fatty acids and most mono- and diacylglycerols are used to resynthesize triacylglycerols as they pass through the intestinal wall, and these triacylglycerols are absorbed into the *lymph system*. Shortly after a meal containing fat, the usually clear lymph fluid takes on a milky appearance due to the presence of tiny droplets composed of triacylglycerols and small amounts of protein. These droplets, called *chylomicra*, enter the blood by way of the *thoracic duct*.

It appears that any phospholipids in the diet are absorbed and reformed before entering the lymph in much the same way as triacylglycerols containing long-chain fatty acids. However, cholesterol requires no digestion, and it is absorbed directly into the lymph system. There some of it combines with fatty acids to form cholesteryl esters; a cholesteryl ester is illustrated in Figure 21-1.

lymph (limf) **system:** a network of vessels and tissue masses that helps maintain fluid balance in blood and tissues.

chylomicra (kye-low-MY-kra): plural of chylomicron.

thoracic (thoh-RAAH-sic) **duct:** the main trunk of the lymph system from which lymph components enter blood; the thoracic duct is located along the spinal column in the chest cavity.

Fig. 21-1 A Cholesteryl Ester (R is the hydrocarbon portion of a fatty acid.)

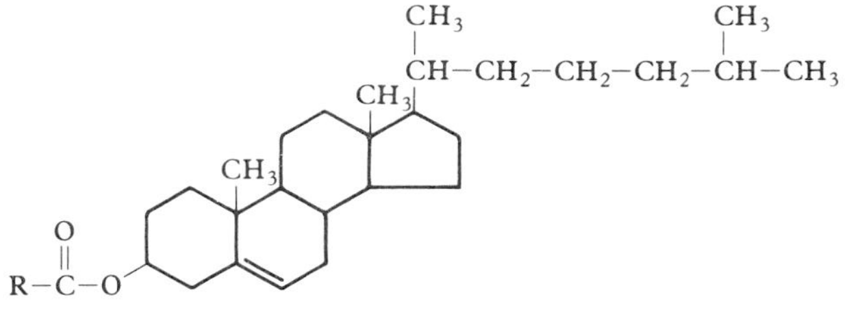

Blood Lipids

Since triacylglycerols and cholesteryl esters have limited solubility in water, an intricate mechanism exists for keeping them dispersed in blood. These molecules form micelles with a special group of plasma proteins, and the lipid-protein aggregates are called *lipoproteins*. The three types of lipoproteins are compared to chylomicra in Table 21-1.

Table 21-1 Major Classes of Human Plasma Lipoproteins

Property	Chylomicra	Very Low-Density (VLDL)	Low-Density (LDL)	High-Density (HDL)
Density, g/mL	less than 0.95	0.95–1.006	1.006–1.063	1.063–1.210
Diameter, nm	30–500	30–75	20–25	10–15
Amount, mg/100 mL plasma	100–250	130–200	210–400	50–130
Approximate composition, %				
Protein	2	9	21	33
Phosphoglyceride	7	18	22	29
Cholesterol:				
Free	2	7	8	7
Ester	6	15	38	23
Triacylglycerol	83	50	10	8
Fatty acids	0	1	1	0

The principal lipoproteins are classified by their densities. The very low-density lipoproteins (VLDLs) contain the largest amount of triacylglycerols. However, most of the cholesterol in the blood of a normal adult after an overnight fast is contained in the low-density lipoproteins (LDLs), formed by the breakdown of VLDLs. The third class of lipoproteins in blood, the high-density lipoproteins (HDLs), contains the most phospholipids. These three types of lipoproteins are responsible for the transport of lipids to tissue cells. There lipoprotein lipases catalyze the hydrolysis of the lipids, and their constituent parts may be metabolized or used to resynthesize triacylglycerols for storage.

The levels of lipids in the blood are often measured as part of a thorough physical examination. However, it is necessary to measure lipid levels after a period of fasting, usually overnight, because they rise and fall in response to diet. Figure 21-2 illustrates this response. Several hours after a meal containing 50–100 g of triacylglycerol, blood lipid levels rise from a normal 500–1000 mg/100 mL to a high level of perhaps over 2000 mg/100 mL. This level gradually falls back to normal within 6–8 hours. A portion of the total blood lipids is due to cholesterol, usually in the range of 130–260 mg/100 mL, and about 70% of the cholesterol is in the form of cholesteryl esters. Another 80–240 mg/100 mL is due to triacylglycerols, while the remainder is phosphoglycerides and fatty acids.

Above-normal levels of serum triacylglycerols and cholesterol have long been associated with the condition known as *atherosclerosis*, a disease in which plaques of fatty deposits form in the lining of the blood vessels. Possible consequences of this disease are high blood pressure, strokes, and heart attacks. However, recent findings suggest that abnormally high levels of cholesterol present as HDL may actually retard the formation of plaque in arterial linings. In addition, it has been found

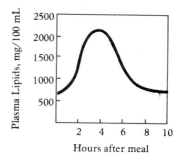

Fig. 21-2 The Rise in Blood Lipids Following a Meal Containing Fat

atherosclerosis: also called hardening of the arteries.

that men with very low levels of cholesterol in their blood have a higher rate of cancer of the colon than men with higher levels of cholesterol.

Phospholipids and cholesterol are not used directly as energy sources. Instead, they are degraded in the liver, and the products are used to synthesize other lipids needed by the body. Triacylglycerols, free fatty acids, and glycerol may be metabolized for energy, or they may be stored in the liver and *adipose* tissue as fat (see Figure 21-3). This reservoir of fat represents a very efficient storage form of energy, as each gram of fat is capable of producing approximately 9 kcal of energy. In contrast, carbohydrates and proteins are each responsible for producing about 4 kcal/g as a result of metabolic oxidations.

adipose: fatty; from the Latin *adipos*, meaning "fat."

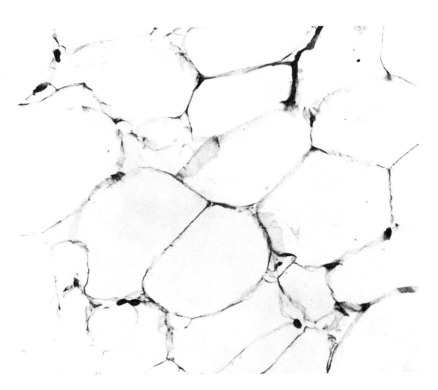

Fig. 21-3 Human Fat Cells Magnified 250 Times. Each cell contains a large globule of fat, with a thin rim of cytoplasm and a slightly bulging nucleus on one side.

Mobilization of Fatty Acids

Although glycogen supplies are able to provide energy needed by the body for a few hours after a meal, fasting depletes these supplies, and the fatty acids stored as triacylglycerols are then called upon. The release of fatty acids from adipose tissue is called *fatty acid mobilization*. In this process, lipases present in fat cells catalyze hydrolysis of triacylglycerols to fatty acids and glycerol in a three-step process summarized by the following equation:

$$\begin{array}{c}
\text{H}_2\text{C}-\text{O}-\overset{\text{O}}{\underset{\|}{\text{C}}}-\text{R} \\
| \\
\text{HC}-\text{O}-\overset{\text{O}}{\underset{\|}{\text{C}}}-\text{R}' + 3\text{H}_2\text{O} \\
| \\
\text{H}_2\text{C}-\text{O}-\overset{\text{O}}{\underset{\|}{\text{C}}}-\text{R}''
\end{array}
\longrightarrow
\begin{array}{c}
\text{H}_2\text{C}-\text{OH} \\
| \\
\text{HC}-\text{OH} \\
| \\
\text{H}_2\text{C}-\text{OH}
\end{array}
+
\begin{array}{c}
\text{R}-\overset{\text{O}}{\underset{\|}{\text{C}}}-\text{OH} \\
+ \\
\text{R}'-\overset{\text{O}}{\underset{\|}{\text{C}}}-\text{OH} \\
+ \\
\text{R}''-\overset{\text{O}}{\underset{\|}{\text{C}}}-\text{OH}
\end{array}$$

The fatty acids released from adipose tissue are transported by serum albumin to other parts of the body. Glycerol formed from hydrolysis of the triacylglycerols in fat cells is converted to dihydroxyacetone phosphate in a two-step process:

$$\begin{array}{c}
\text{H}_2\text{C}-\text{OH} \\
| \\
\text{HC}-\text{OH} \\
| \\
\text{H}_2\text{C}-\text{OH}
\end{array} + \text{ATP} \xrightarrow{\text{glycerol kinase}}
\begin{array}{c}
\text{H}_2\text{C}-\text{OH} \\
| \\
\text{HC}-\text{OH} \\
| \\
\text{H}_2\text{C}-\text{O}-\overset{\text{O}}{\underset{\|}{\text{P}}}-\text{O}^- \\
| \\
\text{O}^-
\end{array} + \text{ADP} + \text{H}^+$$

glycerol　　　　　　　　　　glycerol-3-phosphate

$$\begin{array}{c}
\text{H}_2\text{C}-\text{OH} \\
| \\
\text{HC}-\text{OH} \\
| \\
\text{H}_2\text{C}-\text{O}-\overset{\text{O}}{\underset{\|}{\text{P}}}-\text{O}^- \\
| \\
\text{O}^-
\end{array} + \text{NAD}^+ \xrightarrow{\text{glycerol phosphate dehydrogenase}}
\begin{array}{c}
\text{H}_2\text{C}-\text{OH} \\
| \\
\text{C}=\text{O} \\
| \\
\text{H}_2\text{C}-\text{O}-\overset{\text{O}}{\underset{\|}{\text{P}}}-\text{O}^- \\
| \\
\text{O}^-
\end{array} + \text{NADH} + \text{H}^+$$

glycerol-3-phosphate　　　　　　　dihydroxyacetone phosphate

This compound can then be used in gluconeogenesis or glycolysis to produce glucose or pyruvate, respectively.

The activity of adipose-cell lipase is regulated by four hormones: epinephrine, norepinephrine, glucagon, and adrenocorticotropic hormone (ACTH, or corticotropin). All of these are "cyclic AMP" hormones (see chapter 18), and the increased levels of cAMP activate the lipase indirectly by causing the transfer of phosphate from ATP to an active-site residue in the enzyme. Thus when energy is needed, these hormones act on fat cells to bring about mobilization of fatty acids to be used in energy production.

Because fatty acids are mostly nonpolar structures, they cross cell membranes easily and thus do not need transport systems to enter cells. The ease of entry into cells by fatty acids means that most cells oxidize fatty acids in preference to other molecules, sugars for example, that require transport systems. Two exceptions are red blood cells and brain cells. Red blood cells cannot oxidize fatty acids because fatty acid oxidation is a mitochondrial process, and they have no mitochondria. Brain cells are bathed in cerebrospinal fluid and thus do not obtain nutrients directly from blood. Glucose and many other substances are permitted to enter cerebrospinal fluid, but fatty acids bound to serum albumin are prevented entry. For these reasons, red blood cells and brain cells depend on blood sugar for their primary source of energy.

EXAMPLE 21.1

Why is lymph fluid sometimes cloudy after a meal?

SOLUTION

Lymph fluid will become cloudy after a meal containing fat because chylomicra (tiny droplets of triacylglycerols and protein) are dispersed in the fluid before entry into the blood system.

EXAMPLE 21.2

What mechanism exists for dispersing slightly soluble lipids in blood?

SOLUTION

Lipids having limited solubility are dispersed in blood as lipid-protein aggregates called lipoproteins.

EXAMPLE 21.3

What is the role of adipose-cell lipase in fatty acid mobilization?

SOLUTION

Adipose-cell lipase is the enzyme that catalyzes hydrolysis of stored triacylglycerols to release fatty acids from adipose cells.

Fatty Acid Oxidation

Fatty acids from dietary sources or adipose tissue are oxidized in the mitochondrial matrix by the β-oxidation pathway. This metabolic cycle accepts fatty acid chains attached to coenzyme A and removes two-carbon units from the carboxyl end of the fatty acid chains. Each turn of the cycle produces an acetyl CoA molecule, and the resulting fatty acid chain is recycled until it is eventually degraded entirely into acetyl units attached to coenzyme A. Only fatty acid chains attached to coenzyme A can participate in the β-oxidation pathway, and localization of the pathway within the mitochondria channels the acetyl CoA product directly into the citric acid cycle.

Activation of the Fatty Acid

Fatty acids can diffuse across cell membranes into the cytoplasm, but entry into the β-oxidation pathway requires that they be attached to coenzyme A. In this step, the fatty acid reacts with coenzyme A to form a thioester. This reaction is catalyzed by acyl CoA synthetase, and it is referred to as activation of the fatty acid because the thioester is a high-energy compound. The energy needed for its synthesis is provided by hydrolysis of ATP to AMP and PP_i and subsequent hydrolysis of PP_i,

$$R-CH_2-CH_2-\overset{O}{\underset{\|}{C}}-O^- + \textbf{ATP} + HS-CoA \longrightarrow R-CH_2-CH_2-\overset{O}{\underset{\|}{C}}-S-CoA + AMP + PP_i + H^+$$

$$PP_i + H_2O \longrightarrow 2 P_i$$

The large CoA group prevents the fatty acyl CoA molecules from leaving the cell, but it also blocks their entry into the mitochondria. This problem is overcome by transfer of the acyl units to a carrier which transports them across the mitochondrial membrane, and they are reattached to coenzyme A inside the mitochondria. Typically, the fatty acid will have 16 or 18 carbons.

The β-Oxidation Cycle

After activation, the fatty acyl CoA molecules enter the β-oxidation cycle for degradation. The individual reactions are discussed in the following paragraphs. You will find that the first three steps of the cycle are very similar to the last three reactions of the citric acid cycle (see chapter 19). The β-oxidation cycle is summarized in Figure 21-4, and Table 21-2 lists the steps in the cycle.

Reaction 1. Dehydrogenation. The first step of the cycle involves dehydrogenation of the fatty acid chain. This reaction, which requires FAD as coenzyme, introduces a double bond between the α- and β-carbons. The enzyme is acyl CoA dehydrogenase.

$$R-\overset{\beta}{C}H_2-\overset{\alpha}{C}H_2-\overset{O}{\underset{\|}{C}}-S-CoA + FAD \longrightarrow R-\overset{\beta}{C}H=\overset{\alpha}{C}H-\overset{O}{\underset{\|}{C}}-S-CoA + FADH_2$$

acyl CoA $\qquad\qquad\qquad\qquad\qquad\qquad$ β-enoyl CoA

Table 21-2 Reactions of β-Oxidation

Step	Reaction	Enzyme	Cofactor
1	Dehydrogenation: $RCH_2CH_2\overset{O}{\underset{\|}{C}}-SCoA + FAD \longrightarrow RCH=CH-\overset{O}{\underset{\|}{C}}-SCoA + FADH_2$	Acyl CoA dehydrogenase	FAD
2	Hydration: $RCH=CH-\overset{O}{\underset{\|}{C}}-SCoA + H_2O \rightleftharpoons R\overset{OH}{\underset{\|}{C}}HCH_2\overset{O}{\underset{\|}{C}}-SCoA$	Enoyl CoA hydratase	None
3	Dehydrogenation: $R\overset{OH}{\underset{\|}{C}}HCH_2\overset{O}{\underset{\|}{C}}-SCoA + NAD^+ \rightleftharpoons R\overset{O}{\underset{\|}{C}}CH_2\overset{O}{\underset{\|}{C}}-SCoA + NADH + H^+$	β-Hydroxyacyl CoA dehydrogenase	NAD^+
4	Release of the acetyl unit: $R\overset{O}{\underset{\|}{C}}CH_2\overset{O}{\underset{\|}{C}}-SCoA + HS-CoA \rightleftharpoons R\overset{O}{\underset{\|}{C}}-SCoA + CH_3\overset{O}{\underset{\|}{C}}-SCoA$	β-Ketothiolase	CoA

Fig. 21-4 The β-Oxidation Cycle: Degradation of Palmitate to Acetyl Coenzyme A by the β-Oxidation Pathway

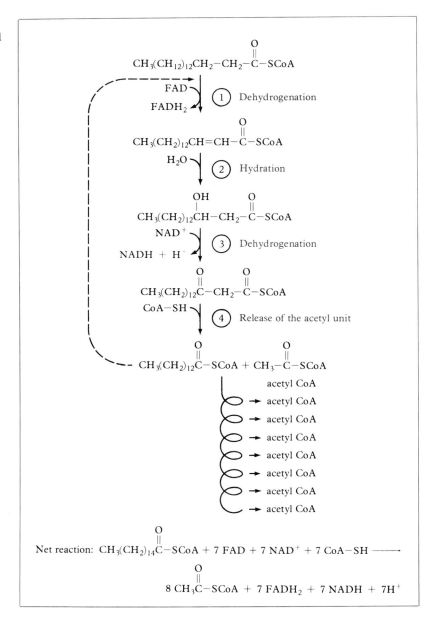

Reaction 2. Hydration. In the second reaction, catalyzed by enoyl CoA hydratase, water is added to the carbon-carbon double bond to produce a secondary alcohol.

$$R-\underset{\beta}{CH}=\underset{\alpha}{CH}-\overset{O}{\underset{\|}{C}}-S-CoA + H_2O \rightleftharpoons R-\underset{\beta}{\overset{OH}{\underset{|}{CH}}}-\underset{\alpha}{CH_2}-\overset{O}{\underset{\|}{C}}-S-CoA$$

β-enoyl CoA   β-hydroxyacyl CoA

Reaction 3. Dehydrogenation. In the third reaction, the β-hydroxy group is oxidized to a β-ketone, with β-hydroxyacyl CoA dehydrogenase as the enzyme and NAD^+ as coenzyme. It is this utilization of the β-carbon that gives the β-oxidation pathway its name.

$$R-\underset{\beta}{CH}(OH)-\underset{\alpha}{CH_2}-C(=O)-S-CoA + NAD^+ \rightleftharpoons R-\underset{\beta}{C}(=O)-\underset{\alpha}{CH_2}-C(=O)-S-CoA + NADH + H^+$$

β-hydroxyacyl CoA   β-ketoacyl CoA

Reaction 4. Release of the Acetyl Unit. In this fourth and final step of the β-oxidation pathway, an acetyl unit is removed from the carboxyl end of the fatty acid chain as acetyl CoA. The reaction, catalyzed by thiolase (more properly called β-ketothiolase), also produces a fatty acyl CoA molecule that has been shortened by two carbon atoms. This molecule can follow the cyclic pathway again and again until it is degraded completely into acetyl CoA units which serve as fuel for the citric acid cycle.

$$R-\underset{\beta}{C}(=O)-CH_2-C(=O)-S-CoA + HS-CoA \rightleftharpoons R-C(=O)-S-CoA + CH_3-C(=O)-S-CoA$$

β-ketoacyl CoA   acyl CoA   acetyl CoA

EXAMPLE 21.4

Which steps in the β-oxidation cycle generate reduced coenzymes?

SOLUTION

Step 1 (dehydrogenation) produces one molecule of $FADH_2$, and step 3 (dehydrogenation) produces one molecule of NADH.

Energetics of Fatty Acid Oxidation

Each turn of the β-oxidation cycle produces one $FADH_2$ and one NADH, which can pass their electrons to the electron transport system to produce a total of five molecules of ATP. In addition, each unit of acetyl CoA fed into the citric acid cycle is ultimately responsible for formation of 12 ATP molecules. With these facts in mind, we can calculate the overall ATP yield from the oxidation of a fatty acid. For example, palmitic acid is a principal fatty acid in our diets. Since there are 16 carbon atoms in palmitic acid, it must pass through the β-oxidation cycle many times. Each turn produces a two-carbon acetyl CoA unit except the last turn, which forms two acetyl CoA units. Thus, a fatty acid containing n carbon atoms passes through $n/2 - 1$ times. For palmitic acid, this amounts to 7 turns of the cycle. Since each turn produces 5 ATP molecules, then β-oxidation is responsible for production of 35 ATP molecules. In addition, 8 acetyl CoA units are produced, each of which is consumed by the citric acid cycle. There each acetyl CoA results ultimately in the formation of 12 ATP molecules. Hence, there will be a total of 96 ATP molecules produced via the citric acid cycle. At this point, we have a total of 35 + 96 ATP molecules, or 131. However, we must not forget that one ATP and one pyrophosphate were used to activate the fatty acid. If we consider these equivalent to 2 ATP molecules, then we must subtract this quantity from 131, giving an overall ATP production of 129 moles from one mole of palmitic acid. These calculations are summarized in Table 21-3. This is equivalent to 129 moles ATP $\times$ 7.3 kcal/mole, or 942 kcal of energy. Oxidation of one mole of palmitic acid by combustion produces 2240 kcal of energy. Thus, our bodies accomplish oxidation of palmitic acid at an efficiency of 42.1% (942 kcal/2240 kcal = 42.1%).

Table 21-3 ATP Production from Complete Oxidation of Palmitic Acid, $CH_3(CH_2)_{14}CO_2H$

7 turns of the β-oxidation cycle (7 $\times$ 5 ATP)	+ 35 ATP
8 turns of the citric acid cycle (8 $\times$ 12 ATP)	+ 96 ATP
activation of palmitic acid	− 2 ATP
Total	+129 ATP

EXAMPLE 21.5

How many moles of ATP are produced by complete oxidation of one mole of stearic acid, the 18-carbon saturated fatty acid?

SOLUTION

Since there are 18 carbons in stearic acid, 8 turns of the β-oxidation cycle ($(18/2) - 1 = 8$) will be needed to degrade it completely to acetyl units. Since each turn of the cycle ultimately results in formation of 5 moles of ATP, then β-oxidation accounts for 40 moles of ATP (8 turns $\times$ 5 ATP/turn). Additionally, 9 acetyl units are produced by β-oxidation, each of which results in 12 moles of ATP *via* the citric acid cycle and oxidative phosphorylation. Thus, oxidation of the acetyl units provides 108 moles of ATP. We now have a total of 148 moles of ATP (40 moles ATP + 108 moles ATP). When we subtract 2 moles of ATP needed for activation, we arrive at a net production of 146 moles of ATP for each mole of stearic acid oxidized:

8 turns of the β-oxidation cycle	+ 40 ATP
9 turns of the citric acid cycle	+108 ATP
activation of stearic acid	− 2 ATP
	+146 ATP

Ketone Bodies

A normal, healthy individual usually consumes a diet that is balanced in carbohydrates and fats as energy sources. Much of the acetyl CoA produced by fatty acids is processed through the citric acid cycle. However, a small amount is used to synthesize cholesterol, steroid hormones, and bile salts, and some is used to produce acetoacetate, a water-soluble transportable form of acetyl units. A summary equation for a series of reactions that results in the formation of acetoacetate can be written:

$$2\ CH_3-\underset{acetyl\ CoA}{\overset{O}{\underset{\|}{C}}-S-CoA} + H_2O \longrightarrow \underset{acetoacetate}{CH_3-\overset{O}{\underset{\|}{C}}-CH_2-\overset{O}{\underset{\|}{C}}-O^-} + 2\ CoA-SH + H^+$$

Under certain conditions, the balance between carbohydrate and fatty acid metabolism becomes disturbed. For example, when an individual fasts for a long period of time, as in starvation, fatty acids from stored fats become the primary energy source. A similar situation exists in untreated diabetes. In these cases, glycolysis is minimal, and oxaloacetate from the citric acid cycle is used for gluconeogenesis to a greater than normal extent as the cells struggle to make their own glucose. Thus, the activity of the citric acid cycle is diminished due to low concentrations of oxaloacetate. These conditions prevail even though fatty acid oxidation will proceed at a normal rate as long as there are fat supplies. The consequence of these events is that more acetyl CoA is produced by fatty acid oxidation than can be used by the citric acid cycle. As the concentration of acetyl CoA builds up, the excess is converted to acetoacetate, which then forms acetone and β-hydroxybutyrate:

$$\underset{acetoacetate}{CH_3-\overset{O}{\underset{\|}{C}}-CH_2-\overset{O}{\underset{\|}{C}}-O^-} + H^+ \longrightarrow \underset{acetone}{CH_3-\overset{O}{\underset{\|}{C}}-CH_3} + CO_2$$

$$\underset{acetoacetate}{CH_3-\overset{O}{\underset{\|}{C}}-CH_2-\overset{O}{\underset{\|}{C}}-O^-} + NADH + H^+ \longrightarrow \underset{\beta\text{-hydroxybutyrate}}{CH_3-\overset{OH}{\underset{|}{C}H}-CH_2-\overset{O}{\underset{\|}{C}}-O^-} + NAD^+$$

Traditionally, acetoacetate, acetone, and β-hydroxybutyrate have been referred to as *ketone bodies* even though one of them is an alcohol instead of a ketone.

These compounds are normally present in blood in small amounts, but when their levels become abnormally high, the condition is called *ketonemia*. Since these ketone bodies are excreted in the urine, *ketonuria* usually accompanies ketonemia. Occasionally, the level of acetone in the blood is high enough that it is expelled through the lungs, and the odor of acetone can be detected in the breath. When ketonemia, ketonuria, and "acetone breath" exist simultaneously, the condition is called *ketosis*.

As the ketone bodies are removed from circulation by the urine, the anions are accompanied by sodium cations, and removal of excessive amounts of these ketone bodies leaves behind large quantities of hydrogen ions in blood. As ketosis becomes gradually worse, a condition of acidosis sets in, and in this case we apply the term *ketoacidosis*. If ketosis is not controlled, the person becomes severely dehydrated because the

ketonemia: abnormally high levels of ketone bodies in the blood.

ketonuria: abnormally high levels of ketone bodies in the urine.

kidneys excrete too much water in attempting to rid the body of excess acid. Prolonged ketosis, as in uncontrolled diabetes, leads to general debilitation, coma, and even death.

EXAMPLE 21.6

During starvation and uncontrolled diabetes, the brain adapts to utilization of acetoacetate for fuel, but not fatty acids. The reason is related to transport. Can you explain why this adaptation takes place?

SOLUTION

Circulating fatty acids are attached to serum albumin, and this complex cannot enter the cerebrospinal fluid surrounding brain cells. However, because acetoacetate is water-soluble, it enters cerebrospinal fluid and hence brain cells.

Fatty Acid Synthesis

Fatty acids and fats are synthesized when an organism takes in more nutrient molecules than are needed to satisfy energy requirements. Most of this synthesis occurs in the liver, adipose tissue, and mammary glands; the mammary glands become especially active in fatty acid synthesis during lactation (milk production).

As was the case with glycolysis and gluconeogenesis, the opposing pathways for fatty acid degradation and fatty acid synthesis are separate from each other. Not only do they utilize different enzyme systems, but they are also located in different cellular compartments. Degradation occurs by the β-oxidation pathway located in mitochondria, whereas biosynthesis of fatty acids occurs in the cytoplasm. Thus, the two pathways can operate at the same time and they can be controlled separately. The major contrasts between fatty acid synthesis and fatty acid degradation are:

1. The intermediates of fatty acid synthesis are attached to sulfhydryl groups of an *acyl carrier protein* (ACP), whereas the intermediates in fatty acid degradation are bonded to coenzyme A.

2. The enzymes of fatty acid synthesis are organized into a multienzyme complex called the *fatty acid synthetase* system. However, the degradative enzymes are not associated.

3. The growing fatty acid chain is elongated by addition of two-carbon units derived from acetyl CoA; however, a molecule called *malonyl-ACP* transfers these units from acetyl CoA to the growing fatty acid chain. Thus, acetyl CoA is only indirectly involved, as contrasted to its direct involvement in fatty acid breakdown.

4. The coenzyme used in the reduction reactions of fatty acid synthesis is the phosphorylated form, NADPH, instead of nonphosphorylated NAD^+ involved in fatty acid degradation.

METABOLISM OF LIPIDS

Table 21-4 Reactions of Fatty Acid Synthesis

Step	Reaction	Enzyme	Cofactor
1	Carbon-carbon bond formation: $CH_3\overset{O}{\overset{\|}{C}}-S-ACP + {}^-O\overset{O}{\overset{\|}{C}}CH_2\overset{O}{\overset{\|}{C}}-S-ACP + H^+ \longrightarrow CH_3\overset{O}{\overset{\|}{C}}CH_2\overset{O}{\overset{\|}{C}}-S-ACP + CO_2 + ACP-SH$	Acyl malonyl-ACP condensing enzyme	None
2	Hydrogenation: $CH_3\overset{O}{\overset{\|}{C}}CH_2\overset{O}{\overset{\|}{C}}-S-ACP + NADPH + H^+ \rightleftharpoons CH_3\overset{OH}{\overset{\|}{C}}HCH_2\overset{O}{\overset{\|}{C}}-S-ACP + NADP^+$	β-Ketoacyl-ACP reductase	NADPH
3	Dehydration: $CH_3\overset{OH}{\overset{\|}{C}}HCH_2\overset{O}{\overset{\|}{C}}-S-ACP \rightleftharpoons CH_3CH=CH\overset{O}{\overset{\|}{C}}-S-ACP + H_2O$	β-Hydroxyacyl-ACP dehydratase	None
4	Hydrogenation: $CH_3CH=CH\overset{O}{\overset{\|}{C}}-S-ACP + NADPH + H^+ \longrightarrow CH_3CH_2CH_2\overset{O}{\overset{\|}{C}}-S-ACP + NADP^+$	Enoyl-ACP reductase	NADPH

Acetyl CoA derived from carbohydrate or amino acid sources is the ultimate precursor of all the carbon atoms of fatty acids. When acetyl CoA produced by glycolysis and/or amino acid degradation is not needed by the citric acid cycle, it is channeled into fatty acid synthesis. Since acetyl CoA is generated inside mitochondria by the pyruvate dehydrogenase complex (chapter 20), it must first be transported into the cytoplasm before it can be used to synthesize fatty acids. Thus, acetyl CoA condenses with oxaloacetate, in the first step of the citric acid cycle (chapter 19) to produce citrate.

$$\text{acetyl CoA} + \text{oxaloacetate} + H_2O \longrightarrow \text{citrate} + \text{CoA-SH}$$

The mitochondrial membrane has a citrate carrier system that allows excess citrate to make its way into the cytoplasm, where it is cleaved to regenerate acetyl CoA and oxaloacetate.

In a preliminary step before actual fatty acid synthesis begins, acetyl CoA in the cytoplasm reacts with CO_2 to form *malonyl CoA*. This reaction is catalyzed by an allosteric enzyme, acetyl CoA carboxylase, which requires the vitamin biotin as cofactor. The enzyme is activated by high levels of citrate, a signal to synthesize fatty acids, and it is inhibited by long-chain acyl CoA molecules, the end products of fatty acid synthesis. The energy for this nonspontaneous reaction is provided by ATP hydrolysis.

$$\underset{\text{acetyl CoA}}{CH_3-\overset{O}{\overset{\|}{C}}-S-CoA} + CO_2 + ATP + H_2O \longrightarrow \underset{\text{malonyl CoA}}{{}^-O_2C-CH_2-\overset{O}{\overset{\|}{C}}-S-CoA} + ADP + P_i + 2H^+$$

After formation of malonyl CoA, fatty acid synthesis occurs in a rather complex series of reactions catalyzed by the multienzyme complex called the *fatty acid synthetase system*. This aggregate contains six enzymes and an additional protein, *acyl carrier protein* (ACP), to which all intermediates are attached by thioester linkage. The four major reactions involved in fatty acid synthesis are listed in Table 21-4, and the summary equation is given below.

$$\underset{\text{acetyl-ACP}}{CH_3-\overset{O}{\underset{\|}{C}}-S-ACP} + \underset{\text{malonyl-ACP}}{^-O_2C-CH_2-\overset{O}{\underset{\|}{C}}-S-ACP} + 2\,NADPH + 3\,H^+ \longrightarrow$$

$$\underset{\text{acyl-ACP}}{CH_3-CH_2-CH_2-\overset{O}{\underset{\|}{C}}-S-ACP} + CO_2 + ACP-SH + 2\,NADP^+ + H_2O$$

The four major reactions of fatty acid synthesis are analogous to those in β-oxidation but in reverse sequence. First, a carbon-carbon single bond is formed. Next, hydrogenation (reduction) occurs, followed by dehydration and a second hydrogenation (reduction). This cycle is repeated over and over again, and each turn results in addition of a 2-carbon unit, obtained from malonyl CoA, to the growing chain. The major product is palmitic acid (16 carbons), although other fatty acids are also formed. The summary equation for synthesis of palmitate from acetyl CoA

$$\underset{\text{acetyl CoA}}{8\,CH_3-\overset{O}{\underset{\|}{C}}-S-CoA} + 14\,NADPH + 6\,H^+ + 7\,ATP + H_2O \longrightarrow$$

$$\underset{\text{palmitate}}{CH_3(CH_2)_{14}CO_2^-} + 8\,CoA-SH + 14\,NADP^+ + 7\,ADP + 7\,P_i$$

demonstrates that a great deal of energy, 7 ATP + 14 NADPH (equivalent to 42 ATP), is required. This large amount of energy stored in fatty acids is what makes it so difficult to lose excess weight due to fat.

In humans and other mammals, most unsaturated fatty acids are those having only one double bond: palmitoleic acid and oleic acid. In these structures, introduction of the double bond occurs after the chain is synthesized. However, linoleic acid (two double bonds) and linolenic acid (three double bonds) in the diet can be converted to other polyunsaturated fatty acids by chain lengthening and introduction of additional double bonds. This utilization of linoleic acid and linolenic acid to form other polyunsaturated fatty acids is the basis of their classification as essential fatty acids for humans (see chapter 14).

EXAMPLE 21.7

How is fatty acid synthesis regulated?

SOLUTION

The preliminary reaction before fatty acid synthesis starts is formation of malonyl CoA from acetyl CoA and CO_2. This reaction is catalyzed by an allosteric enzyme, acetyl CoA carboxylase, which is activated by high levels of citrate and inhibited by long-chain acyl CoA molecules, end products of fatty acid synthesis.

EXAMPLE 21.8

What vitamin is involved in fatty acid synthesis?

SOLUTION

The vitamin biotin is the required cofactor for acetyl CoA carboxylase.

Synthesis of Triacylglycerols and Other Lipids

The body has two sources of fatty acids: diet and internal synthesis. Those fatty acids not needed for membranes and other materials are used to make storage fat, otherwise known as triacylglycerols. The source of the glycerol needed for triacylglycerols is dihydroxyacetone phosphate, the glycolytic intermediate obtained from glucose. Insulin facilitates entry of glucose into fat cells and thus stimulates fat synthesis by formation of both dihydroxyacetone phosphate needed for glycerol and acetyl CoA needed for fatty acid synthesis.

The fatty acyl groups are added one at a time to glycerol by use of three different enzymes. Usually a saturated fatty acyl group is added first, and then two unsaturated fatty acyl groups are added. The triacylglycerols are then stored in fat cells, each of which contains a huge droplet of triacylglycerol (Figure 21-3).

Specialized lipids such as glycolipids and phospholipids are synthesized in the liver by a series of complex reactions. The liver also synthesizes fats, oils, and waxes having special functions, such as ear wax and the oil in skin and hair.

Cholesterol Metabolism

Cholesterol plays an essential role in steroid metabolism, as illustrated in Figure 21-5. All of the carbon atoms in cholesterol are derived from acetyl CoA, as indicated in Figure 21-6. We will not go into cholesterol synthesis in detail, but Figure 21-6 gives structures of important intermediates. Cholesterol then serves as the basis for biosynthesis of the bile acids, vitamin D, and the steroid hormones.

The major site of cholesterol synthesis is the liver. Here regulation occurs at two different steps in the synthetic route. The first control point is at the formation of *mevalonic acid*. Cholesterol itself inhibits the synthesis and the activity of the enzyme that catalyzes this step, so that dietary cholesterol may serve to inhibit cholesterol biosynthesis. However, there is some question as to how efficient dietary restrictions are in lowering total cholesterol. A second control point exists at the cyclization of *squalene* to form *lanosterol*.

Fig. 21-5 The Role of Cholesterol in Steroid Metabolism

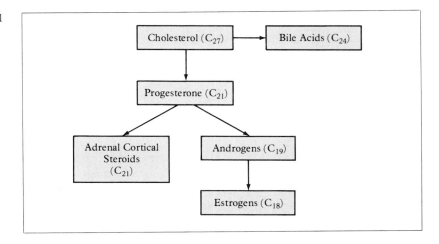

Fig. 21-6 Outline of the Biosynthesis of Cholesterol Showing the Principal Intermediates: Mevalonic Acid, Squalene, and Lanosterol

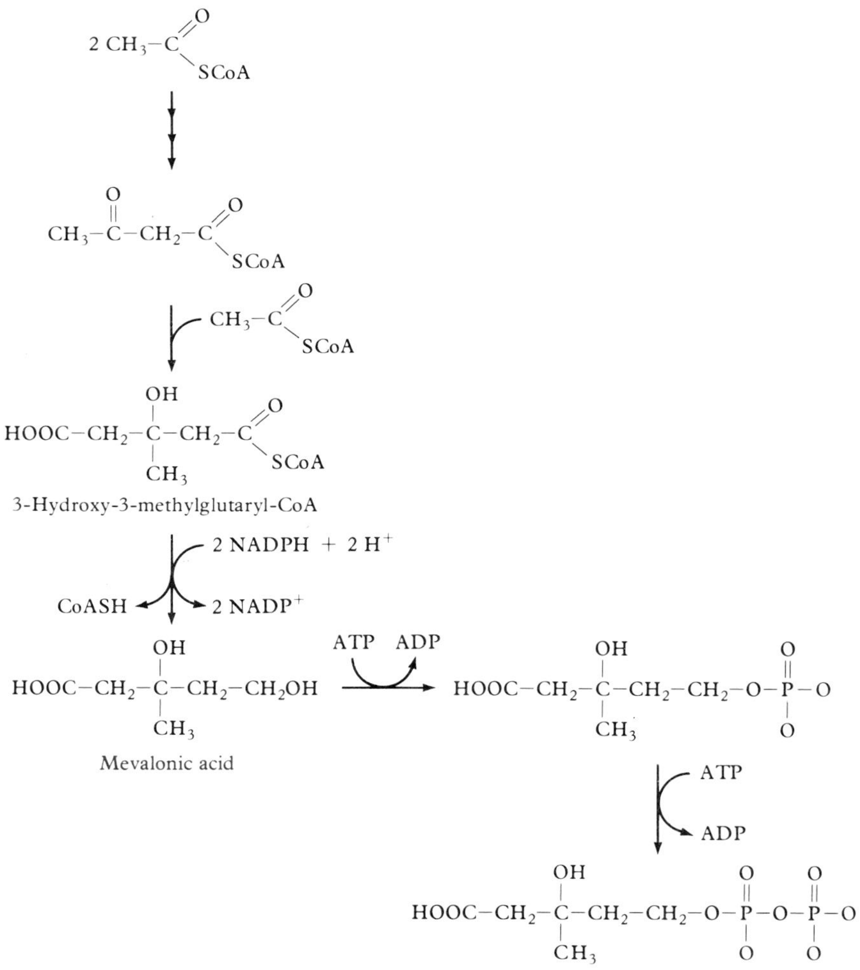

$$2\ CH_3-C\overset{O}{\underset{SCoA}{}}$$

$$CH_3-\overset{O}{\underset{}{C}}-CH_2-C\overset{O}{\underset{SCoA}{}}$$

$$CH_3-C\overset{O}{\underset{SCoA}{}}$$

$$HOOC-CH_2-\overset{OH}{\underset{CH_3}{C}}-CH_2-C\overset{O}{\underset{SCoA}{}}$$

3-Hydroxy-3-methylglutaryl-CoA

2 NADPH + 2 H$^+$

CoASH ⟵ ⟶ 2 NADP$^+$

$$HOOC-CH_2-\overset{OH}{\underset{CH_3}{C}}-CH_2-CH_2OH \quad \xrightarrow{ATP\ \ ADP} \quad HOOC-CH_2-\overset{OH}{\underset{CH_3}{C}}-CH_2-CH_2-O-\overset{O}{\underset{O}{P}}-O$$

Mevalonic acid

ATP ⟶ ADP

$$HOOC-CH_2-\overset{OH}{\underset{CH_3}{C}}-CH_2-CH_2-O-\overset{O}{\underset{O}{P}}-O-\overset{O}{\underset{O}{P}}-O$$

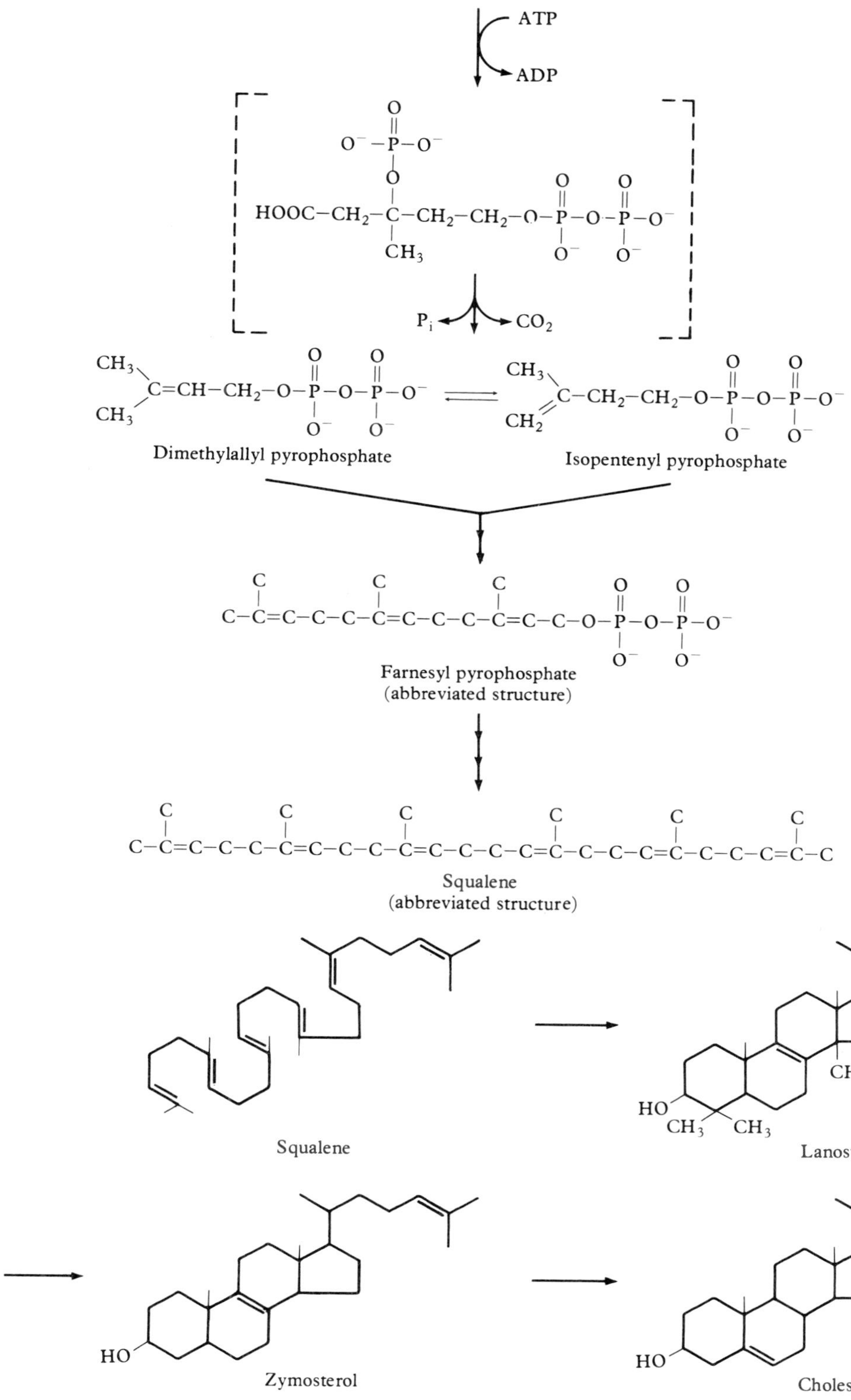

METABOLISM OF LIPIDS

[Figure: Biosynthesis of cholesterol from 3-phospho-5-pyrophosphomevalonate through isoprenoid intermediates]

3-Phospho-5-pyrophosphomevalonate (in brackets, formed from ATP → ADP) loses P_i and CO_2 to yield:

Dimethylallyl pyrophosphate ⇌ **Isopentenyl pyrophosphate**

↓

Farnesyl pyrophosphate (abbreviated structure)

↓

Squalene (abbreviated structure)

Squalene → Lanosterol → Zymosterol → Cholesterol

Another site of cholesterol synthesis is in the intestinal wall; here, formation of mevalonic acid is sensitive not to cholesterol levels but to cholic acid (chapter 18) levels.

An additional regulatory mechanism exists in which changes in the level of blood cholesterol have been observed in response to changes in the amounts of dietary saturated fatty acids. The more saturated fatty acids there are in the diet, the higher is the serum cholesterol level. The explanation for this response is not known.

EXAMPLE 21.9

Describe four ways in which cholesterol biosynthesis is regulated.

SOLUTION

1. The synthesis and activity of the enzyme needed for mevalonic acid synthesis in the liver is inhibited by cholesterol itself.

2. A second control point under the influence of cholesterol in the liver is the cyclization of squalene to form lanosterol.

3. Cholesterol synthesis in the intestinal wall is inhibited by high levels of cholic acid.

4. Saturated fatty acids in the diet stimulate production of cholesterol by an unknown mechanism.

Relationships Between Carbohydrate and Fatty Acid Metabolism

As was mentioned earlier, glucose can be a source of carbon atoms used for fatty acid synthesis. Figure 21-7 illustrates this flow of carbon atoms. However, it should be pointed out that a molecule of glucose (6 carbons) forms two molecules of pyruvate (3 carbons), but each pyruvate molecule loses a carbon atom as CO_2 in forming acetyl CoA. Thus, two of the original six carbons in glucose are lost as two molecules of CO_2. The remaining four atoms from glucose make two molecules of acetyl CoA which then can be used in fatty acid synthesis.

If the carbon atoms derived from fatty acid oxidation are traced through acetyl CoA and glucose synthesis, it can be demonstrated that carbon atoms 1, 2, 5, and 6 of glucose came from fatty acids. However, there is no net formation of new glucose from the two carbon atoms of acetyl CoA. This is because two carbon atoms are lost from the citric acid cycle for each acetyl unit that enters. The oxaloacetate regenerated in each turn of the citric acid cycle contains carbon atoms from the acetyl group, but the input of two carbon atoms into the citric acid cycle is always balanced by the output of two carbon atoms as CO_2.

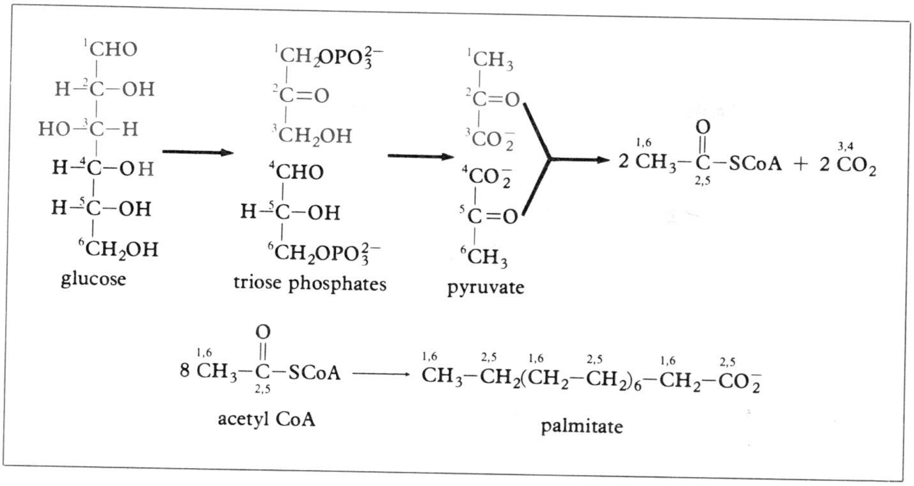

$$\underset{\text{glucose}}{\begin{array}{c}{}^1\text{CHO}\\ \text{H}-{}^2\text{C}-\text{OH}\\ \text{HO}-{}^3\text{C}-\text{H}\\ \text{H}-{}^4\text{C}-\text{OH}\\ \text{H}-{}^5\text{C}-\text{OH}\\ {}^6\text{CH}_2\text{OH}\end{array}} \longrightarrow \underset{\text{triose phosphates}}{\begin{array}{c}{}^1\text{CH}_2\text{OPO}_3^{2-}\\ {}^2\text{C}=\text{O}\\ {}^3\text{CH}_2\text{OH}\\ {}^4\text{CHO}\\ \text{H}-{}^5\text{C}-\text{OH}\\ {}^6\text{CH}_2\text{OPO}_3^{2-}\end{array}} \longrightarrow \underset{\text{pyruvate}}{\begin{array}{c}{}^1\text{CH}_3\\ {}^2\text{C}=\text{O}\\ {}^3\text{CO}_2^-\\ {}^4\text{CO}_2^-\\ {}^5\text{C}=\text{O}\\ {}^6\text{CH}_3\end{array}} \longrightarrow 2\ \overset{1,6}{\text{CH}_3}-\underset{2,5}{\overset{\text{O}}{\overset{\|}{\text{C}}}}-\text{SCoA} + 2\ \overset{3,4}{\text{CO}_2}$$

$$\underset{\text{acetyl CoA}}{8\ \overset{1,6}{\text{CH}_3}-\underset{2,5}{\overset{\text{O}}{\overset{\|}{\text{C}}}}-\text{SCoA}} \longrightarrow \underset{\text{palmitate}}{\overset{1,6}{\text{CH}_3}-\overset{2,5}{\text{CH}_2}(\overset{1,6}{\text{CH}_2}-\overset{2,5}{\text{CH}_2})_6-\overset{1,6}{\text{CH}_2}-\overset{2,5}{\text{CO}_2^-}}$$

Fig. 21-7 The Synthesis of Palmitate from Glucose. The numbers indicate the fates of carbon atoms 1-6 of glucose.

Inborn Errors of Lipid Metabolism

A number of hereditary diseases are caused by excessive accumulation of lipids in tissues and cells. These metabolic defects result from lowered activity or lack of specific enzymes whose normal function is to catalyze degradation of lipids. Table 21-5 lists these hereditary disorders and their corresponding enzyme deficiencies. Except for alleviating the symptoms, there is no specific treatment for any of the lipid storage diseases.

One well-known inborn error of lipid metabolism is *Gaucher's disease*, characterized by enlarged spleen and liver along with other symptoms. A large amount of glucocerebroside is found in fat deposits in these organs. Normally the liver and spleen contain the enzyme *β-glucosidase*, which catalyzes degradation of glucocerebrosides, so that only small amounts of these lipids are usually found in these organs. However, the enzyme activity is impaired in Gaucher's disease, and glucocerebrosides

Table 21-5 Lipid Storage Diseases

Condition	Enzyme Deficiency
Gaucher's disease	β-Glucosidase
Niemann-Pick disease	Sphingomyelinase
Krabbe's disease	β-Galactosidase
Metachromatic leukodystrophy	Aryl sulfatidase
Galactosyl ceramide lipidosis	β-Galactosidase
Fabry's disease	α-Galactosidase
Tay-Sachs disease	Hexosaminidase A
Tay-Sachs variant (Sandhoff's disease)	Hexosaminidase A and B
Generalized gangliosidosis	β-Galactosidase
Fucosidosis	α-Fucosidase

accumulate in the liver and spleen. Most persons with the adult form of the disease do not require treatment; however, the infantile form produces mental retardation and often death. Gaucher's disease and Tay-Sachs disease (also discussed in this section) occur more frequently among individuals of Jewish ancestry than among other populations.

In the *Niemann-Pick disease,* the liver and spleen are also enlarged, but the predominate fatty material is a mixture of phospholipids. In this disease, the defective enzyme is *sphingomyelinase.* This enzyme is normally responsible for catalyzing the degradation of sphingomyelin. This disease occurs in infancy, and the accumulation of sphingomyelin usually causes death at about six months of age.

Tay-Sachs disease results from a deficiency of an enzyme, *hexosaminidase A*, which normally catalyzes catabolism of gangliosides. The accumulated gangliosides in the brain and nerve tissue of victims of this disease result in mental and neurologic disorders. Affected children usually die at about three years of age.

METABOLISM OF AMINO ACIDS

We continuously take proteins in through our diet, and they are then degraded by the digestive process to amino acids. The liver and other body tissues use these amino acids to make new proteins. The need for protein synthesis comes about because all body proteins are in a state of dynamic equilibrium as the old ones are being degraded and the new ones are being synthesized. Liver and blood proteins turn over most rapidly, with *half-lives* on the order of two to ten days. Muscle proteins last considerably longer; their half-lives are about six months.

half-life: the length of time required for one-half of the original quantity to disappear.

The Amino Acid Pool

The average adult takes in about 70 grams of protein each day through diet. After digestion, the constituent amino acids join a circulating pool of amino acids in the blood which can be tapped for synthesis of proteins and other nitrogen compounds or for degradation to produce energy. Amino acids may also join this pool as a result of tissue protein degradation. However, some of the nitrogen from amino acids is lost each day as urea in urine, and thus it is necessary to replace that nitrogen daily through protein in the diet.

Plants and some microorganisms are capable of synthesizing all of the amino acids normally found in proteins. However, animals and most microorganisms cannot synthesize all required carbon skeletons. For this reason, certain amino acids are necessary in the diet of humans in

PERSPECTIVE: Ridding the Body of Toxic Materials

Detoxification is the term used for the chemical methods used by the human body to rid itself of toxic materials. The primary purpose of the detoxification process is to convert the toxic substances to more polar compounds. Even though some "detoxified" products are more toxic than the original substances, their increased polarity makes them more soluble in urine, and thus more easily excreted.

Most detoxifications occur in the liver, and many have elaborate pathways. Some of the reaction types are oxidation, addition of hydroxyl groups, reduction, chemical combination with other molecules, addition of methyl groups, and hydrolysis. Detoxification processes are used by the body to eliminate many steroids and drugs. Some common toxic substances and their "detoxified" products are listed in the accompanying table.

Some Toxic Substances and Their Detoxified Products

Substance	Detoxified Product
benzene	muconic acid + others
chloral	trichloroethanol
p-benzoquinone	hydroquinone
bromobenzene	p-bromophenylmercapturic acid
sulfa drugs	acylated sulfa drugs

order that a complete spectrum of these compounds will be available for protein synthesis. These required amino acids are referred to as *essential amino acids;* those required by humans are listed in Table 21-6.

Amino Acid Degradation

Since the need for dietary protein is essentially a need to replenish amino acids in the circulating pool, much of protein metabolism is really

Table 21-6 Essential and Nonessential Amino Acids for Humans

Essential Amino Acids	Nonessential Amino Acids
Arginine*	Alanine
Histidine**	Asparagine
Isoleucine	Aspartic acid
Leucine	Cysteine
Lysine	Glutamic acid
Methionine	Glutamine
Phenylalanine	Glycine
Threonine	Proline
Tryptophan	Serine
Valine	Tyrosine

*Required by growing children only
**Only recently established as essential for adults

a matter of amino acid metabolism. Degradation of amino acids occurs primarily in the liver. Although each amino acid has its own particular metabolic pathway for degradation, there are two especially important reactions. The first step in the catabolism of any amino acid, removal of nitrogen, is accomplished by either of the two reactions described below.

Transamination

As mentioned in chapter 17, transamination involves transfer of an amino group from an amino acid to an α-keto acid to produce a new amino acid and a new α-keto acid:

$$\underset{\text{amino acid}}{\begin{array}{c} CO_2^- \\ | \\ CH-NH_3^+ \\ | \\ R \end{array}} + \underset{\alpha\text{-keto acid}}{\begin{array}{c} CO_2^- \\ | \\ C=O \\ | \\ R' \end{array}} \rightleftharpoons \underset{\alpha\text{-keto acid}}{\begin{array}{c} CO_2^- \\ | \\ C=O \\ | \\ R \end{array}} + \underset{\text{amino acid}}{\begin{array}{c} CO_2^- \\ | \\ CH-NH_3^+ \\ | \\ R' \end{array}}$$

This process is catalyzed by various transaminases, all of which require pyridoxal phosphate or pyridoxamine phosphate (both derived from vitamin B_6) as coenzyme.

Three different α-keto acids participate in transamination reactions: pyruvate, α-ketoglutarate, and oxaloacetate. When these accept the amino groups from amino acids, the α-keto acids are transformed into alanine, glutamate, and aspartate, respectively:

$$\underset{\text{amino acid}}{\begin{array}{c} CO_2^- \\ | \\ CH-NH_3^+ \\ | \\ R \end{array}} + \underset{\text{pyruvate}}{\begin{array}{c} CO_2^- \\ | \\ C=O \\ | \\ CH_3 \end{array}} \rightleftharpoons \underset{\alpha\text{-keto acid}}{\begin{array}{c} CO_2^- \\ | \\ C=O \\ | \\ R \end{array}} + \underset{\text{alanine}}{\begin{array}{c} CO_2^- \\ | \\ CH-NH_3^+ \\ | \\ CH_3 \end{array}}$$

$$\underset{\text{amino acid}}{\begin{array}{c}CO_2^-\\|\\CH-NH_3^+\\|\\R\end{array}} + \underset{\alpha\text{-ketoglutarate}}{\begin{array}{c}CO_2^-\\|\\C=O\\|\\CH_2\\|\\CH_2\\|\\CO_2^-\end{array}} \rightleftharpoons \underset{\alpha\text{-keto acid}}{\begin{array}{c}CO_2^-\\|\\C=O\\|\\R\end{array}} + \underset{\text{glutamate}}{\begin{array}{c}CO_2^-\\|\\CH-NH_3^+\\|\\CH_2\\|\\CH_2\\|\\CO_2^-\end{array}}$$

$$\underset{\text{amino acid}}{\begin{array}{c}CO_2^-\\|\\CH-NH_3^+\\|\\R\end{array}} + \underset{\text{oxaloacetate}}{\begin{array}{c}CO_2^-\\|\\C=O\\|\\CH_2\\|\\CO_2^-\end{array}} \rightleftharpoons \underset{\alpha\text{-keto acid}}{\begin{array}{c}CO_2^-\\|\\C=O\\|\\R\end{array}} + \underset{\text{aspartate}}{\begin{array}{c}CO_2^-\\|\\CH-NH_3^+\\|\\CH_2\\|\\CO_2^-\end{array}}$$

Thus, in amino acid degradation, amino groups are funneled into the formation of alanine, glutamate, or aspartate. α-Ketoglutarate usually serves as the final collection point, and when the dietary supply of glutamate exceeds need for glutamate, the amino groups are released for excretion. In addition to their involvement in amino acid degradation, transamination reactions also provide a means for readjusting the relative proportions of amino acids to meet the particular needs of an organism.

Transaminases are found in all cells, but their concentrations are especially high in heart and liver. If either of these organs is damaged by trauma or disease, they release abnormally large amounts of transaminases into the blood. Measuring the serum levels of these enzymes can provide information regarding extent of organ damage. In particular, levels of serum glutamate oxaloacetate transaminase, SGOT (also called aspartate transferase, AST), and serum glutamate pyruvate transaminase, SGPT (also called alanine transferase, ALT), are commonly measured for diagnostic purposes, as discussed in chapter 17.

Oxidative Deamination

The net effect of the transamination reactions is to funnel nitrogen from various amino acids to glutamate. The enzyme called glutamate dehydrogenase then catalyzes the removal of the amino group as the ammonium ion, and α-ketoglutarate is regenerated. This reaction is the principal source of NH_4^+ in humans, and since the deamination results in oxidation of glutamate, it is called *oxidative deamination*.

$$\underset{\text{glutamate}}{\begin{array}{c}CO_2^-\\|\\CH-NH_3^+\\|\\CH_2\\|\\CH_2\\|\\CO_2^-\end{array}} + NAD^+ + H_2O \rightleftharpoons \underset{\alpha\text{-ketoglutarate}}{\begin{array}{c}CO_2^-\\|\\C=O\\|\\CH_2\\|\\CH_2\\|\\CO_2^-\end{array}} + NH_4^+ + NADH + H^+$$

The ammonium ions released by the glutamate dehydrogenase reaction are quite toxic, and they must be prevented from accumulating. The liver can serve as a scavenger of NH_4^+ by reversing the glutamate dehydrogenase reaction to resynthesize glutamate. In addition, many tissues, especially the liver and brain, dispose of ammonium ions by formation of glutamine catalyzed by glutamine synthetase:

$$\begin{array}{c} CO_2^- \\ | \\ CH-NH_3^+ \\ | \\ CH_2 \\ | \\ CH_2 \\ | \\ CO_2^- \\ \text{glutamate} \end{array} + NH_4^+ + ATP \xrightarrow[Mn^{2+}]{Mg^{2+} \text{ or}} \begin{array}{c} CO_2^- \\ | \\ CH-NH_3^+ \\ | \\ CH_2 \\ | \\ CH_2 \\ | \\ CONH_2 \\ \text{glutamine} \end{array} + ADP + P_i + H^+$$

However, this reaction results in formation of only a temporary, nontoxic transport form of ammonium ions. In addition, a molecule of ATP is used to provide energy for synthesis of the side-chain amide group of glutamine. The production of urea, as described below, is the ultimate pathway for disposal of ammonium ions.

EXAMPLE 21.10

What are essential amino acids?

SOLUTION

Essential amino acids are those required in the diet because of the inability of an organism to synthesize them internally.

EXAMPLE 21.11

Which α-keto acids participate in transamination reactions?

SOLUTION

Three α-keto acids participate in transamination reactions: pyruvate, α-ketoglutarate, and oxaloacetate.

EXAMPLE 21.12

What is the physiological significance of oxidative deamination?

SOLUTION

Oxidative deamination releases excessive nitrogen as NH_4^+, which then can be excreted in the form of urea in the urine.

The Urea Cycle

Since the synthesis of glutamine represents only a limited storage capacity for ammonium ions, a more important mechanism exists for handling them. The *urea cycle* (Figure 21-8), whose enzymes are present in

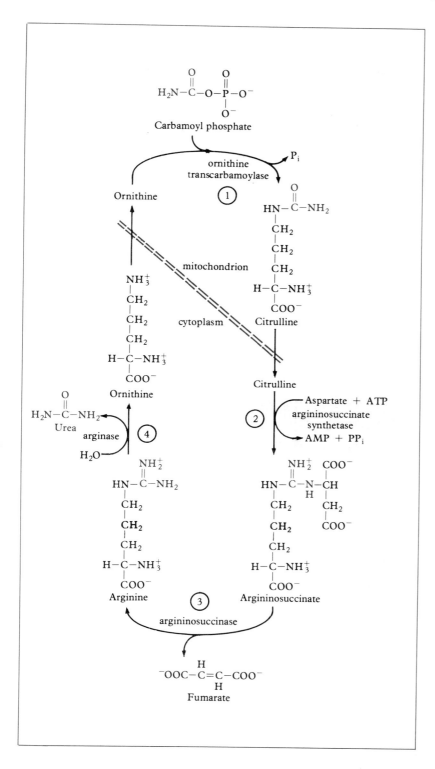

Fig. 21-8 The Urea Cycle

the liver only, converts ammonium ions to urea. This is a less toxic compound which can be allowed to build up until it is convenient to excrete it in urine.

The urea cycle requires as fuel a compound called carbamoyl phosphate. This compound is first synthesized from ammonium ion and carbon dioxide.

$$NH_4^+ + CO_2 + 2\,ATP + H_2O \xrightarrow[\text{synthetase}]{\text{carbamoyl phosphate}} H_2N-\underset{\substack{\|\\O}}{C}-O-\underset{\substack{|\\O^-}}{\overset{\substack{O\\\|}}{P}}-O^- + 2\,ADP + P_i + 3\,H^+$$

carbamoyl phosphate

The energy for its synthesis is derived from two molecules of ATP for each molecule of carbamoyl phosphate formed. Then carbamoyl phosphate, a rather reactive molecule, combines with ornithine to form another amino acid, citrulline, in step 1 of the cycle. In step 2, citrulline combines with aspartate to form argininosuccinate, and this molecule decomposes in step 3 to fumarate and arginine. Then, in step 4, arginine is cleaved to produce urea and regenerate ornithine. Now the cycle is set up to accept another molecule of carbamoyl phosphate and thus dispose of another ammonium ion.

The fumarate formed in step 3 enters the citric acid cycle where it is converted first to malate and then to oxaloacetate. Transamination of oxaloacetate provides another molecule of aspartate which can participate in the urea cycle. Thus, the urea cycle regenerates ornithine, and aspartate is regenerated by the citric acid cycle and transamination.

The summary equation for urea synthesis in the urea cycle is as follows:

$$NH_4^+ + CO_2 + 3\,ATP + \text{aspartate} + 2\,H_2O \longrightarrow$$
$$\text{urea} + 2\,ADP + 2\,P_i + AMP + PP_i + \text{fumarate} + 6\,H^+$$

The pyrophosphate is rapidly hydrolyzed, and thus four high-energy phosphate molecules are used to synthesize one molecule of urea. The preliminary formation of NH_4^+ by the glutamate dehydrogenase reaction, the synthesis of carbamoyl phosphate, and the subsequent formation of citrulline occur in the mitochondrial matrix. The next three reactions in the urea cycle take place in the cytoplasm. Although it is not obvious from the summary equation, two nitrogen groups are used to synthesize one molecule of urea. One comes from the glutamate dehydrogenase reaction (oxidative deamination) and the other from the transamination reaction which produces apartate.

After urea is formed, it diffuses out of liver cells into the blood, and the kidneys filter it out and excrete it into the urine. Usually about 25-30 grams of urea are excreted daily in the urine of a normal adult, although the exact amount varies with the protein content of the diet. The direct excretion of NH_4^+ accounts for a small amount of total

METABOLISM OF AMINO ACIDS **629**

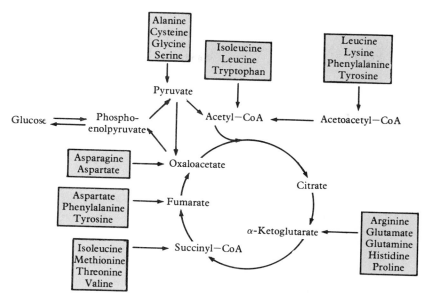

Fig. 21-9 Modes of Entry of the Carbon Atoms from Amino Acids into the Citric Acid Cycle

urinary nitrogen, and this mechanism helps the kidneys control acid-base balance in body fluids. Since ammonia (NH_3) combines with hydrogen ions to form ammonium ions (NH_4^+), the NH_4^+ concentration of urine increases during acidosis and decreases at times of alkalosis.

EXAMPLE 21.13

What is the primary function of the urea cycle?

SOLUTION

The urea cycle converts toxic ammonium ions to less toxic urea, which can be allowed to build up until it can be conveniently excreted in urine.

EXAMPLE 21.14

How are the urea cycle and the citric acid cycle related to each other?

SOLUTION

The urea cycle produces fumarate, a citric acid cycle intermediate. *Via* the citric acid cycle and subsequent transamination, fumarate is used to regenerate the aspartate needed for the urea cycle.

Fates of the Carbon Atoms from Amino Acids

The carbon skeletons derived from amino acids are ultimately disposed of through the citric acid cycle. In addition to the four direct entry points noted in chapter 19, acetyl CoA constitutes a fifth mode of entry into the citric acid cycle for carbon skeletons (see Figure 21-9). The individual pathways leading to the entry point intermediates can be quite complex, but Figure 21-9 illustrates the following points.

1. Eleven amino acid skeletons can enter the acetyl CoA point. Thus, acetyl CoA is the major point of entry into the citric acid cycle for carbon atoms from amino acids.
2. Five amino acid skeletons can enter at the α-ketoglutarate point.
3. Four amino acid skeletons can enter at the succinyl CoA point.
4. Three amino acid skeletons can enter at the fumarate point.
5. Two amino acid skeletons can enter at the oxaloacetate point.

glycogenic: leading to synthesis of glucose or glycogen.

ketogenic: leading to the production of ketone bodies.

As can be seen in Figure 21-9, some carbon skeletons from amino acids can be degraded to pyruvate, α-ketoglutarate, succinyl CoA, and oxaloacetate. Since these intermediates can be used to synthesize phosphoenolpyruvate and then glucose *via* gluconeogenesis, their amino acid precursors are classified as *glycogenic*. These can provide glucose when dietary carbohydrate is low or when supplies are being used rapidly. However, continual use of these amino acids for gluconeogenesis results in wasting away of muscle tissue. In contrast, amino acid skeletons which are degraded to acetyl CoA or acetoacetyl CoA cannot be used for net glucose synthesis from fatty acid carbons, as discussed earlier in this chapter. These amino acids are called *ketogenic* amino acids. You will notice that only one amino acid, leucine, is purely ketogenic. Fourteen amino acids are purely glycogenic, and the remaining five are both ketogenic and glycogenic; this is summarized in Table 21-7. It should be noted that many fad diets involve high protein and low carbohydrate dietary intakes. In these cases, there is a possibility of generating excessive amounts of ketone bodies, resulting in ketosis and possible ketoacidosis.

Amino Acid Synthesis

The liver is a highly active organ in the biochemical sense, and it is responsible for synthesizing most of the amino acids that our body can make. Because we can synthesize these amino acids, they are called *nonessential amino acids* (see Table 21-6). In general, the nonessential amino acids are synthesized by reversals of the transamination and oxidative deamination reactions seen earlier. Several other pathways also exist, but for the sake of brevity we will not go into them.

Inborn Errors of Amino Acid Metabolism

At least 35 different hereditary defects in amino acid metabolism are known. Of these, 29 involve the whole body and thus are classified as systemic. The remaining 6 disorders involve transport of amino acids across membranes of the kidney tubules or the lining of the intestine. We will consider the causes of some of the more well known systemic inborn errors.

Phenylketonuria (PKU), first described in the early 1930s, results from a deficiency of phenylalanine hydroxylase, the enzyme that converts phenylalanine to tyrosine.

Table 21-7 Glycogenic and Ketogenic Amino Acids

Amino Acid	Glycogenic	Ketogenic
Alanine	X	
Arginine	X	
Asparagine	X	
Aspartic acid	X	
Cysteine	X	
Glutamic acid	X	
Glutamine	X	
Glycine	X	
Histidine	X	
Isoleucine	X	X
Leucine		X
Lysine	X	X
Methionine	X	
Phenylalanine	X	X
Proline	X	
Serine	X	
Threonine	X	
Tryptophan	X	X
Tyrosine	X	X
Valine	X	

$$\text{phenylalanine} + O_2 + NADPH + H^+ \longrightarrow$$

$$\text{tyrosine} + NADP^+ + H_2O$$

This defect causes an excessive accumulation of phenylalanine in the blood and spinal fluid. The excessive phenlalanine inhibits tyrosine metabolism, which normally leads to synthesis of melanin, the light brown pigment in skin, hair, and eyes, and thus victims have light-colored skin and hair. More important, however, is that the excessive phenylalanine can produce mental retardation, epileptic seizures, and abnormal brain wave patterns. The urine of newborn infants is routinely tested for levels of phenylpyruvate and phenylacetate ("phenyl ketones"), products of phenylalanine metabolism, and when detected, the defect can be treated by a low-phenylalanine diet.

phenylpyruvate   phenylacetate

Albinism is a general term used to describe a lack of or diminished production of melanin. Individuals with this defect usually have pale skin, white or light blond hair, and pink or bluish eyes. The condition is caused by a deficiency of the enzyme tyrosinase, which is normally involved in the production of melanin from tyrosine. There are usually not any life-threatening consequences of this disorder, and treatment consists of prevention of exposure to sunlight and protection of the eyes by the use of dark glasses.

Alkaptonuria is identified by the presence of arthritis, dark-colored cartilage, and a darkening of the urine upon exposure to air. This disorder results from a deficiency of liver homogentisate oxidase, an enzyme normally involved in tyrosine degradation. When this enzyme is missing, abnormally large amounts of homogentisate are excreted in the urine of affected persons:

$$\text{(OH, HO-substituted benzene)}-CH_2-CO_2^-$$
homogentisate

Oxygen in the air oxidizes homogentisate to a dark-colored material and causes darkening of the urine. Treatment consists of relieving the discomfort of the arthritis.

METABOLIC INTERRELATIONSHIPS

In studying metabolic pathways, it sometimes happens that in focusing on individual pathways, we lose track of how these pathways fit together in an integrated picture. Figure 21-10 illustrates how the major metabolic pathways work together in a living organism.

Glucose metabolism is extremely important for energy production in red blood cells and brain cells. It is the responsibility of the liver to see that blood glucose levels are within normal limits, utilizing glycogen as its source of glucose to furnish to the blood. Most other tissues show a preference for fatty acids or ketone bodies as their energy sources.

When glycogen stores in the liver fall, it uses lactate, citric acid cycle intermediates, or amino acids for gluconeogenesis. (Remember that fatty acids cannot be used for any net synthesis of glucose.) If carbohydrate supplies are severely limited, as in starvation, tissue proteins are degraded to provide carbon atoms for glucose synthesis. This situation results in wasting away of muscle and possible ketoacidosis.

Insulin insufficiency in uncontrolled diabetes causes an increase in liver gluconeogenesis. Protein synthesis slows and amino acid degradation speeds up to provide raw materials for gluconeogenesis. Simultaneous high levels of glucagon cause the liver to send more and more glucose into the blood, and much of it is excreted in urine. (This high

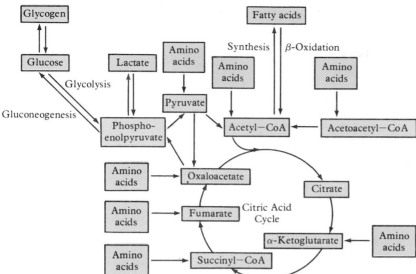

Fig. 21-10 Interrelationships of the Major Metabolic Pathways

load of glucose requires large quantities of urine, hence the large urinary output and increased water intake of uncontrolled diabetics.) In addition, body fat is degraded to provide fatty acid energy sources to compensate for lack of glucose, accounting for the observed loss of weight.

SUMMARY

Metabolism of Lipids: Short-chain fatty acids and glycerol are absorbed directly into the blood stream, but long-chain fatty acids and mono- and diacylglycerols are used to resynthesize triacylglycerols as they pass through the intestinal wall. These then appear in the lymph fluid in the form of chylomicra. When triacylglycerols and cholesterol enter the blood, they are dispersed as lipoproteins. Fatty acid mobilization from fat cells is catalyzed by lipase, and the fatty acids are transported bound to serum albumin. Fatty acids are oxidized in mitochondria by the β-oxidation pathway. Each turn of the β-oxidation cycle results ultimately in production of 5 ATP molecules. The end product of fatty acid oxidation is acetyl CoA, and if levels of this molecule build up, ketone bodies may be produced. Fatty acid synthesis occurs in the cytoplasm, with palmitate as the major product. Fatty acids not needed immediately are used to synthesize triacylglycerols, which are stored in fat cells. Cholesterol synthesis occurs in the liver and intestinal wall. Glucose can serve as a source of carbon atoms for fatty acid synthesis, but there can be no net formation of glucose from fatty acids. Prominent inborn errors of lipid metabolism result in Gaucher's disease, Niemann-Pick disease, and Tay-Sachs disease.

Metabolism of Amino Acids: After digestion, amino acids from protein enter the amino acid pool. Two important reactions in amino acid degradation are transamination and oxidative deamination. Amino groups are removed from amino acids and converted to urea *via* the urea cycle. The carbon skeletons can then enter the citric acid cycle at five points. Glycogenic amino acids are those which can be used for glucose synthesis, while ketogenic amino acids produce acetyl CoA or acetoacetyl CoA. The amino acids which can be synthesized by an organism are called nonessential amino acids, and those required in the diet are essential amino acids. Prominent inborn errors of amino acid metabolism result in phenylketonuria, albinism, and alkaptonuria.

Metabolic Interrelationships: Most tissues prefer fatty acids or ketone bodies as energy sources, but the liver produces glucose from glycogen, lactate, citric acid cycle intermediates, or amino acids for use by red blood cells and brain cells. If liver gluconeogenesis is increased, amino acid degradation speeds up and protein synthesis slows to provide more carbon atoms for gluconeogenesis.

STUDY QUESTIONS AND PROBLEMS

1. Define the following terms:
    (a) chylomicra
    (b) lipoprotein
    (c) VLDL
    (d) LDL
    (e) HDL
    (f) fatty acid mobilization
    (g) β-oxidation pathway
    (h) ketone body
    (i) ketonemia
    (j) ketonuria
    (k) ketosis
    (l) ketoacidosis
    (m) acyl carrier protein
    (n) malonyl-ACP
    (o) essential fatty acid
    (p) mevalonic acid
    (q) Gaucher's disease
    (r) Niemann-Pick disease
    (s) Tay-Sachs disease
    (t) amino acid pool
    (u) essential amino acids
    (v) transamination
    (w) oxidative deamination
    (x) urea cycle
    (y) carbamoyl phosphate
    (z) glycogenic amino acids
    (aa) ketogenic amino acids
    (bb) nonessential amino acids
    (cc) phenylketonuria
    (dd) albinism
    (ee) alkaptonuria

2. How is the absorption of fats different from absorption of sugars and amino acids?
3. Write an equation for hydrolysis of a triacylglycerol to a diacylglycerol and a fatty acid.
4. Write an equation for hydrolysis of a diacylglycerol to a monoacylglycerol and a fatty acid.
5. How does the absorption of short-chain fatty acids differ from that of long-chain fatty acids?
6. What does cloudy lymph fluid signify?
7. Why is fasting necessary prior to having one's blood lipid levels measured?
8. Why do lipids combine with blood proteins to form lipoproteins?
9. Explain how lipoproteins in blood are classified.
10. Give the lipid components of the three types of lipoproteins.
11. What is the principal form of cholesterol in lipoproteins of blood?
12. Eskimos have high levels of high-density lipoproteins.
    (a) How does this information correlate with their diets?
    (b) Would you expect Eskimos to be especially susceptible to atherosclerosis? Explain your answer.
13. Are phospholipids and cholesterol used as direct energy sources? Explain your answer.
14. How many calories are available from one gram of each of the following?
    (a) fat
    (b) carbohydrate
    (c) protein
15. Write an equation for total hydrolysis of a triacylglycerol.
16. List the four hormones that regulate adipose-cell lipase, and explain how they work.
17. How are fatty acids transported in the blood?
18. Why is it that fatty acids do not need membrane transport systems?
19. Why are red blood cells and brain cells unable to oxidize fatty acids?
20. Why is it beneficial that fatty acids are first attached to coenzyme A before oxidation?
21. Write an equation to illustrate fatty acid activation.
22. List the four kinds of reactions involved in β-oxidation of fatty acids, and write an equation to illustrate each.
23. In which steps does β-oxidation produce reduced coenzymes? How many reduced coenzymes are formed in each turn of the cycle?
24. How many moles of ATP are available from β-oxidation of one mole of lauric acid, $CH_3(CH_2)_{10}CO_2H$?
25. Calculate the ATP production from total oxidation of one mole of arachidic acid, $CH_3(CH_2)_{18}CO_2H$, via β-oxidation, the citric acid cycle, and oxidative phosphorylation.
26. Write structures for the three ketone bodies.
27. How does the presence of large amounts of ketone bodies lead to ketoacidosis?
28. Explain why uncontrolled diabetics often have "acetone breath." Are there other conditions which could cause the acetone odor?
29. What are the building blocks used for fatty acid synthesis?
30. How are acetyl units transferred from the mitochondria to the cytoplasm for fatty acid synthesis?

31. What is the role of biotin in fatty acid synthesis?
32. What is the role of malonyl-ACP in fatty acid synthesis?
33. Summarize the differences between fatty acid degradation and fatty acid synthesis.
34. Describe the regulation of fatty acid synthesis.
35. On the basis of your knowledge of fatty acid synthesis, explain why humans do not synthesize fatty acids with an odd number of carbon atoms.
36. Give sources of the molecules needed for synthesis of triacylglycerols.
37. How does insulin influence fat synthesis?
38. What are the building blocks for cholesterol synthesis?
39. Name three compounds synthesized from cholesterol, and draw their structures.
40. Describe the effect of dietary cholesterol on cholesterol biosynthesis.
41. Why do you think cholesterol synthesis is sensitive to cholic acid levels?
42. How does glucose serve as a source of carbon atoms for fatty acid synthesis?
43. Why is it that fatty acids cannot cause any net formation of glucose?
44. List three lipid storage diseases and their corresponding enzyme deficiencies.
45. What is the amino acid pool? What are the sources of amino acids in the pool?
46. What is the difference between essential and nonessential amino acids?
47. Write an equation to illustrate transamination between leucine and pyruvate.
48. Illustrate the reaction catalyzed by serum glutamate oxaloacetate transaminase (SGOT) with an equation.
49. What gets oxidized in oxidative deamination? What gets reduced?
50. Why is the citric acid cycle intermediate, α-ketoglutarate, so important to amino acid metabolism?
51. How does the glutamine synthetase reaction help to lower NH_4^+ levels?
52. Write the summary equation for urea formation *via* the urea cycle. What is the source of energy for urea synthesis?
53. What is the source of amino groups in urea?
54. Differentiate between glycogenic and ketogenic amino acids?
55. How might excessive amino acid degradation, as in uncontrolled diabetes and crash diets, cause acidosis?
56. List three inborn errors of amino acid metabolism and their corresponding enzyme deficiencies.
57. Recent evidence suggests that obese people may be lacking an enzyme, ATPase, that catalyzes the hydrolysis of ATP to produce energy in the form of heat.
    (a) Why would a deficiency of ATPase cause a person to be obese?
    (b) On the basis of the above hypothesis, would you expect the body temperature of an obese person to be normal? Explain your answer.

CHAPTER 22

NUCLEIC ACIDS AND HEREDITY

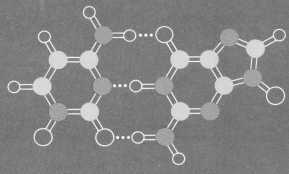

NUCLEIC ACIDS
- Composition of Nucleic Acids
- Nucleosides
- Nucleotides
- The Primary Structure of Nucleic Acids

DNA
- Secondary Structure of DNA
- The Role of DNA
- Replication of DNA
- DNA Repair

RNA
- Messenger RNA
- Transfer RNA
- Ribosomal RNA

PROTEIN BIOSYNTHESIS

GENETIC MUTATIONS

VIRUSES

CARCINOGENS
- Chemical Carcinogens
- Radiation
- Viruses

GENETIC ENGINEERING
- Enzymes Used in Genetic Engineering
- Plasmids
- Recombinant DNA

Everyone is aware of the remarkable resemblance of identical twins and the similarities in appearance among members of a family. We know that certain physical traits are passed from parent to offspring, but it was not known how traits are transmitted until the early nineteenth century. The first glimmering of understanding came when it was learned that eggs and sperm carry hereditary information. Later, in 1865 and 1866, the Austrian monk Gregor Mendel reported results from experiments with plant hybrids that led to the idea that hereditary factors called genes are part of the chromosomes located in cell nuclei. It came to be accepted that the nucleus is the control center of the cell, and early in the twentieth century it became evident that biopolymers called nucleic acids are responsible for the transfer of genetic information. Since that time, understanding of the molecular processes involved in heredity has grown tremendously, to the point where it is now possible to manipulate genes of bacteria by the technology of genetic engineering.

NUCLEIC ACIDS

The *nucleic acids* got their general name because they are acidic substances first isolated from cell nuclei. The two types of nucleic acids are *deoxyribonucleic acid* (DNA) and *ribonucleic acid* (RNA). DNA contains all of the information necessary for a cell to function properly, and RNA translates this information so that proteins can be synthesized. Both types of nucleic acids are polymers with complex structures, and as you shall see, they exist in other places in the cell as well as the nucleus.

Composition of Nucleic Acids

Nucleic acids undergo acid- or enzyme-catalyzed hydrolysis to produce *phosphate*, a *pentose*, and a mixture of *heterocyclic nitrogen bases*. The pentose from DNA is β-deoxyribose, while that from RNA is β-ribose.

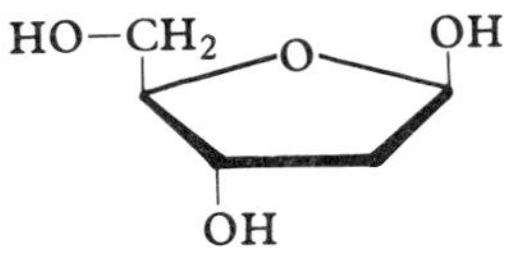

β-deoxyribose
(present in DNA)

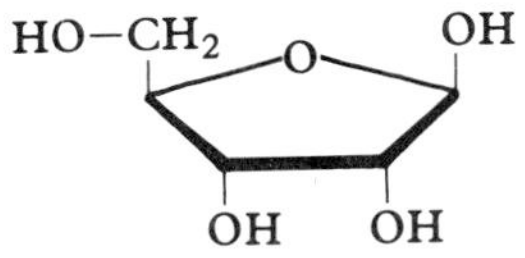

β-ribose
(present in RNA)

Nucleic acids contain two kinds of heterocyclic nitrogen bases, *purines* and *pyrimidines*. Both DNA and RNA contain the purines adenine and guanine; these molecules are considered derivatives of the parent compound purine.

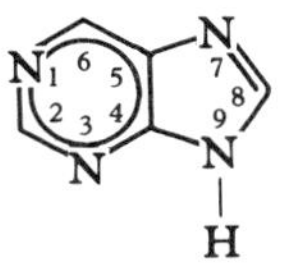

purine

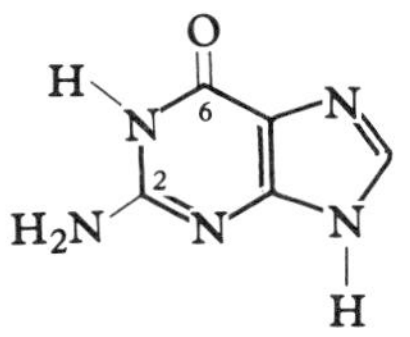

adenine
(present in DNA and RNA)

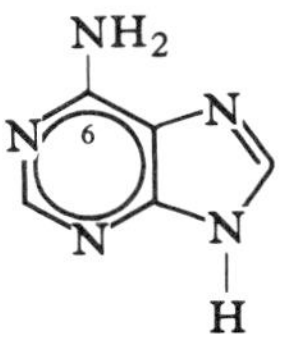

guanine
(present in DNA and RNA)

Both DNA and RNA contain the pyrimidine cytosine, but uracil occurs exclusively in RNA. In place of uracil, DNA contains thymine. (Certain RNA molecules also contain thymine.)

pyrimidine

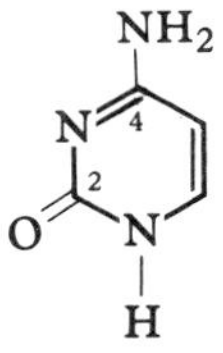

cytosine
(present in DNA and RNA)

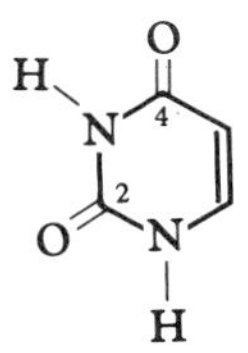

uracil
(present in RNA)

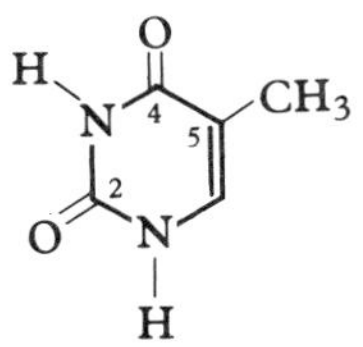

thymine
(present in DNA and, to a small extent, in RNA)

In addition to the five major nitrogen bases shown above, modified forms of these occur to a small extent in some nucleic acids.

Nucleosides

Nucleosides are compounds composed of a purine or a pyrimidine bonded to ribose or deoxyribose in a β-glycosidic linkage. Purines are bonded to the pentose at nitrogen-9 of the base, and pyrimidines are bonded at nitrogen-1. In both cases, the base is attached to the 1'-carbon of the pentose. The carbon atoms of the sugar in a nucleoside are identified by "primes" to distinguish them from atoms in the rings of the nitrogen bases. Examples of two nucleosides follow, and the names of nucleosides derived from nucleic acids are given in Table 22-1.

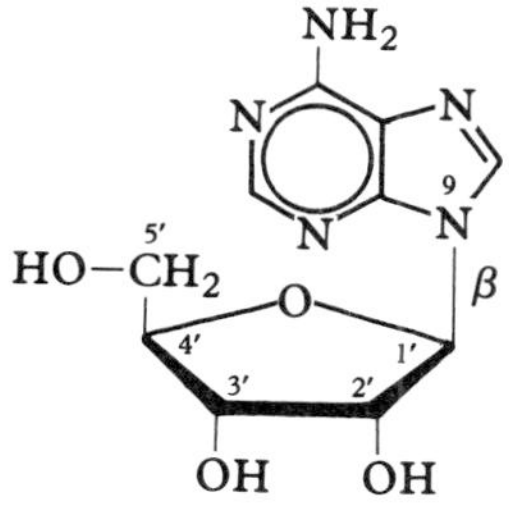

adenosine

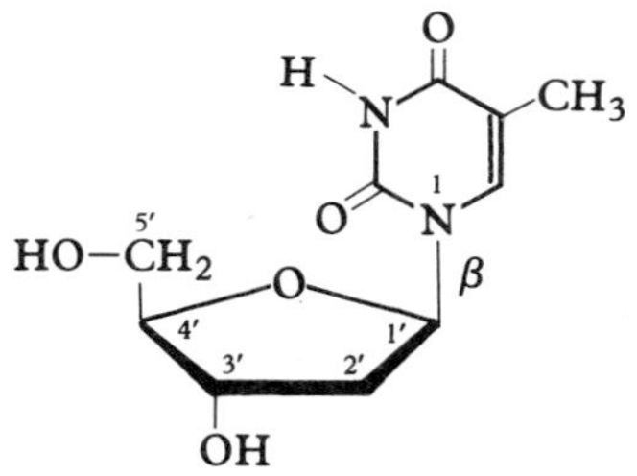

deoxythymidine

Table 22-1 Names of Nucleosides

Base	Ribonucleosides	Deoxyribonucleosides
Adenine	Adenosine	Deoxyadenosine
Guanine	Guanosine	Deoxyguanosine
Uracil	Uridine	Deoxyuridine
Cytosine	Cytidine	Deoxycytidine
Thymine	Ribothymidine or Thymine ribonucleoside	Deoxythymidine or Thymidine

Nucleotides

When a nucleoside reacts with phosphoric acid to form a phosphate ester at the 5'-carbon, the resulting compound is called a *nucleotide*. AMP is an example:

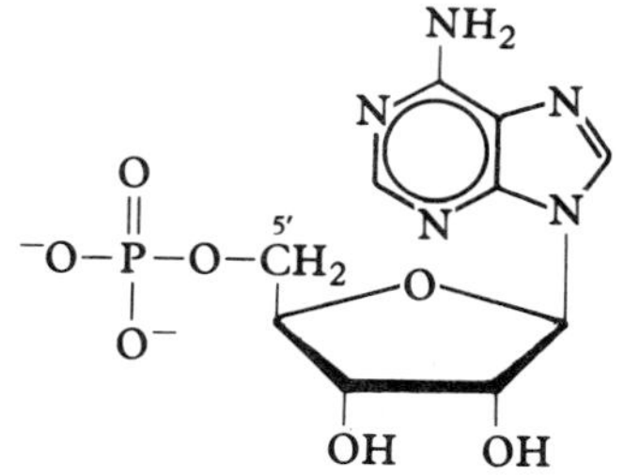

adenosine-5'-monophosphate
(AMP)

Nucleotides are the repeating units of nucleic acids just as amino acids are the repeating units of proteins. The nucleotides present in nucleic acids all contain the phosphate group at the 5'-position of the pentose, and they are named as monophosphate derivatives of nucleosides, as illustrated in Table 22-2.

Table 22-2 Nucleoside-5'-Monophosphates of RNA and DNA

Ribonucleoside-5'-Phosphates	Deoxyribonucleoside-5'-Phosphates
Adenosine-5'-monophosphate, AMP	Deoxyadenosine-5'-monophosphate, dAMP
Guanosine-5'-monophosphate, GMP	Deoxyguanosine-5'-monophosphate, dGMP
Cytidine-5'-monophosphate, CMP	Deoxycytidine-5'-monophosphate, dCMP
Uridine-5'-monophosphate, UMP	Deoxythymidine-5'-monophosphate, dTMP

Free nucleotides are found in all cells, usually with the phosphate group at the 5'-position of the pentose. In addition, a small number of nucleotides are present with the phosphate group attached to the 2'- or 3'-carbon of the pentose. These three types of nucleoside monophosphates differ only by the location of the phosphate group. However, the nucleoside monophosphates can react with phosphoric acid to produce nucleoside di- and triphosphates. Familiar examples are adenosine-5'-diphosphate (ADP) and adenosine-5'-triphosphate (ATP).

adenosine-5'-diphosphate
(ADP)

adenosine-5'-triphosphate
(ATP)

You have learned that nucleoside triphosphates, particularly ATP, serve as energy carriers in cells, but they are also necessary for the synthesis of nucleic acids. Names and abbreviations for nucleoside di- and triphosphates are given in Table 22-3.

Table 22-3 Nucleoside-5'-Di- and Triphosphates

Ribonucleotides	Deoxyribonucleotides
Adenosine-5'-diphosphate, ADP	Deoxyadenosine-5'-diphosphate, dADP
Adenosine-5'-triphosphate, ATP	Deoxyadenosine-5'-triphosphate, dATP
Guanosine-5'-diphosphate, GDP	Deoxyguanosine-5'-diphosphate, dGDP
Guanosine-5'-triphosphate, GTP	Deoxyguanosine-5'-triphosphate, dGTP
Cytidine-5'-diphosphate, CDP	Deoxycytidine-5'-diphosphate, dCDP
Cytidine-5'-triphosphate, CTP	Deoxycytidine-5'-triphosphate, dCTP
Uridine-5'-diphosphate, UDP	Deoxythymidine-5'-diphosphate, dTDP
Uridine-5'-triphosphate, UTP	Deoxythymidine-5'-triphosphate, dTTP

The Primary Structure of Nucleic Acids

Nucleotides are joined together in nucleic acids by phosphate groups connecting the 5'-carbon of one nucleotide to the 3'-carbon of the next. These linkages are called 3', 5'-phosphodiester linkages, and the order of nucleotides in a nucleic acid is its primary structure. Figure 22-1 shows the primary structure of a tetranucleotide from RNA. Like proteins, many nucleic acids have two termini, one called the 5'-terminus and the other the 3'-terminus. As illustrated in Figure 22-1, the ribose at the 5'-terminus has a free C-5' and the ribose at the 3'-terminus has a

Fig. 22-1 A tetranucleotide

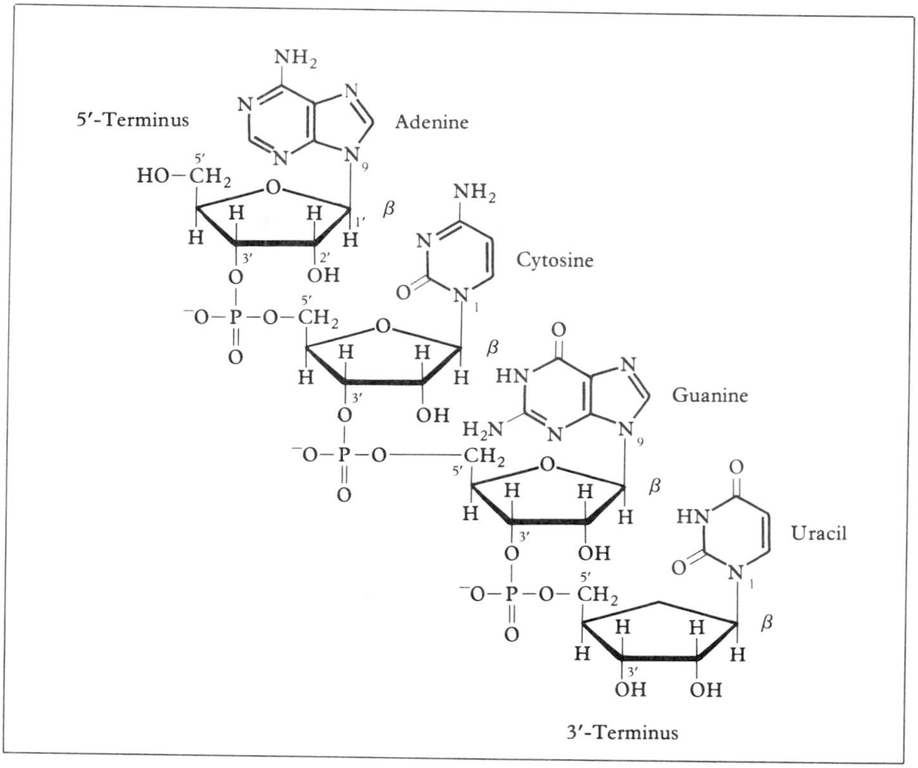

Fig. 22-2 The Synthesis of a Portion of a Nucleic Acid. As two nucleotides are joined together, a pyrophosphate unit is released.

free C-3'. Nucleic acids are also found which have phosphate groups on the 5'-terminus. In addition, there are many examples of circular nucleic acids that have no termini; these are especially plentiful in bacterial cells.

Nucleic acids are synthesized in cells by joining portions of nucleoside-5'-triphosphates; as two nucleotides are joined together, a pyrophosphate unit is released, as illustrated in Figure 22-2. The synthesis of nucleic acid chains is catalyzed by enzymes called polymerases.

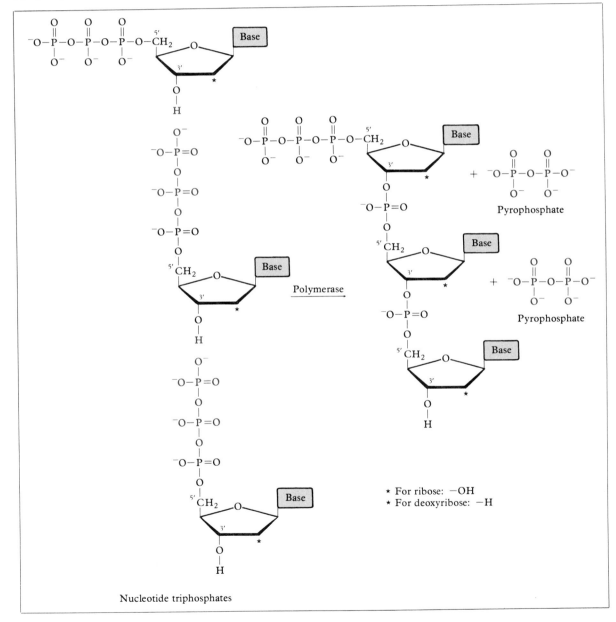

For convenience, the primary structure of a nucleic acid or a polynucleotide is written in abbreviated form. For example, 5' A—C—G—U 3' means a tetranucleotide with the base sequence adenine-cytosine-guanine-uracil, running in the 5' to 3' direction. Since the type of pentose in the nucleic acid is not identified in the abbreviated sequence, the type of nucleic acid must be specified.

EXAMPLE 22.1

Draw a trinucleotide portion of RNA with the following sequence:

5' G—U—C 3'

SOLUTION

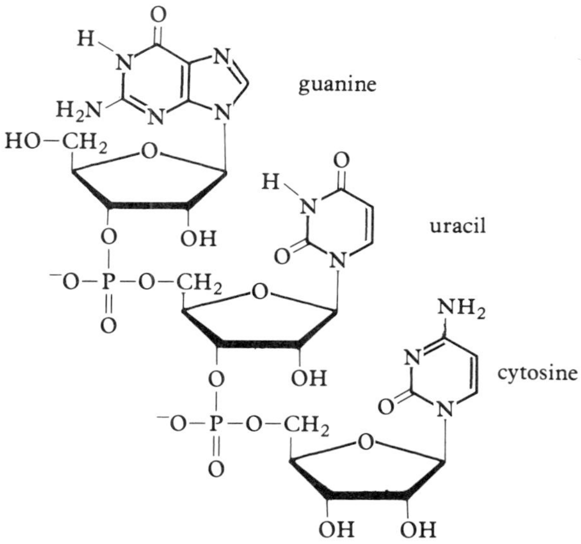

EXERCISE 22.1 Draw a tetranucleotide portion of DNA with the following sequence:

5' A—C—T—G 3'

DNA

In eucaryotic cells, nearly all of the DNA molecules are found in the cell nucleus combined by ionic attraction with basic proteins called *histones*. Small amounts of DNA are also found in mitochondria, accounting for 0.1–0.2% of the total cellular DNA. It is difficult to determine the exact size of intact DNA molecules in eucaryotic cells because the isolation procedures usually result in some degradation of the molecules. The average molecular weight of DNA isolated from such cells is on the order of 10^7 amu, but that probably represents degraded material.

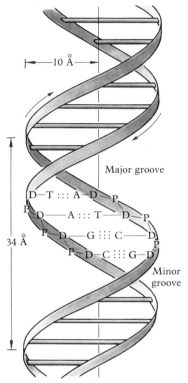

Fig. 22-3 A Diagram of the DNA Double Helix. The two ribbons represent the phosphate (P) and deoxyribose (D) portions of the two DNA strands, and A:::T and G:::C represent adenine-thymine and guanine-cytosine base pairs, respectively.

Secondary Structure of DNA

The determination of the secondary structure of DNA in 1953 by James D. Watson and Francis H. C. Crick (joint Nobel Prize recipients, 1962) was perhaps the greatest discovery of modern biology; it marked the beginning of the field of study called molecular biology. Watson and Crick used data from X-ray diffraction and chemical analysis obtained by others to construct molecular models of DNA. By doing so, they suggested what is now known to be the correct secondary structure, the double-helical model. DNA exists as two helical strands wound together in a double helix. The repeating sugar-phosphate backbone of each DNA strand forms the spiral of the double helix, and the nitrogen bases of the deoxyribonucleotides project in towards the center of the double helix, with the planes of their rings parallel to each other and perpendicular to the double helix axis. A representation of the DNA double helix is shown in Figure 22–3.

The double helix is held together by hydrogen bonds among bases of the two DNA strands. However, the hydrogen bonding is very specific: adenine always forms hydrogen bonds to thymine, and guanine always

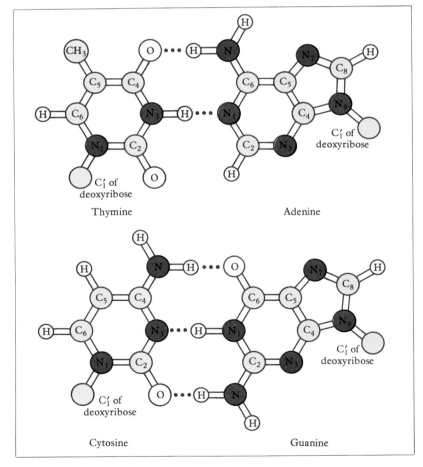

Fig. 22-4 Hydrogen Bonding Between Base Pairs in DNA. An adenine-thymine pair has two hydrogen bonds, and a guanine-cytosine pair has three hydrogen bonds.

forms hydrogen bonds with cytosine. This pattern of base pairing comes about because it represents maximum hydrogen bonding. Models of the base pairs are shown in Figure 22-4.

Several other features of the DNA double helix are noteworthy. The paired bases must lie in the same plane in order to form their hydrogen bonds: two between adenine and thymine and three between guanine and cytosine. The pairing between adjacent complementary bases allows a constant separation of the two DNA strands along the length of the double helix. Also, the two DNA strands run in opposite directions; that is, they are antiparallel. Because of the base pairing, the DNA strands have complementary sequences, meaning that an adenine in one strand is always opposite a thymine in the other strand, and a guanine in one strand is always opposite a cytosine in the other strand.

EXAMPLE 22.2

Write the sequence for a DNA strand that would form a double helix with the following DNA segment:

$$5'\ C\text{—}A\text{—}T\text{—}G\ 3'$$

SOLUTION

Since the second DNA strand will be complementary to the first and the two strands will be antiparallel, the second DNA strand would have the sequence indicated below.

$$\begin{array}{c} 5'\ C\text{—}A\text{—}T\text{—}G\ 3' \\ |\ \ \ |\ \ \ |\ \ \ | \\ 3'\ G\text{—}T\text{—}A\text{—}C\ 5' \end{array}$$

EXERCISE 22.2 Write the sequence for a DNA strand that would form a double helix with the following DNA segment:

$$5'\ T\text{—}G\text{—}A\text{—}T\text{—}C\ 3'$$

The Role of DNA

Nuclei of eucaryotic cells contain several or many *chromosomes*, depending on the species. In human cells, for example, there are 46 chromosomes, arranged as 23 pairs (Figure 22-5). Each chromosome contains one molecule of DNA, coiled tightly about itself in a very compact arrangement. It has been estimated that if the DNA from a single human chromosome were unwound into a linear conformation, it would be about 44 cm (18 inches) long.

The sequence of nucleotides in the DNA of chromosomes carries the information needed by a cell to synthesize all protein molecules necessary for the cell's proper function. Thus, DNA serves as the primary reservoir of genetic information. In humans, one member of each original chromosome pair comes from each parent. When the cell that begins the existence of a human divides for the first time, the two resulting cells each receive a complete set of 46 chromosomes. The cells continue to divide, replicating chromosomes each time, for the lifetime of the human.

chromosome: any of several threadlike bodies found in a cell nucleus that carry the genes in a linear order; chromosomes are composed of DNA and proteins.

Fig. 22-5 Human Chromososome Pairs

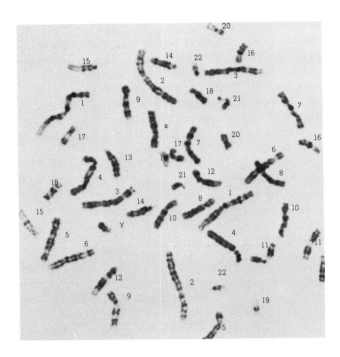

Fig. 22-6 The Distribution of DNA through Two Generations of Cell Division

Original parent molecule

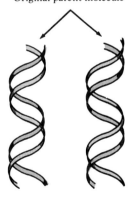

First generation daughter molecules

Second generation daughter molecules

Replication of DNA

When a cell divides, the original DNA in each chromosome is replicated to form DNA for the new cell. In this process, the original double-stranded DNA separates, and a complementary strand is synthesized for each of the original strands. Thus, each daughter DNA molecule contains one newly synthesized strand of DNA and one strand of DNA derived from the parent DNA molecule. Figure 22-6 illustrates the distribution of parent DNA molecules in replicated DNA through two generations of cell division.

The replication process involves several different enzymes, and more details about the process are being discovered almost daily. In particular, double-stranded DNA starts to separate, and at the same time, deoxyribonucleoside triphosphates position themselves by base pairing on the separated DNA strands. Then DNA polymerases catalyze formation of phosphodiester bonds, as shown in Figure 22-7. The PP_i units that are released are hydrolyzed to provide energy for the synthetic process. In this way, single strands of parent DNA are replicated, and the new double-stranded DNA consists of one strand of parent DNA and one strand of newly formed DNA (Figure 22-7).

DNA Repair

Much of our knowledge of nucleic acids comes from studies of bacterial cells; these have simpler structures than animal cells and are thus easier to work with. It has been found that when certain bacteria are exposed to high levels of ultraviolet light, their DNA molecules become damaged. The major form of damage is the production of thymine dimers, shown in Figure 22-8. The covalent linkage of two thymine units creates a distortion in the DNA double helix which subsequently interferes with DNA replication. If the defective DNA is not repaired, cell death is likely to occur. However, some bacterial strains contain enzymes that catalyze removal of the defective portion of DNA and replacement of it by a normal polynucleotide segment. These bacteria are thus resistant to injury by ultraviolet light.

It appears that all cells have DNA repair mechanisms of one type or another that are essential for maintaining the proper structure of the DNA in chromosomes. DNA is damaged by a variety of chemical agents and radiation such as X-rays and ultraviolet light. Nitrogen bases can be altered or destroyed, phosphodiester linkages can be broken, and DNA strands can become cross-linked. Because of the natural repair mechanisms that exist, much of the damage sustained by DNA can be repaired before cell death occurs. However, in a few instances, such as the rare genetic disease called xeroderma pigmentosum, human cells cannot overcome DNA damage. In this case, the cells lack certain repair enzymes, and the skin of an affected person is extremely sensitive to ultraviolet light, including the ultraviolet rays in sunlight. The rapid death of skin cells results in coarse deposits of pigmented tissue, and skin cancer usually develops at several sites. Many patients die before the age of thirty from these spreading malignant skin tumors.

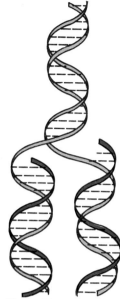

Fig. 22-7 A Simplified Illustration of DNA Replication

RNA

The three major types of ribonucleic acid in cells are called *messenger RNA* (mRNA), *ribosomal RNA* (rRNA), and *transfer RNA* (tRNA). All three types exist as single polyribonucleotide strands, but each has a characteristic range of molecular weight (see Table 22-4). In addition, the three types of RNA occur in multiple molecular forms.

Most cells contain two to eight times as much RNA as DNA. The distribution of RNA in eucaryotic cells differs with cell type. For example, in liver cells, about 11% of the total RNA is in the nucleus, about 15% in the mitochondria, over 50% in the ribosomes, and about 24% in the cytoplasm.

Fig. 22-8 A Thymine Dimer

Table 22-4 Properties of RNAs from *E. coli*

Type	Molecular Weight, amu	Number of Nucleotides	Percent of Total Cell RNA
mRNA	25,000–1,000,000	75–3000	2
tRNA	23,000–30,000	75–90	16
rRNA			
5S	35,000	100	
16S	550,000	1,500	82
23S	1,100,000	3,100	

Messenger RNA

Messenger RNA is synthesized in the nucleus in the process called *transcription*. During synthesis, complementary bases are aligned along one strand of DNA and ribonucleotides are linked together to form the intact mRNA molecule. Small segments of double-stranded DNA separate, and a single-stranded segment thus serves as a *template* for RNA synthesis. Ribonucleoside triphosphates (NTPs) are used as substrates by the enzyme *DNA-directed RNA polymerase* (also called "transcriptase") to bring about RNA synthesis, and all of the adenine-containing nucleotides in DNA give rise to nucleotides containing uracil in RNA. After transcription, the mRNA molecules pass into the cytoplasm and then to the ribosomes where they participate in protein biosynthesis (to be explained later). Messenger RNA molecules appear to have no defined secondary structure.

template: a mold or form used for construction of an object.

Transfer RNA

Transfer RNA molecules are also synthesized in the nucleus using DNA as a template. These relatively small molecules act as carriers of specific individual amino acids during protein biosynthesis on the ribosomes. Each of the 20 amino acids found in proteins has at least one corresponding tRNA, and some have multiple tRNAs.

Transfer RNA molecules have rather well defined secondary structures in which localized hydrogen bonds hold together several intramolecular loops. A simplified illustration of this hydrogen bonding and the resulting L-shape of a tRNA molecule are shown in Figure 22-9.

Ribosomal RNA

Ribosomal RNA, also synthesized in the nucleus, accounts for up to about 65% of the mass of ribosomes. It can be isolated from bacterial ribosomes in the form of linear, single-stranded molecules of three different molecular weights (see Table 22-4), each of which has a different rate of sedimentation when centrifuged. For this reason, the three forms of rRNA from the bacterium *E. coli* are referred to as 5S rRNA, 16S rRNA, and 23S rRNA, where S is a reference to their sedimentation behaviors. In eucaryotic cells, which have larger ribosomes than procaryotic cells, there are four types of rRNA: 5S, 7S, 18S, and 28S.

PROTEIN BIOSYNTHESIS

As mentioned earlier, DNA carries the information necessary for a cell to synthesize all of its proteins. This information is contained in the nucleotide sequence of DNA and is called the *genetic code*. The monumental task of deciphering the genetic code was completed in 1966, and three U.S. scientists—Robert W. Holley, H. Gobind Khorana, and Marshall W. Nirenberg—were awarded the Nobel Prize in 1968 for this

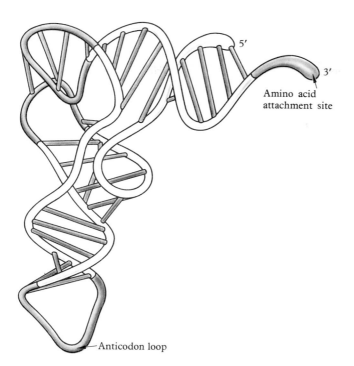

Fig. 22-9 A Diagram of the Three-Dimensional Structure of Transfer RNA (Gray bonds represent hydrogen bonds.)

achievement. The code requires a sequence of three nucleotides in DNA in order to specify one amino acid in each finished polypeptide. The complementary nucleotide triplets in mRNA, called *codons,* occur sequentially along the mRNA chain; overlapping of one codon with another usually does not occur. The segment of DNA that contains the information needed for synthesis of one polypeptide is called a *gene* (a newer term is *cistron*). The genetic code was deciphered between 1961 and 1966, and it was found that there are 64 codons; 61 of these correspond to amino acids and the remaining three serve as signals for terminating the growing polypeptide chain (see Table 22-5). As mentioned earlier, many amino acids are designated by more than one triplet; only tryptophan and methionine are coded by just one triplet. The genetic code as described here is used by all known life forms.

EXAMPLE 22.3

The following segment of mRNA codes for a peptide. What is the amino acid sequence of the peptide? (The code reads from left to right.)

UUACAAGGUCUG

SOLUTION

Since the genetic code usually consists of nonoverlapping nucleotide triplets, the mRNA segment, which contains 12 nucleotides, codes for four amino acids, and a tetrapeptide will be formed with the sequence Leu-Gln-Gly-Leu.

UUA CAA GGU CUG

Leu - Gln - Gly - Leu

Table 22-5 The Genetic Code. The 64 triplet codons are listed in the 5' ⟶ 3' direction in which they are read, and the three termination (Term.) codons are shown in color.

	U	C	A	G
U	UUU Phe UUC Phe UUA Leu UUG Leu	UCU Ser UCC Ser UCA Ser UCG Ser	UAU Tyr UAC Tyr UAA Term. UAG Term.	UGU Cys UGC Cys UGA Term. UGG Trp
C	CUU Leu CUC Leu CUA Leu CUG Leu	CCU Pro CCC Pro CCA Pro CCG Pro	CAU His CAC His CAA Gln CAG Gln	CGU Arg CGC Arg CGA Arg CGG Arg
A	AUU Ile AUC Ile AUA Ile AUG Met	ACU Thr ACC Thr ACA Thr ACG Thr	AAU Asn AAC Asn AAA Lys AAG Lys	AGU Ser AGC Ser AGA Arg AGG Arg
G	GUU Val GUC Val GUA Val GUG Val	GCU Ala GCC Ala GCA Ala GCG Ala	GAU Asp GAC Asp GAA Glu GAG Glu	GGU Gly GGC Gly GGA Gly GGG Gly

EXERCISE 22.3 The following segment of mRNA codes for a peptide. What is the amino acid sequence of the peptide? (The code reads from left to right.)

CAAAUAGUCGAUCGG

When mRNA is synthesized, the coding information in DNA is transcribed in the nucleotide sequence of mRNA; thus, synthesis of mRNA is referred to as transcription. The coding information in DNA is translated into protein synthesis *via* mRNA in a complicated process at the ribosomes. Each tRNA carries a nucleotide triplet called the *anticodon* (see Figure 22-9) that is complementary to one of the amino acid codons listed in Table 22-5. Each of these tRNAs can be joined by an ester linkage to its corresponding amino acid at the 3'-terminus of tRNA by enzymes specific for each amino acid.

EXAMPLE 22.4

A certain protein having only one chain has a molecular weight of 48,000 amu. If the average amino acid residue has a molecular weight of 120 amu, how many nucleotides are there in the mRNA coding for this protein? How many nucleotides are there in the gene for this protein?

SOLUTION

Since the molecular weight of the protein is 48,000, there are 400 amino acid residues in the protein.

$$\frac{48,000 \text{ amu}}{120 \text{ amu/residue}} = 400 \text{ residues}$$

Each amino acid residue requires three nucleotides for its codon in mRNA. Therefore, the mRNA coding will contain 1200 nucleotides.

$$400 \text{ residues} \times \frac{3 \text{ nucleotides}}{\text{residue}} = 1200 \text{ nucleotides in mRNA coding}$$

Since the gene for this protein is double-stranded DNA, the gene will contain twice the number of nucleotides in the mRNA coding, or 2400 nucleotides.

EXERCISE 22.4 A protein having only one chain has a molecular weight of 67,000 amu. If the average amino acid residue has a molecular weight of 120 amu, how many nucleotides are there in the mRNA coding for this protein? How many nucleotides are there in the gene for this protein?

In order for polypeptide synthesis to begin, several units of biological material must come together: an mRNA molecule, a ribosome, certain protein regulatory factors, and a particular initiating tRNA with its attached amino acid. Each ribosome contains one molecule each of the forms of rRNA described earlier, enzymes that catalyze formation of peptide bonds, and other proteins. These components are organized so that each ribosome consists of a large subunit and a small subunit (Figure 22-10).

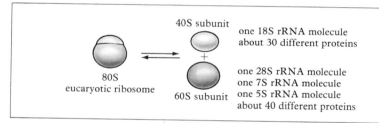

Fig. 22-10 The Subunit Composition of Cytoplasmic Ribosomes from Eucaryotic Cells (During protein synthesis, the participating ribosomes are intact; after a polypeptide is finished, each ribosome dissociates into a large subunit and a small subunit. Subunit compositions are shown on the right of each subunit in the diagram.)

After polypeptide synthesis has been initiated, tRNAs bring in their attached amino acids to the synthetic complex and anticodons of tRNAs associate by hydrogen bonding with complementary codons of mRNA. This association brings appropriate amino acids into position for incorporation into the growing polypeptide chain. The formation of each peptide bond requires energy equivalent to three ATP molecules. Elongation of the chain continues until the ribosome complex encounters a termination codon on mRNA. Since no tRNAs have anticodons that are complementary to termination codons, polypeptide synthesis stops, and the polypeptide chain is hydrolyzed from the last tRNA. At this time, mRNA dissociates from the ribosome. The completed polypeptide folds into its native conformation as it is synthesized, and after it is released from the ribosome, it is free to associate with other polypeptide chains if it is to be a subunit of a protein having quaternary structure.

As illustrated in Figure 22-11, many ribosomes are usually attached to a single mRNA molecule, simultaneously synthesizing each of their growing polypeptide chains. The complex of ribosomes and mRNA is called a *polyribosome* or a *polysome*.

Fig. 22-11 A Simplified Illustration of a Polyribosome. Ribosomes move along the mRNA in the 5' ⟶ 3' direction, functioning independently of each other.

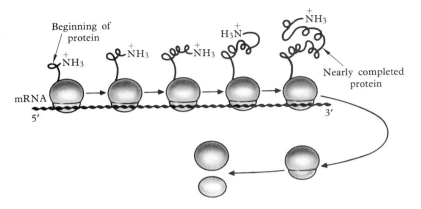

EXAMPLE 22.5

The expression of genetic information involves three major steps: replication, transcription, and translation. What happens in each of these three steps?

SOLUTION

In replication, DNA molecules are synthesized as replicas of the original DNA. In transcription, the genetic code contained in DNA is transcribed into the form of mRNA during synthesis of mRNA. In translation, mRNA carries the genetic message to the ribosomes, where the message is decoded by mRNA serving as a template for directing the amino acid sequence of polypeptides during protein biosynthesis.

EXERCISE 22.5 What substances are ribosomes made of? What is the quaternary structure of ribosomes?

GENETIC MUTATIONS

Any change, even a minor one, in the sequence of nucleotides of DNA brings about a change in the genetic message encoded in DNA. Such an alteration in the genetic message is called a *mutation*. There are several types of mutations: (1) the *substitution* of one or more base pairs for one or more other base pairs; (2) the *deletion* of one or more base pairs; (3) the *insertion* of one or more base pairs. The substitution of one base pair for another is the most common type of mutation.

Mutations occur for many different reasons. Some mutations happen spontaneously as a result of small deviations in base pairing that lead to errors in DNA replication. These are relatively rare, happening about once per offspring per generation. This means that each offspring of a given generation probably contains one gene that is different from those inherited from the parents. Other mutations may be caused by radiation such as X-rays and ultraviolet light. Thus, sunbathing may increase the number of mutations in an individual. A third cause of mutations is the action of chemical agents; some of those suspected are ethanol, caffeine, automobile exhaust, cigarette smoke, nicotine, marijuana, LSD, aspirin, various pollutants, and even some food preservatives.

Some mutations may be more serious than others. For example, if an alteration in DNA results in a change in the amino acid sequence of an enzyme, the change would be relatively harmless if the active site and overall conformation of the enzyme are not affected. However, if the amino acid sequence at the active site is altered, the enzyme would likely be less active or perhaps inactive; a similar result might be observed if the tertiary structure of the enzyme is altered. Such changes might result in serious consequences for the organism, as illustrated by the numerous inborn errors of metabolism discussed in previous chapters.

EXAMPLE 22.6

Classify each of the following mutations as substitution, deletion, or insertion.
A normal segment of a single strand of DNA:

		TTTAAAGATCGA
(a)	Mutation	TTTAAGATCGA
(b)	Mutation	TTTAAAGACCGA
(c)	Mutation	TTTAAAGATACGA

SOLUTION

(a) Deletion
(b) Substitution
(c) Insertion

EXERCISE 22.6 Classify each of the following mutations as substitution, deletion, or insertion.
A normal segment of a single strand of DNA:

		CCAGCCGTGAGA
(a)	Mutation	CCATGCCGTGAGA
(b)	Mutation	CTAGCCGTGAGA
(c)	Mutation	CCAGCCGTGAG

VIRUSES

Viruses (Figure 22-12) are packets of nucleic acid surrounded by protective coats. When outside a living cell, they are inert, but when they infect living cells, they cause the affected cell to manufacture more viruses. Viruses are unable to generate metabolic energy or to synthesize proteins on their own, and there is disagreement among biologists as to whether viruses should be considered living organisms. Viruses range in complexity from some having only four genes to others with about 250 genes.

The sizes and shapes of viruses vary, but all are either rods or spheres (or a combination of these shapes). Each contains one or more molecules of DNA or RNA surrounded by a protein coat or covering. The nucleic acid may be single- or double-stranded, and the protein coat protects

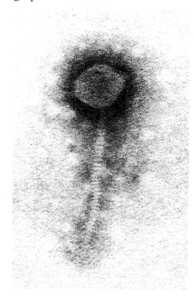

Fig. 22-12 An Electronmicrograph of a Virus.

PERSPECTIVE: Slow Viruses

Slow viruses are said to be the most peculiar of all forms of life, if one considers a virus to be alive. They are characterized by extremely long incubation periods, so long that the infection is not apparent for many years. Slow viruses are almost impossible to kill, resisting such traditional sterilizing procedures as boiling water and exposure to alcohol, formaldehyde, and other disinfectants.

One particular kind of slow virus has been removed from the brain tissue of demented patients in a landlocked tribe in New Guinea, where most members of the group lost their minds by the age of 40 to a disease known as *kuru*. A fact that led to discovery of the virus was that members of the tribe had long practiced ritual cannibalism by eating the brains of the deceased. The virus proved so resistant to sterilization that researchers were able to recover active forms of it from specimens of brain tissue stored in aqueous formaldehyde for at least ten years. When injected into laboratory animals, it produces no evidence of disease for at least 18 months, and then the animal brains undergo rapid destruction. The mysterious virus apparently is too small to be visible under the electron microscope, and scientists have not been able to find any nucleic acid in it, yet it seems to be able to multiply many times over in the affected brain. It has been learned that a disease similar to *kuru* occurs sporadically elsewhere in the world (under the name of Creutzfeldt-Jacob disease), and a similar virus has been recovered from the brain tissue of the victims.

The resemblance of *kuru* and Creutzfeldt-Jacob disease to the senile dementia that afflicts older people is striking. This brain disease, whose major form is known as Alzheimer's disease, affects more than 500,000 people over the age of 50 in the United States, most of them in their seventies and eighties. It accounts for most of the elderly in nursing homes, and the number is likely to grow larger as the proportion of elderly people in the U.S. population increases over the next few decades. Its cause is not currently known.

---

the nucleic acid and determines what type of cell a particular virus can infect. All types of cells seem to be susceptible to viral infections of one sort or another.

When a virus infects a cell, it first attaches to the surface of the cell and then injects its nucleic acid into the cell. The protein coat remains outside the cell, but the viral nucleic acid enters the cell cytoplasm. There it commandeers the synthesizing machinery of the cell, stopping normal synthesis of DNA, RNA, and proteins, and causing synthesis of new viruses instead. Soon after infection, the host-cell DNA is degraded by viral-produced enzymes. After a large number of viruses has been produced, the infected cell releases them to the environment, either by bursting or by allowing the viruses to pass through the cell membrane. The process of viral infection is outlined in Figure 22-13.

One of the reasons that RNA viruses are successful in taking over the operations of host cells is that the RNA viruses contain genetic information for the synthesis of enzymes that direct nucleic acid synthesis from viral RNA templates. An RNA virus can direct synthesis of *RNA-directed RNA polymerase* (also called RNA replicase or RNA synthetase) or of *RNA-directed DNA polymerase* (also called reverse transcriptase). Thus, this new enzyme causes the host cell to either replicate viral RNA or use it as a template for synthesizing new DNA. In either case, the net result is the synthesis and construction of new virus particles.

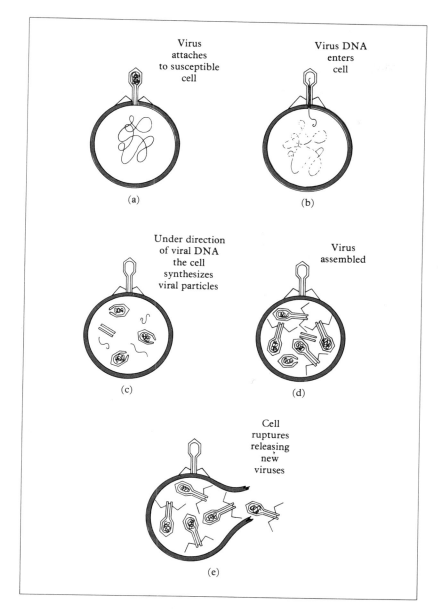

Fig. 22-13 An Outline of the Steps Involved in the Viral Infection of a Cell

In 1979 it was found that a virus contributes to at least one form of diabetes. For the first time, scientists isolated a virus from the pancreas of a patient with juvenile-onset diabetes. This form of diabetes, which strikes less than 0.1% of the U.S. population, is generally more severe than adult-onset diabetes and is believed to have different causes. When the isolated virus was used to inoculate mice, the mice developed diabetes, and the researchers were able to isolate the virus from pancreatic tissue of the diabetic mice. Apparently, the virus causes destruction of pancreatic cells that normally produce insulin. Interestingly, not all of the inoculated mice developed diabetes. It is thought that genetic factors play an important role in susceptibility to diabetes, and in fact, the diabetic patient had relatives with diabetes.

CARCINOGENS

A *carcinogen* is any substance or agent that causes cancer. The malignant process begins as a progressive, unrestrained division of abnormal cells. As the disease progresses, the cells usually become increasingly abnormal in appearance, structure, and function. Cancer grows in all known groups of animals except the lowest forms; even plants develop cancer-like growths. Although there are a few kinds of inherited cancer, such as xeroderma pigmentosum, it is thought that common cancers are not generally inherited. However, the chances of a person getting cancer are increased if there is a family history of cancer. Proven causes of cancer are chemical compounds, radiation, and viruses, as discussed below.

Chemical Carcinogens

Many substances in our environment cause mutations, and they are called *mutagens*. These include such things as certain components of air pollution and cigarette smoke, some herbicides and pesticides, some chemical contaminants in food, and various drugs. Mutations may produce cancer cells, and in these cases, the mutagens are then classified as carcinogens. Hundreds of chemical compounds are known to induce cancer in animals, but fewer, perhaps only dozens, have been proven to be carcinogens in humans.

Part of the problem in identifying carcinogens is that traditional animal tests require an average of three years, and only hundreds of suspected carcinogens can be tested during such a period. It has been discovered recently that benzene may produce leukemia, and benz[a]pyrene (a hydrocarbon present in combustion and distillation products of coal, oil, and other organic materials) has been associated with a high incidence of lung cancer. Other known or suspected chemical carcinogens are trichloromethane (chloroform), tetrachloromethane (carbon tetrachloride), arsenic compounds, asbestos (a form of magnesium silicate), β-naphthylamine, nickel compounds, nitrosamines, various polycyclic aromatic hydrocarbons, degradation products of cyclamates (artificial sweeteners), and aflatoxins (fat-soluble compounds produced by mold growing on improperly processed peanut meal and other food). Structures for some of these follow:

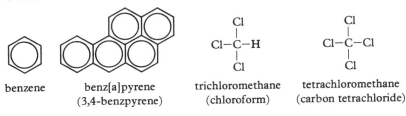

benzene   benz[a]pyrene (3,4-benzpyrene)   trichloromethane (chloroform)   tetrachloromethane (carbon tetrachloride)

β-naphthylamine: aromatic structure with NH_2 group

R_2N-NO — a nitrosamine (formed by reaction of 2° amines with nitrous acid or nitrite ions)

a cyclamate: $(C_6H_{11}-N(H)-SO_3^-)_2 \, Ca^{2+}$

Since chemical carcinogens are usually mutagens, it appears that damage to DNA is fundamental to the occurrence of cancer. An important new test for mutagens was recently devised by Bruce Ames at the University of California, Berkeley. The test is simple to run, very sensitive, and allows for mass screening of suspected mutagens. The Ames test uses a specially grown mutant strain of bacteria that are unable to grow in the absence of histidine because they cannot synthesize this amino acid. The addition of a mutagen to a culture of these bacteria causes many new mutations, and a few of these reverse the original mutation. Some of the new mutants can then grow in the absence of histidine, confirming their exposure to mutagens. This rapid, inexpensive new test will greatly aid in identifying mutagens.

Radiation

Energy from the ultraviolet rays of the sun and X-rays, as well as other types of radiation, causes a variety of cancers. It is thought that DNA damage caused by these agents somehow results in the origin of cancer cells. Pioneer workers with radium and X-rays, Marie Curie and her daughter Irene Joliot-Curie, for example, developed cancer from overexposure to radiation, and uranium miners have been found to have a higher than normal incidence of lung cancer.

Radium is in the same chemical family as calcium and because of their similar properties, radium salts are deposited in the bones, giving rise to bone cancers. In the 1920s, many women were employed to paint radium coatings on watch dials to make them glow in the dark. Unfortunately, some of the painters licked their brushes to make a fine point and thus ingested relatively large amounts of radium. In those cases, there was an unusually high incidence of bone cancer and subsequent death.

Other evidence of radiation-induced cancer comes from the fact that the incidence of skin cancer is related to the amount of sunshine absorbed. In these cases, skin color is important, and skin cancer is most common among light-skinned people. In addition, survivors of the atomic bombs at Hiroshima and Nagasaki have an abnormally high incidence of leukemia and thyroid cancer, and possibly also lung and breast cancer.

Viruses

Over 100 viruses have been discovered that cause cancer in every major group of animals except the lower forms. Although viruses have not been definitely proved to be human carcinogens, the circumstantial evidence suggests that one or more viruses cause cancer in humans. In particular, viruses and viruslike particles have been observed in malignant tumor cells of humans.

Certain RNA and DNA viruses that cause cancer in animals contain only four or five genes, and it is thought that just one or two virus genes are directly involved. When animal cells in special tissue-culture systems are injected with cancer-causing viruses, the transformed animal cells grow continuously and chaotically. These transformed cells contain viral DNA that has been integrated into host-cell genes, and consequently the transformation is then inheritable. In addition, some transformed cells grow into a cancer when they are injected in sufficient number into animals.

The resistance of animal cells to many viruses is greatly increased by a group of small glycoproteins containing about 150 amino acid residues; these proteins have been named *interferons*. Interferons are synthesized and secreted by animal cells in response to virus infection. Interferons bind to cell membranes, and the cells then acquire resistance to a variety of viruses, in contrast to the highly specific immunity brought about by antibody production. It appears that interferons act by inhancing production of certain enzymes that block protein synthesis in infected cells. By an unknown process, interferons also stimulate immune cells that destroy cancer cells.

GENETIC ENGINEERING

During the past decade, scientists have learned an enormous amount of information about the molecular processes of genetics. By 1970, H. Gobind Khorana and his coworkers, using DNA fragments and appropriate enzymes, had assembled a yeast gene, but it did not function. Later they switched their attention to a bacterial gene, and by 1976 they had succeeded in the first laboratory synthesis of a functional gene. Their accomplishment represented a major breakthrough in molecular biology, and it opened the door to the possibility of genetic engineering—the laboratory manipulation of DNA to modify the hereditary mechanisms of microorganisms or cells. However, it was the discovery of restriction enzymes in the 1960s and 1970s by Werner Arbor, Hamilton Smith, and Daniel Nathans (corecipients of the Nobel Prize in 1978) that made genetic engineering possible.

Enzymes Used in Genetic Engineering

Restriction enzymes are found in a wide variety of bacterial cells, and their biological role is to cleave DNA molecules. These enzymes are normally part of a mechanism that allows certain bacteria to protect themselves from invasion of foreign DNA such as that of a virus. In these bacteria, some of the bases in their DNA have methyl groups attached. Examples of such methylated bases follow.

N⁶-Dimethyladenine   1-Methylguanine   5-Methylcytosine

The methylated DNA of these bacteria is not attacked by the restriction enzymes, but foreign DNA, lacking the specific pattern of methyl groups, is rapidly cleaved in both strands and thus becomes nonfunctional. Since there is a "restriction" on the type of DNA that is allowed to be in the bacterial cell, the protective enzymes are called restriction enzymes.

Restriction enzymes act at sites on DNA called *palindromes*. In language, a palindrome is any word or statement that reads the same in either direction, such as "radar" or "Madam, I'm Adam." For double-stranded DNA, a palindrome is a section in which the two strands have the same sequence but running in opposite directions. Examples of DNA palindromes follow; the arrows indicate the points of attack by restriction enzymes.

```
          ↓
5' C—C—G—C—G—G 3'        5' G—G—A—T—C—C 3'
   |  |  |  |  |  |              ↓ |  |  |  |  |  |
3' G—G—C—G—C—C 5'        3' C—C—T—A—G—G 5'
          ↑                               ↑
```

At least 100 restriction enzymes are known, and each catalyzes cleavage of DNA in a specific and predictable way. These enzymes are the tools that are used to take DNA apart and cut it into fragments of known size and nucleotide sequence.

Another set of enzymes, called *DNA ligases*, has been known since 1967. These enzymes normally function to connect DNA fragments during DNA synthesis, and they are used in genetic engineering to put together pieces of DNA produced by the restriction enzymes. By use of the two kinds of enzymes, it is possible to cut apart a long thread of double-stranded DNA and reassemble the pieces to construct a desired gene. However, getting this newly formed gene into a bacterial cell requires the assistance of circular pieces of double-stranded DNA called plasmids.

Plasmids

Plasmids are circular double-stranded DNA molecules ranging in size from two thousand to several hundred thousand nucleotides. They are found in bacteria separate from chromosomes and are dispensible under certain conditions. Plasmids function as accessories to chromosomes by carrying genes for the inactivation of antibiotics, the metabolism of natural products, and the production of toxins. They have the unusual ability to replicate independently of chromosomal DNA. A typical bacterial cell contains about twenty plasmids and one or two chromosomes.

Recombinant DNA

Genetic engineering relies on the use of restriction enzymes and DNA ligases to combine portions of DNA from two quite different sources. In this way, new combinations of unrelated genes can be constructed in the laboratory. The newly tailored DNA, called *recombinant DNA,* can then be inserted into an appropriate cell where it will be reproduced and its genetic information expressed. The major steps in making recombinant DNA and putting it to use are (see Figure 22-14):

1. *Construction of a recombinant DNA molecule.* A DNA fragment of interest is covalently joined to a DNA carrier, usually a plasmid fragment, by use of DNA ligases. In the terminology of genetic

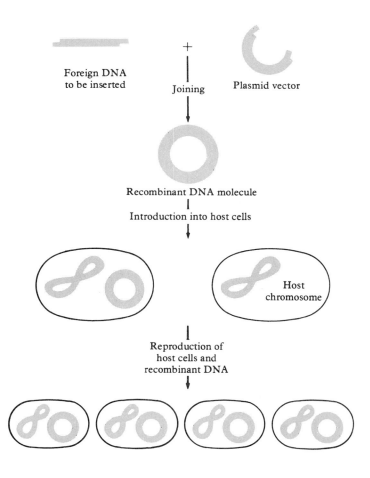

Fig. 22-14 Construction and Replication of Recombinant DNA

engineering, the carrier is called a *vector;* the essential feature of a vector is that it have the capability of replication independent of chromosomal DNA.

2. *Introduction of recombinant DNA into host cells.* Most cells take up free DNA from the surrounding medium. Thus, if appropriate host cells are bathed in a solution containing recombinant DNA plasmids, they will make their way into the cells.

3. *Cell division and replication of recombinant DNA.* After a recombinant DNA plasmid enters a cell, the cell divides, replicating all of its genes including the recombinant DNA plasmid. Subsequently, both the original DNA of the cell and the foreign DNA that has been inserted will be used to make messenger RNA, and the corresponding proteins will be synthesized.

Since the early 1970s, several genetic engineering companies have been established with the goal of production of polypeptides and proteins on a scale far beyond what had ever been achieved. In essence, specially designed recombinant DNA is inserted into bacterial cells, and the large cultures of bacteria synthesize valuable polypeptides and proteins coded for by the foreign DNA. Among the substances recently produced in this way is human insulin. Traditionally, the only convenient source of insulin has been pig and cow pancreases; supplies have barely kept pace with demand, and many diabetics suffer from allergic reactions to these forms of insulin. Human insulin produced by genetic engineering was in the final phase of testing by the Food and Drug Administration in 1981, and it is expected to be approved for use by 1983.

Another substance which is likely to be produced on a large scale is human growth hormone, used to treat dwarfism. The present limited supplies of the hormone, which is usually harvested from pituitary glands of humans soon after death, are insufficient to treat all who might benefit. It appears that genetic engineering can produce human growth hormone on a laboratory scale, and perhaps soon it will be manufactured in large quantities. Other recent achievements of genetic engineering are the large-scale synthesis of human interferon (proposed to be effective in treating cancer) and urokinase (an enzyme used to dissolve blood clots).

Genetic engineering holds great promise for the future in its potential to produce valuable protein molecules otherwise available in only small quantities. Such things as blood clotting factors for treating hemophilia and various polypeptide hormones might well be produced in abundance by using recombinant DNA technology. In fact, it is thought by some that genetic engineering will prove to be the most important technology of the 1980s and perhaps even the late twentieth century. But along with the potential benefits comes the possiblility of future adverse effects that might arise from producing mutant strains of uncontrollable bacteria that are resistant to antibiotics or from accidentally creating a human mutation. These and other concerns, expressed early on by scientists working in the field and later by lay citizens, led to a set of guidelines provided by the National Institutes of Health. Workers in genetic engineering must adhere to these guidelines, which require the use of microorganisms that cannot live outside the sheltered laboratory environment and strict isolation measures for work with unusual microorganisms. These guidelines and the excellent safety record established by genetic engineering have allayed most of the concerns originally held about the work.

SUMMARY

Nucleic Acids: Nucleic acids, acidic substances found in the nucleus and other cell parts, are of two types: DNA and RNA. Nucleic acid hydrolysis products are phosphate, a pentose, and a mixture of the heterocyclic nitrogen bases: adenine, guanine, cytosine, uracil, and thymine. Nucleosides are compounds composed of a nitrogen base bonded to ribose or deoxyribose. If a nucleoside is phosphorylated at the 5'-, 3'-, or 2'-position of the pentose, the product is a nucleotide. Nucleotides also exist as nucleoside di- and triphosphates. Nucleotides, the repeating units of nucleic acids, are joined together by 3', 5'-phosphoester linkages in nucleic acids, and the order of nucleotides in a nucleic acid is its primary structure.

DNA: The secondary structure of DNA is a double helix held together by hydrogen bonds between specific pairs of bases: adenine forms hydrogen bonds with thymine and guanine forms hydrogen bonds with cytosine. The sequence of nucleotides in DNA serves as the primary reservoir of genetic information, and when a cell divides, the original DNA is replicated to form DNA for the new cell. Elongation of DNA utilizes DNA polymerases, of which deoxyribonucleoside triphosphates are substrates. All cells have mechanisms for repairing damaged DNA.

RNA: RNA exists in three types. Messenger RNA (mRNA) is synthesized by the process of transcription, in which a single-stranded segment of DNA serves as the template. Ribonucleoside triphosphates are the substrates for RNA polymerase, the enzyme that brings about RNA synthesis. Transfer RNA (tRNA) molecules, also synthesized in the nucleus, act as carriers of specific individual amino acids during protein biosynthesis. Ribosomal RNA (rRNA) accounts for most of the mass of ribosomes, the sites of protein biosynthesis.

Protein Biosynthesis: The nucleotide sequence of DNA, called the genetic code, consists of sequential, nonoverlapping nucleotide triplets. Each triplet specifies an amino acid in the primary structure of a polypeptide. The genetic code is transferred to mRNA during transcription; mRNA and tRNA with attached amino acids associate with the ribosomes for protein biosynthesis. The anticodons of tRNA bind to codons on mRNA so that polypeptide chains are formed and lengthened, and after completion, the polypeptide is released from the ribosome.

Genetic Mutations: An alteration in the nucleotide sequence of DNA is called a mutation. The three types of mutations are substitution, deletion, and insertion. Mutations are caused by errors in DNA replication, radiation, and chemical agents.

Viruses: Viruses, packets of DNA or RNA surrounded by protein coats, are inert outside of living cells, but when they infect living cells, they cause the cells to manufacture more viruses. Viruses infect cells by injecting viral nucleic acid into the cell.

Carcinogens: A carcinogen is any substance or agent that causes cancer. Carcinogens may be chemical compounds, radiation, or viruses.

Genetic Engineering: The laboratory manipulation of DNA to modify genetic mechanisms of microorganisms or cells is called genetic engineering. Genetic engineering makes use of DNA ligases to combine portions of DNA from different sources. Recombinant DNA is introduced into host cells through the use of vectors, often bacterial plasmids. The host cells then replicate the recombinant DNA, and the corresponding proteins will be synthesized. Genetic engineering has the potential of providing a means of large-scale synthesis of valuable polypeptides and proteins.

STUDY QUESTIONS AND PROBLEMS

1. Define the following terms:
   - (a) nucleic acids
   - (b) DNA
   - (c) RNA
   - (d) purines
   - (e) pyrimidines
   - (f) nucleoside
   - (g) nucleotide
   - (h) nucleoside triphosphate
   - (i) double helix
   - (j) base pairs
   - (k) chromosome
   - (l) DNA polymerases
   - (m) thymine dimer
   - (n) mRNA
   - (o) transcription
   - (p) DNA-directed RNA polymerase
   - (q) tRNA
   - (r) rRNA
   - (s) genetic code
   - (t) codon
   - (u) gene
   - (v) anticodon
   - (w) termination codon
   - (x) polyribosome
   - (y) genetic mutation
   - (z) virus
   - (aa) RNA-directed RNA polymerase
   - (bb) RNA-directed DNA polymerase
   - (cc) carcinogen
   - (dd) mutagen
   - (ee) interferons
   - (ff) genetic engineering
   - (gg) restriction enzymes
   - (hh) palindrome
   - (ii) DNA ligases
   - (jj) plasmid
   - (kk) recombinant DNA

2. What pentose is present in DNA? In RNA?

3. Which heterocyclic nitrogen bases are found in DNA? In RNA?

4. Which of the heterocyclic bases in question 3 are pyrimidines? Which are purines?

5. What is the major difference between a nucleoside and a nucleotide?

6. Draw structures for the following nucleotides:
   - (a) AMP
   - (b) dGTP
   - (c) dTDP
   - (d) GMP
   - (e) dATP
   - (f) UDP

7. Draw structures for the following polynucleotides derived from DNA:
   - (a) 5' A—C—G—T 3'
   - (b) 5' G—A—C—T 3'
   - (c) 5' A—T—C—G—A 3'

8. Draw structures for the following polynucleotides derived from RNA:
   - (a) 5' U—C—G—A 3'
   - (b) 5' C—C—U—G 3'
   - (c) 5' A—G—C—U—C 3'

9. Why are molecular weights for eucaryotic DNA only approximate?

10. Describe the DNA double helix by answering the following questions.
    - (a) What is the backbone of each of the two helices?
    - (b) What forces hold the double helix together?
    - (c) What is the direction of one helix relative to the other?
    - (d) Where are the bases located relative to the helix axis?
    - (e) Why do the paired bases lie in the same plane?
    - (f) Why do the bases pair?
    - (g) Why are paired bases said to be complementary?

11. Draw structures for the two types of base pairs in DNA, and illustrate the hydrogen bonds formed in each pair.

12. In all organisms, DNA contains equal amounts of adenine and thymine and equal amounts of guanine and cytosine. What is the reason for this observation?

13. Write the sequence for a DNA strand that would form a double helix with the following DNA segment:

    5' G—C—A—T—A—C—G—T—A 3'

14. Describe the process by which DNA is replicated.

15. Describe the major form of damage done to DNA by ultraviolet light.

16. What is the molecular defect in the genetic disease xeroderma pigmentosum?

17. Name the three types of RNA, and give their molecular weight ranges.

18. Describe the process of transcription.

19. What is the cellular role of each of the three types of RNA?

20. Each of the following segments of mRNA codes for a peptide. Give the amino acid sequence for each peptide. (The code reads from left to right.)
    - (a) CUAGCACCA
    - (b) AAGCACUCGCUG
    - (c) UUGAUGGUGGCCAAA

21. Horse myoglobin has a molecular weight of 16,900 amu. If the average amino acid residue has a molecular weight of 120 amu, how many nucleotides are there in the mRNA coding for horse myoglobin? How many nucleotides are there in the gene for horse myoglobin?

22. In normal adult human hemoglobin, there are 141 amino acid residues in each α-chain and 146 amino acid residues in each β-chain. How many nucleotides are there in the mRNA coding for each subunit? How many nucleotides are there in the gene for each subunit?

23. Explain how an alteration in the normal nucleotide sequence of DNA can result in synthesis of a defective protein.

24. Classify each of the following mutations as substitution, deletion, or insertion.
    A normal segment of a single strand of DNA:
    AGCAGCGAGCGT
    (a) Mutation: AGTAGCGAGCGT
    (b) Mutation: AGCAGCAGCGT
    (c) Mutation: AGCAGCCGAGCGT

25. In a substitution mutation, the mRNA synthesized on the altered DNA had the triplet UGA substituted for CUG. What will be the effect of the mutation on the polypeptide coded for by the affected portion of mRNA?

26. In sickle-cell hemoglobin (HbS), glutamic acid at position 6 in the β-chain is replaced by valine. What kind of mutation (substitution, deletion, or insertion) does this correspond to? Use the codons for glutamic acid and valine to illustrate the possible alterations in mRNA caused by this mutation.

27. Describe the general structure of a virus.

28. Outline the process by which a virus infects a cell.

29. Why are the enzymes RNA-directed RNA polymerase and RNA-directed DNA polymerase important to RNA viruses?

30. List three proven causes of cancer.

31. Summarize the procedure for the Ames test used to identify mutagens. Why is this test an important new development?

32. What are interferons? Why are they anticipated to be effective in cancer treatment?

33. Describe the actions of the two major enzymes used in genetic engineering.

34. How is recombinant DNA constructed?

35. In genetic engineering, what is a vector? Why do plasmids make good vectors?

36. How is it that genetic engineering can produce large quantities of proteins that are hard to obtain otherwise?

37. List four substances likely to be produced on a large scale by genetic engineering, and give an important use for each.

CHAPTER 23
NUTRITION

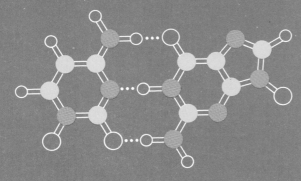

HUMAN NUTRITIONAL REQUIREMENTS
ENERGY PROVIDED BY THE MACRONUTRIENTS
THE MACRONUTRIENTS
- Proteins
- Fats
- Carbohydrates

THE MICRONUTRIENTS
- The Water-Soluble Vitamins
- The Fat-Soluble Vitamins
- Minerals

BALANCED DIETS
DIET THERAPY
- Restorative Diets
- Restrictive Diets

Nutrition is the science of nourishment; it focuses on the study of nutrients and the processes by which organisms utilize them. Nutrition is a relatively young science, having developed mostly in the twentieth century. The term vitamin, derived from the words "vital amine," was used first in 1911, and the isolation, identification, and synthesis of all known vitamins occurred in the present century.

Nutrition is an applied science; that is, its principles are used in actual practice to determine food choices and proper diets. Inorganic, organic, and biochemistry are all important to the study of nutrition, as is physiology. Although coverage of nutrition in this chapter is meant to be only an introduction to human nutrition, you will find many references made to information presented earlier in this book.

HUMAN NUTRITIONAL REQUIREMENTS

The *nutrients* required by humans can be classified as macronutrients and micronutrients. *Macronutrients* are needed in relatively large amounts (gram quantities) to supply energy and building materials for the body; thus the macronutrients are proteins, fats, and carbohydrates. The *micronutrients*, required only in small amounts (milligram or microgram quantities), are vitamins and minerals.

In addition to nutrients, two other nutritional factors are important: water and fiber. Water accounts for 60–70% of human body weight, and cells cannot function properly unless certain of their components are dissolved in water. The loss of 10% of a person's body water poses a serious threat, and the loss of 20% is almost always fatal. Fiber, the indigestible material obtained from plants, is composed mostly of polysaccharides that are resistant to attack by digestive enzymes. Although fiber does not make a chemical contribution to the diet, it does help to

nutrient: any substance that serves to sustain a living organism by promoting growth, replacing loss, or providing energy.

relieve constipation by absorbing water and softening the stool. The amount of fiber needed in the diet and the source of that fiber are matters of controversy at present.

Agencies in a number of countries have established recommendations for amounts of the various nutrients thought to be necessary for maintaining the good health of most of the population. In the United States, the official guidelines (Recommended Dietary Allowances, or RDAs) are established by the Food and Nutrition Board of the National Research Council of the National Academy of Sciences. About every five years, this agency publishes revised guidelines for 17 different population groups classified on the basis of age, sex, height, and weight; these guidelines are given in Table 23-1.

Recommended allowances are sometimes mistaken for *minimal requirements* (the least amount of each nutrient needed to avoid symptoms of deficiency), but it should be emphasized that the quantities in Table 23-1 include a large safety margin. The recommendations are scientific estimates of a reasonable excess of all nutrients for at least 95% of the population. Thus, the recommended allowances are quantities much larger than those experimentally determined as minimal requirements.

ENERGY PROVIDED BY THE MACRONUTRIENTS

The daily intake of calories in a person's diet is generally provided by the macronutrients (proteins, fats, and carbohydrates). Although carbohydrates are the major energy source for most of the world's human population, Americans tend to rely more heavily than other populations on proteins and fats as energy sources.

As emphasized in chapters 19–21, aerobic metabolic processes in the mitochondria oxidize pyruvate (from carbohydrates), amino acid carbon skeletons (from proteins), and fatty acids (from fats). The free energy (ΔG, expressed in calories) released by oxidation of the macronutrients is the source of chemical energy needed by a person to produce ATP necessary for normal life processes. Each gram of fat in the diet results in production of about nine kilocalories, whereas proteins and carbohydrates each produce about four kilocalories per gram. (Remember that one food Calorie is equivalent to one kilocalorie or 1000 calories.) Obesity is generally thought to be a result of a daily intake of calories that exceeds a person's expenditure of calories; such a caloric imbalance causes the excess energy consumed to be stored in the body as fat. (One pound of body fat is roughly equivalent to 3,500 Cal or kcal.) In an effective weight-losing program, the goal is to take in fewer calories, while maintaining a balanced diet, than are expended on a daily basis. Regular exercise is a recommended part of such a program.

The term "empty calories" has become popular in reference to food items that produce energy but contribute none or very little of the micronutrients needed by a person. Sugar is the major food item in this category, and nutritionists are concerned that consumption of such things

Table 23-1 Recommended Daily Dietary Allowances for the United States,[a] Revised 1980

Designed for the maintenance of good nutrition of practically all healthy people in the U.S.A.

	Age (years)	Weight (kg)	Weight (lbs)	Height (cm)	Height (in)	Protein (g)	Fat-Soluble Vitamins			Water-Soluble Vitamins							Minerals					
							Vitamin A (μg R.E.)[b]	Vitamin D (μg)[c]	Vitamin E (mg α T.E.)[d]	Vitamin C (mg)	Thiamin (mg)	Riboflavin (mg)	Niacin (mg N.E.)[e]	Vitamin B$_6$ (mg)	Folacin[f] (μg)	Vitamin B$_{12}$ (μg)	Calcium (mg)	Phosphorus (mg)	Magnesium (mg)	Iron (mg)	Zinc (mg)	Iodine (μg)
Infants	0.0-0.5	6	13	60	24	kg x 2.2	420	10	3	35	0.3	0.4	6	0.3	30	0.5[g]	360	240	50	10	3	40
	0.5-1.0	9	20	71	28	kg x 2.0	400	10	4	35	0.5	0.6	8	0.6	45	1.5	540	360	70	15	5	50
Children	1-3	13	29	90	35	23	400	10	5	45	0.7	0.8	9	0.9	100	2.0	800	800	150	15	10	70
	4-6	20	44	112	44	30	500	10	6	45	0.9	1.0	11	1.3	200	2.5	800	800	200	10	10	90
	7-10	28	62	132	52	34	700	10	7	45	1.2	1.4	16	1.6	300	3.0	800	800	250	10	10	120
Males	11-14	45	99	157	62	45	1000	10	8	50	1.4	1.6	18	1.8	400	3.0	1200	1200	350	18	15	150
	15-18	66	145	176	69	56	1000	10	10	60	1.4	1.7	18	2.0	400	3.0	1200	1200	400	18	15	150
	19-22	70	154	177	70	56	1000	7.5	10	60	1.5	1.7	19	2.2	400	3.0	800	800	350	10	15	150
	23-50	70	154	178	70	56	1000	5	10	60	1.4	1.6	18	2.2	400	3.0	800	800	350	10	15	150
	51+	70	154	178	70	56	1000	5	10	60	1.2	1.4	16	2.2	400	3.0	800	800	350	10	15	150
Females	11-14	46	101	157	62	46	800	10	8	50	1.1	1.3	15	1.8	400	3.0	1200	1200	300	18	15	150
	15-18	55	120	163	64	46	800	10	8	60	1.1	1.3	14	2.0	400	3.0	1200	1200	300	18	15	150
	19-22	55	120	163	64	44	800	7.5	8	60	1.1	1.3	14	2.0	400	3.0	800	800	300	18	15	150
	23-50	55	120	163	64	44	800	5	8	60	1.0	1.2	13	2.0	400	3.0	800	800	300	18	15	150
	51+	55	120	163	64	44	800	5	8	60	1.0	1.2	13	2.0	400	3.0	800	800	300	10	15	150
Pregnant						+30	+200	+5	+2	+20	+0.4	+0.3	+2	+0.6	+400	+1.0	+400	+400	+150	h	+5	+25
Lactating						+20	+400	+5	+3	+40	+0.5	+0.5	+5	+0.5	+100	+1.0	+400	+400	+150	h	+10	+50

[a] The allowances are intended to provide for individual variations among most normal persons as they live in the United States under usual environmental stresses. Diets should be based on a variety of common foods in order to provide other nutrients for which human requirements have been less well defined.

[b] Retinol equivalents. 1 Retinol equivalent = 1 μg retinol or 6 μg carotene.

[c] As cholecalciferol. 10 μg cholecalciferol = 400 I.U. vitamin D.

[d] α-tocopherol equivalents. 1 mg d-α-tocopherol = 1 α T.E.

[e] 1 NE (niacin equivalent) is equal to 1 mg of niacin or 60 mg of dietary tryptophan.

[f] The folacin allowances refer to dietary sources as determined by Lactobacillus *casei* assay after treatment with enzymes ("conjugases") to make polyglutamyl forms of the vitamin available to the test organism.

[g] The RDA for vitamin B$_{12}$ in infants is based on average concentration of the vitamin in human milk. The allowances after weaning are based on energy intake (as recommended by the American Academy of Pediatrics) and consideration of other factors such as intestinal absorption.

[h] The increased requirement during pregnancy cannot be met by the iron content of habitual American diets nor by the existing iron stores of many women; therefore the use of 30-60 mg of supplemental iron is recommended. Iron needs during lactation are not substantially different from those of nonpregnant women, but continued supplementation of the mother for 2-3 months after parturition is advisable in order to replenish stores depleted by pregnancy.

Reproduced from: *Recommended Dietary Allowances*, 9th Edition (1980), with the permission of the National Academy of Sciences, Washington, D.C.

as soft drinks and candy may result in a lowered intake of more nutritious items such as milk and fruits. A diet continuously high in empty calories can lead to dietary deficiencies.

THE MACRONUTRIENTS

Proteins

Nitrogen-balance studies are used to determine minimum protein needs, since proteins are the only significant source of nitrogen for humans. (Other nitrogen compounds in the diet, such as nucleic acids and vitamins, provide only negligible amounts of nitrogen.) Since nitrogen accounts for an average of 16% of protein mass, dietary protein can be weighed and the nitrogen content compared with the nitrogen content of urine, which represents nitrogen removed from amino acids, and that of the feces, which contain any undigested dietary protein. *Positive nitrogen balance* occurs when the nitrogen intake exceeds nitrogen lost in wastes. Such a condition indicates that the body is using nitrogen to synthesize proteins; it is typical of anyone replenishing nitrogen losses due to undernutrition, and it occurs normally for those building protein-containing tissues, such as pregnant women, growing children, and muscle-building athletes. *Negative nitrogen balance* is characterized by nitrogen losses exceeding the amounts ingested and indicates that the body is degrading proteins. This condition may be a result of a dietary deficit of other energy-producing dietary sources, or of such causes as inactivity and the accompanying muscle waste, fever, infection, injury, anxiety, and prolonged stress. *Nitrogen equilibrium* or *zero nitrogen balance* occurs when intake equals output, and the dietary nitrogen intake can be considered a measure of minimal needs to maintain good health.

Protein is necessary in the diet of humans because of the need for amino acids supplied by protein. Table 23-2 lists the amount of each amino acid required for the maintenance of a zero nitrogen balance in young men. The values shown are probably high and provide a large

Table 23-2 Amino Acid Requirements of Young Men

Amino Acid	Amount, mg/kg of Body Weight
Histidine*	not established
Tryptophan	7
Phenylalanine	31
Lysine	23
Threonine	14
Valine	23
Methionine	31
Leucine	31
Isoleucine	20

*Recently shown to be required for adults

margin of safety. In addition, histidine has been recently shown to be required for adults, and arginine is required by growing children. Also, the values for the requirement of a single amino acid are greatly influenced by the composition of the total amino acid mixture. For example, phenylalanine and methionine are used not only for synthesizing protein but also for synthesizing the amino acids tyrosine and cysteine; thus, if tyrosine and cysteine are present in the diet, the dietary need for phenylalanine and methionine is reduced. When only minimal amounts of essential amino acids in the diets of young rats are supplemented by nonessential amino acids, growth is greatly stimulated. Thus, optimal nutrition requires a balanced amino acid mixture. If any one amino acid is supplied in amounts smaller than that needed, it will limit the total amount of protein synthesized from the others. It is important that dietary protein should supply at least the essential amino acids, plus enough of the nonessential ones to allow for normal synthesis of body protein.

Since the amino acid content of one type of protein is likely to be significantly different from that of another protein, it is important to consider the quality of a protein as an amino acid source. A *high-quality protein* is one in which the amino acid proportions approximate very closely those needed by humans. Egg protein has been designated as the reference protein for quality because it most nearly approximates human requirements; other proteins are then compared to egg protein on the basis of their amino acid contents. Such a comparison results in an *amino acid score* for each protein, as shown in Table 23-3. According to this system, meats are next in quality after egg, and then comes milk. Plant proteins (often low in lysine, methionine, and/or tryptophan) usually have one or more amino acid deficiencies, and thus most of these proteins have low amino acid scores.

Another means of rating proteins is based on the *biologic value,* which takes into consideration not only the amino acid content of an individual protein but also the digestibility of the protein. Scores by this rating system, also given in Table 23-3, roughly parallel the amino acid scores.

Table 23-3 The Quality of Proteins of Some Common Foods

Protein Source	Amino Acid Score	Biologic Value	Limiting Amino Acid
Egg	100	94	None
Meat, fish, poultry	66–70	74–76	None
Milk	60	85	None
Wheat	28–44	52–65	Lysine
Corn	41	60	Lysine, tryptophan
Soybeans	47	71	Methionine
Legumes	28–43	55	Methionine, tryptophan

Adapted from Amino-Acid Content of Foods and Biological Data on Proteins, FAO Nutritional Studies No. 24 (Rome, 1970)

Measures of protein quality have the limitation that they are not accurate indicators of protein value in a mixed diet. If two incomplete proteins with offsetting amino acid deficiencies (limiting amounts) such as corn (low in lysine and tryptophan) and legume proteins (low in methionine) are eaten together, a complete essential amino acid supply exists. The two proteins are said to be complementary because they complete each other in the diet. Similarly, peanut butter and whole wheat bread or rice and beans make complete protein combinations.

In the United States, the recommended amount of dietary protein per day for an adult is 0.8 g per kg of body weight. This recommendation is based on the amount of high-quality protein slightly above that necessary to produce negative nitrogen balance. The recommendation also provides a margin of safety to accommodate variations in the needs of individuals. The American diet is typically high in animal protein, and it may seem a bit surprising that a one quarter-pound hamburger patty or the meat and cheese slice of a cheeseburger (not including protein in the bun) adequately supplies half the daily protein needs of an average adult. However, protein deficiencies are common in developing countries in Africa, Asia, and South America.

Perhaps the world's major nutritional problem is the disease known as *kwashiorkor* and the variation of kwashiorkor called *marasmus*. Kwashiorkor typically develops in children between the ages of one and four years who have been weaned and whose diet then becomes deficient in protein. These children fail to grow, and they develop edema. (Edema occurs because depleted blood proteins allow the osmotic flow of water from the blood into the tissue spaces between cells, causing swelling.) Skin rashes, diarrhea, and vomiting are also symptoms of kwashiorkor. Onset of the disease at an early age can produce mental retardation, and the progression of symptoms can ultimately result in death. Marasmus, also called marasmic kwashiorkor or partial starvation, is generally found in infants between six and eighteen months old whose diets contain little more than water. The progression of symptoms is similar to that of kwashiorkor, except that there is no edema. Instead, the water is excreted, and marasmic infants resemble very old, emaciated people. Both kwashiorkor and marasmus are treated by hospitalization, intravenous feeding, and gradual introduction of protein-containing foods, followed by a high-protein diet. The success of treatment depends on the severity of the condition and the child's age at the time of deprivation.

EXAMPLE 23.1

If hamburger meat contains an average of 25% protein, calculate the number of grams of protein in a quarter-pound hamburger patty. (1 lb = 454 g)

SOLUTION

$$(\tfrac{1}{4} \text{ lb}) \left(\frac{454 \text{ g}}{\text{lb}}\right) (0.25) = 28 \text{ g protein}$$

EXERCISE 23.1 Calculate the number of calories available from protein in the hamburger patty described in Example 23.1.

PERSPECTIVE: Liquid Protein Diets

Liquid protein diets have recently become popular as a means of rapid weight loss. They are based on the idea that fasting forces the body to use fat stores as sources of energy. Liquid protein consists mainly of a solution of hydrolyzed gelatin or collagen (both obtained from beef hides), often flavored with cherry- or orange-tasting additives to disguise the taste. The diet program calls for complete abandonment of food, and the dieter drinks liquid protein supplemented by vitamins and a source of potassium. Promoters of the diets maintain that the liquid protein contains all of the amino acids in the optimal proportions needed for synthesis of body proteins. However, animal tests of the liquid protein formulations showed that the protein that they supply is of such low quality that it is not usable. These results suggest that an animal or human on such a diet might degrade some of its own body protein in addition to fat stores.

In one test on obese rats, the Food and Drug Administration found that all of the rats on liquid protein diets died within several months, all victims of irregular heartbeats. In addition, in the summer of 1977, the Food and Drug Administration and the Center for Disease Control began receiving reports that some women who were on liquid protein diets had suddenly died. Most had dieted under medical supervision, and 15 of the 16 victims had died of severe heartbeat irregularity. The middle muscular layer of the heart wall was damaged in all cases, indicating that their bodies had degraded protein from body tissue to satisfy metabolic needs. Preliminary estimates from the Center for Disease Control indicate that women on liquid protein diets have a 16 times greater risk of dying of heartbeat irregularities than does the general population between the ages of 25 and 44.

Fats

Many of the compounds that give foods their flavor and aroma are found in fats and oils. Chicken meat that is skinned before cooking tends to be tasteless and odorless, but the odor of meat fat while it is being cooked whets the appetite. In spite of their universal appeal, there is no RDA for fats even though they are important sources of energy and the essential fatty acids. Since diets low in fat are usually considered unpalatable, the chances of consuming too little fat are slim. However, the National Research Council suggests that 15–25 g daily of well-chosen fat is an appropriate minimum intake. This amount provides approximately 135–225 kcal and would account for less than 10% of the calories in most people's daily consumption.

An interesting observation is that metabolic oxidation of fats produces significantly more water than oxidation of carbohydrates or proteins. This is because fats contain a higher proportion of hydrogen than carbohydrates or proteins do, and upon oxidation, one molecule of water is formed for each two hydrogen atoms present in the original compound. Metabolic studies have demonstrated that the oxidation of 100 g of fat provides about 107 g of water, while oxidation of the same mass of carbohydrates or proteins yields about 55 g or 40 g of water, respectively. Camels can survive for long periods of time without taking in water because metabolic oxidation of fat stored in their humps provides enough water for their needs.

The question of whether excessive dietary cholesterol is harmful is yet to be resolved (see discussions in chapter 21), but it does appear that low dietary intakes reduce serum levels. Consequently, cholesterol restriction, along with limiting total dietary fats, is usually recommended for those in risk of cardiovascular disease. Table 23-4 gives common sources of cholesterol and their cholesterol content. Note that cholesterol is found only in animal sources and is especially abundant in egg yolks, organ meats, and shellfish.

Table 23-4 The Cholesterol Content of Some Common Foods

Food	Amount	Cholesterol, mg
Liver, cooked	3 oz	372
Egg yolk	1	252
Shrimp, canned	1 cup	102
Beef, lean, cooked	3 oz	80
Pork, lean, cooked	3 oz	76
Halibut, cooked	1 fillet	75
Chicken, with skin, cooked	1/2 breast	63
Butter	1 tbsp	35
Cheddar cheese	1 oz	28
Margarine	any amount	0

Adapted from R. M. Feeley, P. E. Criner, and B. K. Watt, Cholesterol Content of Foods, *Journal of the American Dietetic Association* 61:134 (August 1972)

EXAMPLE 23.2

If hamburger meat contains an average of 30% fat, calculate the number of grams of fat in a quarter-pound hamburger patty. (1 lb = 454 g)

SOLUTION

$$(1/4 \text{ lb}) \left(\frac{454 \text{ g}}{\text{lb}}\right) (0.30) = 34 \text{ g fat}$$

EXERCISE 23.2 Calculate the number of calories available from fat in the hamburger patty described in Example 23.2.

Carbohydrates

Dietary carbohydrates provide a source of energy as well as atoms and molecules for synthesis of cell and tissue materials. However, carbohydrates are not absolutely essential in the diet, and animals and humans can be maintained on diets completely devoid of carbohydrates. For most of us, though, carbohydrates constitute an inexpensive source of dietary calories, and carbohydrates usually account for up to 50% of our energy intake. The principal carbohydrates in the diets of humans are starch and sugars.

Although there has been no RDA established for carbohydrates, the National Research Council estimates that a minimum of 50–100 g of carbohydrates will allow normal operation of the citric acid cycle (chapter

> **PERSPECTIVE: Sugar—Is It Really Harmful?**
>
> Sugar is a relatively new addition to the human diet, and it was a scarce luxury until the seventeenth century. In recent years there has been much controversy over the amount of sugar in the diets of Americans; the implication is that perhaps sugar is the cause of many health problems and thus should be minimized or even eliminated from the diet.
>
> One of the health problems long associated with sugar is adult-onset diabetes. The idea has become popular that high levels of sugar in the diet overwork the pancreas to the extent that it stops producing insulin in the necessary amounts. However, genetic factors are involved in diabetes, and the disease can be induced in test animals by feeding them diets high in fat, protein, or sugar. Numerous other studies have shown conflicting results on the involvement of sugar in diabetes, but it seems safe to say that sugar is implicated and that heredity and obesity are major factors in determining susceptibility to diabetes.
>
> Another health problem associated with sugar is the development of dental caries. In this case, the evidence is quite clear that sugar serves as an excellent nutrient for the bacteria that are ultimately responsible for cavities. Thus, diets high in sugar can lead to increased cavities, but other factors such as the length of time the sugar stays in the mouth, whether any other types of food are also present, and proper dental hygiene are equally or perhaps more important.
>
> Sugar, as an energy source, can provide unneeded calories and thus contribute to obesity. This condition is frequently associated with cardiovascular disease and coronary heart disease, but the correlation of poor health with obesity is stronger than the correlation with dietary sugar levels alone. However, the fact that the empty calories of sugar might replace more nutritious energy sources is an important consideration. Also, it has been rumored in recent years that sugar might be a cause of hyperactivity in children; it appears that in some children sugar can produce an allergic reaction that gives rise to behavior similar to that of children diagnosed as being hyperactive. Although many of the studies regarding the effect of sugar on human health are inconclusive, it does seem prudent to limit one's intake of sugar, as recommended by the National Research Council.

19) and thus support normal metabolism. Furthermore, it is recommended that complex carbohydrates and naturally occurring sugars constitute dietary sources of carbohydrates and that refined sugar and processed sugary foods provide 10% or less of the total calorie intake. This latter recommendation is based on the fact that refined sugar contains empty calories and thus is not as conducive to good health as complex carbohydrates and naturally occurring sugars, which also contain other nutrients.

THE MICRONUTRIENTS

The Water-Soluble Vitamins

The water-soluble vitamins include the B-vitamins and vitamin C. Thiamin, the original B-vitamin, was renamed vitamin B_1 when new water-soluble growth factors were discovered. These include niacin, riboflavin (vitamin B_2), vitamin B_6, biotin, pantothenic acid, folic acid (folacin), vitamin B_{12}, and lipoic acid. Since surplus amounts of water-soluble vitamins are excreted in the urine, it is necessary to replenish those supplies frequently. Those water-soluble vitamins having established

RDAs are listed in Table 23-1. As mentioned in chapter 17, each water-soluble vitamin (with the possible exception of vitamin C) is required for the formation of a coenzyme.

Thiamin (vitamin B_1)

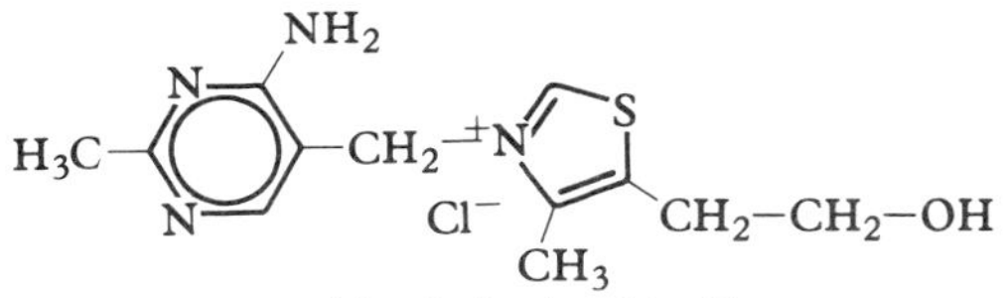

thiamin hydrochloride

The relatively stable form of thiamin is its hydrochloride salt shown here. The vitamin is converted to the active coenzyme, thiamin pyrophosphate (TPP), by attachment of a pyrophosphate group to the side-chain hydroxyl group of thiamin. This coenzyme, essential to numerous enzymes, participates in two major types of reactions: removal of CO_2 (decarboxylation) and transfer of two-carbon units. The pyruvate dehydrogenase reaction is an example of a decarboxylation requiring TPP, while certain reactions of the pentose phosphate pathway require TPP for transfer of two-carbon units from one sugar derivative to another; these reactions were presented in chapter 20.

Thiamin deficiencies produce the disease known as *beriberi*, characterized by fatigue, loss of appetite and weight, intestinal problems, irritability, memory loss, and edema. The condition becomes fatal when the accumulating fluid causes heart failure. Good food sources of thiamin include liver and organ meats, whole-grain and enriched cereals, muscle meats, poultry, fish, eggs, milk, and milk products. However, the vitamin is easily destroyed by prolonged exposure to air (oxidation), heat, and basic pH.

Niacin

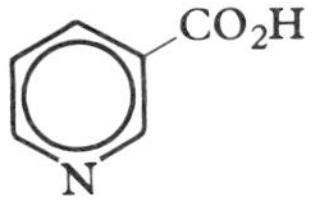

nicotinic acid 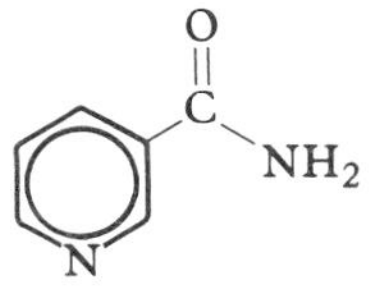   nicotinamide

As mentioned in chapter 17, niacin exists in two forms: nicotinic acid and nicotinamide. These are incorporated into the structures of the coenzymes nicotinamide adenine dinucleotide (NAD^+) and nicotinamide adenine dinucleotide phosphate ($NADP^+$) (Figure 17-9). The function of these coenzymes is to participate in various oxidation-reduction reactions, including a step in the electron transport system.

Because niacin is extremely important in metabolism, the absence of dietary niacin handicaps cells so severely that almost all cells suffer. Deficiencies of niacin produce *pellagra*, characterized by dermatitis, diarrhea, dementia, and, if not corrected, death. The best dietary sources of niacin are the same as those for thiamin. However, niacin is relatively resistant to oxidation, heat, pH extremes, and light—conditions that destroy many other vitamins.

Riboflavin (vitamin B$_2$)

riboflavin (vitamin B$_2$)

Riboflavin, a bright orange yellow solid, derives its name from the Latin word *flavus,* which means "yellow." The vitamin is incorporated into the coenzymes flavin mononucleotide (FMN) and flavin adenine dinucleotide (FAD) (Figure 17-10), which participate in numerous oxidation-reduction reactions. Riboflavin deficiencies produce such symptoms as dermatitis, inflammation of the mucous linings of the mouth, bloodshot eyes, depression, and hysteria. Good dietary sources are the same as those for thiamin and niacin, except that whole-grain cereals lack the vitamin. Riboflavin is easily destroyed by basic pH and exposure to light.

Vitamin B$_6$

pyridoxine   pyridoxal   pyridoxamine

Vitamin B$_6$ is three interconvertible compounds: pyridoxine, pyridoxal, and pyridoxamine. The vitamin is converted to an active coenzyme by attachment of a phosphate group to any one of the three forms at the hydroxyl group shown in color. Enzymes that require vitamin B$_6$ coenzymes are of great importance to amino acid metabolism, particularly in decarboxylations and transaminations of amino acids.

Vitamin B$_6$ deficiencies result in such symptoms as altered brain function, hyperactivity, and convulsions in children, and irritability, weakness, insomnia, nervousness, and anemia in adults. Good sources of vitamin B$_6$ are liver and organ meats, muscle meats, poultry, fish, and eggs. It is a relatively stable vitamin, but it is destroyed by prolonged exposure to air (oxidation) and light.

Biotin

$$\text{biotin structure: imidazolidone ring fused with tetrahydrothiophene ring bearing } (CH_2)_4-CO_2H$$

biotin

Biotin serves as a coenzyme in the addition of carbon dioxide (carboxylation) to substrates. For example, it is involved in the carboxylation of pyruvate to produce oxaloacetate (see chapter 20) and the carboxylation of acetyl CoA to form malonyl CoA in the preliminary step of fatty acid synthesis (chapter 21). In addition to carboxylation reactions, biotin probably has other functions that have not yet been defined in carbohydrate and lipid metabolism. During its participation in carboxylation reactions, biotin is covalently attached to its enzyme by amide linkage between the biotin carboxyl group and the side-chain amino group of a lysine residue of the enzyme.

Biotin deficiencies in humans are rare because intestinal bacteria usually produce the vitamin in sufficient amounts. (It is estimated that 100–200 μg of the vitamin daily is an adequate intake for humans.) However, a protein called *avidin* that is present in raw egg whites binds biotin very tightly and can thus inactivate the vitamin; consequently, eating excessive amounts of raw egg whites can bring on symptoms of deficiency. These include dermatitis, loss of appetite, lethargy, depression, muscle pains, nausea, and possibly anemia and elevation of serum cholesterol. Biotin is widespread in foods, especially liver, organ meats, muscle meats, poultry, fish, eggs, milk, and milk products. However, it is easily destroyed by prolonged exposure to air (oxidation), basic pH, and acid pH.

avidin (AAH-vi-din): so named because it binds biotin avidly (eagerly or greedily).

Pantothenic Acid

$$HO-CH_2-\underset{\underset{CH_3}{|}}{\overset{\overset{CH_3}{|}}{C}}-\underset{\underset{OH}{|}}{CH}-\overset{\overset{O}{\|}}{C}-NH-CH_2-CH_2-CO_2H$$

pantothenic acid

Pantothenic acid is incorporated into the structure of Coenzyme A (Figure 17-11) by cells, and thus the vitamin is essential in the diet of humans to insure adequate supplies of the coenzyme.

Human deficiencies of pantothenic acid are rare, as the vitamin is widely distributed in foods. However, clinically induced symptoms include increased susceptibility to infection, neuromotor disturbances, irritability, fatigue, depression, and digestive difficulties. It is estimated that a safe and adequate daily intake for adults is 4–7 mg of the vitamin.

Especially good sources are liver, organ and muscle meats, poultry, fish, and eggs. The vitamin is inactivated by heat and extreme pH conditions.

Folic Acid (folacin)

$$\text{H}_2\text{N}-\underset{\underset{N}{\|}}{\overset{N}{\underset{\|}{\bigcirc}}}\overset{OH}{\underset{N}{\bigcirc}}-\text{CH}_2-\text{NH}-\bigcirc-\overset{O}{\overset{\|}{C}}-\text{NH}-\underset{\underset{\text{CO}_2\text{H}}{|}}{\text{CH}}-\text{CH}_2-\text{CH}_2-\text{CO}_2\text{H}$$

folic acid (folacin)

Folic acid (also called folacin) was first isolated from spinach leaves, and its name was derived from the Latin word *folium*, which means "leaf." It is converted by reduction into its coenzyme form, called *tetrahydrofolate*, which acts as an intermediate carrier in the enzymatic transfer of one-carbon groups such as $-\text{CHO}$, $-\text{CH}=\text{NH}$, $-\text{CH}_3$, $-\text{CH}_2-$, and $=\text{CH}-$. Folic acid is essential to the synthesis of nucleic acids (DNA and RNA), and folic acid deficiencies limit cell division and thus decrease the rate of formation of new cells. This condition is especially pronounced in the blood, which requires continual production of new red cells. Thus, folic acid deficiencies produce severe anemia. Good sources of the vitamin are liver, organ meats, and green vegetables. It is one of the least stable B-vitamins and is readily inactivated by prolonged exposure to air (oxidation) and light, heat, and extreme pH conditions.

Vitamin B$_{12}$

Vitamin B_{12} has a most unusual structure, as shown in Figure 23-1. Its major feature, the corrin ring system, resembles the porphyrin ring system of heme (Figure 16-10) except that a cobalt ion is bonded to the nitrogen atoms of the four interconnected rings. No other vitamin is known to contain a metal ion. Because vitamin B_{12} is isolated with cyanide ion attached by ionic attraction, the vitamin is also called cyanocobalamin. In the coenzyme form, the cyanide ion is replaced by a 5'-deoxyadenosine group (Figure 23-1).

The way in which coenzyme B_{12} functions is not well understood, but it appears to serve as a carrier for methyl groups and hydrogen atoms. The coenzyme is involved in the biosynthesis of methionine and of nucleic acids.

Vitamin B_{12} is one of the most potent biological compounds known, and only about 3 µg per day are needed by most adults. Dietary deficiencies in humans are rare, but the result is a particularly insidious form of anemia called *pernicious anemia*. This condition can also occur if a mucoprotein, *intrinsic factor*, is not produced in sufficient quantities by the stomach. This protein specifically binds vitamin B_{12} and transports it across the intestinal wall. Thus, individuals who have a lack of intrinsic factor suffer from pernicious anemia. Intrinsic factor deficiency

pernicious (per-NISH-us): ruinous; injurious; deadly.

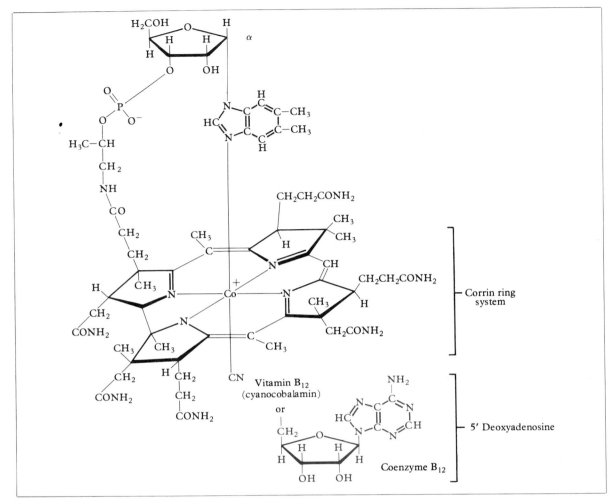

Fig. 23-1 Vitamin B_{12} and Coenzyme B_{12}. (In coenzyme B_{12}, the cyanide ion of vitamin B_{12} is replaced by the 5'-deoxyadenosine group.)

occurs as a genetic abnormality or as a consequence of aging; elderly people often have a diminished production of intrinsic factor and must receive vitamin B_{12} by injection. Good sources of the vitamin are liver, other meats, and egg yolks.

Lipoic Acid

lipoic acid (oxidized form)

lipoic acid (reduced form)

Lipoic acid serves as a coenzyme in oxidation-reduction reactions by being alternately oxidized and reduced itself. It is also involved in the transfer of acyl groups. In particular, it serves as a coenzyme in the α-ketoglutarate dehydrogenase complex (chapter 19) and in the pyruvate dehydrogenase complex (chapter 20). Although lipoic acid was believed to be a vitamin when it was first isolated, it has not been demonstrated to be critical to the growth of either animals or humans. Minute amounts are thought to be needed, but no deficiencies have been observed, and current evidence indicates that animals synthesize the small quantities they need. Because of its coenzyme function, it is often classified as a B-vitamin, but it is more accurately referred to as a *pseudovitamin*.

pseudo (SU-doe): false.

Vitamin C (ascorbic acid)

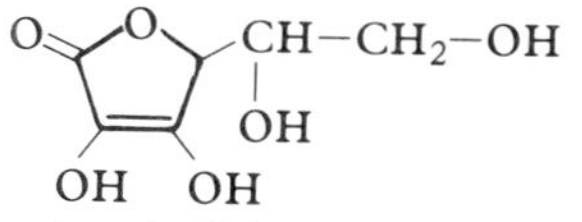

vitamin C (ascorbic acid)

Vitamin C received its chemical name, ascorbic acid, because as little as 10 mg daily will prevent the disease called *scurvy*. Thus, *ascorbic* means "without *scorbutus*" (*scorbutus* is the Latin word for "scaly skin"). In the 1750's, limes were furnished to British sailors as a source of vitamin C, and the sailors came to be known as "limeys." Ascorbic acid is structurally similar to glucose, and many animals can synthesize it from glucose or glucose derivatives. However, humans and a few other animals lack an enzyme essential for its synthesis and therefore must have a dietary source of the vitamin.

The active form of vitamin C is L-ascorbic acid; this compound is a strong reducing agent and readily becomes oxidized to L-dehydroascorbic acid, which also has vitamin activity.

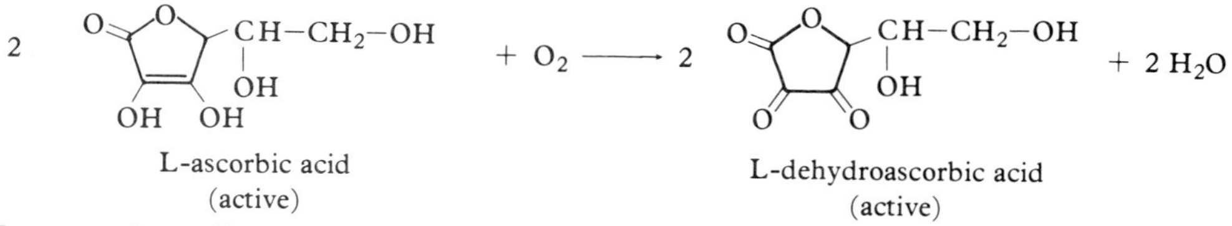

$$2 \text{ L-ascorbic acid} + O_2 \longrightarrow 2 \text{ L-dehydroascorbic acid} + 2 H_2O$$

L-ascorbic acid
(active)

L-dehydroascorbic acid
(active)

However, the cyclic ester of L-dehydroascorbic acid is easily hydrolyzed, and the L-diketogulonic acid produced is not active as a vitamin. Thus, oxidation of ascorbic acid leads to inactivation of the vitamin.

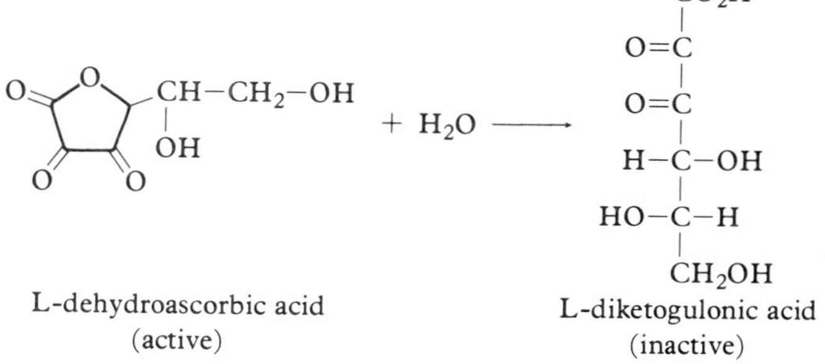

L-dehydroascorbic acid
(active)

L-diketogulonic acid
(inactive)

Ascorbic acid is said to be the first vitamin ever to be used specifically for the purpose of preventing disease. It is ironic that vitamin C is still a subject of controversy. Although no coenzyme form of the vitamin has ever been detected, ascorbic acid is thought to be a requirement for the addition of hydroxyl groups to proline residues after they have been incorporated into collagen. Hydroxyproline residues are essential for proper function of collagen in connective tissue, and ascorbic acid deficiency causes formation of abnormal collagen associated with structural weaknesses in bones, walls of blood vessels, and other tissues. Ascorbic acid may also be involved in the addition of hydroxyl groups to tryptophan to produce 5-hydroxytryptophan, an intermediate in the biosynthesis of serotonin.

serotonin (5-hydroxytryptamine)

Serotonin is involved in impulse transmission between nerve endings in the brain and is thus critical to normal mental processes. Still another possible role of ascorbic acid is to participate in the hydroxylation of tyrosine to form an intermediate in the synthesis of norepinephrine (chapter 15), a hormone secreted by the adrenal gland; norepinephrine also serves as a neurotransmitter.

In addition to the specific metabolic roles proposed above for ascorbic acid, in recent years Linus Pauling and others have supported use of the vitamin in gram quantities for prevention and treatment of common colds and for treating cancer. Unfortunately, the studies to date on these uses of ascorbic acid have been judged by many to be incomplete or invalid, and medical scientists have not reached agreement on whether large doses of ascorbic acid are effective in such treatment.

Deficiencies of vitamin C result in a number of physiological changes. Skin rashes and discolorations are the first symptoms, and if the deficiency continues, scurvy will develop. This disease is characterized by swollen and bleeding gums, dry and itchy skin, loose teeth, fatigue, painful joints, faulty bone growth in children, and hysteria. All of these symptoms respond quickly to the ingestion of ascorbic acid, but if the deficiency is not corrected, they culminate in death.

Vitamin C is widely distributed in foods, and the recommendation of 60 mg per day for adults is usually satisfied easily. The best sources are the rapidly growing fruits and vegetables that are exposed to sunlight, such as broccoli, brussels sprouts, parsley, green peppers, and citrus fruits. The vitamin is inactivated by oxygen, alkali, heat, light, and some metals, but its stability is prolonged in acid solutions.

The Fat-Soluble Vitamins

The four fat-soluble vitamins are A, D, E, and K, and they are usually regarded as lipids. Vitamins A, E, and K are terpenes, while D is a sterol synthesized from cholesterol. (Terpenes and sterols were discussed in chapter 14.) Unlike the water-soluble vitamins, fat-soluble vitamins can be stored in the body because they dissolve in lipid substances instead of body fluids. Since the body does not rid itself of excesses of the fat-soluble vitamins, excessive intakes can result in the toxic condition called *hypervitaminosis*.

hypervitaminosis: a toxic condition caused by an excessive intake of vitamins.

Vitamin A (retinol)

vitamin A (retinol)

Vitamin A serves as a source of retinal needed by the retina for vision processes in the rods. Within the retina, all-*trans*-retinal is converted to 11-*cis*-retinal,

rods: the photoreceptors in the eyes that are sensitive to dim light; rods do not contribute to color vision.

all-*trans*-retinal   11-*cis*-retinal

which binds to a protein called *opsin*. The complex of 11-*cis*-retinal bound to opsin is called *rhodopsin*. When light strikes rhodopsin, the energy converts 11-*cis*-retinal back to all-*trans*-retinal, which then dissociates from opsin because the shape of all-*trans*-retinal is not compatible with the binding site on opsin. When dissociation occurs, nerve impulses travel to the brain to produce an image, and all-*trans*-retinal is ready to be converted back to 11-*cis*-retinal, a process that occurs in the dark. As the two forms of retinal are continuously interconverted, some is degraded, and dietary supplies of vitamin A are needed for replenishing the retinal that is lost. A dietary deficiency of vitamin A results in short lapses of vision due to the delayed adjustment from bright light to dim. This condition is called *night blindness*.

In addition to its role in vision, vitamin A appears to be necessary for the normal production of mucus. Deficiencies can result in diminished mucus secretion by the tear glands, causing hardening of the cornea and, eventually, blindness. Vitamin A may also be involved in the transport of calcium ions across membranes.

The recommended allowances of vitamin A are not difficult to obtain in the United States and other developed countries because of abundant food supplies, but dietary lack of vitamin A is a serious problem world wide. It is estimated that at least one million cases of blindness occur every year in certain parts of the Middle East and Asia because of vitamin A deficiencies.

The richest source of vitamin A is liver, because it is the storage organ for the vitamin in animals. It is also present in eggs, whole milk and its products, butter, and meats. β-Carotene, a precursor form of the vitamin, is widespread in the yellow, orange, and dark green fruits and vegetables. After ingestion, this *provitamin* is cleaved into two molecules of vitamin A by an enzymatic reaction in the intestinal wall and the liver.

provitamin: an inactive precursor of a vitamin.

(cleavage here produces two vitamin A molecules)

β-Carotene

Overconsumption of vitamin A can result in hypervitaminosis. Although it is almost impossible to get too much of the vitamin from natural sources, children have been known to ingest accidentally large quantities of vitamin tablets, and adults sometimes mistakenly take large quantities of vitamin supplements. Common symptoms of such overdoses include loss of appetite and weight, nausea, vomiting, irritability, abdominal pain, pain in the joints and bones, fragile bones, and abnormal fetal development (in pregnant women). *Hypercarotenosis* (high levels of β-carotene) can develop as a result of eating large quantities of vegetables containing β-carotene. The condition imparts a yellow cast to the skin, but is not regarded as dangerous.

Vitamin D

The vitamin D of animals is vitamin D_3 (cholecalciferol). The provitamin 7-dehydrocholesterol is produced in the liver and intestine from cholesterol and then converted in skin by the ultraviolet rays of sunlight to vitamin D_3 (Figure 23-2). In plants, a parallel reaction occurs in which ergosterol is converted to vitamin D_2 (ergocalciferol). These two forms of the vitamin are quite similar in structure, and both exist in foods and show vitamin activity.

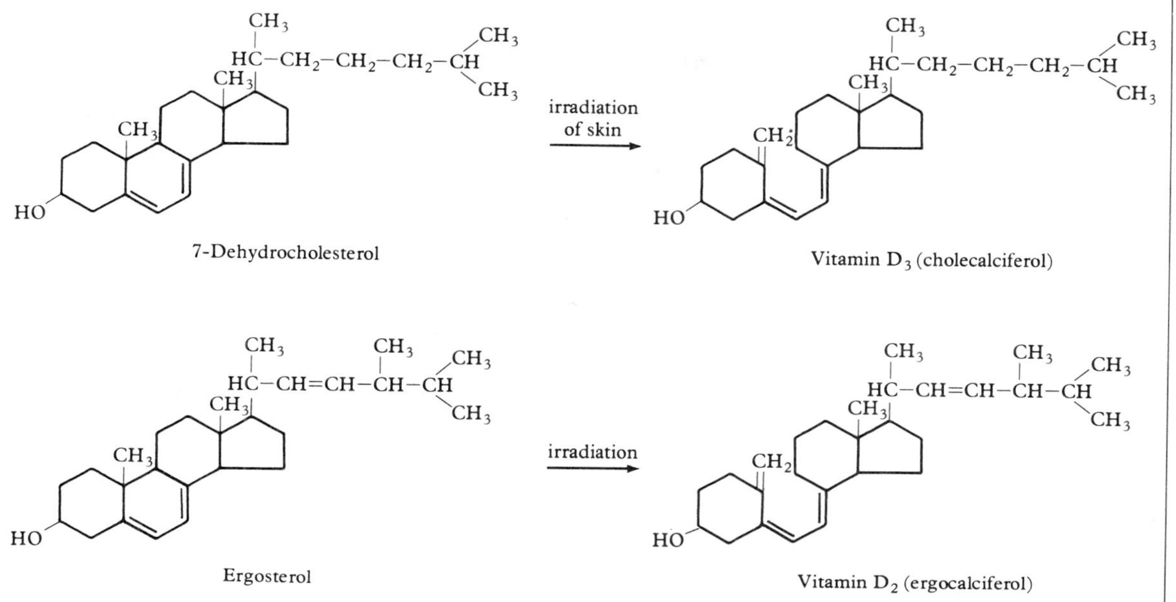

Fig. 23-2 Formation of Vitamins D_3 and D_2

After synthesis in the skin, vitamin D_3 is converted to 1,25-dihydroxycholecalciferol by the successive addition of a hydroxyl group at position 25 in the liver and at position 1 in the kidneys. The dihydroxy compound appears to be the most active form, and it regulates the absorption of calcium ions, blood calcium levels, calcium retention in bones, and calcium excretion in urine. It has recently been proposed that the dihydroxy form of vitamin D_3 is actually a hormone because it is produced by one organ (kidneys) and transported by the blood to its target tissues (intestinal wall and bones).

Vitamin D deficiency in childhood produces *rickets*, a condition in which the legs become bowed. It is brought on by the downward force of the body weight on legs containing inadequately calcified cartilage and bone. Rickets was once quite prominent in England, where there is so little sunlight for much of the year; however, it was learned that the condition could be prevented by generous doses of cod liver oil, a good source of vitamin D.

Today, vitamin D is considered a dietary essential only for growing children and pregnant and lactating women. Others usually can make enough of the vitamin in their skin if they have sufficient exposure to sunlight. Fish liver oils are especially good sources of vitamin D. In addition, milk is fortified with vitamin D to provide a convenient source of the vitamin.

Excessive amounts of vitamin D, as in the indiscriminate use of vitamin supplements, can result in hypervitaminosis, with symptoms of brittle bones, irritability, vomiting, and weight loss. Studies have shown that as little as six to ten times the RDA, ingested regularly, can be toxic for small children, and twice those amounts are dangerous for adults. The possible consequences of prolonged overdosage are irreversible kidney damage, due to precipitation of insoluble calcium salts, and calcification of major arteries.

Vitamin E

Vitamin E is a family of compounds called *tocopherols,* so named because they cure infertility in rats grown on a diet of milk. Tocopherols differ from each other in the arrangement of methyl groups and in the presence of double bonds in the side chain, but the most abundant and active form is α-tocopherol:

tocopherols (toe-KOF-fer-roles): from the Greek word *tokos,* meaning "to bear young."

α-tocopherol

The tocopherols are oxidized especially easily, and the hydrogen shown in color in α-tocopherol is lost as a result. This ease of oxidation causes vitamin E to oxidize in preference to other oxidizable substances present, thereby protecting the other substances from oxidative changes. For this reason, vitamin E is said to be an *antioxidant*.

antioxidant: a substance that retards or prevents oxidation of other substances

Since the discovery of vitamin E, its role in human nutrition has proved elusive, but it is thought to protect the double bonds in unsaturated fatty acids from oxidation and thus also prevent damage to other substances from reactive intermediates that would be produced by such oxidation. Because of the uncertainties of the role of vitamin E, many popular beliefs have arisen. For example, vitamin E is said to increase sexual potency, make wrinkles disappear, retard aging, and prevent cancer. Indeed, it does cure infertility in rats grown on a diet of milk, but any effect on human sexual performance is yet to be proved. The other popular beliefs are likely conceptual extensions of its antioxidant properties.

Vitamin E deficiencies are seldom observed in adults, but some newborn infants are nearly deficient in vitamin E due to poor placental transport of the vitamin. In these cases, early deterioration of red blood cell membranes produces hemolytic anemia. The vitamin is quite prevalent in foods, especially vegetable oils and their products, such as margarines and shortenings.

Vitamin K

Vitamin K was named for the Danish word *koagulation;* as implied by this word, vitamin K is needed for normal blood clotting. Specifically, it is essential for the synthesis of prothrombin and several other clotting factors (see chapter 18). Two active forms of the vitamin are known: vitamin K_2 and vitamin K_3 (menadione). Vitamin K_2 is the biologically produced form, while vitamin K_3 is a water-soluble, synthetically produced form that is commercially available.

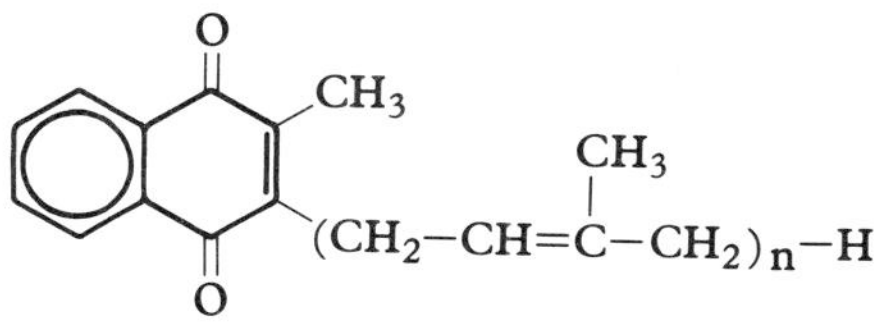

vitamin K_2
(n is 6–10)

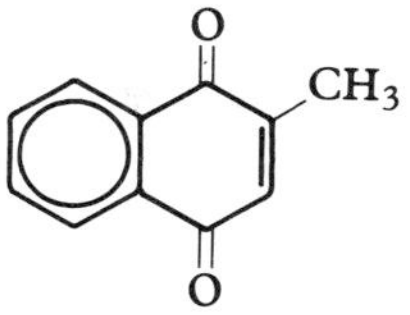

vitamin K_3
(menadione)

Dicoumarol, which can be isolated from moldy clover hay, and warfarin, a synthetic analog,

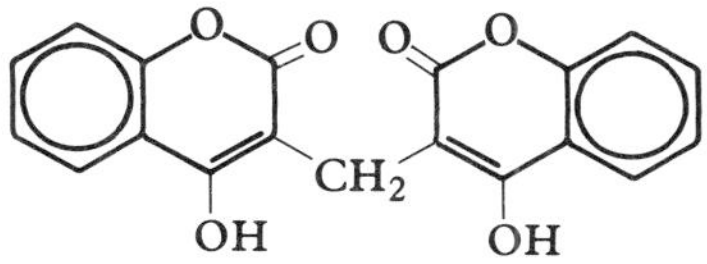

dicoumarol

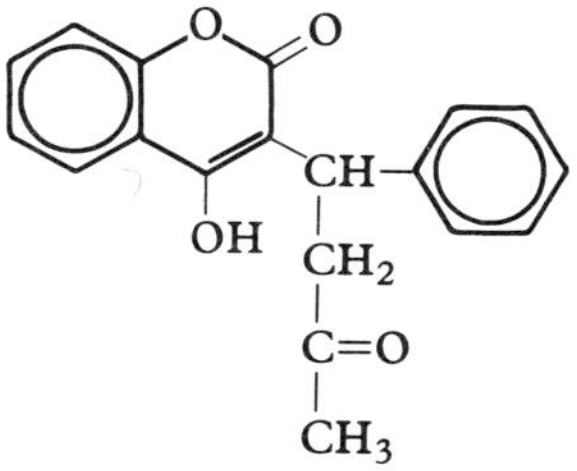

warfarin

interfere with the action of vitamin K and thus prevent blood clotting. Dicoumarol is now synthesized and administered to patients to prevent unwanted blood clots. (In the event of dicoumarol overdosage, vitamin K_3 can be administered to encourage clotting.) Warfarin got its name from the initials of the Wisconsin Alumni Research Foundation, the agency that sponsored research on the compound. It is used as an ingredient in rat poisons to permit unchecked internal hemorrhaging.

Deficiencies of vitamin K are seldom observed in humans, as intestinal bacteria usually produce enough to satisfy normal needs, and no dietary standards have been established.

Minerals

A *nutritional mineral* is an inorganic substance needed in small quantities for life processes. Those minerals needed by humans are given in Table 23–5.

The minerals needed in greatest amounts, the *major minerals,* are calcium, phosphorus, potassium, sulfur, sodium, chlorine, and magnesium. Calcium is required for formation of bones and teeth, which also serve as reserves of calcium used in other processes, such as the activation of enzymes, the blood clotting process, and muscle contraction. The best sources of calcium are milk and milk products, with the exception of butter. Phosphorus is also needed for formation of bones and

Table 23-5 Minerals Essential to Humans

Classification	Mineral Element	Physiological Role
Major Minerals	Calcium (Ca)	Formation of bones and teeth; enzyme activation; blood clotting; muscle contraction
	Phosphorus (P)	Formation of bones and teeth; synthesis of nucleotides, nucleic acids, phosphorylated sugars, and phospholipids; blood buffer
	Potassium (K)	Electrolyte
	Sulfur (S)	Formation of sulfur proteins, mucopolysaccharides, and certain lipids
	Sodium (Na)	Electrolyte
	Chlorine (Cl)	Electrolyte
	Magnesium (Mg)	Cofactor in ATP reactions and in synthesis of nucleic acids
Trace Minerals	Iron (Fe)	Formation of heme and certain proteins
	Zinc (Zn)	Structural ingredient of many enzymes
	Manganese (Mn)	Cofactor for many enzymes
	Selenium (Se)	Helps to protect cell membranes against oxidative damage
	Copper (Cu)	Cofactor for many enzymes; involved in oxygen transport and formation of heme
	Iodine (I)	Formation of thyroid hormones
	Cobalt (Co)	Structural component of vitamin B_{12}
	Molybdenum (Mo)	Enzyme cofactor
	Chromium (Cr)	Essential to normal metabolism of glucose
	Fluorine (F)	Stabilizes the structure of teeth and bones
	Silicon (Si)	Undetermined
	Vanadium (V)	Undetermined
	Nickel (Ni)	Undetermined
	Tin (Sn)	Undetermined
	Arsenic (As)	Undetermined

teeth; in addition, it is a structural ingredient of nucleotides and nucleic acids, phosphorylated sugars, and phospholipids, and it supplies the inorganic phosphate used as one of the blood buffers. The major food sources are milk, milk products, meats, fish, poultry, and eggs. Sulfur is used by the body in the synthesis of sulfur-containing proteins and as a structural ingredient of polysaccharides in mucus and of certain lipids. Primary food sources are meats, fish, poultry, eggs, milk, and legumes. Magnesium ions are essential cofactors in all ATP reactions and in the synthesis of nucleic acids. Food sources of magnesium are quite diverse, and a typical mixed diet supplies adequate amounts.

Potassium, sodium, and chlorine furnish ions for the major electrolytes in the body. The electrolytes are important for proper nerve impulse transmission and maintenance of blood pressure and body fluid distribution. These minerals are so readily available in a variety of foods that deficiencies almost never occur. However, electrolyte depletion becomes a threat if unusually great losses, as in heavy and prolonged perspiration or in severe vomiting, are not replenished.

Aside from the major minerals, fifteen others are required by the human body in trace amounts and are thus referred to as *trace minerals*. The most prominent member of this group is iron, needed for the heme in hemoglobin and the cytochromes and as a functional part of certain other proteins. The body has an intricate mechanism for conserving iron so that the only losses occur through bleeding. However, failure to replenish such losses or inadequate intake can result in iron-deficiency anemia. The richest food source of iron is liver; other organ meats and oysters also contain generous amounts.

Another prominent trace mineral is iodine, needed for synthesis of thyroxine and related thyroid hormones. If maternal iodine supplies are insufficient, the infant is likely to be born with *cretinism*, a condition characterized by arrested physical and mental development and severe mental retardation. Dietary deficiencies in adults produce *goiter*, in which the thyroid becomes enlarged to cause a swelling in the neck area. The enlargement occurs as a result of cellular growth caused by increased thyroid activity in an attempt to produce more of the deficient thyroid hormones. The number of cases of iodine deficiency has declined dramatically in those countries that produce iodized salt, in which small amounts of soluble iodide salts (usually KI) are added to table salt (NaCl). Since iodide salts are especially plentiful in seawater, seafoods are a good source of the mineral, and even vegetables and fruit grown in coastal regions provide iodine in the diet.

Additional trace minerals required by humans are listed with their roles in Table 23-5. Many of the specific functions of trace minerals are poorly defined, and no dietary standards have been set for most of them. In addition, knowledge of their distribution in foods is incomplete. However, more information regarding trace minerals and their physiological roles should become available as research on them continues.

BALANCED DIETS

The nutrient content of individual foods and the nutritional requirements of humans (Table 23-1) are the basis for the categorization of food into four basic groups, as summarized in Table 23-6. These groups, called the *Basic Four*, are recommended by American nutritionists to serve as a guide to food selection and daily meal planning.

The major advantage of the Basic Four is that it provides a reasonably adequate diet for most people. In addition, the guide is simple, easy to learn, and easy to remember. However it is difficult to estimate serving size accurately, and some foods, such as pizza, are hard to categorize. There is no guidance on consumption of fat, cholesterol, sugar, or fiber, and the caloric intake can vary over quite a range, although about 1200-1800 kcal is probably typical of most people.

Table 23-6 A Summary of the Basic Four Food Guide

Group	Foods Included	Amounts Recommended Daily	Major Nutrients Provided
MILK	Milk (whole milk, 2%-fat milk, skimmed milk, buttermilk, yogurt, other milk products except butter)	Children, 2–4 cups Teenagers, 4 cups or more Adults, 2 cups or more Pregnant women, 3 cups or more Lactating women, 4 cups or more	Calcium Phosphorus Complete protein Riboflavin Niacin Vitamin D (if fortified milk is used)
MEAT	Muscle meats (veal, beef, pork, lamb, mutton, venison) Fish Poultry	2 or more 2–3 oz servings	Complete protein Iron Thiamin Riboflavin Niacin Vitamin B_{12}
VEGETABLES/ FRUIT	Vegetables and fruits	4 or more servings	Vitamin C Vitamin A Some iron and calcium
BREADS/ CEREALS	Whole-grain and enriched breads and cereals	4 or more servings	Thiamin Niacin Riboflavin (if enriched) Iron (if enriched) Incomplete protein

The Basic Four guide can be interpreted very narrowly and an unbalanced diet will result. However, the best nutritional advice is to eat as wide a variety of food as possible; by doing so, a balanced diet will be the result, and there should be no need for vitamin or mineral supplements for a healthy person. In addition, though, it is wise to analyze one's activities, so that caloric intake is balanced by caloric expenditure.

DIET THERAPY

A person's diet can be an effective tool in the treatment of various health problems, and in some cases, diet therapy is essential for maintenance of good health. Therapeutic diets may be designed to restore losses resulting from wasting disease (*restorative diets*), or they may be designed to remove excess weight or to reduce the strain on an organ or tissue damaged by or susceptible to disease (*restrictive diets*).

Restorative Diets

Wasting of the body is likely to occur after any severe infectious disease, trauma from injury, or major surgery. After the infection is overcome or the injury repaired, a person's appetite is often impaired, and medical assistance is necessary for proper nutrition.

When normal feeding is possible, foods should be chosen that are rich in protein and energy. In these cases, such items as dried milk, fat emulsions, and sugar can be used to enrich foods, and supplements of vitamins and minerals are often recommended.

When normal feeding is not possible, as when patients are too ill to eat properly, are unconscious, or have diseases affecting the mouth or digestive tract, other means of providing nutrients are necessary. Feeding can be accomplished via a tube passed into the stomach through the nose, a tube surgically inserted into the stomach or intestine, or by intravenous feeding. In intravenous feeding, mixtures of amino acids, fat emulsions, simple sugars, vitamins, and other substances may be used. A patient may receive nourishment by these methods for many weeks while recovering from previous disease or injury.

Restrictive Diets

Low-Energy Diets

As more and more people turn their attention to the health problems brought on by obesity, various low-energy diets gain popularity as ways of losing weight. Anyone who considers a weight-loss program should always consult a physician to make sure that there are no previously undetected medical problems that could be aggravated by a low-calorie diet. A reducing diet should always contain sufficient quantities of the four food groups to insure that needs for protein, minerals, and vitamins are met. Those who are extremely overweight and who wish to embark on diets severely low in calories should do so only under a physician's supervision. Overweight patients with diabetes or high blood pressure are usually advised to restrict the calories in their diets. For many of these, such a diet may give pronounced relief from symptoms, and in some cases, no other therapy is needed.

Low-Protein Diets

Patients suffering from kidney disease are often placed on low-protein diets to reduce the load on the kidneys for excreting urea and other nitrogen waste products. The extent to which dietary protein is reduced for these patients depends on the amount of kidney damage and the availability of hemodialysis facilities.

Low-Salt Diets

Edema often occurs in patients suffering from heart failure or kidney or liver disease. Reduction or elimination of salt in the diet will help to reduce the edema, as salt tends to hold water in the tissues. An added benefit is that restricting the salt intake tends to lower blood pressure, even for a healthy person, and most patients with high blood pressure are advised to either reduce salt levels or eliminate salt from their diets.

Low-Fat Diets

Since bile and pancreatic juice are needed for digestion of fats, diseases of the pancreas, liver, and bile duct may impair digestion of fats. In these cases, patients are often advised to restrict their fat intake. When a patient is on a low-fat diet, a supplement of fat-soluble vitamins is usually given to replace those eliminated from the diet.

Special Diets for Inborn Errors of Metabolism

For individuals suffering from certain inborn errors of metabolism, various special diets have been designed which eliminate foods containing the intolerable ingredient. For example, children with phenylketonuria (chapter 21) or galactosemia (chapter 20) can be raised successfully by use of synthetic diets that contain little or no phenylalanine or galactose, respectively. Commercial preparations are available for such use.

SUMMARY

Human Nutritional Requirements: Nutrients required by humans can be classified as macronutrients and micronutrients. The macronutrients are proteins, fats, and carbohydrates, while the micronutrients are vitamins and minerals. Recommended dietary allowances (RDAs) are established as guidelines for intake of nutrients.

Energy Provided by the Macronutrients: The free energy released by oxidation of macronutrients provides chemical energy needed by the body. Each gram of fat in the diet produces 9 kcal, whereas proteins and carbohydrates each produce 4 kcal per gram. The term "empty calories" refers to food items that produce energy but contribute none or very little of the micronutrients.

The Macronutrients: Nitrogen-balance studies are used to determine minimum protein needs. Positive nitrogen balance occurs when the nitrogen intake exceeds nitrogen lost in wastes, and negative nitrogen balance is characterized by nitrogen losses exceeding nitrogen intake. Nitrogen equilibrium (zero nitrogen balance) occurs when nitrogen intake equals nitrogen losses. Proteins, the only significant dietary source of nitrogen, are needed because of the amino acids they contain. A high-quality protein is one in which the amino acid proportions approximate very closely those needed by humans. The quality of a protein may be expressed by its amino acid score or by its biological value. Protein deficiencies in infants and children can result in the diseases kwashiorkor and marasmus.

Fats and oils account for much of the flavor and aroma of foods. It is recommended that fats contribute no more than 10% of daily calories. Dietary carbohydrates usually contribute about 50% of daily energy intake, and it is recommended that most dietary carbohydrates be complex carbohydrates and naturally occurring sugars.

The Micronutrients: The water-soluble vitamins include the B-vitamins and vitamin C. The B-vitamins are thiamin (vitamin B_1), niacin, riboflavin (vitamin B_2), vitamin B_6, biotin, pantothenic acid, folic acid (folacin), vitamin B_{12}, and lipoic acid. All are needed by the body as sources of coenzymes. Vitamin C (ascorbic acid) is a controversial vitamin that appears to be required for adding hydroxyl groups to proline residues in collagen. Vitamin C deficiencies produce the disease known as scurvy.

The fat-soluble vitamins are A, D, E, and K. Vitamin A is a source of retinal needed for vision processes in the rods and for proper mucus secretions. Vitamin D, synthesized in the skin, is a precursor of 1,25-dihydroxycholecalciferol, which regulates calcium metabolism. Vitamin E serves as an antioxidant in cells, and vitamin K is essential for the synthesis of prothrombin and several other blood clotting factors.

Nutritional minerals are inorganic substances needed in small quantities for life processes. The major minerals, those needed in greatest amounts, are calcium, phosphorus, potassium, sulfur, sodium, chorine, and magnesium. In addition, there are 15 other minerals needed by humans in trace amounts.

Balanced Diets: The four basic food groups (Basic Four) are recommended as a guide to food selection and meal planning. A wide variety of foods from the Basic Four represents a balanced diet, in which there should be no need normally for vitamin or mineral supplements.

Diet Therapy: The diet can be used as effective therapy for various health problems. A restorative diet is one rich in protein and energy designed to restore health after disease, injury, or surgery. A restrictive diet is one that restricts food intake so that harmful substances are not consumed. Examples of restrictive diets are low-energy diets, low-protein diets, low-salt diets, low-fat diets, and diets designed especially to treat inborn errors of metabolism.

STUDY QUESTIONS AND PROBLEMS

1. Define the following terms:

    (a) nutrient
    (b) macronutrient
    (c) micronutrient
    (d) dietary fiber
    (e) RDA
    (f) minimal dietary requirement
    (g) empty calories
    (h) positive nitrogen balance
    (i) negative nitrogen balance
    (j) nitrogen equilibrium (zero nitrogen balance)
    (k) high-quality protein
    (l) amino acid score
    (m) biologic value
    (n) kwashiorkor
    (o) marasmus
    (p) water-soluble vitamins
    (q) thiamin
    (r) niacin
    (s) pellagra
    (t) riboflavin
    (u) vitamin B_6
    (v) biotin
    (w) avidin
    (x) pantothenic acid
    (y) folic acid
    (z) vitamin B_{12}
    (aa) pernicious anemia
    (bb) intrinsic factor
    (cc) lipoic acid
    (dd) pseudovitamin
    (ee) vitamin C
    (ff) scurvy
    (gg) fat-soluble vitamins
    (hh) hypervitaminosis
    (ii) vitamin A
    (jj) rhodopsin
    (kk) night blindness
    (ll) β-carotene
    (mm) provitamin
    (nn) hypercarotenosis
    (oo) vitamin D
    (pp) rickets
    (qq) vitamin E
    (rr) antioxidant
    (ss) vitamin K
    (tt) nutritional mineral
    (uu) major mineral
    (vv) trace mineral
    (ww) goiter
    (xx) Basic Four
    (yy) diet therapy

2. Why do humans need a certain amount of fiber in their diets? What is the source of fiber?

3. How many extra calories are needed per day by a person wishing to gain 5 pounds of weight over a period of a month?

4. A person who consumed an average of 2,800 kcal per day for 30 days gained 2.0 pounds. How many excess calories in the diet accounted for the weight gain? By how much would the person have to reduce his daily caloric intake to lose the 2.0 pounds in two weeks?

5. Running is a good form of exercise, and it is often used by those wishing to lose weight. Running expends an average of 210 cal/min per kg of body weight. How long would a 59-kg (130-lb) person have to run to expend calories equivalent to 10 pounds of body weight?

6. A glass of orange juice contains about the same number of calories as a glass of root beer. Why is the orange juice more nutritious?

7. How are nitrogen-balance tests performed?

8. How does positive nitrogen balance differ from negative nitrogen balance? What kinds of conditions might bring on each type of nitrogen balance?

9. Why is a balanced supply of amino acids necessary in dietary protein?

10. Why do plant proteins usually have low amino acid scores?

11. Distinguish between amino acid score and biologic value for a protein.

12. How much dietary protein does a 150-lb person need per day? (1 lb = 454 g)

13. What are the possible metabolic consequences of a diet high in proteins and low in carbohydrates and fats? (Hint: Consider the effects of metabolizing stored body fat and ketogenic amino acids.)

14. Suggest reasons for holding fat intake to 10% or less of dietary calories.

15. Carbohydrates are not absolutely essential in the diet, yet they usually account for about 50% of a person's caloric intake. Why?

16. Suggest reasons for limiting one's intake of refined sugar.

17. A 1-cup serving of ice cream contains 6 g of protein, 14 g of fat, and 28 g of carbohydrate. Calculate the number of calories available from the ice cream.

18. A 12-ounce serving of lean, broiled steak contains 108 g of protein, 24 g of fat, and no carbohydrate. Calculate the number of calories available from the steak.

19. Does the steak described in question 18 satisfy the protein needs for one day for a 60-kg (132-lb) person? Explain your answer.
20. A 50-g carrot contains 88% water, 2% protein, no fat, and 10% digestible carbohydrate. How many calories are available from the carrot?
21. A large hamburger from a popular fast food restaurant contains 48% water, 14% protein, 17% fat, and 21% carbohydrate. How many calories are available from one of these hamburgers (typical weight 187 g)?
22. Many who go on weight-reduction diets experience a temporary weight plateau after about three weeks, and then weight loss resumes. The weight plateau results from a combined loss of fat tissue and an increase in body water. Why does body water increase?
23. Make a list of the B-vitamins and name each of their corresponding coenzymes.
24. Which of the B-vitamins serve as sources of coenzymes needed for oxidation-reduction reactions?
25. How are the functions of the coenzymes formed from thiamin, lipoic acid, pantothenic acid, riboflavin, and niacin involved in processing acetyl units through the citric acid cycle?
26. How does amino acid metabolism depend on vitamin B_6?
27. Overconsumption of raw eggs results in symptoms of biotin deficiency, but consumption of large amounts of cooked eggs shows no such effect. Why?
28. Why does a deficiency in vitamin B_{12} or folic acid cause anemia?
29. What vitamin deficiency is caused by insufficient quantities of intrinsic factor? How can this condition be alleviated?
30. Describe the physiological role of vitamin C. Why are humans unable to synthesize this vitamin?
31. Why is it that excessive intakes of fat-soluble vitamins can result in toxic conditions?
32. Why does vitamin A alleviate night blindness?
33. What is the provitamin for vitamin A? What are good dietary sources of the provitamin?
34. β-Carotene is added to margarines to provide the yellow color. Does this addition improve the nutritive value of margarines? Explain your answer.
35. Why is cholesterol important to vitamin D production?
36. Why might lack of exposure to sunlight result in vitamin D deficiency?
37. Why is vitamin D important to calcium metabolism? In what specific ways does it affect calcium metabolism?
38. What is the apparent role of vitamin E? How does this role serve to protect membrane structure?
39. Why is vitamin K needed for normal blood clotting?
40. Name the vitamin deficiency that produces each of the following disorders.
    (a) pernicious anemia
    (b) beriberi
    (c) night blindness
    (d) pellagra
    (e) scurvy
    (f) rickets
41. Suggest natural food sources which would provide the deficient vitamins in your answer for question 40.
42. Name the major minerals, give a food source for each, and explain why each is needed in the body.
43. Why does iron deficiency lead to anemia? What is the best dietary source of iron?
44. How does iodine deficiency cause goiter? What are good dietary sources of iodine?
45. What is the best way to obtain a balanced diet?
46. Explain what is meant by diet therapy.
47. Distinguish between restorative diets and restrictive diets.
48. Describe five types of restrictive diets, and give reasons for using each.

CHAPTER 24
CONSUMER PRODUCTS

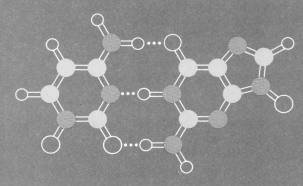

FOOD PRESERVATION AND FOOD PRODUCTS
- Food Preservation
- Dairy Products
- Fats and Oils
- Food Additives

DRUGS
- Drug Names
- Drugs for Treating Symptoms
- Drugs for Curing Disease

HOME PRODUCTS
- Dental Chemistry and Fluoride Treatment
- Soaps and Detergents
- Drain Cleaners and Oven Cleaners

In the economic sense, a consumer is one who uses a commercial product or service. By this definition, we are all consumers, and the array of consumer products increases daily. It is up to each of us to choose products that are suitable for our needs and to use them wisely. To do so, we must educate ourselves about the products. This chapter is intended to provide a basis for self-education about products that are chemical in nature, and although the treatment is not comprehensive in scope, it should serve as a starting point for a learning process that will continue as you go through life.

FOOD PRESERVATION AND FOOD PRODUCTS

Since prehistory, humans have survived on a diet consisting of relatively few kinds of plants and animals. We will examine the chemical nature of the preservation and production of some of those foods in the following sections.

Food Preservation

As soon as food is harvested or slaughtered, it begins to deteriorate. Although heat, light, oxygen, and moisture play roles in food deterioration, the major cause is decomposition by bacteria, yeasts, and molds. These food spoilage microorganisms are found everywhere: in soil, air, water, and in the food itself. Various methods have been devised for preserving food; some of these methods have been used for hundreds of years and thus can be classified as traditional, while other methods of food preservation are relatively modern developments.

Traditional Methods

The simplest traditional method of preserving food is refrigeration, a method used by early hunters and gatherers who stored food in caves and other cool places. Low temperatures help to preserve food by decreasing the rates of both deteriorative chemical reactions in the food itself and metabolic reactions needed for growth of microorganisms.

Sugar (sucrose) has been used for centuries in preserving fruits, and its preservative effect results from the high osmotic pressure of a concentrated sugar solution. Microorganisms cannot live in a concentrated sugar solution because of the osmotic flow of water out of microorganism cells into the sugar solution. Salt (NaCl) has the same effect, and it is used to preserve meat in the process known as curing. Curing is accomplished by packing the meat in salt crystals or by injecting it with a salt solution. Bacon, ham, and corned beef are examples of cured meats.

Smoking and drying are also effective means of preserving food. Smoking is generally used to enhance the preservation of cured meats. It is believed that the smoke dehydrates the surface of the meat and deposits a surface coating of formaldehyde and other chemical compounds that prevent growth of microorganisms. Drying can be used for fruits, fish, and meats, and it prevents growth of microorganisms by removing the water necessary for their life processes.

Although the goal of food preservation is to prevent the growth of microorganisms, controlled growth of certain microorganisms can be beneficial in food production. For example, fermentation of grains and fruit juices by yeast results in production of beer and wine, respectively, and bacterial fermentation of milk is used to produce yogurt. Pickling, in which fruits and vegetables (typically cucumbers) are soaked in a concentrated salt solution, is another well-known fermentation process. The salt suppresses growth of most bacteria, but certain bacteria survive. These produce lactic acid in the low-oxygen environment of the pickling container, and the salt and lactic acid are then capable of preserving the pickles for several years. Before the pickles are eaten, they are soaked in warm water to remove excess salt and packaged in vinegar, which then serves as the main preservative. Both lactic acid and vinegar function as preservatives by lowering the pH to a range unsuitable for bacterial growth.

Vinegar itself is a bacteria product; it is formed when the ethanol of fermented fruit juice or grain is acted on by bacterial enzymes in the presence of oxygen to produce acetic acid:

$$\underset{\text{ethanol}}{CH_3-CH_2-OH} + O_2 \longrightarrow \underset{\text{acetic acid}}{CH_3-CO_2H} + H_2O$$

The resulting mixture is usually distilled, and the final product is an aqueous solution typically containing 4–12% acetic acid and small amounts of esters and other carboxylic acids derived from the fermented source material. The acids and esters are responsible for the flavors and aromas of the various kinds of vinegar.

Modern Methods

One of the methods developed in modern times for preserving food is the process called *pasteurization,* in which the food is heated to destroy harmful microorganisms. All food canned commercially is heated to temperatures in the range of 100–125°C, and sealing the can prevents future contamination. Milk is heated to lower temperatures, typically 63–85°C, if it is to be used fresh. Higher temperatures cook milk and cause coagulation of milk proteins, but the more gentle heating destroys most microorganisms. However, since there are a few left, fresh milk must be kept refrigerated and used within a few days.

pasteurization: a heat treatment that destroys pathogenic microorganisms in foods and beverages; named for Louis Pasteur (1822–1895), the French chemist and microbiologist who originated the process.

Modern refrigeration systems have replaced traditional cold storage methods, so that now food can be stored for a matter of days at temperatures just above 0°C, or for weeks or months when frozen and stored at temperatures in the range of −10 to −20°C. Before vegetables are frozen, mild heat treatment (blanching) is used to denature enzymes that would catalyze reactions contributing to loss of texture and development of inferior flavor.

Other modern methods of food preservation include freeze-drying (discussed in chapter 6) and vacuum drying, in which moisture is removed under vacuum with gentle heating. Both of these modern methods of dehydration avoid undesirable changes that would be brought about by strong heating.

Although the use of chemical agents (sugar, salt, vinegar) to preserve food originated in early civilizations, most chemical preservatives have been developed in modern times. For example, trace amounts of sodium benzoate and other soluble salts of benzoic acid are used to preserve acidic foods; these salts form benzoic acid at low pH,

$$\text{Ph–}CO_2^- Na^+ + H^+ \rightleftharpoons \text{Ph–}CO_2H + Na^+$$

sodium benzoate   benzoic acid

and the benzoic acid inhibits the growth of yeasts and bacteria. The ability of cranberries to resist rapid deterioration is a result of their naturally high content of benzoic acid. Other chemical compounds used to preserve acidic foods are sulfur dioxide (a gas) and various salts of sulfurous acid; these agents form sulfurous acid, a weak acid that is un-ionized at low pH and which inhibits growth of yeasts, molds, and bacteria.

$$SO_2 + H_2O \longrightarrow H_2SO_3$$
sulfur dioxide   sulfurous acid

$$Na_2SO_3 + 2\,H^+ \longrightarrow H_2SO_3 + 2\,Na^+$$
sodium sulfite   sulfurous acid

Some additional examples of chemical preservatives are vanillic acid esters, monochloroacetic acid, propionic acid salts, and glycols; others are discussed as food additives later in this chapter.

Dairy Products

Milk

Whole cow's milk contains approximately 87% water, 5% carbohydrate (lactose), 4% fat (called butterfat), 3% protein, and 1% minerals (the most well known is calcium). The fat exists as tiny droplets that will rise to the top of the liquid to form a cream layer unless the milk is homogenized. In *homogenization,* milk is forced through a nozzle at high pressure and the fat droplets are reduced in size so that they remain suspended. If the cream is allowed to rise and then skimmed off, it can be used to make butter. Churning causes the droplets to coalesce into large particles that coagulate to form butter, which is about 80% fat. The amount of fat in milk and certain milk products determines their flavor and physical characteristics. The fat content of milk and various milk products is given in Table 24–1.

casein (CASE-een)

About 80% of the protein in milk is *casein*. The rest of milk protein is a mixture of albumins and globulins known as *whey proteins*. Casein itself is actually a group of four closely related proteins referred to as α-, β-, γ-, and κ-casein. (κ is the Greek letter *kappa*.) The α- and β-caseins are rich in phosphate, which is present mainly as phosphoserine residues:

$$-NH-CH-\overset{\overset{O}{\|}}{C}-$$
$$|$$
$$CH_2$$
$$|$$
$$O$$
$$|$$
$$^-O-P=O$$
$$|$$
$$O^-$$

a phosphoserine residue

Cheese

chymosin (kye-MOE-sin)

If cow's milk is acidified to pH 4.7, casein precipitates as the insoluble calcium salt. In fact, this reaction is the basis for the curdling of spoiled milk, in which bacterial fermentation produces the lactic acid responsible for curd formation. The fourth stomach of ruminants contains a protease called *chymosin* (formerly known as rennin), which causes milk curds to form at pH 7. In this reaction, a small peptide is released from κ-casein, resulting in lower solubility of the remaining protein. The liquid left after curd formation, the whey, contains the whey proteins.

Curd formation in milk is the basis of cheese making. Rather than

Table 24-1 The Fat Content of Milk and Milk Products

Milk or Milk Product	% (w/w) Fat
Butter	80.6
Whipping cream	36.0
Cheddar cheese	32.2
Medium cream	30.0
Light cream	18.0
Half-and-half	11.5
Sweetened condensed milk	8.1
Evaporated milk	7.9
Whole milk	3.9
Yogurt	1.7
Skimmed milk	0.1
Cultured buttermilk	0.1

waiting for bacterial fermentation to make curds, it is common practice to add a crude preparation of chymosin obtained from the gastric juices of calves. This crude extract is known as *rennet*. Legend has it that the first person to make cheese was an Arabian merchant who traveled across the desert on a hot day with milk in a pouch made from a sheep's stomach. When he stopped to drink his milk, he discovered curds (cheese) and whey, presumably formed by the chymosin from the sheep's stomach.

Curds contain most of the fat, casein, and other dissolved or suspended substances originally in milk. After formation, the curds are cut, pressed, and heated mildly to remove most of the whey, and the net result is reduction of about ten volumes of milk to one volume of cheese. Hence, cheese making is a way of preserving many of the nutritious ingredients of milk. Hundreds of varieties of cheese are made from the milk of cows, goats, sheep, water buffalo, mares, llamas, and yaks. The products vary according to the type of milk and its treatment, adjustment of its fat content, heating or pasteurization, and addition of enzymes or cultures of bacteria, molds, or yeasts.

Some cheeses are ripened (or aged) while others are not. Before ripening, cheese is said to be fresh or green. Cottage cheese and cream cheese are examples of fresh cheeses, which are usually soft and mild tasting. Ripened cheeses are characterized by semihard to hard texture and distinctive flavor, and the method of ripening determines the type of cheese that forms. A variety of bacteria and molds are applied to the surface of the curd, injected into the curd, or added to the milk before curd formation. One of the first chemical changes during ripening is conversion of lactose trapped in the curds to lactic acid. Many other chemical changes also occur, and these account for the character of a particular cheese. For example, the mold *Penicillium roquefortii* is used to inoculate the curds of Roquefort cheese, producing the distinctive blue veins and sharp flavor. In this cheese, and in most others, mold or bacterial enzymes catalyze hydrolysis of triacylglycerols to glycerol and fatty acids, many of which have distinctive flavors and aromas. In addition, some of the fatty acids are converted to other substances that contribute flavor and aroma, such as 2-heptanone, an important ingredient of Roquefort and blue cheeses:

$$CH_3-CH_2-CH_2-CH_2-CH_2-\overset{\overset{\displaystyle O}{\|}}{C}-CH_3$$
2-heptanone

Table 24–2 gives characteristics of many common cheeses.

Table 24-2 Characteristics of Some Common Cheeses

Class	Milk Used	Flavor	Processing
Very Hard			
Parmesan	Partly skimmed cow's milk	Sharp	Bacteria-ripened, 16 mos to 3 yrs; rubbed with oil containing burnt umber or other coloring matter
Romano	Ewe's, cow's, or goat's milk	Sharp	Bacteria-ripened, 5 mos to 1 yr
Hard			
Cheddar	Cow's milk	Mildly acidic	Ripened 2 mos to 2 yrs
Swiss	Cow's milk	Nutlike	Bacteria-ripened, 3–6 mos; one of the most difficult cheeses to make (large holes formed by escaping CO_2)
Gruyere	Cow's milk	Mild	Bacteria-ripened
Semisoft			
Brick	Cow's milk	Mild	Bacteria-ripened, 3 mos
Muenster	Cow's milk	Mild to sharp	Bacteria-ripened, 1–8 wks or longer
Limburger	Cow's milk	Strong	Bacteria-ripened, 1–2 mos
Port du Salut	Cow's milk	Mild	Bacteria-ripened, 6–8 wks
Roquefort	Ewe's milk	Sharp	Mold-ripened, 2–5 mos
Blue	Cow's or goat's milk	Sharp	Mold-ripened for several months
Gorgonzola	Cow's milk	Sharp	Mold-ripened, 3–6 mos
Stilton	Cow's milk with cream added	Sharp	Mold-ripened, 4–6 mos or longer
Soft			
Bel Paese	Cow's milk	Mild	Ripened 6–9 wks
Camembert	Cow's milk	Mild to sharp	Mold-ripened (applied to surface only), 60 days or longer
Neufchatel	Cow's milk	Mild	Unripened
Cottage	Skimmed cow's milk	Mild	Unripened
Cream	Cow's milk and cream	Mild	Unripened

Yogurt

Yogurt is produced by allowing the fermentation of skimmed milk by lactic acid–producing bacteria specially cultured for this purpose. The milk is prepared by heating to boiling, to kill all microorganisms initially present, and then a small amount of the special bacteria is added to the warm milk. As the mixture is warmed for a number of hours, it takes on a thick, creamy texture as lactic acid is produced by the bacteria. The final product derives its tart taste from the lactic acid. The yogurt itself may be used as a source of bacteria for subsequent batches of yogurt.

yogurt bacteria: bacteria that use lactose as their nutrient and form lactic acid under anaerobic conditions.

Sour Cream and Buttermilk

Sour cream is produced by bacterial fermentation of cream, and the final product contains about 18% fat. Buttermilk is the liquid remaining after milk or cream is churned to make butter. Commercial buttermilk is often cultured with special strains of bacteria, and both types of buttermilk have a low fat content. Both sour cream and buttermilk derive their sour taste from lactic acid.

Fats and Oils

Vegetable oils (see Table 14-7) are obtained by pressing oilseeds and extracting the residue with liquid hydrocarbons to obtain the last traces of oil. Animal fats are isolated by the process called *rendering*, in which fatty tissues are cut into small pieces and heated. The major food products of fats and oils are discussed in the following sections.

Margarine and Butter

Margarine and butter are familiar fat products that contain about 80% fat. Butter derives its fat from milk (see the preceding section on milk); its other constituents are water (16%) and small amounts of lactose, casein, salt, and other minerals. The fat in margarines is derived almost totally from vegetable oils, and federal law requires conventional margarines to contain a minimum of 80% fat. Margarine also contains about 16% water and small amounts of salt and added nonfat milk solids. Various additives are also present to improve appearance and stability. Diet margarines, also called imitation margarines, may be marketed with as little as 40% fat.

Margarine provides a reasonably inexpensive alternative to butter; it also has the advantage of containing fewer saturated fats and no cholesterol, since it is a product of vegetable oils. Margarine is prepared from vegetable oils containing a relatively high proportion of long-chain fatty acids and unsaturated fatty acids. For example, butterfat has an iodine number of 25-42, whereas margarines typically have iodine numbers in the range of 78-90. Some of the original double bonds in vegetable oils are hydrogenated in the production of margarine, and the result is an increase in melting point. (Iodine numbers and hydrogenation of double bonds in food oils were discussed in chapter 14.) Thus, margarines have much higher melting points than oils and are solids if kept refrigerated. β-Carotene (see chapter 23) is added to margarines to provide the yellow color, giving them an appearance similar to that of butter.

Salad Oils, Cooking Oils, and Shortening

Salad oils, cooking oils, and shortening are products that are almost pure triacylglycerols. Salad oils and cooking oils are purified vegetable oils that contain small amounts of additives to improve appearance, ease of handling, and stability. Shortening is made primarily from hydrogenated vegetable oils, but sometimes animal fat is blended into the product. Shortening is more highly saturated than margarine so that it can be stored at room temperature without melting.

Food Additives

Substances are often added to processed foods to enhance their quality or to improve processing, packaging, or storage. Such substances are called food additives. Laws require that additives be proven safe for their intended use, and in general, there is a greater body of scientific knowledge about food additives than about the natural components of foods. The major classes of food additives are discussed below.

Preservatives

Salt (NaCl), the oldest additive known, has been used for centuries to preserve meat. Cured meats are treated with salt containing sodium nitrate ($NaNO_2$), which protects against the growth of botulism bacteria and also reacts with myoglobin and moisture in the meat to form nitrosomyoglobin, the pink color of cured ham and bacon and of corned beef. There has been some concern about adding sodium nitrite to meat because of the likelihood of formation of nitrosamines (see chapter 22). These are known to cause cancer in laboratory animals, but apparently humans ingest nitrite-forming compounds from such a wide variety of natural sources that nitrites contributed by food additives may not be significant. Other examples of antimicrobial food additives are sodium benzoate, propionic acid, and calcium propionate.

Other preservatives are antioxidants, which prevent oxidation that causes rancidity in oils and fats and the browning of fruits. In oils and fats, oxidation of double bonds in the unsaturated fatty acids produces aldehydes and carboxylic acids, both of which have bad odors and tastes. Common antioxidants include vitamin E, citric acid, ascorbic acid, butylated hydroxytoluene (BHT), and butylated hydroxyanisole (BHA).

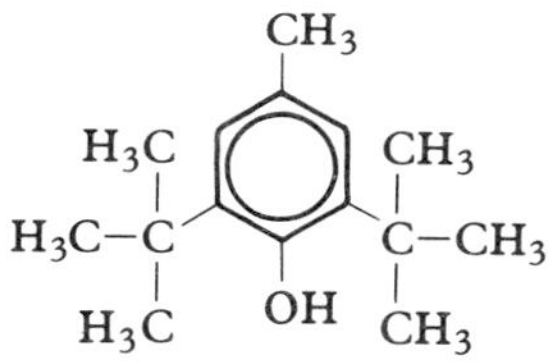

BHT

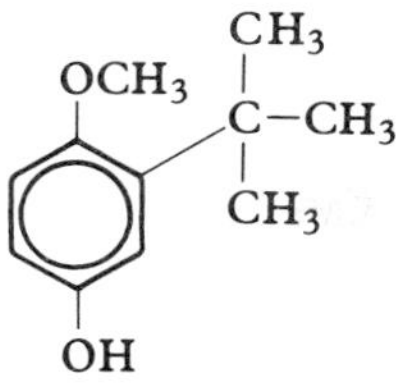

BHA

Additives That Maintain Consistency or Texture

Emulsifiers, stabilizers, and thickeners affect food consistency and/or texture. *Emulsifiers* encourage the suspension of one fluid in another, as in the mixture of oil and water in margarine, shortening, and salad dressing. Examples of emulsifying additives are lecithin, mono- and diacylglycerols, and propylene glycol monostearate.

Stabilizers are additives that maintain the emulsions promoted by emulsifiers. Examples are gum arabic, carrageenan, and dextrin, all polysaccharides. *Thickeners* are used to give body to sauces and other liquid products. Examples are the polysaccharides agar (obtained from seaweed), gelatin, pectin, and gum arabic.

Acids and Bases

Acids and bases are used as additives to control the pH of food; many act as buffers or as neutralizing agents. Examples are citric acid, sodium bicarbonate, lactic acid, adipic acid, and potassium hydrogen tartrate.

$$HO_2C-(CH_2)_4-CO_2H$$

adipic acid

$$HO_2C-\underset{OH}{CH}-\underset{OH}{CH}-CO_2^-K^+$$

potassium hydrogen tartrate ("cream of tartar")

Nutrient Supplements

Nutrient supplements are vitamins and minerals added to foods to increase nutritive value and sometimes to provide specific nutrients in which populations are deficient. Flour and bread products are often enriched with iron, thiamin, riboflavin, and niacin, and citrus fruit beverages which naturally contain vitamin C may be fortified with additional vitamin C. The addition of iodide salts to table salt has eliminated goiter in certain areas, and vitamin D may be added to baby food to combat rickets.

Flavors and Flavor Enhancers

Both natural and synthetic agents are added to foods to give a wide variety of flavors. Such additives are widely used in processed foods, including dairy products, soft drinks, processed fruits and vegetables, prepared meats, baked goods, and confections. Examples of flavoring additives are natural lemon and orange flavors, dried garlic, herbs, spices, hydrolyzed vegetable protein, vanillin (flavor of vanilla), and artificial flavors (mainly fruit flavors).

Flavor enhancers are used to intensify existing flavors without adding any of their own. The best known flavor enhancer is monosodium glutamate (MSG), a sodium salt of the amino acid glutamic acid:

$$^-O_2C-CH_2-CH_2-CH-CO_2^-Na^+$$
$$|$$
$$NH_3^+$$

monosodium glutamate (MSG)

It is especially effective in bringing out the flavor of meats, and it is used in frozen meat, fish, and chicken dinners, dry soup mixes, canned stews, sauces, and meat spreads. Glutamic acid is a precursor of γ-aminobutyrate, a neurotransmitter,

$$^-O_2C-CH_2-CH_2-CH-CO_2^- + H^+ \longrightarrow$$
$$|$$
$$NH_3^+$$

glutamate

$$^-O_2C-CH_2-CH_2-CH_2-NH_3^+ + CO_2$$

γ-aminobutyrate

and it is theorized that MSG works by increasing the sensitivity of the taste buds or by stimulating saliva flow, which would help to release flavors from foods. Ingestion of large amounts of MSG has been associated with "Chinese restaurant syndrome," named for the relatively heavy use of MSG in Chinese food. Symptoms include pain and a burning sensation in the neck, forearms, chest, and head. Other flavor enhancers are inosine monophosphate (IMP, an intermediate in the synthesis of AMP), guanosine monophosphate (GMP), maltol, and ethyl maltol.

Coloring Agents

coal tar: the residue left from heating coal in the absence of air.

Coloring agents include those from natural sources (usually plants), inorganic pigments, and synthetic coal-tar products. Used to enhance food appearance, they may be added to orange skins, sausage casings, baked goods, confections, soft drinks, gelatin desserts, and many other food products. Examples of coloring agents are β-carotene, caramel color, beet powder, and artificial colors.

Miscellaneous Food Additives

Many food additives do not fit into the preceding categories and thus are classified as miscellaneous additives. These include such things as anticaking agents (calcium silicate, calcium stearate, sodium aluminosilicate), antifoaming agents (methyl silicone), bleaching agents, and nonnutritive sweeteners.

Nonnutritive sweeteners provide sweetness with few or no calories and they are valuable to those with weight problems or diabetes. Calcium cyclamate was once a popular artificial sweetener, but it was banned in 1969 because its degradation products were found to be animal carcinogens. Saccharin has a long history of use as a nonnutritive sweetener; it is about 450 times as sweet as sucrose.

calcium cyclamate   saccharin

The recent history of saccharin has been controversial, as it is also thought to be an animal carcinogen. However, test results have not provided unequivocal evidence of its carcinogenicity, and it still appears on the market. Perhaps the most promising artificial sweetener is one derived from two amino acids, aspartic acid and phenylalanine. It is called aspartame, and it is about 200 times as sweet as sucrose.

PERSPECTIVE: The Chemistry of Leavening Agents

Leavening agents are used to cause expansion of doughs and batters by the release of gases within these mixtures, thus producing baked products with porous structure. The most popular leavening agents are bakers' yeast, bacteria, and chemical compounds.

Bakers' yeast is used to leaven most types of breads, rolls, and sweet-dough products such as coffee cakes, raised doughnuts, and Danish pastry. This strain of yeast converts sugars (glucose, fructose, maltose, and sucrose) in the dough or batter to carbon dioxide, water, and ethanol. As the carbon dioxide forms and escapes, the dough or batter expands. The ethanol produced is an important component of the aroma of freshly baked bread.

Bakers' yeast cannot act on lactose, the predominant sugar in milk, or other carbohydrates.

Sourdough breads are leavened by special strains of sugar-fermenting bacteria that produce carbon dioxide, water, and lactic acid. The carbon dioxide causes the dough to expand, and the lactic acid is responsible for the slightly sour taste of sourdough breads.

Layer cakes, cookies, biscuits, and many other baked goods are leavened by carbon dioxide from sodium bicarbonate (baking soda). This leavening agent requires the presence of added acid ingredients such as honey, sour cream, sour milk, molasses, or cocoa, because acid is needed to produce carbon dioxide from sodium bicarbonate:

$$NaHCO_3 + H^+ \longrightarrow H_2CO_3 + Na^+$$
$$H_2CO_3 \rightleftharpoons H_2O + CO_2$$

Instead of adding baking soda and acids separately, many bakers use baking powder, a mixture of baking soda and a solid acid. In the dough or batter, the solid acid dissolves and reacts with the dissolved baking soda to produce carbon dioxide. Double-acting baking powder contains two solid acids; one of these reacts with sodium bicarbonate as soon as the baking powder is added to dough or batter, but the other requires the high temperatures of the baking oven for reaction. In this way, carbon dioxide is released first in the freshly made dough or batter and then again during the actual baking.

$$^+H_3N-CH-\overset{\overset{O}{\|}}{C}-NH-CH-\overset{\overset{O}{\|}}{C}-OCH_3$$
$$||$$
$$CH_2CH_2$$
$$||$$
$$CO_2^-C_6H_5$$

aspartame (aspartylphenylalanine methyl ester)

In quantities needed for sweetening, aspartame provides almost no calories. It was approved in July, 1981, for use in tablet form or as a free-flowing tabletop sweetener, and it may be approved as an additive to processed foods in the future. Since aspartame provides the amino acid phenylalanine on digestion, the marketed product will carry a warning label for persons with phenylketonuria.

DRUGS

Drugs are chemical substances that affect the functions of living things. The experimental study of drugs and drug actions dates from the mid-nineteenth century, and since then drugs have been used to treat, prevent, and diagnose diseases.

Drug Names

The names used to identify drugs are a growing concern of consumers. Three types of names can be used to specify a drug. One is its *systematic name,* the name derived for the drug by the IUPAC system of nomenclature. Systematic names are seldom used except in technical literature, and most of us are more familiar with generic and brand names. *Generic names* are rather technical but unsystematic names assigned to drugs while they are being tested. *Brand names* (or trademark names) are those adopted by a drug company for its own particular products. A drug may have only one generic name, but it can have several brand names, depending on the number of different companies that produce it.

Brand names are capitalized and they frequently appear with the sign ® or ™ to indicate the name is a registered trademark and its use restricted to the owner of the name. Generic names are not capitalized. Examples of generic and brand names of two drugs are:

$$HO-C_6H_4-NH-\overset{\overset{O}{\|}}{C}-CH_3$$

acetaminophen (Tylenol, Tempra, Datril, Liquiprin, Trilium)

$$\text{H}_2\text{N}-\overset{\overset{\text{O}}{\|}}{\text{C}}-\text{O}-\text{CH}_2-\overset{\overset{\text{CH}_3}{|}}{\underset{\underset{\underset{\text{CH}_3}{|}}{\underset{\text{CH}_2}{|}}}{\underset{\text{CH}_2}{|}}}{\text{C}}-\text{CH}_2-\text{O}-\overset{\overset{\text{O}}{\|}}{\text{C}}-\text{NH}_2$$

meprobamate (Equanil, Miltown)

Drugs with brand names are frequently more expensive than their counterparts with generic names. This cost difference is thought by many to be a result of expenses incurred in advertising the brand names. Drugs marketed by their generic names are held by some to be equivalent to their brand-name counterparts, and there is no doubt that the main ingredients are chemically identical. However, the two may have different additives that affect drug performance. Even so, documented differences between brand-name drugs and generic-name drugs are rare.

Drugs for Treating Symptoms

Most drugs used by consumers are for treating symptoms such as headache, fever, and nervous tension. Many of these drugs are available over the counter, but some require a physician's or dentist's prescription.

Analgesics

There is a wide array of analgesics (pain relievers) available, both over the counter and by prescription. Aspirin is by far the most popular over-the-counter drug, and it is available as aspirin tablets and in combination with cold remedies and other pain relievers. After ingestion, aspirin (acetylsalicylic acid) is hydrolyzed to salicylic acid:

acetylsalicylic acid (aspirin) + H_2O ⟶ salicylic acid + CH_3-CO_2H (acetic acid)

Salicylic acid, the actual pain reliever, acts by inhibiting prostaglandin synthesis, but the compound acetylsalicylic acid is marketed because it is less irritating to the stomach (see chapter 14). However, aspirin still causes irritation for some, and it also has a pronounced anticoagulation effect. For those wishing an alternative analgesic, acetaminophen (illustrated in the preceding section) is available. Aspirin is especially effective in relieving symptoms of arthritis, as it is also an antipyretic (lowers fever) and an antiinflammatory agent (reduces swelling and inflammation).

When pain is severe, more potent pain relievers may be prescribed by a physician or dentist. The use of these analgesics is restricted because most are addictive. One that is very popular is propoxyphene, marketed as the hydrochloride salt and sold under its generic name (propoxyphene) and the brand names of Darvon, Darvocet-N (in combination with acetaminophen), and Darvon Compound 65 (in combination with aspirin, caffeine, and phenacetin).

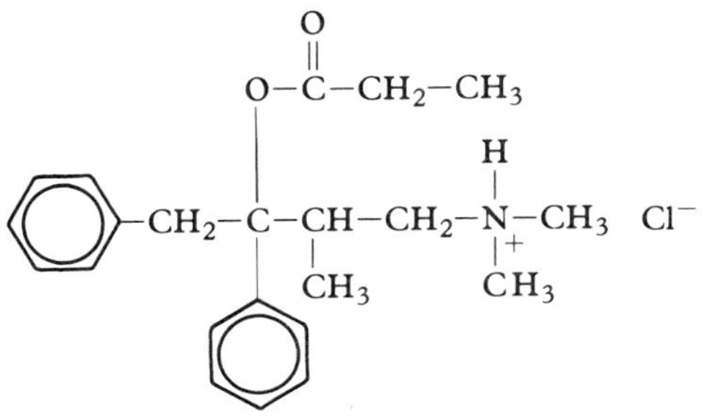

propoxyphene hydrochloride

The drug has a rather low degree of potential for addiction, but studies have shown that its usual dose is not as effective as aspirin or acetaminophen in relieving pain.

The *opiates* are a group of highly addictive pain relievers available only by prescription. They include morphine and heroin (see chapter 14), codeine, meperidine, and methadone.

opiates: drugs isolated from the opium poppy (morphine and codeine) and other drugs that have similar narcotic properties (heroin, meperidine, and methadone).

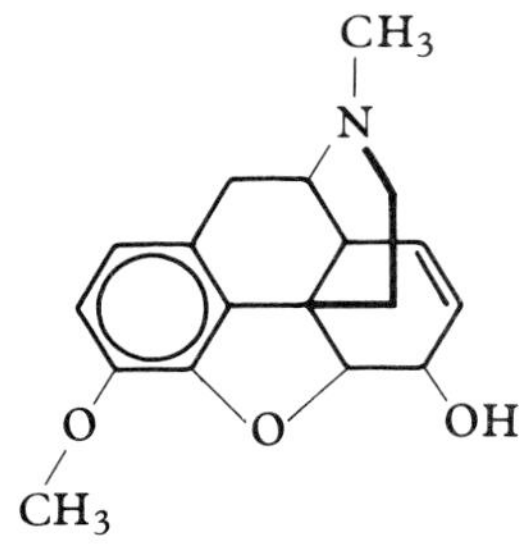

codeine

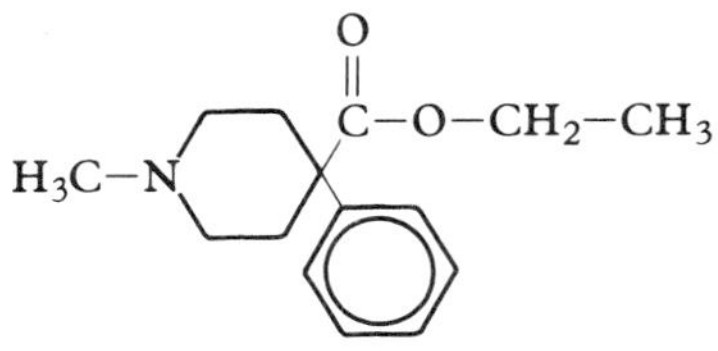

meperidine (Demerol)

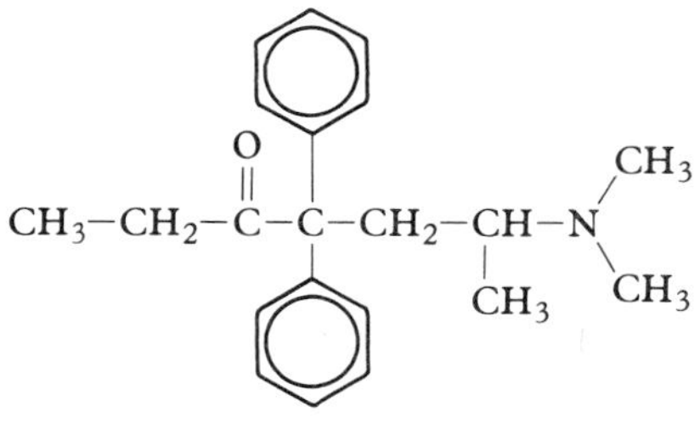

methadone

Anesthetics

Diethyl ether, first used as an anesthetic in 1847, is an effective anesthetic, analgesic, and muscle relaxant, with a wide range between an effective dose and the lethal dose. However, its extreme flammability and explosive nature have caused its use to diminish. Chloroform (trichloromethane) also was first used as an anesthetic in 1847, but even though it is nonflammable, it causes liver damage, and there is a narrow margin between an effective dose and the lethal dose. It is seldom used any more for surgery.

Presently, the primary drugs used for surgery are Sodium Pentothal, nitrous oxide (N_2O, sometimes called "laughing gas"), enflurane, and halothane.

$$\text{Sodium Pentothal} \qquad \text{enflurane} \qquad \text{halothane}$$

Sodium Pentothal rapidly induces unconsciousness, but it is not effective enough as an analgesic and muscle relaxant to be used alone, and its administration is frequently followed by that of enflurane or halothane. These are nonflammable substances that are quite safe to use, and enflurane is preferred because it is a very good analgesic and muscle relaxant.

For very minor surgery, a local anesthetic may be used. Two popular ones are procaine (Novocaine) and lidocaine (Xylocaine), which were discussed in chapter 15. Either can be injected in preparation for surgery. Another local anesthetic, ethyl chloride (CH_3-CH_2-Cl), is sprayed on the skin surface to achieve numbness. It has a low boiling point (12°C, or 54°F) and a very high molar heat of vaporization (see chapter 6), so that as it evaporates, it cools the skin to the point of freezing. It is often sprayed on a wound to relieve pain, and it is used for minor skin surgery.

Drugs for Curing Disease

Although most drugs are used for treating the symptoms of disease, some are known to be highly effective in curing disease. Drugs that act against bacterial infections may be classified as bacteriostatic or bacteriocidal. A *bacteriostatic drug* inhibits microbial growth, usually by interfering with protein synthesis, whereas a *bacteriocidal drug* kills microorganisms. Although bacteriocidal drugs are more desirable in treating diseases, bacteriostatic drugs are quite effective because while they are inhibiting the growth of microorganisms, the normal body defenses (the immune system) are allowed to develop.

The Sulfa Drugs

The first sulfa drug was discovered in 1932 by a German chemist who noticed that a sulfur-containing dye he had synthesized had antibacterial action. It was found that the sulfur compound is converted to

sulfanilamide
(sul-fah-NIL-ah-mide)

sulfanilamide in the body and that this compound also has antibacterial action. Since then, a series of biologically active sulfanilamide derivatives have been synthesized, and these are called the sulfa drugs:

$$H_2N-\underset{\text{sulfanilamide}}{\bigcirc}-\underset{O}{\overset{O}{\underset{\|}{\overset{\|}{S}}}}-NH_2 \qquad H_2N-\underset{\text{sulfathiazole}}{\bigcirc}-\underset{O}{\overset{O}{\underset{\|}{\overset{\|}{S}}}}-NH-\bigcirc$$

$$H_2N-\underset{\text{sulfadiazine}}{\bigcirc}-\underset{O}{\overset{O}{\underset{\|}{\overset{\|}{S}}}}-NH-\bigcirc \qquad H_2N-\underset{\text{sulfisoxazole}}{\bigcirc}-\underset{O}{\overset{O}{\underset{\|}{\overset{\|}{S}}}}-NH-\bigcirc$$

The sulfa drugs block bacterial synthesis of folic acid (see chapter 23), essential for the survival of bacteria, but if the bacteria accumulate folic acid from their surroundings, they will continue to live. Thus, sulfa drugs exert both bacteriostatic and bacteriocidal actions. Because of their disruption of bacterial metabolism, the sulfa drugs are called *antimetabolites*.

The sulfa drugs were once used to treat a variety of infections, but their use has diminished because of their numerous side effects, including kidney damage and hypersensitivity reactions of the skin and bone marrow. Presently, they are used mainly for treating urinary infections, since they are efficiently absorbed and excreted in the urine.

Antibiotics

An *antibiotic* is any substance produced by one microorganism that kills or inhibits the growth of other microorganisms. The penicillins constitute a large group of antibiotics obtained from natural sources (molds) and by chemical modification of the natural penicillins. Table 24–3 gives features of the penicillins.

The penicillins are the classic example of antibiotics for which microorganisms have developed resistance. A *resistant organism* is usually a mutant that has developed a way of living in the presence of an antibiotic. In the case of the penicillins, microorganisms that produce the enzyme *penicillinase* can develop over several generations. This enzyme catalyzes hydrolysis of the amide bond in the four-membered ring of the penicillins, thus inactivating the antibiotics. As microorganisms developed resistance to a form of penicillin, new forms were developed which are less susceptible to attack by penicillinase, as indicated in Table 24–3. However, some penicillins are unstable in the acid environment of the stomach and thus are not given orally.

Table 24-3 The Penicillins

Core structure: R—C(=O)—NH—CH—CH—S—C(CH₃)₂ ... (β-lactam ring with C(=O)—N) ... C—COOH

(Penicillinase catalyzes cleavage here — at the C—N bond of the β-lactam)

Name	Side Chain (R—)	Stability in Acid	Sensitivity to Penicillinase
Penicillin G	C₆H₅—CH₂—	Poor	Sensitive
Phenoxymethyl penicillin (Penicillin V)	C₆H₅—OCH₂—	Good	Sensitive
Methicillin	2,6-dimethoxyphenyl— (OCH₃ groups at both ortho positions)	Poor (not given orally)	Resistant
Oxacillin	3-phenyl-5-methylisoxazol-4-yl—	Good	Resistant
Dicloxacillin	3-(2,6-dichlorophenyl)-5-methylisoxazol-4-yl—	Good	Resistant
Cloxacillin	3-(2-chlorophenyl)-5-methylisoxazol-4-yl—	Good	Resistant
Nafcillin	2-ethoxynaphth-1-yl— (OC₂H₅)	Poor	Resistant
Ampicillin	C₆H₅—CH(NH₂)—	Good	Sensitive

Table 24-3 The Penicillins (continued)

Name	Side Chain (R—)	Stability in Acid	Sensitivity to Penicillinase
Carbenicillin	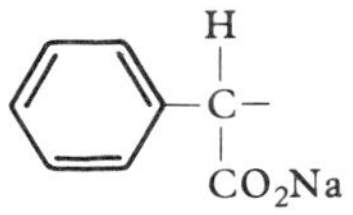 phenyl—CH(CO₂Na)—	(not given orally)	Sensitive
Amoxicillin	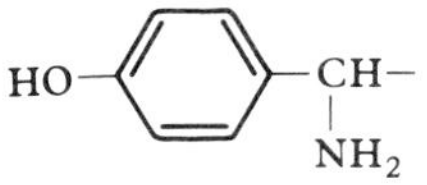 HO—C₆H₄—CH(NH₂)—	Good	Sensitive
Ticarcillin	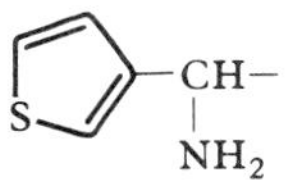 thienyl—CH(NH₂)—	(not given orally)	Sensitive

In bacteria, a cell wall surrounds each cell, and a cell membrane lies beneath the cell wall. The cell wall gives structural support to the cell and is responsible for the distinctive shapes of bacterial cells. The membrane of a bacterial cell is not strong enough to withstand the unusually high osmotic pressures inside a bacterial cell. The penicillins exhibit bacteriocidal action by reacting with an enzyme that is normally involved in synthesis of bacterial cell walls. The reaction with the penicillins occurs at the active site of the enzyme, and a penicillin molecule becomes covalently attached to the enzyme, preventing it from acting as a catalyst. This action weakens the cell wall, and the cell bursts. Animal cells do not have cell walls, and thus the penicillins do not harm animal cells.

The *tetracyclines* (see Figure 24-1) are bacteriostatic antibiotics because they interfere with protein synthesis in microorganisms by bind-

Fig. 24-1 Some of the Tetracyclines

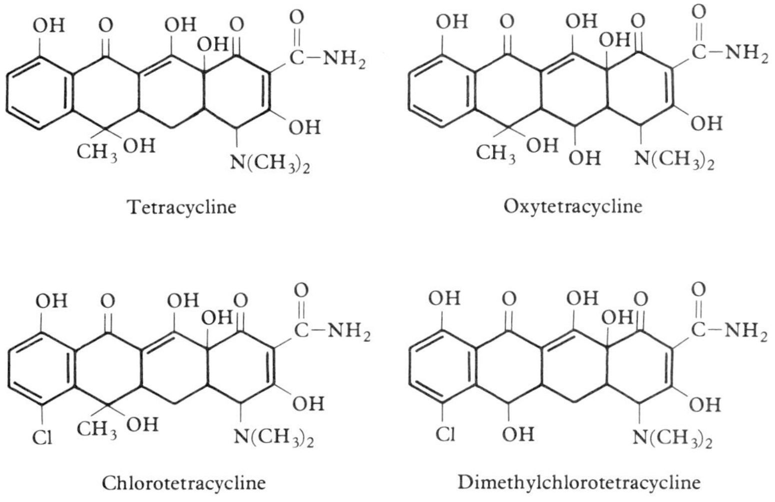

Tetracycline

Oxytetracycline

Chlorotetracycline

Dimethylchlorotetracycline

ing to the ribosomes responsible for protein synthesis. They are derived from several species of bacteria found in soil, and they have a broad spectrum of antibiotic activity that overlaps some of the penicillins. The tetracyclines usually are effective against microorganisms that are resistant to other drugs. The tetracyclines are absorbed from the stomach and small intestine, but in general, absorption is impaired by milk, milk products, aluminum hydroxide (an ingredient in some antacids), and salts of calcium and magnesium.

Antiviral Agents

The development of drugs for treating viral diseases has presented difficult problems because any agents that are harmful to viruses are also likely to injure host cells. Thus far, the drugs developed show narrow activities, limited to only one or a few specific viruses. Most antiviral agents are obtained by chemical synthesis.

Idoxuridine is an antiviral agent that acts on DNA viruses. It does so by being converted to its triphosphate derivative,

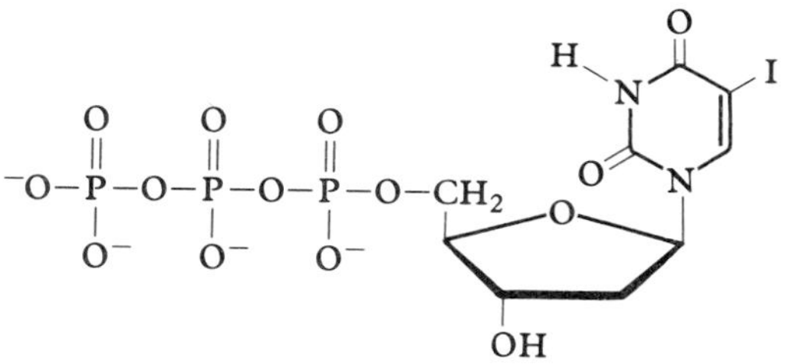

idoxuridine triphosphate

whose structure resembles that of deoxythymidine triphosphate (dTTP, see Table 22-3), one of the deoxynucleoside triphosphates needed for synthesizing DNA. Idoxuridine triphosphate is then used to synthesize viral DNA. Incorporation of idoxuridine monophosphate into DNA makes DNA more fragile, and it alters viral proteins due to faulty RNA synthesis. The primary use of the drug has been in the treatment of herpes simplex keratitis, a severe eye infection that can lead to blindness.

Amantadine is effective in preventing infection by different strains of Asian influenza viruses. Although its mechanism of action is not well understood, it somehow prevents penetration into the host cell by the virus.

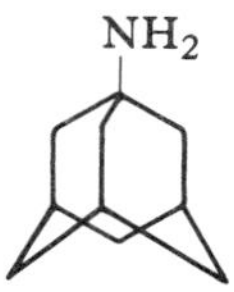

amantadine

Cytarabine, a structural analog of deoxycytidine (see Table 22-1), is an antiviral agent that acts to inhibit DNA polymerase (chapter 22) and thus DNA synthesis. Like idoxuridine, the drug acts on DNA viruses.

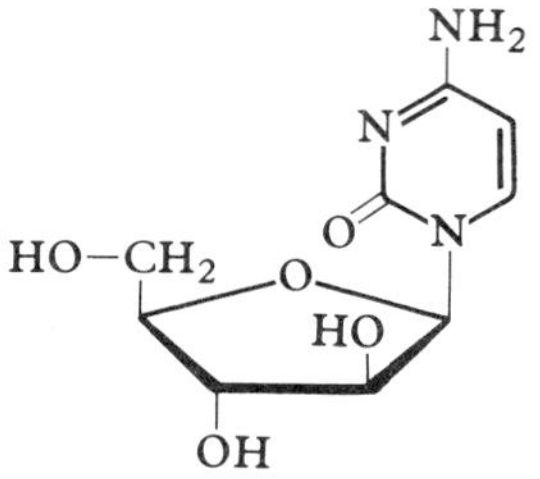

cytarabine

Anticancer Drugs

Although there have been no spectacular breakthroughs in cancer chemotherapy, much basic knowledge has been acquired since the mid-1960s, and at least 30 anticancer drugs are now known.

One of the earliest known groups of anticancer drugs is known as the *alkylating agents*. These substances react with DNA to add alkyl groups to adenine, cytosine, and guanine. This action results in disruption of base pairing in DNA and leads to faulty synthesis of DNA and RNA. However, the alkylating agents are highly reactive, and a wide variety of other chemical reactions probably occur that can cause a number of harmful effects on cell function. Some examples of alkylating agents are given in Figure 24-2.

Another group of anticancer drugs act as *antimetabolities* by severely inhibiting a variety of enzymes involved in cell metabolism. The consequences of enzyme inhibition are experienced most dramatically by cells undergoing rapid cell division. Thus, normal cells are also harmed by these agents, but not to the same extent as cancer cells. The classic example of an anticancer antimetabolite is 5-fluorouracil, which acts as an inhibitor in the synthesis of deoxthymidine monophosphate (dTMP, see Table 22-2), thereby slowing DNA synthesis.

Fig. 24-2 Some Alkylating Agents Used in Cancer Chemotherapy

Mechlorethamine

Cyclophosphamide

Uracil Mustard

Carmustine

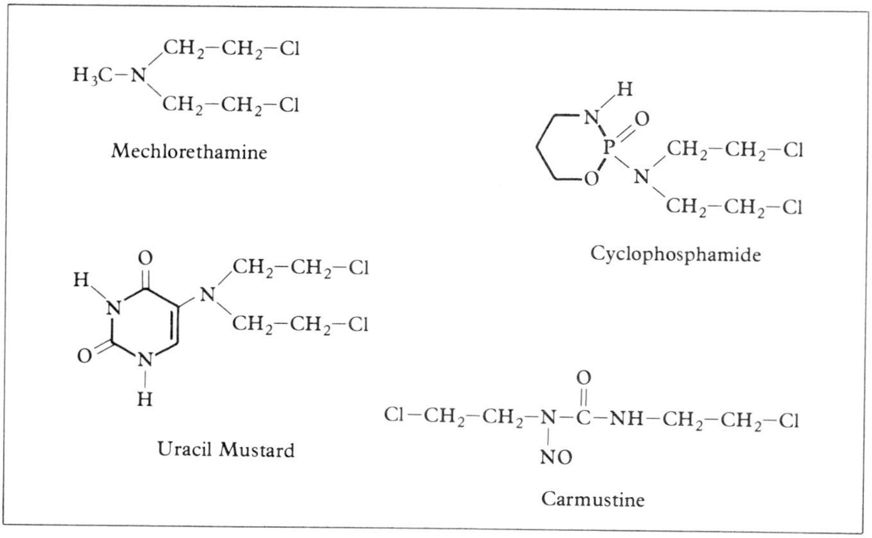

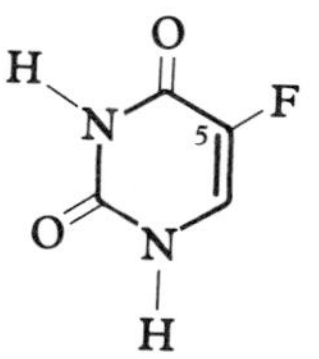

5-fluorouracil

Various *natural products,* including some antibiotics, constitute a third group of anticancer drugs. One of the more successful antibiotics used in cancer chemotherapy is Adriamycin (generic name doxorubicin), a fermentation product of a certain fungus.

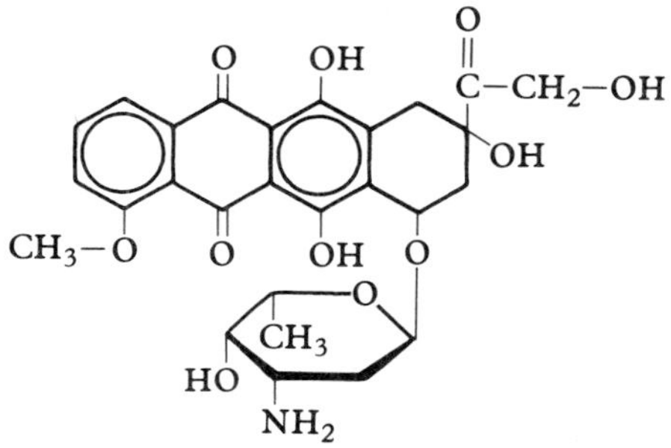

Adriamycin
(doxorubicin)

Even as a relatively new drug, it has an impressive record of activity against a wide spectrum of tumors. The drug appears to inhibit DNA replication by binding tightly to the interior of DNA, fitting itself between the planes of adjacent base pairs in a DNA double helix.

HOME PRODUCTS

Many home products have been developed on the basis of known chemical principles; some of the more familiar ones are discussed in this section.

Dental Chemistry and Fluoride Treatment

Tooth enamel consists mainly of hydroxyapatite crystals, $Ca_5(PO_4)_3OH$, with some protein integrated into the enamel structure. Although hydroxyapatite is considered to be water insoluble, in the mouth it exists in equilibrium with a small number of its ions. The forward reaction is called demineralization, and the reverse reaction is called remineralization:

$$Ca_5(PO_4)_3OH \underset{\text{remineralization}}{\overset{\text{demineralization}}{\rightleftharpoons}} 5\ Ca^{2+} + 3\ PO_4^{3-} + OH^-$$

If bacteria are allowed to remain in the vicinity of teeth for long periods of time, and if there is sugar present to serve as a nutrient for them, they will metabolize the sugar and produce lactic acid. The accumulation of lactic acid lowers the pH in the vicinity of the teeth and serves to neutralize the OH^- ions produced by demineralization. This action then shifts the equilibrium (by LeChatelier's principle, chapter 8) to the right, encouraging demineralization and, in effect, degradation of tooth enamel. Such degradation leads to tooth decay.

Fluoride ions have proved to be effective in preventing tooth decay. Research indicates that fluoride ions become incorporated into the structure of tooth enamel by replacing hydroxide ions in hydroxyapatite:

$$5\ Ca^{2+} + 3\ PO_4^{3-} + F^- \rightleftharpoons Ca_5(PO_4)_3F$$

However, fluoride does not replace all of the OH^- ions, and a mixture of $Ca_5(PO_4)_3F$ and $Ca_5(PO_4)_3OH$ then exists in tooth enamel. The important point, however, is that fluoridated hydroxyapatite is even less soluble in water (and saliva) than normal hydroxyapatite and thus more resistant to decay. Because of the favorable effects of fluoride ions on tooth enamel, some communities add sodium fluoride, NaF, to their drinking water. In addition, many toothpaste manufacturers have incorporated fluoride into their toothpastes as stannous fluoride (SnF_2), sodium fluoride (NaF), or monofluorophosphate (MFP, PO_3F^{2-}) salts. All of these provide fluoride ions for incorporation into tooth enamel.

Soaps and Detergents

As discussed in chapter 14, the cleaning action of soaps is attributed to the formation of micelles which suspend dirt and grease particles in water. However, soaps do not work well in hard water, because the Ca^{2+} and Mg^{2+} in hard water form insoluble salts with the fatty acid anions of soaps. Precipitation of these salts removes some of the soap from solution and leaves scum deposits. Detergents were developed in response to the hard water problem.

Detergents are soluble salts of sulfonic acids; these form micelles and have cleaning action similar to that of soaps. Sulfonate anions of detergents form compounds with hard-water ions (Ca^{2+} and Mg^{2+}), but the compounds are water soluble and thus do not form scums. The first detergents developed were alkylbenzenesulfonates (ABS), one of which is shown here.

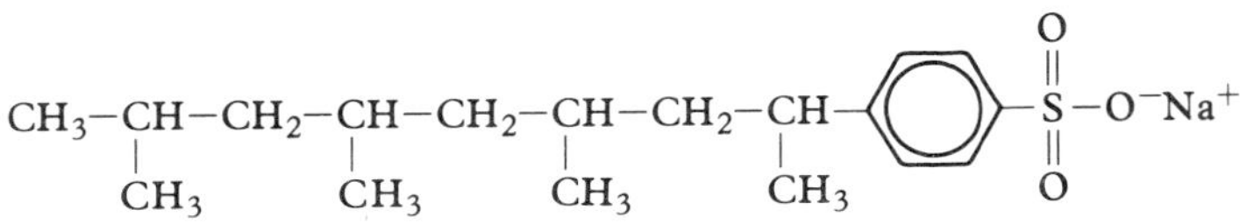

$$CH_3-\underset{\underset{CH_3}{|}}{CH}-CH_2-\underset{\underset{CH_3}{|}}{CH}-CH_2-\underset{\underset{CH_3}{|}}{CH}-CH_2-\underset{\underset{CH_3}{|}}{CH}-\underset{}{\bigcirc}-\underset{\underset{O}{\|}}{\overset{\overset{O}{\|}}{S}}-O^-Na^+$$

a sodium alkylbenzenesulfonate (ABS)

However, it was found that microorganisms decompose ABS detergents extremely slowly, if at all, in sewage-treatment plants, and thus these detergents remain in the water supply indefinitely. In other words, ABS detergents are not biodegradable. Consequently, by 1965, the ABS detergents were replaced by linear alkylbenzenesulfonate (LAS) detergents, which are biodegradable by virtue of their unbranched hydrocarbon chains.

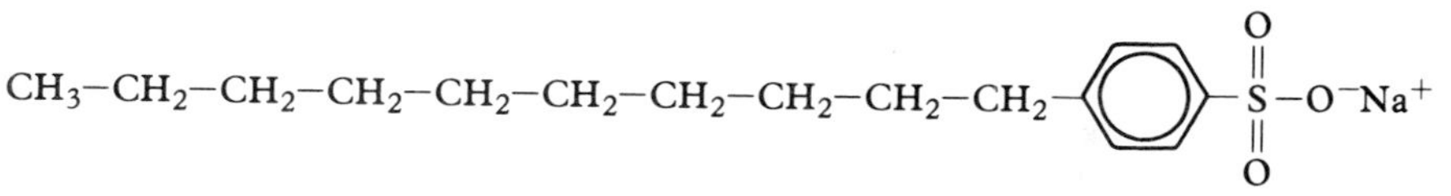

$$CH_3-CH_2-CH_2-CH_2-CH_2-CH_2-CH_2-CH_2-CH_2-CH_2-C_6H_4-\overset{O}{\underset{O}{\overset{\|}{\underset{\|}{S}}}}-O^-Na^+$$

a linear (sodium) alkylbenzenesulfonate (LAS)

Another problem associated with detergents arose in the 1960s. Despite the fact that detergents form soluble compounds with hard-water ions, these compounds show no cleaning action, and their formation detracts from detergent effectiveness. Because of this, manufacturers began adding "builders" to detergents. A builder is a substance that chemically softens water by forming soluble compounds with hard-water ions, thereby removing them from water and increasing the effectiveness of the accompanying detergent. The substances used most often as builders are sodium pyrophosphate and sodium tripolyphosphate; the anions of these compounds are shown here in combination with hard-water ions, represented by M^{2+}:

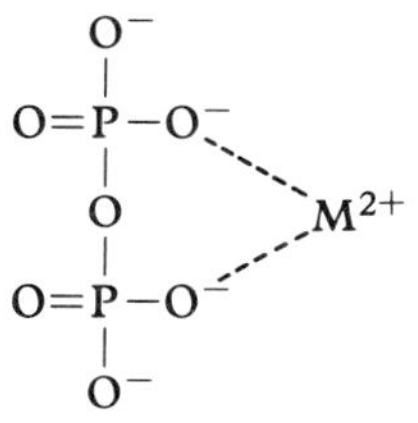

calcium or magnesium pyrophosphate

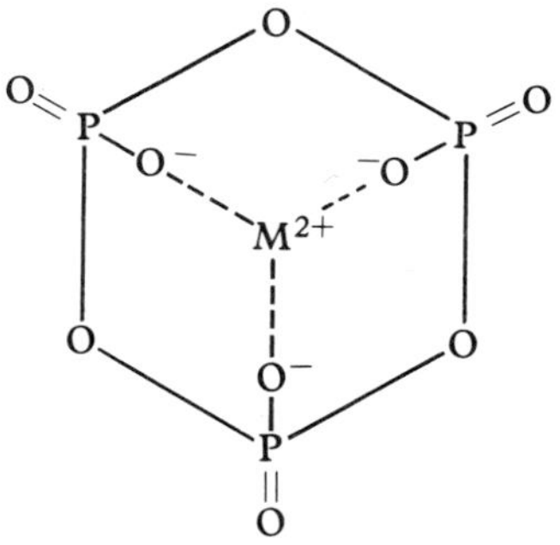

calcium or magnesium tripolyphosphate

The problem with phosphate builders is that they provide too much phosphorus nutrient to algae in lakes and other waterways, and as phosphates accumulate, a lake may be overrun with algae. This destroys the ecological balance of the lake to the point where much of the aquatic life dies. To alleviate this problem, manufacturers replaced many of the phosphate builders in detergents with other compounds that also remove hard-water ions. Unfortunately, the substitutes are not as effective as phosphates in softening water.

Drain Cleaners and Oven Cleaners

Both drain cleaners and oven cleaners have the same active ingredient—sodium hydroxide (NaOH). Since drains usually become clogged by fat, sodium hydroxide works well as a drain cleaner for two reasons: (1) the

NaOH partially saponifies the fat (see chapter 14), converting some of it to soluble fatty acid salts and glycerol; (2) dissolving solid NaOH or diluting a concentrated NaOH solution liberates much heat that serves to melt some of the fat. Oven cleaners also contain NaOH, and they are usually brushed or sprayed onto oven walls. In this case also, the NaOH dissolves the greasy residue on the oven surface by saponification.

If sodium hydroxide comes into contact with the skin, serious burns can result; however, the eyes can be damaged much more severely, and blindness can be caused by NaOH in the eyes. For these reasons, it is not only prudent but essential that skin and eye protection be worn when working with drain cleaners and oven cleaners.

SUMMARY

Food Preservation and Food Products: Decomposition of food by bacteria, yeasts, and molds is the major cause of food spoilage. Traditional methods used for preserving food include refrigeration, addition of sugar or salt, smoking, drying, and pickling. Modern methods are heat treatment, canning, cold storage, freezing, freeze-drying, vacuum drying, and the use of chemical agents.

Whole cow's milk contains 87% water, 5% carbohydrates, 4% fat, 3% protein, and 1% minerals. Churning coagulates the fat droplets in milk to make butter, whereas homogenization helps to suspend the fat droplets. Acidifying cow's milk or adding rennet causes formation of curds composed of mostly fat and casein. Cheese is produced by cutting the curds and pressing the whey from them. Cheeses are ripened by treatment with bacteria or molds to develop distinctive textures, flavors, and aromas. Yogurt and sour cream are made by bacterial fermentation of skimmed milk and cream, respectively. Buttermilk is the liquid remaining after butter is made from milk.

Vegetable oils are obtained by pressing oilseeds and extracting the residue with liquid hydrocarbons. Margarine and shortening are made from hydrogenated vegetable oils, while salad oils and cooking oils are purified vegetable oils.

Food additives are substances added to processed foods to enhance their quality or to improve processing, packaging, or storage. Food additives may be classified as preservatives, additives that maintain consistency or texture, acids and bases, nutrient supplements, flavors and flavor enhancers, coloring agents, and miscellaneous additives.

Drugs: Drugs are chemical substances that affect the functions of living things. Drugs have three types of names: systematic names, generic names, and brand names. Drugs with brand names are frequently more expensive than their counterparts with generic names, but the two usually function the same. Of the over-the-counter analgesics, aspirin is the most popular, but acetaminophen is also available. A popular prescription analgesic is propoxyphene, but it does not seem to be as effective as aspirin. For severe pain, physicians and dentists sometimes prescribe one of the opiates. The primary drugs presently used to induce general anesthesia are Sodium Pentothal, nitrous oxide, enflurane, and halothane. The two popular local anesthetics are procaine and lidocaine.

Some drugs are known to be effective in curing diseases. The sulfa drugs are used for bacterial urinary infection, while the penicillins and tetracyclines are used for many types of bacteral infections. Some effective antiviral agents are idoxuridine, amantadine, and cytarabine. Alkylating agents, some antimetabolites, and some natural products are effective in treating cancer.

Home Products: Fluoride treatment of teeth results in incorporation of fluoride ions into the enamel structure to make the enamel less soluble in water and more resistant to tooth decay. The cleaning action of soaps occurs because of micelle formation; however, soaps precipitate in hard water. Detergents are soluble salts of sulfonic acids which also form micelles and exhibit cleaning action. Modern detergents are linear alkylsulfonates, which are biodegradable. The amount of phosphate builders in detergents has been reduced due to overabundant growth of algae in waterways. Drain cleaners and oven cleaners contain NaOH that saponifies fats, causing them to dissolve in water. Skin and eye protection should be worn when handling these products because of the harmful effects of NaOH on living tissues.

STUDY QUESTIONS AND PROBLEMS

1. Define the following terms:

    (a) pasteurization
    (b) homogenization
    (c) casein
    (d) chymosin
    (e) rennet
    (f) rendering
    (g) butter
    (h) margarine
    (i) shortening
    (j) food additive
    (k) emulsifier
    (l) stabilizer
    (m) thickener
    (n) flavor enhancer
    (o) MSG
    (p) nonnutritive sweetener
    (q) drug
    (r) systematic drug name
    (s) generic drug name
    (t) brand name (for a drug)
    (u) analgesic
    (v) opiates
    (w) bacteriostatic drug
    (x) bacteriocidal drug
    (y) sulfa drugs
    (z) antimetabolite
    (aa) antibiotic
    (bb) penicillins
    (cc) resistant microorganisms
    (dd) penicillinase
    (ee) tetracyclines
    (ff) antiviral agent
    (gg) alkylating agents
    (hh) hydroxyapatite
    (ii) demineralization
    (jj) remineralization
    (kk) monofluorophosphate
    (ll) soap
    (mm) detergent
    (nn) detergent builder

2. Name six traditional methods of preserving food, and explain why each is successful.

3. Suggest an easy way of converting wine to vinegar, and illustrate with a chemical equation.

4. Name six modern methods of preserving food, and explain why each is successful.

5. Sodium benzoate can be used to preserve acidic foods, but it is not effective in preserving foods in which the pH is basic. Why?

6. Why is sulfur dioxide a good preservative for acidic foods?

7. What are the major components of cow's milk? In what amounts are they present?

8. How would you go about making skimmed milk?

9. Describe two methods of precipitating casein from milk.

10. Why is rennet important to cheese making?

11. Explain why cheese making is a way of preserving the nutritious ingredients of milk.

12. What are fresh cheeses? What are their qualities? Give two examples of fresh cheeses.

13. Explain what is meant by "ripening" cheese. How is Roquefort cheese ripened? What is the origin of the blue veins in Roquefort cheese?

14. What carbohydrate is used by the bacteria in making yogurt? What is the major metabolic pathway involved in making yogurt?

15. What would happen to the lactic acid component of yogurt if too much oxygen were present during fermentation?

16. How do butter and margarine differ in composition?

17. Why does margarine have a higher iodine number than butter?

18. Why is cholesterol found in butter but not in margarine?

19. How do margarine and shortening differ in composition? Why do the two have different colors?

20. Cottonseed oil (iodine number 97–112) is more desirable for use in frying foods than safflower oil (iodine number 140–150). Why?

21. Give five examples of antioxidants used as food additives, and explain how they help to prevent rancidity in fats and oils.

22. In what ways do acidic and basic food additives control the pH of foods?

23. Give five examples of nutrient supplements in foods.

24. Distinguish between a flavoring agent and a flavor enhancer.

25. How is MSG thought to work as a flavor enhancer?

26. Why is it that individuals with phenylketonuria should not use aspartame?

27. Why do nonnutritive sweeteners contribute few or no calories to the diet?

28. What three kinds of names are used to identify drugs? How do these names differ from each other?

29. What is the actual pain reliever in aspirin? Why is this substance not marketed as an analgesic?

30. Name four opiates, and explain why they are available only by prescription.

31. What are the advantages and disadvantages of each of the following general anesthetics?

    (a) diethyl ether
    (b) chloroform
    (c) Sodium Pentothal
    (d) enflurane

32. Name three local anesthetics, and explain how each is used.
33. Draw structures and give names of four sulfa drugs. Why are sulfa drugs both bacteriostatic and bacteriocidal?
34. Why is the use of sulfa drugs restricted mainly to treatment of urinary infections?
35. How do microorganisms develop resistance to the penicillins?
36. Explain how the penicillins exhibit bacteriocidal action.
37. How do the tetracyclines exhibit bacteriostatic action?
38. Why are tetracyclines sometimes preferred over the penicillins as antibiotics?
39. Name three antiviral agents, and explain how each works.
40. Name three general classes of anticancer drugs, explain how each works, and give an example for each class.
41. Explain how fluoride treatment helps to prevent tooth decay.
42. Why is it that soaps do not clean well in hard water?
43. What is the advantage of a detergent over a soap?
44. What are ABS and LAS detergents? Give a structural example of each.
45. Explain how phosphate builders enhance the cleaning action of detergents. Why has the amount of phosphate builders in detergents been reduced?
46. What is the active ingredient in drain cleaners and oven cleaners? How do these products work?

CHAPTER 25

RADIOACTIVITY AND NUCLEAR PROCESSES

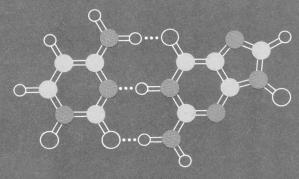

RADIOACTIVITY
- Natural Radioactivity
- Artificial Radioactivity

RADIATION: DETECTION AND MEASUREMENT
- Detection of Radiation
- Half-Life
- Measurement of Radiation

RADIATION SAFETY
- The Basis for Biological Damage
- Radiation Exposure
- Protection Against Radiation

APPLICATIONS OF RADIOCHEMISTRY
- Archeological Dating
- Isotopic Tracers
- Radiation Therapy
- Medical Diagnosis

NUCLEAR POWER
- Nuclear Binding Energy
- Nuclear Fission
- Nuclear Fusion

The discovery of radioactivity by the French physicist Henri Becquerel in 1896 came only a few months after the discovery in Germany of X-rays. These two events led to an unprecedented period of exciting scientific advances in the study of atomic structure. For example, J. J. Thomson discovered the electron in 1897, the three different types of radioactive decay were observed in 1900, and in 1902 it was recognized that nuclear changes accompany radioactive decay. The most significant event of all, Einstein's theory of the equivalence of mass and energy (E = mc^2), came about in 1905. The twentieth century has witnessed an unraveling of the structure of the atom and the unleashing of nuclear fury at Hiroshima and Nagasaki. An understanding of the fundamental principles of radioactivity and nuclear processes is important so that we can safely utilize the beneficial applications of radiochemistry.

RADIOACTIVITY

Radioactivity is defined as the spontaneous emission of energy and/or subatomic particles by certain types of matter called radioactive matter. There are two general classes of radioactivity: natural and artificial.

Natural Radioactivity

Approximately one-third of the elements have natural radioactive isotopes. Most of these are very slow to decay and have existed since the earth was formed. Natural radioisotopes stabilize themselves by the three nuclear decay processes discussed in chapter 2: α-, β-, and γ-decay.

In addition to slowly decaying natural radioisotopes, others with shorter half-lives, such as tritium (hydrogen-3) and carbon-14, are continuously formed by bombardment from *cosmic rays*. This radiation consists of particles streaming into the earth's atmosphere from the sun and outer space. The incoming particles are mostly protons, with some electrons, α-particles, and nuclei of higher elements. When these *primary cosmic rays* collide with gas atoms and molecules in the upper atmosphere, they produce *secondary cosmic rays* composed of all of the known elementary particles—electrons, protons, neutrons, positrons, and other less familiar particles. The heavier nuclei present in primary cosmic rays are broken up by collisions. It is probably the secondary cosmic rays which generate hydrogen-3 and carbon-14.

When a radioisotope undergoes decay, a new radioisotope is often produced. This means that the decay process did not produce a stable product. Instead, the product itself will decay to another product, which may or may not be stable. In this way, a whole succession of radioactive disintegrations may take place, ultimately forming a stable isotope. Naturally occurring radioisotopes of atomic number greater than 81 decay by one of three *radioactive disintegration series*, one of which is illustrated in Figure 25-1. All of the series produce stable isotopes of lead as end products.

Artificial Radioactivity

Nuclear reactions can be used to produce artificial radioisotopes. If a stable nucleus is bombarded with α-particles, neutrons, or other sub-

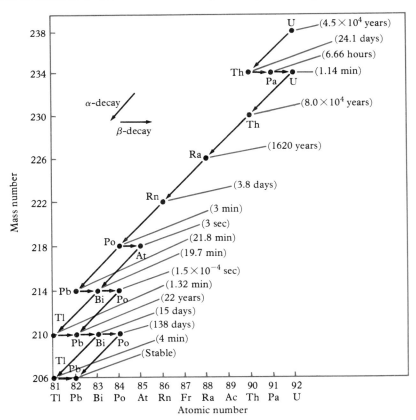

Fig. 25-1 The Uranium-238 Disintegration Series. (The half-lives of the intermediates are given in parentheses.)

atomic particles, the first atom may be changed into a new atom that is unstable. The resulting radioactive decay is often called *artificial radioactivity*. The synthesis of elements beyond uranium in the periodic table (transuranium elements) has been accomplished by this bombardment technique, as illustrated by the following equations:

$$^{238}_{92}U + ^{1}_{0}n \longrightarrow ^{239}_{93}Np + ^{0}_{-1}\beta$$
$$^{238}_{92}U + ^{2}_{1}H \longrightarrow ^{238}_{93}Np + 2^{1}_{0}n$$
$$\hookrightarrow ^{238}_{94}Pu + ^{0}_{-1}\beta$$

As these equations illustrate, nuclear bombardment brings about *transmutation*, the conversion of one element into another. In fact, natural radioactivity also results in transmutation. However, it was through artificial transmutation that Ernest Rutherford proved the existence of protons in 1919:

$$^{14}_{7}N + ^{4}_{2}He \longrightarrow ^{17}_{8}O + ^{1}_{1}H$$

Another artificial transmutation led to the discovery of the neutron by James Chadwick in 1932:

$$^{9}_{4}Be + ^{4}_{2}He \longrightarrow ^{12}_{6}C + ^{1}_{0}n$$

RADIATION: DETECTION AND MEASUREMENT

Radiation is energy and/or particles emitted by matter. Radioactive matter gives off radiation in the form of α-particles, β-particles, or γ-rays. In addition, other types of radiation can be created artificially; in general these are protons, neutrons, and X-rays. Certain forms of radiation cause ionization in substances that they strike; Table 25-1 lists the types of ionizing radiation along with the properties and sources of each.

X-rays are high-energy ultraviolet light given off when a nucleus captures an electron. Energy is released as an outer electron falls into the space created by the electron capture, as illustrated in Figure 25-2.

Table 25-1 The Major Types of Ionizing Radiation

Type	Symbol	Mass (amu)	Charge	Source
Alpha	$^{4}_{2}\alpha$	4	+2	Spontaneous radioactive decay (primarily from heavy atoms)
Beta	$^{0}_{-1}\beta$	0	−1	Spontaneous radioactive decay
Beta (positron)	$^{0}_{1}\beta$	0	+1	Spontaneous radioactive decay
Proton	$^{1}_{1}H$	1	+1	Artificially produced by nuclear reactors
Neutron	$^{1}_{0}n$	1	0	Artificially produced by nuclear reactors
Gamma	γ	0	0	Spontaneous radioactive decay
X-Ray	—	0	0	X-Ray machines

Fig. 25-2 X-Ray Emission Following Electron Capture

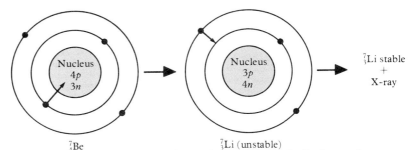

X-rays can be generated by a high-voltage electrical discharge in a vacuum tube; this design allows high-speed electrons to bombard a target metal and thus produce X-rays.

X-rays have a high energy content and pass through soft tissues to varying degrees. They are used in medicine and dentistry to examine internal organs and the structure of bone and teeth. Figure 25-3 shows an X-ray machine. After passing through the soft tissue, the X-rays expose a photographic film placed on the other side of the body. Bones and other hard substances absorb the X-rays, leaving white images on the photographic film; an example is shown in Figure 25-4.

Fig. 25-3 A Commercial X-Ray Machine

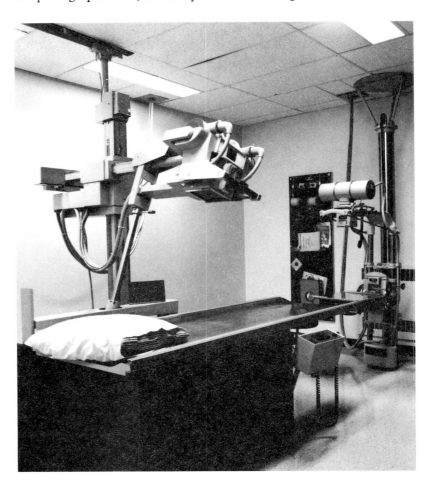

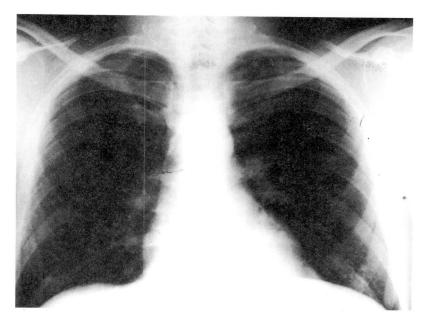

Fig. 25-4 A Developed X-Ray of Healthy, Normal Lungs

Detection of Radiation

Many forms of radiation produce molecular fragments and ions when they interact with matter. Such *ionizing radiation* can be detected by the *Geiger-Muller* counter (Geiger counter is a shortened version of the name). This device contains an unreactive gas in a closed chamber (see Figure 25-5). The window of the chamber is constructed of a material that allows high-energy radiation, such as β-, γ-, and X-rays, to pass through into the chamber. Once inside, the radiation produces ions from some of the gas molecules, and the ions move to electrodes, completing an electrical circuit. The intensity of the radiation is registered by a meter and a clicking sound. The speed of the clicking indicates the intensity of the radiation. A Geiger counter is shown in Figure 25-6.

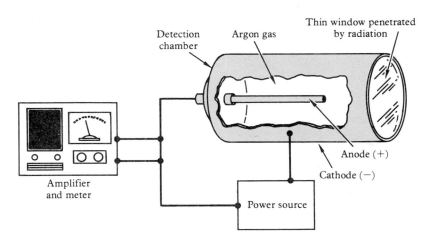

Fig. 25-5 Schematic Diagram of a Geiger Counter

Fig. 25-6 A Geiger Counter (The detection chamber is the cylinder attached to the top of the meter.)

For detection of low-energy radiation, such as α- and some β-radiation, a device called a *scintillation counter* is used. In this instrument special compounds that emit bursts of light (scintillations) when they absorb radiation are exposed to the material being tested. These light pulses can then be counted by extremely sensitive detectors, converted to electrical signals, and registered on a meter. In this way, very low levels of radiation can be detected and measured. This method is particularly useful for detecting the low-energy β-radiation of carbon-14, a radioisotope used widely in research.

$$^{14}_{6}C \longrightarrow ^{14}_{7}N + ^{0}_{-1}\beta$$

Radioactivity produces radiation that has many of the same qualities as light. In this respect, radiation will expose photographic film. Thus, people who work around radiation often wear *film badges* (Figure 25–7), sometimes called dosimeters, to record the amount of radiation exposure they receive. These badges contain several layers of film, each shielded from light but not from higher-energy radiation. The extent of darkening of the film can be translated into the amount of radiation exposure received by the wearer. In most cases, cumulative records are kept to insure that workers are not exposed to unsafe amounts of radiation.

Fig. 25-7 A Radiation Film Badge. On each film badge, the worker's name is written on the white plastic holder (*lower left*) and the date (*upper center*) is printed on paper that encloses the radiation-sensitive film.

EXAMPLE 25.1

Why do X-rays provide images of bones and other hard tissue?

SOLUTION

X-rays have enough energy to pass through soft tissue and expose a photographic film, but they are absorbed by hard tissue. This absorption leaves a white image on the film corresponding to the shape of the hard tissue.

EXERCISE 25.1 Name three devices which can be used to detect radiation.

Half-Life

Each different radioisotope decays at its own characteristic rate. Thus, it takes some radioisotopes a long time to decay to a significant degree, whereas others decay rapidly. In order to compare rates of decay, we use the quantity called the *half-life*. This is the length of time required for half of the original radioactive atoms to decay. For example, mercury-203 has a half-life of 47 days, while carbon-14 has a half-life of 5730 years. This means that half of the radioactive Hg-203 atoms will have decayed after just 47 days, but it will take almost 6000 years for half of the radioactive C-14 atoms to decay. Furthermore, even though more radiation is emitted when a large number of radioactive atoms is present, the half-life of a radioisotope is always the same, regardless of how much was present originally. Thus, after the first 47 days, Hg-203 will have lost half its radioactivity; after the second 47 days it will have lost half again, and so on. Table 25-2 illustrates how the relative amount of radioactivity diminishes with each half-life. Note that the radioactivity will never completely disappear, but will diminish more and more, approaching a value of zero. This behavior is illustrated graphically in Figure 25-8. Note the constant value for the half-life.

The half-life of a radioactive material is an important consideration in the use and storage of radioisotopes. For example, it is desirable that radioisotopes used in the body for medical diagnosis have half-lives of a few days or less in order to minimize the patient's radiation exposure. Examples of some radioisotopes used for diagnostic purposes are given in Table 25-3. Radioisotopes with such short half-lives may be allowed to decay to a low level of radioactivity in a shielded container before conventional disposal. In contrast, radioisotopes with long half-lives are disposed of by burial in locations approved by the Nuclear Regulatory Commission (NRC).

EXAMPLE 25.2

The half-life of Na-24, a gamma emitter used in medicine, is 15.0 hours. What proportion of Na-24 atoms will have decayed in a sample after 45.0 hours?

SOLUTION

The length of time, 45.0 hours, corresponds to 3 half-lives (45.0 hours/15.0 hours per half-life = 3.0 half-lives). After the first half-life, half the radioactive

Table 25-2 The Radioactive Decay of Hg-203

Time (number of half-lives)	Percent of Hg-203 Remaining
0	100
1 × 47 days (1 half-live)	50
2 × 47 days (2 half-lives)	25
3 × 47 days (3 half-lives)	12.5
4 × 47 days (4 half-lives)	6.25
5 × 47 days (5 half-lives)	3.125
10 × 47 days (10 half-lives)	0.098

Fig. 25-8 Radioactive Decay of Mercury-203

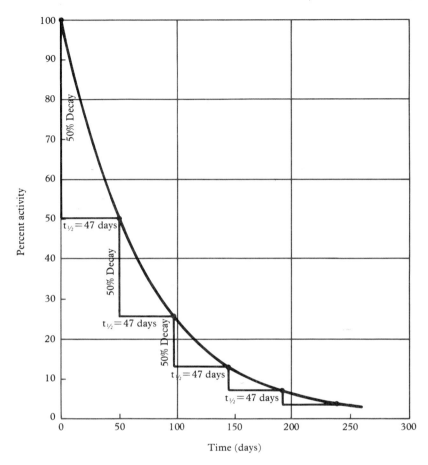

atoms will have decayed. After the second half-life, another half of the atoms will have decayed, leaving one-fourth of the original number ($½ \times ½ = ¼$); and after the third half-life, ⅛ of the original radioactive atoms will remain ($¼ \times ½ = ⅛$). Thus ⅞ of the original radioactive atoms of Na-24 will have decayed after 45.0 hours.

Measurement of Radiation

Radiation is measured by methods that rely on the ionization of gases caused by radiation, the emission of light from radiation-sensitive materials, or the blackening of photographic film by radiation. Thus, the detection devices mentioned earlier are also used to measure levels of radiation.

Radiation measurements are reported in a variety of units, depending on the reason for each measurement. For example, it might be necessary for a hospital to purchase a radiation source. In that case, it would be necessary to know the *activity* of the sample—that is, the number of radioactive disintegrations over a given period of time. It would also be important to know the *dosage* amount delivered by the sample so that it could be used properly. In addition, health care facilities must know the

Table 25-3 Half-Lives of Some Radioisotopes Used in Medical Diagnosis

Radioisotope	Half-Life
Calcium-47	4.5 days
Chromium-51	27.8 days
Cobalt-57	270 days
Gold-198	2.7 days
Iodine-125	60.2 days
Iodine-131	8.1 days
Iron-59	45.6 days
Krypton-79	1.5 days
Mercury-197	2.7 days
Phosphorus-32	14.3 days
Strontium-87	2.8 hours
Technetium-99m	6.0 hours
Xenon-133	5.3 days

energy level of X-rays produced by their machines. Radiation measurements can be made and expressed for all three purposes, as the following sections will describe; the three types of radiation measurement are summarized in Table 25–4.

Activity

A sample of radioactive material is never pure, because its atoms are continuously breaking down into new atoms. Thus, it is often necessary to measure the number of nuclear disintegrations per second in order to characterize the sample accurately. For a long time the standard unit of measurement has been the *curie*, abbreviated as Ci. This unit, named for Marie Curie, is defined as 3.7×10^{10} disintegrations per second. As shown below, disintegrations can also be expressed in millicuries, microcuries, and picocuries.

$$1 \text{ millicurie} = 10^{-3} \text{Ci} = 3.7 \times 10^{7} \text{ disintegrations/sec}$$
$$1 \text{ microcurie} = 10^{-6} \text{Ci} = 3.7 \times 10^{4} \text{ disintegrations/sec}$$
$$1 \text{ picocurie} = 10^{-12} \text{Ci} = 3.7 \times 10^{-2} \text{ disintegrations/sec}$$

Table 25-4 Radiation Quantities and Units

Measurement	Radiation Type	Effect Measured	Units
Activity	All types	Nuclear disintegrations	curie (Ci) becquerel (Bq)
Dosage	X- or γ-radiation	Ionization	roentgen (R)
	All types	Energy absorbed	rad; gray (Gy)
	All types	Biological damage	rem
Energy	All types	Energy emitted	electron-volt (eV)

becquerel (beck-where-REHL)

With the introduction of SI units, the becquerel was devised. Named for Henri Becquerel, the *becquerel* (Bq) is defined as 1 disintegration per second. Thus, 1 Ci = 3.7×10^{10} Bq.

Dosage

roentgen (RENT-gehn)

It is also necessary to measure radiation in terms of exposure, the dose delivered by X-rays or γ-rays. The traditional unit is named for Wilhelm Roentgen, who was awarded a Nobel Prize in 1901 for his discovery of X-rays. One *roentgen* (R) is that amount of X- or γ-radiation which will produce ions totaling 2.1×10^9 units of electrical charge in 1 cubic centimeter of dry air at standard temperature and pressure. Thus, the higher the energy of X- or γ-radiation, the more ions it produces, both in air and in any other matter it encounters, such as biological tissue.

A more general and more accurately defined measure of exposure to all types of radiation is the *rad*, the *r*adiation *a*bsorbed *d*ose. This is a measure of the radiation dose absorbed in tissue itself. It is defined as the dosage equivalent to 100 ergs of energy absorbed per gram of tissue. An erg is a very small amount of energy, roughly the energy expended when a mosquito does one push-up. However, it is not the absorbed energy that damages tissue; it is the highly reactive ions created by the radiation that can cause harm. The SI units for absorbed dosage are named for the British radiobiologist Louis Harold Gray. One *gray* (Gy) is defined as 1 joule (J) of energy absorbed per kilogram of tissue. Since 1 J = 10^7 erg, then 1 Gy = 100 rad.

Absorption of different kinds of radiation results in different biological responses even if the dosage in rads is the same. For this reason, it really is not informative to compare the rad dosage of α-radiation to the rad dosage of β-radiation. To settle this dilemma, another unit has been devised. One *rem* of any kind of radiation is defined as that quantity of radiation that causes damage equivalent to the absorption of 1 roentgen. Rem stands for *r*oetgen *e*quivalent *m*an. If we use these units, then the dosages of several different exposures can be added together, regardless of the type of radiation for each exposure.

Energy of Radiation

Another unit, the *electron-volt* (eV), is used to describe the energy of radiation emitted by radioactive materials or by X-ray generators. This unit is that quantity of energy acquired by an electron accelerating through a voltage change of one volt. This is an extremely small quantity of energy, equal to 1.602×10^{-19} J. Typically, radioactive materials emit energy on the order of a million electron-volts (MeV), while X-rays have energies of about 0.09–0.25 MeV.

EXAMPLE 25.3

List seven units of radiation measurement and indicate what is measured by each.

SOLUTION

curie:	radioactive disintegrations
becquerel:	radioactive disintegrations
roentgen:	dosage level
rad:	dosage level
gray:	dosage level
rem:	dosage level
electron-volt:	energy of radiation

RADIATION SAFETY

We are all aware of the need to avoid excessive dosages of radiation because it is damaging to biological tissue. However, the reasons for biological damage are not so well known. In the following sections you will learn how tissue damage occurs, the typical responses to various levels of exposure, and the precautions necessary for observing proper radiation safety.

The Basis for Biological Damage

Nuclear radiation and X-rays are harmful to living tissues because of the chemical changes they cause. As mentioned earlier, nuclear radiation and X-rays interact with molecules by stripping away electrons to form ions and by fragmenting molecules to form reactive structures called *free radicals*. Figure 25-9 illustrates the effects of radiation on water molecules, the most plentiful compound present in living tissue.

Generally speaking, the more massive a particle is, the less deeply it penetrates into tissue. Thus α-particles travel only short distances through tissue, on the order of 0.05 mm. β-Particles, having smaller mass, penetrate about 4 mm into tissue. X- and γ-rays have no mass, and thus they are the most penetrating forms of radiation, passing through as much as 50 cm of tissue. You can understand these relative penetrating abilities by imagining that radiation travels through tissue by passing between the atoms present. The larger the radiation particle, the greater chance it has of bumping into an atom and being stopped. In fact, bumping into atoms is what brings about formation of ions and free radicals.

The least penetrating forms of radiation cause the greatest biological damage. This is because the large particles do not travel very far in the tissue, but they transfer all of their energy to a rather small volume of tissue. The smaller particles as well as X- and γ-radiation travel over longer distances inside the tissue, and they release their energy all along the path, thus doing less harm.

Ions and free radicals are formed when water in the tissue reacts with radiation, and these structures then react with whatever is nearby. In a cell, this means an assortment of biological molecules, but the most harmful effects come from reactions with proteins, enzymes, and genetic material. These reactions can denature proteins and inactivate enzymes,

free radical: a highly reactive structure having one or more unpaired electrons.

Fig. 25-9 The Effects of Ionizing Radiation on Water Molecules. After the radiation is absorbed, a water molecule can (a) split into H^+ and OH^-; (b) lose an electron and become a positively charged ion, H_2O^+; or (c) split into two free radicals

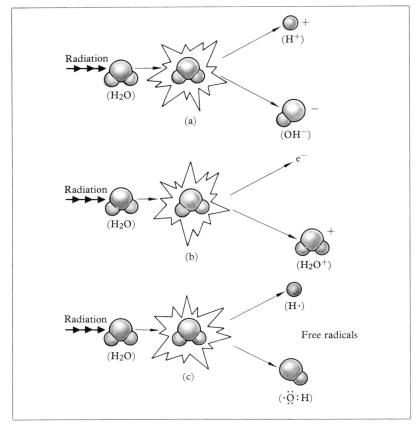

thus interfering with normal metabolic processes. But these effects will be temporary provided that DNA damage is minimal. If DNA can still function to direct protein synthesis, then the damaged proteins and enzymes can be replaced, although in some cases, cellular control mechanisms may be impaired. Cells can withstand some damage to DNA because they have enzymes whose function is to repair damaged DNA. In fact, small amounts of damage to DNA occur naturally. But if the DNA is damaged extensively, then the repair enzymes, already smaller in number, cannot overcome the damage, and the cell dies.

Cells undergoing rapid division are more susceptible to radiation damage. For this reason, lymph tissue, the intestinal tract, bone marrow, the gonads, and developing embryos are especially vulnerable. For the same reason, radiation is a powerful form of treatment for cancer.

Radiation Exposure

All of us are exposed to a certain amount of radiation each day arising from natural sources such as cosmic rays, deposits of radioactive ore, and radioisotopes in building materials and the air around us. It is estimated that each U.S. citizen receives an average of about 100–125 millirems of radiation exposure each year from these natural sources. In addition, medical procedures, the atomic energy industry, and radioactive fallout account for 60–70 millirems, on the average. Thus, each of us receives about 150–200 millirems of radiation each year during normal existence. Indeed it appears that humans can withstand a single dose of 25 rems without any immediate bad effects, but the possibility of long-term effects such as tumors, birth defects, cancer, and mutations has not been evaluated. As a matter of precaution, it is recommended that the maximum annual exposure due to man-made sources be limited to 170 millirems (0.170 rem). One reason for this is that there is evidence that exposure has a cumulative effect; thus, even exposure to very low levels of radiation over a long period of time may be as serious a health threat as a short-term exposure to a very high level.

In short-term doses of 25–100 rems, a person may not experience any symptoms of illness, but anemia and low white blood cell counts may occur. For higher short-term exposure, 150–600 rems, survival is still possible, but recovery may take weeks or even a year. Symptoms are nausea, vomiting, weakness, anemia, low white blood cell count, internal hemorrhaging, susceptibility to infection, hair loss, emaciation, and fever.

Persons receiving in excess of 600 rems in a single dose of radiation are not likely to survive. In these cases, painful, ugly, and repulsive symptoms occur prior to a delayed death. Early symptoms are loss of appetite, nausea, vomiting, diarrhea, fever, anemia, low white blood cell count, and hemorrhaging. Complete loss of hair within ten days of exposure has been taken as an indicator of a lethal dosage.

The effects of short-term radiation doses are summarized in Table 25-5.

Protection Against Radiation

In light of the hazards of nuclear radiation and X-rays, we have learned to treat them with caution and respect. In this way, we can derive many

Table 25-5 Probable Effects of Short-Term Radiation Doses

Dose (rems)	Probable Effect
0–25	No noticeable effects
25–100	Slight blood changes
100–200	Some vomiting within hours, with fatigue and loss of appetite; except for the blood-forming system, recovery will be complete within a few weeks.
200–600	For doses of 300 rems and more, vomiting will occur in two hours or less; severe blood changes will be accompanied by hemorrhage and infection; recovery will occur in 20–100% of all cases, usually within one month to one year.
600–1000	Vomiting within one hour; severe blood changes, hemorrhaging, and infection; loss of hair; 80–100% of exposed individuals will die within two months; survivors will be convalescent over a long period of time.

benefits from applications of nuclear processes, as will be discussed shortly. But the basis of any safety precautions lies in the properties of the various forms of radiation.

Since we know that radiation moves in straight lines from the origin, we can deduce that it streams out in all directions from a radioactive sample, much as light waves spread from a light source. Thus, in order to have sufficient protection from radiation, we must have a shield present between ourselves and the emitting source. Since α- and β-particles have the least penetrating abilities, as indicated in Table 25-6, these forms of radiation may be stopped by paper, wood, or cardboard. However protection against γ- and X-radiation requires use of very dense materials; lead is the most common substance used. Thus, radiologists spread lead-impregnated cloths over parts of the body not intended to receive radiation. It is also important that radiologists and associated workers take extra precautions to shield themselves from radiation, as they have far more opportunities for exposure than most people.

Table 25-6 Penetrating Abilities of Common Forms of Radiation

Radiation Type	APPROXIMATE DEPTH OF PENETRATION OF RADIATION INTO:		
	Dry Air	Tissue	Lead
Alpha	4 cm	0.05 mm	0
Beta	6–300 cm	0.06–4 mm	0.005–0.3 mm
	THICKNESS NEEDED TO REDUCE INITIAL INTENSITY BY 10%		
Gamma	400 meters	50 cm	30 mm
X-Rays	120–240 meters	15–30 cm	0.3–1.5 mm

A second way of protecting one's self from radiation is to stay far away from the radioactive source. The rays spread out in the shape of a cone, and less radiation will strike a given amount of surface area as it is moved away from the source. In a vacuum, the radiation intensity striking a given amount of surface area decreases with the square of the distance from the source. Another way of stating this is that radiation intensity in a vacuum is inversely proportional to the square of the distance from the source:

$$\text{radiation intensity} \propto \frac{1}{d^2}$$

Although the relationship is not strictly true in air, it is a very good approximation. Estimations of radiation intensities at new distances can be made on the basis of previously measured intensities by using the following equation:

$$\frac{\text{intensity at } d_1}{\text{intensity at } d_2} = \frac{d_2^2}{d_1^2}$$

EXAMPLE 25.4

A radioactive material produced a radiation intensity of 100 millirems at a distance of 10 feet. What is the intensity at a distance of 5 feet?

SOLUTION

Using the relationship

$$\frac{\text{intensity at } d_1}{\text{intensity at } d_2} = \frac{d_2^2}{d_1^2}$$

we can solve for intensity at d_2:

$$\text{intensity at } d_2 = \frac{d_1^2}{d_2^2} \times \text{intensity at } d_1$$

$$= \frac{(10 \text{ ft})^2}{(5 \text{ ft})^2} \times 100 \text{ millirems}$$

$$= 400 \text{ millirems}$$

In addition to proper shielding and distance from the source, it is best to use a low-intensity source whenever possible and to keep the exposure time to a minimum. These precautions help to minimize dosage levels and prevent unnecessary exposure. Above all, radioactive materials should not be ingested except under the direction of a physician. This means that workers should never eat, drink, or smoke in the vicinity of radioactive substances.

APPLICATIONS OF RADIOCHEMISTRY

Since radioactivity was discovered, many practical uses have been found for radioisotopes and X-rays. As more and more applications are developed, the number of people working around radiation also increases. Thus, it is important that each of us has an appreciation for the precautions necessary for proper handling and storage of radioactive samples. In this way, radiochemistry will continue to be of great benefit to mankind in the uses discussed below and others yet to be discovered.

Archeological Dating

As mentioned earlier, secondary cosmic radiation is responsible for producing carbon-14 in our atmosphere:

$$^{14}_{7}\text{N} + ^{1}_{0}n \longrightarrow ^{14}_{6}\text{C} + ^{1}_{1}\text{H}$$

This radioisotope of carbon makes up a fraction of a percent of all carbon on earth. Once C-14 is formed, it makes its way into live plants and animals as they exchange carbon compounds with their environment. Thus, any living creature gives off and takes in C-14 until an equilibrium is reached, and then the level of C-14 within the organism remains constant.

As fast as new C-14 is formed, some is lost by radioactive decay; its β-radiation is easily detected.

$$^{14}_{6}C \longrightarrow \, ^{14}_{7}N + \, ^{0}_{-1}\beta$$

Because of its continuous formation and decomposition, the amount of C-14 in the environment is also constant. In fact, the level of C-14 in a living creature is equal to the level of the radioisotope in the surroundings. However, when death occurs, exchange of C-14 with the environment ceases, but the decay of the previously incorporated radioisotope continues. After 5730 years, the half-life of C-14, one-half of the normal level is found in the organic matter that was once part of a living organism. Thus, the amount of C-14 continuously diminishes in nonliving organic matter. Measuring relative amounts of this radioisotope in organic matter allows archeologists to estimate the age of a sample. This measurement is based, of course, on the known half-life of C-14. In practice, the carbon content of a sample is converted to CO_2 by burning, and the level of β-radiation is measured. Some examples of objects dated with carbon-14 are listed in Table 25-7.

Table 25-7 Approximate Ages of Some Objects Dated with Carbon-14

Object	Age (years)
Franchthi Cave artifacts (Greece)	22,000
Lescaux Cave wall paintings (France)	15,000
Bristlecone pine trees (U.S. southwest)	7,000
Crater Lake, Oregon	6,500
Egyptian tombs	4,900
Stonehenge (England)	3,700
California giant sequoia	2,900
Dead Sea Scrolls	1,900

Isotopic Tracers

The incorporation of radioisotopes into chemical compounds allows scientists to monitor chemical processes which otherwise might be quite complicated to observe. Biochemical research has depended heavily on the use of "labeled" compounds to determine the metabolic pathways. For example, the use of $^{14}CO_2$ allowed Melvin Calvin (Nobel Prize in chemistry, 1961) to explain the process of photosynthesis. He illuminated green algae in the presence of $^{14}CO_2$ for short time periods and then analyzed the cells for the presence of C-14. One of the earliest radioactive compounds formed by the algae was 3-phosphoglycerate, labeled at the 1 position:

$$O = \overset{*}{C} = O \qquad\qquad \begin{array}{l} \overset{*}{C}O_2^- \\ | \\ CH-OH \\ | \\ CH_2OPO_3^{2-} \end{array}$$

carbon dioxide   3-phosphoglycerate

(The asterisk represents the C-14 label.)

Calvin subsequently found that plants use 3-phosphoglycerate to make glucose-6-phosphate, which then produces cellulose. In essence, Calvin allowed the plant to do the chemistry for him and then analyzed the results. Similar studies have been done on other major metabolic pathways.

Radiation Therapy

Radiation is used to treat cancer and other abnormal growths because rapidly multiplying cells are more sensitive to radiation damage. Alpha- and beta-radiation are not normally used for therapy because they have such low penetrating abilities. However, γ- and X-radiation readily penetrate tissues, and they are used extensively. The most common examples are machine-generated X-rays and γ-rays from cobalt-60 and cesium-137. These radiations may be shielded and guided so that they are directed at only the cancer site (see Figure 25-10). In this way, cancer cells are preferentially destroyed; unfortunately, some normal cells are also destroyed and the patient is likely to suffer the effects of radiation sickness during therapy. However, radiation is one of the more effective ways of treating cancer.

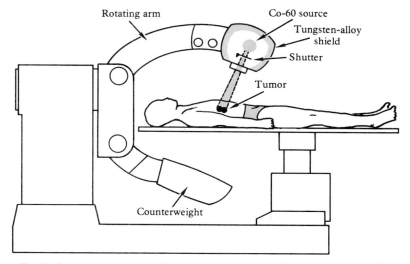

Fig. 25-10 Radiation Therapy for Cancer with Cobalt-60

Radiation therapy can utilize external sources of radiation, as outlined above, or it can employ sources located inside the body. This latter treatment is possible because certain elements are concentrated in specific parts of the body. For example, the thyroid gland concentrates iodide ions for the purpose of synthesizing thyroid hormone (thyroxine):

$$HO-\underset{I}{\overset{I}{\bigcirc}}-O-\underset{I}{\overset{I}{\bigcirc}}-CH_2-\underset{NH_3^+}{\overset{}{CH}}-CO_2^-$$

thyroxine
(3, 5, 3′, 5′-tetraiodothyronine)

leukemia: a form of cancer characterized by proliferation of white blood cells.

extracorporeal: literally, outside of the body; corporeal comes from the Latin *corpore,* meaning "of the body."

Overactive thyroid gland is often treated by administering orally a solution of sodium iodide (NaI) that contains a small amount of radioactive iodine-131. The radioisotope becomes concentrated in the thyroid gland where radiation destroys cells. Since thyroid overactivity can have a number of causes, including cancer, cell destruction helps to reduce thyroxine production to normal. The short half-life of I-131, 8.1 days, allows a large dose to be used, because the radioactivity dies away rather quickly. If the cause for overactivity was cancer, and if the cancer cells have broken off and moved to other parts of the body, they will still concentrate the radioactive iodide and suffer the consequences.

Phosphorus-32 incorporated into phosphate has been used for treatment of *leukemia.* Since phosphate is a common chemical component of all cells, it circulates freely throughout the body by way of the bloodstream. Thus, the blood cells receive a high dose of radiation compared to other cells. In addition, the radioactive phosphate is incorporated into bones, where it inhibits white blood cell production in the bone marrow.

Other radioisotopes have been shaped into metal wires and needles that can be embedded into cancer sites. These include Cr-51, Y-90, Au-198, Pd-109, and Sr-90.

An interesting alternative to internal radiation therapy for certain illnesses is the use of *extracorporeal irradiation* (Figure 25-11). In this technique, an external plastic tube is spliced in the blood flow of the patient, so that the blood is circulated externally before it reenters the body. While it is in the plastic tube, the blood is irradiated with γ-radiation. In this way the blood receives a predetermined dose of radiation, while other body parts are not exposed.

Fig. 25-11 Extracorporeal Irradiation

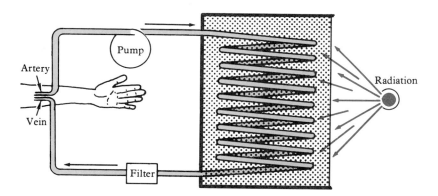

Medical Diagnosis

Nuclear medicine is the rapidly growing field in which radioisotopes are used to diagnose disease. Radioisotopes can be introduced into the body by injection or ingestion. As they move through the body, they continue to emit radiation, usually γ-radiation, which can be detected by scanning with a scintillation camera (Figure 25-12). This device is similar to a scintillation counter in that tiny flashes of light are produced when

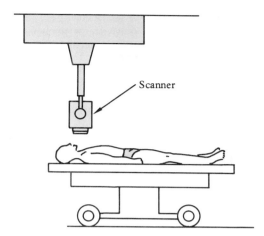

Fig. 25-12 A Scanning Scintillation Camera

radiation strikes crystals inside the camera. These light flashes are then used to expose photographic film. The developed film resembles an X-ray photograph, except that images of certain body parts arise from radiation coming from them. In this way, organ and body scans are obtained.

Bleomycin, a compound produced by certain bacteria, is selectively absorbed and degraded by tumors. It is possible to incorporate indium-111 into the structure of bleomycin. Since indium-111 is a γ-ray emitter, the accumulation of indium-bleomycin in tumors can be visualized with the scintillation camera, and a scan of the tumor is obtained. This compound can be used to detect a variety of tumors.

Other examples of radioisotopes used for organ scans are mercury-203 (brain scans), iodine-131 (thyroid scans), and strontium-85 (bone scans). The basis for the use of these isotopes is that mercury is concentrated in brain tumors, iodine accumulates in the thyroid gland, and strontium, which is in the same chemical family as calcium, tends to become localized in bone. An example of an organ scan is shown in Figure 25-13.

Newer methods employ *technetium-99m.* This radioisotope decays by γ-radiation only; since there are no particles emitted, there is little tissue damage. The intensity of the γ-radiation makes it easy to detect, and the relatively short half-life of 61 days minimizes patient exposure to radiation. These and other diagnostic uses of radioisotopes are summarized in Table 25-8.

Fig. 25-13 Results from a Brain Scan. The large white spot in the upper center part of the brain is a tumor

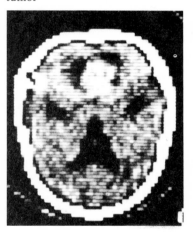

technetium-99m: the symbol "m" (metastable) means that the nucleus is somewhat unstable and will become stable by γ-emission; Tc-99m is one of the few isotopes that release γ-rays without any other type of accompanying radioactive decay.

EXAMPLE 25.5

Why does radioactive iodine accumulate in the thyroid gland?

SOLUTION

The thyroid gland accumulates iodine, including radioactive iodine, in order to synthesize thyroxine.

EXAMPLE 25.6

How are diagnostic uses of radioisotopes different from therapeutic uses of radioisotopes and radiation?

Table 25-8 Some Diagnostic Uses of Radioisotopes

Part of Body	Isotope	Use
Blood	chromium-51	Determine blood volume and red blood cell lifetime
	krypton-79	Study cardiovascular system and blood flow
	iodine-125	Determine concentration of protein hormones in blood
	iron-59	Study iron metabolism in red blood cells
Bone	barium-131 and technetium-99m	Detection of bone tumors and sites of rheumatoid arthritis
Brain	iodine-131	Detection of fluid buildup on brain and assessment of flow of cerebrospinal fluid
	mercury-197	Brain scans
	technetium-99m	Detection of brain tumors, blood clots, and hemorrhages
Heart	technetium-99m	Measurement of cardiac function, size and shape of heart; location of cardiac infarcts and holes between chambers
Kidney	chromium-51, iodine-131, and technetium-99m	Assessment of kidney function; location of cysts
Liver/spleen	copper-64	Aids in diagnosis of Wilson's disease (copper storage disease)
	gold-198	Assessment of liver function
	mercury-197	Assessment of spleen function
	technetium-99m	Measurement of size and shape of liver and spleen; location of tumors
Lung	iodine-131 and technetium-99m	Location of blood clots
	xenon-133	Measurement of lung volume; assessment of respiratory function; location of tumors
Pancreas	selenium-75	Measurement of size and shape of pancreas; location of abnormal masses
Thyroid	iodine-131	Measurement of rate of iodine uptake by thyroid; measurement of size and shape of thyroid
	technetium-99m	Measurement of size and shape of thyroid; location of abnormal masses

SOLUTION

For purposes of diagnosis, radioisotopes are used to detect abnormalities. Therapy involves use of radioisotopes and radiation to treat abnormal conditions.

NUCLEAR POWER

It is a well-known fact that nuclear changes usually result in liberation of energy. The theoretical basis for these energy changes lies in the equation worked out by Albert Einstein in 1905.

$$E = mc^2$$

The now famous equation states that energy (in ergs) is equal to mass (in grams) multiplied by the square of the speed of light (in cm/sec). In essence, Einstein's equation says that energy and matter are interchangeable, as related by the square of the speed of light. Einstein's ideas were far ahead of his time, and it took other scientists 40 years to verify his reasoning. However, the validity of his equation is now readily accepted. Since the speed of light is such a large number (3×10^{10} cm/sec), a very small amount of matter can release a fantastic amount of energy.

Nuclear Binding Energy

Einstein's equation does not tell us the whole story about the origin of nuclear energy. We can learn more if we examine the mass of a nucleus, helium-4 for example. If we add up the masses of the component parts of the helium nucleus, we get a value of 4.031882 amu:

Two protons	2×1.007276 amu =	2.014552 amu
Two neutrons	2×1.008665 amu =	2.017330 amu
	Total mass =	4.031882 amu

But experimentally we find the mass of the helium nucleus to be 4.001506 amu. Thus, some mass calculated to be in the helium nucleus is not actually present. This deficiency in mass is called the *mass defect;* for helium it is 0.030376 amu. If we use Einstein's equation, we can calculate the energy equivalent for this mass, which is 28 MeV (million electron-volts). The term used for the energy equivalent of the mass defect is *nuclear binding energy*. It is defined as the energy released in the combination of particles to form a nucleus. Indeed, all atomic nuclei weigh less than the sum of their counterparts; this mass defect is considered to be equivalent to the energy holding atomic nuclei together.

We can calculate binding energies for all atomic nuclei as we did for helium. If we divide nuclear binding energy by the number of particles in a particular nucleus, we get the *average binding energy* per particle. A plot of average binding energy as a function of mass number is given in Figure 25-14. As you can see, average binding energies are very low in the region of low atomic mass; however, average binding energy rises to a maximum around mass number 56, corresponding to iron. Iron is one of the most abundant elements in the universe, and its high average binding energy, and consequently its high nuclear stability, are factors in determining its abundance.

As atomic mass increases beyond iron, average binding energy begins to fall again. You can see that elements having intermediate atomic masses are the most stable. Thus, it would appear that converting elements of small mass number to elements in the middle of the curve would result in release of energy and greater stability. Indeed, this is

mass defect: the mass equivalent of the energy released in the formation of a nucleus.

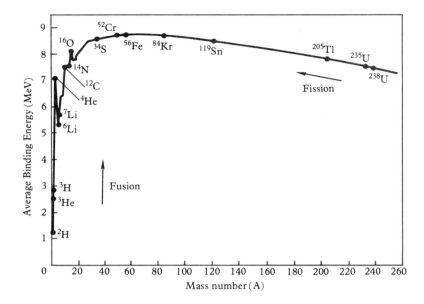

Fig. 25-14 A Plot of Average Binding Energy as a Function of Mass Number

what happens in *nuclear fusion*, the process of combining two light nuclei to produce a heavier nucleus of intermediate atomic mass. Similarly, if we could split the very heavy nuclei into lighter ones, we could also achieve release of energy and greater stability. Again, such processes really do occur; *nuclear fission* is the name given to the process of splitting a heavy nucleus into two lighter nuclei of intermediate atomic mass. In both fusion and fission, the total mass of the nuclei involved decreases, and energy is released.

In radioactive decay, small amounts of mass are converted to energy. In these cases, the energy liberated appears mostly as kinetic energy of the particles and nuclei produced by the reaction. These particles and nuclei collide with atoms and molecules in their surroundings, gradually losing their excess energy by transfer to other particles. Most of this lost energy is ultimately converted to heat.

Nuclear Fission

When certain large atoms are bombarded with particles emitted by a nuclear source, smaller atoms are produced. In the fission process, the binding energies of the parent nucleus and the smaller fragments are sufficiently different that large amounts of energy are liberated. One of the most common examples is the fission of uranium-235 that results from bombardment with neutrons:

$$^{235}_{92}U + ^{1}_{0}n \longrightarrow ^{135}_{53}I + ^{97}_{39}Y + 4\,^{1}_{0}n$$
$$^{235}_{92}U + ^{1}_{0}n \longrightarrow ^{139}_{56}Ba + ^{94}_{36}Kr + 3\,^{1}_{0}n$$
$$^{235}_{92}U + ^{1}_{0}n \longrightarrow ^{103}_{42}Mo + ^{131}_{50}Sn + 2\,^{1}_{0}n$$
$$^{235}_{92}U + ^{1}_{0}n \longrightarrow ^{139}_{54}Xe + ^{95}_{38}Sr + 2\,^{1}_{0}n$$

Only a few of the many decomposition routes are shown, but in each case, more neutrons are produced than are consumed. Thus, fission of U-235 can become self-sustaining since the neutrons produced can serve as bombarding neutrons for fission of more U-235. Two well-known applications of nuclear fission are found in nuclear power plants and the atomic bomb.

Nuclear Power Plants

It is possible to control the rate of the chain reaction of nuclear fission so that energy can be produced slowly enough to be used constructively. In this case, the equipment in which the fission is carried out is called a *nuclear reactor*.

Most nuclear power plants, such as the one shown in Figure 25-15, utilize fission reactors that require U-235 as fuel. However, since this isotope accounts for only 0.7% of natural uranium, there is a limit to the amount of fuel that will be available for nuclear power generation in the future. Furthermore, naturally occurring uranium must be enriched so that the U-235 isotope amounts to 3% of the total. This is both a costly and an energy-consuming process.

Fig. 25-15 A Nuclear Power Plant

The Atomic Bomb

The violent release of energy is what distinguishes an atomic bomb (Figure 25-16) from a nuclear reactor. This characteristic results from two factors. One is that the fissionable material (either U-235 or Pu-239) is highly concentrated, on the order of 97%. The other factor is related to what is called the *critical mass*, the smallest amount of fissionable isotope required for a self-sustaining nuclear chain reaction. If the fissionable isotope is present in amounts less than the critical mass, too many of the neutrons produced will escape to the surroundings and not be available to sustain the chain reaction (see Figure 25-17).

Fig. 25-16 The Atomic Explosion at Nagasaki, Japan, in 1945

Fissionable isotopes must be handled in amounts less than their critical masses during construction and transportation of bombs. Thus it is necessary to have methods of assembling a supercritical mass at the time that the bomb is to be exploded. Two such designs have been used. In one (see Figure 25–18), a small nonnuclear explosive propels a subcritical mass at one end of a barrel into a subcritical mass at the other end,

PERSPECTIVE: Development of the Atomic Bomb

Our present knowledge of nuclear power is the result of developments in the 1930s in Europe. The Italian physicist Enrico Fermi was the first to study the bombardment of uranium with neutrons, and two Austrians, Lise Meitner and her nephew Otto Frisch, calculated the energy that could be derived from U-235 fission. When Nazi Germany invaded Austria in 1938, Meitner fled to Sweden and conveyed her research findings to the Danish physicist Niels Bohr. Bohr then brought the information to the attention of Albert Einstein, who had already taken refuge in the United States, and he convinced President Franklin D. Roosevelt of the importance of the discovery. Thus, in 1942 the United States began a massive research project on atomic energy under the code name Manhattan Project.

Some of the most illustrious scientists the world has known gathered to work on the Manhattan Project. On December 2, 1942, Enrico Fermi achieved the first self-sustaining nuclear chain reaction in a laboratory hastily constructed under the bleachers of the football field at the University of Chicago. In less than three years, knowledge gained from this work resulted in the atomic bombs used at Hiroshima and Nagasaki.

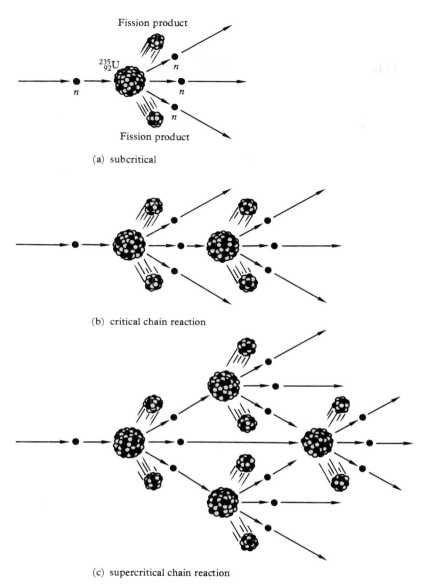

Fig. 25-17 Mass Considerations in a Nuclear Chain Reaction Initiated by Neutron Bombardment of Uranium-235 (a) In a subcritical mass, the number of neutrons released that cause further fission averages less than one per reaction, and a chain reaction will not be sustained. (b) In a critical mass, the number of neutrons released that cause further fission averages one per reaction, and the reaction will be self-sustaining. (c) In a supercritical mass, the number of neutrons released that cause further fission averages more than one per reaction, and the accelerating chain reaction can lead to an explosion

producing a supercritical mass that will explode upon exposure to a neutron source. This is called the *gun method* of assembly. In the second method, called the *implosion design*, a number of nonnuclear explosive charges are placed on the inner surface of a sphere. These fire many subcritical pieces of fissionable material into one common ball at the center of the sphere, thus creating the supercritical mass by implosion. The bomb can then be triggered by a neutron source. Both atomic bomb designs were used in 1945, the gun assembly at Hiroshima and the implosion assembly at Nagasaki.

Fig. 25-18 Designs of Atomic Bombs. Design I is the "gun" design and Design II is the implosion design

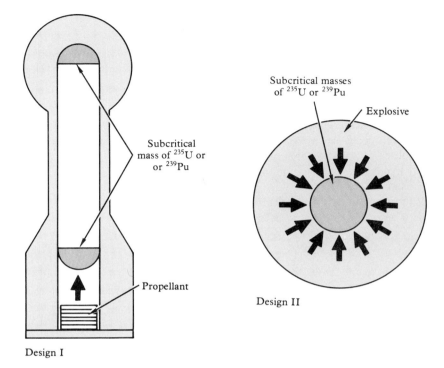

Nuclear Fusion

The sun constitutes the only known continuous nuclear fusion reactor. In fact, it is the fusion of hydrogen atoms to produce helium that is the main source of energy in the sun. The reaction below is a summary of several intermediate reactions in the sun.

$$_1^1H + {}_1^2H \longrightarrow {}_2^3He$$

It is important to remember that even though nuclear fusion produces large quantities of energy, very high temperatures are required to sustain the process. This is because the nuclei must overcome the electrical repulsions between their positive charges in order to fuse together. Thus, at the temperatures in the sun (about 15 million degrees Celsius), the light nuclei are moving fast enough to collide and stay fused.

It once seemed impossible that we would ever be able to achieve the temperatures needed for fusion reactions on earth. But by 1952, scientists had developed the *hydrogen bomb*. A hydrogen bomb can be described as a fission bomb surrounded by the compound formed between deuterium (hydrogen-2) and lithium-6; this compound has the formula ${}_3^6Li{}_1^2H$. When the fission bomb releases its large amounts of heat and neutrons, fusion occurs as illustrated in the following sequence.

$$_3^6Li + {}_0^1n \longrightarrow {}_2^4He + {}_1^3H$$
$$_1^2H + {}_1^3H \longrightarrow {}_2^4He + {}_0^1n$$

As a result of the current energy shortage, much interest has developed in the possibility of putting nuclear fusion to practical use. The reaction between deuterium and tritium could be used as a source of power, since both reactants can be obtained rather easily. However, two major problems remain to be solved. One is the matter of containment. Fusion occurs in the sun because the reactants are held close together by the sun's large gravitational forces. However, there are no such powerful forces on earth, and there are no materials known which can withstand the temperatures of fusion and thus serve as a container. A second problem is that the temperatures necessary for fusion have been achieved in hydrogen bombs but not in the controlled systems necessary for realistic energy production. However, much progress has been made in solving these problems, and there is optimism that the fusion process can be harnessed for the benefit of mankind.

SUMMARY

Radioactivity: Most of the naturally occurring radioisotopes decay very slowly and have existed since formation of the earth. Others are formed by bombardment from secondary cosmic rays. Naturally occurring radioisotopes of atomic number greater than 81 decay by one of three radioactive disintegration series. Stable nuclei can be bombarded with subatomic particles to produce new, unstable nuclei. In this way, nuclear bombardment brings about transmutation.

Radiation: Detection and Measurement: Radiation can be detected and measured by Geiger counters, scintillation counters, and film badges. Since each radioisotope decays at its own rate, the half-life is used to compare rates of decay. The units of radioactive disintegration are curies and becquerels, while those for dosage are roentgens, rads, grays, and rems. Electron-volts are the units for energy of radiation.

Radiation Safety: Nuclear radiation and X-rays form ions and free radicals in living tissues that denature proteins, inactivate enzymes, and damage DNA. Cells can withstand small radiation dosages, but high levels of radiation damage DNA to such an extent that cell death occurs. It is recommended that the maximum annual exposure from man-made sources be limited to 170 millirems. Those working with radiation should observe all safety precautions.

Applications of Radiochemistry: Knowledge of the half-life of C-14 allows estimation of the age of archeological specimens. Isotopic tracers are used to elucidate metabolic pathways and other complex biological systems. Radiation therapy is a powerful treatment for cancer; normally γ- and X-radiation are used for this purpose. Radiation therapy can utilize external or internal sources of radiation. Radioisotopes injected or ingested into the body can be used to diagnose disease. The pattern of radiation can be used to diagnose tumors and other internal abnormalities.

Nuclear Power: The Einstein equation indicates that matter and energy are interchangeable. The relationship can be used to calculate nuclear binding energy from mass defect. In nuclear fusion, two light atomic nuclei are combined to produce a heavier nucleus of greater binding energy, and the excess energy is liberated to the surroundings. Energy is also released when a heavy nucleus is split into two lighter nuclei in the fission process. Nuclear power plants take advantage of the fission process first utilized in atomic bombs. Continuous nuclear fusion occurs in the sun, but we are only beginning to develop the process on earth. Fusion power is one possible solution to the current energy shortage.

STUDY QUESTIONS AND PROBLEMS

1. Define the following terms:
   (a) primary cosmic rays
   (b) secondary cosmic rays
   (c) radioactive disintegration series
   (d) artificial radioactivity
   (e) transmutation
   (f) X-rays
   (g) ionizing radiation
   (h) Geiger counter
   (i) scintillation counter
   (j) film badge
   (k) radioactive half-life
   (l) curie
   (m) roentgen
   (n) rad
   (o) rem
   (p) electron-volt
   (q) free radical
   (r) carbon dating
   (s) isotopic tracer
   (t) extracorporeal irradiation
   (u) Einstein equation
   (v) mass defect
   (w) nuclear binding energy
   (x) average binding energy
   (y) nuclear fusion
   (z) nuclear fission
   (aa) nuclear reactor
   (bb) atomic bomb
   (cc) critical mass
   (dd) hydrogen bomb

2. Suggest how tritium might be formed by cosmic rays. Write a balanced equation to illustrate.

3. Suggest how each of the following transuranium elements might be synthesized by bombardment of each precursor with a single particle.
   (a) $^{240}_{95}$Am from $^{239}_{94}$Pu
   (b) $^{242}_{96}$Cm from $^{239}_{94}$Pu
   (c) $^{243}_{97}$Bk from $^{241}_{95}$Am
   (d) $^{245}_{98}$Cf from $^{242}_{96}$Cm

4. Irene and Frederick Joliot-Curie first observed induced radioactivity in 1934 by bombarding $^{27}_{13}$Al with α-particles. The products were a neutron and $^{30}_{15}$P which subsequently decayed by positron emission. Write balanced equations for this two-step process.

5. Describe the machine-generation of X-rays.

6. Explain how a Geiger counter works.

7. Explain how a scintillation counter works.

8. What percentage of radioactivity will remain in a sample of radioisotope after a period of four half-lives?

9. Why is it desirable for radioisotopes used in medical diagnosis to have short half-lives?

10. Why is the half-life an important consideration in disposal of radioactive waste?

11. What information is conveyed by each of the following units for measuring radiation?
    (a) curie
    (b) roentgen
    (c) rad
    (d) rem
    (e) electron-volt

12. If a person receives 10 rems of radiation each year for 60 years, there might not be any noticeable effects. However, if the same person receives the same total of 600 rems in one dose, the exposure is likely to be fatal. Explain why the single dose is so much more harmful than the multiple doses.

13. Both Marie Curie and her daughter Irene died of leukemia, probably caused by years of exposure to radiation. Explain how radiation might produce cancer. (Leukemia is a form of cancer characterized by proliferation of white blood cells.)

14. Give four ways of minimizing exposure for those who must work with radiation.

15. Explain the basis of radiocarbon dating.

16. How did Melvin Calvin use radioisotope tracers to learn about photosynthesis?

17. Why does radiation sickness usually accompany radiation treatment for cancer?

18. Explain how I-131 can be used to treat overactive thyroid.

19. Explain how P-32 can be used to treat leukemia.

20. It is ironic that radiation can be both a cause and a cure for cancer. Suggest an explanation for this seeming contradiction.

21. What is the advantage of extracorporeal irradiation?

22. Describe how an organ scan is performed.

23. How are mass defect and nuclear binding energy important to energy produced by nuclear processes?

24. Distinguish between nuclear fission and nuclear fusion.

25. Give two major differences between a nuclear reactor and an atomic bomb.

26. How can a nuclear reaction become a chain reaction?

27. Describe two methods of assembling an atomic bomb.
28. Use an equation to summarize the nuclear fusion process which provides the sun's energy.
29. Distinguish between an atomic bomb and a hydrogen bomb.
30. What are the two major drawbacks in developing nuclear fusion as a practical energy source?

APPENDIX A
BASIC MATHEMATICS

EXPONENTS AND SCIENTIFIC NOTATION

When numbers are either very large or very small, they can be expressed conveniently by using *exponents*. An exponent is the power to which a number is raised. For example, 1 million (1,000,000) can be expressed as 10^6, where 6 is the exponent and 10 is the base. A *positive exponent* indicates how many times the base must be multiplied by itself to give the original number, while a *negative exponent* tells how many times 1 must be divided by the base to give the original number.

Scientific notation is the use of powers (exponents) of 10 to express values. In this system, exponents describe the location of the decimal point. The general form for scientific notation is

$$N \times 10^{\text{exponent}},$$

where N is usually a number between 1 and 10, and the exponent is a whole number. In using scientific notation, N must contain the correct number of significant figures. Table A-1 illustrates the use of scientific notation to express very large and very small numbers to two significant figures.

Table A-1 Scientific Notation

Number	Scientific Notation
1,000,000	1.0×10^6
100,000	1.0×10^5
10,000	1.0×10^4
1,000	1.0×10^3
100	1.0×10^2
10	1.0×10^1
1	1.0×10^0
0.1	1.0×10^{-1}
0.01	1.0×10^{-2}
0.001	1.0×10^{-3}
0.0001	1.0×10^{-4}
0.00001	1.0×10^{-5}
0.000001	1.0×10^{-6}

Note that 10^0 is equal to 1.

As mentioned, exponents of ten describe the location of the decimal point. For example, the number 1 million (1,000,000) is written in scientific notation by moving the decimal point six places to the left and using the exponent 6. In this case, N is 1, and it is assigned the correct number of significant figures by placing zeroes after the decimal point. For two significant figures, 1 million is expressed as 1.0×10^6.

$$1 \text{ million} = 1{,}000{,}000 = 1.0 \times 10^6$$

move decimal 6 places to left

The number one millionth, 1/1,000,000, can also be expressed in scientific notation by moving the decimal point six places to the right and using the exponent -6.

$$1 \text{ millionth} = 1/1{,}000{,}000 = 0.000001 = 1.0 \times 10^{-6}$$

move decimal 6 places to right

Thus, a positive exponent signifies that the number is greater than 1 and that the decimal point has been moved to the left. A negative exponent signifies that the number is less than 1 and that the decimal point has been moved to the right. To convert a number expressed in scientific notation to its ordinary form, simply reverse the operations just described. A positive exponent signifies that the decimal point must be moved to the right, and a negative exponent signifies that the decimal point must be moved to the left.

EXAMPLE A.1

Express 35,000 in scientific notation with three significant figures.

SOLUTION

The goal is to express 35,000 in the form

$$N \times 10^{\text{exponent}},$$

where N is between 1 and 10. To do this, we must move the decimal point four places to the left and use the exponent 4. We must also have three significant figures in our answer. Thus, the correct answer is 3.50×10^4.

A-1

EXERCISE A.1 Express each of the following numbers in scientific notation with the number of significant figures indicated.
(a) 125,000 (three significant figures)
(b) 4,694 (two significant figures)
(c) 7,430,000 (four significant figures)
(d) 97,600 (two significant figures)

SOLUTION

(a) 1.25×10^5
(b) 4.7×10^3
(c) 7.430×10^6
(d) 9.8×10^4

EXAMPLE A.2
Express 0.0000493 in scientific notation with two significant figures.

SOLUTION

To give N a value between 1 and 10, we must move the decimal point five places to the right and use the exponent -5. Thus, the correct answer is 4.9×10^{-5}.

EXERCISE A.2 Express each of the following numbers in scientific notation with the number of significant figures indicated.
(a) 0.00243 (two significant figures)
(b) 0.0001057 (three significant figures)
(c) 0.0752 (three significant figures)
(d) 0.00000023 (two significant figures)

SOLUTION

(a) 2.4×10^{-3}
(b) 1.06×10^{-4}
(c) 7.52×10^{-2}
(d) 2.3×10^{-7}

To *add* or *subtract* numbers expressed in scientific notation, adjust the decimal point in N and change the exponents of the tens so that all exponents are alike. Then proceed with addition or subtraction.

To *multiply* numbers expressed in scientific notation, use the following three steps:

1. Multiply the numbers (N) standing before the tens.
2. Add the exponents of the tens.
3. Adjust the exponent in the answer so that N is between 1 and 10.

EXAMPLE A.3
Multiply 7.4×10^2 by 1.92×10^{-4}.

SOLUTION

$$(7.4 \times 10^2)(1.92 \times 10^{-4}) = 7.4 \times 1.92 \times 10^{(2-4)}$$
$$= 14 \times 10^{-2}$$
$$= 1.4 \times 10^{-1}$$

EXERCISE A.3 Find answers for the following multiplications.
(a) $(2.5 \times 10^{-2})(3.0 \times 10^5)$
(b) $(7.23 \times 10^4)(1.62 \times 10^{-2})$
(c) $(3.89 \times 10^{-3})(4.7 \times 10^{-2})$
(d) $(5.62 \times 10^6)(1.0 \times 10^{-5})(2.3 \times 10^{-23})$

SOLUTION

(a) 7.5×10^3
(b) 1.17×10^3
(c) 1.8×10^{-4}
(d) 1.3×10^{-21}

To *divide* numbers expressed in scientific notation, use the following three steps:

1. Perform the indicated division with the numbers (N) standing before the tens.
2. Subtract the exponent(s) of the divisor(s) from the exponent(s) of the number(s) being divided.
3. Adjust the exponent in the answer so that N is between 1 and 10.

EXAMPLE A.4
Divide 7.21×10^5 by 8.46×10^{-2}.

SOLUTION

$$\frac{7.21 \times 10^5}{8.46 \times 10^{-2}} = 0.852 \times 10^{[5-(-2)]}$$
$$= 0.852 \times 10^{[5+2]}$$
$$= 0.852 \times 10^7$$
$$= 8.52 \times 10^6$$

EXERCISE A.4 Perform the following divisions.
(a) $5.41 \times 10^{-2} \div 1.22 \times 10^4$
(b) $2.36 \times 10^5 \div 1.82 \times 10^2$
(c) $3.57 \times 10^{-2} \div 3.65 \times 10^{-4}$
(d) $8.96 \times 10^5 \div 4.72 \times 10^{-2}$

SOLUTION

(a) 4.43×10^{-6}
(b) 1.30×10^3
(c) 9.78×10^1
(d) 1.90×10^7

EXAMPLE A.5

Find the answer for the following operation.

$$\frac{(7.36 \times 10^{-2})(2.49 \times 10^5)}{6.65 \times 10^{-3}}$$

SOLUTION

$$\frac{(7.36 \times 10^{-2})(2.49 \times 10^5)}{6.65 \times 10^{-3}} = 2.76 \times 10^{[-2+5-(-3)]}$$
$$= 2.76 \times 10^{[-2+5+3]}$$
$$= 2.76 \times 10^6$$

EXERCISE A.5 Find the answers for the following operations.

(a) $\dfrac{(2.43 \times 10^{-2})(7.25 \times 10^{-3})}{1.65 \times 10^{-4}}$

(b) $\dfrac{1.74 \times 10^5}{(3.59 \times 10^{-2})(6.83 \times 10^{-4})}$

(c) $\dfrac{(4.71 \times 10^{-4})(5.22 \times 10^5)}{(6.79 \times 10^{-2})(2.46 \times 10^3)}$

(d) $\dfrac{(1.17 \times 10^3)(2.66 \times 10^{-2})(4.95 \times 10^5)}{(6.74 \times 10^2)(4.85 \times 10^4)}$

SOLUTION

(a) 1.07 (c) 1.47
(b) 7.10×10^9 (d) 4.71×10^{-1}

LOGARITHMS

A *logarithm* is the exponent of a number. Most work is done with *common logarithms* (abbreviated log), in which the base 10 is used, and the log of 10^{exponent} is the exponent.

$$\log 10^{\text{exponent}} = \text{exponent}$$

Thus, the logarithm of 10 raised to a power is simply the power of 10, and the log can be determined by inspection. Remember that 10^0 is always 1, and thus the log of 1 is 0.

Logarithms of numbers between 1 and 10 cannot be determined by inspection, but they can be obtained from the table of logarithms in Appendix B (or from certain pocket calculators). The numbers in the table do not have decimal points, but each of the numbers listed is assumed to have a decimal point following the first digit, and each of the logarithms listed is assumed to have a decimal point immediately preceding the value given. Thus, the logarithm of 2.720 is 0.4346.

The logarithm of a number greater than 10 or less than 1 is obtained by the following steps:

1. Express the number in scientific notation. For example, 1750 expressed in scientific notation is 1.75×10^3 (to three significant figures).

2. Since logarithms are exponents, and since exponents are added when numbers are multiplied, then

$$\log 1.75 \times 10^3 = \log 1.75 + \log 10^3$$

Determine the logarithm of each number and add the logarithms together. (The number of significant figures in each number determines the number of significant figures in each logarithm; whole-number exponents are considered to be exact numbers.) Thus,

$$\log 1.75 \times 10^3 = 0.243 + 3.000$$
$$= 3.243$$

EXAMPLE A.6

Find the logarithm of 0.0257.

SOLUTION

$$0.0257 = 2.57 \times 10^{-2}$$
$$\log 2.57 \times 10^{-2} = \log 2.57 + \log 10^{-2}$$
$$= 0.410 + (-2.000)$$
$$= -1.590$$

EXERCISE A.6 Find the logarithm of each number.
(a) 7.29×10^4
(b) 2.36×10^{-3}
(c) 85,100 (three significant figures)
(d) 0.000295 (three significant figures)

SOLUTION

(a) 4.863 (c) 4.930
(b) −2.627 (d) −3.530

Sometimes it is necessary to find an *antilogarithm*; this is the number that corresponds to a given logarithm. The given logarithm is composed of two parts: a decimal fraction (the *mantissa*) and a positive or negative whole number (the *characteristic*).

For example, in the logarithm 2.959, 0.959 is the mantissa and +2 is the characteristic. The antilogarithm (abbreviated antilog) is found by reversing the procedure for finding a logarithm.

1. Using the table in Appendix B (or a pocket calculator), find the antilogarithm of the mantissa (0.959).

$$\text{antilog } 0.959 = 9.10$$

2. Then, by inspection, find the antilogarithm of the characteristic (+2).

$$\text{antilog } 2 = 10^2$$

3. Then, since logs are added when numbers are multiplied, show the total antilog in scientific notation.

$$\begin{aligned}\text{antilog } 2.959 &= \text{antilog } 0.959 + \text{antilog } 2 \\ &= 9.10 \times 10^2 \\ &= 9.10 \times 10^2\end{aligned}$$

EXAMPLE A.7

Find the antilogarithm of 3.841.

SOLUTION

$$\begin{aligned}\text{antilog } 3.841 &= \text{antilog } 0.841 + \text{antilog } 3 \\ &= 6.93 \times 10^3\end{aligned}$$

EXERCISE A.7 Find the antilogarithm of each of the following logarithms:
(a) 0.845
(b) 2.480
(c) 5.781
(d) 1.891

SOLUTION

(a) 7.00
(b) 3.02×10^2
(c) 6.04×10^5
(d) 7.78×10^1

All the mantissas in a table of logarithms are positive, and this fact must be taken into account when an antilogarithm of a negative number is found. For example, to take the antilogarithm of -5.802, we must write the logarithm in such a way that the mantissa is positive. Thus,

$$\begin{aligned}\text{antilog } (-5.802) &= \text{antilog } (+0.198 - 6.000) \\ &= 1.58 \times 10^{-6}\end{aligned}$$

or $\quad -5.802 = \log (1.58 \times 10^{-6})$

EXAMPLE A.8

Find the antilogarithm of -3.850.

SOLUTION

$$\begin{aligned}\text{antilog } (-3.850) &= \text{antilog } (+0.150 - 4.000) \\ &= 1.41 \times 10^{-4}\end{aligned}$$

EXERCISE A.8 Find the antilogarithm of each of the following logarithms:
(a) -2.493
(b) -0.225
(c) -4.959
(d) -10.873

SOLUTION

(a) 3.21×10^{-3}
(b) 5.96×10^{-1}
(c) 1.10×10^{-5}
(d) 1.34×10^{-11}

SIMPLE ALGEBRA

Simple algebra involves formulating equations and solving them for a given quantity. One example is the formulation of a *proportionality*, a statement of how one quantity varies with another quantity. For example, Charles' Law states that the volume of a gas is directly proportional to its absolute temperature; in other words, the volume of a gas increases or decreases as its absolute temperature increases or decreases. Thus, we can write

$$V \propto T,$$

where "$\propto$" means "is proportional to." To convert the proportionality into an algebraic equation, we write k (a proportionality constant) in place of "$\propto$." Then,

$$V = kT,$$

and the value of k is constant and is determined by the values of V and T.

It is often necessary to solve an algebraic equation for a quantity of interest. For example, in the equation below, suppose we wanted to solve for x.

$$x + 2 = y$$

To do so, we must rearrange the equation so that only x is left on one side of the equation. In this case, we want to move 2 from the left side to the right side of the equation. Since 2 is added to the left side, we can move it to the right side by the reverse mathematical operation, *subtracting* it from the left side. But we must also subtract it from the right side, because adding or subtracting the same term from both sides of an equation does not alter the equality of the two sides. Thus,

$$x + 2 - 2 = y - 2$$

and

$$x = y - 2$$

In another example, to solve

$$x - 12 = y$$

for x, we must *add* 12 to both sides of the equation. Thus,

$$x - 12 + 12 = y + 12$$

and

$$x = y + 12$$

EXAMPLE A.9

Solve the following equation for x.

$$x - 20 + a = y$$

SOLUTION

In this case, to solve for x we must add 20 to both sides of the equation and subtract a from both sides.

$$x - 20 + 20 + a - a = y + 20 - a$$
$$x = y + 20 - a$$

EXERCISE A.9 Solve each of the following equations for x.
(a) $x - z = 4$
(b) $x + 9 = y$
(c) $x + a - y = 0$
(d) $x - b - c + z = 1$

SOLUTION

(a) $x = 4 + z$
(b) $x = y - 9$
(c) $x = y - a$
(d) $x = 1 + b + c - z$

An equation can also be rearranged by *multiplying* or *dividing* both sides by the same term, in effect reversing the operation indicated in the original equation. Thus, in the Charles' law equation,

$$V = kT,$$

we can solve for k by dividing both sides by T. This operation reverses the original multiplication by T.

$$V = kT$$
$$\frac{V}{T} = \frac{kT}{T}$$
$$= k$$

or

$$k = \frac{V}{T}$$

Note, however, that any necessary multiplication or division should be performed *after* any rearrangement made by addition or subtraction.

EXAMPLE A.10

Solve the following equation for x.

$$10x + 20 = 40$$

SOLUTION

First we must subtract 20 from both sides of the equation.

$$10x + 20 - 20 = 40 - 20$$
$$10x = 20$$

Then, since x is multiplied by 10, dividing both sides by 10 does not change the equality but does isolate x to one side of the equation.

$$\frac{10x}{10} = \frac{20}{10}$$
$$x = 2$$

EXERCISE A.10 Solve each of the following equations for x.
(a) $5x = 25$
(b) $2x + 12 = 20$
(c) $3x - 6 = 12$
(d) $4x - 10 + b = 15$

SOLUTION

(a) $x = 5$
(b) $x = 4$
(c) $x = 6$
(d) $x = \dfrac{25 - b}{4}$

EXAMPLE A.11

Solve the following equation for x.

$$\frac{x}{2} + 4 = 6$$

SOLUTION

$$\frac{x}{2} + 4 - 4 = 6 - 4$$

$$\frac{x}{2} = 2$$

$$(2)\left(\frac{x}{2}\right) = (2)(2)$$

$$x = 4$$

EXERCISE A.11 Solve each of the following equations for x.

(a) $\frac{x}{3} - 7 = 10$

(b) $\frac{x}{5} + 2 = y$

(c) $\frac{x}{10} = y$

(d) $\frac{x}{7} - b + c = y$

SOLUTION

(a) $x = 51$ (c) $x = 10y$
(b) $x = 5y - 10$ (d) $x = 7y + 7b - 7c$

In the event that both multiplication and division are needed to rearrange an equation, multiplication is usually done first, followed by division. The following example uses the equation describing a gas at initial conditions P_i, V_i, and T_i, and final conditions P_f, V_f, and T_f:

$$\frac{P_i V_i}{T_i} = \frac{P_f V_f}{T_f}$$

To solve for P_i, first we multiply both sides by T_i,

$$(\cancel{T_i})\left(\frac{P_i V_i}{\cancel{T_i}}\right) = \left(\frac{P_f V_f}{T_f}\right)(T_i)$$

and then we divide both sides by V_i:

$$\frac{P_i \cancel{V_i}}{\cancel{V_i}} = \frac{P_f V_f T_i}{T_f V_i}$$

$$P_i = \frac{P_f V_f T_i}{T_f V_i}$$

EXAMPLE A.12

Solve the following equation for V_i.

$$\frac{P_i V_i}{T_i} = \frac{P_f V_f}{T_f}$$

SOLUTION

$$(\cancel{T_i})\left(\frac{P_i V_i}{\cancel{T_i}}\right) = \left(\frac{P_f V_f}{T_f}\right)(T_i)$$

$$\frac{\cancel{P_i} V_i}{\cancel{P_i}} = \frac{P_f V_f T_i}{T_f P_i}$$

$$V_i = \frac{P_f V_f T_i}{T_f P_i}$$

EXERCISE A.12 Solve the following equation for each of the terms indicated.

$$\frac{P_i V_i}{T_i} = \frac{P_f V_f}{T_f}$$

(a) P_f
(b) V_f
(c) T_i
(d) T_f

SOLUTION

(a) $P_f = \dfrac{P_i V_i T_f}{T_i V_f}$ (c) $T_i = \dfrac{P_i V_i T_f}{P_f V_f}$

(b) $V_f = \dfrac{P_i V_i T_f}{T_i P_f}$ (d) $T_f = \dfrac{P_f V_f T_i}{P_i V_i}$

APPENDIX B

TABLE OF LOGARITHMS
(See Appendix A for instructions on use.)

N	0	1	2	3	4	5	6	7	8	9
10	0000	0043	0086	0128	0170	0212	0253	0294	0334	0374
11	0414	0453	0492	0531	0569	0607	0645	0682	0719	0755
12	0792	0828	0864	0899	0934	0969	1004	1038	1072	1106
13	1139	1173	1206	1239	1271	1303	1335	1367	1399	1430
14	1461	1492	1523	1553	1584	1614	1644	1673	1703	1732
15	1761	1790	1818	1847	1875	1903	1931	1959	1987	2014
16	2041	2068	2095	2122	2148	2175	2201	2227	2253	2279
17	2304	2330	2355	2380	2405	2430	2455	2480	2504	2529
18	2553	2577	2601	2625	2648	2672	2695	2718	2742	2765
19	2788	2810	2833	2856	2878	2900	2923	2945	2967	2989
20	3010	3032	3054	3075	3096	3118	3139	3160	3181	3201
21	3222	3243	3263	3284	3304	3324	3345	3365	3385	3404
22	3424	3444	3464	3483	3502	3522	3541	3560	3579	3598
23	3617	3636	3655	3674	3692	3711	3729	3747	3766	3784
24	3802	3820	3838	3856	3874	3892	3909	3927	3945	3962
25	3979	3997	4014	4031	4048	4065	4082	4099	4116	4133
26	4150	4166	4183	4200	4216	4232	4249	4265	4281	4298
27	4314	4330	4346	4362	4378	4393	4409	4425	4440	4456
28	4472	4487	4502	4518	4533	4548	4564	4579	4594	4609
29	4624	4639	4654	4669	4683	4698	4713	4728	4742	4757
30	4771	4786	4800	4814	4829	4843	4857	4871	4886	4900
31	4914	4928	4942	4955	4969	4983	4997	5011	5024	5038
32	5051	5065	5079	5092	5105	5119	5132	5145	5159	5172
33	5185	5198	5211	5224	5237	5250	5263	5276	5289	5302
34	5315	5328	5340	5353	5366	5378	5391	5403	5416	5428
35	5441	5453	5465	5478	5490	5502	5514	5527	5539	5551
36	5563	5575	5587	5599	5611	5623	5635	5647	5658	5670
37	5682	5694	5705	5717	5729	5740	5752	5763	5775	5786
38	5798	5809	5821	5832	5843	5855	5866	5877	5888	5899
39	5911	5922	5933	5944	5955	5966	5977	5988	5999	6010
40	6021	6031	6042	6053	6064	6075	6085	6096	6107	6117
41	6128	6138	6149	6160	6170	6180	6191	6201	6212	6222
42	6232	6243	6253	6263	6274	6284	6294	6304	6314	6325
43	6335	6345	6355	6365	6375	6385	6395	6405	6415	6425
44	6435	6444	6454	6464	6474	6484	6493	6503	6513	6522
45	6532	6542	6551	6561	6571	6580	6590	6599	6609	6618
46	6628	6637	6646	6656	6665	6675	6684	6693	6702	6712
47	6721	6730	6739	6749	6758	6767	6776	6785	6794	6803
48	6812	6821	6830	6839	6848	6857	6866	6875	6884	6893
49	6902	6911	6920	6928	6937	6946	6955	6964	6972	6981

(continued on next page)

TABLE OF LOGARITHMS (continued)

N	0	1	2	3	4	5	6	7	8	9
50	6990	6998	7007	7016	7024	7033	7042	7050	7059	7067
51	7076	7084	7093	7101	7110	7118	7126	7135	7143	7152
52	7160	7168	7177	7185	7193	7202	7210	7218	7226	7235
53	7243	7251	7259	7267	7275	7284	7292	7300	7308	7316
54	7324	7332	7340	7348	7356	7364	7372	7380	7388	7396
55	7404	7412	7419	7427	7435	7443	7451	7459	7466	7474
56	7482	7490	7497	7505	7513	7520	7528	7536	7543	7551
57	7559	7566	7574	7582	7589	7597	7604	7612	7619	7627
58	7634	7642	7649	7657	7664	7672	7679	7686	7694	7701
59	7709	7716	7723	7731	7738	7745	7752	7760	7767	7774
60	7782	7789	7796	7803	7810	7818	7825	7832	7839	7846
61	7853	7860	7868	7875	7882	7889	7896	7903	7910	7917
62	7924	7931	7938	7945	7952	7959	7966	7973	7980	7987
63	7993	8000	8007	8014	8021	8028	8035	8041	8048	8055
64	8062	8069	8075	8082	8089	8096	8102	8109	8116	8122
65	8129	8136	8142	8149	8156	8162	8169	8176	8182	8189
66	8195	8202	8209	8215	8222	8228	8235	8241	8248	8254
67	8261	8267	8274	8280	8287	8293	8299	8306	8312	8319
68	8325	8331	8338	8344	8351	8357	8363	8370	8376	8382
69	8388	8395	8401	8407	8414	8420	8426	8432	8439	8445
70	8451	8457	8463	8470	8476	8482	8488	8494	8500	8506
71	8513	8519	8525	8531	8537	8543	8549	8555	8561	8567
72	8573	8579	8585	8591	8597	8603	8609	8615	8621	8627
73	8633	8639	8645	8651	8657	8663	8669	8675	8681	8686
74	8692	8698	8704	8710	8716	8722	8727	8733	8739	8745
75	8751	8756	8762	8768	8774	8779	8785	8791	8797	8802
76	8808	8814	8820	8825	8831	8837	8842	8848	8854	8859
77	8865	8871	8876	8882	8887	8893	8899	8904	8910	8915
78	8921	8927	8932	8938	8943	8949	8954	8960	8965	8971
79	8976	8982	8987	8993	8998	9004	9009	9015	9020	9025
80	9031	9036	9042	9047	9053	9058	9063	9069	9074	9079
81	9085	9090	9096	9101	9106	9112	9117	9122	9128	9133
82	9138	9143	9149	9154	9159	9165	9170	9175	9180	9186
83	9191	9196	9201	9206	9212	9217	9222	9227	9232	9238
84	9243	9248	9253	9258	9263	9269	9274	9279	9284	9289
85	9294	9299	9304	9309	9315	9320	9325	9330	9335	9340
86	9345	9350	9355	9360	9365	9370	9375	9380	9385	9390
87	9395	9400	9405	9410	9415	9420	9425	9430	9435	9440
88	9445	9450	9455	9460	9465	9469	9474	9479	9484	9489
89	9494	9499	9504	9509	9513	9518	9523	9528	9533	9538
90	9542	9547	9552	9557	9562	9566	9571	9576	9581	9586
91	9590	9595	9600	9605	9609	9614	9619	9624	9628	9633
92	9638	9643	9647	9652	9657	9661	9666	9671	9675	9680
93	9685	9689	9694	9699	9703	9708	9713	9717	9722	9727
94	9731	9736	9741	9745	9750	9754	9759	9763	9768	9773
95	9777	9782	9786	9791	9795	9800	9805	9809	9814	9818
96	9823	9827	9832	9836	9841	9845	9850	9854	9859	9863
97	9868	9872	9877	9881	9886	9890	9894	9899	9903	9908
98	9912	9917	9921	9926	9930	9934	9939	9943	9948	9952
99	9956	9961	9965	9969	9974	9978	9983	9987	9991	9996

APPENDIX C
DECIMAL pH VALUES

Decimal pH values are calculated by use of logarithms, as outlined in Appendix A. For example, let's calculate the pH of a solution containing $2.3 \times 10^{-4} M$ H^+.

$$\begin{aligned} pH &= -\log [H^+] \\ &= -\log (2.3 \times 10^{-4}) \\ &= -[\log 2.3 + \log 10^{-4}] \\ &= -[0.36 + (-4.00)] \\ &= -[0.36 - 4.00] \\ &= -[-3.64] \\ &= 3.64 \end{aligned}$$

EXAMPLE C.1

Calculate the pH of a solution containing $4.5 \times 10^{-8} M$ H^+.

SOLUTION

$$\begin{aligned} pH &= -\log [H^+] \\ &= -\log (4.5 \times 10^{-8}) \\ &= -[\log 4.5 + \log 10^{-8}] \\ &= -[0.65 + (-8.00)] \\ &= -[0.65 - 8.00] \\ &= -[-7.35] \\ &= 7.35 \end{aligned}$$

EXERCISE C.1 Calculate the pH corresponding to each of the following H^+ concentrations.

(a) $7.93 \times 10^{-2} M$
(b) $3.79 \times 10^{-10} M$
(c) $5.38 \times 10^{-7} M$
(d) $9.85 \times 10^{-3} M$

SOLUTION

(a) 1.101
(b) 9.421
(c) 6.269
(d) 2.007

Hydrogen ion concentrations are calculated from decimal pH values by finding antilogarithms as outlined in Appendix A. For example, let's calculate $[H^+]$ for a solution having a pH of 7.45.

$$\begin{aligned} pH &= 7.45 \\ -\log [H^+] &= 7.45 \\ \log [H^+] &= -7.45 \\ [H^+] &= \text{antilog}(-7.45) \\ &= \text{antilog}(0.55) + \text{antilog}(-8.00) \\ &= 3.5 \times 10^{-8} \end{aligned}$$

EXAMPLE C.2

Calculate $[H^+]$ for a solution having a pH of 3.38.

SOLUTION

$$\begin{aligned} pH &= 3.38 \\ -\log [H^+] &= 3.38 \\ \log [H^+] &= -3.38 \\ [H^+] &= \text{antilog}(-3.38) \\ &= \text{antilog}(0.62) + \text{antilog}(-4.00) \\ &= 4.2 \times 10^{-4} \end{aligned}$$

EXERCISE C.2 Calculate $[H^+]$ for each of the following pH values.

(a) 6.29
(b) 9.35
(c) 7.05
(d) 4.47

SOLUTION

(a) 5.1×10^{-7}
(b) 4.5×10^{-10}
(c) 8.9×10^{-8}
(d) 3.4×10^{-5}

APPENDIX D
SOLUTIONS TO EXERCISES AND ANSWERS TO SELECTED STUDY QUESTIONS AND PROBLEMS

CHAPTER 1

Exercises
1.1
(a) $17 \text{ mL} \times \dfrac{1 \text{ L}}{1000 \text{ mL}} = 0.017 \text{ L}$

(b) $10 \text{ dg} \times \dfrac{1 \text{ g}}{10 \text{ dg}} = 1.0 \text{ g}$

(c) $2 \text{ L} \times \dfrac{1000 \text{ mL}}{1 \text{ L}} = 2000 \text{ mL}$

(d) $1.68 \text{ kg} \times \dfrac{1000 \text{ g}}{1 \text{ kg}} = 1680 \text{ g}$

1.2
(a) $4230 \text{ ft} \times \dfrac{1 \text{ mi}}{5280 \text{ ft}} \times \dfrac{1 \text{ km}}{0.621 \text{ mi}} = 1.29 \text{ km}$

(b) $3.65 \text{ kg} \times \dfrac{1000 \text{ g}}{1 \text{ kg}} \times \dfrac{10^6 \text{ }\mu\text{g}}{1 \text{ g}} = 3.65 \times 10^9 \text{ }\mu\text{g}$

(c) $4.30 \text{ m} \times \dfrac{39.37 \text{ in}}{1 \text{ m}} \times \dfrac{1 \text{ ft}}{12.0 \text{ in}} = 14.1 \text{ ft}$

(d) $1.67 \text{ m}^3 \times \dfrac{1 \text{ L}}{0.001 \text{ m}^3} \times \dfrac{1000 \text{ mL}}{1 \text{ L}} = 1.67 \times 10^6 \text{ mL}$

1.3
(a) 3 (b) 4 (c) 2 (d) 2 (e) 4 (f) 4

1.4
(a) 2080 (b) 1.89×10^5 (c) 1660 (d) 0.0358
(e) 1.99 (f) 3.33

1.5
(a) 20,000 (b) 7.1 (c) 10 (d) 1.29 (e) 15,000
(f) 2400

1.6
$d = \dfrac{m}{v} = \dfrac{21.7 \text{ g}}{21.2 \text{ mL}} = 1.02 \text{ g/mL}$

1.7
$v = \dfrac{m}{d} = \dfrac{27.6 \text{ g}}{1.02 \text{ g/mL}} = 27.1 \text{ mL}$

1.8
$m = dv = (2.70 \text{ g/mL})(38 \text{ mL}) = 100 \text{ g}$

1.9
specific gravity $= \dfrac{1.02 \text{ g/mL}}{1.0000 \text{ g/mL}} = 1.02$

1.10
(a) $°C = (°F - 32)(5/9) = (17.3 - 32)(5/9) = -8.17$
(b) $K = °C + 273 = 25 + 273 = 298$
(c) $°F = 9/5\ °C + 32 = (9/5)(43) + 32 = 109$
(d) $°C = (°F - 32)(5/9) = (0 - 32)(5/9) = -18$
(e) $K = °C + 273 = 0 + 273 = 273$
(f) $K = °C + 273 = (°F - 32)(5/9) + 273 =$
$\qquad (78 - 32)(5/9) + 273 = 299$

1.11
cal = sp heat $\times$ g $\times$ °C = $(0.217 \text{ cal/g °C})(10.0 \text{ g})(13°C)$
    = 28 cal

Study Questions and Problems
2. (a) 3 (b) 2 (c) 3 (d) 5 (e) 4 (f) 4 (g) 2
(h) 4 (i) 3 (j) 4

4. (a) 3.9 (b) 2.00 (c) 4.7 (d) 73 (e) 130
(f) 83.0 (g) 0.2167 (h) 51.74 (i) 107

6. (a) 4.399×10^{-1} m (b) 2.349×10^1 L
(c) 1.5×10^1 in (d) 1.19×10^1 m (e) 1.03×10^2 mi
(f) 3.20×10^{-1} lb (g) 2.64×10^{-1} qt (h) 8.63 ft
(i) 3.20×10^1 cm (j) 1.56 L

8. 0.9981 g/mL **10.** 5.5 g/mL **12.** 86 g **16.** 1.1
18. 1.02 **22.** 7500 cal **24.** 1300 cal **26.** 2 kg
28. 2.1°C **30.** 1300 cal **32.** 5600 mg/wk **34.** 7 g

CHAPTER 2

Exercises
2.1
(a) oxygen (b) chlorine (c) sodium (d) aluminum
(e) neon

2.2
80.30 atoms $\times$ 11 amu/atom = 883.3 amu
19.70 atoms $\times$ 10 amu/atom = 197.0 amu
total mass of 100.00 atoms = 1080.3 amu

atomic weight = average mass per atom = $\dfrac{1080.3 \text{ amu}}{100.00 \text{ atoms}}$
$\qquad = 10.803$ amu

2.3
$1s^2 2s^2 2p^6 3s^2 3p^3$

2.4
(a) $^{13}_{7}\text{N} \longrightarrow\ ^{0}_{1}\beta +\ ^{13}_{6}\text{C}$ (b) $^{198}_{84}\text{Po} \longrightarrow\ ^{4}_{2}\alpha +\ ^{194}_{82}\text{Pb}$

2.5
(a) $^{190}_{78}\text{Pt} \longrightarrow\ ^{4}_{2}\alpha +\ ^{186}_{76}\text{Os}$ (b) $^{210}_{82}\text{Pb} \longrightarrow\ ^{0}_{-1}\beta +\ ^{210}_{83}\text{Bi}$

2.6
(a) $^{99}_{46}\text{Pd} \longrightarrow\ ^{0}_{1}\beta +\ ^{99}_{45}\text{Rh}$ (b) $^{192}_{78}\text{Pt} \longrightarrow\ ^{4}_{2}\alpha +\ ^{188}_{76}\text{Os}$

Study Questions and Problems
8. (a) iron (b) barium (c) gold (d) phosphorus
(e) krypton (f) mercury (g) uranium (h) chlorine
(i) lead (j) silver

10. 20.17 amu **12.** 32.09 amu

16. (a) 1 (b) 8 (c) 3 (d) 4 (e) 2 (f) 5
(g) 7 (h) 6

20. (a) Li, B, C (b) Li, N, F (c) Rb, Sb, I
(d) Na, Si, Cl (e) Ba, Tl, Bi (f) K, Ga, Se
(g) Sr, Sb, Te (h) Na, Al, P (i) Ba, Pb, Po (j) Be, N, O

22. (a) nonmetal (b) metal (c) metal (d) nonmetal
(e) metal (f) nonmetal (g) metal (h) nonmetal
(i) metal (j) nonmetal

38.
(a) $^{205}_{84}\text{Po} \longrightarrow\ ^{4}_{2}\alpha +\ ^{201}_{82}\text{Pb}$ (b) $^{56}_{27}\text{Co} \longrightarrow\ ^{0}_{1}\beta + \gamma +\ ^{56}_{26}\text{Fe}$
(c) $^{223}_{88}\text{Ra} \longrightarrow\ ^{4}_{2}\alpha +\ ^{219}_{86}\text{Rn}$ (d) $^{238}_{93}\text{Np} \longrightarrow\ ^{0}_{-1}\beta +\ ^{238}_{94}\text{Pu}$
(e) $^{237}_{94}\text{Pu} +\ ^{0}_{-1}\beta \longrightarrow\ ^{237}_{93}\text{Np}$ (f) $^{6}_{2}\text{He} \longrightarrow\ ^{0}_{-1}\beta +\ ^{6}_{3}\text{Li}$
(g) $^{8}_{3}\text{Li} \longrightarrow\ ^{0}_{-1}\beta +\ ^{8}_{4}\text{Be}$ (h) $^{10}_{6}\text{C} \longrightarrow\ ^{0}_{1}\beta +\ ^{10}_{5}\text{B}$
(i) $^{18}_{7}\text{N} \longrightarrow\ ^{0}_{-1}\beta +\ ^{18}_{8}\text{O}$ (j) $^{190}_{77}\text{Ir} +\ ^{0}_{-1}\beta \longrightarrow\ ^{190}_{76}\text{Os}$

CHAPTER 3
Exercises
3.1
(a) H—C(H)(H)—H   (b) H—S̈—H   (c) H—C(H)(H)—C(H)(H)—Ö—H

3.2
(a) carbon disulfide (b) hydrogen iodide
(c) phosphorus trichloride (d) carbon tetrafluoride

3.3
(a) $[Na]^+ [:\ddot{C}l:]^-$ (b) $[Li]^+ [:\ddot{F}:]^-$
(c) $[K]^+ [:\ddot{B}r:]^-$ (d) $[Rb]^+ [:\ddot{I}:]^-$

3.4
(a) $[:\ddot{B}r:]^- [Mg]^{2+} [:\ddot{B}r:]^-$
(b) $[:\ddot{C}l:]^- [Ba]^{2+} [:\ddot{C}l:]^-$
(c) $[:\ddot{I}:]^- [Ca]^{2+} [:\ddot{I}:]^-$
(d) $[:\ddot{F}:]^- [Sr]^{2+} [:\ddot{F}:]^-$

3.5
Na $1s^2 2s^2 2p^6 3s^1$ Na⁺ $1s^2 2s^2 2p^6$
Cl $1s^2 2s^2 2p^6 3s^2 3p^5$ Cl⁻ $1s^2 2s^2 2p^6 3s^2 3p^6$

3.6
(a) K_2S (b) NaI (c) MgF_2 (d) CaO

3.7
(a) Na_2S, sodium sulfide (b) Li_2CO_3, lithium carbonate
(c) $Al(NO_3)_3$, aluminum nitrate (d) BaF_2, barium fluoride
(e) $Mg_3(PO_4)_2$, magnesium phosphate
(f) K_2HPO_4, potassium hydrogen phosphate

3.8
(a) tetrahedron (b) pyramid (c) bent (H—S̈—H)

Study Questions and Problems
2.
(a) :N̈e: (b) ·Al· (c) ·N̈· (d) ·Mg· (e) K·
(f) ·Be· (g) ·C̈· (h) ·P̈· (i) H· (j) :C̈l·

4. (a) carbon monoxide (b) hydrogen sulfide
(c) carbon tetraiodide (d) phosphorus trichloride
(e) hydrogen bromide (f) dinitrogen oxide
(g) carbon dioxide (h) sulfur trioxide
(i) boron trifluoride (j) diphosphorus pentoxide

8. (a) $1s^2 2s^2 2p^6$ (b) $1s^2 2s^2 2p^6$
(c) $1s^2 2s^2 2p^6 3s^2 3p^6$ (d) $1s^2 2s^2 2p^6 3s^2 3p^6$

10. (a) sodium bromide (b) magnesium oxide
(c) aluminum sulfide (d) potassium oxide
(e) calcium chloride (f) aluminum fluoride

14. (a) bent shape (b) pyramid (c) tetrahedron

16. (a) nonpolar covalent (b) ionic (c) polar covalent
(d) ionic (e) polar covalent (f) polar covalent
(g) nonpolar covalent (h) polar covalent (i) ionic
(j) polar covalent

18. (a) nonpolar (b) polar (c) nonpolar (d) nonpolar

20.
(a) Ö=C=Ö   linear; bond angles of 180°
(b) :N≡N: linear
(c) H—C≡N: linear; bond angles of 180°
(d) H—C(=O)—H   planar; bond angles of 120°
(e) H—C≡C—H   linear; bond angles of 180°
(f) H₂C=N—H   planar; bond angles of 120°

CHAPTER 4
Exercises
4.1
total mass of reactants = total mass of products
5.0 g C + 13.3 g O_2 = 18.3 g CO_2

4.2
total mass of reactants = total mass of products
8.50 g S + 8.50 g O_2 = 17.0 g SO_2

4.3
(a) $2S + 3O_2 \longrightarrow 2SO_3$
(b) $2C_2H_6 + 7O_2 \longrightarrow 4CO_2 + 6H_2O$
(c) $2Na + Cl_2 \longrightarrow 2NaCl$
(d) $4Na + O_2 \longrightarrow 2Na_2O$

4.4
(a) $MW = (2 \text{ atoms C} \times \frac{12.0 \text{ amu}}{\text{atom C}})$
$+ (6 \text{ atoms H} \times \frac{1.0 \text{ amu}}{\text{atom H}})$
$+ (1 \text{ atom O} \times \frac{16.0 \text{ amu}}{\text{atom O}})$
$= 46.0 \text{ amu}$

(b) $MW = (2 \text{ atoms C} \times \frac{12.0 \text{ amu}}{\text{atom C}})$
$+ (4 \text{ atoms H} \times \frac{1.0 \text{ amu}}{\text{atom H}})$
$= 28.0 \text{ amu}$

(c) $MW = (1 \text{ atom C} \times \frac{12.0 \text{ amu}}{\text{atom C}})$
$+ (1 \text{ atom O} \times \frac{16.0 \text{ amu}}{\text{atom O}})$
$= 28.0 \text{ amu}$

(d) $MW = (2 \text{ atoms N} \times \frac{14.0 \text{ amu}}{\text{atom N}})$
$+ (3 \text{ atoms O} \times \frac{16.0 \text{ amu}}{\text{atom O}})$
$= 76.0 \text{ amu}$

(e) $MW = (1 \text{ atom C} \times \frac{12.0 \text{ amu}}{\text{atom C}})$
$+ (4 \text{ atoms Cl} \times \frac{35.5 \text{ amu}}{\text{atom Cl}})$
$= 154.0 \text{ amu}$

(f) $MW = (2 \text{ atoms N} \times \frac{14.0 \text{ amu}}{\text{atom N}})$
$+ (4 \text{ atoms H} \times \frac{1.0 \text{ amu}}{\text{atom H}})$
$= 32.0 \text{ amu}$

4.5
(a) $FW = (1 \text{ atom Ca} \times \frac{40.1 \text{ amu}}{\text{atom Ca}})$
$+ (2 \text{ atoms O} \times \frac{16.0 \text{ amu}}{\text{atom O}})$
$+ (2 \text{ atoms H} \times \frac{1.0 \text{ amu}}{\text{atom H}})$
$= 74.1 \text{ amu}$

(b) $FW = (1 \text{ atom Al} \times \frac{27.0 \text{ amu}}{\text{atom Al}})$
$+ (1 \text{ atom P} \times \frac{31.0 \text{ amu}}{\text{atom P}})$
$+ (4 \text{ atoms O} \times \frac{16.0 \text{ amu}}{\text{atom O}})$
$= 122.0 \text{ amu}$

(c) $FW = (2 \text{ atoms Na} \times \frac{23.0 \text{ amu}}{\text{atom Na}})$
$+ (1 \text{ atom S} \times \frac{32.1 \text{ amu}}{\text{atom S}})$
$+ (4 \text{ atoms O} \times \frac{16.0 \text{ amu}}{\text{atom O}})$
$= 142.1 \text{ amu}$

(d) $FW = (1 \text{ atom Al} \times \frac{27.0 \text{ amu}}{\text{atom Al}})$
$+ (3 \text{ atoms O} \times \frac{16.0 \text{ amu}}{\text{atom O}})$
$+ (3 \text{ atoms H} \times \frac{1.0 \text{ amu}}{\text{atom H}})$
$= 78.0 \text{ amu}$

(e) $FW = (3 \text{ atoms Mg} \times \frac{24.3 \text{ amu}}{\text{atom Mg}})$
$+ (2 \text{ atoms P} \times \frac{31.0 \text{ amu}}{\text{atom P}})$
$+ (8 \text{ atoms O} \times \frac{16.0 \text{ amu}}{\text{atom O}})$
$= 262.9 \text{ amu}$

(f) $FW = (1 \text{ atom Na} \times \frac{23.0 \text{ amu}}{\text{atom Na}})$
$+ (1 \text{ atom H} \times \frac{1.0 \text{ amu}}{\text{atom H}})$
$+ (1 \text{ atom C} \times \frac{12.0 \text{ amu}}{\text{atom C}})$
$+ (3 \text{ atoms O} \times \frac{16.0 \text{ amu}}{\text{atom O}})$
$= 84.0 \text{ amu}$

4.6
$\% \text{ Mg} = \frac{\text{mass of Mg in Mg(OH)}_2}{\text{FW of Mg(OH)}_2} \times 100 = \frac{24.3 \text{ amu}}{58.3 \text{ amu}} \times 100$
$= 41.7$

$\% \text{ O} = \frac{\text{mass of O in Mg(OH)}_2}{\text{FW of Mg(OH)}_2} \times 100 = \frac{32.0 \text{ amu}}{58.3 \text{ amu}} \times 100$
$= 54.9$

$\% \text{ H} = \frac{\text{mass of H in Mg(OH)}_2}{\text{FW of Mg(OH)}_2} \times 100 = \frac{2.0 \text{ amu}}{58.3 \text{ amu}} \times 100$
$= 3.4$

4.7
(a) $MW = 2 \text{ atoms H} \times \frac{1.0 \text{ amu}}{\text{atom H}} = 2.0 \text{ amu}$
Thus, weight of 1 mole $H_2 = 2.0 \text{ g}$

(b) $MW = (1 \text{ atom C} \times \frac{12.0 \text{ amu}}{\text{atom C}})$
$+ (2 \text{ atoms O} \times \frac{16.0 \text{ amu}}{\text{atom O}})$
$= 44.0 \text{ amu}$
Thus, weight of 1 mole of $CO_2 = 44.0 \text{ g}$

(c) $FW = (1 \text{ atom Mg} \times \frac{24.3 \text{ amu}}{\text{atom Mg}})$
$+ (1 \text{ atom S} \times \frac{32.1 \text{ amu}}{\text{atom S}})$
$+ (4 \text{ atoms O} \times \frac{16.0 \text{ amu}}{\text{atom O}})$
$= 120.4 \text{ amu}$
Thus, weight of 1 mole of $MgSO_4 = 120.4 \text{ g}$

(d) $FW = 1 \text{ K}^+ \text{ ion} \times \frac{39.1 \text{ amu}}{\text{K}^+ \text{ ion}} = 39.1 \text{ amu}$
Thus, weight of 1 mole of $K^+ = 39.1 \text{ g}$

4.8
(a) The simplest mole ratio of Cl_2 to Na is 1 : 2.
(b) $1 \text{ mole Cl}_2 \times \frac{71.0 \text{ g}}{\text{mole Cl}_2} = 71.0 \text{ g}$
$2 \text{ moles Na} \times \frac{23.0 \text{ g}}{\text{mole Na}} = 46.0 \text{ g}$
Mass ratio = 71.0 : 46.0
Simplest mass ratio $= \frac{71.0}{46.0} : \frac{46.0}{46.0} = 1.54 : 1.00$

4.9
(a) moles O_2 = ¼ moles Na = $(¼)(7.69 \text{ g} \times \frac{1 \text{ mole}}{23.0 \text{ g}})$
$= 0.0836 \text{ mole}$
$\text{g } O_2 = 0.0836 \text{ mole} \times \frac{32.0 \text{ g}}{\text{mole}} = 28.6 \text{ g}$

(b) moles $Na_2O = 2 \times$ moles $O_2 = (2)(0.0836 \text{ mole}) = 0.167 \text{ mole}$
$\text{g } Na_2O = 0.167 \text{ mole} \times \frac{62.0 \text{ g}}{\text{mole}} = 10.4 \text{ g}$

4.10
(a) combination (b) isomerization (c) single replacement
(d) decomposition (e) double replacement

4.11
K   Mn O_4
+1 +7   −2

4.12
(a) Zn is oxidized; H is reduced. (b) Br is oxidized; Cl_2 is reduced. (c) C_2H_4O is oxidized; O_2 is reduced. (d) C_6H_{12} is simultaneously oxidized to C_6H_6 and reduced to H_2.

Study Questions and Problems

6. (a) 410.7 amu (b) 189.3 amu (c) 208.1 amu
(d) 392.3 amu (e) 235.9 amu (f) 187.5 amu

(g) 295.5 amu (h) 109.9 amu (i) 142.0 amu
(j) 223.3 amu
8. 3.2×10^8 moles   **10.** 2.8×10^{-3} mole
12. (a) 69.9% Fe; 30.1% O (b) 36.1% Ca; 69.3% Cl
(c) 15.9% C; 2.2% H; 18.5% N; 63.4% O (d) 15.8% Al; 28.1% S; 56.1% O (e) 43.5% Ca; 26.1% C; 30.4% N
14. 2.5 g   **16.** 8.00 moles CO_2; 4.00 moles H_2O **18.** 22.3 g
20. 0.65 g   **22.** 21.5 g   **24.** 0.0161 mole   **26.** 2.17 g
28.
(a) single replacement and oxidation-reduction
(b) single replacement and oxidation-reduction
(c) oxidation-reduction   (d) double replacement
(e) isomerization   (f) combination
(g) decomposition and oxidation-reduction
(h) single replacement and oxidation-reduction
(i) decomposition and oxidation-reduction
(j) double replacement
30.

	Oxidizing agent	Reducing agent
(a)	H_2SO_4	Ca
(b)	K_2O	K_2O
(c)	C_2H_4	H_2
(d)	O_2	Na
(e)	HCl	Fe
(f)	O_2	CH_3CHO
(g)	H_2O	Mg
(h)	O_2	CH_3OH
(i)	Fe_2O_3	H_2
(j)	O_2	S

CHAPTER 5

Exercises

5.1
$V_f = 75.0 \text{ mL} \times \dfrac{720 \text{ torr}}{700 \text{ torr}} = 77.1 \text{ mL}$

5.2
$P_f = 760 \text{ torr} \times \dfrac{140 \text{ mL}}{200 \text{ mL}} = 532 \text{ torr}$

5.3
$T_i = °C + 273 = 100 + 273 = 373 \text{ K}$
$T_f = 373 \text{ K} \times \dfrac{8.00 \text{ mL}}{10.0 \text{ mL}} = 298 \text{ K}$
$T_f = \text{K} - 273 = 298 - 273 = 25°C$

5.4
$T_i = °C + 273 = 25 + 273 = 298 \text{ K}$
$T_f = °C + 273 = 50 + 273 = 323 \text{ K}$
$V_f = 1.00 \text{ L} \times \dfrac{323 \text{ K}}{298 \text{ K}} = 1.08 \text{ L}$

5.5
$T_i = °C + 273 = 100 + 273 = 373 \text{ K}$
$T_f = °C + 273 = 50 + 273 = 323 \text{ K}$
$P_f = 2.50 \text{ atm} \times \dfrac{323 \text{ K}}{373 \text{ K}} = 2.16 \text{ atm}$

5.6
$T_i = °C + 273 = 150 + 273 = 423 \text{ K}$
$T_f = °C + 273 = 50 + 273 = 323 \text{ K}$
$V_f = 1.50 \text{ L} \times \dfrac{800 \text{ torr}}{700 \text{ torr}} \times \dfrac{323 \text{ K}}{423 \text{ K}} = 1.31 \text{ L}$

5.7
$T_i = °C + 273 = 25 + 273 = 298 \text{ K}$
$T_f = °C + 273 = 50 + 273 = 323 \text{ K}$
$P_f = 760 \text{ torr} \times \dfrac{323 \text{ K}}{298 \text{ K}} \times \dfrac{250 \text{ mL}}{200 \text{ mL}} = 1030 \text{ torr}$

5.8
$T_i = °C + 273 = 20 + 273 = 293 \text{ K}$
$T_f = 293 \text{ K} \times \dfrac{10.0 \text{ L}}{7.50 \text{ L}} \times \dfrac{650 \text{ torr}}{720 \text{ torr}} = 353 \text{ K}$
$T_f = \text{K} - 273 = 353 - 273 = 80°C$

5.09
$\text{K} = °C + 273 = 30 + 273 = 303 \text{ K}$
$n = \dfrac{PV}{RT} = \dfrac{(2 \text{ atm})(1.25 \text{ L})}{(0.0820 \text{ L atm/mole K})(303 \text{ K})} = 0.1 \text{ mole}$

5.10
$\text{K} = °C + 273 = 20 + 273 = 293 \text{ K}$
$n = \dfrac{PV}{RT} = \dfrac{(1.50 \text{ atm})(80.1 \text{ L})}{(0.0820 \text{ L atm/mole K})(293 \text{ K})} = 5.00 \text{ moles}$
weight per mole $= \dfrac{220 \text{ g}}{5.00 \text{ mole}} = 44.0 \text{ g}$

5.11
$T = \dfrac{PV}{nR} = \dfrac{(1.2 \text{ atm})(140 \text{ L})}{(7.25 \text{ moles})(0.0820 \text{ L atm/mole K})} = 280 \text{ K}$
$°C = \text{K} - 273 = 280 - 273 = 7$

5.12
$T_i = °C + 273 = 0 + 273 = 273 \text{ K}$
$T_f = \dfrac{P_f V_f T_i}{P_i V_i} = \dfrac{(1.50 \text{ atm})(2.00 \text{ L})(273 \text{ K})}{(2.00 \text{ atm})(1.35 \text{ L})} = 303 \text{ K}$
$°C = \text{K} - 273 = 303 - 273 = 30°C$

5.13
$P_{HCl} = P_{total} - P_{H_2O} = 750 \text{ torr} - 24 \text{ torr} = 726 \text{ torr}$

Study Questions and Problems

12. 270 mL
14. (a) 0.800 atm (b) 2.00 atm (c) 0.200 atm
16. 68.2 mL   **18.** 106°C   **20.** 439 mL   **22.** 2640 torr
24. 2.5 L
26. $P_f = 2.15$ atm   Yes, it would be safe because the pressure inside the can would not exceed 3.0 atm at 120°F.
28. 0.893 atm   **30.** 2.88 moles, 80.6 g   **32.** 0.0155 L
34. 41.9 amu   **36.** 9.00 g/L
38. 13.2% O_2; 5.3% CO_2; 6.2% H_2O; 75.4% N_2

CHAPTER 6

Exercises

6.1
(a) $\delta+ \text{C}\equiv\text{O} \ \delta-$ CO is a polar molecule; it will exert London dispersion forces and dipolar attractions.

(b) $\overset{\delta-}{:\ddot{\text{S}}:}$ with $\underset{\delta+}{\text{H}}$ $\underset{\delta+}{\text{H}}$ H_2S is a polar molecule, but hydrogen is not bonded to N, O, or F, so no hydrogen bonds are formed; therefore, H_2S will exert London dispersion forces and dipolar attractions.

(c) H—H H_2 is a nonpolar molecule and will exert only London dispersion forces.

(d) $\delta+H-F\delta-$ HF is a polar molecule which will exert London dispersion forces and form hydrogen bonds.

6.2
heat = moles Hg $\times \Delta H_{vap}$
= $(2.50\text{ g})(\frac{1\text{ mole}}{200.6\text{ g}}) \times 13.6$ kcal/mole
= 0.169 kcal or 169 cal

6.3
heat = moles benzene $\times \Delta H_{fus}$
= $(15.0\text{ g})(\frac{1\text{ mole}}{78.1\text{ g}}) \times 2.35$ kcal/mole
= 0.451 kcal or 451 cal

6.4
Heating the ice from $-5°C$ to $0°C$:
 heat = sp heat $\times$ g $\times$ temperature change
 = $(0.492\text{ cal/g °C})(10.0\text{ g})(5°)$ = 24.6 cal
Converting the ice to water at $0°C$:
 heat = moles ice $\times \Delta H_{fus}$
 = $(10.0\text{ g} \times \frac{1\text{ mole}}{18.0\text{ g}})(1.44$ kcal/mole$)$
 = 0.800 kcal or 800 cal
Heating the water from $0°C$ to $100°C$:
 heat = sp heat $\times$ g $\times$ temperature change
 = $(1.00\text{ cal/g °C})(10.0\text{ g})(100°C)$ = 1000 cal
Converting the water to steam at $100°C$:
 heat = moles water $\times \Delta H_{vap}$
 = $(10.0\text{ g} \times \frac{1\text{ mole}}{18.0\text{ g}})(9.72$ kcal/mole$)$
 = 5.40 kcal or 5400 cal
Total heat required = 25 cal + 800 cal + 1000 cal + 5400 cal
 = 7225 cal

Study Questions and Problems

16. Perspiration evaporates from the skin, absorbing large quantities of heat as it turns from liquid to vapor; the loss of body heat lowers body temperature.
18. 4840 cal   **32.** 2.35 kcal/mole   **34.** 660 cal
38. 9960 cal
40. 10,890 cal liberated from steam; 1170 cal liberated from water

CHAPTER 7

Exercises

7.1
g glucose = $(0.5\text{ mole})(\frac{180.0\text{ g}}{1\text{ mole}})$ = 90 g

7.2
moles = $(1.50\text{ moles/L})(0.250\text{ L})$ = 0.375 mole
 g = $(0.375\text{ mole})(262.9\text{ g/mole})$ = 98.6 g

7.3
moles = $(26.2\text{ g})(\frac{1\text{ mole}}{68.9\text{ g}})$ = 0.380 mole
 L = $\frac{0.380\text{ mole}}{3.70\text{ mole/L}}$ = 0.103 L

7.4
%(w/v) = $\frac{25.0\text{ g}}{200\text{ mL}} \times 100$ = 12.5

7.5
mg% = $\frac{23\text{ mg}}{125\text{ mL}} \times 100$ = 18

7.6
%(v/v) = $\frac{11.5\text{ mL}}{80.0\text{ mL}} \times 100$ = 14.4

7.7
%(w/w) = $\frac{0.085\text{ g}}{8.36\text{ g}} \times 100$ = 1.02

7.8
$\text{conc}_f = (1.75M)(\frac{250\text{ mL}}{800\text{ mL}})$ = 0.547M

7.9
$V_f = (100\text{ mL})(\frac{2.00M}{0.100M})$ = 2000 mL

Study Questions and Problems

18. (a) 90 g (b) 4.7 g (c) 17 g (d) 77.0 g (e) 4.8 g (f) 112 g (g) 10.6 g (h) 439 g (i) 24 g (j) 12 g
20. (a) 11.3% (w/v) (b) 2.43% (w/v) (c) 3.97% (w/v) (d) 2.92% (w/v) (e) 4.11% (w/v) (f) 4.88% (w/v) (g) 2.66% (w/v) (h) 6.98% (w/v) (i) 7.52% (w/v) (j) 21.2% (w/v)
22. 20 g   **24.** 0.88% (w/v)
26. (a) 11.3% (v/v) (b) 9.88% (v/v) (c) 6.24% (v/v) (d) 4.99% (v/v) (e) 22.7% (v/v)
28. (a) 11.3% (w/w) (b) 2.76% (w/w) (c) 25.1% (w/w) (d) 15.9% (w/w) (e) 1.96% (w/w)
30. (a) 37.5 mL (b) 179 mL (c) 100 mL (d) 43.8 mL (e) 12.5 mL
36. 0.15M MgCl$_2$, 0.15M NaCl, 0.10M LiBr, 0.15M C$_2$H$_5$OH, 0.05M KNO$_3$
38. Eating salty food causes a buildup of salt in body fluids; the increase in osmotic pressure in body fluids causes less water to be excreted, thus increasing fluid volume.
40. Water moves from the pickles into the hypertonic salt solution, causing shrinkage and shriveling of the cucumbers.
42. Homogenized milk is more opaque than nonfat milk because homogenized milk contains more fat droplets; these are dispersed as a colloid and thus scatter visible light.

CHAPTER 8

Exercises

8.1
$K_{eq} = \frac{[NH_3]^2}{[N_2][H_2]^3}$

8.2
$K_{eq} = \frac{[NH_3]^2}{[N_2][H_2]^3} = \frac{(0.0434\text{ mole/L})^2}{(0.9783\text{ mole/L})(0.935\text{ mole/L})^3}$
= 2.36×10^{-3} L^2/mole^2

8.3
(a) $K_{eq} = \frac{[NO_2]^2}{[N_2O_4]}$ [N$_2$O$_4$] will increase, and the equilibrium will shift to the left.

(b) $K_{eq} = \frac{[CO]^2[O_2]}{[CO_2]^2}$ [CO] and [O$_2$] will increase, and the equilibrium will shift to the right.

(c) $K_{eq} = \frac{[H_2]^2[O_2]}{[H_2O]^2}$ [H$_2$] and [O$_2$] will decrease, and the equilibrium will shift to the right.

8.4
(a) Nonspontaneous reaction because ΔG is positive.
(b) Reaction is at equilibrium because $\Delta G = 0$.
(c) Spontaneous reaction because ΔG is negative.

8.5
(a) Since $K_{eq} > 1$, $\Delta G°$ is negative and the position of equilibrium lies on the right, favoring products.
(b) Since $K_{eq} < 1$, $\Delta G°$ is positive and the position of equilibrium lies on the left, favoring reactants.
(c) Since $K_{eq} > 1$, $\Delta G°$ is negative and the position of equilibrium lies on the right, favoring products.

Study Questions and Problems

2.
(a) $K_{eq} = \dfrac{[NO]^4[H_2O]^6}{[NH_3]^4[O_2]^5}$
(b) $K_{eq} = \dfrac{[N_2O_4]}{[NO_2]^2}$
(c) $K_{eq} = \dfrac{[SO_2]^2[O_2]}{[SO_3]^2}$
(d) $K_{eq} = \dfrac{[PCl_3][Cl_2]}{[PCl_5]}$
(e) $K_{eq} = \dfrac{[HCl]^2}{[H_2][Cl_2]}$

4. 1 **6.** 4.2×10^{-2} mole/L

10. (a) The concentrations of products will increase.
(b) The concentrations of reactants will increase.
(c) The concentrations of products will decrease.
(d) The concentrations of reactants will decrease.

12. For an exothermic reaction, a temperature increase will cause K_{eq} to decrease.

14. (a) $[H_2O]$ will increase. (b) $[H_2O]$ will decrease.
(c) $[H_2O]$ will decrease.

20. (a) spontaneous (b) nonspontaneous
(c) at equilibrium (d) spontaneous (e) at equilibrium

22. (a) $T\Delta S$ (b) positive (c) negative

30. rate = $k[A][B]$
(a) The rate will be twice the original rate.
(b) The rate will be twice the original rate.
(c) The rate will be four times the original rate.

32. 4.2×10^{-4} L/mole sec

34. (a) increased by a factor of 1000 (b) no effect
(c) no effect

38. Since K_{eq} is a ratio of forward and reverse rate constants, a change in temperature affects the rates, and thus the rate constants, to different extents; thus, K_{eq} will be changed.

CHAPTER 9

Exercises

9.1

	Conjugate base		Conjugate acid
(a)	CO_3^{2-}	(d)	$HClO_4$
(b)	HSO_4^-	(e)	H_2CO_3
(c)	F^-	(f)	NH_4^+

9.2
$HA(aq) + H_2O(l) \longrightarrow H_3O^+(aq) + A^-(aq)$

$K_a = \dfrac{[H_3O^+][A^-]}{[HA]} = \dfrac{(2.0 \times 10^{-3}M)(2.0 \times 10^{-3}M)}{0.10M}$
$= 4.0 \times 10^{-5}$

9.3
$[H_3O^+] = \sqrt{K_a[HCN]} = \sqrt{(4.9 \times 10^{-10})(0.15)} = 8.6 \times 10^{-6}M$

9.4
$K_b = \dfrac{[BH^+][OH^-]}{[B]} = \dfrac{(7.7 \times 10^{-8})(7.7 \times 10^{-8})}{0.40}$
$= 1.5 \times 10^{-14}$

9.5
$N = $ eq/L $= (120$ g/L$)(1$ mole$/40.0$ g$)(1$ eq/mole$) = 3.00$ eq/L

9.6
g/L = (eq/L)(g/eq) = (eq/L)(mole/eq)(g/mole)
$= (6.00$ eq/L$)(1$ mole$/3$ eq$)(98.0$ g/mole$) = 196$ g/L
for 500 mL,
g = $(196$ g/L$)(0.500$ L$) = 98.0$ g

9.7
meq/mL = $(0.100$ eq/L$)\left(\dfrac{1 \text{ L}}{1000 \text{ mL}}\right)\left(\dfrac{1 \text{ meq}}{0.001 \text{ eq}}\right)$
$= 0.100$ meq/mL
meq/20.0 mL = $(20.0$ mL$)(0.100$ meq/mL$) = 2.00$ meq

9.8
$N = (0.02$ mole/L$)(2$ eq/mole$) = 0.04$ eq/L

9.9
$N_a = \dfrac{V_b \times N_b}{V_a} = \dfrac{(26.2 \text{ mL})(0.100 \text{ eq/L})}{20.0 \text{ mL}} = 0.131$ eq/L

9.10
$V_b = \dfrac{V_a \times N_a}{N_b} = \dfrac{(25.0 \text{ mL})(0.250 \text{ eq/L})}{0.150 \text{ eq/L}} = 41.7$ mL

Study Questions and Problems

6. (a) $Na_2O(s) + H_2O(l) \longrightarrow 2\,NaOH(s)$
(b) $CaO(s) + H_2O(l) \longrightarrow Ca(OH)_2(s)$
(c) $K_2O(s) + H_2O(l) \longrightarrow 2\,KOH(s)$
(d) $MgO(s) + H_2O(l) \longrightarrow Mg(OH)_2(s)$

8. $6.0M$ **10.** 4.9×10^{-10} **12.** $5.8 \times 10^{-3}M$

14. (a) diprotic (b) triprotic (c) monoprotic
(d) diprotic

16.
$H_2CO_3(aq) + H_2O(l) \longrightarrow H_3O^+(aq) + HCO_3^-(aq)$
acid$_1$ base$_1$ acid$_2$ base$_2$

$HCO_3^-(aq) + H_2O(l) \longrightarrow H_3O^+(aq) + CO_3^{2-}(aq)$
acid$_1$ base$_1$ acid$_2$ base$_2$

22. 1.7×10^{-5}

24. Lemon juice is an acid solution; washing with lemon juice neutralizes the amines responsible for fish odor and forms odorless salts.

26. $2.00\,N$ **28.** 202 g **30.** 7.50 meq **32.** $3.00\,N$
34. $0.75\,N$ **36.** $2.14\,N; 2.14\,M$

38. Since NO_2 is an acid anhydride, any NO_2 in automobile exhaust fumes will react with moisture in the atmosphere to form acid:
$3\,NO_2(g) + H_2O(l) \longrightarrow 2\,HNO_3(aq) + NO(g)$
However, N_2 produced by catalytic converters is highly unreactive and does not pollute the atmosphere, which already contains about 80% N_2.

42.
(a) $Ca^{2+}(aq) + 2\,OH^-(aq) + 2\,H^+(aq) + SO_4^{2-}(aq) \longrightarrow CaSO_4(s) + 2\,H_2O(l)$
(b) Same as (a)

CHAPTER 10

Exercises

10.1
pH = $-\log[H^+] = -\log(1 \times 10^{-9}) = 9$
The solution is basic.

10.2
$\log[H^+] = -pH = -6.0; [H^+] = 1 \times 10^{-6} M$

10.3
pOH = 14 − pH = 14 − 11 = 3

10.4
pOH = 14 − pH = 14 − 5 = 9
$\log[OH^-] = -pOH = -9; [OH^-] = 1 \times 10^{-9} M$

10.5
pH = $-\log[H^+] = -\log(1 \times 10^{-3}) = 3$

10.6
pOH = $-\log[OH^-] = -\log(1 \times 10^{-2}) = 2$
pH = 14 − pOH = 14 − 2 = 12

10.7
pH = $pK_a + \log([A^-]/[HA])$
  = 4.76 + log (0.1M/0.2M) = 4.5

10.8
$\log([A^-]/[HA]) = pH - pK_a = 3.50 - 3.75 = -0.25$
$[A^-]/[HA] = 0.56$

10.9
$\log([A^-]/[HA]) = pH - pK_a = 3.75 - 3.75 = 0$
$[A^-]/[HA] = 1.00$
moles HA + moles A^- = 0.200 mole
  x + x = 0.200 mole
  $2x$ = 0.200 mole
  x = 0.100 mole = moles HA = moles A^-
Thus, to make 1 liter of buffer, 0.100 mole of HCO$_2$H and 0.100 mole of HCO$_2$Na are required.

Study Questions and Problems

6. (a) $1.00 \times 10^{-2} N$ (b) 1.00 meq (c) 1.00 meq
8. (a) Na and I$_2$ (b) Li and Br$_2$ (c) Mg and Cl$_2$
(d) Al and F$_2$ **12.** $1 \times 10^{-11} M$
14. (a) 13 (b) 2 (c) 10 (d) 5 (e) 14 (f) 8
16. (a) $1 \times 10^{-3} M$ (b) $1 \times 10^{-11} M$ (c) 11 (d) 3
18. When CO$_2$ dissolves in water, the weak acid carbonic acid is formed, and the added hydrogen ions make the solution slightly acidic, with a pH less than 7.
CO$_2$(g) + H$_2$O(l) $\rightleftharpoons$ H$_2$CO$_3$(aq)
$\rightleftharpoons$ H$^+$(aq) + HCO$_3^-$(aq)
20. $[H_3O^+] = 1.56 \times 10^{-7} M = [OH^-]$ **24.** 3.63
26. $[HCO_2^-]/[HCO_2H] = 1.8$
28. Yes, it is possible to make the buffer. First, make an aqueous solution of acetic acid, and then add sodium hydroxide to neutralize part of the acetic acid and produce ionized sodium acetate:
CH$_3$CO$_2$H(aq) + NaOH(aq) $\longrightarrow$ CH$_3$CO$_2$Na(aq) + H$_2$O(l)
The amount of sodium hydroxide added determines the amount of acetate anion formed and thus the pH of the resulting solution.
30. 0.448 mole HCO$_2$H; 0.251 mole HCO$_2$Na

CHAPTER 11

Exercises

11.1

(a) H$_3$C—CH$_2$—(C)—CH$_2$—C (carbonyl group) (carboxyl group, OH)

(b) H$_3$C—(S—S)—CH$_2$CH$_3$
  disulfide group

(c) H$_3$C—CH$_2$—(Cl) chloro group

(d) H$_3$C—(N)—CH$_2$—CH$_3$
        |
        H   amino group

(e) H$_3$C—(C)—CH$_2$—(OH) hydroxyl group
       ‖
       O carbonyl group

11.2
(a) CH$_3$—CH$_2$—CH—CH$_3$ 3-methylpentane
                |
                CH$_2$
                |
                CH$_3$

(b)   CH$_3$
           |
    CH$_3$—CH$_2$—C—CH$_2$—CH$_3$ 3,3-dimethylpentane
           |
           CH$_3$

(c) CH$_3$—CH—CH$_2$—CH—CH$_2$—CH$_3$
          | |
          CH$_2$ CH$_3$ 3,5-dimethylheptane
          |
          CH$_3$

11.3
(a) CH$_3$—CH=CH—CH$_3$ 2-butene

(b) CH$_3$—CH—CH=CH—CH—CH$_3$
         | |
         CH$_3$ CH$_2$
                    |
                    CH$_3$

  2,5-dimethyl-3-heptene

(c) CH$_2$=CH—CH=CH—CH—CH$_3$
                     |
                     CH$_3$

  5-methyl-1,3-hexadiene

11.4

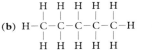

(a)
```
    H  H  H
    |  |  |
H — C — C — C — H
    |  |  |
    H  Br H
```

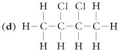

(c)
```
    H  H  OH H
    |  |  |  |
H — C — C — C — C — H
    |  |  |  |
    H  H  |  H
          H — C — H
              |
              H
```

(b)
```
    H  H  H  H  H
    |  |  |  |  |
H — C — C — C — C — C — H
    |  |  |  |  |
    H  H  H  H  H
```

(d)
```
    H  Cl Cl H
    |  |  |  |
H — C — C — C — C — H
    |  |  |  |
    H  H  H  H
```

11.5

(a) H−C(H)(H)−C(H)(H)−C(H)=C−C(H)(H)−H + H−Cl

(b) H−C(H)(H)−C(H)=C−C(H)(H)−H + H−Cl with H−C−H branch

(c) H−C(H)(H)−C(H)(H)−C(=CH₂)−C(H)(H)−H + H−Br

Study Questions and Problems

2.

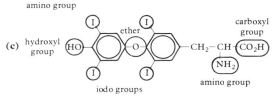

(a) ring structure with double bonds and hydroxyl group circled

(b) pyridine ring with CO₂H (carboxyl group) and N (amino group) circled

(c) structure with hydroxyl group, ether, iodo groups, carboxyl group, amino group labeled

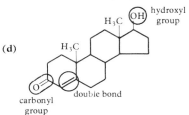

(d) steroid structure with hydroxyl group, double bond, carbonyl group labeled

(e) hydroxyl groups (HO), amino group (NH) circled on structure with CH−CH₂−NH−CH₃ and OH

(f) CH₃−C(=O)−O−CH₂−CH₂−CH₂−CH (ester group circled)

4. (a) 2-methylpropane (b) 2,4,4,6-tetramethyloctane (c) 2-methylpentane (d) 3-methyl-5-propyloctane (e) 2,2,4-trimethylpentane

6.

(a) CH₃−CH(CH₃)−CH₂−CH₂−CH₃

(b) CH₃−C(CH₃)(CH₃)−CH₂−CH₃

(c) CH₃−CH₂−C(CH₃)(CH₂−CH₃)−CH₂−CH(CH₃)−CH₂−CH₂−CH₂−CH₂−CH₃

(d) CH₃−CH₂−CH−CH(CH₃)−CH(CH₂CH₂CH₃)−CH₂−CH₂−CH₂−CH₃ with CH₂−CH₃ and CH₂−CH₃ branches

(e) CH₃−CH(CH₂CH₂CH₃)−CH₂−CH(CH₂CH₂CH₃)−CH₂−CH₂−CH₂−CH₃

14. (a) propene (b) 4-methyl-1-pentene (c) 3-methyl-2-pentene (d) 3-heptene (e) 6-methyl-3-heptene

30.

Br−CH₂−CH₂−CH₂−CH₃ | CH₃−C(CH₃)(Br)−CH₃
1-bromobutane | 2-bromo-2-methylpropane

CH₃−CH(Br)−CH₂−CH₃ | CH₃−CH(CH₃)−CH₂−Br
2-bromobutane | 1-bromo-2-methylpropane

32. (a) 3-methyl-1-butyne (b) methylcyclohexane (c) bromocyclopentane (d) 1-chloro-3-hydroxybenzene (e) cyclohexene (f) methylbenzene (g) 2-pentyne (h) 1-amino-4-iodobenzene (i) 2-bromobutane

36. $(CH_3CH_2)_4Pb + 14\,O_2 \longrightarrow 8\,CO_2 + 10\,H_2O + PbO_2$ (lead dioxide)

CHAPTER 12

Exercises

12.1
(a) tertiary (b) secondary (c) primary (d) secondary (e) primary

12.2
(a) 2,4-dimethyl-1-pentanol (b) cyclohexanol (c) 2-methylcyclopentanol (d) 2-methyl-2-propanol (e) 2-butanol

12.3
(a) cyclohexene + H₂O

(b) CH₃−CH₂−CH₂−O−CH₂−CH₂−CH₃ + H₂O

(c) $CH_3-CH_2-CH=CH_2 + H_2O$
(d) $CH_2=C-CH_3 + H_2O$
        |
        CH_3
(e) $CH_3-CH=CH-CH_3$ (major product) $+ H_2O$;
    $CH_2=CH-CH_2-CH_3$ (minor product) $+ H_2O$

12.4
(a) $4 CO_2 + 5 H_2O$
(b) $CH_3-\underset{\underset{CH_3}{|}}{\overset{\overset{CH_3}{|}}{C}}-CH_2-C\overset{\diagup O}{\underset{\diagdown H}{}}$
(c) $CH_3-CH-CH_2-CH_2-CH_3$
        |
        C=O
        |
        CH_3
(d) NR
(e) $CH_3-CH_2-CH_2-CH_2-C\overset{\diagup O}{\underset{\diagdown OH}{}}$

12.5
(a) dibutyl ether   (b) ethyl methyl ether
(c) cyclohexyl propyl ether

Study Questions and Problems

2. (a) secondary (b) tertiary (c) secondary
(d) secondary (e) secondary (f) primary
(g) primary (h) tertiary (i) primary (j) secondary

4.
(a) $CH_3-CH-CH_2-CH_2-CH_2-CH_3$
          |
          OH
(b) $HO-CH_2-CH-CH_3$ (c) $CH_3-CH-CH_2-CH-CH_3$
                | | |
                CH_3 OH CH_3

(d) $HO-CH_2-\underset{\underset{CH_3}{|}}{\overset{\overset{CH_3}{|}}{C}}-CH_2-CH_2-CH_3$

(e) $CH_3-\underset{\underset{OH}{|}}{CH}-\underset{\underset{CH_3}{|}}{C}-CH_3$ (f) $HO-CH_2-\underset{\underset{OH}{|}}{CH}-CH_2-Cl$

6.
(a) $CH_3-\underset{\underset{}{|}}{\overset{\overset{CH_3}{|}}{C}}=O$
(b) $CH_3-CH_2-CH_2-O-CH_2-CH_2-CH_3 + H_2O$
(c) ⌬$-CH=CH_2 + H_2O$ (d) NR
(e) $CH_3-\underset{\underset{CH_2}{||}}{C}-CH_3 + H_2O$
(f) $CH_3-CH_2-C\overset{\diagup O}{\underset{\diagdown OH}{}}$
(g) $6 CO_2 + 8 H_2O$
(h) $CH_3-C\overset{\diagup O}{\underset{\diagdown O-CH_2-CH_3}{}} + H_2O$

(i) $CH_3-\overset{\overset{CH-CH_3}{||}}{\underset{\underset{CH_2-CH_3}{|}}{C}}-CH_3$ (major product) $+ H_2O$;
    $CH_2=C-CH_3$ (minor product) $+ H_2O$
            |
            CH_2-CH_3

(j) $CH_3-CH_2-CH_2-CH_2-C\overset{\diagup O}{\underset{\diagdown OH}{}}$

14. $2 R-SH \xrightarrow{[O]} R-S-S-R$

16. (a) ether (b) disulfide (c) aldehyde
(d) carboxylic acid (e) alcohol (f) ketone
(g) phenol (h) ester (i) alcohol (j) thiol
(k) aldehyde (l) ketone (m) disulfide
(n) carboxylic acid (o) ether

22. 40%

CHAPTER 13

Exercises

13.1

	Common Name	IUPAC Name
(a)	methyl ethyl ketone	butanone
(b)	formaldehyde	methanal
(c)	valeraldehyde	pentanal
(d)	acetophenone or methyl phenyl ketone	1-phenylethanone
(e)	ethyl cyclopentyl ketone	1-cyclopentylpropanone

13.2
(a) $H-C\overset{\diagup O}{\underset{\diagdown OH}{}}$ (b) $6 CO + 6 H_2O$

(c) CH_3-CH_2-OH (d) $CH_3-\underset{\underset{OCH_3}{|}}{\overset{\overset{OH}{|}}{C}}-H$

(e) $H-\underset{\underset{OH}{|}}{\overset{\overset{OH}{|}}{C}}-H$ (f) CH_3-OH

Study Questions and Problems

6. a, e, b, h, c, d, f, g

8.
(a) $H-C\overset{\diagup O}{\underset{\diagdown H}{}}$ (b) ⌬$-C\overset{\diagup O}{\underset{\diagdown H}{}}$

(c) $CH_3-CH_2-CH_2-CH_2-C\overset{\diagup O}{\underset{\diagdown CH_2-CH_2-CH_2-CH_2-CH_2-CH_3}{}}$

(d) $I-CH_2-CH_2-C\overset{\diagup O}{\underset{\diagdown H}{}}$

(e) $CH_3-CH_2-CH_2-CH_2-C\overset{\diagup O}{\underset{\diagdown H}{}}$

(f) $CH_3-C\overset{\diagup O}{\underset{\diagdown CH_2-CH_3}{}}$ (g)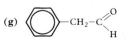

(h) $CH_3-CH_2-CH_2-\underset{\underset{OH}{|}}{CH}-CH_2-\overset{O}{\underset{H}{C}}$

10.
(a) $CH_3-\overset{O}{\underset{H}{C}} + H_2O \rightleftharpoons CH_3-\underset{\underset{OH}{|}}{\overset{\overset{OH}{|}}{C}}-H$

(b) $CH_3-\overset{O}{\underset{H}{C}} + CH_3-OH \overset{H^+}{\rightleftharpoons} CH_3-\underset{\underset{O-CH_3}{|}}{\overset{\overset{OH}{|}}{C}}-H$

(c) $CH_3-\overset{O}{\underset{H}{C}} + 2\,CH_3-OH \overset{H^+}{\rightleftharpoons} CH_3-\underset{\underset{O-CH_3}{|}}{\overset{\overset{O-CH_3}{|}}{C}}-H + H_2O$

12.
$CH_3-CH_2-CH_2-CH_2-CH_2-CH_2-CH_2-CH_2-CH_2-CH_2-CH_2-OH$

16. (a) aldotetrose (b) ketopentose (c) ketohexose (d) aldopentose

18.
$$\begin{array}{c} CH_2OH \\ | \\ C=O \\ | \\ HO-C^*-H \\ | \\ H-C^*-OH \\ | \\ H-C^*-OH \\ | \\ CH_2OH \end{array}$$
D-fructose

30. Fructose can be used as a low-calorie sweetener because it is approximately 70% sweeter than table sugar (sucrose), and thus less fructose than table sugar is needed to achieve the same degree of sweetness in foods.

32.

α-D-mannose ⇌ D-mannose ⇌ β-D-mannose

34. (a) glucose (b) galactose and glucose (c) glucose and fructose
36. (a) disaccharide (b) α-1,1-linkages (c) no reaction
42. (a) glucose (b) glucose (c) glucose

CHAPTER 14

Exercises

14.1

	Common Name	IUPAC Name
(a)	γ-bromocaproic acid	4-bromohexanoic acid
(b)	butyric acid	butanoic acid
(c)	lactic acid	2-hydroxypropanoic acid
(d)	m-chlorobenzoic acid	3-chlorobenzoic acid
(e)	cyclohexanecarboxylic acid	cyclohexanecarboxylic acid

14.2
(a) $CH_3-CH_2-CO_2H(aq) + OH^-(aq) \longrightarrow CH_3-CH_2-CO_2^-(aq) + H_2O(l)$

(b) $C_6H_5-CO_2H(aq) + OH^-(aq) \longrightarrow C_6H_5-CO_2^-(aq) + H_2O(l)$

(c) $CH_3-CH_2-CH_2-CO_2H(aq) + NH_3(aq) \longrightarrow CH_3-CH_2-CH_2-CO_2^-(aq) + NH_4^+(aq)$

14.3
(a) $CH_3-CH_2-CO_2H + CH_3-OH \overset{H^+}{\rightleftharpoons} CH_3-CH_2-\underset{O-CH_3}{\overset{O}{C}} + H_2O$

(b) $CH_3-CO_2H + C_6H_5-OH \overset{H^+}{\rightleftharpoons} CH_3-C(=O)-O-C_6H_5 + H_2O$

(c) $C_6H_5-CO_2H + CH_3-\underset{CH_3}{CH}-OH \overset{H^+}{\rightleftharpoons} C_6H_5-\overset{O}{C}-O-\underset{CH_3}{CH}-CH_3 + H_2O$

14.4

	Common Name	IUPAC Name
(a)	ethyl acetate	ethyl ethanoate
(b)	phenyl propionate	phenyl propanoate
(c)	butyl benzoate	butyl benzoate

14.5
(a) $CH_3-CH_2-\overset{O}{C}-O-CH_2-CH_2-CH_2-CH_3$

(b) $CH_3-\overset{O}{C}-O-CH_2-CH_2-CH_3$

(c) $CH_3-CH_2-\underset{Br}{CH}-CH_2-\overset{O}{C}-O-CH_3$

(d) $C_6H_5-\overset{O}{C}-O-CH_2-CH_3$

14.6
(a) $CH_3-\underset{CH_3}{CH}-CO_2H + HO-C_6H_5$

(b) $CH_3-CH_2-CH_2-CO_2H + HO-CH_3$

(c) $C_6H_5-CO_2H + HO-C_6H_5$

(d) $CH_3-CO_2H + HO-CH_2-C_6H_5$

Study Questions and Problems

2.

	Common Name	IUPAC Name
(a)	butyric acid	butanoic acid
(b)	α,α-dimethylpropionic acid	2,2-dimethylpropanoic acid
(c)	p-bromobenzoic acid	4-bromobenzoic acid
(d)	α-phenylacetic acid	2-phenylethanoic acid
(e)	α-chloroacetic acid	2-chloroethanoic acid
(f)	lactic acid	2-hydroxypropanoic acid

8.

	Common Name	IUPAC Name
(a)	sodium acetate	sodium ethanoate
(b)	potassium propionate	potassium propanoate
(c)	sodium lactate	sodium 2-hydroxypropanoate
(d)	trisodium citrate	sodium 2-hydroxypropan-1,2,3-tricarboxylate

10. When the carboxylic acid comes into contact with aqueous base, the carboxylic acid is converted to its salt, which is more soluble in water than in the nonpolar solvent; thus, the carboxylic acid salt dissolves in the aqueous base, removing the carboxylic acid from the nonpolar solvent.

14.

(a) C$_6$H$_{11}$—CO$_2$H + CH$_3$—OH $\xrightleftharpoons[\Delta]{H^+}$

C$_6$H$_{11}$—C(=O)—O—CH$_3$ + H$_2$O

(b) CH$_3$—CO$_2$H + CH$_3$—CH$_2$—OH $\xrightleftharpoons[\Delta]{H^+}$

CH$_3$—C(=O)—O—CH$_2$—CH$_3$ + H$_2$O

(c) CH$_3$—CO$_2$H + C$_6$H$_5$—OH $\xrightleftharpoons[\Delta]{H^+}$

CH$_3$—C(=O)—O—C$_6$H$_5$ + H$_2$O

(d) H$_2$N—C$_6$H$_4$—CO$_2$H + CH$_3$—CH(CH$_3$)—CH$_2$—OH $\xrightleftharpoons[\Delta]{H^+}$

H$_2$N—C$_6$H$_4$—C(=O)—O—CH$_2$—CH(CH$_3$)—CH$_3$ + H$_2$O

(e) CH$_3$—CH=CH—CO$_2$H + C$_6$H$_5$—OH $\xrightleftharpoons[\Delta]{H^+}$

CH$_3$—CH=CH—C(=O)—O—C$_6$H$_5$ + H$_2$O

16.

(a) CH$_3$—C(=O)—O—CH$_2$—CH$_2$—CH$_2$—CH$_2$—CH$_2$—CH$_2$—CH$_3$

(b) C$_6$H$_5$—C(=O)—O—CH$_2$—CH$_2$—CH$_3$

(c) CH$_3$—C(=O)—O—CH$_2$—CH$_3$

(d) CH$_3$—CH$_2$—C(=O)—O—C$_6$H$_5$

(e) CH$_3$—CH(OH)—C(=O)—O—CH$_3$

18. In considering molecular weights and structures, the compounds show the following order of decreasing water solubility.

CH$_3$—CHO, CH$_3$—CO$_2$H, and CH$_3$—CH(OH)—CH$_2$—OH are nearly equally soluble;

CH$_3$—CH$_2$—C(=O)—CH$_3$;

CH$_3$—CH$_2$—CH$_2$—CH$_3$; CH$_3$—CH$_2$—O—CH$_2$—CH$_3$;

C$_6$H$_{11}$—OH; C$_6$H$_5$—CO$_2$—CH$_2$—CH$_3$.

20.

(a) CH$_3$—C(=O)—O—(CH$_2$)$_6$—CH$_3$ + H$_2$O $\xrightleftharpoons[\Delta]{H^+}$

CH$_3$—CO$_2$H + HO—(CH$_2$)$_6$—CH$_3$

(b) C$_6$H$_5$—C(=O)—O—CH$_2$—CH$_2$—CH$_3$ + H$_2$O $\xrightleftharpoons[\Delta]{H^+}$

C$_6$H$_5$—CO$_2$H + HO—CH$_2$—CH$_2$—CH$_3$

(c) CH$_3$—C(=O)—O—CH$_2$—CH$_3$ + H$_2$O $\xrightleftharpoons[\Delta]{H^+}$

CH$_3$—CO$_2$H + HO—CH$_2$—CH$_3$

(d) CH$_3$—CH$_2$—C(=O)—O—C$_6$H$_5$ + H$_2$O $\xrightleftharpoons[\Delta]{H^+}$

CH$_3$—CH$_2$—CO$_2$H + HO—C$_6$H$_5$

(e) CH$_3$—CH(OH)—C(=O)—O—CH$_3$ + H$_2$O $\xrightleftharpoons[\Delta]{H^+}$

CH$_3$—CH(OH)—CO$_2$H + HO—CH$_3$

22.

o-HO—C$_6$H$_4$—C(=O)—O—CH$_3$ + H$_2$O $\xrightleftharpoons[]{H^+}$

o-HO—C$_6$H$_4$—CO$_2$H + HO—CH$_3$

A-22 APPENDIX D

32.

H₂C—O—C(=O)—(CH₂)₁₆—CH₃
|
HC—O—C(=O)—(CH₂)₁₆—CH₃ + 3 H₂O →(lipase)
|
H₂C—O—C(=O)—(CH₂)₁₆—CH₃

H₂C—OH
|
HC—OH + 3 CH₃—(CH₂)₁₆—CO₂H
|
H₂C—OH

CHAPTER 15

Exercises

15.1
(a) primary (b) secondary (c) secondary (d) tertiary

15.2

	Common Name	IUPAC Name
(a)	diethylamine	N-ethylaminoethane
(b)	o-hydroxyaniline or 2-hydroxyaniline	1-amino-2-hydroxybenzene
(c)	aminoacetic acid	aminoethanoic acid
(d)	1,3-propanediamine	1,3-diaminopropane
(e)	—	4-(N,N-dimethylamino)-pentanal

15.3
(a) CH₃—CH(+NH₃)—CH₃ HSO₄⁻ (aq)

(b) CH₃—N⁺(CH₃)(CH₂—CH₃)—CH₃ Cl⁻

(c) CH₃—C(=O)—NH—CH₃ + H₂O

(d) C₆H₅—C(=O)—NH₂ + H₂O

15.4
Epinephrine and *norepinephrine* are hormones secreted by the adrenal gland that raise blood pressure, increase alertness, and serve as neurotransmitters. The *amphetamines* are synthetic drugs that stimulate the central nervous system. *p-Aminobenzoic acid* is required by several types of microorganisms to synthesize the vitamin folic acid. *Esters* of *p-aminobenzoic acid* act as local anesthetics. *Lidocaine* is a local anesthetic. (See structures given in chapter 15.)

15.5
(a) CH₂(+NH₃)—CO₂H, CH₂(+NH₃)—CO₂⁻, CH₂(NH₂)—CO₂⁻

(b) C₆H₅—CH₂—CH(+NH₃)—CO₂H, C₆H₅—CH₂—CH(+NH₃)—CO₂⁻,
    C₆H₅—CH₂—CH(NH₂)—CO₂⁻

(c) H₂N—C(+NH₂)=NH—(CH₂)₃—CH(+NH₃)—CO₂H,

H₂N—C(+NH₂)=NH—(CH₂)₃—CH(+NH₃)—CO₂⁻,

H₂N—C(+NH₂)=NH—(CH₂)₃—CH(NH₂)—CO₂⁻,

H₂N—C(NH₂)=NH—(CH₂)₃—CH(NH₂)—CO₂⁻

15.6

	Common Name	IUPAC Name
(a)	butyramide	butanamide
(b)	N-ethylformamide	N-ethylmethanamide
(c)	N,N-dimethylacetamide	N,N-dimethylethanamide

15.7

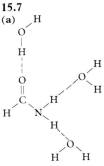

(a), (b), (c) hydrogen bonding diagrams for acetamide, N-methylbenzamide, and N,N-dimethylacetamide with water.

15.8
(a) C₆H₅—CO₂⁻ + NH₄⁺

(b) CH₃—CO₂⁻ + CH₃—CH₂—NH₃⁺

(c) C₆H₅—CO₂⁻ + CH₃—CH₂—NH₂⁺—CH₃

Study Questions and Problems

2. (a) tertiary (b) primary (c) primary (d) secondary
(e) secondary (f) tertiary (g) tertiary (h) secondary

4.
(a) CH₃—N(CH₃)—CH₂—CH₃   (b) H₂N—CH₂—CH₂—OH

(c) CH₃—CH(NH₂)—CH₂—CO₂H   (d) CH₃—NH—CH₂—CO₂H

(e) C₆H₄(Cl)—NH₂ (chloro ortho to NH₂)

(f) CH₃—CH(NH₂)—CH₂—CH(CH₃)—CH₂—CH₂—CH₃

SOLUTIONS AND ANSWERS FOR CHAPTER 16   A-23

10. Diethylamine is a weak base, and thus its aqueous solutions turn red litmus blue.

12.
(a) CH₃—CH—C(=O)—NH—CH₃ + H₂O
          |
          CH₃

(b) C₆H₅—⁺NH₃Cl⁻

(c) C₆H₅—C(=O)—NH₂ + H₂O

(d) C₆H₅—CH₂—⁺NH₂—CH₃ Br⁻

(e) CH₃—C(=O)—NH—CH₃ + H₂O

(f) CH₃—CH₂—C(=O)—NH—C₆H₅ + HCl

18. Acidic side chains and basic side chains, neutral polar side chains, nonpolar side chains.

20.
$H_3\overset{+}{N}$—CH₂—CO₂⁻ + OH⁻ ⟶ H₂N—CH₂—CO₂⁻ + H₂O
$H_3\overset{+}{N}$—CH₂—CO₂⁻ + H⁺ ⟶ $H_3\overset{+}{N}$—CH₂—CO₂H

22. (a) 0 (b) 0 (c) 0 (d) − (e) +

30.
	Common Name	IUPAC Name
(a)	N,N-dimethylformamide	N,N-dimethylmethanamide
(b)	propionamide	propanamide
(c)	N-methylbenzamide	N-methylbenzamide
(d)	benzamide	benzamide
(e)	N-ethyl-N-propylacetamide	N-ethyl-N-propylethanamide

32.
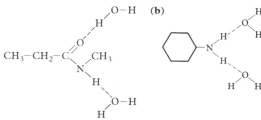

(e) piperidine N—H⋯O(H)H

CHAPTER 16

Exercises

16.1

$H_3\overset{+}{N}$—CH—C(=O)—NH—CH₂—CO₂⁻ , $H_3\overset{+}{N}$—CH₂—C(=O)—NH—CH—CO₂⁻
             | |
             CH₂—C₆H₅   CH₂—C₆H₅

16.2

$H_3\overset{+}{N}$—CH—C(=O)—NH—CH—C(=O)—NH—CH—C(=O)—N—CH—CO₂⁻
         | | | (proline ring)
         (CH₂)₂   (CH₂)₄   CH₂
         | | |
         CO₂⁻   NH₃⁺   C₆H₄—OH

glutamic acid, lysine, tyrosine, proline

16.3
(a) collagen triple helix (b) α-helix (c) β-pleated sheet

16.4
(a) An ionic attraction will exist.

R—NH₃⁺ - - - ⁻O₂C—R
lysine   glutamic acid

(b) A hydrophobic attraction will exist.

R—CH—CH₂—CH₃
    |
    CH₃
isoleucine   tryptophan

(c) A hydrogen bond will form.

R—O—H⋯O⁻—C(=O)—R
serine   aspartic acid

(d) A hydrophobic attraction will exist.

R—CH—CH₃   C₆H₅—CH₂—R
    |
    CH₃
valine   phenylalanine

Study Questions and Problems

4.
(a) $H_3\overset{+}{N}$—CH₂—C(=O)—NH—CH—CO₂⁻ ,
                                      |
                                      CH₂OH

$\text{H}_3\overset{+}{\text{N}}-\text{CH}-\overset{\text{O}}{\overset{\|}{\text{C}}}-\text{NH}-\text{CH}_2-\text{CO}_2^-$
$\quad\quad\quad |$
$\quad\quad\text{CH}_2\text{OH}$

(b) $\text{H}_3\overset{+}{\text{N}}-\text{CH}-\overset{\text{O}}{\overset{\|}{\text{C}}}-\text{NH}-\text{CH}-\overset{\text{O}}{\overset{\|}{\text{C}}}-\text{NH}-\text{CH}-\text{CO}_2^-$,
with side chains CH_3, $\text{CH}_2\text{CH}(\text{CH}_3)\text{CH}_3$, $(\text{CH}_2)_4\text{NH}_3^+$

$\text{H}_3\overset{+}{\text{N}}-\text{CH}-\overset{\text{O}}{\overset{\|}{\text{C}}}-\text{NH}-\text{CH}-\overset{\text{O}}{\overset{\|}{\text{C}}}-\text{NH}-\text{CH}-\text{CO}_2^-$,
with side chains CH_3, $(\text{CH}_2)_4\text{NH}_3^+$, $\text{CH}_2\text{CH}(\text{CH}_3)\text{CH}_3$

$\text{H}_3\overset{+}{\text{N}}-\text{CH}-\overset{\text{O}}{\overset{\|}{\text{C}}}-\text{NH}-\text{CH}-\overset{\text{O}}{\overset{\|}{\text{C}}}-\text{NH}-\text{CH}-\text{CO}_2^-$,
with side chains $\text{CH}_2\text{CH}(\text{CH}_3)\text{CH}_3$, CH_3, $(\text{CH}_2)_4\text{NH}_3^+$

$\text{H}_3\overset{+}{\text{N}}-\text{CH}-\overset{\text{O}}{\overset{\|}{\text{C}}}-\text{NH}-\text{CH}-\overset{\text{O}}{\overset{\|}{\text{C}}}-\text{NH}-\text{CH}-\text{CO}_2^-$,
with side chains $\text{CH}_2\text{CH}(\text{CH}_3)\text{CH}_3$, $(\text{CH}_2)_4\text{NH}_3^+$, CH_3

$\text{H}_3\overset{+}{\text{N}}-\text{CH}-\overset{\text{O}}{\overset{\|}{\text{C}}}-\text{NH}-\text{CH}-\overset{\text{O}}{\overset{\|}{\text{C}}}-\text{NH}-\text{CH}-\text{CO}_2^-$,
with side chains $(\text{CH}_2)_4\text{NH}_3^+$, CH_3, $\text{CH}_2\text{CH}(\text{CH}_3)\text{CH}_3$

$\text{H}_3\overset{+}{\text{N}}-\text{CH}-\overset{\text{O}}{\overset{\|}{\text{C}}}-\text{NH}-\text{CH}-\overset{\text{O}}{\overset{\|}{\text{C}}}-\text{NH}-\text{CH}-\text{CO}_2^-$,
with side chains $(\text{CH}_2)_4\text{NH}_3^+$, $\text{CH}_2\text{CH}(\text{CH}_3)\text{CH}_3$, CH_3

(c) $\text{H}_3\overset{+}{\text{N}}-\text{CH}-\overset{\text{O}}{\overset{\|}{\text{C}}}-\text{NH}-\text{CH}-\text{CO}_2^-$,
with side chains $(\text{CH}_2)_3\text{NH}-\text{C}(=\text{NH}_2^+)\text{NH}_2$, $\text{CH}_2\text{C}(=\text{O})\text{NH}_2$

$\text{H}_3\overset{+}{\text{N}}-\text{CH}-\overset{\text{O}}{\overset{\|}{\text{C}}}-\text{NH}-\text{CH}-\text{CO}_2^-$,
with side chains $\text{CH}_2\text{C}(=\text{O})\text{NH}_2$, $(\text{CH}_2)_3\text{NH}-\text{C}(=\text{NH}_2^+)\text{NH}_2$

(d) $\text{H}_3\overset{+}{\text{N}}-\text{CH}-\overset{\text{O}}{\overset{\|}{\text{C}}}-\text{N}-\text{CH}-\overset{\text{O}}{\overset{\|}{\text{C}}}-\text{NH}-\text{CH}-\text{CO}_2^-$,
(Glu)–(Pro)–(Phe)

$\text{H}_3\overset{+}{\text{N}}-\text{CH}-\overset{\text{O}}{\overset{\|}{\text{C}}}-\text{NH}-\text{CH}-\overset{\text{O}}{\overset{\|}{\text{C}}}-\text{N}-\text{CH}-\text{CO}_2^-$,
(Glu)–(Phe)–(Pro)

$\text{H}_2\overset{+}{\text{N}}-\text{CH}-\overset{\text{O}}{\overset{\|}{\text{C}}}-\text{NH}-\text{CH}-\overset{\text{O}}{\overset{\|}{\text{C}}}-\text{NH}-\text{CH}-\text{CO}_2^-$,
(Pro)–(Glu)–(Phe)

$\text{H}_2\overset{+}{\text{N}}-\text{CH}-\overset{\text{O}}{\overset{\|}{\text{C}}}-\text{NH}-\text{CH}-\overset{\text{O}}{\overset{\|}{\text{C}}}-\text{NH}-\text{CH}-\text{CO}_2^-$,
(Pro)–(Phe)–(Glu)

$\text{H}_3\overset{+}{\text{N}}-\text{CH}-\overset{\text{O}}{\overset{\|}{\text{C}}}-\text{NH}-\text{CH}-\overset{\text{O}}{\overset{\|}{\text{C}}}-\text{N}-\text{CH}-\text{CO}_2^-$,
(Phe)–(Glu)–(Pro)

$\text{H}_3\overset{+}{\text{N}}-\text{CH}-\overset{\text{O}}{\overset{\|}{\text{C}}}-\text{N}-\text{CH}-\overset{\text{O}}{\overset{\|}{\text{C}}}-\text{NH}-\text{CH}-\text{CO}_2^-$,
(Phe)–(Pro)–(Glu)

8. α-Keratin can be stretched when heated by steam because heat energy from the steam ruptures the hydrogen bonds that hold together the α-helix of α-keratin; this allows α-keratin to be stretched into the more extended β-pleated sheet structure.

10. (a) The small size of the glycine residues repeating at regular intervals allows the chains to intertwine to form a triple helix. **(b)** Hydrogen bonds are the principal forces holding the triple helix together.
12. (a) α-helix **(b)** collagen triple helix **(c)** α-helix **(d)** β-pleated sheet **(e)** α-helix **(f)** α-helix and collagen triple helix
14.
(a) A hydrogen bond will form.

glutamic acid

(b) An ionic attraction will exist.

$$R-NH-C\begin{matrix}NH_2^+ --- O^- \\ NH_2 \end{matrix}\begin{matrix} \\ O \end{matrix}C-R$$

arginine   aspartic acid

(c) A hydrophobic attraction will exist.

leucine   isoleucine

(d) An ionic attraction will exist.

$$R-NH_3^+ --- O^- \\ O=C-R$$

lysine

               aspartic acid

(e) A hydrophobic attraction will exist.

R—CH$_2$—⟨C$_6$H$_5$⟩   CH$_3$—CH—R
                                  |
                                 CH$_3$
phenylalanine   valine

30. HbS can be separated from HbA by electrophoresis because HbS has a valine residue instead of a glutamic acid residue at position 6 in each β-chain; thus, at neutral pH, HbS has a greater positive charge per molecule than HbA and moves faster toward the cathode.

CHAPTER 17

Exercises

17.1

	Class	Recommended Name
(a)	transferase	glutamate-oxaloacetate transaminase
(b)	oxido-reductase	malate dehydrogenase
(c)	hydrolase	protease
(d)	lyase	alanine decarboxylase

17.2
(a) decrease in enzyme activity   **(b)** increase in enzyme activity   **(c)** decrease in enzyme activity   **(d)** decrease in enzyme activity   **(e)** variable effect, depending on the pH

optimum for each particular enzyme; extreme pH values usually denature enzymes.

17.3
(a) folic acid   **(b)** lipoic acid   **(c)** riboflavin   **(d)** niacin   **(e)** vitamin B$_{12}$

Study Questions and Problems

6. Enzymes exhibit specificity in both the type of reaction to be catalyzed and the particular substrate to be acted upon.

8. The equation represents the reversible binding of substrate (S) to enzyme (E) in an enzyme-catalyzed reaction to form an intermediate, the Michaelis complex (ES), which is then converted to free enzyme (E) and product (P).

10. The zinc ion exerts an electrical attraction for the carbonyl oxygen of the substrate; the attraction probably imposes strain on the substrate, making it more susceptible to hydrolysis.

12. (a) Cofactors are necessary for the activity of some enzymes; thus, the amount of cofactor available determines the extent of enzyme activity. **(b)** Enzyme activity increases with substrate concentration in a nonlinear fashion until the enzyme is saturated with substrate; under these conditions, the enzyme shows maximum activity. **(c)** Enzyme activity is directly proportional to enzyme concentration. **(d)** Enzyme activity increases with temperature in a nonlinear fashion until a temperature is reached that denatures the enzyme; then enzyme activity drops. **(e)** Every enzyme has a pH optimum, and its activity is lower at other pH values; in general, most enzymes are denatured by extreme pH values and thus show no activity under these conditions.

14. The pH optimum of pepsin is near pH 2, corresponding roughly to the pH of gastric juice. Since the pH of intestinal juice is quite different from pepsin's pH optimum, pepsin is not active in the intestine.

16. $NAD^+ + H^- \rightleftharpoons NADH$
$FAD + H^+ + H^- \rightleftharpoons FADH_2$

22. Paint that contains lead is poisonous because lead ions inactivate enzymes by forming covalent bonds with essential thiol groups, thus removing the catalysts needed for metabolism and other essential functions, such as nerve transmission.

CHAPTER 18

Exercises

18.1
(a) salivary amylase, mucins, inorganic ions   **(b)** hydrochloric acid, pepsin   **(c)** bicarbonate ions and enzymes, some initially present as zymogens: trypsin, chymotrypsin, elastase, carboxypeptidase, phospholipase, lipase, pancreatic amylase, deoxyribonuclease, ribonuclease   **(d)** digestive enzymes: enteropeptidase, aminopeptidases, dipeptidases, maltase, sucrase, lactase, intestinal lipase   **(e)** inorganic salts, mucins, bile salts, bile pigments

18.2
In the lungs, hemoglobin binds O_2 and liberates CO_2 and H^+. The reverse process occurs in actively metabolizing tissue, where hemoglobin unloads O_2 and forms bonds to CO_2 and H^+.

18.3
Factor X_a is a protease which catalyzes the formation of thrombin from prothrombin. Thrombin then catalyzes formation of

Study Questions and Problems

2. Chewing grinds up food, thus increasing its surface area and its exposure to the action of digestive enzymes.

8. When chyme enters the duodenum, the small intestine releases the polypeptide hormone enterocrinin, which stimulates the intestine to release intestinal juice.

10. The bile salts aid the function of pancreatic lipase by emulsifying lipids; they also help to keep cholesterol in solution and aid in the intestinal absorption of fat-soluble vitamins.

14.
(a) $R-CO_2H$,

$$H_2C-OH$$
$$|$$
$$HC-O-\overset{O}{\underset{||}{C}}-R$$
$$|$$
$$H_2C-O-\overset{O}{\underset{||}{C}}-R,$$

$$H_2C-OH$$
$$|$$
$$HC-O-\overset{O}{\underset{||}{C}}-R$$
$$|$$
$$H_2C-OH$$

where R represents the hydrocarbon chain of any fatty acid.

(b)

$$\text{(glucose pyranose structure with CH}_2\text{OH, OH, HO, OH, OH)}$$

(plus other monosaccharides)

(c) $R-\underset{\underset{NH_3^+}{|}}{CH}-CO_2^-$, where R represents a side chain of any amino acid.

22. The low P_{50} of myoglobin relative to that of hemoglobin means that myoglobin has a greater oxygen affinity than hemoglobin. The higher oxygen affinity of myoglobin allows for efficient transfer of oxygen from hemoglobin to myoglobin for use in the tissues.

26. Lowered blood pH causes hemoglobin to bind carbon dioxide more tightly; thus, carbon dioxide transport is enhanced.

32. Vitamin K is needed by the liver for synthesis of prothrombin. If vitamin K transport to the liver is inhibited, prothrombin synthesis will also be inhibited, impairing the body's ability to form blood clots.

34.

(a) $P-NH-\underset{\underset{R}{|}}{CH}-\overset{O}{\underset{||}{C}}-NH-P' + H_2O \longrightarrow$

$P-NH-\underset{\underset{R}{|}}{CH}-CO_2^- + H_3\overset{+}{N}-P'$

where P and P′ represent the rest of the protein molecule and R represents the side chain of an aromatic amino acid, leucine, methionine, aspartic acid, or glutamic acid.

(b) $P-NH-\underset{\underset{R}{|}}{CH}-\overset{O}{\underset{||}{C}}-NH-P' + H_2O \longrightarrow$

$P-NH-\underset{\underset{R}{|}}{CH}-CO_2^- + H_3\overset{+}{N}-P'$

where P and P′ represent the rest of the protein molecule and R represents the side chain of lysine or arginine.

(c) $P-\overset{O}{\underset{||}{C}}-NH-\underset{\underset{R}{|}}{CH}-CO_2^- + H_2O \longrightarrow$

$P-CO_2^- + H_3\overset{+}{N}-\underset{\underset{R}{|}}{CH}-CO_2^-$

where P represents the rest of the protein molecule and R represents any amino acid side chain.

(d) $P-NH-\underset{\underset{R}{|}}{CH}-\overset{O}{\underset{||}{C}}-NH-CH_2-\overset{O}{\underset{||}{C}}-NH-P' + H_2O \longrightarrow$

$P-NH-\underset{\underset{R}{|}}{CH}-CO_2^- + H_3\overset{+}{N}-CH_2-\overset{O}{\underset{||}{C}}-NH-P'$

where P and P′ represent the rest of the protein molecule and R represents an arginine side chain.

(e)

(disaccharide hydrolysis reaction showing two pyranose/furanose sugar rings + H_2O → two separate sugar rings)

(f)

$$H_2C-O-\overset{O}{\underset{||}{C}}-R$$
$$|$$
$$HC-O-\overset{O}{\underset{||}{C}}-R + H_2O \longrightarrow$$
$$|$$
$$H_2C-O-\overset{O}{\underset{||}{C}}-R$$

$$H_2C-OH$$
$$|$$
$$HC-O-\overset{O}{\underset{||}{C}}-R + R-CO_2H$$
$$|$$
$$H_2C-O-\overset{O}{\underset{||}{C}}-R$$

$$H_2C-OH$$
$$|$$
$$HC-O-\overset{O}{\underset{||}{C}}-R + H_2O \longrightarrow$$
$$|$$
$$H_2C-O-\overset{O}{\underset{||}{C}}-R$$

$$H_2C-OH$$
$$|$$
$$HC-O-\overset{O}{\underset{||}{C}}-R + R-CO_2H$$
$$|$$
$$H_2C-OH$$

where R represents the hydrocarbon chain of any fatty acid.

(g) $H_3\overset{+}{N}-\underset{\underset{R}{|}}{CH}-\overset{O}{\underset{||}{C}}-NH-\underset{\underset{R'}{|}}{CH}-CO_2^- + H_2O \longrightarrow$

$H_3\overset{+}{N}-\underset{\underset{R}{|}}{CH}-CO_2^- + H_3\overset{+}{N}-\underset{\underset{R'}{|}}{CH}-CO_2^-$

where R and R′ represent side chains of any amino acids.

(h) [structural formulas of sugars] + H$_2$O →

38. The body responds by releasing aldosterone from the adrenal gland. Aldosterone acts on the kidneys to cause them to retain Na$^+$ and some water.

40. The body responds by releasing the enzyme renin from the kidneys; renin catalyzes formation of angiotensin II, a hormone that causes constriction of blood vessels and thus raises blood pressure.

CHAPTER 19

Study Questions and Problems

4. (a) duplication of DNA and transcription of the genetic code of DNA into RNA in prepartation for protein synthesis (b) oxidation of derivatives of carbohydrates, fats, and amino acids, and synthesis of ATP (c) transport of substances to and from cell nucleus, and synthesis of proteins (d) degradation of cellular substances

8. Due to hydrophobic attractions, inhaled hydrocarbon vapors dissolve to some extent in lung tissue membranes, disrupting membrane structure and damaging cells.

10. Salt solutions dislodge peripheral proteins by disrupting the ionic attractions and hydrogen bonds that normally keep the peripheral proteins attached to membranes.

12. Membrane proteins stabilize membrane structure, participate in transport processes, catayze reactions, and participate in intercellualr recognition.

14. The fluid mosaic model suggests that membranes are fluid solutions of globular proteins and lipids; the lipid bilayer serves as both a permeability barrier and a solvent for integral proteins, and the proteins are distributed through the membrane in a mosaic pattern.

22. The products of digestion of the three major types of food are degraded to acetyl units attached to coenzyme A; acetyl CoA then serves as fuel for the citric acid cycle.

24. NADH and FADH$_2$ are reoxidized by the electron transport system, located within the inner membrane of mitochondria.

26. In the cytochromes, iron cycles back and forth between the +2 and +3 oxidation states; in hemoglobin, iron is always in the +2 oxidation state.

28. Cyanide is highly toxic because cyanide ions form coordinate covalent bonds to Fe^{3+} in cytochrome oxidase, thus blocking the final step of the electron transport system.

34. 72 ATP

36. The enzyme catalyzing step 1 of the cycle is inhibited by ATP and NADH and activated by ADP. The enzyme catalyzing step 4 is inhibited by NADH and activated by ADP. The enzyme catalyzing step 6 is inhibited by succinyl CoA, NADH, and ATP. Thus, high energy levels in the cell result in high ATP, NADH, and succinyl CoA levels that inhibit the citric acid cycle, and low energy levels result in high ADP levels that activate the citric acid cycle.

CHAPTER 20

Study Questions and Problems

2. During digestion, carbohydrates are degraded to glucose, fructose, and galactose. After absorption of these simple sugars, fructose and galactose are converted to glucose or other compounds that are metabolized by the same pathway as glucose.

8. They are usually separate systems so that each pathway, with its own energy requirements, can be controlled separately.

10. Glycogenolysis is restricted to the liver, kidney, and intestine because only these cells contain glucose-6-phosphatase, the enzyme that catalyzes formation of glucose from glucose-6-phosphate.

14. The coenzyme is NAD$^+$; its vitamin source is niacin.

16. Fructose can be converted enzymatically to fructose-6-phosphate.

fructose + ATP ⟶ fructose-6-phosphate + ADP + H$^+$

18. Reduction of pyruvate to lactate regenerates NAD$^+$ needed for glycolysis.

20. Reduction of acetaldehyde to ethanol regenerates NAD$^+$ needed for glycolysis.

22.
Cofactor	Vitamin Source
coenzyme A	pantothenic acid
NAD$^+$	niacin
thiamin pyrophosphate	thiamin
lipoic acid	lipoic acid
FAD	riboflavin

24. The reducing power of cytoplasmic NADH is transferred to mitochondria by the glycerol phosphate shuttle, in which a hydride ion from NADH is transported across the mitochondrial membrane by glycerol-3-phosphate. Once inside the mitochondrion, glycerol-3-phosphate is oxidized to dihydroxyacetone phosphate and FAD is reduced to FADH$_2$. Since each FADH$_2$ molecule is ultimately responsible for production of 2 ATP molecules, the net cost of the shuttle system is one ATP per hydride ion transported.

30. The primary control point is step 3, the conversion of fructose-6-phosphate to fructose-1,6-diphosphate. It is responsive to cellular needs in that the enzyme, phosphofructokinase, is activated by ADP and AMP and inhibited by ATP and citrate. Thus, when the energy level of the cell is low, glycolysis is stimulated, and when the energy level of the cell is high, glycolysis is inhibited.

32. Their actions are similar in that both stimulate glycogen breakdown and retard glycogen synthesis. Their actions are different from each other in that epinephrine acts primarily on muscle cells, while glucagon acts on liver cells.

34. It is important because adenyl cyclase catalyzes formation of cyclic AMP, and increased levels of cyclic AMP indirectly activate glycogen phosphorylase and indirectly inactivate glycogen synthesis.

38. Hexokinase deficiency results in lowered production of DPG, leading to increased oxygen affinity by hemoglobin.

40. Since metabolism of both alcohols utilizes the same enzyme, high levels of ethanol would complex a large amount of the enzyme, decreasing the amount of toxic formaldehyde formed over a period of time and allowing time for the kidneys to excrete methanol unchanged.

CHAPTER 21

Study Questions and Problems

6. Cloudy lymph signifies the presence of chylomicra, the tiny droplets composed of triacylglycerols and proteins; chylomicra appear shortly after eating a meal containing fat.

8. The lipoprotein complex is more soluble in water than the free lipids would be; thus, combination with blood proteins helps to solubilize the lipids.

12. (a) Eskimos eat a lot of animal fat and thus would be expected to have high levels of blood lipids. The fact that their high density lipoprotein levels are elevated instead of their very low density lipoprotein levels suggests that their diets contain an abundance of short-chain fatty acids, which would not appear as lipoproteins in the blood. (b) No, Eskimos should not be especially susceptible to atherosclerosis, because this disease is correlated with high blood levels of very low density lipoproteins and low density lipoproteins.

16. The four hormones are epinephrine, norepinephrine, glucagon, and adrenocorticotropic hormone. All of these are "cyclic AMP" hormones, and the increased levels of cAMP in fat cells indirectly activate adipose-cell lipase.

24. 25 ATP

28. In uncontrolled diabetes, the utilization of glucose for energy production is impaired, and the body must rely on fatty acid oxidation for energy production to an abnormally large extent. In addition, oxaloacetate is used more than normally for gluconeogenesis, thus reducing the activity of the citric acid cycle. Under these circumstances, levels of acetyl CoA increase, and more and more acetyl CoA is converted to ketone bodies, including acetone. Exceedingly high levels of acetone in the blood lead to "acetone breath."

30. Acetyl units are transferred from the mitochondria to the cytoplasm by the citrate carrier system, in which acetyl CoA reacts with oxaloacetate to produce citrate. The citrate carrier system in the mitochondrial membrane transports citrate to the cytoplasm, where it is cleaved to regenerate acetyl CoA and oxaloacetate.

32. Malonyl-ACP serves to transfer acetyl units from acetyl CoA to growing fatty acid chains during fatty acid synthesis.

34. The preliminary reaction of fatty acid synthesis, formation of malonyl CoA, is catalyzed by an enzyme that is activated by citrate and inhibited by long-chain acyl CoA molecules, end products of fatty acid synthesis.

36. Fatty acids are obtained from diet and by internal synthesis, and glycerol is produced from dihydroxyacetone phosphate, a glycolytic intermediate.

40. In the liver, dietary cholesterol inhibits the synthesis of mevalonic acid and the cyclization of squalene to lanosterol; thus, dietary cholesterol inhibits cholesterol biosynthesis in the liver.

42. The product of glycolysis, pyruvate, can be converted to acetyl CoA by the pyruvate dehydrogenase system, and the acetyl units can then be used for fatty acid synthesis.

46. Essential amino acids cannot be synthesized at a rate adequate to meet metabolic needs, while nonessential amino acids can.

48.

$$\begin{array}{c} CO_2^- \\ | \\ CH-NH_3^+ \\ | \\ (CH_2)_2 \\ | \\ CO_2^- \end{array} + \begin{array}{c} CO_2^- \\ | \\ C=O \\ | \\ CH_2 \\ | \\ CO_2^- \end{array} \rightleftharpoons \begin{array}{c} CO_2^- \\ | \\ C=O \\ | \\ (CH_2)_2 \\ | \\ CO_2^- \end{array} + \begin{array}{c} CO_2^- \\ | \\ CH-NH_3^+ \\ | \\ CH_2 \\ | \\ CO_2^- \end{array}$$

50. α-Ketoglutarate can react with NH_4^+ to form glutamate in one of the few reactions that convert inorganic nitrogen to organic nitrogen. In addition, α-ketoglutarate serves to funnel amino groups from various amino acids into glutamate, which can then lose an amino group as NH_4^+ through oxidative deamination. In this way, excess nitrogen of amino acids is converted to NH_4^+ for processing through the urea cycle.

54. Glycogenic amino acids have carbon skeletons that are degraded to intermediates used in gluconeogenesis, while ketogenic amino acids have carbon skeletons that are degraded to acetyl CoA or acetoacetyl CoA, compounds leading to formation of ketone bodies.

CHAPTER 22

Exercises

22.1

[structural formula of a trinucleotide containing adenine, cytosine, thymine, and guanine linked by phosphodiester bonds]

22.2

5' T—G—A—T—C 3'
   | | | | |
3' A—C—T—A—G 5'

22.3

CAA AUA GUC GAU CGG

Gln-Ile-Val-Asp-Arg

22.4

67,000 amu/(120 amu/residue) = 560 residues
(560 residues)(3 nucleotides/residue) =
        1680 nucleotides in mRNA segment
nucleotides in gene = (2)(1680 nucleotides in mRNA segment)
        = 3360 nucleotides

22.5

Ribosomes are composed of each of the three types of RNA and 30–40 different proteins. Each ribosome has two subunits.

22.6

(a) insertion (b) substitution (c) deletion

Study Questions and Problems

12. The constant ratios of adenine/thymine and guanine/cytosine reflect the fact that each of these pairs is a hydrogen-bonded base pair.

14. Double-stranded DNA starts to separate and complementary deoxyribonucleoside triphosphates align themselves by base pairing on the separated DNA strands. Then DNA polymerases catalyze phosphodiester bond formation among the deoxyribonucleoside triphosphate units to form new DNA and release PP_i units.

16. There is a deficiency of certain DNA repair enzymes, and skin cells cannot repair DNA that has been damaged by ultraviolet light.

18. Double-stranded DNA separates in small segments, and complementary ribonucleoside triphosphates align themselves by base pairing on a single strand of DNA. Then DNA-directed RNA polymerase catalyzes phosphodiester bond formation among ribonucleoside triphosphates to form mRNA and release PP_i units.

20. (a) CUA GCA CCA
Leu-Ala-Pro

(b) AAG CAC UCG CUG
Lys-His-Ser-Leu

(c) UUG AUG GUG GCC AAA
Leu-Met-Val-Ala-Lys

22. For α-chain: 423 nucleotides in mRNA segment; 846 nucleotides in gene   For β-chain: 438 nucleotides in mRNA segment; 876 nucleotides in gene

24. (a) substitution (b) deletion (c) insertion

26. The alteration corresponds to substitution of U for A in the codon for glutamic acid or substitution of UU or UC for AA or AG in the codon for glutamic acid.

codons for glutamic acid: GAA GAG
codons for valine: GUA GUG GUU GUC

32. Interferons are small glycoproteins that are synthesized and secreted by animal cells in response to viral infections; they increase cellular resistance to viruses. Interferons are expected to be effective in cancer treatment because of their antiviral actions and because they stimulate immune cells that destroy cancer cells.

34. DNA fragments are obtained by use of restriction enzymes, and a DNA fragment is then covalently joined to a carrier, usually a plasmid fragment, by use of DNA ligases.

36. Using genetic engineering techniques, it is possible to grow large cultures of bacteria that have been programmed to produce large amounts of polypeptides and proteins normally produced in only very small quantities by animal cells.

CHAPTER 23

Exercises
23.1
(28 g)(4 kcal/g) = 100 kcal
23.2
(34 g)(9 kcal/g) = 300 kcal

Study Questions and Problems
2. Humans need dietary fiber because it helps to prevent or relieve constipation by absorbing water and softening the stool. Sources are fruits, vegetables, and grains.

4. 7000 kcal excess; 500 kcal/day reduction

6. The orange juice is more nutritious because it contains micronutrients not found in the root beer.

10. Plant proteins usually have low amino acid scores because they usually are deficient in one or more amino acids.

12. 50 g

14. Prevention of ketosis and control of cholesterol and saturated fat intake are two reasons for limiting fat intake.

18. 600 kcal   **20.** 24 kcal

22. Body water increases because increased fatty acid oxidation produces an abnormally large amount of body water.

24. The B-vitamins that provide redox coenzymes are niacin, riboflavin, and lipoic acid (a pseudo vitamin).

26. Vitamin B_6 provides interconvertible coenzymes that participate in amino acid decarboxylations and transminations.

28. Both of these vitamins provide coenzymes that participate in the synthesis of nucleic acids; deficiencies of these vitamins cause a decrease in formation of new cells, resulting in fewer red blood cells and thus a state of anemia.

30. Vitamin C is thought to participate in the addition of hydroxyl groups to proline residues after they have been incorporated into collagen. Humans are unable to synthesize vitamin C because they lack an enzyme needed for its synthesis.

38. Vitamin E is an antioxidant; it protects membrane structure by preventing oxidation of double bonds in fatty acid components of membranes.

40. (a) vitamin B_{12} (b) thiamin (c) vitamin A
(d) niacin (e) vitamin C (f) vitamin D

CHAPTER 24

Study Questions and Problems
6. Sulfur dioxide, an acid anhydride, forms undissociated sulfurous acid, the effective agent, at low pH.

8. Skimmed milk is easily prepared by skimming the fat from the top of whole milk that has been allowed to separate.

14. The bacteria convert lactose to lactic acid. The major metabolic pathway involved is glycolysis.

16. Butter is composed of milk fat, while margarine is composed of triacylglycerols from vegetable oils.

18. Cholesterol is found in butter but not in margarine because butter is an animal product and margarine is a vegetable product. Only animals synthesize cholesterol.

20. The higher iodine number of safflower oil means that it has more carbon-carbon double bonds than cottonseed oil. Highly unsaturated oils do not make good cooking oils because oxidation of their many double bonds under the high-temperature conditions of cooking produces an abundance of bad-tasting and bad-smelling aldehydes and carboxylic acids. More highly saturated oils do not form as many oxidation products and thus do not become rancid as easily as highly saturated oils do.

26. During digestion, aspartame is hydrolyzed to aspartic acid, phenylalanine, and methanol. Phenylalanine must be eliminated from the diet of those with phenylketonuria.

34. The use of sulfa drugs is restricted mainly to treatment of urinary infections because of the numerous side effects of the drugs and the fact that they are efficiently absorbed and excreted in urine.

36. The penicillin molecules become covalently attached to and thus inactivate an enzyme needed to synthesize bacterial cell walls. This action weakens the cell walls, and the cells burst.

CHAPTER 25

Exercises

25.1 Geiger counters, scintillation counters, and film badges can be used to detect radiation.

Study Questions and Problems

2. Tritium (H-3) could be formed by the bombardment of deuterium (H-2) with neutrons:
$${}^{2}_{1}H + {}^{1}_{0}n \longrightarrow {}^{3}_{1}H$$

4. ${}^{27}_{13}Al + {}^{4}_{2}\alpha \longrightarrow {}^{30}_{15}P + {}^{1}_{0}n$

${}^{30}_{15}P \longrightarrow {}^{30}_{14}Si + {}^{0}_{1}\beta$

8. 6.25%

12. A single large dose of radiation is more likely to damage cellular DNA so extensively that it cannot be repaired.

16. Calvin illuminated green algae in the presence of $^{14}CO_2$ for short periods and then analyzed the cells for the presence of C-14. Appearance of the C-14 label in metabolic intermediates and products allowed Calvin to explain the process of photosynthesis.

20. Damage to normal DNA produced by radiation can produce cancer due to loss of normal cellular control mechanisms, while damage to abnormal DNA, as exists in cancer, can result in death of the cancer cells.

22. Radioisotopes are introduced into the body by injection or ingestion, and as they become concentrated in body parts, the emitted radiation is detected by scanning with a scintillation camera.

26. A nuclear reaction can become a chain reaction if a sufficient number of the neutrons produced bombard unreacted fissionable material to cause a self-sustaining reaction.

28. ${}^{1}_{1}H + {}^{2}_{1}H \longrightarrow {}^{3}_{2}He$

30. The two major drawbacks are containment of the extremely hot mixture of components and achieving the extremely high temperatures under the controllable conditions needed for energy production.

GLOSSARY

Accuracy: the amount of error in a measurement.

Acetal: the compound formed by reaction of an alcohol molecule with a hemiacetal.

Acid: a substance that produces H^+ in water (Arrhenius), or a substance that can donate a proton to another substance (Bronsted-Lowry).

Acid anhydride: a nonmetallic oxide that reacts with water to form an acid.

Acid-base indicator: a substance that is itself an acid or base and that is one color in acidic solution and another color in basic solution.

Acid chloride: a carboxylic acid derivative having the general formula $R-CO-Cl$.

Acidosis: a condition in which blood pH is less than 7.35.

Activated complex: the unstable arrangement of atoms that exists in a reaction pathway and possesses the activation energy; also called the transition state.

Activation energy (E_a): the energy required to reach the energy maximum in a reaction pathway.

Active metals: the elements of groups IA and IIA.

Active site: the location on an enzyme where substrate is bound and catalysis occurs.

Active transport: a type of facilitated transport in which movement of transported substances occurs against a concentration gradient and which requires an input of free energy.

Acyl carrier protein (ACP): the protein to which acyl intermediates are bound during fatty acid synthesis.

Adrenal glands: a pair of endocrine glands located just above the kidneys.

Alchemy: an ancient art concerned principally with finding an elixir of life and discovering a method of transmuting base metals into gold.

Alcohol: an organic compound having one or more hydroxyl groups, $-OH$, bonded to alkyl groups.

Aldehyde: an organic compound containing the aldehyde group, $-CHO$.

Aldose: a monosaccharide containing an aldehyde group.

Aliphatic hydrocarbon: a hydrocarbon that is not aromatic.

Alkali metals: elements of group IA.

Alkaline earth metals: elements of group IIA.

Alkaloid: a member of a class of weakly basic organic compounds having rings containing nitrogen atoms.

Alkalosis: a condition in which blood pH is greater than 7.45.

Alkane: a hydrocarbon containing only single bonds; also called paraffins.

Alkene: a hydrocarbon having one or more double bonds; also called olefins.

Alkyl group: a saturated hydrocarbon substituent in a molecule.

Alkyl halide: a halogen derivative of an aliphatic hydrocarbon.

Alkyne: a hydrocarbon containing one or more triple bonds.

Allosteric enzyme: an enzyme involved in metabolic regulation that exhibits allosteric interactions.

Allosteric interaction: an interaction between spatially different sites on a protein molecule.

Allosteric protein: a protein having two or more spatially different sites that interact with small molecules or ions.

Alpha (α) decay: the emission of alpha particles by radioactive nuclei.

Alpha (α) helix: the right-handed helical structure that is the secondary structure of many proteins.

Alpha (α) particle: a particle of atomic number 2 and mass number 4 emitted by a radioactive nucleus; an alpha particle is represented by $^{4}_{2}\alpha$ and is the same as a helium nucleus.

Amide: an organic compound having the general formula $R-CO-NR'_2$, where R' may be hydrogen or one or two alkyl or aromatic groups.

Amine: an organic compound derived by replacing one or more of the hydrogen atoms of ammonia with alkyl or aromatic groups, as in RNH_2, R_2NH, and R_3N.

Amino acid: an organic compound containing an amino group and a carboxyl group.

Amino acid pool: the circulating supply of amino acids in the blood.

Amino acid residue: an amino acid incorporated into the structure of a protein.

Amorphous solid: a solid composed of disordered particles having no particular geometric pattern.

Amphoteric substance: any substance that can behave as both an acid and a base.

Anabolism: the collection of metabolic reactions that deal with synthesis of biological molecules.

Analgesic: a drug that relieves pain.

Anesthetic: a drug that produces loss of the senses of feeling, such as pain, heat, cold, and touch.

Anion: a negatively charged ion.

Anode: an electrode that loses electrons and takes on a positive charge.

Anomers: a pair of cyclic stereoisomers of monosaccharides whose members differ from each other by the orientation of the hydroxyl group formed by ring closure.

Antibiotic: a substance produced by one microorganism that kills or inhibits the growth of other microorganisms.

Antibodies: highly specific proteins that combine with viruses, bacteria, and other foreign substances that make their way into the blood or tissues; also called immunoglobulins.

Anticodon: a nucleotide triplet in tRNA that is complementary to one of the amino acid codons in mRNA.

Antimetabolite: a substance that interferes with the normal metabolism of an organism.

Antioxidant: a substance that retards or prevents oxidation of other substances.

Antipyretic agent: a drug that reduces fever.

Antiseptic: a substance used on or in the body to prevent the growth of infectious microorganisms.

Apoenzyme: a catalytically inactive protein formed by removal of the cofactor from a holoenzyme.

Apoprotein: a protein without its characteristic prosthetic group.

Aromatic hydrocarbon: a cyclic, unsaturated hydrocarbon having hybrid bonds intermediate between single and double bonds and possessing resonance.

Artificial radioactivity: radioactive decay induced by bombardment of atoms with subatomic particles.

Atom: the smallest unit of an element that has properties of the element.

Atomic hypothesis: the hypothesis that the smallest unit of an element is an atom.

Atom mass unit (amu): a relative unit used to express masses of atoms, ions, and molecules; an amu is 1/12 the mass of an atom of the carbon isotope of mass number 12.

Atomic nucleus: the dense core of an atom that contains principally protons and neutrons.

Atomic number: the number of protons (also the number of electrons) in an atom.

Atomic orbital: any one of several regions in an atom where an electron is most likely to be found.

Atomic weight: the average mass, expressed in amu, of all isotopes of an element as they occur in nature.

Average binding energy: the binding energy of a nucleus divided by the number of protons and neutrons in the nucleus.

Avidin: the protein in raw egg white that inactivates the vitamin biotin by binding tightly to it.

Avogadro's hypothesis: the hypothesis that equal volumes of any gases at the same temperature and pressure contain equal numbers of molecules.

Bacteriocidal drug: a drug that kills bacteria.

Bacteriostatic drug: a drug that inhibits the growth of bacteria, usually by interfering with protein synthesis.

Base: a substance that produces OH^- in water (Arrhenius), or a substance that accepts a proton from another substance (Bronsted-Lowry).

Basic anhydride: a metallic oxide that reacts with water to form a base.

Basic Four: the four basic food groups recommended by nutritionists to serve as a guide to food selection and daily meal planning.

Beta (β) decay: a type of nuclear decay involving electron emission, positron emission, or electron capture.

Beta (β) oxidation cycle: the metabolic pathway in which fatty acids are oxidized and converted to acetyl CoA units.

Beta (β) particle: an electron emitted by a radioactive nucleus, represented by $_{-1}^{0}\beta$.

Beta (β) pleated sheet: the extended, zigzag sheet structure that is the secondary structure of some proteins.

Bile: the fluid produced by the liver and stored in the gall bladder until needed in the intestine to participate in digestion.

Bile pigments: the hemoglobin degradation products bilirubin and biliverdin that give bile its yellow green color.

Bile salts: salts of glycocholic and taurocholic acids that are present in bile and that aid digestion by pancreatic lipase, help to keep cholesterol in solution, and aid in the intestinal absorption of fat-soluble vitamins.

Biomolecular reaction: a reaction in which two reactant molecules are needed to form the activated complex.

Binary compound: a compound composed of two elements.

Bioenergetics: the study of the production and use of energy by living organisms.

Biologic value: the nutritional value of a protein, based on the protein's amino acid content and its digestibility.

Biological cell: a functional unit of biological activity surrounded by a membrane and capable of reproduction.

Blood gases: the gases (CO_2 and O_2) transported by blood.

Blood plasma: the liquid portion of blood.

Blood serum: the liquid left after blood clots.

Boiling point: the temperature at which the vapor pressure of a liquid is equal to atmospheric pressure.

Boyle's law: the pressure and volume of a fixed sample of any gas are inversely proportional at constant temperature.

Brownian motion: the constant, random motion of colloidal particles.

Buffer: a solution that is able to resist pH change when small amounts of acid or base are added.

Calorie (cal): the quantity of energy that will raise the temperature of 1 gram of water from 14.5°C to 15.5°C.

Capillary action: the rise of a liquid within small-diameter tubes.

Carbohydrate: an organic compound containing one or more hydroxyl groups and either an aldehyde or a ketone group.

Carbonyl group: the functional group of aldehydes and ketones, $-CO-$.

Carboxyl group: the functional group of carboxylic acids, $-CO_2H$.

Carboxylic acid: an organic acid containing the carboxyl group.

Carcinogen: any substance or agent that causes cancer.

Catabolism: the collection of metabolic reactions that deal with the degradation of biological molecules.

Catalyst: a substance that affects the rate of a reaction by its presence in the reaction mixture but does not undergo permanent chemical change itself; most catalysts increase reaction rates.

Catecholamine: any of the biologically active amino derivatives of catechol.

Cathode: an electrode that gains electrons and takes on a negative electrical charge.

Cation: a positively charged ion.

Cell membrane: the lipid bilayer that surrounds a biological cell and contains some proteins.

Cell nucleus: the central feature of a biological cell, where DNA is duplicated and the genetic code is transcribed.

Celsius temperature scale: the temperature scale that sets the freezing point of water at 0° and the boiling point of water at 100°; formerly called the centigrade scale.

Charles's law: the volume of a fixed sample of any gas at constant pressure varies directly with its absolute temperature; alternately, the pressure of a fixed sample of gas at constant volume varies directly with its absolute temperature.

Chemical bond: an attractive force that joins two atoms in a compound; the two types of chemical bonds are covalent and ionic.

Chemical equation: a written description of chemical change showing formulas for all reactants and products.

Chemical family: a vertical group of elements in the periodic table.

Chemical period: a horizontal row of elements in the periodic table.

Chemical reaction: a transformation of one or more substances; a change in the composition of matter.

Chirality: the quality possessed by an organic compound having an asymmetric carbon atom.

Chromosomes: threadlike bodies found in a cell nucleus that carry the genes in a linear order and are composed of DNA and proteins.

Chylomicra: tiny droplets suspended in lymph fluid and composed of triacylglycerols and small amounts of protein.

Chyme: the semifluid food mass that leaves the stomach and passes into the small intestine.

Codon: a nucleotide triplet in mRNA that specifies one amino acid in the sequence of a finished polypeptide synthesized by an organism.

Coenzyme: an enzyme cosubstrate that is regenerated in other enzymatic reactions; coenzymes are derived from vitamins.

Cofactor: a nonprotein molecule or ion required by an enzyme for catalytic activity.

Collagen triple helix: the triple helical arrangement of protein strands that constitutes the secondary structure of collagen.

Colligative properties: those properties of solutions that depend only on the number of dissolved solute particles and are independent of solute identity; colligative properties include vapor-pressure lowering, boiling-point elevation, freezing-point lowering, and osmotic pressure.

Colloid: a homogeneous mixture in which relatively large particles are dispersed in a solid, liquid, or gas.

Combination reaction: a chemical reaction in which two or more reacting ions, atoms, or molecules combine to form one product.

Combined gas equation: the equation that describes the behavior of ideal gases, $PV = nRT$; also called the ideal gas equation.

Combustion: the process of burning.

Common pathway: the pathway responsible for the final stages of blood clot formation; the intrinsic and extrinsic pathways merge to form the common pathway.

Competitive inhibitor: an enzyme inhibitor that competes with substrate for binding at the active site of the enzyme.

Complete ionic equation: an equation that shows all ions and molecules in a reaction mixture.

Complex ion: an ion composed of a metal ion and two or more other ions or groups attached to the metal ion by coordinate covalent bonds.

Complex lipid: a lipid containing one or more long-chain carboxylic acids joined to alcohol groups by ester linkages.

Compound: a pure substance composed of two or more elements and having constant composition.

Concentration: the amount of solute in a given quantity of solution.

Concentration gradient: a graded difference in concentration.

Configuration: the arrangement of atoms about an asymmetric carbon atom.

Conformation: the shape of a flexible molecule.

Conjugate acid: the acid formed when a base accepts a proton.

Conjugate base: the base formed when an acid loses a proton.

Conjugated proteins: proteins that have prosthetic groups.

Cooperative effect: the cooperation among spatially different sites on a protein molecule to bring about a biological effect.

Coordinate covalent bond: a covalent bond in which both bonding electrons are contributed by one atom.

Cori cycle: the cycle by which the liver provides glucose for use by other tissues, especially muscle, and the lactate formed from glucose is transported by blood back to the liver, where it can be used for making glucose again.

Cosmic rays: the radiation consisting of particles streaming into the earth's atmosphere from the sun and outer space.

Covalent bond: a chemical bond composed of a shared pair of electrons.

Covalent crystal: a network of atoms in a crystal lattice held together by covalent bonds.

Crenation: the collapse of red blood cells in a hypertonic solution.

Critical mass: the smallest amount of fissionable isotope needed to produce a self-sustaining nuclear fission chain reaction.

Crystal: a bit of solid matter having a regular geometric shape and composed of ions, atoms, or molecules arranged in an orderly array.

Crystal lattice: the orderly array of particles in a crystal.

Curie (Ci): a unit of radioactivity equal to 3.7×10^{10} disintegrations per second.

Cytoplasm: the colorless, semifluid substance inside a cell, exclusive of cellular organelles; also called protoplasm.

Decomposition reaction: a chemical reaction in which a reactant is decomposed into two or more products.

Dehydration: the loss of water.

Dehydrogenase: an enzyme that catalyzes transfer of electrons and hydrogen atoms.

Denaturation: any physical change that alters the native conformation of a protein molecule to render it biologically inactive.

Density: the amount of mass in a volume of 1 unit, usually 1 mL; $d = m/v$.

Deoxyribonucleic acid (DNA): the nucleic acid that contains the genetic information of an organism.

Detergents: soluble salts of sulfonic acids that have cleaning properties similar to those of soaps.

Dextrins: a heterogeneous group of polymers of D-glucose, intermediate in size between starch and maltose.

Dialysis: the separation process in which small molecules and ions diffuse through a semipermeable membrane while large molecules and ions are not allowed to pass across.

Diet therapy: the use of diet in treating health problems.

Diffusion: the spontaneous mixing of particles to form a homogeneous mixture.

Digestion: the degradation of food macromolecules to small molecules capable of being absorbed into the circulatory system.

Dilution factor: the factor by which a solution concentration is reduced by dilution with solvent; also the ratio of initial solution volume to final solution volume after a dilution.

Dimensional analysis: the use of the dimensions (units) associated with a quantity as an aid in setting up the solution to a problem.

Dimer: a pair of chemical units held together by covalent bonds, hydrogen bonds, or other attractive forces.

Dipolar attractions: attractive forces between polar molecules.

Dipole moment: a measurable physical property possessed by an electric dipole.

Disaccharide: a carbohydrate composed of two monosaccharide units joined by acetal or ketal linkage.

Dissolution: the process of dissolving, in which solute particles become evenly distributed through the solvent in a solution.

Distillation: a separation method in which a liquid solution is boiled; solids and materials with high boiling points are left behind as residue, while the volatile components boil off in increasing order of their boiling points.

Disulfide: a compound containing the disulfide group, $-S-S-$.

Diuresis: the increased discharge of urine.

Diuretic: a substance that causes an increase in the quantity of urine produced.

DNA: see deoxyribonucleic acid.

Double bond: a covalent bond composed of two pairs of shared electrons; alternately, two covalent bonds between two atoms.

Double helix: the secondary structure of DNA in which two antiparallel strands of DNA are wound together in a double helix.

Double replacement reaction: a reaction in which parts of two different reacting compounds replace each other, $AB + CD \longrightarrow AC + BD$; also called double displacement or metathesis.

Drug: a chemical substance that affects the functions of living things.

Dynamic equilibrium: a state of rest or balance in which there is a constant interchange of particles; the opposing changes occur at the same time and the same rate.

Edema: the condition in which fluid collects in the tissue spaces between cells and causes swelling.

Electric dipole: a localized separation of positive and negative electrical charge.

Electrode: a metal rod or strip immersed in a solution that allows the flow of electricity through the solution to another electrode.

Electrolysis: the process of decomposing compounds by the use of electricity.

Electrolyte: a substance that conducts electricity when melted or dissolved in water.

Electron: a subatomic particle found outside the nucleus of an atom and having a -1 charge and negligible mass.

Electron configuration: the arrangement of electrons in atomic orbitals in an atom or ion.

Electron transport system (ETS): the mitochondrial pathway in which electrons are transferred spontaneously from reduced coenzymes through a system of carriers to O_2 to produce H_2O; the energy released by the pathway is used to synthesize ATP from ADP and P_i; also called the respiratory chain.

Electron-volt (eV): a unit describing energy of radiation, defined as that quantity of energy acquired by an electron accelerating through a voltage change of one volt.

Electronegativity: the ability of an atom to draw electrons in a covalent bond toward itself.

Electronic formula: a formula which shows covalent bonds, nonbonded valence electrons, and the arrangement in which atoms are bonded together in a molecule or ion; also called electron dot structure or Lewis structure.

Electrophoresis: an electrical method used to separate charged particles in solution.

Element: a pure substance that cannot be separated into simpler substances by ordinary processes.

Elimination reaction: a reaction in which an organic molecule loses the elements of a small molecule and forms a carbon-carbon multiple bond.

Emulsifying agent: a substance that promotes the suspension (emulsion) of one liquid in another.

Emulsion: a heterogeneous mixture in which droplets of one liquid are suspended in another liquid.

Enantiomers: isomeric pairs of chiral molecules whose members are mirror images of each other and thus nonsuperimposable; also called optical isomers.

Endergonic process: a process that absorbs (or requires) free energy.

Endocrine glands: ductless glands that secrete hormones directly into the bloodstream.

Endoplasmic reticulum: the system of flattened, interconnected channels in eucaryotic cells that serves to transport substances to and from the nucleus.

Endothermic: heat absorbing.

Energy: the capacity to do work, to move an object over a distance.

Enthalpy change (ΔH): the heat change that accompanies a chemical reaction.

Entropy: a measure of disorder in a system; since a disordered system is more probable than an ordered system, entropy is also a measure of probability.

Enzyme: a protein catalyst that increases the rate of a biochemical reaction.

Enzyme specificity: the ability of an enzyme to choose both the type of reaction to be catalyzed and the particular substrate to be reacted.

Equilibrium constant (K_{eq}): the constant ratio of product concentrations and reactant concentrations for a reversible reaction at equilibrium at a specific temperature.

Equivalence point: the end point of a titration in which the number of equivalents of acid or base added are equal to the original number of equivalents of the substance being titrated.

Equivalent: the weight of an acid that neutralizes one mole of OH^- or the weight of a base that neutralizes one mole of H^+; the weight of one mole of an ion divided by the absolute value of its charge.

Essential amino acids: amino acids that cannot be synthesized by an organism at a rate adequate to meet metabolic needs and that are thus essential in the diet of the organism.

Essential fatty acids: fatty acids that cannot be synthesized by an organism and thus are essential in the diet of the organism; for humans, linoleic and linolenic acids are essential fatty acids.

Ester: an organic compound with the functional group $-CO_2-$.

Ether: an organic molecule containing an oxygen atom attached to two alkyl or aromatic groups, as in $R-O-R$.

Eucaryotic cell: a biological cell that has nuclear material concentrated within a nuclear membrane.

Exergonic process: a process that releases free energy.

Exothermic: heat releasing.

Extracellular fluid: body fluid found outside of cells.

Extracorporeal irradiation: a form of radiation therapy in which blood is irradiated as it is circulated outside of the body.

Extrinsic pathway: one of two pathways for initiating blood clot formation; the extrinsic pathway is activated by the addition of substances not normally present in blood.

Facilitated diffusion: a type of facilitated transport in which movement of transported substances occurs from a region of high concentration to a region of low concentration.

Facilitated transport: a type of membrane transport in which membrane proteins serve as carriers of transported substances.

Fahrenheit temperature scale: the temperature scale that sets the freezing point of water at 32° and the boiling point of water at 212°.

Fat: a triacylglycerol that is a solid at room temperature.

Fat-soluble vitamins: vitamins that are insoluble in water but soluble in fatty body tissues.

Fatty acid: a long-chain carboxylic acid, usually having an even number of carbon atoms, that can be isolated from fats by hydrolysis.

Fatty acid mobilization: the release of fatty acids from adipose tissue.

Feedback inhibition: a type of metabolic regulation in which the end product of a sequence of enzyme-catalyzed reactions allosterically inhibits the enzyme catalyzing the first reaction in the sequence.

Fermentation: the anaerobic utilization of glucose by microorganisms to form either ethanol and carbon dioxide or lactic acid.

Fibrous proteins: protein molecules that have long, rodlike or stringlike shapes.

Fischer projection: a method of drawing structures of open-chain chiral molecules in which horizontal bonds are projected toward the viewer and vertical bonds are projected away from the viewer.

Flavor enhancer: a substance that enhances or intensifies flavors existing in foods without adding any flavor of its own.

Fluid: any substance having the ability to flow and thus assume the shape of its container.

Fluid mosaic model: the current model of membrane structure that suggests that membranes are fluid solutions of globular proteins distributed in a mosaic pattern through lipid molecules.

Food additives: substances added to processed foods to enhance their quality or to improve processing, packaging, or storage.

Formed elements (of blood): the red blood cells, white blood cells, and platelets, all of which are suspended in blood.

Formula unit: a unit of a compound corresponding to the formula of the compound.

Formula weight: the sum of atomic weights of all of the elements in a formula unit of a compound; if two or more atoms of the same element are present in the formula unit, the atomic weight of the element is added as many times as atoms of the element appear in the formula unit; usually applied to ionic compounds.

Free energy change (ΔG): the amount of usable, or free, energy that is made available or required by a process.

Free radical: a highly reactive chemical structure having one or more unpaired electrons.

Functional group: a distinctive atomic grouping in an organic compound that is responsible for many of the physical and chemical properties of the compound.

Fusion: melting.

Gamma (γ) decay: a type of nuclear decay in which energy in the form of gamma rays is released, usually accompanied by other types of radiation.

Gamma (γ) ray: a highly penetrating form of energy emitted by a radioactive nucleus.

Gas: the state of matter in which particles are far away from each other and moving rapidly and randomly; a gas has neither fixed volume nor shape, and it expands or contracts to fill containers of all shapes and sizes.

Gastric juice: the fluid secreted by the stomach.

Gel filtration chromatography: a type of liquid chromatography that utilizes porous beads as column packing material and separates mixtures of molecules on the basis of their sizes; also called gel permeation chromatography.

Gene: the segment of DNA that contains the information needed for synthesis of one polypeptide; a newer term is cistron.

Genetic code: the code composed of triplet nucleotide sequences in DNA that specifies amino acid sequences for all proteins synthesized by an organism.

Genetic engineering: the laboratory manipulation of DNA to modify the hereditary mechanisms of microorganisms or cells.

Geometric isomers: isomers of alkenes in which identical substituents on each of the carbons connected by the double bond can be on the same side of the double bond (*cis*) or on opposite sides of the double bond (*trans*); also called *cis-trans* isomers.

Germicide: a substance that destroys microorganisms.

Globular proteins: protein molecules that have spherical or nearly spherical shapes.

Gluconeogenesis: the metabolic synthesis of glucose from noncarbohydrate precursors.

Glycogenesis: the metabolic synthesis of glycogen.

Glycogenic amino acids: amino acids whose carbon skeletons can be converted metabolically to intermediates used in the synthesis of glucose.

Glycogenolysis: the metabolic degradation of glycogen to glucose.

Glycol: a common name for diols that is also used specifically for 1,2-ethanediol.

Glycolysis: the metabolic pathway that degrades glucose to pyruvate.

Glycoproteins: protein molecules that have carbohydrate groups attached.

Glycosidic linkage: the acetal or ketal linkages in sugar derivatives and in di-, oligo-, and polysaccharides.

Halogenation: reaction of an organic compound with a halogen so that one or more halogen atoms are introduced into the compound.

Halogens: elements of group VIIA.

Haworth structure: a way of drawing cyclic structures of monosaccharides that illustrates the location of hydroxyl substituents relative to the ring.

Heat: the form of energy related to the average kinetic energy of the particles in a sample of matter; heat flows from a region of higher temperature to a region of lower temperature.

Hemiacetal: the compound formed by addition of an alcohol molecule to the carbonyl group of an aldehyde.

Hemiketal: the compound formed by addition of an alcohol molecule to the carbonyl group of a ketone.

Hemolysis: red blood cell destruction, usually caused by rupturing.

Hemoproteins: proteins that contain the heme prosthetic group.

Hemostasis: the natural system in animals that prevents blood loss.

Henderson-Hasselbalch equation: the equation that describes buffer action in terms of pH, pK_a, and log $[A^-]/[HA]$; $pH = pK_a + \log [A^-]/[HA]$.

Heterocyclic compound: a cyclic organic compound having other atoms in the ring(s) in addition to carbon atoms.

High-quality protein: a protein having amino acid proportions similar to those needed by humans.

Histones: basic proteins found in the cell nucleus combined in ionic attraction with DNA.

Holoenzyme: an enzyme-cofactor complex.

Homogeneous mixture: a mixture of two or more different substances that is uniform throughout, with no detectible boundaries separating the substances.

Homogenization: the process in which milk is forced through a nozzle at high pressure and the fat droplets are reduced in size so that they remain suspended.

Hormone: a chemical messenger produced by one tissue or cell type and transported by the blood to act on another tissue or cell type.

Hydrate: a crystalline compound that contains a fixed number of water molecules within its crystals, or an organic molecule formed by addition of a water molecule to the carbonyl group of an aldehyde or ketone.

Hydration: the process in which water molecules surround solute particles in an aqueous solution, or the addition of a water molecule to a multiple bond in an organic compound.

Hydrocarbons: binary molecular compounds composed of hydrogen and carbon.

Hydrogen bond: an attractive force between molecules in which a hydrogen atom bonded to a highly electronegative atom such as N, O, or F is shared between two molecules.

Hydrogenation: the addition of hydrogen to one or more multiple bonds in an organic compound.

Hydrolases: a class of enzymes that catalyze hydrolysis of esters, acetals, and amides.

Hydrolysis: a reaction with water in which the water molecule is split into $-H$ and $-OH$.

Hydronium ion: the ion H_3O^+ that is formed when a water molecule forms a coordinate covalent bond to a proton.

Hydrophilic: water loving, or attracted to water.

Hydrophobic: water hating, or having very little attraction for water.

Hypercarotenosis: a condition characterized by abnormally high levels of β-carotene in the body, usually caused by eating large quantities of vegetables containing β-carotene.

Hyperglycemia: a condition characterized by an abnormally high level of blood glucose.

Hypertonic solution: a solution having a higher osmotic pressure than a reference solution.

Hypervitaminosis: a toxic condition caused by an excessive intake of vitamins.

Hypoglycemia: a condition in which the blood glucose concentration is abnormally low.

Hypotonic solution: a solution having a lower osmotic pressure than a reference solution.

Inborn error of metabolism: a disease in which a genetic defect causes a deficiency of a particular protein, often an enzyme.

Induced-fit hypothesis: the theory of enzyme action that suggests that the binding of substrate induces a conformational change in the enzyme structure that brings about catalysis.

Inhibitor: a substance that slows the rate of a catalyzed reaction.

Inorganic chemistry: the chemistry of nonliving substances.

Integral protein: a protein that is partially or totally buried in a biological membrane and can only be removed from the membrane by using detergents or organic solvents.

Interstitial fluid: body fluid found in the tissue spaces (interstices) between cells.

Intracellular fluid: fluid found inside biological cells.

Intrinsic factor: the protein responsible for transporting vitamin B_{12} across the intestinal wall.

Intrinsic pathway: one of two pathways for initiating blood clot formation; the intrinsic pathway is activated when blood encounters an abnormal or foreign surface.

Iodine number: the number of grams of I_2 that would react with 100 g of unsaturated triacylglycerol.

Ion: an atom or group of bonded atoms with an electrical charge.

Ion-exchange chromatography: a type of liquid chromatography that utilizes charged packing material in the column and thus separates mixtures of ions on the basis of their charges.

Ion product constant (K_w): the product of molar concentrations of H^+ and OH^- in water; $K_w = 1 \times 10^{-14}$ at 25°C.

Ionic bond: a chemical bond formed by the attraction of two oppositely charged ions.

Ionic compound: a compound composed of ions.

Ionization constant (K_a or K_b): the ratio of the product of molar concentrations of dissociated ions to the molar concentration of undissociated acid (K_a) or base (K_b); also called acid dissociation constant or base dissociation constant.

Ionizing radiation: radiation that produces molecular fragments and ions when it interacts with matter.

Isoelectric point (P_i): the pH at which molecules of an ionizable compound such as an amino acid have zero average charge.

Isomerases: a class of enzymes that catalyze isomerization reactions.

Isomerization reaction: a chemical reaction in which a compound is converted to its isomer.

Isomers: compounds having the same numbers and kinds of atoms arranged in different structural forms.

Isotonic solution: a solution having the same osmotic pressure as a reference solution.

Isotopes: atoms of the same element (same atomic number) that have different mass numbers.

Jaundice: a condition that exists when excessive amounts of bile pigments make their way into the skin, giving a yellow color to skin and the whites of the eyes.

Kelvin temperature scale: the temperature scale derived from the theoretical behavior of gases; the freezing point of water on the Kelvin scale is 273 and the boiling point of water is 373; also called the absolute temperature scale.

Ketal: the compound formed by reaction of an alcohol molecule with a hemiketal.

Ketoacidosis: a condition characterized by the simultaneous existence of ketosis and acidosis.

Ketogenic amino acids: amino acids whose carbon skeletons can be converted metabolically to acetyl CoA or acetoacetyl CoA, which can then be used to synthesize ketone bodies.

Ketone: an organic compound containing an oxygen atom double bonded to a carbon atom carrying two alkyl or aromatic substituents, as in R—CO—R.

Ketone bodies: acetone, acetoacetic acid, and β-hydroxybutyric acid, compounds that are formed in higher than normal amounts as a result of increased fatty acid oxidation.

Ketonemia: a condition characterized by abnormally high levels of ketone bodies in the blood.

Ketonuria: a condition characterized by abnormally high levels of ketone bodies in the urine.

Ketose: a monosaccharide containing a ketone group.

Ketosis: a condition characterized by the simultaneous existence of ketonemia, ketonuria, and acetone odor of the breath.

Kinetic energy: energy of motion.

Kinetic molecular theory: the explanation for gaseous behavior that suggests that gas particles move freely and rapidly along straight lines, with frequent collisions causing variation in velocity and direction.

Law of conservation of energy: energy cannot be created or destroyed, but it can be changed from one form to another.

Law of conservation of mass: matter is neither created nor destroyed in a chemical reaction.

Law of partial pressures: each component in a gas mixture exerts its own partial pressure, and the total pressure of the mixture is the sum of the partial pressures.

LeChatelier's principle: a chemical system in equilibrium will respond to a disturbance by offsetting the disturbance partially and restoring equilibrium.

Ligases: a class of enzymes that catalyze the formation of covalent bonds, usually by joining two molecules to form a larger one.

Lipase: an enzyme that catalyzes hydrolysis of triacylglycerols.

Lipid: a member of a heterogeneous group of organic compounds characterized by a greasy feeling, insolubility in water, and solubility in nonpolar solvents.

Lipid bilayer: a double layer of lipid molecules.

Lipoproteins: protein molecules that have lipid groups attached.

Liquid: the condensed state of matter in which particles are in contact with each other but moving freely throughout the sample of matter; a liquid has a definite volume but assumes the shape of its container.

Liquid chromatography: a method used for separating mixtures of substances in solution; it utilizes a glass or metal column filled with water-insoluble material that has an affinity for the molecules or ions to be separated.

Liquid crystal: a state of matter intermediate between liquid and solid that possesses some properties of liquid matter and others of solid matter.

Litmus: an acid-base indicator that is red in acidic solution and blue in basic solution.

Lock-and-key theory: the theory of enzyme specificity that suggests that a substrate has a shape complementary to that of the enzyme's active site, and that a substrate fits the active site like a key fits a lock.

London dispersion forces: weak attractive forces between instantaneous dipoles in atoms and molecules.

Lyases: a class of enzymes that catalyze the addition of water, ammonia, and carbon dioxide to compounds; they also catalyze the removal of these substances to create double bonds.

Lysosomes: cellular organelles that contain hydrolytic enzymes necessary for degradation of various substances.

Macronutrient: a nutrient needed in relatively large amounts (gram quantities).

Major minerals: the nutritional minerals needed in greatest amount by humans.

Markovnikov's rule: the pattern of addition of hydrogen-containing molecules to carbon-carbon double bonds; the added hydrogen bonds to the carbon in the double bond that originally had the greater number of hydrogen atoms, and the rest of the adding molecule bonds to the other carbon in the double bond.

Mass: the amount of matter contained in an object.

Mass defect: the deficiency in mass that results from combining subatomic particles to form an atomic nucleus.

Mass number: the sum of protons and neutrons in an atom.

Matter: anything that occupies space and has mass.

Maximal velocity (V_{max}): the theoretical maximum rate of an enzyme-catalyzed reaction that would occur when all enzyme molecules present are converted to Michaelis complexes.

Melting: the conversion of a solid to a liquid.

Melting point: the temperature at which a solid substance is in equilibrium with its corresponding liquid.

Membrane fluidity: the fluid, constantly changing structure of biological membranes.

Messenger RNA (mRNA): the type of RNA that carries genetic information needed for protein synthesis from DNA to the ribosomes.

Metabolic acidosis: acidosis brought on by metabolic or physiological disorders other than respiratory disorders.

Metabolic alkalosis: alkalosis brought on by metabolic or physiological disorders other than respiratory disorders.

Metabolism: the network of chemical reactions responsible for extracting energy from food and synthesizing needed cellular materials.

Metalloids: elements having some properties of metals and others of nonmetals; metalloids border the zigzag line on the periodic table that separates metals from nonmetals.

Metalloproteins: protein molecules that have metal ions attached.

Metals: elements that generally have hard, lustrous surfaces and that can be pounded into sheets and drawn into wires; metals are usually good conductors of heat and electricity.

Metric system: the system of measurement based on the decimal system.

Micelle: a spherical cluster of molecules or ions in which hydrophobic parts of the molecules or ions are arranged in the micelle interior and hydrophilic parts of the molecules or ions are located on the micelle surface.

Michaelis complex: the complex formed when an enzyme binds its substrate.

Micronutrient: a nutrient required in relatively small amounts (mg or μg quantities).

Milligram percent (mg%): solution concentration expressed as a percentage based on mg of solute per mL of solution.

Minerals: naturally occurring substances found in the earth's crust; nutritional minerals are inorganic substances needed in small quantities for life processes.

Mitochondria: the distinctive, cigar-shaped organelles found in most animal cells; mitochondria are the site of oxidative reactions and ATP production.

Mixture: a collection of two or more substances that are not chemically united and that do not exist in fixed proportions to each other.

Molar heat of fusion (ΔH_{fus}): the amount of heat required to melt one mole of crystalline solid at its normal melting point.

Molar heat of solution ($\Delta H_{solution}$): the heat change that occurs when one mole of solute is dissolved in a particular solvent.

Molar heat of vaporization (ΔH_{vap}): the amount of heat required to vaporize one mole of liquid at its normal boiling point.

Molarity (M): solution concentration expressed as moles of solute per liter of solution (moles/L).

Mole: 6.02×10^{23} particles; the mass contained in the molecular weight of a compound expressed in grams.

Molecular compound: a compound composed only of molecules.

Molecular formula: a formula which gives the number of atoms of each element in one molecule of a substance.

Molecular weight: the sum of atomic weights of all the elements in a molecular element or compound; if two or more atoms of the same element are present in the molecule, the atomic weight of the element is added as many times as atoms of the element appear in the formula.

Molecule: an uncharged group of two or more atoms connected by covalent bonds.

Monosaccharide: a carbohydrate that cannot be broken down into simpler molecules by acid hydrolysis.

Mutagen: any substance or agent that causes a genetic mutation; mutagens are often carcinogens.

Mutation: an alteration in the nucleotide sequence of DNA that brings about a change in the genetic message contained in DNA.

Narcotic: a substance that produces a state of sleep, drowsiness, or unconsciousness.

Native conformation (of a protein): the predominating three-dimensional structure of a protein that exists under natural conditions and possesses normal biological activity.

Net ionic equation: an equation showing only ions and molecules that undergo change in a chemical reaction.

Neutralization reaction: the reaction between an acid and a base to produce a salt.

Neutron: a subatomic particle, found in the nucleus of an atom, having a mass of 1 amu and no electrical charge.

Nitrogen balance: a comparison of nitrogen intake (as proteins for humans) and nitrogen losses (via urine and feces for humans).

Noble gases: elements of group VIIIA; also called rare gases and inert gases.

Noncompetitive inhibitor: an enzyme inhibitor that binds to the enzyme at a location other than the active site.

Nonelectrolyte: a substance that produces no ions when dissolved in water.

Nonessential amino acids: amino acids that can be synthesized by an organism at a rate adequate to meet metabolic needs and thus are not essential in the diet of the organism.

Nonmetals: elements that generally crumble easily, if they are solids, and are poor conductors of heat and electricity.

Nonnutritive sweetner: a substance that provides a sweet taste but few or no calories.

Normality (N): solution concentration expressed as number of equivalents per liter of solution.

Nuclear binding energy: the energy released when subatomic particles combine to form an atomic nucleus.

Nuclear fission: the process of splitting an atomic nucleus into two smaller atomic nuclei.

Nuclear fusion: the process of combining two atomic nuclei to produce a larger atomic nucleus.

Nuclear reactor: the equipment in which controlled nuclear fission is carried out.

Nucleic acids: the acidic substances that can be isolated from cell nuclei (DNA and RNA).

Nucleoside: a compound composed of a purine or pyrimidine bonded to ribose or deoxyribose in a β-glycosidic linkage.

Nucleotide: a compound that produces a pentose, a heterocyclic nitrogen base, and phosphoric acid on hydrolysis; nucleotides are the repeating units of nucleic acids.

Nutrient: any substance that serves to sustain a living organism by promoting growth, replacing loss, or providing energy.

Octet rule: the statement that an atom is least reactive when it contains eight electrons in its highest occupied principal energy level.

Oil: a viscous liquid hydrocarbon or hydrocarbon mixture isolated from petroleum, or a triacylglycerol that is a liquid at room temperature.

Oligosaccharide: a carbohydrate composed of three to ten monosaccharide units joined by acetal or ketal linkage.

Opiates: drugs isolated from the opium poppy, and other drugs that have similar narcotic properties.

Optical isomers: members of a pair of enantiomers; one member rotates plane-polarized light to a certain extent in one direction, and the other member rotates plane-polarized light to an equal extent in the opposite direction.

Organic chemistry: the study of covalent carbon compounds.

Osmosis: the movement of water across a semipermeable membrane from a region of high water concentration to one of low water concentration.

Osmotic membrane: a membrane that allows only water molecules to pass across.

Osmotic pressure: the minimum amount of pressure that must be applied to the surface of a solution to prevent the flow of water across an osmotic membrane into the solution.

Oxidation: an increase in oxidation number, a loss of electrons, a gain of oxygen atoms, or a loss of hydrogen atoms.

Oxidation number: a signed whole number assigned to an element in a compound to indicate the element's relative state of oxidation.

Oxidative deamination: the oxidative removal of an amino group from an amino acid to form NH_4^+ and an α-keto acid.

Oxidative phosphorylation: the mitochondrial synthesis of ATP from ADP and P_i using free energy released by the electron transport system.

Oxidizing agent: an element or compound responsible for oxidation.

Oxido-reductases: a class of enzymes catalyzing oxidation-reduction reactions involving transfer of electrons and hydrogen atoms.

Oxyacid: an acid in which one or more OH groups and possibly additional oxygen atoms are bonded to a central atom.

Partial pressure: the pressure exerted by an individual gas in a mixture of gases.

Pasteurization: a heat treatment that destroys pathogenic microorganisms in foods and beverages.

Pentose phosphate pathway: the metabolic pathway responsible for the interchange of carbon atoms among 3-, 4-, 5-, 6-, and 7-carbon phosphate derivatives of carbohydrates.

Peptide: a molecule composed of 2–20 amino acids joined by amide linkage.

Peptide bond: the amide linkage that joins two amino acids.

Peptide group: the atoms involved in a peptide bond, $-CO-NH-$.

Periodic law: the periodic recurrence of similar physical and chemical properties of elements as a function of atomic number.

Periodic table: the table that contains all of the elements listed in order of atomic number and arranged in horizontal rows called periods and vertical columns called groups.

Peripheral protein: a membrane protein that is attached to the membrane surface and can easily be removed from the membrane.

Permeability: the extent to which a membrane allows passage of particles.

Peroxide: a compound having two oxygen atoms bonded together, as in $H-O-O-H$ and $R-O-O-R$.

pH: the negative logarithm of the hydrogen ion concentration in water; $pH = -\log [H^+]$.

pH meter: a meter that measures pH.

pH scale: a scale of 0–14 used for expressing the level of acidity in aqueous solutions.

Phenol: an organic compound that contains one or more $-OH$ groups attached to an aromatic ring, as in $Ar-OH$; phenol is also the name used specifically for the compound having one benzene ring and one $-OH$ group.

Phosphoglyceride: a complex lipid molecule containing a phosphodiester group.

Pituitary: a small oval endocrine gland attached to the base of the brain; also called the hypophysis.

pK_a: the negative logarithm of an acid ionization constant; $pK_a = -\log K_a$; also, the pH at which a weak acid is half-neutralized.

Plasmid: a circular double-stranded DNA molecule found in bacteria separate from chromosomes.

pOH: the negative logarithm of the hydroxide ion concentration in water; $pOH = -\log [OH^-]$.

Polar covalent bond: a covalent bond in which the shared pair of electrons is shifted toward an atom of high electronegativity; also called a dipolar bond.

Polyatomic ion: an ion composed of two or more atoms.

Polycyclic aromatic hydrocarbon: a hydrocarbon molecule composed of two or more aromatic rings fused together.

Polymer: a very large molecule composed of many repeating units.

Polypeptide: a molecule composed of 20–50 amino acids joined by amide linkage.

Polyprotic acid: an acid that is capable of donating more than one proton from each formula unit.

Polyribosome: the complex of ribosomes and mRNA that forms during protein synthesis; also called polysome.

Polysaccharide: a polymeric carbohydrate containing more than ten monosaccharide units joined by acetal or ketal linkage.

Positron: a subatomic particle having a $+1$ charge and negligible mass emitted by a radioactive nucleus; positrons, represented by $^0_1\beta$, are similar to electrons but have a positive charge.

Potential energy: the energy stored in an object because of the object's position, condition, or composition.

Precision: the fineness of measurement and the extent of reproducibility of a measurement.

Pressure: a force that acts on a given surface area; common units of pressure are pounds per square inch (psi), torrs, and atmospheres (atm).

Primary (1°) alcohol: an alcohol in which the carbon carrying the —OH is attached to only one alkyl or aromatic group, as in R—CH$_2$—OH; methanol, the simplest alcohol, is also classified as a primary alcohol.

Primary (1°) amine: an amine having one alkyl or aromatic group bonded to nitrogen, as in R—NH$_2$.

Primary protein structure: the amino acid sequence of a protein.

Principal energy level: a region of space in an atom occupied by one or more atomic orbitals having similar energy content.

Procaryotic cell: a biological cell that has nuclear material concentrated in a small section but lacks a nuclear membrane.

Prohormone: an inactive precursor of a polypeptide hormone.

Prostaglandins: a group of biologically active, naturally occurring cyclic derivatives of the 20-carbon fatty acid prostanoic acid; the prostaglandins are tissue hormones.

Prosthetic group: a nonpolypeptide chemical unit attached tightly to a protein.

Protease: an enzyme that catalyzes hydrolysis of proteins; also called a proteolytic enzyme.

Protein: a polymeric biological molecule composed of at least 50 amino acids joined by amide linkage.

Proton: a subatomic particle found in an atomic nucleus; it has a +1 charge and a mass of 1 amu; a proton is the same as a hydrogen ion (H$^+$).

Protoplasm: the viscous, jellylike fluid inside a living cell; also called cytoplasm.

Provitamin: an inactive precursor of a vitamin.

Ptomaines: a group of diamino compounds formed during putrefaction of protein-containing substances.

Quaternary ammonium ion: a positively charged ion in which four alkyl or aromatic groups are bonded to nitrogen, as in R$_4$N$^+$.

Quaternary ammonium salt: a salt composed of a quaternary ammonium ion and an anion, as in R$_4$N$^+$Cl$^-$.

Quaternary protein structure: the arrangement of subunits in a protein.

Racemic mixture: a mixture of equal parts of any pair of enantiomers.

Rad: a unit of radiation dosage equivalent to 100 ergs of energy absorbed per gram of tissue.

Radiation: energy and/or particles emitted by matter.

Radioactive decay: the change in the nucleus of an atom resulting from radioactivity.

Radioactivity: the spontaneous emission of particles and/or energy, or the capture of electrons, by unstable nuclei.

Radioisotope: a radioactive isotope.

Rate constant: the proportionality constant that relates a reaction rate to concentration(s) of reactant(s).

Reaction pathway: the hypothetical route whereby reactants are converted to products in a chemical reaction.

Reagent: a reactive substance used in the laboratory for chemical synthesis or analysis.

Recombinant DNA: DNA formed by genetic engineering techniques and composed of portions of DNA from different sources.

Reducing agent: an element or compound responsible for reduction.

Reducing sugar: any sugar capable of reducing Cu^{2+} to Cu$^+$.

Reduction: a decrease in oxidation number, a gain of electrons, a loss of oxygen atoms, or a gain of hydrogen atoms.

Regulatory factors: polypeptide hormones secreted by the hypothalamus that act on the pituitary to stimulate or inhibit synthesis and release of other hormones.

Regulatory site: a site other than the active site where an inhibitor or activator can bind to an allosteric enzyme and thus regulate its activity.

Rem: a unit of radiation dosage defined as that amount of radiation that causes damage equivalent to the absorption of 1 roentgen.

Representative elements: elements in which the outermost p orbitals are being filled (groups IIIA–VIIA).

Resistant organism: a microorganism, usually a mutant, that has developed resistance to an antibiotic.

Resonance: the quality possessed by a molecule or polyatomic ion having resonance forms; resonance imparts high stability and low chemical reactivity to a molecule or polyatomic ion.

Resonance forms: two or more alternative electronic formulas for a molecule or polyatomic ion differing only in the placement of electrons.

Respiratory acidosis: acidosis brought on by respiratory disorders.

Respiratory alkalosis: alkalosis brought on by respiratory disorders.

Restorative diet: a diet intended to restore good health after disease, injury, or surgery.

Restriction enzymes: certain enzymes found in bacterial cells that catalyze cleavage of foreign DNA.

Restrictive diet: a diet that restricts intake of one or more nutrients designed to remove previous excess weight or to reduce the strain on an organ or tissue.

Reversible reaction: a chemical reaction that proceeds forwards and backwards at the same time.

Ribonucleic acid (RNA): the nucleic acid that transcribes and translates the genetic information contained in DNA; there are three types of RNA: mRNA, tRNA, and rRNA.

Ribosomal RNA (rRNA): the type of RNA that is incorporated into the structure of ribosomes.

Ribosomes: tiny granules, composed of RNA and proteins, attached to portions of the endoplasmic reticulum in cells; ribosomes are the sites of protein synthesis.

RNA: see ribonucleic acid.

Roentgen (R): a unit of radiation dosage equal to the amount of X- or γ-radiation that produces ions totaling 2.1×10^9 units of electrical charge in one cubic centimeter of dry air at normal temperature and pressure.

Salt: an ionic compound composed of one or more cations other than H^+ and one or more anions other than O^{2-} or OH^-; acid-base neutralization reactions produce salts.

Salt bonds: ionic attractions between or within protein molecules.

Saponification: the process of soap making, in which triacylglycerols are reacted with NaOH or KOH to produce glycerol and Na^+ or K^+ salts of fatty acids.

Saturated hydrocarbon: a hydrocarbon having only single bonds.

Saturated solution: a solution in which dissolved and undissolved solute particles are in dynamic equilibrium with each other.

Saytzeff's rule: when HX is eliminated from an alkyl halide (or H_2O from an alcohol), hydrogen is lost from the carbon adjacent to C—X (or C—OH) that had the fewest hydrogens originally.

Second messenger: the term applied to cyclic AMP because it conveys the influence of certain hormones to the interior of cells after the hormones bind to protein receptors on the external surfaces of the target cell membranes.

Secondary (2°) alcohol: an alcohol in which the carbon carrying the —OH group is attached to two alkyl or aromatic groups, as in R_2CH—OH.

Secondary (2°) amine: an amine having two alkyl or aromatic groups bonded to nitrogen, as in R_2NH.

Secondary protein structure: the spatial arrangement of the polypeptide backbone of a protein.

Semipermeable membrane: a membrane that selectively allows only certain molecules to pass across.

Significant figures: the digits in a number that have physical meaning.

Simple lipid: a lipid that does not contain carboxylic acid esters.

Single replacement reaction: a reaction in which an ion, atom, or group of atoms replaces an ion, atom, or group of atoms in a compound, as in A + BC $\longrightarrow$ AC + B; also called substitution or single displacement.

Soap: a Na^+ or K^+ salt of a fatty acid.

Solid: the condensed state of matter in which particles are in contact with each other and occupy fixed positions in space; a solid has a definite volume and fixed shape.

Solubility: the maximum amount of a solute that will dissolve in a given quantity of solvent at a particular temperature.

Solute: a minor component that is dissolved in the solvent to form a solution.

Solution: a homogeneous mixture of two or more components.

Solvation: the process in which solvent molecules surround solute particles in a solution.

Solvent: the component present in greatest amount in a solution.

Specific gravity: the density of a material relative to that of water at 4°C.

Specific heat: the amount of heat, measured in calories, required to raise the temperature of 1 g of a substance 1°C.

Spectator ion: an ion that does not undergo change in a reaction mixture.

Sphingolipid: a complex lipid containing sphingomyelin or a related base.

Standard free energy change ($\Delta G°$): a reference free energy change for a chemical reaction under standard conditions; $\Delta G° = -2.303\, RT \log K_{eq}$

Standard temperature and pressure (STP): 273 K and 1 atmosphere pressure.

Stereoisomers: isomers having the same structure but different arrangements of atoms about an asymmetric carbon atom.

Steric hindrance: the prevention or retardation of a chemical reaction because of the presence of bulky groups.

Steroid: a member of a class of organic compounds having a characteristic pattern of four fused nonaromatic rings, three composed of six carbon atoms and one composed of five carbon atoms.

Sterol: a member of the subgroup of steroids having a hydroxyl group at carbon 3 in ring A.

Strong acid: an acid having an ionization constant greater than 1 ($K_a > 1$).

Strong base: a base that completely ionizes in water to form metal cations and OH^- ions.

Strong electrolyte: a substance that dissociates completely into ions when dissolved in water.

Structural formula: a formula which indicates the geometry of a molecule or polyatomic ion.

Structural isomers: compounds having the same molecular formula but different sequences of atoms bonded together.

Substrate: the substance that undergoes a chemical change catalyzed by an enzyme.

Substrate-level phosphorylation: the synthesis of ATP or similar high-energy compounds utilizing free energy released by a substrate reaction rather than the electron transport system.

Sugar: a sweet-tasting, water-soluble carbohydrate.

Surface tension: the elasticlike force on the surface of a liquid that tends to minimize the liquid's surface area.

Suspension: a heterogeneous mixture in which relatively large particles are temporarily scattered through a solid, liquid, or gas.

Temperature: a measure of the warmth of an object.

Terminal amino group: the free α-amino group at one end of a peptide or protein chain; also called N-terminus.

Terminal carboxyl group: the free α-carboxyl group at one end of a peptide or protein chain; also called C-terminus.

Terpene: a hydrocarbon synthesized from isoprene units by plants and animals.

Tertiary (3°) alcohol: an alcohol in which the carbon carrying the $-OH$ is attached to three alkyl or aromatic groups, as in R_3C-OH.

Tertiary (3°) amine: an amine having three alkyl or aromatic groups bonded to nitrogen, as in R_3N.

Tertiary protein structure: the bending and folding of protein helices and pleated sheets to make the three-dimensional shape of a protein molecule.

Tetrahedron: a four-sided geometric solid.

Thiol: an organic compound containing the thiol group, $-SH$.

Tissue hormone: a hormone produced by one cell type in a tissue that acts on another cell type in the same tissue.

Titration: a laboratory procedure for determining concentration of an acid or base on the basis of equivalents required for neutralization.

Toxin: a natural substance that is poisonous to plants or animals.

Trace minerals: nutritional minerals required by the human body in trace amounts.

Transamination: the reversible metabolic transfer of an amino group from an amino acid to an α-keto acid.

Transfer RNA (tRNA): the type of RNA that carries amino acids to the ribosomes in preparation for protein synthesis.

Transferases: a class of enzymes that catalyze the transfer of groups of atoms.

Transition elements: the elements whose d-orbitals are incomplete; transition elements are usually designated by the letter B following the group number.

Transmutation: the conversion of one element to another.

Triacylglycerol: a fatty acid triester of glycerol; also called a triglyceride or a neutral fat.

Triple bond: a covalent bond composed of three pairs of shared electrons; alternately, three covalent bonds between two atoms.

Tyndall effect: the visible scattering of light as it passes through a colloid.

Unimolecular reaction: a reaction in which only one molecule of reactant is needed to form the activated complex.

Unit-factor method: a method of dimensional analysis using unit factors for making conversions and solving problems.

Unsaturated hydrocarbon: a hydrocarbon having one or more double or triple bonds.

Unshared electron pair: a pair of valence electrons of an atom that are not involved in covalent bond formation; also called nonbonding electrons and lone pairs.

Urea cycle: the metabolic pathway responsible for converting NH_4^+ to urea.

Urine: the solution of metabolic waste products and other substances excreted by the body via the kidneys.

Valence: the number of bonds an atom can form.

Valence electrons: electrons occupying the highest principal energy level of an atom.

Vapor pressure: the pressure exerted by a vapor in equilibrium with its corresponding liquid.

Virus: a packet of nucleic acid surrounded by a protective coat of protein and capable of infecting a cell.

Viscosity: a liquid's resistance to flow.

Vitamin: an organic substance essential to the diet in small quantities, needed for normal metabolism, and found in foodstuffs or produced synthetically.

Volume-to-volume (v/v) percentage concentration: solution concentration expressed as a percentage based on mL of solute per mL of solution.

Water of hydration: the fixed amount of water associated with certain crystalline solids called hydrates.

Water-soluble vitamins: vitamins that are soluble in water and from which coenzymes are derived.

Wax: a solid ester of a high molecular weight fatty acid and either a long-chain alcohol or a sterol.

Weak acid: an acid having an ionization constant (K_a) of 10^{-1} or less.

Weak base: a base having an ionization constant (K_b) of 10^{-1} or less.

Weak electrolyte: a substance for which dissociation into ions is incomplete when the substance is dissolved in water.

Weight: the force exerted on the mass of an object by the earth's gravitational attraction.

Weight-to-volume (w/v) percentage concentration: solution concentration expressed as a percentage based on grams of solute per mL of solution.

Weight-to-weight (w/w) percentage concentration: solution concentration expressed as a percentage based on grams of solute per gram of solution.

X-ray diffraction: a technique used to study the structure of proteins and other crystals; a narrow beam of X-rays strikes the crystal and is scattered into a pattern that can be analyzed.

X-rays: high energy ultraviolet light given off when a nucleus captures an electron.

Zwitterion: an ion carrying both a positive and a negative charge; a dipolar ion.

Zymogen: an inactive precursor of an enzyme.

ILLUSTRATION SOURCES

The author is indebted to the following for permission to use copyrighted materials.

Art Credits

Fig. 1-7 © 1980 by Sidney Harris/*Science 80*

Fig. 16-6 Jerome Gross, Developmental Biology Laboratory, Massachusetts General Hospital, Boston.

Fig. 16-12 Reprinted by permission of Academic Press and Dr. Richard E. Dickerson from H. Neurath, ed., *The Proteins*, vol. 2, 2d ed., (New York: Academic Press, 1964), pg. 634.

Fig. 16-15 From Dickerson and Geis, *Hemoglobin: Structure, Function, Evolution, and Pathology,* Benjamin/Cummings, Menlo Park, Calif., 1982. Illustration copyright by Irving Geis. Reprinted by permission of Irving Geis.

Fig. 16-16 From "The Hemoglobin Molecule" by M.F. Perutz. Copyright © 1964 by Scientific American, Inc. All rights reserved.

Page 476 Illustration courtesy of Richard Kilker, Jr., Drew University; adapted with permission from *Chemistry 51* (6), 7 (1978). Copyright 1978 American Chemical Society.

Fig. 16-20 Phillips Electronic Instruments, Inc., Mahwah, New Jersey.

Fig. 19-13 After S.J. Singer and G.L. Nicolson, *Science* **175**:723, 1972. Copyright 1972 by the American Association for the Advancement of Science.

Fig. 22-5 Reproduced by permission of the Cytogenetics Laboratory, University of California, San Francisco, CA.

Fig. 22-9 After a drawing by Dr. Alexander Rich. Reproduced by permission of Dr. Alexander Rich.

Table 23-1 Reproduced from: Recommended Dietary Allowances, Ninth Edition (1980), with the permission of the National Academy of Sciences, Washington, D.C.

Photo Credits

Fig. 1-1, Reproduction on pg. 4 Bettmann Archive. *1-3* © 1980 Richard Wood/Taurus Photos. *1-5* Monkmeyer Press Photo Service. *Reproductions on pg. 29* Bettmann Archive. *2-3* Courtesy of Albert V. Crewe, University of Chicago. *Photos on pgs. 53 and 80* Bettmann Archive. *Photo on pg. 82* © 1981 Martin M. Rotker/Taurus Photos. *5-1* Monkmeyer Press Photo Service. *5-3* Bettmann Archive. *5-6* © 1981 Martin M. Rotker/Taurus Photos. *6-12* Charles Farrow. *6-13* © 1981 Martin M. Rotker/Taurus Photos. *Photo on pg. 155* Monkmeyer Press Photo Service. *Photo on pg. 230* Paul Conklin from Monkmeyer Press Photo Service. *9-4* © 1981 Martin M. Rotker/Science Photo Library International. *10-6* © 1981 Martin M. Rotker/Taurus Photos. *Photo on pg. 295* Monkmeyer Press Photo Service. *Photos on pgs. 342, 402, 531* Bettmann Archive. *20-3* Courtesy of Albert L. Jones and Ira S. Goldman. *21-3* © 1981 Martin M. Rotker/Taurus Photos. *22-12* Photo reprinted by permission of Dr. M.J. Nadakavukaren. *25-3* Russell A. Thompson/Taurus Photos. *25-4* © 1981 Martin M. Rotker/Taurus Photos. *25-6* Baird Corporation. *25-7, 25-13* © 1981 Martin M. Rotker/Taurus Photos. *25-15* Mimi Forsyth/Monkmeyer Press Photo Service. *25-16* USAF/Monkmeyer Press Photo Service.

INDEX

Page numbers in bold face type refer to a principal presentation or to the structure of a compound.

A

Absolute temperature scale. *See* Kelvin temperature scale
Absolute zero, 120
Accuracy, **12**
Acetaldehyde, **330**, **333**, 355, 358
Acetals: from monosaccharides, 374; from simple hemiacetals, 356-57
Acetaminophen, **708**, 709, 710
Acetic acid, 227, **228**, 330, 333, 392
Acetoacetate, **613**
Acetoacetic acid, **523**
Acetone, 355, 358, **523**, 613
Acetylcholinesterase, 505
Acetyl CoA. *See* Acetyl coenzyme A
Acetyl CoA carboxylase, 615
Acetyl coenzyme A, 547, 608, 611, 612, 613, 615, 617, 618, 620
Acetylene. *See* Ethyne
Acetylsalicylic acid (aspirin), 399, **709**
Acid anhydride, 228
Acid-base indicator, 243
Acid-base neutralization, 226, 243, 245-46
Acid-base theories, **221-24**
Acid-base titration, **243-45**
Acid chloride, 429
Acid ionization constant (K_a), **230-33**, 267, 268
Acidosis, **271-73**, 613; metabolic, 272-73, 588; respiratory, 272, 273
Acids, **221-36**; Arrhenius, 221; Bronsted-Lowry, 222-24; carboxylic, 235-36, **385-94**; commercially available, 227; common, **226-28**; conjugate, 224, 232; as food additives, **705**; naming, **225**; normality of, **238-42**; oxy-, 235; polyprotic, 234-35; properties of, **225-26**; strong, **231**, 233-34; weak, **231**, 234-36
Aconitase, 564-65
cis-Aconitate, **563**, **564**, 565
Actin, 456
Activated complex, 212
Activation energy (E_a), **211-12**, 214
Active metals, 44
Active site, 493
Active transport, 554-55
Activity of radioactive material, 732, **733-34**

Acyl carrier protein (ACP), 614, 616
Acyl CoA, 609
Acyl CoA dehydrogenase, 609
Acyl CoA synthetase, 608
Acyl group, 391
Addition reactions: of aldehydes and ketones, **354-56**; of alkenes, **298-301**
Adenine, **449**, 638
Adenosine, 639
Adenosine diphosphate. *See* ADP
Adenosine monophosphate. *See* AMP
Adenosine triphosphate. *See* ATP
Adenyl cyclase, 596
Adipic acid, **705**
Adipose tissue, 456, 606
ADP (adenine disphophate), 489, **457**, **640**; hydrolysis of, 558
Adrenal glands, 169, 430, 534, 535, 537
Adrenocorticotropic hormone (ACTH), 536, 607
Adriamycin, **717**
Aflatoxins, 656
Alanine, 433, **434**
β-Alanine, **441**
Albinism, 632
Albumin, 474, 522; serum, 456, 607
Alchemy, 3
Alcohol dehydrogenase, 486, 508, 589, 594
Alcohols, **319-34**; chemical properties of, **325-31**; ester formation from, **331**, 391-92, 394; hydrogen bonding in, **320**; nomenclature of, **321-23**; physical properties of, **323-25**; sources of, **331-32**; structure of, **319-21**; subclasses of, **320-21**; uses of, **332-34**
Aldehydes, 328, **349-59**; chemical properties of, **354-57**; nomenclature of, **351-53**; sources and uses of, **358-59**; structure and physical properties of, **350-51**
Aldolase, 583
Aldoses, 360
Aldosterone, **533**, 537
Aliphatic hydrocarbons, 284; cyclic, **304-5**; *See also* Alanes; Alkenes; Alkynes
Alkali metals, 45
Alkaline earth metals, 45
Alkaloids, 401
Alkalosis, **271-73**; metabolic, 273; respiratory, 273
Alkanes, **284-93**; chemical properties of, **291-92**; nomenclature of, **286-90**; physical properties of, **291**; sources

and uses of, **293**; structure of, **284-86**
Alkaptonuria, 632
Alkenes, **293-301**; chemical properties of, **298-301**; nomenclature of, **294-97**; physical properties of, **297-98**; sources and uses of, **301**; structure of, **293-94**
Alkoxy group, 391
Aklylating agents, **716**
Alkybenzenesulfonates (ABS), **718**
Alkyl groups, 289
Alkyl halides, **307-12**
Alkynes, **301-4**; chemical properties of, **303-4**; nomenclature of, **303**; physical properties of, 303; sources and uses of, **304**; structure of, **302**
Allosteric enzymes, 505
Allosteric interactions, 474
Alpha (α) decay, **51**
Alpha- (α-) helix, **463-65**; hydrogen bonding in, 464
Alpha- (α-) keratin, 464
Alpha (α) particles, 51, 735, 738
Amantadine, **715**
Ames, Bruce, 657
Ames test, 657
Amide (amido) group, 283, **442**
Amides, 428-29, 440-41, **442-47**; chemical properties of, 446-47; examples of, **447**; formation of, 428-29, 440-41; hydrogen bonding in, 444-45; hydrolysis of, 446-47; neutrality of, **446**; nomenclature of, **442-44**; physical properties of, **444-46**; structure of, **442**
Amine salts, **427-28**
Amines, **421-32**; amide formation from, **428-29**; aromatic, **425-26**; basicity of, **427-28**; chemical properties of, **427-29**; examples of, **430-32**; hydrogen bonding in, 425; nomenclature of, **423-25**; physical properties of, **425-27**; structure of, **422-23**; subclasses of, **422**
Amino acid composition, 460
Amino acid degradation, **623-26**; oxidative deamination in, **625-26**; transamination in, 624-25
Amino acid metabolism **622-32**; fates of carbon atoms in, 629-30; inborn errors in, **630-32**; urea cycle in, **626-29**
Amino acid pool, 622-23
Amino acid score, 671
Amino acid synthesis, **630**

I-1

Amino acids, **432-42**, 603; abbreviations of, 433, **434-35**; acidic (negatively charged) side chains of, **435, 436**; alpha-, 432; amphoteric nature of, **437-39**; basic (positively charged) side chains of, **435, 436**; chemical properties of, **437-41**; chirality of, 436; classification of, **433-36**; degradation of, **623-26**; essential, 623, 624, 671; families of, 436-37; fates of carbon atoms in, **629-30**; formation of amides from, **440-41**; glycogenic, 630, 631; human requirements for, 670-71; ionic forms of, **437-38**; ketogenic, 630, 631; limiting, 671, 672; metabolism of, **622-32**; neutral (uncharged) polar side chains of, **433, 434-35**; nonessential, 624, 630, 671; nonpolar (hydrophobic) side chains of, **433, 434**; optical activity of, **436-37**; physical properties of, **436-37**; reaction of, with ninhydrin, 440; side chains of, 432-33; structure of, **432**; symbols of, 433, 434-35; synthesis of, **630**; unusual, **441-42**. *See also names of individual amino acids*
p-Aminobenzoic acid (PABA), **431**
p-Aminobenzoic acid esters, **431-32**
γ-Aminobutyric acid, **441, 706**
Amino group, 283, **421**
Ammonia, 75, 229
Ammonium hydroxide, 227, **229**
Ammonium ions, 71, 236, **239**; physiological disposal of, **626-29**
Amorphous solids, **154-56**
AMP (adenosine monophosphate), **557, 639**
Amphetamines, **430-31**
Ampholytes, 437
Amphoteric, 437
Amylase: pancreatic, 516, 517; salivary, 514
Amylopectin, **378**
Amylose, **378**
Amytal, 560
Anabolism, 546
Analgesics, 337, **709-10**
Androgens, **538**
Anemia, 737; hemolytic, 686; pernicious, 679; sickle-cell, 104, 480-81
Anesthetics, 337, **710-11**
Angiotensin II, 533-34, 537
Angstrom (Å), 43
Anhydride linkage, 556
Anhydrides: acid, 228; basic, 229
Anions, 66; polyatomic, 71
Anode, 256
Anomers, 368
Antacids, **240**

Antibiotics, **712-15**
Antibodies, 456-522
Anticaking agents, 707
Anticancer drugs, **716-17**
Anticoagulants, 393, 522
Anticodons, 650-51
Antifoaming agents, 707
Antiinflammatory agents, 709
Antimetabolites, 712, 716
Antioxidants, 686, 704
Antipyretic agents, 394, 709
Antiseptic, 334
Antiviral agents, **715**
Apoenzyme, 491
Aqueous, 170
Archeological dating, **739-40**
Arginine, 435, **436**
Argininosuccinate, **627**, 628
Aristotle, 30
Aromatic hydrocarbons, 284, **305-7**; benzene, **306-7**; polycyclic, **307**
Arrhenius, Svante, 221
Arrhenius acids and bases, 221-22
Aryl halides, 309
Ascorbic acid (Vitamin C), **400**, 675, **681-82**, 704
Asparagine, 433, **435**
Aspartame, 707-8
Aspartate, 624, 625
Aspartic acid, **435, 436**, 477; ionic forms of, 438
Aspirin (acetylsalicylic acid), **399, 709**, 710
Asymmetric carbon atoms, 361
Atherosclerosis, 605
Atom, 29-30
Atomic bomb, **747-50**
Atomic crystals, **154**
Atomic hypothesis, 30
Atomic mass units, **35**
Atomic nucleus, 33
Atomic number, **34**
Atomic orbitals: d-orbital, 40, 41; f-orbital, 40; p-orbital, 36-37, 40, 41; s-orbital, 36-37, 40, 41
Atomic size, 43
Atomic structure, **33-41**
Atomic weight, **36**
ATP (adenosine triphosphate), 489, 545, **556-58, 640**; hydrolysis of, 557-58; production of, by citric acid cycle, 569; stability of, 557; yield of, from complete oxidation of glucose, 591
ATPase, 555, 556
Atropine, **401-2**
Average binding energy, 745
Avidin, 678
Avogadro, Amedeo, 95, 127-28
Avogadro's hypothesis, **127-28**, 131

Avogadro's number, 95, 128

B

Balanced diets, **689-90**
Barbital, **447**
Base ionization constant (K_b), **236-38**
Base pairing (in DNA), 644-45
Bases, **221-29, 236-38**; Arrhenius, 222; Bronsted-Lowry, 222-24; commercially available, 227; common, **229**; conjugate, 224, 232; as food additives, **705**; normality of, **238-43**; properties of, **226**; strong, 236; weak, **236-38**
Basic anhydride, 229
Basic Four food guide, 689-90
Becquerel (Bq), 733, 734
Becquerel, Henri, 50, 725, 734
Benedict's test, 371
Benz [a] pyrene, **307, 656**
Benzedrine (amphetamine), 430-31
Benzene, **306-7**
Benzocaine (ethyl p-aminobenzoate), **431**
Benzoic acid, **393, 699**
Benzoic acid salts, 699
3,4-Benzpyrene, **307, 656**
Beriberi, 676
Beta (β) decay, 51, **52-53**; electron capture in, 51, **52**; electron emission in, 51, **52**; positron emission in, 51, **52**
Beta- (β-) oxidation pathway, 608, **609-12**; cellular location of, 607, 608, 614; energetics of, **611-12**; reactions of, **609-11**
Beta (β) particle, 51, 52, 735, 738
Beta- (β-) pleated sheet, **465-66**; hydrogen bonding in, 465
BHA (butylated hydroxyanisole), 704-5
BHT (butylated hydroxytoluene), 704-5
Bile, **517-20**
Bile acids, 618
Bile pigments, **519-20**
Bile salts, **517-19**, 613
Bilirubin, **519**, 524
Biliverdin, **519**
Bimolecular reactions, 214
Binary ionic compounds, **70-71**; rules for naming, 70
Binary molecular compounds, **64-65**; rules for naming, **64**
Binding energy, nuclear, **745-46**
Bioenergetics, **555-62**
Biological damage by radiation, **735-36**
Biologic value, 671
Biosphere, 210
Biotin, 615, 675, **678**
Bleaching agents, 707

Bleomycin, 743
Blood, **520-30**; composition of, **521-22**; functions of, 520-21; hemoglobin in, **524-28**
Blood cells, 521
Blood clotting, **528-30**, 686, 687; extrinsic pathway of, **530**; factors of, 528, 686-87; formation of clot fibers (the common pathway) in, **529**; intrinsic pathway of, **529**
Blood gases, 134
Blood lipids, **604-6**
Blood plasma, 521
Blood serum, 370, 521
Blood sugar, **576-77**
Body fluids, 190, **513-34**; bile, **517-20**; blood, **520-30**; in digestion, **513-20**; extracellular, 513; gastric juice, **514-16**; interstitial, 513; intestinal juice, **516-17**; intracellular, 513; pancreatic juice, **516**; saliva, **514**; urine, **530-34**
Bohr, Niels, 748
Bohr effect, 527
Boiling-point elevation (of solutions), **185**
Boiling points, **147-48**
Bonds. *See* Chemical bonds
Boyle, Robert, 27, 29, 116
Boyle's law, **116-17**
Bradykinin, **461**
Brand drug names, 708-9
Bromelain, 506
Bronsted, Johannes, 222
Bronsted-Lowry acids and bases, 222-24
Brown, Robert, 188
Brownian motion, 188
Buffer pairs, 265-66, 271; carbonic acid/bicarbonate, 271; dihydrogen phosphate/hydrogen phosphate, 271; table of, 265
Buffers, **265-73**, 705; in blood, **271**; concentrations of, 270
Builders' detergent, 719
Butane, **285**, 286, 293
Butene, 295, 301
Butesin (butyl *p*-aminobenzoate), **431**
Butter, **703-4**
Buttermilk, **703**
Butyl *p*-aminobenzoate (Butesin), **431**
Butylated hydroxyanisole (BHA), 704-5
Butylated hydroxytoluene (BHT), 704-5
B-vitamins, **675-81**. *See also names for individual B-vitamins*

C

Caffeine, 710
Calcium cyclamate, **707**
Calories (cal), 6, 18; empty, 668; food, 18, 668

Calorimeter, 206, 207
Calvin, Melvin, 740, 741
cAMP. *See* Cyclic AMP
Capillary action, 167
Capillary blood vessels, 191
Carbamoyl phosphate, **627**, **628**
Carbamoyl phosphate synthetase, 628
Carbohydrate hydrolase reaction, 488
Carbohydrate metabolism, **575-99**; energetics of glycolysis in, **586-87**; fates of pyruvate in, **587-90**; glycogen breakdown in, **578-80**; glycogen synthesis in, **577-78**; glycolysis in, **580-87**; gluconeogenesis in, **592-93**; inborn errors in, **580**, **598-99**; pentose phosphate pathway of, **593-94**; regulation of, **594-97**
Carbohydrates, **359-79**; caloric content of, 603, 606, 668; disaccharide, 360, **374-77**; membrane, 551; metabolism of, **575-99**; in milk, 700; monosaccharide, 359, **366-74**; as nutrients, **674-75**; optical isomerism in, **361-66**; polysaccharide, 360, **377-79**; recommended dietary allowances of, 669, 674
Carbolic acid, 340, 342
Carbon dating, **739-40**
Carbon dioxide, 228, 292; in acidosis, 272, 273; in alkalosis, 273; formation of, by citric acid cycle, 562, 563, 565, 566; in leavening, **707**
Carbonic acid, **228-29**
Carbon monoxide, 292, 354; binding of, to hemoglobin, 526
Carbon tetrachloride, 79, 656
Carbonyl group, 282, 349; polarity of, **350-51**
Carboxylase/decarboxylase reaction, 488
Carboxylate ions, 390; as conjugate bases, 390
Carboxylation, 678
Carboxyl group, 282, **385**
Carboxylic acid dimers, 386
Carboxylic acid salts, 390-91; nomenclature of, 390-91
Carboxylic acids, 235, 329, **385-94**; acidity of, **389-90**; chemical properties of, **389-91**; ester formation from, 331, **391-92**, 394; examples of, **392-94**; hydrogen bonding of, 386; neutralization reactions of, 390; nomenclature of, **386-88**; physical properties of, **388-89**; structure of, **386**
Carboxypeptidase, 495
Carcinogens, **656-58**; chemical, **656-57**; radiation as, **657**; viruses as, **657-58**
Cardiac glycosides, 555
Carnosine, **461**

Carmustine, **716**
β-Carotene, **413**, **684**, 704, 706
Carotenoids, 413
Casein, 456, 700, 701
Catabolism, 546
Catalysis, **214-15**
Catalysts, **214-15**
Catechol, **430**
Catecholamines, 430
Cathode, 256
Cation, 66
Cells, **543-46**; eucaryotic, **543-46**, 643, 645, 647; membrane of, 544, **548-55**; nucleus of, **545**; procaryotic, 543-44; wall of, 714
Cellulose, **379**, 575, 741
Celsius temperature scale, 19-20
Cerebrosides, **412**, 548, 549
Ceruloplasmin, 475
Chadwick, James, 727
Charles, J.A.C., 119
Charles's law, **119-20**, 122
Cheeses, **700-2**; characteristics of, 702; ripening (aging) of, 701
Chemical bonds, **59-83**; coordinate covalent, 223; covalent, 59, **60-61**; ionic, 59, **66-67**
Chemical carcinogens, **656-57**
Chemical equations, **88-89**; balancing, **90-93**
Chemical equilibrium, **197-211**, 216; dynamic aspects of, **197-99**; factors affecting, **202-5**; and LeChatelier's principle, **202-5**, 216; position of, **201-2**, 209-210; and reaction rates, **215-16**; and reversible reactions, **197-211**; shift in, **202-5**. *See also* Equilibrium
Chemical family, 32
Chemical group, 32
Chemical period, 32
Chemical periodicity, **41-43**
Chemical preservatives, 699, 704-5
Chemical reactions, **87-108**; types of, **101-8**
Chemical symbols, 32
Chiral center, 363
Chirality, **362-63**
Chloral hydrate, 355-56
Chloroform, 310-12, 656, 711
Chlorotetracycline, **714**
Cholecalciferol. *See* Vitamin D_3
Cholecystokinin, 516, 537
Cholesterol, **414**, 518, 524, 604, 605-6, 613, **618-619**, 684; dietary, 674; synthesis of, 617, 618
Cholesterol metabolism, **618-19**
Cholesteryl esters, **604**
Cholic acid, 517-18, 618
Choline, **410**

Chromatography: gel filtration (gel permeation), 478; ion-exchange, 478; liquid, **477-79**
Chromosomes, 637, 645
Chylomicra, 604, 605
Chyme, 514
Chymosin, 700, 701
Chymotrypsin, 492, 503
Chymotrypsinogen, 502-3
Cistron, 649
Citrate, 562-**63**, **564**, 565. *See also* Citric acid
Citrate carrier system, 615
Citrate synthetase, 564-568
Citric acid, 393, **562**, 704, 705. *See also* Citrate
Citric acid cycle, 393, **562-70**; cellular location of, 562; energetics in, **569-70**; entry points for amino acid skeletons in, **629-30**; reactions in, **562-67**; regulation of, **568**; summary and major features of, **569**
Citrulline, 441, 627, 628
Clark, Leland, C., Jr., 312
Clinitest tablets, 371
Clostridium botulinum, 457
Clotting factors, 528, 686-87
Cobalt ion, 679, 680
Codeine, **710**
Codons, 649-51
Coefficients (in chemical equations), 88
Coenzyme A, **501**, 608, 678
Coenzyme B$_{12}$, 679, **680**
Coenzyme Q (CoQ), 559-60
Coenzymes, **497-502**; flavin, **499-500**; pyridine, **497-99**. *See also* names for *individual coenzymes*
Cofactors, 491
Collagen, 441, 456, **466**, 471, 475, 682
Collagen triple helix, **466**
Colligative properties: boiling-point elevation, **185**; freezing-point lowering, **185**; osmotic pressure, **186-87**; vapor-pressure lowering, **184**
Colloidal dispersion. *See* Colloids
Colloids, **187-89**; types of (table), 189
Coloring agents, **706**
Combination reactions, **102**
Combined (ideal) gas equation, 128
Combustion, 89, 105; of alcohols, 328; of aldehydes and ketones, 354; of alkanes, **291-92**; of alkenes, 298; of alkynes, 303; of ethers, 337
Common pathway, **529**
Competitive inhibition, 504
Complementary proteins, 672
Complex ionic compounds, **71-72**
Complex ions, 393
Complex lipids, 403; phosphoglycerides,

409-11; sphingolipids, **411-14**; triacylglycerols, **405-9**; waxes, **405**
Compounds, 28-29
Concentration gradient, 185
Concentrations: effect of, on reaction rates, **213-14**; of solutions, **176-80**
Condensed structural formulas, 286
Condensing agents, 440
Configuration, 361
Conformation, 470
Conjugate acid-base pairs, 224, 232
Conjugated proteins, 475
Consumer products: drugs, **708-17**; food preservation and food products, **697-708**; home products, **717-20**
Controlled oxidation: of primary alcohols, **328-29**; of secondary alcohols, 329
Conversions: of measurement systems, **7-11**; of temperature scales, 18-21; unit-factor method of, **8-11**
Cooking oils, **704**
Cooperative effect, 474
Coordinate covalent bonds, 223
Corey, Robert, 463, 464, 465
Cori cycle, 593
Corrin ring, 679
Cosmic rays, 725-26
Covalent bonds, 59, **60-61**; coordinate, 223; multiple, **80-81**
Covalent crystals, **153**
Covalent modification, **504**
Creatine phosphokinase (CPK), 506
Crenation, 187
Cresols, **341**
Crick, Francis H.C., 644
Critical mass, 747, 748, 749
Crystal lattice, 68-69
Crystalline solids, *See* Crystals
Crystals, 68-69, **150-54**, 160; atomic, **154**; covalent, **153**; ionic, **151-52**; liquid, 160; metallic, **154**; molecular, **152-53**
C-terminus, 459
Curie (Ci), 733, 734
Curie, Marie, **53**, 657, 733
Cyanide ions, 71; reaction of, with cytochrome oxidase, 505, 560; reaction of, with enzymes, 491, 505; in vitamin B$_{12}$, 679, 680
Cyclamates, 656, **657**, 707
Cyclic adenosine monophosphate (cAMP). *See* Cyclic AMP
Cyclic aliphatic hydrocarbons, **304-5**
Cyclic AMP (cAMP, cyclic adenosine monophosphate), **539**, 597, 607
Cyclohexane, 305
Cyclohexene, **305**
Cyclopentane, **305**

Cyclophosphamide, **716**
Cyclopropane, 305
Cysteine, 433, **435**
Cystine, **433**, 459
Cytochrome oxidase, 505, 560
Cytochromes, 475, 560
Cytoplasm, 545
Cytosine, **449**, 638

D

Dairy products, **700-3**
Dalton, John, 27, 132
Darvocet-N, 710
Darvon, 710
Darvon Compound 65, 710
Datril, 708
Deaminase reaction, 488
Deamination, **625-26**
Decarboxylation, 676
Decomposition reactions, **102**
Dehydration: in fatty acid synthesis, 616; of humans, **169**; of primary alcohols, **325-26**; of secondary and tertiary alcohols, **326-27**
L-Dehydroascorbic acid, **681**
7-Dehydrocholesterol, 684, **685**
Dehydrogenases, 584
Dehydrogenation, 486, **609-10**, **611**
Demerol, 710
Demineralization, 717
Democritus, 29-30
Denaturation, 475-77
Density, **15-16**; of liquids, **144-45**; values of (table), 15
Dental chemistry, **717-18**
Deoxyribose, **366-69**, **638**
Deoxythymidine, **639**
Detection of radiation, **729-30**
Detergent builders, 719
Detergents, **718-19**
Detoxification, 623
Deuterium, 35, 750, 751
Dexedrine, 430-**31**
Dextrins, 514
Dextroamphetamine. *See* Dexedrine
Dextrorotatory, 364
Diabetes insipidus, 533
Diabetes mellitus, 533, 597, 613, 632, 655-56
Diacylglycerols, 603, 604, 705
Dialysis, **189-90**
Dicoumarol, **687**
Diet: balanced, **689-90**; restorative, **690-91**; restrictive, 690, **691-92**
Diethyl ether, 337, 710-11
Diet therapy, **690-92**
Diffusion: facilitated, 554-55, 577; of gases, **116**, 207, 208

Digestion, 491, **513-20**; of fats, 516-17, 603; of polysaccharides, 514, 515-17; of proteins, 516-17
Dihydroxyacetone phosphate, **584**, 590, 607, 617
1,25-Dihydroxycholecalciferol, 685
L-Diketogulonic acid, **681**
Dilution factor, 181
Dilutions, **180-82**
Dimensional analysis, 7
Dimers, 386
N^6-Dimethyladenine, **659**
Dimethylchlorotetracycline, **714**
2,3-Diphosphoglycerate (DPG), **526**-27, **585**, 599
1,3-Diphosphoglyceric acid, **584**, **585**
Diphosphoglyceromutase, 585
Dipolar attractions, **140-41**; in tertiary protein structure, 470
Dipolar bond, 78
Dipolar ions (zwitterions), 437
Dipole moments, 78
Disaccharides, 360, **374-77**. See also lactose; maltose; sucrose
Dissolution, **171-72**
Distal histidine, 469
Distillation, 183
Disulfide, 343
Disulfide bond linkages, 343, 459, 470
Disulfide group, 283, 343, 459, 470
Diuresis, 532
Diuretics, **532**
DNA, 637, 638, **643-47**, 659, 660, 715, 716, 717; base pairing of, 644-45; damage to, 647, 736; molecular weight of, 643; recombinant, **659-62**; repair of, **646-47**, 736; replication of, **646**; role of, in cell, **645**; secondary structure of, **644-45**
DNA-directed RNA polymerase, 648
DNA ligases, 659, 660
DNA polymerases, 642, 646, 715
DNA repair, **646-47**, 736
Dosage of radiation, 732, **734**
Dosimeter, 730
Double bond, **81**
Double helix, **644-45**
Double replacement reactions, 101
Doxorubicin, **717**
DPG (2,3-diphosphoglycerate), **526**-27, **585**, 599
Drain cleaners, **719-20**
Drug names, **708-9**
Drugs, **708-17**; bacteriocidal, 711; bacteriostatic, 711; for curing disease, **711-17**; names of, **708-9**; sulfa, **711-12**; for treating symptoms, **709-11**. *See also individual drug names*
Drying, 698, 699

Duodenum, 516
Dynamic equilibrium, **146-47**; of liquid-solid mixtures, 156; of liquid-vapor mixtures, **146-47**; of saturated solutions, 173
Dynamite, **402**

E

Edema, 532, 672, 676, 691
Einstein, Albert, 745, 748
Einstein equation, 725, 745
Elastin, 456, 475
Electric dipole, 78
Electrodes, 255
Electrolysis, **255-57**
Electrolytes, **253-57**, 688-89; concentrations of, 255; examples of (table), 255; in humans (table), 256; non-, **254-55**; strong, **254**; weak, 254
Electron capture, 51, **52**
Electron configurations, **36-41**; of ions, 67; steps for determining, 40-41; table of first thirty-six elements, 39
Electron dot structure. *See* Electronic formulas
Electronegativity, **77-78**
Electron emission, 51, **52**
Electron formulas, 61-64; rules for writing, 62
Electron spin, 37-38
Electron transport system, **559-60**
Electron-volt (eV), 733, 734
Electrons, 33, 35, **36-41**, 51-**52**; valence, 42
Electrophoresis: of amino acids, **439**; of proteins, 479
Elemental composition of the human body, 49
Elements, 27-29, 32, 44-50; ancient Greek, 27; first list of, 28; important to life, **48-50**; and life, **47-48**; inner transition, 45; representative, 44-45; transition, 44
Elements important to life: carbon, **49**; hydrogen **48-49**; nitrogen, **49-50**; oxygen, **49**; table of, 48
Elimination reactions, 309
Embden-Meyerhof pathway. *See* Glycolysis
Empty calories, 668, 675
Emulsifiers. *See* Emulsifying agents
Emulsifying agents, 411, 705
Enantiomers, 363
Endergonic, 208
Endocrine glands, 534
Endoplasmic reticulum, **545-46**
Endorphins, **462**
Endothermic, 18, 173, 206

Energetics: of the citric acid cycle, **569-70**; of fatty acid oxidation, **611-12**; of glycolysis, **586-87**
Energy, **17-18**; and the biosphere, **210-11**; from carbohydrates, 603, 608, 668; and chemical change, **205-10**; classes of, 17; from fats, 603, 608, 668; from food, **546-47**; forms of, 18; free, 208, 210-11; from macronutrients, **668-70**; from proteins, 603, 608, 668; of radiation, 733, **734-35**; units of, 6, 18
Energy conservation from glucose oxidation, 591
Energy of activation (E_a). *See* Activation energy
Energy of radiation, 733, **734-35**
Enflurane, 711
Enkephalins, 462
Enolase, 586
Enolpyruvate, **586**
Enoyl CoA hydratase, 610
Enterocrinin, 516, 537
Enthalpy change (ΔH), 205-8
Entropy, 208, 211
Entropy change (ΔS), 208
Enzyme activity, **495-97**; effect of co-factor concentration on, **495**; effect of enzyme concentration on, **496**; effect of pH on, **496**; effect of substrate concentration on, **495-96**; effect of temperature on, **496**; regulation of, **502-6**
Enzyme catalysis, **492-94**; rates and equilibria in, **492**; theories of, **493-94**
Enzyme inhibition, **504-5**
Enzyme inhibitors, 215, **504-5**
Enzyme specificity, **492-93**
Enzyme-substrate complex. *See* Michaelis complex
Enzymes, 215, 456, **485-97**, **502-8**, 735, 736; factors affecting activity of, **495-97**; in genetic engineering, **658-59**; names and classes of, **485-91**; properties and functions of, **491**; restriction, 658-59; uses of, **506-8**. *See also specific names of enzymes*
Epidermin, 465
Epinephrine, **430**, 539, **596**, 597, 607
Equanil, **709**
Equilibrium, **146-47**; chemical, **197-211**, 216; dynamic, **146-47**, **197-99**; glucose-fructose, 197-99; of liquid-solid mixtures, 156; of liquid-vapor mixtures, **146-47**; position of, **201-2**, 209-10; of reversible reactions, **197-211**; in saturated solutions, 173
Equilibrium constant (K_{eq}), **199-205**; calculation of, 199-201;

Equilibrium constant (K_{eq}) (cont.)
and estimation of product formation, 201-2; expressions for, 199-201; and rate constants, 215-16; and standard free energy change, 209-10; utility of, **201-2**
Equivalence point, 243
Equivalents, 239-43; milli-, 242
Erg, 6, 734
Ergocalciferol. *See* Vitamin D_2
Erythrocytes. *See* Red blood cells
Essential amino acids, 623, 624
Essential fatty acids, 405, 616, 673
Ester froup, 282, 394
Esters, 331, 391, **394-402**; chemical properties of, **397-98**; examples of, **399-402**; formation of, 331, 391-92, 394; hydrolysis of, **397-98**; nomenclature of, **395-96**; physical properties and uses of, **396-97**
Estrogen, 537, **538**
Ethane, **285**
1,2-Ethanediol (ethylene glycol, glycol), **334**
Ethanol, 330, **333-34**, 508, 589, 594
Ethanolamine, **410**
Ethene, **81**, 293, **294**, 301
Ether group, 282, **334**
Ethers, **334-37**; chemical properties of, **337**; nomenclature of, **335**; physical properties of, **336-37**; sources and uses of, **337**; structure of, **335**
Ethyl *p*-aminobenzoate (Benzocaine), **431**
Ethyl chloride, 711
Ethyne (acetylene), **81**, 302, 304
Eucaryotic cells, **543-46**, 643, 645, 647
Evaporation, 146
Exergonic, 208
Exothermic, 18, 173, 174, 206
Exposure to radiation, 732, 734, **736-37**
Extracellular fluid, 513
Extracorporeal irradiation, 742
Extrinsic pathway, 530

F

Facilitated diffusion, 554-55, 577
Facilitated transport, 554
FAD (flavin adenine dinucleotide, oxidized), **500**, 547, 677
$FADH_2$ (flavin adenine dinucleotide, reduced), **500**, 547, 559, 560, 562
Fahrenheit temperature scale, 18-20
Fats, 405, 406, 617, 719, 720; absorption of, **603-4**; caloric content of, 603, 608, 668; caloric equivalent of, in weight gain and loss, 668; digestion of, 516-17; metabolic water from, 673; and oils, **703-4**; recommended dietary allowances of, 669, 673
Fat-soluble vitamins, **683-87**; absorption of, 519. *See also individual names of fat-soluble vitamins*
Fatty acid oxidation, **608-12**; cellular location of, 607, 608, 614; energetics of, **611-12**
Fatty acids, **404-5**, 603, 605, 606, 617; activation of, 608; essential, 405, 616, 673; mobilization of, **606-8**
Fatty acid synthesis, **614-17**; cellular location of, 614
Fatty acid synthetase, 614, 616
Feedback inhibition, 505
Fehling's test, 371
Fermentation, 332, 589, 698, 700, 701, 703
Fermi, Enrico, 748
Ferritin, 456
Fiber, dietary, 667-8
Fibrin, 465, 529
Fibrinogen, 457, 521, 529
Fibrous protein, 475
Film badge, 730
Fischer, Emil, 363, 364, 494
Fischer projection, **363**
Fission, nuclear, 746-50
Flavin adenine dinucleotide, oxidized. *See* FAD
Flavin adenine dinucleotide, reduced. *See* $FADH_2$
Flavin coenzymes, **499-500**
Flavin mononucleotide, oxidized. *See* FMN
Flavin mononucleotide, reduced. *See* $FMNH_2$
Flavor enhancers, **706**
Flavoring agents, **705-6**
Fluid mosaic model, 553
Fluids, 139. *See also* Body fluids
Fluoride ions, 718
Fluoride treatment, 718
FMN (flavin mononucleotide, oxidized), **500**, 559, 607
$FMNH_2$ (flavin mononucleotide, reduced), **500**, 559, 560
Folacin. *See* Folic acid
Folic acid (folacin), **431**, 675, **679**
Food additives, **704-8**; for consistency or texture, **705**; miscellaneous, **706-8**; as nutrient supplements, **705**;
Food preservation, **697-708**; modern methods of, **699-700**; traditional methods of, 698
Food preservatives, 698, 699-700, **704-5**
Food products: dairy products, **700-3**; fats and oils, **703-4**
Formaldehyde, 328, 333, **350**, 358
Formed elements (in blood), 521
Formic acid, **392**
Formula unit, 93
Formula weight (FW), **93**
Formulas, chemical, **61-64**, **66-67**, **69-72**
Free energy, 208, 210-11
Free energy change (ΔG), 208; standard (ΔG°), 209-10, 557
Free radicals, 735
Freeze-drying, 699
Freezing-point lowering (of solutions), **185**
Freons, **310**
Frisch, Otto, 748
Fructose, 198, 199, 204, 205, **373**, 575, 598-99
Fructose-1,6-diphosphatase, 592
Fructose-1,6-diphosphate, **583**, 584
Fructose-6-phosphate, **583**
Fructosemia, 598-99
Fumarase, 567
Fumarate, **563**, 566-67
Functional groups, 280, **282-83**. *See also names of individual functional groups*
Fusion: nuclear, 746, **750-51**; of solids, **156-57**

G

Galactose, **371-72**, 575
Galactosemia, 372, 598, 692
Gamma (γ) decay, 51, **52-53**
Gamma (γ) rays, 51, 52, 735, 738, 741
Gangliosides, **412**, 548, 549, 622
Gas mixtures, **132-34**
Gases, 5, **113-35**; blood, 134; ideal, 134; interpretation of behavior of, **127-32**; partial pressure of, 132-34; properties of, **113-16**; real, **134-35**; relationships of volume, temperature, and pressure of, **116-27**
Gastric juice, **514-16**
Gastric juice enzymes, 516
Gastrins, 514, 537
Gastrointestinal tract, 513-14
Gastrointestinal X-ray, 246
Gay-Lussac, Joseph L., 119
Geiger-Muller counter, 729
Gel filtration (gel permeation) chromatography, 478
Gene, 649
Generic drug names, 708-9
Genetic code, **648-50**
Genetic engineering, **658-62**; enzymes in, **658-59**; plasmids and, **659-60**, 661; recombinant DNA and, **660-62**
Genetic mutations, **652-53**
Geometric (*cis-trans*) isomers, 294
Geraniol, **413**
Germicide, 334
Glass, 154-55

Globular proteins, 474
Globulins, 474, 522
Glucagon, **596**, 597, 607
Glucocerebrosides, 622
Gluconeogenesis, **592-93**, 596
Glucose, 198, 199, 204, 205, **307-71**, 507, **575-77**; Benedict's test for, 371; in blood, **576-77**; Fehling's test for, 371; o-toluidine test for, 370; in urine, 576
Glucose oxidase, 507
Glucose-6-phosphatase, 579, 580, 592
Glucose-1-phosphate, 577, 578, 579
Glucose-6-phosphate, 577, 579, 580, **583**, 741
Glucose tolerance test, 576
β-Glucosidase, 622
Glucosuria, 576
Glutamate, **565**, 624, 625. *See also* Glutamic acid
Glutamate dehydrogenase, 625, 626, 628
Glutamate oxaloacetate transaminase, 625
Glutamate pyruvate transaminase, 625
Glutamic acid, **435**, 436, 477, 706. *See also* Glutamate
Glutamine, 433, **435**
Glutamine synthetase, 626
Gluteraldehyde, 381
Glutathione, **461**
Glyceraldehyde, **361**
Glyceraldehyde-3-phosphate, **584**, 589, 590
Glyceraldehyde-3-phosphate dehydrogenase, 584
Glycerol (1,2,3-propanetriol), **334**, 603, 604, 606, 607
Glycerol kinase, 607
Glycerol-3-phosphate, 607
Glycerol phosphate dehydrogenase, 607
Glycerol phosphate shuttle, 590-91
Glycine, 433, **434**, 517-18
Glycocholic acid, 517-**18**
Glycogen, **378-79**, **577-80**; breakdown of, **578-80**; synthesis of, **577-78**
Glycogenesis, **577-78**
Glycogen granules, 578
Glycogenic amino acids, 630, 631
Glycogen metabolism, **577-80**
Glycogenolysis, **578-80**
Glycogen phosphorylase, 504, 539, 579, 596, 597; regulation of, 504, 539, 596, 597
Glycogen storage disease I (von Gierke's disease), 580
Glycogen synthesis, **577-78**
Glycogen synthetase, 578, 596, 597
Glycolipids, 548, 549, 551, 617
Glycols, 334, 700

Glycolysis, 575, **580-87**, 594; aerobic, 580; anaerobic, 580-81; cellular location of, 580; energy yield of, **586-87**; reactions in, **581-86**
Glycosides, 375
Glycosidic linkages, 375, 639
Glycoproteins, 475, **476**, 551
GMP (guanosine monophosphate), 706
Goiter, 689
Gramicidin S, **482**
Gray (Gy), 731, 734
Gray, Louis Harold, 734
Greek prefixes, 64
Growth hormone, 457; human, 661
GTP (guanosine triphosphate), 566, 569
Guanidino group, **436**
Guanine, **449**, **638**
Guanosine monophosphate. *See* GMP
Guanosine triphosphate. *See* GTP
Gun method of assembly, 749-50

H
Half-life: of proteins, 622; of radioisotopes, **731-32**
Hall, Charles Martin, **258**
Halogenated hydrocarbons, **307-12**; in the environment, **311**
Halogenation: of alkanes, **292**; of alkenes, 298; of alkynes, 304; of unsaturated triacylglycerols, 409
Halogen derivatives of hydrocarbons, **307-12**
Halogens, 45
Halothane, **312**, 711
Hard water, 718-19
Hard-water ions, 718-19
Hasselbalch, Karl, 267
Haworth structures, 369
Heat, **18**; changes of, in chemical reactions, 205-8; of fusion, **156-57**; of solution, 173-74; of vaporization, **148-49**
Heating curve, 158
Heat of fusion (ΔH_{fus}), **156-57**
Heat of solution ($\Delta H_{solution}$), 173-74
Heat of vaporization (ΔH_{vap}), **148-50**
Heavy metal ions: reactions of, with enzymes, 491, 505; reaction of, with proteins, 477
Heme, **448**, 467-**68**, 469, 524
Hemiacetals: from monosaccharides, 368-72; from simple aldehydes, **356**
Hemiketals: from monosaccharides, 373; from simple ketones, **356**
Hemodialysis, **190**
Hemoglobin, 456, 471-**74**, 475, 521, **524-28**; adult, 524, 527; allosteric interactions of, 474; cooperative effect of, 474; fetal, 524, 527; heme in, 473;

interaction of, with DPG, **526-27**; oxygen affinity of, **525-27**; oxygen binding in, 473-74; primary structure of, 473; quaternary structure of, 472; regulation of oxygen binding in, **526-27**; tertiary structure of, 473; transport of carbon dioxide by, **527-28**; transport of oxygen by, **525-26**; transport of protons by, 526, 527; types of, **524**
Hemolysis, **187**
Hemophilia, 530, 531
Hemoproteins, 475
Hemostasis, 528
Henderson-Hasselbalch equation, **267-70**
Henderson, Lawrence, 267
2-Heptanone, **702**
Heroin, **402-3**, 710
Heterocyclic compounds, 448
Heterocyclic nitrogen compounds, **448-49**; with five-membered rings, **448**; fused-ring, **449**; with six-membered rings, **449**
Heterogeneity, 165
Hexachlorophene, **340**
Hexokinase, 582
Hexokinase deficiency, 599
Hexosaminidase A, 622
Hexyl resorcinol, **340**
High-quality proteins, 671
Histidine, **435**, 436
Histones, 643
Holley, Robert W., 648
Holoenzyme, 491
Home products: dental chemistry and fluoride treatment, **717-18**; drain cleaners and oven cleaners, **719-20**; soaps and detergents, **718-19**
Homocysteine, 441
Homogeneity, 165, 479; of solutions, 183
Homogeneous, 116
Homogenization, 700
Homogentisate, **632**
Homogentisate oxidase, 632
Homoserine, **441**
Hormones, 169, 416, 457, **534-39**, 685; amine, 537, 538, 539; classes of, **537-38**; polypeptide, 537, 538, 539; regulatory, 535; steroid, 537, 538, 539; theories of molecular action of, **538-39**; tissue, 416; trophic, 535. *See also names of individual hormones*
Human growth hormone, 661
Hydrase/dehydrase reaction, 488
Hydrates: inorganic, **167-68**; organic, 355
Hydration: of aldehydes and ketones, 355-56; of alkenes, 299, 332;

Hydration (cont.)
of alkynes, 304; of ions and molecules, 171-72; in the β-oxidation cycle, **609-10**

Hydrocarbons, 80-81, 102-4, **281-84**; occurrence and general uses of, **281**; physical properties of, **281-84**; table of, 103; types of, **284**. *See also* Alkanes; Alkenes; Alkynes; Aromatic hydrocarbons

Hydrochloric acid, 226-27; in gastric juice, 514

Hydrogenation: of aldehydes and ketones, 354-55; of alkenes, 298; of alkynes, 304; in fatty acid synthesis, 616; of unsaturated triacylglycerols, 409, 704

Hydrogen bomb, 750, 751

Hydrogen bonding: in alcohols, **320**; in amides, 444-45; in amines, 425; in base pairs, 644-45; in carboxylic acids, 386; in the DNA double helix, 644-45; effect of, on boiling points (table), 142; in ice, **153**, **166**; in the α-helix, 464; in liquid water, **143**, **153**, **166**; in the β-pleated sheet, 465; in tertiary protein structure, 470; in tRNA, 648

Hydrogen bonds, **142-44**, 172; effect of, on boiling points (table), 142

Hydrogen phosphate ion (inorganic phosphate, P_i), 234, 235, 557

Hydrolases, 487-88

Hydrolysis: of ADP, 558; of alkyl halides, 332; of ATP, 557-58; of esters, **397-98**; of pyrophosphate, 558; of triacylglycerols, **406-7**

Hydronium ions, 223

Hydrophilic, 175-76

Hydrophobic, 175-76

Hydrophobic forces, 470

β-Hydroxyacyl CoA dehydrogenase, 611

Hydroxyapatite, 152, 717, 718

β-Hydroxybutyrate, **613**. *See also* β-Hydroxybutyric acid

β-Hydroxybutyric acid, **523**. *See also* β-hydroxybutyrate

Hydroxyl group, 282, **319**

Hydroxylysine, **441**

Hydroxyproline, **441**, 682

Hypercarotenosis, 684

Hyperglycemia, 576

Hypertonic solutions, 187

Hypervitaminosis, 684, 685

Hypoglycemia, 370, 576, 580

Hypotonic solutions, 187

I

Ideal gas, 134

Ideal gas equation, 128

Idoxuridine, 715

Idoxuridine triphosphate, **715**

Imidazole, **448**

Immunoglobulins, 522

IMP (inosine monophosphate), 706

Implosion design, 749-50

Inborn errors: of amino acid metabolism, **630-32**; of carbohydrate utilization and glycolysis, 588-89; of glycogen metabolism, 580; of lipid metabolism, 621-22; of metabolism, 580, 692

Indicator, acid-base, 243

Indole, **449**

Induced-fit hypothesis, **494**

Inert gases. *See* Noble gases

Ingram, Vernon, 479

Inner transition elements, 45

Inorganic phosphate (P_i), 504, **557**

Inosine monophosphate. *See* IMP

Insulin, 535, 536, 596, 597-98, 617, 632, 661

Integral protein, 551

Interferons, 658, 661

Intermolecular forces: dipolar attractions, **140-41**; hydrogen bonds, **142-44**; London dispersion forces, **141-42**

International Union of Pure and Applied Chemistry (IUPAC), 288

Intestinal juice, **516-17**

Intracellular fluid, 513

Intrinsic factor, 503, 679-80

Intrinsic pathway, **529**

Iodide ions, 741-42

Iodine numbers, 409, 703-4

Iodized salt, 689

Ion-exchange chromatography, 478

Ionic attractions, 470

Ionic bonds, 59, **66-67**

Ionic compounds, **68-74**; binary, **70-71**; complex, **71-72**; naming, **70-72**; properties of, **72-74**

Ionic crystals, **151-52**

Ionic equations, 245-46

Ionization constants: acid (K_a), 230-33, 267, 268; base (K_b), 236-37

Ionizing radiation, 727, 729

Ion-product constant (K_w), 258-59

Ions, 66; electron configurations of, **67**; in hard water, 718-19; nitrite, **82**; physiologically important (table), 73; polyatomic, 71. *See also names of the individual ions*

Isoalloxazine, **449**

Isocitrate, **563**, **564**, **565**

Isocitrate dehydrogenase, 565, 568

Isoelectric point (pI), 439

Isoleucine, 433, **434**

Isomers, **102**; of alkanes (table), 287-88; geometric (*cis-trans*), 294; optical, 362, 364; stereo-, 361; structural, 286, 362

Isomerases, **488-89**

Isomerization reactions, **102-5**; glucose-fructose, 197-99, 215; glucose-6-phosphate/fructose-6-phosphate, 583; of triose phosphates, **584**

Isotonic solutions, 187

Isotopes, 35

Isotopic tracers, **740-41**

IUPAC (International Union of Pure and Applied Chemistry), 288

J

Jaundice, 519-20

Jenson, Elwood V., 539

Joliot-Curie, Irene, 657

Joule, 6, 18

K

Kelvin (absolute) temperature scale, 20-21; and gaseous behavior, 120

Kendrew, John, 467, 468, 472

Keratins, 475; α-, **456**, 464, 465, 466; β-, 465, 466

Ketals: from monosaccharides, 374; from simple hemiketals, **356-57**

Ketoacidosis, 613, 630

β-Keto acids, 624, 625

Ketogenic amino acids, 630, 631

α-Ketoglutarate, **563**, **565**, **566**, 624, 625

α-Ketoglutarate dehydrogenase complex, 566, 568, 681

Ketone bodies, **523**, **612-14**, 630

Ketonemia, 613

Ketones, 329, **349-59**; chemical properties of, **354-57**, nomenclature of, **351-53**; sources and uses of, **358-59**; structure and physical properties of, **350-51**

Ketonuria, 613

Ketoses, 360

Ketosis, 613, 630

β-Ketothiolase, 611

Khorana, H. Gobind, 648

Kidneys: in acid-base balance, **533**; in electrolyte balance, **533**; in regulation of blood pressure, **533-34**

Kilogram, 6, 7

Kinetic energy, 17; of gaseous particles, 132; of liquid particles, 140; relationship to temperature, 132, 140

Kinetic molecular theory, **131-32**; application of, to liquids, 140

Koshland, Daniel E., Jr., 494

Krebs cycle. *See* Citric acid cycle

Krebs, Hans A., 393, 562

Kwashiorkor, 672

L

Lactate, **587**. *See also* Lactic acid
Lactate dehydrogenase (LDH), 506, 587
Lactic acid, **392**, **523**, 698, 701, 703, 705, 718. *See also* Lactate
Lactose, **375-76**, 575, 700, 701
Lactose intolerance, **376**
Lanosterol, 618, **620**
Lavoisier, Antoine Laurent, **29**
Laws: Boyle's, **116-17**; Charles's, **119-20**, **122**; of conservation of energy, 18; of conservation of mass, **89**; of partial pressures, 132; periodic, 43
Leavening agents, **707**
LeChatelier, Henri Louis, 202
LeChatelier's principle, **202-5**, 216; and concentration changes, 202-5; and reaction rates, 216; and temperature changes, 205
Lecithin, **410**, 705. *See also* Phosphatidyl choline
Leucine, 433, **434**
Leukemia, 742
Leukocytes. *See* White blood cells
Levorotatory, 364
Lewis structures. *See* Electronic formulas
Lidocaine (Xylocaine), **432**, **447**, 711
Ligases, **489**, 659, 660
Limiting amino acids, 671, 672
Limonene, **413**
Linear alkylbenzenesulfonates (LAS), **719**
Linoleic acid, 404, 405, 616
Linolenic acid, 404, 405, 616
Lipase reaction, 487
Lipases, 406; adipose-cell, 606, 607; lipoprotein, 605; pancreatic, 603
Lipid bilayers, **548-50**, 553
Lipid metabolism, **603-22**; absorption of fats in, **603-4**; blood lipids in, **604-6**; cholesterol metabolism and, **618**; fatty acid oxidation in, **608-12**; fatty acid synthesis in, **614-17**; inborn errors of, **621-22**; ketone bodies in, **612-14**; mobilization of fatty acids in, **606-8**; relationships between carbohydrate and lipid metabolism, **632-33**; synthesis of triacylglycerols and other lipids in, **617-18**
Lipids, **403-16**; biological roles of, 403; blood, **604-6**; classification of, **403**; complex, 403, **405-12**; fatty-acid, **404-5**; membrane of, **548**; simple, 403, **412-16**
Lipid storage diseases, 621-22
Lipoic acid, 589, 675, **680-81**
Lipoproteins, 475, **604-5**

Liquid chromatography, **477-79**
Liquid crystals, **160**
Liquids, 5, 139-140, **144-50**
Liquiprin, 708
Lister, Joseph, **342**
Liter, 6, 7
Lithium, 68
Litmus, 225-26
Lock-and-key theory, **494**
London, Fritz, 142
London dispersion forces, **141-42**
Lone (electron) pair, 61
Low-energy diets, **691**
Low-protein diets, **692**
Lowry, Thomas, 222
Low-salt diets, **691**
Lyases, **488**
Lymph system, 604
Lysine, **435**, 436; ionic forms of, 438
Lysosomes, **546**

M

McArdle's disease, 580
Macronutrients, 667, 668-70; carbohydrates; **674-75**; fats, **673-74**; proteins, **670-73**
Magdeburg hemispheres, 114
Major minerals, 687-89
L-Malate, **563**, 567
Malate dehydrogenase, 567
Malonyl-ACP, 614, **618**
Malonyl CoA, **615**, 616
Maltose, **375**
Marasmus, 672
Margarine, 409, **703-4**
Markovnikov's rule, 299
Mass, 6, 7, **14-15**
Mass defect, 745
Mass number, **34**
Mass ratio (of reacting substances), 98
Mass units, 6, 7
Matter, 4-5; theory of continuous, 30; theory of discontinuous, 29-30. *See also* Gases; Liquids; Solids
Maximal velocity (V_{max}), 496
Measurement: of radiation, 732-35. *See also* Systems of measurement
Mechloroethamine, **716**
Meitner, Lise, 748
Melanin, 631, 632
Melting point, **157**
Membranes, 544, **548-55**; carbohydrates of, 551; fatty acid composition of, 552; fluidity of, **551-52**; fluid mosaic model of, **553**; lipid bilayers of, **548-50**; lipids of, **548**; mitochondrial, 545, 550, 559; permeability of, **550**; proteins of, **550-51**
Membrane transport, **553-55**

Mendel, Gregor, 637
Menthol, **413**
Meperidine, **710**
Meprobamate, **709**
Messenger RNA. *See* mRNA
Metabolic acidosis, 272-73, 588
Metabolic reactions, 197. *See also names of individual metabolic pathways*
Metabolic water, 673
Metabolism: of amino acids, **622-32**; of carbohydrates, **575-99**; of cholesterol, **617-20**; of glycogen, **577-80**; interrelationships among pathways of, **632-33**; of lipids, **603-22**; overview of, **546-47**
Metallic crystals, **154**
Metalloids, 45
Metalloproteins, 475
Metals, 44-45
Meter, 6, 7
Methadone, **710**
Methamphetamine (Methedrine), **431**
Methane, 75, 284, **285**, 293
Methanol, 333
Methedrine (methamphetamine), 430-31
Methionine, 433, **434**
Methylated bases (in DNA), 658-59
5-Methylcytosine, **659**
1-Methylguanine, **659**
Methyl salicylate (oil of wintergreen), **399**
Mevalonic acid, 618, **619**
MFP (monofluorophosphate), 718
Micelles, 408, 409, 549, 550, 718
Michaelis complex, 493, 495
Micronutrients, 667; fat-soluble vitamins, **683-87**; minerals, **687-89**; water-soluble vitamins, **675-82**
Milliequivalent (meq), 242, 255
Milk, **700-2**
Miltown, **709**
Minerals, 247; major, 687-89; in milk, 700; nutritional, **687-89**; trace, 688, 689
Minimal dietary requirements, 668
Mitochondria, 543, **545**, 559, 607
Mitochondrial matrix, 545, 562
Mitochondrial membranes, 545, 550, 559, 589
Mixture, 29
Molar heat of fusion (ΔH_{fus}), 156
Molar heat of solution ($\Delta H_{solution}$), 173-74
Molar heat of vaporization (ΔH_{vap}), **148-49**
Molarity, **176-78**
Mole, **95-98**
Molecular biology, 644
Molecular compounds, **61-66**;

Molecular compounds (cont.)
  binary, **64-65**; naming, **64-65**;
  properties of, **65-66**
Molecular crystals, **152-53**
Molecular formulas, 62, 285
Molecular shapes, **74-77**; of ammonia,
  75; of carbon tetrachloride, **79**; of
  methane, **75**; of water, **76**
Molecular weight (MW), **93**
Molecule, 60
Mole ratio (of reacting substances), 98
Monoacylglycerol, 603, 604, 705
Monofluorophosphate (MFP), 718
Monosaccharides, **359-60, 366-74**;
  families of, 366; subclasses of,
  359-60. *See also names of individual
  monosaccharides*
Monosodium glutamate (MSG), **706**
Morphine, **402-3**, 710
mRNA, 647, **648**
MSG (monosodium glutamate), **706**
Mucins, 514, 515
Multiple covalent bonds, **80-81**
Mutagens, 656, 657
Mutations, genetic, **652-53**
Myelin, 550
Myoglobin, **467-70**, 475, 704; heme in,
  467-68, 469; oxygen binding by,
  469-70; primary structure of, **467**;
  secondary structure of, 468; tertiary
  structure of, 468-69
Myosin, 456, 464

N

NAD^+, **498-99**, 547, 559, 614, 676
NADH, **498-99**, 547, 559, 560, 561, 562
NADH dehydrogenase, 559
$NADP^+$, **498-99**, 676
NADPH, **498-99**, 594, 614
β-Naphthylamine, **426**, 656, **657**
Narcotics, 402
Native conformation, 470
Natural gas, 293
Nerve gas, 505
Neutralization, acid-base, 226, 243,
  245-46
Neutron, 33, 35
Neutron/proton ratio, 50-51
Niacin (vitamin B_3), **449**, 497, 675, **676**
Nicolson, Garth, 553
Nicotinamide, **449**, 497, **676**
Nicotinamide adenine dinucleotide
  (NAD^+, NADH), **498-99**
Nicotinamide adenine dinucleotide,
  oxidized. *See* NAD^+
Nicotinamide adenine dinucleotide,
  reduced. *See* NADH
Nicotinamide adenine dinucleotide phos-
  phate ($NADP^+$, NADPH), **498-99**

Nicotinamide adenine dinucleotide
  phosphate, oxidized. *See* $NADP^+$
Nicotinamide adenine dinucleotide
  phosphate, reduced. *See* NADPH
Nicotine, **448**
Nicotinic acid, **497**, 676
Niemann-Pick disease, 622
Night blindness, 683
Ninhydrin, **440**
Nirenberg, Marshall W., 648
Nitrate ion, 82, **83**
Nitric acid, **227-28**
Nitrite ion, 82, **83**
Nitrogen balance, 670, 672
Nitroglycerin, **400-1**, 402
Nitro group, 283
Nitrosamines, 656, **657**, 704
Nobel, Alfred, **402**
Nobel Prizes, 402
Noble (rare, inert) gases, 45
Nomenclature: of acids, 225; of alcohols,
  321-23; of aldehydes and ketones,
  351-53; of alkanes, **286-90**; of alkenes,
  294-97; of alkynes, **303**; of amides,
  442-44; of amines, **423-25**; of benzene
  derivatives, 307, 308; of binary ionic
  compounds, **70-71**; of binary mole-
  cular compounds, **64**; of carboxylic
  acids, **386-88**; of carboxylic acid salts,
  390-91; of complex ionic compounds,
  71-72; of esters, **395-96**; of ethers,
  335; of phenols, **338**
Nonbonding electrons, 61
Noncompetitive inhibition, 505
Nonelectrolytes, **254**
Nonnutritive sweeteners, 707-8
Norepinephrine, **430**, 607, 682
Normality, **238-43**
Novocaine (procaine), 431-**32**, 711
N-terminus, 459
Nuclear binding energy, **745-46**
Nuclear fission, **746-50**
Nuclear fusion, 746, **750-51**
Nuclear medicine, **742-44**
Nuclear power, **744-51**
Nuclear power plants, **747**
Nuclear reactions, **53-55**
Nuclear reactors, 747
Nuclear stability, 745-46
Nucleic acids, **637-62**; composition of,
  638; DNA, 637, 638, **643-47**, 659-62;
  genetic engineering and, **658-62**;
  nucleosides, **639**; nucleotides, **639-41**;
  primary structure of, **641-43**; RNA,
  637, 638, **647-48**
Nucleosides, **639**
Nucleoside diphosphate, 640
Nucleoside monophosphate, 640
Nucleoside triphosphate, 640, 648

Nucleotides, 498, **639-41**
Nucleotide triplets, 649-50
Nucleus: atomic, 33; cell, **545**
Nutrients, 667; macro-, 667, 668-70,
  670-74; micro-, 667, **675-89**
Nutrient supplements, **705**
Nutrition, **667-92**; and balanced diets,
  689-90; and diet therapy, **690-92**;
  human requirements for, 667-68; and
  macronutrients, 667, 668-70, **670-74**;
  and micronutrients, 667, **675-89**
Nutritional minerals, **687-89**
Nylon, **447**, 448

O

Obesity, 668
Octane scale, 314
Octet rule, 45
Oils, 295, 405, 406, 617; cooking, **704**;
  food, **703-4**
Oleic acid, 404, 616
Oligosaccharides, 360
Opiates, 710
Opsin, 683
Optical activity: of α-amino acids,
  436-37; of monosaccharides, **363-66**
Optical isomerism, **361-66**
Optical isomers, 362, 364
Optical rotation, 364-66
Orbitals, 36-38, 40, 41
Organelles, cellular, **544-46**
Organic chemistry, **279-81**
Organ scan, 743
Ornithine, **441**, **627**, 628
Osmosis, **185-87**
Osmotic membranes, 185
Osmotic pressure, **186-87**, 714
Ouabain, **555**
Ovalbuman, 456
Oven cleaners, **719-20**
Oxaloacetate, **563**, 567, 624, 625
Oxalosuccinate, **563**, 565
Oxidant. *See* Oxidizing agent
Oxidation, **107-8**; of alcohols, **328-30**;
  of aldehydes, 354; of alkanes, **291-92**;
  of alkenes, 298; of alkynes, 303-4;
  of foods, 547, 570; of glucose, **590-91**;
  of hydrocarbons, 332
Oxidation numbers, 105
Oxidation process, 105
Oxidation-reduction reactions, **105-8**
Oxidative deamination, **625-26**
Oxidative phosphorylation, 547, **560-62**;
  energy conservation and, 561
Oxidizing agent, 107
Oxido-reductases, **486-87**
Oxyacids, 235
Oxygen affinity of hemoglobin, **525-27**;
  regulation of, by diphosphoglycerate

(DPG), **526-27**
Oxygen dissociation curve, 525
Oxytetracycline, **714**
Oxytocin, **460**, **461**, **535**, 536
Ozone, **82**

P

Palindrome, 659
Palmitic acid, 404, 611, 612
Palmitoleic acid, 404, 616
Pancreatic juice, **516**
Pantothenic acid, 501, 675, **678-79**
Papain, 506
Paracelsus, 4
Partial pressure, **132-33**
Pasteurization, 699
Pauling, Linus, 77, **80**, 463, 464, 465 479, 480
Pellagra, 676
Penetrating abilities (of forms of radiation), 735, 738
Penicillins, **712-14**
Penicillinase, 712, 713
Pentose phosphate pathway, **593-94**
Pepsin, 496, 516
Pepsinogen, 516
Peptides, 458-59
Peptide bonds, 457
Peptide group, **463**
Peptide hormones, 458, 461, 535, 536, 537, 539
Percentage concentrations of solutions, **178-80**
Perfluorocarbons, 312
Periodic law, 43
Periodic table, **31-33**, **44-45**
Peripheral protein, 551
Peroxides, 336-37
Pernicious anemia, 503, 679
Perutz, Max, 472, 473
Petroleum, 281, 293, **295**
pH: concept of, **259-62**; measurement of, **263-64**
pH meter, 264
pH scale, **262-63**
Phase changes, **158-60**
Phenacetin, 710
Phenols, **337-42**; acidity of, **338-39**; properties of, **338-39**; sources and uses of, **341**; structure and nomenclature of, **338**
Phenoxide ion, 339
Phenylacetate, **631**
Phenylalanine, 433, **434**
Phenylalanine hydroxylase, 631
Phenylketonuria (PKU), 506, 630, 692, 708
Phenylpyruvate, **631**
Phenylsalicylate (Salol), **400**

Phosphatase reaction, 487
Phosphate builders, **719**
Phosphate diesters, 409, 410-11; acidity of, 410-11
Phosphate monoesters, 409
Phosphatidyl choline (lecithin), **410**, 411, 524, **548**. See also Lecithin
Phosphatidyl ethanolamine, **410**,524, **548**
Phosphatidyl serine, **410**, **548**
Phosphodiester linkages, 641
Phosphoenolpyruvate, **586**
Phosphoenolpyruvate carboxykinase, 592
Phosphofructokinase, 583, 594
Phosphoglucoisomerase, 583
Phosphoglucomutase, 577, 579
2-Phosphoglycerate, **585**, **586**
3-Phosphoglycerate, **585**, 740, 741
Phosphoglycerate kinase, 585
Phosphoglycerides, **409-11**, 605
Phosphoglyceromutase, 585
Phospholipids, 548, 549, 551, 604, 605, 606, 617
Phosphoric acid, 227, **228**, 557
Phosphorylation, 558; oxidative, 547, 560-62; substrate-level, 566, 585
Phosphoserine residues, **700**
Phosphotransferases, 487
P$_i$. See Inorganic phosphate
Picric acid, **341**
Pituitary gland, 169, 534, 535, 536
pK_a, 267-70; table of values of, 268
Plane-polarized light, 363-64
Plasma, blood, 521; inorganic constituents of, **523-24**; organic constituents of, other than proteins, 523; proteins in, **522**
Plasmids, **659-60**, 661
Platelets, 521, 545
Plunkett, Roy, 312
Polar covalent bonds, **78-79**
Polar molecules, 78-79
Polyatomic anions, 71
Polyatomic ions (table), 71
Polycyclic aromatic hydrocarbons, **307**, 656
Polyethylene, 300-**1**
Polymers, 155-56; table of, 302
Polymerases, 642, 646
Polymerization of alkenes, **300-1**
Polypeptide hormones, 458, 461, 535, 536, 537, 539
Polypeptides, 459
Polypropylene, **301**
Polyprotic acids, 234-35
Polyribosome, 651-52
Polysaccharides, 332, 360, **377-79**; digestion of, 514, 515-17; glycogen, 378-79; starch, **378**. See also Cellulose
Polysome, 651-52
Pompe's disease, 580
Position of equilibrium 201-2, 209-10
Positron, 51, 52
Positron emission, 51, **52**
Potassium hydrogen tartrate, **705**
Potential energy, 17
PP$_i$. See Pyrophosphate
Precision, **11-12**
Pressure, **114-15**, 116-118, 122; effect of, on boiling point, **147-48**; effect of, on solubility, **174**; osmotic, **186-87**; partial, 132-34; units of, 114-15
Primary (1°) alcohols, **320-21**; dehydration of, **325-26**; oxidation of, **328-29**
Primary (1°) amines, **422**
Primary cosmic rays, 726
Primary structure: of nucleic acids, **641-43**; of proteins, **457-63**
Principal energy levels. 37
Procaine (Novocaine), **531-32**, 711
Prohormones, **538**
Proline, 433, **434**
Propane, **285**, 293
1,2,3,-Propanetriol (glycerol), **334**
2-Propanol (isopropyl alcohol), **334**
Propene, **301**
Propoxyphene, 710
Propoxyphene hydrochloride, **710**
Prostaglandins, **416**, 535, 709
Prostanoic acid, **416**
Prosthetic group, 491
Protease reaction, 488
Proteases, 446
Protein biosynthesis, **648-52**
Proteins, **455-81**, **550-51**, 735, 736; amino acid score, 671; biologic value of, 671; biosynthesis of, **648-52**; caloric content of, 603, 606, 668; classes of, **474-75**; complementary, 672; digestion of, 516-17; functions of, **455-57**; membrane, **550-51**; in milk, 700; as nutrients, **670-73**; properties of, **475-77**; quality, 671; recommended dietary allowances of, 669; separation of, **447-81**; structure of, **457-63**; whey, 700. See also names of specific proteins
Proteolytic enzymes, 446
Prothrombin, 529, 686
Protium, 35
Proton, 33, 35
Protoplasm, 165, 188
Provitamin, 684
Proximal histidine, 469
Pseudovitamin, 681
Ptomaines, 425
Purines, **449**, **638**, 639

Pyridine, 449
Pyridine coenzymes, **497-99**
Pyridoxal, **677**
Pyridoxal phosphate, 624
Pyrodoxamine, **677**
Pyridoxamine phosphate, 624
Pyridoxine, **677**
Pyrimidines, **449**, **638**, 639
Pyrophosphate (PP_i), **557-58**
Pyrophosphate linkages, 556-57
Pyrophosphoric acid, **557**
Pyrrole, **448**
Pyrrolidine, 433, **448**
Pyruvate, 586, **587-90**, 624; conversion of, to acetyl CoA, **589-90**; reduction of, to ethanol, **588-90**; reduction of, to lactate, **587-88**
Pyruvate carboxylase, 592, 596
Pyruvate decarboxylase, 588
Pyruvate dehydrogenase complex, 589, 676, 681
Pyruvate fates, **587-90**
Pyruvate kinase, 586; deficiency in, 599

Q

Quaternary ammonium salts, 428
Quaternary structure of proteins, **471-74**

R

Racemic mixture, 364
Rad, 733, 734
Radiation, 657, **727-39**, 741-42; basis for biological damage, **735-36**; as a carcinogen, **657**; detection and measurement of, **727-35**; dosage of, 732, 733, **734**; energy of, 733, **734-35**; exposure to, 730, 732, **736-37**; ionizing, 727; protection against, **737-39**; safety and, **735-39**; shielding from, 738
Radiation measurement, **732-35**
Radiation safety, **735-39**
Radiation therapy, **741-42**
Radioactive decay, **50-53**, 725, 726, 727, 731; types of, **51-53**
Radioactivity, **50-55**, **725-27**; artificial, **726-27**; natural, **725-26**
Radiocarbon dating, **739-40**
Radiochemistry applications, **739-44**; archeological dating, **739-40**; isotopic tracers, **740-41**; medical diagnosis, **742-44**; radiation therapy, **741-42**
Radioisotopes, 50; in archeological dating, **739-40**; in medical diagnosis, 731, 733, **742-44**; in medicine (table), 50; in radiation therapy, **741-42**; as tracers, **740-41**
Rare gases. *See* Noble gases
Rate constants, 214, 216

Rates of chemical reactions, **211-15**; and chemical equilibrium, **215-16**; factors affecting, **213-15**; and LeChatelier's principle, 216; and rate constants, 214
Rates of dissolution, **172**
RDA. *See* Recommended Dietary Allowances
Reaction pathway, 211-12
Reaction rates. *See* Rates of chemical reactions
Reactions, **101-8**
Reagents, 180, 507
Real gases, **134-35**
Recombinant DNA, 660-61
Recommended Dietary Allowances (RDAs), 668, 669
Recommended enzyme name, 486
Red blood cells (erythrocytes), 521, **524**, 545, 607
Reducing agent, 107
Reducing sugars, 371
Reductant. *See* Reducing agent
Reduction, **107-8**; of aldehydes and ketones, 354-55; of organic compounds, 332; of pyruvate, **587-90**
Refrigeration, 698, 699
Regulatory site, 505
Rem, 733, 734
Remineralization, 717
Rendering, 703
Renin, 533-34
Rennet, 701
Rennin. *See* Chymosin
Repair enzymes, 647, 736
Representative elements, 44-45
Residues, 459
Resistant organism, 712
Resonance, **81-83**
Resonance forms, 82
Respiratory chain. *See* Electron transport system
Restorative diets, **690-91**
Restriction enzymes, 658-59, 660
Restrictive diets, 690, **691-92**
Retinal, **683**
Retinol. *See* Vitamin A
Reversible reactions, **197-99**
Rhodopsin, 457, 683
Riboflavin (vitamin B_2), **449**, 475, **499**, 675, **677**
Ribonucleoside triphosphates, 648
Ribose, **366-69**, 594, **638**
Ribosomal RNA. *See* rRNA
Ribosomes, 545-46, 650, 651, 715
Rickets, 685
RNA, 637, 638, **647-48**, 715, 716
RNA-directed DNA polymerase, 654
RNA-directed RNA polymerase, 654
Roentgen (R), 733, 734

Roentgen, Wilhelm, 734
Rotenone, 560
Rounding off, 14
rRNA, 647, **648**
Rubber, 413, **414**
Rutherford, Ernest, 51, 727

S

Saccharin, **707**
Salad oil, **704**
Salicylic acid, **393-94**, 399, **709**
Saliva, **514**
Salt (sodium chloride): in food preservation, 698, 699, 704; iodized, 689
Salts, **245-47**; from amines, 427-28; solubility guidelines for, 247
Saponification, **408**, 720
Saran, **312**
Saturated solutions, 173
Saytzeff's rule, 309, 327
Scintillation counter, 730
Scurvy, 681, 682
Seawater, 170
Secondary (2°) alcohols, **320-21**; dehydration of, **326-27**; oxidation of, **329**
Secondary (2°) amines, **422**
Secondary cosmic rays, 726
Secondary structure: of DNA, **644-45**; of proteins, **463-66**; of tRNA, 648, 649
Secretin, 516, 536
Semipermeable membranes, 185
Serine, **410**, 433, **434**
Serotonin, **682**
Serum, blood, 370, 521
Serum albumin, 456, 607
Serum glutamate-oxaloacetate transaminase (SGOT), 506
Shielding, radiation, 738
Shortening, 409, **704**
Sickle-cell anemia, 104, 480-81
Sickle-cell hemoglobin (HbS), 479-81
Significant figures, **12**; rules for, **13-14**
Simple lipids, 403; prostaglandins, **416**; steroids, **414-15**; terpenes, **412-14**
Singer, S. Jonathan, 553
Single replacement reactions, **101**
Smoking, 698
Soaps, 408, **718-19**
Sodium benzoate, **699**, 704
Sodium bicarbonate, 705; in leavening agents, 707
Sodium fluoride, 718
Sodium hydroxide, **229**, 719, 720
Sodium nitrite, **82**, 704
Sodium Pentothal, **711**
Sodium pump, 555
Solids, 5, **139-140**; amorphous, **154-56** crystalline, **150-54**

Solubility, **172-76**; of common substances in water (table), 173; factors affecting, **174-76**; of salts, 247
Solute particles, **183**
Solutes, 170
Solutions, **169-87**; colligative properties of, **184-87**; concentrations of, **176-82**; dilutions of, **180-82**; formation of, **169-76**; homogeneity of, **183**; hypertonic, 187; hypotonic, 186-87; isotonic, 186; preparation of, **176-82**; properties of, **182-87**; saturated, 173; types of (table), 170
Solvation, 171
Solvents, 167, 170
Sour cream, **703**
Specific gravity, **17**
Specific heat, **21-23**; values of (table), 22
Spectator ions, 245
Sphingolipids, **411-12**
Sphingomyelins, **411**, 548, 622
Sphingomyelinase, 622
Sphingosine, **411**
Spontaneous changes, 205-10, 211
Squalene, **413**, 618, **620**
Stabilizers, 705
Standard free energy change ($\Delta G°$), 209-10; of ADP hydrolysis, 558; of ATP hydrolysis, 557, 558; for conversion of glucose to lactate, 558; in the electron transport system, 561; of pyrophosphate hydrolysis, 558
Standard temperature pressure (STP), 128
Stannous fluoride, 718
Starch, **378**, 575
States of matter: gaseous, **113-35**; liquid, **144-50**; properties of, 139-40; solid, **150-56**
Stereoisomers, 361
Steroid hormones, 613, 618
Steroid metabolism, 617, 618
Steroids, **414-15**
Sterols, 415
Structural formulas, 79, 286
Structural isomers, 286, 362
Subatomic particles, **33**; table of, 35
Substrate-level phosphorylation, 566, 585
Substrates, 486, 491, 492, 493, 494, 495, 497, 502, 504, 505
Subtilisin, 494
Succinate, **563**, 566, 567
Succinate dehydrogenase, 566
Succinyl CoA, **563**, **566**
Succinyl CoA synthetase, 566
Sucrose, **377**, 575
Sugar, 359, 366-77, 717, 718; in blood, **576-77**; in the diet, 668-69, 675; in food preservation, 698; sweetness scale for, 377
Sulfadiazine, **712**
Sulfa drugs, 711-**12**
Sulfanilamide, **712**
Sulfathiazole, **712**
Sulfisoxazole, **712**
Sulfonic acids, 718
Sulfur dioxide, 699
Sulfuric acid, **227**
Sulfurous acid, 699
Sulfurous acid salts, 699
Surface tension, **145**
Suspensions, **189**
Sutherland, Earl, 539
Synthetases, 489
Systematic drug names, 708
Systematic enzyme names, 486
Systems of measurement, **5-7**; units in, 6

T

Taurine, 517-**18**
Taurocholic acid, 517-**18**
Tay-Sachs disease, 622
Teflon, **312**
Temperature, **18-21**; effect of, on chemical equilibrium, 205; effect of, on gas pressure, **122**; effect of, on gas volume, **119-20**; effect of, on rates of dissolution, 172; effect of, on reaction rates, **213**; effect of, on solubility, 174; relationship of, to kinetic energy, 132, 140
Temperature conversions, 19-21
Temperature scales, 18-21
Template, 649
Tempra, **708**
Terminal amino group (N-terminus), 459
Terminal carboxyl group (C-terminus), 459
Terpenes, **412-14**
Tertiary (3°) alcohols, **320-21**; dehydration of, **326-27**; oxidation of, **329**
Tertiary (3°) amines, **422**
Tertiary structure (of proteins), **467-71**
Tetracyclines, **714**-15
Tetraethyl lead, **316**
Tetrahedron, 75
Tetrahydrofolate, 679
Thiamin (vitamin B_1), 588, 589, 675, **676**
Thiamin pyrophosphate (TPP), 588, 589, 676
Thickeners, 705
Thioesters, 608
Thiolase, 611
Thiol group, 283, **342**
Thiols, **342-43**
Thomson, J.J., 725
Thoracic duct, 604
Threonine, 433, **434**
Thrombin, 457, 529
Thromboplastin, 530
Thymine, **449**, **638**
Thymine dimers, **647**
Thymol, **340**
Thyroid gland, 689
Thyroid hormones, 689, 741
Thyrotropic releasing factor, **461**
Thyrotropin, 457
Thyroxine, **741**
Tissue factor, 530
Tissue hormones, 416
Titration, acid-base, **243-45**
Tocopherols, **686**
Torr, 115
Torricelli, Evangelista, 114-15
Toxins, 457
Trace minerals, 688, 689
Transaminases, 487, 624, 625
Transaminations, 487, **624-25**
Transcription, 648, 650
Transferases, 487
Transferrin, 456
Transfer RNA. *See* tRNA
Transition elements, 44
Transition state, 212
Translation, 650-51
Transmutation, 3, 727
Transport, membrane. *See* Membrane transport
Trehalose, 382
Triacylglycerols, **405-9**, 603, 605, 606; caloric content of, 603, 606, 668; chemical properties of, **407-9**; digestion of, 603-4; functions of, 406-7; nomenclature of, 406; synthesis of, **617**
Tricarboxylic acid cycle. *See* Citric acid cycle
Triglycerides. *See* Triacylglycerols
Triose phosphate isomerase, 584
Triose phosphates, **583-84**
Triple bond, **81**
Triple helix, **466**
Triplet, nucleotide, 649-50
Tritium, 35, 708, 725, 751
tRNA, 647, **648**
Tropomysin, 464-65
Trypsin, 503
Tryptophan, 433, **434**
Two-messenger hypothesis, 539
Tylenol, **708**
Tyndall effect, 188, 475
Tyrocidin A, **461**
Tyrosinase, 632
Tyrosine, 433, **435**

U

UDP, 577, 578
UDP-glucose, 577, 578
UDP-glucose pyrophosphorylase, 577
Unimolecular reactions, 214
Unit conversions. *See* Conversions
Unit-factor method, **8-11**
Unit factors, **8**
Units: of energy of radiation, 731, **734-35**; in measurement systems, 6; of radiation dosage, 733, **734**; of radiation measurements, **733-35**; of radioactivity, **733-35**
Unshared (electron) pair, 61
Uracil, **449**, **638**
Uracil mustard, **716**
Urea, **447**, **507**, **622**, **626-29**; adult daily excretion of, 628
Urea cycle, **626-29**; cellular location of, 628
Urease, 507
Uridine diphosphate. *See* UDP
Uridine triphosphate. *See* UTP
Urine, **530-34**; hormones affecting composition of, **532-33**
Urokinase, 506, 661
UTP, 577, 578

V

Valence, 59
Valence electrons, 42
Valine, **433**, **434**
Vanillic acid esters, 700
Vaporization, **146-50**
Vapor pressure, 147
Vapor-pressure lowering (of solutions), **184**
Vasopressin, **461**, **532**, 533, 535, 536
Vector, 661
Vinegar: in food preservation, 698; formation of, 698
Virus infections, 654-55, 658
Viruses, **653-56**; as carcinogens, **657-58**; slow, 654
Viscosity, **145**
Vitamins, **675-87**. *See also names of individual vitamins*
Vitamin A (retinol), 413, **414**, **683-84**
Vitamin B, 520
Vitamin B_1. *See* Thiamin
Vitamin B_2. *See* Riboflavin
Vitamin B_3. *See* Niacin
Vitamin B_6, 624, 675, **677**
Vitamin B_{12}, 675, **679-80**
Vitamin C. *See* Ascorbic acid
Vitamin D, 3, 617, **684-86**
Vitamin D_2 (ergocalciferol), 684, **685**
Vitamin D_3 (cholecalciferol), **684-86**
Vitamin E, 413, **414**, **686**, 704
Vitamin K, 413, **414**, 519, 520, 529, **686-87**
Vitamin K_2, **687**
Vitamin K_3 (menadione), **687**
Von Gierke's disease (glycogen storage disease I), 580
Von Guericke, Otto, 114

W

Warfarin, **687**
Water, 29, **165-69**; density of, 166; distribution of, in the human body (table), 168; hard, 718-19; in the human body, **168-69**, 667; of hydration, 168; hydrogen bonding in, **143**, **153**, **166**; ionization of, **258-59**; ion-product constant (K_w) of, **258-59**; liquid structure of, **143**, **153**, **165-66**; metabolic, 673; in milk, 700; molecular structure of, **76**; physical properties of (table), 166; solid structure of, **153**, **166**
Water-soluble vitamins, **675-82**. *See also names of individual water-soluble vitamins*
Watson, James D., 644
Waxes, **405**, 617
Weight, **15**
Whey, 700, 701
Whey proteins, 700
White blood cells (leukocytes), 521
Wöhler, Friedrich, 279

X

Xeroderma pigmentosum, 647, 656
X-ray diffraction, 468, 472, 644
X-rays, 723, **727-28**, 735, 738, 741
Xylocaine (lidocaine), **432**, **447**, 711

Y

Yogurt, **703**

Z

Zwitterions, 437
Zymogens, **502-3**